Consciousness and the Universe

Quantum Physics, Evolution, Brain & Mind

Editors
Sir Roger Penrose
University of Oxford, Oxford, United Kingdom

Stuart Hameroff, M.D.
University of Arizona, Arizona,

Subhash Kak, Ph.D.
Oklahoma State University, Oklahoma

Consciousness and the Universe

Quantum Physics, Evolution, Brain & Mind

Copyright © 2009, 2010, 2011, 2017, Cosmology Science Publishers, Cosmology.com

Published by: Cosmology Science Publishers, Cambridge, MA

All rights reserved. This book is protected by copyright. No part of this book may be reproduced in any form or by any means, including photocopying, or utilized in any information storage and retrieval system without permission of the copyright owner.

The publisher has sought to obtain permission from the copyright owners of all materials reproduced. If any copyright owner has been overlooked please contact: Cosmology Science Publishers at Editor@Cosmology.com, so that permission can be formally obtained.

Editor-in-Chief, Rudolf E. Schild, Ph.D., Center for Astrophysics, Harvard-Smithsonian, Cambridge, MA

Consciousness and the Universe
Quantum Physics, Evolution, Brain and Mind

ISBN: 978-1938024306

1. Consciousness, 2. Cosmology 3. Quantum Physics, 4. Evolution,
5. Neuroscience, 6. Universe, 7. Brain, 8. Mind

Acknowledgments: All of the chapters in this book have been peer reviewed.

Consciousness and the Universe

Quantum Physics, Evolution, Brain & Mind

Roger Penrose, Stuart Hameroff, Chris King, Walter J. Christensen Jr., Edgar D. Mitchell, Robert Staretz, Menas Kafatos, Rudolph E. Tanzi, Deepak Chopra, Ezequiel Morsella, Tiffany Jantz, Rhawn Joseph, Andrea Nani, Andrea E. Cavanna, Walter J. Freeman, Giuseppe Vitiello, Michael A. Persinger, William James, L. Dossey B. Greyson, P.A. Sturrock, Etienne Vermeersch, Ernest Lawrence Rossi, Kathryn Lane Rossi, Bruce J. MacLennan, GianCarlo Ghirardi, Ellen Langer, Michel Cabanac, Rémi Cabanac, Harold T. Hammel, Riccardo Manzotti, Steven Bodovitz, Samanta Pino, Ernesto Di Mauro, Chris J. S. Clarke, David Deamer, Thomas Suddendorf, Lori Marino, Jennifer Mather, A. N. Mitra, G. Mitra-Delmotte, Martin Lockley, Ian Tattersall, Arnold Trehub, Hans Liljenström, Tom Lombardo, Ellert R.S. Nijenhuis, Michel Bitbol, Pier-Luigi Luisi, Bruce Greyson, Jean-Pierre Jourdan, Kevin R. Nelson, Etzel Cardeña, Gordon Globus, Michael B. Mensky, Shan Gao, Fred Kuttner, Bruce Rosenblum, Michael Nauenberg, Antonella Vannini, Ulisse Di Corpo, Don N. Page, Henry P. Stapp

Contents

1. Consciousness in the Universe: Neuroscience, Quantum Space-Time Geometry and Orch OR Theory 8

2. Cosmological Foundations of Consciousness 48

3. Does the Universe have Cosmological Memory? Does This Imply Cosmic Consciousness? 68

4. The Quantum Hologram And the Nature of Consciousness 82

5. How Consciousness Becomes the Physical Universe 115

6. Conscious States Are a Crosstalk Mechanism for Only a Subset of Brain Processes 126

7. The Neuroanatomy of Free Will: Loss of Will, Against the Will, "Alien Hand" 138

8. Brain, Consciousness, and Causality 168

9. The Dissipative Brain and Non-Equilibrium Thermodynamics 181

10. Electromagnetic Bases of the Universality of the Characteristics of Consciousness: Quantitative Support 189

11. Does 'Consciousness' Exist? 197

12. Consciousness -- What Is It? 211

13. Consciousness: Solvable and Unsolvable Problems 228

14. Decoding the Chalmers Hard Problem of Consciousness: Qualia of the Molecular Biology of Creativity and Thought   235

15. Protophenomena and their Physical Correlates   254

16. The Macro-Objectification Problem and Conscious Perceptions   266

17. A Mindful Alternative to the Mind/Body Problem   285

18. Consciousness: The Fifth Influence 292

19. The Spread Mind: Seven Steps to Situated Consciousness   308

20. Consciousness Vectors 330

21. Gaia Universalis   337

22. What Consciousness Does: A Quantum Cosmology of Mind 346

23. The Stream of Consciousness 354

24. The Quest for Animal Consciousness   371

25. Consciousness and Intelligence in Mammals: Complexity thresholds   385

26. Mirror Self-Recognition on Earth   393

27. Consciousness in Cephalopods?   416

28. Origins of Thought: Consciousness, Language, Egocentric Speech and the Multiplicity of Mind   429

29. Consciousness: A Direct Link to Life's Origins? 456

30. The Evolution of Human Consciousness: Reflections on the Discovery of Mind and the Implications for the Materialist Darwinian Paradigm   466

31. Evolution of Paleolithic Cosmology and Spiritual Consciousness, and the Temporal and Frontal Lobes   475

32. Evolution of Modern Human Consciousness   530

33. Evolution's Gift: Subjectivity and the Phenomenal World   540

34. Intention and Attention in Consciousness: Dynamics and Evolution   552

35. The Ecological Cosmology of Consciousness   564

36. Consciousness, Dissociation and Self-Consciousness   575

37. Science and the Self-Referentiality of Consciousness   592

38. Cosmological Implications of Near-Death Experiences   608

39. Near Death Experiences and the 5th Dimensional Spatio-Temporal Perspective   622

40. In the Borderlands of Consciousness and Dreams: Spirituality Rising from Consciousness in Crisis   649

41. Dreams and Hallucinations: Lifting the Veil to Multiple Perceptual Realities   667

42. Altered Consciousness Is A Many Splendored Thing   703

43. Quantum Physics and the Multiplicity of Mind: Split-Brains, Fragmented Minds, Dissociation, Quantum Consciousness   715

44. Consciousness and Quantum Physics: A Deconstruction of the Topic   747

45. Logic of Quantum Mechanics and Phenomenon of Consciousness   754

46. A Quantum Physical Effect of Consciousness   767

47. The Conscious Observer in the Quantum Experiment   776

48. Does Quantum Mechanics Require A Conscious Observer?   784

49. Quantum Physics, Advanced Waves and Consciousness   790

50. Consciousness and the Quantum   803

51. Quantum Reality and Mind   812

1. Consciousness in the Universe: Neuroscience, Quantum Space-Time Geometry and Orch OR Theory

Roger Penrose, PhD, OM, FRS[1], and Stuart Hameroff, MD[2]

[1]Emeritus Rouse Ball Professor, Mathematical Institute, Emeritus Fellow, Wadham College, University of Oxford, Oxford, UK
[2]Professor, Anesthesiology and Psychology, Director, Center for Consciousness Studies, The University of Arizona, Tucson, Arizona, USA

Abstract

The nature of consciousness, its occurrence in the brain, and its ultimate place in the universe are unknown. We proposed in the mid 1990's that consciousness depends on biologically 'orchestrated' quantum computations in collections of microtubules within brain neurons, that these quantum computations correlate with and regulate neuronal activity, and that the continuous Schrödinger evolution of each quantum computation terminates in accordance with the specific Diósi–Penrose (DP) scheme of 'objective reduction' of the quantum state (OR). This orchestrated OR activity (Orch OR) is taken to result in a moment of conscious awareness and/or choice. This particular (DP) form of OR is taken to be a quantum-gravity process related to the fundamentals of spacetime geometry, so Orch OR suggests a connection between brain biomolecular processes and fine-scale structure of the universe. Here we review and update Orch OR in light of criticisms and developments in quantum biology, neuroscience, physics and cosmology. We conclude that consciousness plays an intrinsic role in the universe.

KEY WORDS: Consciousness, microtubules, OR, Orch OR, quantum computation, quantum gravity

1. Introduction: Consciousness, Brain and Evolution

Consciousness implies awareness: subjective experience of internal and external phenomenal worlds. Consciousness is central also to understanding, meaning and volitional choice with the experience of free will. Our views of reality, of the universe, of ourselves depend on consciousness. Consciousness defines our existence.

Three general possibilities regarding the origin and place of consciousness in the universe have been commonly expressed.

(A) Consciousness is not an independent quality but arose as a natural evolutionary consequence of the biological adaptation of brains and nervous systems. The most popular scientific view is that consciousness emerged as a property of complex biological computation during the course of evolution. Opinions vary as to when, where and how consciousness appeared, e.g. only recently in humans, or earlier in lower organisms. Consciousness as evolutionary adaptation is commonly assumed to be epiphenomenal (i.e. a secondary effect without independent influence), though it is frequently argued to confer beneficial advantages to conscious species (Dennett, 1991; 1995; Wegner, 2002).

(B) Consciousness is a quality that has always been in the universe. Spiritual and religious approaches assume consciousness has been in the universe all along, e.g. as the 'ground of being', 'creator' or component of an omnipresent 'God'. Panpsychists attribute consciousness to all matter. Idealists contend consciousness is all that exists, the material world an illusion (Kant, 1781).

(C) Precursors of consciousness have always been in the universe; biology evolved a mechanism to convert conscious precursors to actual consciousness. This is the view implied by Whitehead (1929; 1933) and taken in the Penrose-Hameroff theory of 'orchestrated objective reduction' ('Orch OR'). Precursors of consciousness, presumably with proto-experiential qualities, are proposed to exist as the potential ingredients of actual consciousness, the physical basis of these proto-conscious elements not necessarily being part of our current theories of the laws of the universe (Penrose and Hameroff, 1995; Hameroff and Penrose, 1996a; 1996b).

2. Ideas for how consciousness arises from brain action

How does the brain produce consciousness? An enormous amount of detailed knowledge about brain function has accrued; however the mechanism by which the brain produces consciousness remains mysterious (Koch, 2004).

The prevalent scientific view is that consciousness somehow emerges from complex computation among simple neurons which each receive and integrate synaptic inputs to a threshold for bit-like firing. The brain as a network of 10^{11} 'integrate-and-fire' neurons computing by bit-like firing and variable-strength chemical synapses is the standard model for computer simulations of brain function, e.g. in the field of artificial intelligence ('AI').

The brain-as-computer view can account for non-conscious cognitive functions including much of our mental processing and control of behavior. Such non-conscious cognitive processes are deemed 'zombie modes', 'auto-pilot', or 'easy problems'. The 'hard problem' (Chalmers, 1996) is the question of how cognitive processes are accompanied or driven by phenomenal conscious experience and subjective feelings, referred to by philosophers as 'qualia'. Other issues also suggest the brain-as-computer view may be incomplete, and that other approaches are required. The conventional brain-as-computer view fails to account for:

The 'hard problem' Distinctions between conscious and non-conscious processes are not addressed; consciousness is assumed to emerge at a critical level (neither specified nor testable) of computational complexity mediating otherwise non-conscious processes.

'Non-computable' thought and understanding, e.g. as shown by Gödel's theorem (Penrose, 1989; 1994).

'Binding and synchrony', the problem of how disparate neuronal activities are bound into unified conscious experience, and how neuronal synchrony, e.g. gamma synchrony EEG (30 to 90 Hz), the best measurable correlate of consciousness does not derive from neuronal firings.

Causal efficacy of consciousness and any semblance of free will. Because measurable brain activity corresponding to a stimulus often occurs after we've responded (seemingly consciously) to that stimulus, the brain-as-computer view depicts consciousness as epiphenomenal illusion (Dennett, 1991; 1995; Wegner, 2002).

Cognitive behaviors of single cell organisms. Protozoans like Paramecium can swim, find food and mates, learn, remember and have sex, all without synaptic computation (Sherrington, 1957).

In the 1980s Penrose and Hameroff (separately) began to address these issues, each against the grain of mainstream views.

3. Microtubules as Biomolecular Computers

Hameroff had been intrigued by seemingly intelligent, organized activities inside cells, accomplished by protein polymers called microtubules (Hameroff and Watt, 1982; Hameroff, 1987). Major components of the cell's structural cytoskeleton, microtubules also accounted for precise separation of chromosomes in cell division, complex behavior of Paramecium, and regulation of synapses within brain neurons (Figure 1). The intelligent function and periodic lattice structure of microtubules suggested they might function as some type of biomolecular computer.

Microtubules are self-assembling polymers of the peanut-shaped protein dimer tubulin, each tubulin dimer (110,000 atomic mass units) being composed of an alpha and beta monomer (Figure 2). Thirteen linear tubulin chains ('protofilaments') align side-to-side to form hollow microtubule cylinders (25 nanometers diameter) with two types of hexagonal lattices. The A-lattice has multiple winding patterns which intersect on protofilaments at specific intervals matching the Fibonacci series found widely in nature and possessing a helical symmetry (Section 9), suggestively sympathetic to large-scale quantum processes.

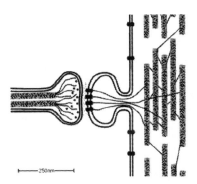

Figure 1. Schematic of portions of two neurons. A terminal axon (left) forms a synapse with a dendritic spine of a second neuron (right). Interiors of both neurons show cytoskeletal structures including microtubules, actin and microtubule-associated proteins (MAPs). Dendritic microtubules are arrayed in mixed polarity local networks, interconnected by MAPs. Synaptic inputs are conveyed to dendritic microtubules by ion flux, actin filaments, second messengers (e.g. CaMKII, see Hameroff et al, 2010) and MAPs.

Along with actin and other cytoskeletal structures, microtubules establish cell shape, direct growth and organize function of cells including brain neurons. Various types of microtubule-associated proteins ('MAPs') bind at specific lattice sites and bridge to other microtubules, defining cell architecture like girders and beams in a building. One such MAP is tau, whose displacement from microtubules results in neurofibrillary tangles and the cognitive dysfunction of Alzheimer's disease (Brunden et al, 2011). Motor proteins (dynein, kinesin) move rapidly along microtubules, transporting cargo molecules to specific locations.

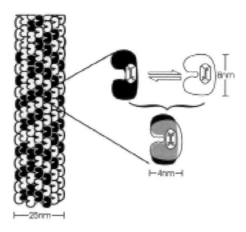

Figure 2. Left: Portion of single microtubule composed of tubulin dimer proteins (black and white) in A-lattice configuration. Right, top: According to pre-Orch OR microtubule automata theory (e.g. Hameroff and Watt, 1982; Rasmussen et al, 1990), each tubulin in a microtubule lattice switches between alternate (black and white) 'bit' states, coupled to electron cloud dipole London forces in internal hydrophobic pocket. Right, bottom: According to Orch OR, each tubulin can also exist as quantum superposition (quantum bit, or 'qubit') of both states, coupled to superposition of London force dipoles in hydrophobic pocket.

Microtubules also fuse side-by-side in doublets or triplets. Nine such doublets or triplets then align to form barrel-shaped mega-cylinders called cilia, flagella and centrioles, organelles responsible for locomotion, sensation and cell division. Either individually or in these larger arrays, microtubules are responsible for cellular and intra-cellular movements requiring intelligent spatiotemporal organization. Microtubules have a lattice structure comparable to computational systems. Could microtubules process information?

The notion that microtubules process information was suggested in general terms by Sherrington (1957) and Atema (1973). With physicist colleagues through the 1980s, Hameroff developed models of microtubules as information processing devices, specifically molecular ('cellular') automata, self-organizing computational devices (Figure 3). Cellular automata are computational systems in which fundamental units, or 'cells' in a grid or lattice can each exist in specific states, e.g. 1 or 0, at a given time (Wolfram, 2002). Each cell interacts with its neighbor cells at discrete, synchronized time steps, the state of each cell at any particular time step determined by its state and its neighbor cell states at the previous time step, and rules governing the interactions. In such ways, using simple neighbor interactions in simple lattice grids, cellular automata can perform complex computation and generate complex patterns.

Cells in cellular automata are meant to imply fundamental units. But biological cells are not necessarily simple, as illustrated by the clever Paramecium. Molecular automata are cellular automata in which the fundamental units, bits or cells are states of molecules, much smaller than biological cells. A dynamic, interactive molecular grid or lattice is required.

Microtubules are lattices of tubulin dimers which Hameroff and colleagues modeled as molecular automata. Discrete states of tubulin were suggested to act as bits, switching between states, and interacting (via dipole-dipole coupling) with neighbor tubulin bit states in 'molecular automata' computation (Hameroff and Watt, 1982; Rasmussen et al., 1990; Tuszynski et al., 1995). The mechanism for bit-like switching at the level of each tubulin was proposed to depend on the van der Waals–London force in non-polar, water-excluding regions ('hydrophobic pockets') within each tubulin.

Proteins are largely heterogeneous arrays of amino acid residues, including both water-soluble polar and water-insoluble non-polar groups, the latter including phenylalanine and tryptophan with electron resonance clouds (e.g. phenyl and indole rings). Such non-polar groups coalesce during protein folding to form homogeneous water-excluding 'hydrophobic' pockets within which instantaneous dipole couplings between nearby electron clouds operate. These are London forces which are extremely weak but numerous and able to act collectively in hydrophobic regions to influence and determine protein state (Voet and Voet, 1995).

London forces in hydrophobic pockets of various neuronal proteins are the mechanisms by which anesthetic gases selectively erase consciousness (Franks and Lieb, 1984). Anesthetics bind by their own London force attractions with electron clouds of the hydrophobic pocket, presumably impairing normally-occurring London forces governing protein switching required for consciousness (Hameroff, 2006).

In Figure 2, and as previously used in Orch OR, London forces are illustrated in cartoon fashion. A single hydrophobic pocket is depicted in tubulin, with portions of two electron resonance rings in the pocket. Single electrons in each ring repel each other, as their electron cloud net dipole flips (London force oscillation). London forces in hydrophobic pockets were used as the switching mechanism to distinguish discrete states for each tubulin in microtubule automata. In recent years tubulin hydrophobic regions and switching in the Orch OR proposal that we describe below have been clarified and updated (see Section 8).

To synchronize discrete time steps in microtubule automata, tubulins in microtubules were assumed to oscillate synchronously in a manner proposed by Fröhlich for biological coherence. Biophysicist Herbert Fröhlich (1968; 1970; 1975) had suggested that biomolecular dipoles constrained in a common geometry and voltage field would oscillate coherently, coupling, or condensing to a common vibrational mode. He proposed that biomolecular dipole lattices could convert ambient energy to coherent, synchronized dipole excitations, e.g. in the gigahertz (10^9 s^{-1}) frequency range. Fröhlich coherence or condensation can be either quantum coherence (e.g. Bose-Einstein condensation) or classical synchrony (Reimers et al., 2009).

In recent years coherent excitations have been found in living cells emanating from microtubules at 8 megahertz (Pokorny et al., 2001; 2004). Bandyopadhyay (2011) has found a series of coherence resonance peaks in single microtubules ranging from 12 kilohertz to 8 megahertz.

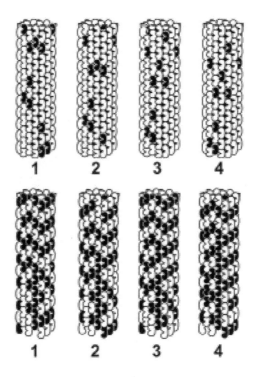

Figure 3. Microtubule automata (Rasmussen et al, 1990). Top: 4 time steps (e.g. at 8 megahertz, Pokorny et al, 2001) showing propagation of information states and patterns ('gliders' in cellular automata parlance). Bottom: At different dipole coupling parameter, bi-directional pattern movement and computation occur.

Rasmussen et al (1990) applied Fröhlich synchrony (in classical mode) as a clocking mechanism for computational time steps in simulated microtubule automata. Based on dipole couplings between neighboring tubulins in the microtubule lattice geometry, they found traveling gliders, complex patterns, computation and learning. Microtubule automata within brain neurons could potentially provide another level of information processing in the brain.

Approximately 10^8 tubulins in each neuron switching and oscillating in the range of 10^7 per second (e.g. Pokorny 8 MHz) gives an information capacity at the microtubule level of 10^{15} operations per second per neuron. This predicted capacity challenged and annoyed AI whose estimates for information processing at the level of neurons and synapses were virtually the same as this single-cell value, but for the entire brain (10^{11} neurons, 10^3 synapses per neuron, 10^2 transmissions per synapse per second = 10^{16} operations per second). Total brain capacity when taken at the microtubule level (in 10^{11} neurons) would potentially be 10^{26} operations per second, pushing the goalpost for AI brain equivalence farther into the future, and down into the quantum regime.

High capacity microtubule-based computing inside brain neurons could account for organization of synaptic regulation, learning and memory, and perhaps act as the substrate for consciousness. But increased brain information capacity per se didn't address most unanswered questions about consciousness (Section 2). Something was missing.

4. Objective Reduction (OR)

In 1989 Penrose published The Emperor's New Mind, which was followed in 1994 by Shadows of the Mind. Critical of AI, both books argued, by appealing to Gödel's theorem and other considerations, that certain aspects of human consciousness, such as understanding, must be beyond the scope of any computational system, i.e. 'non-computable'. Non-computability is a perfectly well-defined mathematical concept, but it had not previously been considered as a serious possibility for the result of physical actions. The non-computable ingredient required for human consciousness and understanding, Penrose suggested, would have to lie in an area where our current physical theories are fundamentally incomplete, though of important relevance to the scales that are pertinent to the operation of our brains. The only serious possibility was the incompleteness of quantum theory—an incompleteness that both Einstein and Schrödinger had recognized, despite quantum theory having frequently been argued to represent the pinnacle of 20th century scientific achievement. This incompleteness is the unresolved issue referred to as the 'measurement problem', which we consider in more detail below, in Section 5. One way to resolve it would be to provide an extension of the standard framework of quantum mechanics by introducing an objective form of quantum state reduction—termed 'OR' (objective reduction), an idea which we also describe more fully below, in Section 6.

In Penrose (1989), the tentatively suggested OR proposal would have its onset determined by a condition referred to there as 'the one-graviton' criterion. However, in Penrose (1995), a much better-founded criterion was used, now sometimes referred to as the Diósi–Penrose proposal (henceforth 'DP'; see Diósi 1987, 1989, Penrose 1993, 1996, 2000, 2009). This is an objective physical threshold, providing a plausible lifetime for quantum-superposed states. Other such OR proposals had also been put forward, from time to time (e.g.

Kibble 1981, Pearle 1989, Pearle and Squires 1994, Ghirardi et al., 1986, 1990; see Ghirardi 2011, this volume) as solutions to the measurement problem, but had not originally been suggested as having anything to do with the consciousness issue. The Diósi-Penrose proposal is sometimes referred to as a 'quantum-gravity' scheme, but it is not part of the normal ideas used in quantum gravity, as will be explained below (Section 6). Moreover, the proposed connection between consciousness and quantum measurement is almost opposite, in the Orch OR scheme, to the kind of idea that had frequently been put forward in the early days of quantum mechanics (see, for example, Wigner 1961) which suggests that a 'quantum measurement' is something that occurs only as a result of the conscious intervention of an observer. This issue, also, will be discussed below (Section 5).

5. The Nature of Quantum Mechanics and its Fundamental Problem

The term 'quantum' refers to a discrete element of energy in a system, such as the energy E of a particle, or of some other subsystem, this energy being related to a fundamental frequency ν of its oscillation, according to Max Planck's famous formula (where h is Planck's constant):

$$E = h\nu.$$

This deep relation between discrete energy levels and frequencies of oscillation underlies the wave/particle duality inherent in quantum phenomena. Neither the word "particle" nor the word "wave" adequately conveys the true nature of a basic quantum entity, but both provide useful partial pictures.

The laws governing these submicroscopic quantum entities differ from those governing our everyday classical world. For example, quantum particles can exist in two or more states or locations simultaneously, where such a multiple coexisting superposition of alternatives (each alternative being weighted by a complex number) would be described mathematically by a quantum wavefunction. We don't see superpositions in the consciously perceived world; we see objects and particles as material, classical things in specific locations and states.

Another quantum property is 'non-local entanglement,' in which separated components of a system become unified, the entire collection of components being governed by one common quantum wavefunction. The parts remain somehow connected, even when spatially separated by significant distances (e.g. over 10 kilometres, Tittel et al., 1998). Quantum superpositions of bit states (quantum bits, or qubits) can be interconnected with one another through entanglement in quantum computers. However, quantum entanglements cannot, by themselves, be used to send a message from one part of an entangled system to another; yet entanglement can be used in conjunction with classical signaling to achieve strange effects—such as the strange phenomenon referred to as quantum teleportation—that classical signalling cannot achieve by itself (e.g. Bennett and Wiesner, 1992; Bennett et al., 1993; Bouwmeester et al., 1997; Macikic et al., 2002).

The issue of why we don't directly perceive quantum superpositions is a manifestation of the measurement problem referred to in Section 4. Put more precisely, the measurement problem is the conflict between the two fundamental procedures of quantum mechanics. One of these procedures, referred to as unitary evolution, denoted here by U, is the continuous deterministic evolution of the quantum state (i.e. of the wavefunction of the entire system) according to the fundamental Schrödinger equation, The other is the procedure that is adopted whenever a measurement of the system—or observation—is deemed to have taken place, where the quantum state is discontinuously and probabilistically replaced by another quantum state (referred to, technically, as an eigenstate of a mathematical operator that is taken to describe the measurement). This discontinuous jumping of the state is referred to as the reduction of the state (or the 'collapse of the wavefunction'), and will be denoted here by the letter R. The conflict that is termed the measurement problem (or perhaps more accurately as the measurement paradox) arises when we consider the measuring apparatus itself as a quantum entity, which is part of the entire quantum system consisting of the original system under observation together with this measuring apparatus. The apparatus is, after all, constructed out of the same type of quantum ingredients (electrons, photons, protons, neutrons etc.—or quarks and gluons etc.) as is the system under observation, so it ought to be subject also to the same quantum laws, these being described in terms of the continuous and deterministic U. How, then, can the discontinuous and probabilistic R come about as a result of the interaction (measurement) between two parts of the quantum system? This is the measurement problem (or paradox).

There are many ways that quantum physicists have attempted to come to terms with this conflict (see, for example, Bell 1966, Bohm 1951, Rae 1994, Polkinghorne 2002, Penrose, 2004). In the early 20th century, the Danish physicist Niels Bohr, together with Werner Heisenberg, proposed the pragmatic 'Copenhagen interpretation', according to which the wavefunction of a quantum system, evolving according to U, is not assigned any actual physical 'reality', but is taken as basically providing the needed 'book-keeping' so that eventually probability values can be assigned to the various possible outcomes of a quantum measurement. The measuring device itself is explicitly taken to behave classically and no account is taken of the fact that the device is ultimately built from quantum-level constituents. The probabilities are calculated, once the nature of the measuring device is known, from the state that the wavefunction has U-evolved to at the time of the measurement. The discontinuous "jump" that the wavefunction makes upon measurement, according to R, is attributed to the change in 'knowledge' that the result of the measurement has on the observer. Since the wavefunction is not assigned physical reality, but is considered to refer merely to the observer's knowledge of the quantum system, the jumping is considered simply to reflect the jump in the observer's knowledge state, rather than in the quantum system under consideration.

Many physicists remain unhappy with such a point of view, however, and regard it largely as a 'stop-gap', in order that progress can be made in applying the quantum formalism, without this progress being held up by a lack of a serious quantum ontology, which might provide a more complete picture of what is actually going on. One may ask, in particular, what it is about a measuring device that allows one to ignore the fact that it is itself made from quantum constituents and is permitted to be treated entirely classically. A good many proponents of the Copenhagen standpoint would take the view that while the physical measuring apparatus ought actually to be treated as a quantum system, and therefore part of an over-riding wavefunction evolving according to U, it would be the conscious observer, examining the readings on that device, who actually reduces the state, according to R, thereby assigning a physical reality to the particular observed alternative resulting from the measurement. Accordingly, before the intervention of the observer's consciousness, the various alternatives of the result of the measurement including the different states of the measuring apparatus would, in effect, still coexist in superposition, in accordance with what would be the usual evolution according to U. In this way, the Copenhagen viewpoint puts consciousness outside science, and does not seriously address the nature and physical role of superposition itself nor the question of how large quantum superpositions like Schrödinger's superposed live and dead cat (see below) might actually become one thing or another.

A more extreme variant of this approach is the 'multiple worlds hypothesis' of Everett (1957) in which each possibility in a superposition evolves to form its own universe, resulting in an infinite multitude of coexisting 'parallel' worlds. The stream of consciousness of the observer is supposed somehow to 'split', so that there is one in each of the worlds—at least in those worlds for which the observer remains alive and conscious. Each instance of the observer's consciousness experiences a separate independent world, and is not directly aware of any of the other worlds.

A more 'down-to-earth' viewpoint is that of environmental decoherence, in which interaction of a superposition with its environment 'erodes' quantum states, so that instead of a single wavefunction being used to describe the state, a more complicated entity is used, referred to as a density matrix. However decoherence does not provide a consistent ontology for the reality of the world, in relation to the density matrix (see, for example, Penrose 2004, Sections 29.3-6), and provides merely a pragmatic procedure. Moreover, it does not address the issue of how R might arise in isolated systems, nor the nature of isolation, in which an external 'environment' would not be involved, nor does it tell us which part of a system is to be regarded as the 'environment' part, and it provides no limit to the size of that part which can remain subject to quantum superposition.

Still other approaches include various types of objective reduction (OR) in which a specific objective threshold is proposed to cause quantum state reduction (e.g. Kibble 1981; Pearle 1989; Ghirardi et al., 1986; Percival, 1994; Ghirardi, 2011). The specific OR scheme that is used in Orch OR will be described in Section 6.

The quantum pioneer Erwin Schrödinger took pains to point out the difficulties that confront the U-evolution of a quantum system with his still-famous thought experiment called 'Schrödinger's cat'. Here, the fate of a cat in a box is determined by magnifying a quantum event (say the decay of a radioactive atom, within a specific time period that would provide a 50% probability of decay) to a macroscopic action which would kill the cat, so that according to Schrödinger's own U-evolution the cat would be in a quantum superposition of being both dead and alive at the same time. If this U-evolution is maintained until the box is opened and the cat observed, then it would have to be the conscious human observing the cat that results in the cat becoming either dead or alive (unless, of course, the cat's own consciousness could be considered to have already served this purpose). Schrödinger intended to illustrate the absurdity of the direct applicability of the rules of quantum mechanics (including his own U-evolution) when applied at the level of a cat. Like Einstein, he regarded quantum mechanics as an incomplete theory, and his 'cat' provided an excellent example for emphasizing this incompleteness. There is a need for something to be done about quantum mechanics, irrespective of the issue of its relevance to consciousness.

6. The Orch OR Scheme

Orch OR depends, indeed, upon a particular OR extension of current quantum mechanics, taking the bridge between quantum- and classical-level physics as a 'quantum-gravitational' phenomenon. This is in contrast with the various conventional viewpoints (see Section 5), whereby this bridge is claimed to result, somehow, from 'environmental decoherence', or from 'observation by a conscious observer', or from a 'choice between alternative worlds', or some other interpretation of how the classical world of one actual alternative may be taken to arise out of fundamentally quantum-superposed ingredients.

It must also be made clear that the Orch OR scheme involves a different interpretation of the term 'quantum gravity' from what is usual. Current ideas of quantum gravity (see, for example Smolin, 2002) normally refer, instead, to some sort of physical scheme that is to be formulated within the bounds of standard quantum field theory—although no particular such theory, among the multitude that has so far been put forward, has gained anything approaching universal acceptance, nor has any of them found a fully consistent, satisfactory formulation. 'OR' here refers to the alternative viewpoint that standard quantum (field) theory is not the final answer, and that the reduction R of the quantum state ('collapse of the wavefunction') that is adopted in standard quantum mechanics is an actual physical phenomenon which is not part of the conventional unitary formalism U of quantum theory (or quantum field theory) and does not arise as some kind of convenience or effective consequence of environmental decoherence, etc., as the conventional U formalism would seem to demand. Instead, OR is taken to be one of the consequences of melding together the principles of Einstein's general relativity with those of the conventional unitary quantum formalism U, and this demands a departure from the strict rules of U. According to this OR viewpoint, any quantum measurement—whereby the quantum-superposed alternatives produced in accordance with the U formalism becomes reduced to a single actual occurrence—is real objective physical phenomenon, and it is taken to result from the mass displacement between the alternatives being sufficient, in gravitational terms, for the superposition to become unstable.

In the DP (Diósi–Penrose) scheme for OR, the superposition reduces to one of the alternatives in a time scale τ that can be estimated (for a superposition of two states each of which can be taken to be stationary on its own) according to the formula

$$\tau \approx \hbar/E_G.$$

Here $\hbar$ (=h/2π) is Dirac's form of Planck's constant h and E_G is the gravitational self-energy of the difference between the two mass distributions of the superposition. (For a superposition for which each mass distribution is a rigid translation of the other, E_G is the energy it would cost to displace one component of the superposition in the gravitational field of the other, in moving it from coincidence to the quantum-displaced location; see Disói 1989, Penrose 1993, 2000, 2009).

According to Orch OR, the (objective) reduction is not the entirely random process of standard theory, but acts according to some non-computational new physics (see Penrose 1989, 1994). The idea is that consciousness is associated with this (gravitational) OR process, but occurs significantly only when the alternatives are part of some highly organized structure, so that such occurrences of OR occur in an extremely orchestrated form. Only then does a recognizably conscious event take place. On the other hand, we may consider that any individual occurrence of OR would be an element of proto-consciousness.

The OR process is considered to occur when quantum superpositions between slightly differing space-times take place, differing from one another by an integrated space-time measure which compares with the fundamental and extremely tiny Planck (4-volume) scale of space-time geometry. Since this is a 4-volume Planck measure, involving both time and space, we find that the time measure would be particularly tiny when the space-difference measure is relatively large (as with Schrödinger's cat), but for extremely tiny space-difference measures, the time measure might be fairly long, such as some significant fraction of a second. We shall be seeing this in more detail shortly, together with its particular relevance to microtubules. In any case, we recognize that the elements of proto-consciousness would be intimately tied in with the most primitive Planck-level ingredients of space-time geometry, these presumed 'ingredients' being taken to be at the absurdly tiny level of 10^{-35}m and 10^{-43}s, a distance and a time some 20 orders of magnitude smaller than those of normal particle-physics scales and their most rapid processes. These scales refer only to the normally extremely tiny differences in space-time geometry between different states in superposition, and OR is deemed to take place when such space-time differences reach the Planck level. Owing to the extreme weakness of gravitational forces as compared with those of the chemical and electric

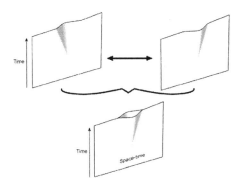

Figure 4. From Penrose, 1994 (P. 338). With four spatiotemporal dimensions condensed to a 2-dimensional spacetime sheet, mass location may be represented as a particular curvature of that sheet, according to general relativity. Top: Two different mass locations as alternative spacetime curvatures. Bottom: a bifurcating spacetime is depicted as the union ("glued together version") of the two alternative spacetime histories that are depicted at the top of the Figure. Hence a quantum superposition of simultaneous alternative locations may be seen as a separation in fundamental spacetime geometry.

forces of biology, the energy E_G is liable to be far smaller than any energy that arises directly from biological processes. However, E_G is not to be thought of as being in direct competition with any of the usual biological energies, as it plays a completely different role, supplying a needed energy uncertainty that then allows a choice to be made between the separated space-time geometries. It is the key ingredient of the computation of the reduction time τ. Nevertheless, the extreme weakness of gravity tells us there must be a considerable amount of material involved in the coherent mass displacement between superposed structures in order that τ can be small enough to be playing its necessary role in the relevant OR processes in the brain. These superposed structures should also process information and regulate neuronal physiology. According to Orch OR, microtubules are central to these structures, and some form of biological quantum computation in microtubules (most probably primarily in the more symmetrical A-lattice microtubules) would have to have evolved to provide a subtle yet direct connection to Planck-scale geometry, leading eventually to discrete moments of actual conscious experience.

The degree of separation between the space-time sheets is mathematically described in terms of a symplectic measure on the space of 4-dimensional metrics (cf. Penrose, 1993). The separation is, as already noted above, a space-time separation, not just a spatial one. Thus the time of separation contributes as well as the spatial displacement. Roughly speaking, it is the product of the temporal separation T with the spatial separation S that measures the overall degree of separation, and OR takes place when this overall separation reaches a critical amount. This critical amount would be of the order of unity, in absolute units, for which the Planck-Dirac constant $\hbar$, the gravitational constant G, and the velocity of light c, all take the value unity, cf. Penrose, 1994 - pp. 337-339. For small S, the lifetime $\tau \approx T$ of the superposed state will be large; on the other hand, if S is large, then τ will be small.

To estimate S, we compute (in the Newtonian limit of weak gravitational fields) the gravitational self-energy E_G of the difference between the mass distributions of the two superposed states. (That is, one mass distribution counts positively and the other, negatively; see Penrose, 1993; 1995.) The quantity S is then given by:

$$S \approx E_G$$

and $T \approx \tau$, whence

$$\tau \approx \hbar/E_G, \text{ i.e. } E_G \approx \hbar/\tau.$$

Thus, the DP expectation is that OR occurs with the resolving out of one particular space-time geometry from the previous superposition when, on the average, $\tau \approx \hbar/E_G$. Moreover, according to Orch OR, this is accompanied by an element of proto-consciousness.

Environmental decoherence need play no role in state reduction, according to this scheme. The proposal is that state reduction simply takes place spontaneously, according to this criterion. On the other hand, in many actual physical situations, there would be much material from the environment that would be entangled with the quantum-superposed state, and it could well be that the major mass displacement—and therefore the major contribution to E_G—would occur in the environment rather than in the system under consideration. Since the environment will be quantum-entangled with the system, the state-reduction in the environment will effect a simultaneous reduction in the system. This could shorten the time for the state reduction R to take place very considerably. It would also introduce an uncontrollable random element into the result of the reduction, so that any non-random (albeit non-computable, according to Orch OR) element influencing the particular choice of state that is actually resolved out from the superposition would be completely masked by this randomness. In these circumstances the OR-process would be indistinguishable from the R-process of conventional quantum mechanics. If the suggested non-computable effects of this OR proposal are to be laid bare, if E_G is to be able to evolve and be orchestrated for conscious moments, we indeed need significant isolation from the environment.

As yet, no experiment has been refined enough to determine whether this (DP) OR proposal is actually respected by Nature, but the experimental testing of the scheme is fairly close to the borderline of what can be achieved with present-day technology (see, for example, Marshall et al. 2003). One ought to begin to see the effects of this OR scheme if a small object, such as a 10-micron cube of crystalline material could be held in a superposition of two locations, differing by about the diameter of an atomic nucleus, for some seconds, or perhaps minutes.

A point of importance, in such proposed experiments, is that in order to calculate E_G it may not be enough to base the calculation on an average density of the material in the superposition, since the mass will be concentrated in the atomic nuclei, and for a displacement of the order of the diameter of a nucleus, this inhomogeneity in the density of the material can be crucial, and can provide a much larger value for E_G than would be obtained if the material is assumed to be homogeneous. The Schrödinger equation (more correctly, in the zero-temperature approximation, the Schrödinger–Newton equation, see Penrose 2000; Moroz et al. 1998) for the static unsuperposed material would have to be solved, at least approximately, in order to derive the expectation value of the mass distribution, where there would be some quantum spread in the locations of the particles constituting the nuclei.

For Orch OR to be operative in the brain, we would need coherent superpositions of sufficient amounts of material, undisturbed by environmental entanglement, where this reduces in accordance with the above OR scheme in a rough time scale of the general order of time for a conscious experience to take place. For an ordinary type of experience, this might be say about $\tau = 10^{-1}$s which concurs with neural correlates of consciousness, such as particular frequencies of electroencephalograhy (EEG).

Penrose (1989; 1994) suggested that processes of the general nature of quantum computations were occurring in the brain, terminated by OR. In quantum computers (Benioff 1982, Deutsch 1985, Feynman 1986), information is represented not just as bits of either 1 or 0, but also as quantum superposition of both 1 and 0 together (quantum bits or qubits) where, moreover, large-scale entanglements between qubits would also be involved. These qubits interact and compute following the Schrödinger equation, potentially enabling complex and highly efficient parallel processing. As envisioned in technological quantum computers, at some point a measurement is made causing quantum state reduction (with some randomness introduced). The qubits reduce, or collapse to classical bits and definite states as the output.

The proposal that some form of quantum computing could be acting in the brain, this proceeding by the Schrödinger equation without decoherence until some threshold for self-collapse due to a form of non-computable OR could be reached, was made in Penrose 1989. However, no plausible biological candidate for quantum computing in the brain had been available to him, as he was then unfamiliar with microtubules.

7. Penrose-Hameroff Orchestrated Objective Reduction ('Orch OR')

Penrose and Hameroff teamed up in the early 1990s. Fortunately, by then, the DP form of OR mechanism was at hand to be applied to the microtubule-automata models for consciousness as developed by Hameroff. A number of questions were addressed.

How does $\tau \approx \hbar/E_G$ relate to consciousness? Orch OR considers consciousness as a sequence of discrete OR events in concert with neuronal-level activities. In $\tau \approx \hbar/E_G$, τ is taken to be the time for evolution of the pre-conscious quantum wavefunction between OR events, i.e. the time interval between conscious moments, during which quantum superpositions of microtubule states evolve according to the continuous Schrödinger equation before reaching (on the average) the $\tau \approx \hbar/E_G$ OR threshold in time τ, when quantum state reduction and a moment of conscious awareness occurs (Figure 5).

The best known temporal correlate for consciousness is gamma synchrony EEG, 30 to 90 Hz, often referred to as coherent 40 Hz. One possible viewpoint might be to take this oscillation to represent a succession of 40 or so conscious moments per second ($\tau=25$ milliseconds). This would be reasonably consistent with neuroscience (gamma synchrony), with certain ideas expressed in philosophy (e.g. Whitehead 'occasions of experience'), and perhaps even with

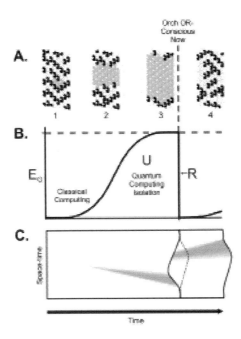

Figure 5. Three descriptions of an Orch OR conscious event by $E_G = \hbar/\tau$. A. Microtubule automata. Quantum (gray) tubulins evolve to meet threshold after Step 3, a moment of consciousness occurs and tubulin states are selected. For actual event (e.g. 25 msec), billions of tubulins are required; a small number is used here for illustration. B. Schematic showing U-like evolution until threshold. C. Space-time sheet with superposition separation reaches threshold and selects one reality/spacetime curvature.

ancient Buddhist texts which portray consciousness as 'momentary collections of mental phenomena' or as 'distinct, unconnected and impermanent moments which perish as soon as they arise.' (Some Buddhist writings quantify the frequency of conscious moments. For example the Sarvaastivaadins, according to von Rospatt 1995, described 6,480,000 'moments' in 24 hours—an average of one 'moment' per 13.3 msec, ~75 Hz—and some Chinese Buddhism as one "thought" per 20 msec, i.e. 50 Hz.) These accounts, even including variations in frequency, could be considered to be consistent with Orch OR events in the gamma synchrony range. Accordingly, on this view, gamma synchrony, Buddhist 'moments of experience', Whitehead 'occasions of experience', and our proposed Orch OR events might be viewed as corresponding tolerably well with one another.

Putting $\tau=25$ msec in $E_G \approx \hbar/\tau$, we may ask what is E_G in terms of superpositioned microtubule tubulins? E_G may be derived from details about the superposition separation of mass distribution. Three types of mass separation were considered in Hameroff–Penrose 1996a for peanut-shaped tubulin proteins of 110,000 atomic mass units: separation at the level of (1) protein spheres, e.g.

by 10 percent volume, (2) atomic nuclei (e.g. carbon, ~ 2.5 Fermi length), (3) nucleons (protons and neutrons). The most plausible calculated effect might be separation at the level of atomic nuclei, giving E_G as superposition of 2 x 10^{10} tubulins reaching OR threshold at 25 milliseconds.

Brain neurons each contain roughly 10^8 tubulins, so only a few hundred neurons would be required for a 25msec, gamma synchrony OR event if 100 percent of tubulins in those neurons were in superposition and avoided decoherence. It seems more likely that a fraction of tubulins per neuron are in superposition. Global macroscopic states such as superconductivity ensue from quantum coherence among only very small fractions of components. If 1 percent of tubulins within a given set of neurons were coherent for 25msec, then 20,000 such neurons would be required to elicit OR. In human brain, cognition and consciousness are, at any one time, thought to involve tens of thousands of neurons. Hebb's (1949) 'cell assemblies', Eccles's (1992) 'modules', and Crick and Koch's (1990) 'coherent sets of neurons' are each estimated to contain some 10,000 to 100,000 neurons which may be widely distributed throughout the brain (Scott, 1995).

Adopting $\tau \approx \hbar/E_G$, we find that, with this point of view with regard to Orch-OR, a spectrum of possible types of conscious event might be able to occur, including those at higher frequency and intensity. It may be noted that Tibetan monk meditators have been found to have 80 Hz gamma synchrony, and perhaps more intense experience (Lutz et al. 2004). Thus, according to the viewpoint proposed above, where we interpret this frequency to be associated with a succession of Orch-OR moments, then $E_G \approx \hbar/\tau$ would appear to require that there is twice as much brain involvement required for 80 Hz than for consciousness occurring at 40 Hz (or $\sqrt{2}$ times as much if the displacement is entirely coherent, since then the mass enters quadratically in E_G). Even higher (frequency), expanded awareness states of consciousness might be expected, with more neuronal brain involvement.

On the other hand, we might take an alternative viewpoint with regard to the probable frequency of Orch-OR actions, and to the resulting frequency of elements of conscious experience. There is the possibility that the discernable moments of consciousness are events that normally occur at a much slower pace than is suggested by the considerations above, and that they happen only at rough intervals of the order of, say, one half a second or so, i.e. ~500msec, rather than ~25msec. One might indeed think of conscious influences as perhaps being rather slow, in contrast with the great deal of vastly faster unconscious computing that might be some form of quantum computing, but without OR. At the present stage of uncertainty about such matters it is perhaps best not to be dogmatic about how the ideas of Orch OR are to be applied. In any case, the numerical assignments provided above must be considered to be extremely rough, and at the moment we are far from being in a position to be definitive about the precise way in which the Orch-OR is to operate. Alternative possibilities will need to be considered with an open mind.

How do microtubule quantum computation avoid decoherence? Technological quantum computers using e.g. ion traps as qubits are plagued by decoherence, disruption of delicate quantum states by thermal vibration, and require extremely cold temperatures and vacuum to operate. Decoherence must be avoided during the evolution toward time τ ($\approx \hbar/E_G$), so that the non-random (non-computable) aspects of OR can be playing their roles. How does quantum computing avoid decoherence in the 'warm, wet and noisy' brain?

It was suggested (Hameroff and Penrose, 1996a) that microtubule quantum states avoid decoherence by being pumped, laser-like, by Fröhlich resonance, and shielded by ordered water, C-termini Debye layers, actin gel and strong mitochondrial electric fields. Moreover quantum states in Orch OR are proposed to originate in hydrophobic pockets in tubulin interiors, isolated from polar interactions, and involve superposition of only atomic nuclei separation. Moreover, geometrical resonances in microtubules, e.g. following helical pathways of Fibonacci geometry are suggested to enable topological quantum computing and error correction, avoiding decoherence perhaps effectively indefinitely (Hameroff et al 2002) as in a superconductor.

The analogy with high-temperature superconductors may indeed be appropriate, in fact. As yet, there is no fully accepted theory of how such superconductors operate, avoiding loss of quantum coherence from the usual processes of environmental decoherence. Yet there are materials which support superconductivity at temperatures roughly halfway between room temperature and absolute zero (He et al., 2010). This is still a long way from body temperature, of course, but there is now some experimental evidence (Bandyopadhyay 2011) that is indicative of something resembling superconductivity (referred to as 'ballistic conductance'), that occurs in living A-lattice microtubules at body temperature. This will be discussed below.

Physicist Max Tegmark (2000) published a critique of Orch OR based on his calculated decoherence times for microtubules of 10^{-13} seconds at biological temperature, far too brief for physiological effects. However Tegmark didn't include Orch OR stipulations and in essence created, and then refuted his own quantum microtubule model. He assumed superpositions of solitons separated from themselves by a distance of 24 nanometers along the length of the microtubule. As previously described, superposition separation in Orch OR is at the Fermi length level of atomic nuclei, i.e. 7 orders of magnitude smaller than Tegmark's separation value, thus underestimating decoherence time by 7 orders of magnitude, i.e. from 10^{-13} secs to microseconds at 10^{-6} seconds. Hagan et al (2001) used Tegmark's same formula and recalculated microtubule decoherence times using Orch OR stipulations, finding 10^{-4} to 10^{-3} seconds, or longer due to topological quantum effects. It seemed likely biology had evolved optimal information processing systems which can utilize quantum computing, but there was no real evidence either way.

Beginning in 2003, published research began to demonstrate quantum coherence in warm biological systems. Ouyang and Awschalom (2003) showed that quantum spin transfer through phenyl rings (the same as those in protein hydrophobic pockets) is enhanced at increasingly warm temperatures. Other studies showed that quantum coherence occurred at ambient temperatures in proteins involved in photosynthesis, that plants routinely use quantum coherence to produce chemical energy and food (Engel et al, 2007). Further research has demonstrated warm quantum effects in bird brain navigation (Gauger et al, 2011), ion channels (Bernroider and Roy, 2005), sense of smell (Turin, 1996), DNA (Rieper et al., 2011), protein folding (Luo and Lu, 2011), biological water (Reiter et al., 2011) and microtubules.

Recently Anirban Bandyopadhyay and colleagues at the National Institute of Material Sciences in Tsukuba, Japan have used nanotechnology to study electronic conductance properties of single microtubules assembled from porcine brain tubulin. Their preliminary findings (Bandyopadhyay, 2011) include: (1) Microtubules have 8 resonance peaks for AC stimulation (kilohertz to 10 megahertz) which appear to correlate with various helical conductance pathways around the geometric microtubule lattice. (2) Excitation at these resonant frequencies causes microtubules to assemble extremely rapidly, possibly due to Fröhlich condensation. (3) In assembled microtubules AC excitation at resonant frequencies causes electronic conductance to become lossless, or 'ballistic', essentially quantum conductance, presumably along these helical quantum channels. Resonance in the range of kilohertz demonstrates microtubule decoherence times of at least 0.1 millisecond. (4) Eight distinct quantum interference patterns from a single microtubule, each correlating with one of the 8 resonance frequencies and pathways. (5) Ferroelectric hysteresis demonstrates memory capacity in microtubules. (6) Temperature-independent conductance also suggests quantum effects. If confirmed, such findings would demonstrate Orch OR to be biologically feasible.

How does microtubule quantum computation and Orch OR fit with recognized neurophysiology? Neurons are composed of multiple dendrites and a cell body/soma which receive and integrate synaptic inputs to a threshold for firing outputs along a single axon. Microtubule quantum computation in Orch OR is assumed to occur in dendrites and cell bodies/soma of brain neurons, i.e. in regions of integration of inputs in integrate-and-fire neurons. As opposed to axonal firings, dendritic/somatic integration correlates best with local field potentials, gamma synchrony EEG, and action of anesthetics erasing consciousness. Tononi (2004) has identified integration of information as the neuronal function most closely associated with consciousness. Dendritic microtubules are uniquely arranged in local mixed polarity networks, well-suited for integration of synaptic inputs.

Membrane synaptic inputs interact with post-synaptic microtubules by activation of microtubule-associated protein 2 ('MAP2', associated with learning), and calcium-calmodulin kinase II (CaMKII, Hameroff et al, 2010). Such inputs were suggested by Penrose and Hameroff (1996a) to 'tune', or 'orchestrate' OR-mediated quantum computations in microtubules by MAPs, hence 'orchestrated objective reduction', 'Orch OR'.

Proposed mechanisms for microtubule avoidance of decoherence were described above, but another question remains. How would microtubule quantum computations which are isolated from the environment, still interact with that environment for input and output? One possibility that Orch OR suggests is that perhaps phases of isolated quantum computing alternate with phases of classical environmental interaction, e.g. at gamma synchrony, roughly 40 times per second. (Computing pioneer Paul Benioff suggested such a scheme of alternating quantum and classical phases in a science fiction story about quantum computing robots.)

With regard to outputs resulting from processes taking place at the level of microtubules in Orch-OR quantum computations, dendritic/somatic microtubules receive and integrate synaptic inputs during classical phase. They then become isolated quantum computers and evolve to threshold for Orch OR at which they reduce their quantum states at an average time interval τ (given by by $\tau \approx \hbar/E_G$). The particular tubulin states chosen in the reduction can then trigger axonal firing, adjust firing threshold, regulate synapses and encode memory. Thus Orch OR can have causal efficacy in conscious actions and behavior, as well as providing conscious experience and memory.

Orch OR in evolution In the absence of Orch OR, non-conscious neuronal activities might proceed by classical neuronal and microtubule-based computation. In addition there could be quantum computations in microtubules that do not reach the Orch OR level, and thereby also remain unconscious.

This last possibility is strongly suggested by considerations of natural selection, since some relatively primitive microtubule infrastructure, still able to support quantum computation, would have to have preceded the more sophisticated kind that we now find in conscious animals. Natural selection proceeds in steps, after all, and one would not expect that the capability of the substantial level of coherence across the brain that would be needed for the non-computable OR of human conscious understanding to be reached, without something more primitive having preceded it. Microtubule quantum computing by U evolution which avoids decoherence would well be advantageous to biological processes without ever reaching threshold for OR.

Microtubules may have appeared in eukaryotic cells 1.3 billion years ago due to symbiosis among prokaryotes, mitochondria and spirochetes, the latter the apparent origin of microtubules which provided movement to previously immobile cells (e.g. Margulis and Sagan, 1995). Because Orch OR depends on

$\tau \approx \hbar/E_G$, more primitive consciousness in simple, small organisms would involve smaller E_G, and longer times τ to avoid decoherence. As simple nervous systems and arrangements of microtubules grew larger and developed anti-decoherence mechanisms, inevitably a system would avoid decoherence long enough to reach threshold for Orch OR conscious moments. Central nervous systems around 300 neurons, such as those present at the early Cambrian evolutionary explosion 540 million years ago, could have τ near one minute, and thus be feasible in terms of avoiding decoherence (Hameroff, 1998d). Perhaps the onset of Orch OR and consciousness with relatively slow and simple conscious moments, precipitated the accelerated evolution.

Only at a much later evolutionary stage would the selective advantages of a capability for genuine understanding come about. This would require the non-computable capabilities of Orch OR that go beyond those of mere quantum computation, and depend upon larger scale infrastructure of efficiently functioning microtubules, capable of operating quantum-computational processes. Further evolution providing larger sets of microtubules (larger EG) able to be isolated from decoherence would enable, by $\tau \approx \hbar/E_G$, more frequent and more intense moments of conscious experience. It appears human brains could have evolved to having Orch OR conscious moments perhaps as frequently as every few milliseconds.

How could microtubule quantum states in one neuron extend to those in other neurons throughout the brain? Assuming microtubule quantum state phases are isolated in a specific neuron, how could that quantum state involve microtubules in other neurons throughout the brain without traversing membranes and synapses? Orch OR proposes that quantum states can extend by tunneling, leading to entanglement between adjacent neurons through gap junctions.

Gap junctions are primitive electrical connections between cells, synchronizing electrical activities. Structurally, gap junctions are windows between cells which may be open or closed. When open, gap junctions synchronize adjacent cell membrane polarization states, but also allow passage of molecules between cytoplasmic compartments of the two cells. So both membranes and cytoplasmic interiors of gap-junction-connected neurons are continuous, essentially one complex 'hyper-neuron' or syncytium. (Ironically, before Ramon-y-Cajal showed that neurons were discrete cells, the prevalent model for brain structure was a continuous threaded-together syncytium as proposed by Camille Golgi.) Orch OR suggests that quantum states in microtubules in one neuron could extend by entanglement and tunneling through gap junctions to microtubules in adjacent neurons and glia (Figure 6), and from those cells to others, potentially in brain-wide syncytia.

Open gap junctions were thus predicted to play an essential role in the neural correlate of consciousness (Hameroff, 1998a). Beginning in 1998, evidence began to show that gamma synchrony, the best measureable correlate

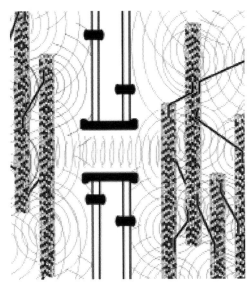

Figure 6. Portions of two neurons connected by a gap junction with microtubules (linked by microtubule-associated proteins, 'MAPs') computing via states (here represented as black or white) of tubulin protein subunits. Wavy lines suggest entanglement among quantum states (not shown) in microtubules.

of consciousness, depended on gap junctions, particularly dendritic-dendritic gap junctions (Dermietzel, 1998; Draguhn et al, 1998; Galaretta and Hestrin, 1999). To account for the distinction between conscious activities and non-conscious 'auto-pilot' activities, and the fact that consciousness can occur in various brain regions, Hameroff (2009) developed the "Conscious pilot' model in which syncytial zones of dendritic gamma synchrony move around the brain, regulated by gap junction openings and closings, in turn regulated by microtubules. The model suggests consciousness literally moves around the brain in a mobile synchronized zone, within which isolated, entangled microtubules carry out quantum computations and Orch OR. Taken together, Orch OR and the conscious pilot distinguish conscious from non-conscious functional processes in the brain.

Libet's backward time referral In the 1970s neurophysiologist Benjamin Libet performed experiments on patients having brain surgery while awake, i.e. under local anesthesia (Libet et al., 1979). Able to stimulate and record from conscious human brain, and gather patients' subjective reports with precise timing, Libet determined that conscious perception of a stimulus required up to 500 msec of brain activity post-stimulus, but that conscious awareness occurred at 30 msec post-stimulus, i.e. that subjective experience was referred 'backward in time'.

Bearing such apparent anomalies in mind, Penrose put forward a tentative suggestion, in The Emperor's New Mind, that effects like Libet's backward time referral might be related to the fact that quantum entanglements are not mediated in a normal causal way, so that it might be possible for conscious experience not to follow the normal rules of sequential time progression, so long as this does not lead to contradictions with external causality. In Section 5, it was pointed out that the (experimentally confirmed) phenomenon of 'quantum teleportation' (Bennett et al., 1993; Bouwmeester et al., 1997; Macikic et al., 2002) cannot be explained in terms of ordinary classical information processing, but as a combination of such classical causal influences and the acausal effects of quantum entanglement. It indeed turns out that quantum entanglement effects—referred to as 'quantum information' or 'quanglement' (Penrose 2002, 2004)—appear to have to be thought of as being able to propagate in either direction in time (into the past or into the future). Such effects, however, cannot by themselves be used to communicate ordinary information into the past. Nevertheless, in conjunction with normal classical future-propagating (i.e. 'causal') signalling, these quantum-teleportation influences can achieve certain kinds of 'signalling' that cannot be achieved simply by classical future-directed means.

The issue is a subtle one, but if conscious experience is indeed rooted in the OR process, where we take OR to relate the classical to the quantum world, then apparent anomalies in the sequential aspects of consciousness are perhaps to be expected. The Orch OR scheme allows conscious experience to be temporally non-local to a degree, where this temporal non-locality would spread to the kind of time scale τ that would be involved in the relevant Orch OR process, which might indeed allow this temporal non-locality to spread to a time $\tau=500$ms. When the 'moment' of an internal conscious experience is timed externally, it may well be found that this external timing does not precisely accord with a time progression that would seem to apply to internal conscious experience, owing to this temporal non-locality intrinsic to Orch OR.

Measurable brain activity correlated with a stimulus often occurs several hundred msec after that stimulus, as Libet showed. Yet in activities ranging from rapid conversation to competitive athletics, we respond to a stimulus (seemingly consciously) before the above activity that would be correlated with that stimulus occurring in the brain. This is interpreted in conventional neuroscience and philosophy (e.g. Dennett, 1991; Wegner, 2002) to imply that in such cases we respond non-consciously, on auto-pilot, and subsequently have only an illusion of conscious response. The mainstream view is that consciousness is epiphenomenal illusion, occurring after-the-fact as a false impression of conscious control of behavior. We are merely 'helpless spectators' (Huxley, 1986).

However, the effective quantum backward time referral inherent in the temporal non-locality resulting from the quanglement aspects of Orch OR, as suggested above, enables conscious experience actually to be temporally non-that are driving the physical (chemical, electronic) processes of relevance. In a clear sense E_G is, instead, an energy uncertainty—and it is this uncertainty that allows quantum state reduction to take place without violation of energy conservation. The fact that E_G is far smaller than the other energies involved in the relevant physical processes is a necessary feature of the consistency of the OR scheme. It does not supply the energy to drive the physical processes involved, but it provides the energy uncertainty that allows the freedom for processes having virtually the same energy as each other to be alternative actions. In practice, all that E_G is needed for is to tell us how to calculate the lifetime τ of the superposition. E_G would enter into issues of energy balance only if gravitational interactions between the parts of the system were important in the processes involved. (The Earth's gravitational field plays no role in this either, because it cancels out in the calculation of E_G.) No other forces of nature directly contribute to E_G, which is just as well, because if they did, there would be a gross discrepancy with observational physics.

Tegmark, 2000. Physicist Max Tegmark (2000) confronted Orch OR on the basis of decoherence. This was discussed at length in Section 7.

Koch and Hepp, 2006. In a challenge to Orch OR, neuroscientists/physicists Koch and Hepp published a thought experiment in Nature, describing a person observing a superposition of a cat both dead and alive with one eye, the other eye distracted by a series of images (binocular rivalry). They asked 'Where in the observer's brain would reduction occur?', apparently assuming Orch OR followed the Copenhagen interpretation in which conscious observation causes quantum state reduction. This is precisely the opposite of Orch OR in which consciousness is the orchestrated quantum state reduction given by OR.

Orch OR can account for the related issue of bistable perceptions (e.g. the famous face/vase illusion, or Necker cube). Non-conscious superpositions of both possibilities (face and vase) during pre-conscious quantum superposition then reduce by OR at time τ to conscious perception of one or the other, face or vase. The reduction would occur among microtubules within neurons interconnected by gap junctions in various areas of visual and pre-frontal cortex and other brain regions.

Reimers et al (2009) described three types of Fröhlich condensation (weak, strong and coherent, the first classical and the latter two quantum). They validated 8 MHz coherence measured in microtubules by Pokorny (2001; 2004) as weak condensation. Based on simulation of a 1-dimensional linear chain of tubulin dimers representing a microtubule, they concluded only weak Fröhlich condensation occurs in microtubules. Claiming Orch OR requires strong or coherent Fröhlich condensation, they concluded Orch OR is invalid. However Samsonovich et al (1992) simulated a microtubule as a 2-dimensional lattice plane with toroidal boundary conditions and found Fröhlich resonance maxima at discrete locations in super-lattice patterns on the simulated micro-local, thus providing a means to rescue consciousness from its unfortunate characterization as epiphenomenal illusion. Accordingy, Orch OR could well enable consciousness to have a causal efficacy, despite its apparently anomalous relation to a timing assigned to it in relation to an external clock, thereby allowing conscious action to provide a semblance of free will.

8. Orch OR Criticisms and Responses

Orch OR has been criticized repeatedly since its inception. Here we review and summarize major criticisms and responses.

Grush and Churchland, 1995. Philosophers Grush and Churchland (1995) took issue with the Gödel's theorem argument, as well as several biological factors. One objection involved the microtubule-disabling drug colchicine which treats diseases such as gout by immobilizing neutrophil cells which cause painful inflammation in joints. Neutrophil mobility requires cycles of microtubule assembly/disassembly, and colchicine prevents re-assembly, impairing neutrophil mobility and reducing inflammation. Grush and Churchland pointed out that patients given colchicine do not lose consciousness, concluding that microtubules cannot be essential for consciousness. Penrose and Hameroff (1995) responded point-by-point to every objection, e.g. explaining that colchicine does not cross the blood brain barrier, and so doesn't reach the brain. Colchicine infused directly into the brains of animals does cause severe cognitive impairment and apparent loss of consciousness (Bensimon and Chemat, 1991).

Tuszynski et al, (1998). Tuszynski et al (1998) questioned how extremely weak gravitational energy in Diósi-Penrose OR could influence tubulin protein states. In Hameroff and Penrose (1996a), the gravitational self-energy E_G for tubulin superposition was calculated for separation of tubulin from itself at the level of its atomic nuclei. Because the atomic (e.g. carbon) nucleus displacement is greater than its radius (the nuclei separate completely), the gravitational self-energy E_G is given by: $E_G = Gm^2/a_c$, where a_c is the carbon nucleus sphere radius equal to 2.5 Fermi distances, m is the mass of tubulin, and G is the gravitational constant. Brown and Tuszynski calculated E_G (using separation at the nanometer level of the entire tubulin protein), finding an appropriately small energy E of 10^{-27} electron volts (eV) per tubulin, infinitesimal compared with ambient energy kT of 10^{-4}eV. Correcting for the smaller superposition separation distance of 2.5 Fermi lengths in Orch OR gives a significantly larger, but still tiny 10^{-21}eV per tubulin. With 2×10^{10} tubulins per 25msec, the conscious Orch OR moment would be roughly 10^{-10}eV (10^{-29} joules), still insignificant compared to kT at 10^{-4}eV.

All this serves to illustrate the fact that the energy E_G does not actually play a role in physical processes as an energy, in competition with other energies

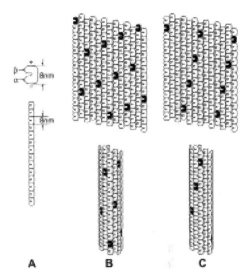

Figure 7. Simulating Fröhlich coherence in microtubules. A) Linear column of tubulins (protofilament) as simulated by Reimers et al (2010) which showed only weak Fröhlich condensation. B) and C) 2-dimensional tubulin sheets with toroidal boundary conditions (approximating 3-dimensional microtubule) simulated by Samsonovich et al (1992) shows long range Fröhlich resonance, with long-range symmetry, and nodes matching experimentally-observed MAP attachment patterns.

tubule surface which precisely matched experimentally observed functional attachment sites for microtubule-associated proteins (MAPs). Further, Bandyopadhyay (2011) has experimental evidence for strong Fröhlich coherence in microtubules at multiple resonant frequencies.

McKemmish et al (2010) challenged the Orch OR contention that tubulin switching is mediated by London forces, pointing out that mobile π electrons in a benzene ring (e.g. a phenyl ring without attachments) are completely delocalized, and hence cannot switch between states, nor exist in superposition of both states. Agreed. A single benzene cannot engage in switching. London forces occur between two or more electron cloud ring structures, or other non-polar groups. A single benzene ring cannot support London forces. It takes two (or more) to tango. Orch OR has always maintained two or more non-polar groups are necessary (Figure 8). McKemmish et al are clearly mistaken on this point.

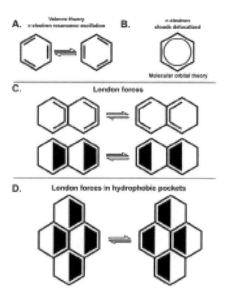

Figure 8. A) Phenyl ring/benzene of 6 carbons with three extra π electrons/double bonds which oscillate between two configurations according to valence theory. B) Phenyl ring/benzene according to molecular orbital theory in which π electrons/double bonds are delocalized, thus preventing oscillation between alternate states. No oscillation/switching can occur. C) Two adjacent phenyl rings/benzenes in which π electrons/double bonds are coupled, i.e. van der Waals London (dipole dispersion) forces. Two versions are shown: In top version, lines represent double bond locations; in bottom version, dipoles are filled in to show negative charge locations. D) Complex of 4 rings with London forces.

McKemmish et al further assert that tubulin switching in Orch OR requires significant conformational structural change (as indicated in Figure 2), and that the only mechanism for such conformational switching is due to GTP hydrolysis, i.e. conversion of guanosine triphophate (GTP) to guanosine di-phosphate (GDP) with release of phosphate group energy, and tubulin conformational flexing. McKemmish et al correctly point out that driving synchronized microtubule oscillations by hydrolysis of GTP to GDP and conformational changes would be prohibitive in terms of energy requirements and heat produced. This is agreed. However, we clarify that tubulin switching in Orch OR need not actually involve significant conformational change (e. g. as is illustrated in Figure 2), that electron cloud dipole states (London forces) are sufficient for bit-like switching, superposition and qubit function. We acknowledge tubulin conformational switching as discussed in early Orch OR publications and illustrations do indicate significant conformational changes. They are admittedly, though unintentionally, misleading.

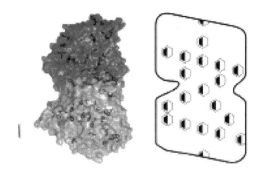

Figure 9. Left: Molecular simulation of tubulin with beta tubulin (dark gray) on top and alpha tubulin (light gray) on bottom. Non-polar amino acids phenylalanine and tryptophan with aromatic phenyl and indole rings are shown. (By Travis Craddock and Jack Tuszynski.) Right: Schematic tubulin with non-polar hydrophobic phenyl rings approximating actually phenyl and indole rings. Scale bar: 1 nanometer.

The only tubulin conformational factor in Orch OR is superposition separation involved in E_G, the gravitational self-energy of the tubulin qubit. As previously described, we calculated E_G for tubulin separated from itself at three possible levels: 1) the entire protein (e.g. partial separation, as suggested in Figure 2), 2) its atomic nuclei, and 3) its nucleons (protons and neutrons). The dominant effect is 2) separation at the level of atomic nuclei, e.g. 2.5 Fermi length for carbon nuclei (2.5 femtometers; 2.5×10^{-15} meters). This shift may be accounted for by London force dipoles with Mossbauer nuclear recoil and charge effects (Hameroff, 1998). Tubulin switching in Orch OR requires neither GTP hydrolysis nor significant conformational changes.

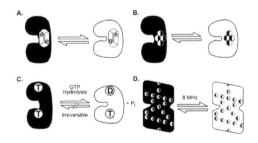

Figure 10. Four versions of the schematic Orch OR tubulin bit (superpositioned qubit states not shown). A) Early version showing conformational change coupled to/driven by single hydrophobic pocket with two aromatic rings. B) Updated version with single hydrophobic pocket composed of 4 aromatic rings. C) McKemmish et al (2009) mis-characterization of Orch OR tubulin bit as irreversible conformational change driven by GTP hydrolysis. D) Current version of Orch OR bit with no significant conformational change (change occurs at the level of atomic nuclei) and multiple hydrophobic pockets arranged in channels.

Schematic depiction of the tubulin bit, qubit and hydrophobic pockets in Orch OR has evolved over the years. An updated version is described in the next Section.

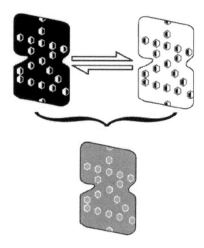

Figure 11. 2011 Orch OR tubulin qubit. Top: Alternate states of tubulin dimer (black and white) due to collective orientation of London force electron cloud dipoles in non-polar hydrophobic regions. There is no evident conformational change as suggested in previous versions; conformational change occurs at the level of atomic nuclei. Bottom: Depiction of tubulin (gray) superpositioned in both states.

9. Topological Quantum Computing in Orch OR

Quantum processes in Orch OR have consistently been ascribed to London forces in tubulin hydrophobic pockets, non-polar intra-protein regions, e.g. of π electron resonance rings of aromatic amino acids including tryptophan and phenylalanine. This assertion is based on (1) Fröhlich's suggestion that protein states are synchronized by electron cloud dipole oscillations in intra-protein non-polar regions, and (2) anesthetic gases selectively erasing consciousness by London forces in non-polar, hydrophobic regions in various neuronal proteins (e.g. tubulin, membrane proteins, etc.). London forces are weak, but numerous and able to act cooperatively to regulate protein states (Voet and Voet, 1995).

The structure of tubulin became known in 1998 (Nogales et al, 1998), allowing identification of non-polar amino acids and hydrophobic regions. Figure 9 shows locations of phenyl and indole π electron resonance rings of non-polar aromatic amino acids phenylalanine and tryptophan in tubulin. The ring locations are clustered along somewhat continuous pathways (within 2 nanometers) through tubulin. Thus, rather than hydrophobic pockets, tubulin may have within it quantum hydrophobic channels, or streams, linear arrays of electron resonance clouds suitable for cooperative, long-range quantum London forces. These quantum channels within each tubulin appear to align with those in adjacent tubulins in microtubule lattices, matching helical winding patterns (Figure 12). This in turn may support topological quantum computing in Orch OR.

Quantum bits, or qubits in quantum computers are generally envisioned as information bits in superposition of simultaneous alternative representations, e.g. both 1 and 0. Topological qubits are superpositions of alternative pathways, or channels which intersect repeatedly on a surface, forming 'braids'. Quasiparticles called anyons travel along such pathways, the intersections forming logic gates, with particular braids or pathways corresponding with particular information states, or bits. In superposition, anyons follow multiple braided pathways simultaneously, then reduce, or collapse to one particular pathway and functional output. Topological qubits are intrinsically resistant to decoherence.

An Orch OR qubit based on topological quantum computing specific to microtubule polymer geometry was suggested in Hameroff et al. (2002). Conductances along particular microtubule lattice geometry, e.g. Fibonacci helical pathways, were proposed to function as topological bits and qubits. Bandyopadhyay (2011) has preliminary evidence for ballistic conductance along different, discrete helical pathways in single microtubules

As an extension of Orch OR, we suggest topological qubits in microtubules based on quantum hydrophobic channels, e.g. continuous arrays of electron resonance rings within and among tubulins in microtubule lattices, e.g. following Fibonacci pathways. Cooperative London forces (electron cloud dipoles) in quantum hydrophobic channels may enable long-range coherence and topological quantum computing in microtubules necessary for optimal brain function and consciousness.

10. Conclusion: Consciousness in the Universe

Our criterion for proto-consciousness is OR . It would be unreasonable to refer to OR as the criterion for actual consciousness, because, according to the DP scheme, OR processes would be taking place all the time, and would be providing the effective randomness that is characteristic of quantum measurement. Quantum superpositions will continually be reaching the DP threshold for OR in non-biological settings as well as in biological ones, and usually take place in the purely random environment of a quantum system under measurement. Instead, our criterion for consciousness is Orch OR, conditions for which are fairly stringent: superposition must be isolated from the decoherence effects of the random environment for long enough to reach the DS

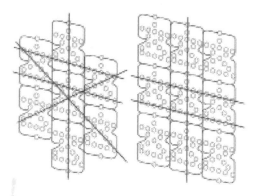

Figure 12. Left: Microtubule A-lattice configuration with lines connecting proposed hydrophobic channels of near-contiguous (<2 nanometer separation) electron resonance rings of phenylalanine and tryptophan. Right: Microtubule B-lattice with fewer such channels and lacking Fibonacci pathways. B-lattice microtubules have a vertical seam dislocation (not shown).

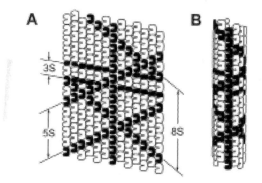

Figure 13. Extending microtubule A-lattice hydrophobic channels (Figure 12) results in helical winding patterns matching Fibonacci geometry. Bandyopadhyay (2011) has evidence for ballistic conductance and quantum inteference along such helical pathways which may be involved in topological quantum computing. Quantum electronic states of London forces in hydrophobic channels result in slight superposition separation of atomic nuclei, sufficient E_G for Orch OR. This image may be taken to represent superposition of four possible topological qubits which, after time T=tau, will undergo OR, and reduce to specific pathway(s) which then implement function.

threshold. Small superpositions are easier to isolate, but require longer reduction times τ. Large superpositions will reach threshold quickly, but are intrinsically more difficult to isolate. Nonetheless, we believe that there is evidence that such superpositions could occur within sufficiently large collections of microtubules in the brain for τ to be some fraction of a second.

Very large mass displacements can also occur in the universe in quantum-mechanical situations, for example in the cores of neutron stars. By OR, such superpositions would reduce extremely quickly, and classically unreasonable superpositions would be rapidly eliminated. Nevertheless, sentient creatures might have evolved in parts of the universe that would be highly alien to us. One possibility might be on neutron star surfaces, an idea that was developed ingeniously and in great detail by Robert Forward in two science-fiction stories (Dragon's Egg in 1980, Starquake in 1989). Such creatures (referred to as 'cheelas' in the books, with metabolic processes and OR-like events occurring at rates of around a million times that of a human being) could arguably have intense experiences, but whether or not this would be possible in detail is, at the moment, a very speculative matter. Nevertheless, the Orch OR proposal offers a possible route to rational argument, as to whether life of a totally alien kind such as this might be possible, or even probable, somewhere in the universe.

Such speculations also raise the issue of the 'anthropic principle', according to which it is sometimes argued that the particular dimensionless constants of Nature that we happen to find in our universe are 'fortuitously' favorable to human existence. (A dimensionless physical constant is a pure number, like the ratio of the electric to the gravitational force between the electron and the proton in a hydrogen atom, which in this case is a number of the general order of 10^{40}.) The key point is not so much to do with human existence, but the existence of sentient beings of any kind. Is there anything coincidental about the dimensionless physical constants being of such a nature that conscious life is possible at all? For example, if the mass of the neutron had been slightly less than that of the proton, rather than slightly larger, then neutrons rather than protons would have been stable, and this would be to the detriment of the whole subject of chemistry. These issues are frequently argued about (see Barrow and Tipler 1986), but the Orch OR proposal provides a little more substance to these arguments, since a proposal for the possibility of sentient life is, in principle, provided.

The recently proposed cosmological scheme of conformal cyclic cosmology (CCC) (Penrose 2010) also has some relevance to these issues. CCC posits that what we presently regard as the entire history of our universe, from its Big-Bang origin (but without inflation) to its indefinitely expanding future, is but one aeon in an unending succession of similar such aeons, where the infinite future of each matches to the big bang of the next via an infinite change of scale. A question arises whether the dimensionless constants of the aeon prior to ours, in the CCC scheme, are the same as those in our own aeon, and this relates to the question of whether sentient life could exist in that aeon as well as in our own. These questions are in principle answerable by observation, and again they would have a bearing on the extent or validity of the Orch OR proposal. If Orch OR turns out to be correct, in it essentials, as a physical basis for consciousness, then it opens up the possibility that many questions may become answerable, such as whether life could have come about in an aeon prior to our own, that would have previously seemed to be far beyond the reaches of science.

Moreover, Orch OR places the phenomenon of consciousness at a very central place in the physical nature of our universe, whether or not this 'universe' includes aeons other than just our own. It is our belief that, quite apart from detailed aspects of the physical mechanisms that are involved in the production of consciousness in human brains, quantum mechanics is an incomplete theory. Some completion is needed, and the DP proposal for an OR scheme underlying quantum theory's R-process would be a definite possibility. If such a scheme as this is indeed respected by Nature, then there is a fundamental additional ingredient to our presently understood laws of Nature which plays an important role at the Planck-scale level of space-time structure. The Orch OR proposal takes advantage of this, suggesting that conscious experience itself plays such a role in the operation of the laws of the universe.

Acknowledgment We thank Dave Cantrell, University of Arizona Biomedical Communications for artwork.

References

Atema, J. (1973). Microtubule theory of sensory transduction. Journal of Theoretical Biology, 38, 181-90.

Bandyopadhyay A (2011) Direct experimental evidence for quantum states in microtubules and topological invariance. Abstracts: Toward a Science of Consciousness 2011, Sockholm, Sweden, HYPERLINK "http://www.consciousness.arizona.edu"www.consciousness.arizona.edu

Barrow, J.D. and Tipler, F.J. (1986) The Anthropic Cosmological Principle (OUP, Oxford).

Bell, J.S. (1966) Speakable and Unspeakable in Quantum Mechanics (Cambridge Univ. Press, Cambridge; reprint 1987).

Benioff, P. (1982). Quantum mechanical Hamiltonian models of Turing Machines. Journal of Statistical Physics, 29, 515-46.

Bennett C.H., and Wiesner, S.J. (1992). Communication via 1- and 2-particle operators on Einstein-Podolsky-Rosen states. Physical Reviews Letters, 69, 2881-84.

Bensimon G, Chemat R (1991) Microtubule disruption and cognitive defects: effect of colchicine on teaming behavior in rats. Pharmacol. Biochem. Behavior 38:141-145.

Bohm, D. (1951) Quantum Theory (Prentice–Hall, Englewood-Cliffs.) Ch. 22, sect. 15-19. Reprinted as: The Paradox of Einstein, Rosen and Podolsky, in Quantum Theory and Measurement, eds., J.A. Wheeler and W.H. Zurek (Princeton University Press, Princeton, 1983).

Bernroider, G. and Roy, S. (2005) Quantum entanglement of K ions, multiple channel states and the role of noise in the brain. SPIE 5841-29:205–14.

Bouwmeester, D., Pan, J.W., Mattle, K., Eibl, M., Weinfurter, H. and Zeilinger, A. (1997) Experimental quantum teleportation. Nature 390 (6660): 575-579.

Brunden K.R., Yao Y., Potuzak J.S., Ferrer N.I., Ballatore C., James M.J., Hogan A.M., Trojanowski J.Q., Smith A.B. 3rd and Lee V.M. (2011) The characterization of microtubule-stabilizing drugs as possibletherapeutic agents for Alzheimer's disease and related taupathies. Pharmacological Research, 63(4), 341-51.

Chalmers, D. J., (1996). The conscious mind - In search of a fundamental theory. Oxford University Press, New York.

Crick, F., and Koch, C., (1990). Towards a neurobiological theory of consciousness. Seminars in the Neurosciences, 2, 263-75.

Dennett, D.C. (1991). Consciousness explained. Little Brown, Boston. MA.

Dennett, D.C. (1995) Darwin's dangerous idea: Evolution and the Meanings of Life, Simon and Schuster.

Dermietzel, R. (1998) Gap junction wiring: a 'new' principle in cell-to-cell communication in the nervous system? Brain Research Reviews. 26(2-3):176-83.

Deutsch, D. (1985) Quantum theory, the Church–Turing principle and the universal quantum computer, Proceedings of the Royal Society (London) A400, 97-117.

Diósi, L. (1987) A universal master equation for the gravitational violation of quantum mechanics, Physics Letters A 120 (8):377-381.

Diósi, L. (1989). Models for universal reduction of macroscopic quantum fluctuations Physical Review A, 40, 1165-74.

Draguhn A, Traub RD, Schmitz D, Jefferys (1998). Electrical coupling underlies high-frequency oscillations in the hippocampus in vitro. Nature, 394(6689), 189-92.

Eccles, J.C. (1992). Evolution of consciousness. Proceedings of the National Academy of Sciences, 89, 7320-24.

Engel GS, Calhoun TR, Read EL, Ahn T-K, Mancal T, Cheng Y-C, Blankenship RE, Fleming GR (2007) Evidence for wavelike energy transfer through quantum coherence in photosynthetic systems. Nature 446:782-786.

Everett, H. (1957). Relative state formulation of quantum mechanics. In Quantum Theory and Measurement, J.A. Wheeler and W.H. Zurek (eds.) Princeton University Press, 1983; originally in Reviews of Modern Physics, 29, 454-62.

Feynman, R.P. (1986). Quantum mechanical computers. Foundations of Physics, 16(6), 507-31.

Forward, R. (1980) Dragon's Egg. Ballentine Books.

Forward, R. (1989) Starquake. Ballentine Books.

Fröhlich, H. (1968). Long-range coherence and energy storage in biological systems. International Journal of Quantum Chemistry, 2, 641-9.

Fröhlich, H. (1970). Long range coherence and the actions of enzymes. Nature, 228, 1093.

Fröhlich, H. (1975). The extraordinary dielectric properties of biological materials and the action of enzymes. Proceedings of the National Academy of Sciences, 72, 4211-15.

Galarreta, M. and Hestrin, S. (1999). A network of fast-spiking cells in the neocortex connected by electrical synapses. Nature, 402, 72-75.

Gauger E., Rieper E., Morton J.J.L., Benjamin S.C., Vedral V. (2011) Sustained quantum coherence and entanglement in the avian compass http://arxiv.org/abs/0906.3725.

Ghirardi, G.C., Rimini, A., and Weber, T. (1986). Unified dynamics for microscopic and macroscopic systems. Physical Review D, 34, 470.

Ghirardi, G.C., Grassi, R., and Rimini, A. (1990). Continuous-spontaneous reduction model involving gravity. Physical Review A, 42, 1057-64.

Grush R., Churchland P.S. (1995), 'Gaps in Penrose's toilings', J. Consciousness Studies, 2 (1):10-29.

Hagan S, Hameroff S, and Tuszynski J, (2001). Quantum Computation in Brain Microtubules? Decoherence and Biological Feasibility, Physical Review E, 65, 061901.

Hameroff, S.R., and Watt R.C. (1982). Information processing in microtubules. Journal of Theoretical Biology, 98, 549-61.

Hameroff, S.R.(1987) Ultimate computing: Biomolecular consciousness and nanotechnology. Elsevier North-Holland, Amsterdam.

Hameroff, S.R., and Penrose, R., (1996a). Orchestrated reduction of quantum coherence in brain microtubules: A model for consciousness. In: Toward a Science of Consciousness ; The First Tucson Discussions and Debates. Hameroff, S.R., Kaszniak, and Scott, A.C., eds., 507-540, MIT Press, Cambridge MA, 507-540. Also published in Mathematics and Computers in Simulation (1996) 40:453-480.

Hameroff, S.R., and Penrose, R. (1996b). Conscious events as orchestrated spacetime selections. Journal of Consciousness Studies, 3(1), 36-53.

Hameroff, S. (1998a). Quantum computation in brain microtubules? The Penrose-Hameroff "Orch OR" model of consciousness. Philosophical Transactions of the Royal Society (London) Series A, 356, 1869-1896.

Hameroff, S. (1998b). 'Funda-mentality': is the conscious mind subtly linked to a basic level of the universe? Trends in Cognitive Science, 2, 119-127.

Hameroff, S. (1998c). Anesthesia, consciousness and hydrophobic pockets – A unitary quantum hypothesis of anesthetic action. Toxicology Letters, 100, 101, 31-39.

Hameroff, S. (1998d). HYPERLINK "http://www.hameroff.com/penrose-hameroff/cambrian.html"Did consciousness cause the Cambrian evolutionary explosion? In: Toward a Science of Consciousness II: The Second Tucson Discussions and Debates. Eds. Hameroff, S.R., Kaszniak, A.W., and Scott, A.C., MIT Press, Cambridge, MA.

Hameroff, S., Nip, A., Porter, M., and Tuszynski, J. (2002). Conduction pathways in microtubules, biological quantum computation and microtubules. Biosystems, 64(13), 149-68.

Hameroff S.R., & Watt R.C. (1982) Information processing in microtubules. Journal of Theoretical Biology 98:549-61.

Hameroff, S.R. (2006) The entwined mysteries of anesthesia and consciousness. Anesthesiology 105:400-412.

Hameroff, S.R, Craddock TJ, Tuszynski JA (2010) Memory 'bytes' – Molecular match for CaMKII phosphorylation encoding of microtubule lattices. Journal of Integrative Neuroscience 9(3):253-267.

He, R-H., Hashimoto, M., Karapetyan. H., Koralek, J.D., Hinton, J.P., Testaud, J.P., Nathan, V., Yoshida, Y., Yao, H., Tanaka, K., Meevasana, W., Moore, R.G., Lu, D.H.,Mo, S-K., Ishikado, M., Eisaki, H., Hussain, Z., Devereaux, T.P., Kivelson, S.A., Orenstein, Kapitulnik, J.A., Shen, Z-X. (2011) From a Single-Band Metal to a High Temperature Superconductor via Two Thermal Phase Transitions. Science, 2011;331 (6024): 1579-1583.

Hebb, D.O. (1949). Organization of Behavior: A Neuropsychological Theory, John Wiley and Sons, New York.

Huxley TH (1893; 1986) Method and Results: Essays.

Kant I (1781) Critique of Pure Reason (Translated and edited by Paul Guyer and Allen W. Wood, Cambridge University Press, 1998).

Kibble, T.W.B. (1981). Is a semi-classical theory of gravity viable? In Quantum Gravity 2: a Second Oxford Symposium; eds. C.J. Isham, R. Penrose, and D.W. Sciama (Oxford University Press, Oxford), 63-80.

Koch, C., (2004) The Quest for Consciousness: A Neurobiological Approach, Englewood, CO., Roberts and Co.

Koch C, Hepp K (2006) Quantm mechanics in the brain. Nature 440(7084):611.

Libet, B., Wright, E.W. Jr., Feinstein, B., & Pearl, D.K. (1979) Subjective referral of the timing for a conscious sensory experience. Brain 102:193-224.

Luo L, Lu J (2011) Temperature dependence of protein folding deduced from quantum transition. http://arxiv.org/abs/1102.3748

Lutz A, Greischar AL, Rawlings NB, Ricard M, Davidson RJ (2004) Long-term meditators self-induce high-amplitude gamma synchrony during mental practice The Proceedings of the National Academy of Sciences USA 101(46)16369-16373.

Macikic I., de Riedmatten H., Tittel W., Zbinden H. and Gisin N. (2002) Long-distance teleportation of qubits at telecommunication wavelengths Nature 421, 509-513.

Margulis, L. and Sagan, D. 1995. What is life? Simon and Schuster, N.Y.

Marshall, W, Simon, C., Penrose, R., and Bouwmeester, D (2003). Towards quantum superpositions of a mirror. Physical Review Letters 91, 13-16; 130401.

McKemmish LK, Reimers JR, McKenzie RH, Mark AE, Hush NS (2009) Penrose-Hameroff orchestrated objective-reduction proposal for human consciousness is not biologically feasible. Physical Review E. 80(2 Pt 1):021912.

Moroz, I.M., Penrose, R., and Tod, K.P. (1998) Spherically-symmetric solutions of the Schrödinger–Newton equations:. Classical and Quantum Gravity, 15, 2733-42.

Nogales E, Wolf SG, Downing KH. (1998) HYPERLINK "http://dx.doi.org/10.1038/34465"Structure of the αβ-tubulin dimer by electron crystallography. Nature. 391, 199-203.

Ouyang, M., & Awschalom, D.D. (2003) Coherent spin transfer between molecularly bridged quantum dots. Science 301:1074-78.

Pearle, P. (1989). Combining stochastic dynamical state-vector reduction with spontaneous localization. Physical Review A, 39, 2277-89.

Pearle, P. and Squires, E.J. (1994). Bound-state excitation, nucleon decay experiments and models of wave-function collapse. Physical Review Letters, 73(1), 1-5.

Penrose, R. (1989). The Emperor's New Mind: Concerning Computers, Minds, and the Laws of Physics, Oxford University Press, Oxford.

Penrose, R. (1993). Gravity and quantum mechanics. In General Relativity and Gravitation 13. Part 1: Plenary Lectures 1992. Proceedings of the Thirteenth International Conference on General Relativity and Gravitation held at Cordoba, Argentina, 28 June - 4 July 1992. Eds. R.J.Gleiser, C.N.Kozameh, and O.M.Moreschi (Inst. of Phys. Publ. Bristol and Philadelphia), 179-89.

Penrose, R. (1994). Shadows of the Mind; An Approach to the Missing Science of Consciousness. Oxford University Press, Oxford.

Penrose, R. (1996). On gravity's role in quantum state reduction. General Relativity and Gravitation, 28, 581-600.

Penrose, R. (2000). Wavefunction collapse as a real gravitational effect. In Mathematical Physics 2000, Eds. A.Fokas, T.W.B.Kibble, A.Grigouriou, and B.Zegarlinski. Imperial College Press, London, 266-282.

Penrose, R. (2002). John Bell, State Reduction, and Quanglement. In Quantum Unspeakables: From Bell to Quantum Information, Eds. Reinhold A. Bertlmann and Anton Zeilinger , Springer-Verlag, Berlin, 319-331.

Penrose, R. (2004). The Road to Reality: A Complete Guide to the Laws of the Universe. Jonathan Cape, London.

Penrose, R. (2009). Black holes, quantum theory and cosmology (Fourth International Workshop DICE 2008), Journal of Physics, Conference Series 174, 012001.

Penrose, R. (2010). Cycles of Time: An Extraordinary New View of the Universe. Bodley Head, London.

Penrose R. and Hameroff S.R. (1995) What gaps? Reply to Grush and Churchland. Journal of Consciousness Studies.2:98-112.

Percival, I.C. (1994) Primary state diffusion. Proceedings of the Royal Society (London) A, 447, 189-209.

Pokorný, J., Hasek, J., Jelínek, F., Saroch, J. & Palan, B. (2001) Electromagnetic activity of yeast cells in the M phase. Electro Magnetobiol 20, 371–396.

Pokorný, J. (2004) Excitation of vibration in microtubules in living cells. Bioelectrochem. 63: 321-326.

Polkinghorne, J. (2002) Quantum Theory, A Very Short Introduction. Oxford University Press, Oxford.

Rae, A.I.M. (1994) Quantum Mechanics. Institute of Physics Publishing; 4th edition 2002.

Rasmussen, S., Karampurwala, H., Vaidyanath, R., Jensen, K.S., and Hameroff, S. (1990) Computational connectionism within neurons: A model of cytoskeletal automata subserving neural networks. Physica D 42:428-49.

Reimers JR, McKemmish LK, McKenzie RH, Mark AE, Hush NS (2009) Weak, strong, and coherent regimes of Frohlich condensation and their applications to terahertz medicine and quantum consciousness Proceedings of the National Academy of Sciences USA 106(11):4219-24

Reiter GF, Kolesnikov AI, Paddison SJ, Platzman PM, Moravsky AP, Adams MA, Mayers J (2011) Evidence of a new quantum state of nano-confined water http://arxiv.org/abs/1101.4994

Rieper E, Anders J, Vedral V (2011) Quantum entanglement between the electron clouds of nucleic acids in DNA. http://arxiv.org/abs/1006.4053.

Samsonovich A, Scott A, Hameroff S (1992) Acousto-conformational transitions in cytoskeletal microtubules: Implications for intracellular information processing. Nanobiology 1:457-468.

Sherrington, C.S. (1957) Man on His Nature, Second Edition, Cambridge University Press.

Smolin, L. (2002). Three Roads to Quantum Gravity. Basic Books. New York.

Tegmark, M. (2000) The importance of quantum decoherence in brain processes. Physica Rev E 61:4194-4206.

Tittel, W, Brendel, J., Gisin, B., Herzog, T., Zbinden, H., and Gisin, N. (1998) Experimental demonstration of quantum correlations over more than 10 km, Physical Reiew A, 57:3229-32.

Tononi G (2004) An information integration theory of consciousness BMC Neuroscience 5:42.

Turin L (1996) A spectroscopic mechanism for primary olfactory reception Chem Senses 21(6) 773-91.

Tuszynski JA, Brown JA, Hawrylak P, Marcer P (1998) Dielectric polarization, electrical conduction, information processing and quantum computation in microtubules. Are they plausible? Phil Trans Royal Society A 356:1897-1926.

Tuszynski, J.A., Hameroff, S., Sataric, M.V., Trpisova, B., & Nip, M.L.A. (1995) Ferroelectric behavior in microtubule dipole lattices; implications for information processing, signaling and assembly/disassembly. Journal of Theoretical Biology 174:371–80.

Voet, D., Voet, J.G. 1995. Biochemistry, 2nd edition. Wiley, New York.

von Rospatt, A., (1995) The Buddhist Doctrine of Momentariness: A survey of the origins and early phase of this doctrine up to Vasubandhu (Stuttgart: Franz Steiner Verlag).

Wegner, D.M. (2002) The illusion of conscious will Cambridge MA, MIT Press.

Whitehead, A.N., (1929) Process and Reality. New York, Macmillan.

Whitehead, A.N. (1933) Adventure of Ideas, London, Macmillan.

Wigner E.P. (1961). Remarks on the mind-body question, in The Scientist Speculates, ed. I.J. Good (Heinemann, London). In Quantum Theory and Measurement, eds., J.A. Wheeler and W.H. Zurek, Princeton Univsity Press, Princeton, MA. (Reprinted in E. Wigner (1967), Symmetries and Reflections, Indiana University Press, Bloomington).

Wolfram, S. (2002) A New Kind of Science. Wolfram Media incorporated.

2. Cosmological Foundations of Consciousness
Chris King, Ph.D.

Emeritus, Mathematics Department,
University of Auckland, New Zealand

Abstract

This paper explores the cosmological foundations of subjective consciousness in the biological brain, from cosmic-symmetry-breaking, through biogenesis, evolutionary diversification and the emergence of metazoa, to humans, presenting a new evolutionary perspective on the potentialities of quantum interactions in consciousness, and the ultimate relationship of consciousness with cosmology.

KEY WORDS: cosmology, consciousness, evolution, neurodynamics, quantum theory, chaos

1: Introduction: Scope and Design

This overview explores the cosmological foundations of consciousness as evidenced in current research and uses this evidence to present a radical view of what subjective consciousness is, how it evolved, and how it might be supported through quantum processes in the biological brain.

To do full justice to this very broad topic within the confines of the special issue and its planned book edition, I have prepared this paper as a short review article, referring to the full research monograph (King 2011b), as supporting online material, containing all the detailed references, a more complete explanation of the ideas and the ongoing state of the research in the diverse areas covered.

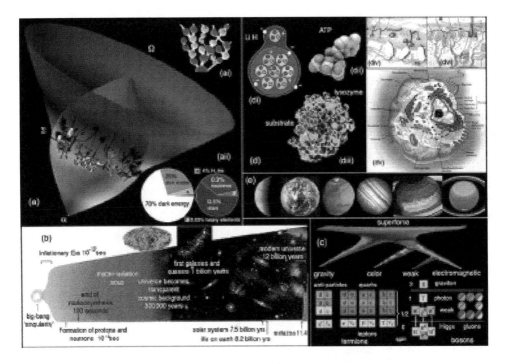

Fig 1: Cosmic symmetry-breaking and its interactive fractal and chaotic effects leading to biogenesis. (a) Life is the consummation of interactive complexity (Σ) resulting from symmetry-breaking of the fundamental force of nature in the big-bang (α), whatever ultimate fate is in store (&Omega). Inset (i) fractal inflation model, (ii) the distribution of dark energy and matter and the matter of stars and planets. (b) Logarithmic time scale of cosmological events showing life on earth existing for a third of the universe's lifetime. (c) Symmetry-breaking of the forces of nature results in the color and weak forces, generating 100 atomic nuclei, while gravity and electromagnetism govern long-range structure determining biogenesis, from fractal chemical bonding, to solar systems capable of photosynthetic life in the goldilocks zone of liquid water. (d) Interactive effects of cosmic symmetry-breaking lead to hierarchical interaction of the forces, generating hadrons, atomic nuclei and molecules (i). Non-linear energetics of chemical bonding lead to a cascade of cooperative weak-bonding effects, which generate fractal molecular complexity, from the molecular orbitals of simple molecules (ii), through the 3D structures of complex proteins and nucleic acids (iii) to supra-molecular cell organelles (iv), cells (v), and tissues (vi) and organisms. (e) Chaotic effects of gravity as a non-linear force, results in extreme planetary variation, generating a diversity of potential conditions for biogenesis, similar to dynamic variations surrounding the Mandelbrot set.

2: Non-linear Quantum and Cosmological Foundations of Biogenesis

While it is well understood that the fundamental forces of nature appear to have differentiated from a super-force in a founding phase of cosmic inflation, the interactive implications of cosmic symmetry-breaking for the chemical basis of life and its evolution into complex sentient organisms are equally as striking, and central to our existence. Cosmic symmetry-breaking and the ensuing preponderance of matter over anti-matter results in the hierarchi cal arrangement of quarks into neutrons and positively charged protons and then the 100 or so stable atomic nuclei, through the interaction of the strong and weak nuclear forces with electromagnetic charge, providing a rich array of stable, electromagnetically polarized, atoms with graduated energetics.

The non-linear molecular orbital charge energetics that results in strong covalent and ionic bonds also leads to a cascade of successively weaker bonding effects from H-bonds, to van-der-Waal's interactions, whose globally cooperative nature is responsible for the primary, secondary, and tertiary structures of proteins and nucl eic acids, and in a fractal manner to quaternary supra-molecul ar assemblies, cell organelles, cells, tissues and organisms. Thus, although genetic coding is a necessary condition for the development of cell organelles and organismic tissues, this is possible only because the symmetry-broken laws of nature can give rise to such dynamical structures. In this sense, tissue, culminating in the sentient brain, is the natural interactive full-complexity product of cosmic symmetry-breaking. Despite the periodi c quantum properties of the s, p, d and f-orbitals, which form the basis of the table of the elements, successive rows have non-periodic trends because of non-linear charge interactions, which result in a symmetrybreaking determining the bio-elements pivotal to biogenesis. Life as we know it is based on the strong covalent bonding of first row elements C, N and O in relation to H, stemming from the optimally strong multiple -CN, -CC-, and >CO bonds, which are cosmically abundant in forming star systems and readily undergo polymerization to heterocyclic molecules, including the nucleic acid bases A, U, G, C and a variety of amino acids, as well as optically active cofactors such as porphyrins.

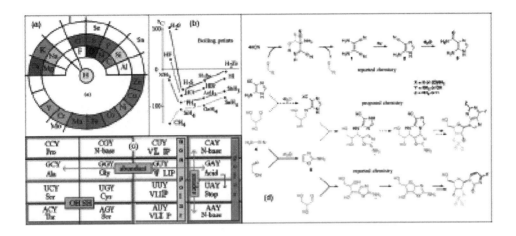

Fig2: (a) Symmetry-breaking quasi-periodic table of the bioelements displays covalent optimality. (b) Optimality of H^2O in terms of internal weak-bonding expressed in its high boiling point. (c) Evidence for a symmetry-breaking origin of the genetic code. (d) Realized and proposed direct synthesis paths from primordial precursors such as HCN to nucleotides (Powner et. al. (2010).

This interactive symmetry-breaking continues in a cascade. As we trend from C > N > O the electronegativity increases from non-polar C-H, to highly electronegative O, resulting in H_2O having extreme optimal properties as a polar hydride, bifurcating molecular dynamics into polar and non-polar phases, in addition to pH, and H-bonding effects, which define the aqueous structures and dynamics of proteins, nucleic acids, lipid membranes, ion and electron transport. Following on are secondary properties of S in lower energy -SH and -SS- bonds and the role of P as oligomeric phosphates in the energetics of biogenesis, cellular metabolism, dehydration polymerization and the nucleic acid backbone. We then have bifurcations of ionic properties K_+/Na_+ and Ca_{++}/Mg_{++} and finally the catalytic roles of transition elements as trace ingredients.

This does not imply that this is the only elemental arrangement possible for life, as organisms claimed to be adapted to use arsenic in the place of phosphorus (Wolfe-Simon et. al. 2010) suggest, but it does confirm that life as we know it has optimal symmetry-breaking properties cosmologically. Many of the fundamental molecules associ ated with membrane excitation, including lipids such as phosphatidyl choline and amine-based neurotransmitters, also have potentially primordial status (King 1996). Effects of symmetry-breaking may also extend to the genetic code (King 1982). Recent research has begun to elucidate a plausible 'one-pot' rout e (Powner et. al. 2010) from simple cosmically abundant molecules such as HCN and HCHO to the nucleotide units making up RNA, giving our genetic origin a potentially cosmological status. There have also been advances with inducing selected RNAs to self-assemble from precursors and assume catalytic functions (see King 2011b).

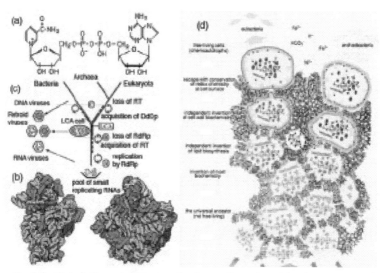

Fig 3: (a) Catalytic nicotine-adenine dinucleotide is essential in respiration. (b) Large and small subunits of the ribosome are centrally and functionally RNA [pink] (c) Molecular fossil evidence for a viral-based cellular transition from the RNA world to DNA based chromosomes, through cellular cooption of viral RNA-directed RNA-polymerase, followed by reverse transcriptase and finally DNA-dependent DNA-polymerase. (d) Independent evolution of archaean and bacterial cellular life from a non-cellular form of life at the interface of olivine and acid, iron-rich sea water forming 'lost city' undersea vents able to solve the concentration and encapsulation problems (Martin and Russell 2003).

3: Emergence of the Excitable Cell: From Universal Common Ancestor to Eucaryotes

Looking back at the universal common ancestor of life, likewise indicates a transition through an era in which RNA functioned as both catalyst and replicator, through the establishment of the genetic code, whose ribosomal protein translation units are still RNA-based, to the eventual emergence of DNA-based life, probably through viral genes (King 2011a). However the genetic picture of cell wall proteins is consistent with independent cellular origins of bacteria and archaea, implying more than one evolution of cellular life from a protected environment conducive to naked nucleotide replication (Martin and Russell 2003; Russell 2011).

Nevertheless, once the branches of cellular life evolved, excitability based on ion channels and pumps rapidly became universal. It has been reported that as early as 3.3 billion years ago there was a massive genetic expansion, which may have contributed to the genes common to all forms of life (David and Alm 2010) facilitated by high levels of horizontal gene transfer, promoted by viruses (Dagan et. al. 2006; Wickramasinghe 2011).

Estimates of the adaptive computational power of the collective bacterial and archaean genome (King 2011a) give a presentation rate of new combinations of up to 10^{30} bits per second, compared with the current fastest computer at about 10^{17} bit ops per second. Corresponding rates for complex life forms are much lower, around 10^{17} per second, because they are fewer in total number and have lower reproduction rates and longer generation times. This picture of bit rates coincides closely with the Archaean expansion scenario and suggests that evolution has been a two-phase process of genetic algorithm super-computation, which arrived at a global solution to the notoriously intractable protein-folding problems of the central metabolic and electro-chemical pathways, which are later capitalized on by eukaryotes and metazoa.

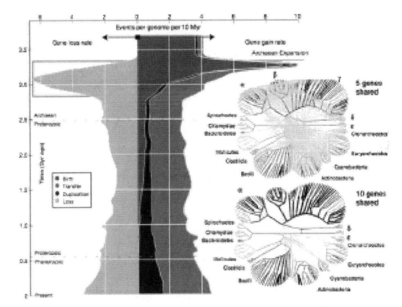

Fig 4: (Left) Archaean genetic expansion around 3.3 billion years ago generated most critical genes common to life (David and Alm 2010) (Right) Evidence of ubiquitous horizontal transfer of genes between bacterial species at different trigger levels (Dagan et. al. 2006).

Horizontal gene transfer, endosymbiosis and gene fusion may have led to a situation where sexuality and excitability, along with all the critical components for neural dynamics including ion-channels specific for Ca^{++}, K^+ and Na^+, G-protein linked receptors, microtubules, and fast action potential became common among eukaryotes (Wickramasinghe 2011). Ion channel structure appears to have been established during the soup of lateral gene transfers that drove the evolution of eukaryotes. This means we should find neurotransmitter receptors from GABA a, b, and glutamate, through opioid, to dopamine, epinephrine, serotonin and melatonin in all multi-cellular eukaryotes. This universality would have continued up the evolutionary tree, implying that the very different nervous system designs of arthropods and vertebrates mask a deeper common neurodynamic and genetic basis.

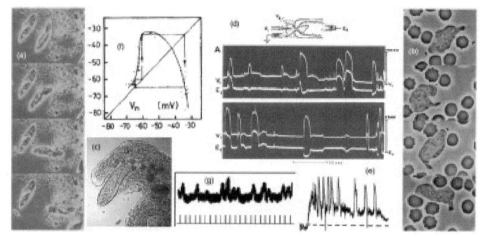

Fig 5: Real-time purposive behavior in single cells (a) Paramecium reverses, turns right and explores a cul-de-sac. (b) Human neutrophil chases an escaping bacterium (black), before engulfing it. (c) Chaos chaos engulfs a paramecium. Action potentials in Chaos chaos (d) and paramecium (e). Period 3 perturbed excitations in alga Nitella indicate chaos. (g) Frog retinal rod cells are sensitive to single quanta in an ultra-low intensity beam.

The evolutionary key to sentient consciousness may lie in the survival advantage it could provide in anticipating threats and strategic opportunities. Since key genes for the brain evolved even before the Cambrian radiation (Wickramasinghe 2011), the key to the emergence of conscious sentience may be sourced in the evolution of excitable single cells. Chaotic excitation provides a eukaryote cell with a generalized quantum sense organ. Sensitive dependence would give a cell feedback about its external environment, perturbed by a variety of quantum modes - chemically through molecular orbital interaction, electromagnetically through photon emission and absorption, electrochemically through the perturbations of the fluctuating fields generated by the excitations, and through acoustic, mechanical and osmotic interaction.

As we move to founding metazoa, we find Hydra, which supports only a primitive diffuse neural net, in continuous transformation and reconstruction, has a rich repertoire of up to 12 forms of 'intuitive' locomotion from snail-like sliding to somersaulting (King 2008), as well as coordinated tentacle movements. This is consistent with much of the adaptive capacity of nervous systems arising from cellular complexity, rather than neural net design alone. Pyramidal neurons for example engage up to 10^4 synaptic junctions, having a diversity of excitatory and inhibitory synaptic inputs involving up to five types of neurotransmitter, with differing effects depending on receptor types, and their location on dendrites, cell body, or axons.

In the complex central nervous systems of vertebrates, we see the same dynamical features, now expressed in whole system excitations, such as the EEG, in which interacting excitatory and inhibitory neurons provide a basis for broadspectrum oscillation, phase coherence and chaos in the global dynamics, with the synaptic organization enabling the dynamics to resolve complex context-sensitive decision-making problems. Nevertheless the immediate decisionmaking situations around which life or death results, in the theatre of conscious attention are qualitatively similar to those made by single celled organisms, based strongly on sensory input, and short term anticipation of immediate existential threats and opportunities, in a context of remembered situations that bear upon the current experience.

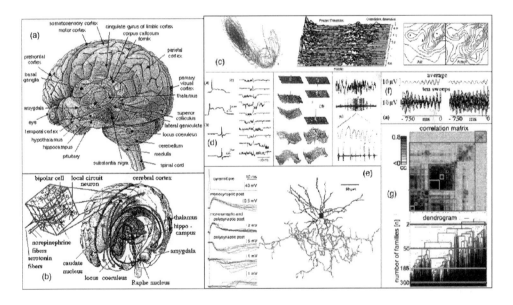

Fig 6a,b,c,d,e,f: Structural overview of the brain as a dynamical organ. (a) Major anatomical features including the cerebral cortex, its underlying driving centres in the thalamus, and surrounding limbic regions involving emotion and memory, including the cingulate cortex, hippocampus and amygdala. (b) Conscious activity of the cortex is maintained through the activity of ascending pathways from the thalamus and brain stem, including the reticular activating system and serotonin and nor-adrenaline pathways involved in light and dreaming sleep. Processes which enable global dynamics to be affected by small perturbations. (c) Evidence for dynamical chaos includes modulated strange attractors (Freeman 1991), and broad spectrum excitations with moderate fractal (correlation) dimensions (Basar et. al. 1989). These dynamics are complemented by holographic processing across the cortex illustrated in an experimental representation of olfactory excitations corresponding to recognized odors (Skarda and Freeman 1987). (d) Stochastic resonance enables fractal instabilities to grow from ion channel to neuron to hippocampal excitation (Liljenström and Uno 2005). (e) Chandelier cells can facilitate an spreading of excitation to many pyramidal cells (Molnar et. al. 2008, Woodruff and Yuste 2008). (f) Wave front coherence in processing becomes manifest when a cue is recognized by the subject (left) (g) Correlation matrix and dendrogram of cortical slice is consistent with fractal self-organized criticality (Beggs and Plenz 2004).

4: A Dynamical View of the Conscious Brain

Although long distance axons involve pulse coded action potentials, the brain appears to utilize dynamic processing involving broad-spectrum oscillations, rather than discrete signals. Unlike the digital computer, the human brain is a massively parallel organ with perhaps the order of 10 synapses between input and output, despite having an estimated 10^{10} neurons and 10^{14} synapses. Such design is essential to enable quick reactions to complex stimuli in real time and avoid the intractability problem of serial computers, which neural nets and genetic algorithms do solve effectively.

The cerebral cortex consists of six layers of cells organized in a sheet of functional columns about 1mm square. These have a fractal modular architecture, with each column representing one aspect of experience, from primary processing of lines at given angles, color, motion and auditory tones, through to cells recognizing individual faces. Major areas of the cortex also follow a modular pattern centered on the primary senses and our coordinated motor responses to our ongoing situation. Frontal areas are involved in abstractions of motor events, strategic planning and execution, parietal areas between touch and visual cortices are involved in spatial abstractions, with the temporal lobes extending laterally beyond visual and auditory areas representing attributes with specific meaning, such as specific faces and complex melodies, semantic and symbolic process, such as language, and the temporal relationships between experiences. This is consistent with a 'holographic' model – each experience being represented collectively, like a Fourier transform, in terms of its attributes – consistently with the many-to-many connections neurons provide.

No single cortical area has been identified as the seat of consciousness (Joseph 2009). One proposal (Ananthaswamy 2009, 2010) is that conscious processes correspond to the coordinated activity of the whole brain engaging active communication in 'working memory' between the frontal cortex and major sensory and association areas, while activity confined to regional processing is subconscious. This tallies with Bernard Baars' (1997) model of the Cartesian theatre of consciousness as 'global workspace'.

While major input and output pathways pass through thalamic nuclei underlying the cortex, two other systems modulate the dynamics of brain activity. The cortex is energized by ascending pathways from the brain stem, involving the reticular activating system, and dopamine, nor-adrenalin and serotonin pathways, fanning out across wide areas of the cortex, modulating active wakefulness, dreaming and sleep. Our emotional experiences are modulated through the limbic system, a lateral circuit, passing through the hypothalamus regulating internal and hormonal processes, the cingulate cortex dealing with emotional representations, and the hippocampus and amygdala, setting down sequential memories and dealing with flight and fight survival.

There is also evidence active conscious processing corresponds to (30-80 Hz) EEG oscillations in the gamma band, driven by mutual feedback between excitatory and inhibitory neurons in the cortex, and that phase coherence distinguishes 'in-synch' neuronal assemblies forming conscious thought process from peripheral pre-processing (Basar et. al. 1989, Crick & Koch 1992).

While the brain may be 'holographic' spatially, it appears to use phases of dynamical chaos in the time domain. Modulated transitions at the edge of chaos can explain many phenomena, from perception to insight learning in a 'eureka' brain wave. In olfactory perception, the brain appears to enter high energy chaos, which frees the dynamic from getting inappropriately locked-in, as annealing does in formal networks, fully-exploring dynamical space, followed by a reduction of energy, causing the dynamic to fall, either into a recognized state, represented by a strange attractor, or to form a new attractor through an adaptive change in the potential energy landscape, through learning (Skarda and Freeman 1987). The same idea fits with the 'eureka' of insight, where an unstable dynamic generated by the problem is resolved in a single bifurcation from chaotic instability into lucidity.

Non-linear mode-locking, common to oscillating chaotic systems, has the potential to facilitate the coherent excitations that characterize coupled neurosystems, going a good way towards resolving the 'binding' problem – how the brain 'brings it all back together'. By modulating the coupling between oscillating neurosystems, mode-locking could selectively bring related systems into phase coherence, just as the heartbeat is mode-locked to its local and brain pacemakers.

Chaos also makes the brain state arbitrarily sensitive to small perturbations, which is essential for a dynamical brain to be sensitive to small changes in its environment, and to its local instabilities. If the global state is critically poised at a tipping point, an unstable chaotic dynamic could become sensitive to perturbations at the level of the cell, synapse, or ion channel. There are several additional ways in which such sensitivity could come about. Stochastic resonance has been demonstrated to facilitate sensitivity, from ion channel, to cell, to global dynamic (Liljenström and Uno 2005). Fractal self-organized criticality has been found in cortical slices (Beggs and Plenz 2004). Chandelier cells have been shown to facilitate lateral spreading of local excitations to multiple pyramidal cells (Molnar et. al. 2008, Woodruff and Yuste 2008).

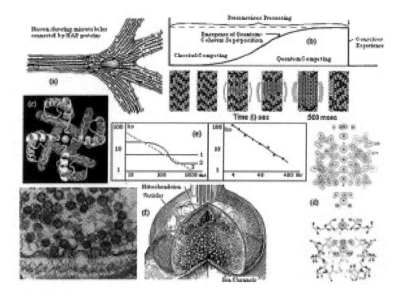

Fig 7: Features of quantum processing in proposed models. (a) Microtubule MAP proteins as envisaged in the OOR model (Hameroff and Penrose 2003). (b) The ensuing relationship between classical and quantum computing and consciousness. (c, d) gated K+ ion channels from MacKinnon's group (Zhou et. al. 2001). (e) Fractal kinetics in the channels (Liebovitch et. al.) (f) Synaptic junction may invoke uncertainty of position of the vesicle.

5: Quantum Dynamics and Conscious Anticipation

The two key questions confounding science about the brain are (1) how and why brain function generates subjective experience, and (2) whether there is any basis for our subjective conscious intentions having physical consequences in 'free-will' (Joseph 2009, 2011).

We thus explore how central to neurodynamic processes might exploit quantum effects to enhance survival prospects of the organism. To develop a realistic quantum theory of consciousness, we have to consider how whole brain states might become capable of quantum interaction (Joseph 2009) and how this could arise from neurophysiological processes common to excitable cells.

We have seen that various forms of global instability, from chaos, through tipping points to self-organized criticality could make the global brain state ultimately sensitive to change at the cellular, molecular or quantum level. Ion channels, such as for acetyl-choline display non-linear (quadratic) concentration dynamics, being excited by two molecules. Many aspects of synaptic release are also highly non-linear, due to biochemical feedback loops. A single vesicle excites up to 2000 ion channels, providing extreme amplification of a potentially quantum event. In addition to being candidates for quantum coherence, voltage gated ion channels display fract al kinetics (Liebovitch 1987).

How interacting systems respond to the quantum suppression of chaos, in processes such as scarring of the wave function (Gutzwiller 1992), received clarification (Chaudhury et. al. 2009, Steck 2009), when it was discovered that an electron in an orbit around a Cs atom in a classically chaotic regime enters into entanglement with nuclear spin. This illustrates how the chaotic 'billiards' of molecular kinetics, and chaotic membrane excitation, might become entangled with other states at the quantum level. One characteristic of time-dependent quantum 'chaos' is transient chaotic behavior ending up in a periodic orbital scar as wave spreading occurs. This would suggest that chaotic sensitivity, with an increasing dominance by quantum uncertainty over time, would contribute to which entanglements ultimately occur in a given kinetic encounter.

The evolutionary argument is a potent discriminator of models of consciousness. Quantum attributes making subjective consciousness possible need to evolve in confluence with essential physiological processes, thus potentially dating back to the epoch when the central components of modulated excitability evolved. Many theories of consciousness have been devised invoking quantum processes which emphasize unusual interpretations of physics, esoteric forms of quantum computation invoking properties extraneous to the known physiological functions of biological organelles, or hypothetical fields in addition to known physiology, raising questions as to whether they pass the evolutionary test. One of the most famous is Hameroff and Penrose's (2003) OOR theory combining objective reduction of the wave function with hypothetical forms of quantum computing on microtubules, which might be extended between cells through gap junctions. These are extensively discussed in the supporting online material, (King 2011b).

One idea fitting closely with neurophysiology is Bernroider's (2003, 2005) proposal that quantum coherence may be sustained in ion channels long enough to be relevant for neural processes and that the channels could be entangl ed with surrounding lipids and proteins and with other channels in the same membrane. He suggests that the ion channel functions through quantum coherence. MacKinnon's group (Zhou et. al. 2001) have shown that the K+-specific ion channel filter works by holding two K^+ ions bound to water structures induced by protein side chains. These have similarities to models of quantum computing using ion traps. The solitonic nature of action potentials could provide such entangled connectivity between channels.

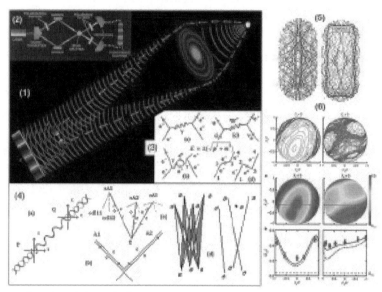

Fig 8: Wheeler delayed choice experiment (1) shows that a decision can be made after a photon from a distant quasar has traversed a gravitationally lensing galaxy by deciding whether to detect which way the photon traveled or to demonstrate it went both ways by sampling interference. The final state at the absorber thus appears to be able to determine past history of the photon. Quantum erasure (2) likewise enables a distinction already made, which would prevent interference, to be undone after the photon is released. Feynman diagrams (3) show similar time-reversible behavior. In particular time reversed electron scattering (d) is identical to positron creation-annihilation. (4a) In the transactional interpretation (Cramer 1983), a single photon exchanged between emitter and absorber is formed by constructive interference between a retarded offer wave (solid) and an advanced confirmation wave (dotted). (b) Experiments of quantum entanglement involving pair-splitting are resolved by combined offer and confirmation waves, because confirmation waves intersect at the emission point. Contingent absorbers of an emitter in a single passage of a photon (c). Collapse of contingent emitters and absorbers in a transactional match-making (d). (5) Scarring of the wave function of the quantum stadium along repelling orbits (Gutzwiller 1992). (6) Generation of quantum entanglement by quantum chaos in the quantum kicked top (Chaudhury et. al. 2009, Steck 2009).

While decoherence theories and objective reduction do not provide an active role for will, several physicists have suggested consciousness could play a part in the way the wave function representing a superposition of states, collapses to one real instance of the particle. Quantum theory predicts Schrödinger's cat subjected to cyanide if a radioactive scintillation occurs, is in a shadowy superposition - both alive and dead. In our conscious experience of the real world, we find the cat is either alive or dead. This suggests subjective consciousness could play an intervening role within quantum reality, reducing the superabundance of quantum probability multiverses to the historical process we experience. If so, consciousness may have a direct window on the entangled sub-quantum realm (Joseph 2009). We thus explore a model of quantum anticipation, which could extend back to single celled evolution.

Feynman diagrams of quantum interactions show that the quantum interaction is time-reversible. The diagram for electron scattering, when the scattered electron path is time-reversed, becomes positron creation and annihilation. Moreover in real quantum experiments, such as quantum erasure and the Wheeler delayed-choice experiment, it is possible to change how an intervening wave-particle behaves by making different measurements after the wave-particle has passed through the 'apparatus'. All forms of quantum entanglement possess this time-symmetric property. John Cramer (1983) incorporated time-symmetry into the 'transactional interpretation' of quantum mechanics, in which space-time handshaking between the future and past becomes the basis of each real quantum interaction. The emitter of a particle sends out an offer wave forwards and backwards in time, whose energies cancel. The prospective absorbers respond with confirmation waves, and the real quantum exchange arises from constructive interference between the retarded component of the chosen emitter's offer wave and the advanced, time-reversing component of the chosen absorber's confirmation wave. The boundary conditions defining the exchange thus involve both past and future states of the universe. Upon wave function collapse, the exchanged real particle traveling from the emitter to the absorber is identical with its negative energy anti-particle traveling backwards in time.

The transactional interpretation is a heuristic device, which is not essential to the argument, since its predictions coincide, largely, or exclusively with conventional quantum mechanics, but it does highlight future boundary conditions, which could play a part in conscious anticipation. Regardless of the interpretation of quantum mechanics we use, an exchanged particle has a wave function existing throughout the space-time interval in which it exists, so any process involving collapse of a wave function has boundary conditions extending in principle throughout space-time, involving future prospective absorbers. Advanced entanglement becomes clear in experiments creating two entangled particles (Aspect 1981), where subsequent measurement of the polarization of one photon immediately results in the other having complementary polarization, although neither had a defined polarization beforehand. The only way this correlation can be maintained within quantum reality is through a wave function extending back to the creation event of the pair and forward again in time to the other particle.

If subjective consciousness has a complementary role to brain function, correlated with entangled, quanta emitted and absorbed by the biological brain, it is then correlated with a superposition of possible states in the brain's future, as well as having access to memories of the past. In pair-splitting experiments, the boundary conditions do not permit a classically-causal exploitation. This does not result in a contradiction here, because the brain state is quantum indeterminate and the conscious experience corresponding to the entangled collapse provides an intuitive 'hunch', not a causal deduction.

A possible basis for the emergence of subjective consciousness, which could also be pivotal in explaining the source of free-will, is thus that the excitable cell gained a fundamental form of anticipation of threats to survival. These cells also evolved the ability to perceive strategic opportunities, through anticipatory quantum non-locality induced by chaotic excitation of the cell membrane, in which the cell becomes both an emitter and absorber of its own excitations. Non-locality in space-time is a fundamental quantum property shared by all physical systems, including macroscopic systems with coherent resonance.

The coherent global excitations in the gamma range associated with conscious states, could thus be the 'excitons' in such a quantum model. Unlike quantum computing, which depends on not being disturbed by decoherence caused by interaction with other quanta. Stringent requirements, avoiding decoherence, may not apply to transactions, where real particle exchange occurs even under scattering.

Quantum phenomena abound in biological tissues. Entanglement has been observed in healthy tissues in quantum coherence MRI imaging and bird navigation has been suggested to use entangled electrons. Excitations in photosynthetic antennae have also been shown to perform spatial quantum computing. Enzyme activation energy transition states and synaptic transmission also use quantum tunneling.

By making the organism sensitive to a short envelope of time, extending into the immediate future, as well as the past, subjective consciousness could thus gain an evolutionary advantage, making the organism sensitive to anticipated threats to survival as well as hunting and foraging opportunities. It is these primary needs, guided by the nuances of hunch and familiarity, rather than formal calculations, that the central nervous systems of vertebrates have evolved to successfully handle. Such temporal anticipation need not be of causal efficacy but just provide a small statistical advantage, complemented by computational brain processes associated with learning, which edge-of-chaos wave processing is ideally positioned to do.

These objectives are shared in precisely the same way by single-celled organisms and complex nervous systems. Because of the vastly longer evolutionary time since the Archaean expansion, than the Cambrian metazoan radiation, and the fact that all the components of neuronal excitability were already present when the metazoa emerged, quantum anticipation could have become an evolutionary feature of single celled eukaryot es, before metazoa evolved.

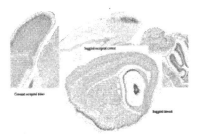

Fig 9: Expression of rhodopsin in the CNS shows both strong selective neuronal activity and a focal expression in the occipital cortex consistent with function in primary visual areas (King 2007).

6: Quantum Sensitivity, Sensory Transduction and Subjective Experience

One of the mysteries that distinguish the richness of subjective conscious experience from the colorless logic of electrodynamics is that sensory experiences of vision, sound, smell and touch are richly and qualitatively so different that it is difficult to see how mere variations in neuronal firing organization can give rise to such qualitatively different subjective affects. How is it that when dreaming, or in a psychedelic reverie, we can experience ornate visions, hear entrancing music, or smell fragrances as rich, real, intense and qualitatively diverse as those of waking life?

Since the senses are actually fundamental quantum modes by which biological organisms can interact with the physical world, this raises the question whether subjective sensory experience is in some way related to the quantum modes by which the physical senses communicate with the world (Joseph 2009). Clearly our senses are sensitive to the quantum level.

Individual frog rod cells have been shown to respond to individual photons, the quietest sound involves movements in the inner ear of only the radius of a hydrogen atom and single molecules are sufficient to excite pheromonal receptors. Many genes we associate with peripheral sensory transduction in several senses are also expressed in the mouse brain (King 2007) at least in the form of RNA transcripts, including stomatin-like protein 3 associated with touch, epsin, otocadherin and otoferlin associated with hearing, and several types of opsin, including rhodopsin and encephalopsin. This suggests the brain could harbour an 'internal sensory system' which might play a role in generating the 'internal model of reality', although these ideas are speculative and it is a major challenge to see how such processes could be activated reversibly in the CNS.

Several researchers (Pocket 2000, McFadden 2002) have proposed that neural excitation is associated with electromagnetic fields, which might play a formative role in brain dynamics. Attention has recently been focused on biophotons as a possible basis of processing in the visual cortex based on quantum releases in mitochondrial redox reactions (Rahnama et. al. 2010, Bókkon et. al. 2010). Microtubules have also been implicated (Cifra et. al. 2010).

All excitable cells have ion channels, which undergo conformation changes associated with voltage, and orbital or 'ligand'-binding, both of internal effectors such as G-proteins and externally via neurotransmitters, such as acetylcholine. They also have osmotic and mechano-receptive activation, as in hearing, and in some species can be also activated directly and reversibly by photoreception. Conformation changes of ion channels are capable of exchanging photons, phonons, mechano-osmotic effects and orbital perturbations, representing a form of quantum synesthesia. Since the brain uses up to 40% of our metabolic energy for functions with little or no direct energy output, it is plausible that some of the 'dissipated' energy could be generating novel forms of interaction.

7: Complementarity, Symmetry-breaking, Subjective Consciousness, and Cosmology

This leads to the most perplexing chasm facing the scientific description of reality. What is the existential nature of subjective consciousness, from waking life, through dreaming to psychedelic and mystical experience, and does it have cosmological status in relation to the physical universe?

The key entities forming the physical universe manifest as symmetry-broken complementarities. Quanta are waveparticles, with complementary discrete particle and continuous wave aspects. The fundamental forces are symmetry broken in a manner that results in complementary force-radiation bearing bosons and matter forming fermions. In the standard model these have symmetry broken properties, with differing collections of particles. Supersymmetry proposes each boson has a fermion partner to balance their positive and negative energy contributions, but E8's 112 'bosonic' and 128 'fermionic' root vectors, suggest symmetry-breaking could be fundamental (Fielder and King 2010).

Further symmetry-broken complementarities apply to the biological world, where the dyadic sexes of complex organisms and many eukaryotes are both complementary and symmetry broken, with themes of discreteness and continuity even more obviously expressed at the level of sperm and ovum than in our highly symmetry-broken human bodily forms, involving pregnancy, live birth and lactation.

The relationship between subjective consciousness and the physical universe displays a similar complementarity, with profound symmetry breaking. The 'hard problem of consciousness research' (Chalmers 1995) underlines the fundamental differences between subjective 'qualia' and the participatory continuity of the Cartesian theatre on the one hand, and the objective, analyzable properties of the physical world around us.

Although we depend on a pragmatic acceptance of the real world, knowing we will pass out if concussed and could die if we cut our veins, from birth to death, the only veridical reality we experience is the envelope of subjective conscious experience. It is only through the consensual regularities of subjective consciousness that we come to know the real world and discover its natural and scientific properties. As pointed out by Indian philosophy, this suggests that mind is more fundamental than matter. The existential status of subjective consciousness thus also has a claim to cosmological status.

A further cosmological interpretation of consciousness we have noted in association with the cat paradox is that it may function to solve the problem of super-abundance, by reducing probability multiverses to the unique course of history we know and witness. This view of consciousness in shaping the universe is consistent with several of the conclusions of biocentrism (Lanza 2009).

The lessons of quantum and fundamental particle complementarity and symmetry-breaking, sexuality and the Yin-Yang complementarity of the Tao and of Shakti-Shiva in Tantric mind-world cosmologies, lead to a cosmology of consciousness, as symmetry-broken complement to the physical universe.

References

Ananthaswamy, A. (2009) Whole brain is in the grip of consciousness New Scientist 18 March.

Ananthaswamy, A. (2010), Firing on all neurons: Where consciousness comes from, New Scientist, 22 March.

Aspect, A., Grangier P., Roger G. (1981), Phys. Rev. Lett. 47, 460; (1982) Phys. Rev. Lett. 49, 1804; 49, 91.

Baars, B. (1997) In the Theatre of Consciousness: Global Workspace Theory, A Rigorous Scientific Theory of Consciousness. Journal of Consciousness Studies, 4/4 292-309.

Basar E., Basar-Eroglu J., Röschke J., Schütt A., (1989) The EEG is a quasi-deterministic signal anticipating sensory-cognitive tasks, in Basar E., Bullock T.H. eds. Brain Dynamics Springer-Verlag, 43-71.

Beggs J, Plenz D. (2004) Neuronal Avalanches Are Diverse and Precise Activity Patterns That Are Stable for Many Hours in Cortical Slice Cultures Journal of Neuroscience, 24, 5216-9.

Bernroider, G. (2003) Quantum neurodynamics and the relation to conscious experience Neuroquantology, 2, 163–8.

Bernroider, G., Roy, S. (2005) Quantum entanglement of K ions, multiple channel states and the role of noise in the brain SPIE 5841/29 205–214.

Bókkon I, Salari V, Tuszynski J, Antal I (2010) Estimation of the number of biophotons involved in the visual perception of a singleobject image: Biophoton intensity can be considerably higher inside cells than outside http://arxiv.org/abs/1012.3371

Chalmers D. (1995) The Puzzle of Conscious Experienceeс, Scientific American Dec. 62-69.

Chaudhury S, Smith A, Anderson B, Ghose S, Jessen P (2009) Quantum signatures of chaos in a kicked top, Nature 461 768-771.

Cifra M, Fields J, Farhadi A (2010) Electromagnetic cellular interactions Progress in Biophysics and Molecular Biology doi:10.1016/j.pbiomolbio.2010.07.003

Cramer J.G. (1983) The Transactional Interpretation of Quantum Mechanics, Found. Phys. 13, 887.

Crick F, Koch C. (1992) The Problem of Consciousness, Sci. Am. Sep. 110-117.

Dagan T, Artzy-Randrup Y, Martin W (2006) Modular networks and cumulative impact of lateral transfer in prokaryote genome evolution, PNAS 105/29, 10039-10044.

Darwin C. (1871) The Descent of Man and Selection in Relation to Sex, John Murray, London.

David L, Alm E (2010) Rapid evolutionary innovation during an Archaean genetic expansion Nature doi:10.1038/nature09649

Fielder Christine and King Chris (2004) Sexual Paradox: Complementarity, Reproductive Conflict and Human Emergence Lulu Press ISBN141165532X (2010) edition: http://www.sexualparadox.org

Gutzwiller, M.C. (1992) Quantum Chaos, Scientific American 266, 78 - 84.

Hameroff, Stuart, Penrose, Roger (2003) Conscious Events as Orchestrated Space-Time Selections, NeuroQuantology; 1, 10-35.

Hauser M. (2009) Origin of the Mind, Scientific American, Sept, 44-51.

Joseph, R. (2009). Quantum Physics and the Multiplicity of Mind: Split-Brains, Fragmented Minds, Dissociation, Quantum Consciousness, Journal of Cosmology, 3, 600-640.

Joseph, R. (2011). The neuroanatomy of free will. Loss of will, against the will, "alien hand". Journal of Cosmology, 14, 6000-6045.

King C.C. (1982) A Model for the Development of Genetic Translation, Origins of Life, 12 405-417.

King C.C, (1996) Fractal Neurodynamics and Quantum Chaos : Resolving the Mind-Brain Paradox through Novel Biophysics, in Advances in Consciousness Research, The Secret Symmetry : Fractals of Brain Mind and Consciousness (eds.) E. Mac Cormack and M. Stamenov, John Benjamin.

King C.C. (2007) Sensory Transduction and

Subjective Experience Nature Preceedings hdl:10101/npre.2007.1473.1 2009 edition: Activitas Nervosa Superior, 51/1, 45-50. http://www.dhushara.com/lightf/light.htm

King C.C. (2008) The Central Enigma of Consciousness Nature Preceedings hdl:10101/npre.2008.2465.1 2010 edition: http://www.dhushara.com/enigma/enigma.htm

King C. C. (2011a) The Tree of Life: Tangled Roots and Sexy Shoots: Tracing the genetic pathway from the Universal Common Ancestor to Homo sapiens, DNA Decipher J., 1. http://www.dhushara.com/book/unraveltree/unravel.htm

King C. C., (2011b) Cosmological Foundations of Consciousness http://www.dhushara.com/cosfcos/cosfcos2.html

Lanza, Robert and Berman, Bob (2009) Biocentrism: How Life and Consciousness are the Keys to Understanding the True Nature of the Universe, BenBella, ISBN 978-1933771694

Liebovitch L.S., Sullivan J.M., (1987) Fractal analysis of a voltage-dependent potassium channel from cultured mouse hippocampal neurons, Biophys. J., 52, 979-988.

Liljenström Hans, Svedin Uno (2005) Micro-Meso-Macro: Addressing Complex Systems Couplings, Imperial College Press.

Martin, W. and Russell, M. J. (2003) On the origins of cells: a hypothesis for the evolutionary chemoautotrophic transitions from abiotic geochemistry to prokaryotes, and from prokaryotes to nucleated cells, Phil. Trans. R. Soc. Lond. B 358, 59-85.

McFadden J (2002) The Conscious Electromagnetic Information (Cemi) Field Theory: The Hard Problem Made Easy? Journal of Consciousness Studies, 9/8, 45-60. http://www.surrey.ac.uk/qe/pdfs/mcfadden_JCS2002b.pdf

Molnar, G et. al. (2008) Complex Events Initiated by Individual Spikes in the Human Cerebral Cortex, PLOS Biology, 6/9 222. Pockett, Susan (2000) The Nature of Consciousness, ISBN 0595122159.

Powner M., Sutherland J., Szostak J. (2010) Chemoselective Multicomponent One-Pot Assembly of Purine Precursors in Water, J. Am. Chem. Soc., 132, 16677-16688.

Rahnama M, Bókkon I, Tuszynski J, Cifra M, Sardar P, Salari V (2010) Emission of Biophotons and Neural Activity of the Brain, http://arxiv.org/abs/1012.3371

Russell, M. (2011). Origins, abiogenesis, and the search for life. Cosmology Science Publishers, Cambridge.

Skarda C.J., Freeman W.J., (1987) How brains make chaos in order to make sense of the world, Behavioral and Brain Sciences, 10, 161-195.

Steck D (2009) Passage through chaos, Nature, 461, 736-7.

Wickramasinghe, C. (2011) The Biological Big Bang. Cosmology Science Publishers. Cambridge.

Woodruff, A and Yuste R (2008) Of Mice and Men, and Chandeliers, PLOS Biology, 6/9, 243.

Zhou, Y., Morais-Cabral, A., Kaufman, A. & MacKinnon, R. (2001) Chemistry of ion coordination and hydration revealed in K+ channel-Fab complex at 2.0 A resolution, Nature, 414, 43-48.

3. Does the Universe have Cosmological Memory? Does This Imply Cosmic Consciousness?
Walter J. Christensen Jr.

Physics Department, Cal Poly Pomona University, 3801 W. Temple Ave, Pomona CA 91768

Abstract

Does the universe have cosmological memory? If so, does this imply cosmic consciousness? In this paper a cosmological model is proposed similar in structure to the famous thought experiment presented by James Clerk Maxwell, in which a "demon" tries to violate the Second Law of Thermodynamics. In such a proposed cosmological scenario, if the Second Law of Thermodynamics is to be preserved, it implies the existence of cosmological memory. Since consciousness and memory are intimately linked and a demon-like (intelligent) creature permits only fast moving particles to pass through a cosmic gate, it is argued the universe is necessarily conscious.

KEY WORDS: Consciousness, Cosmology, Quantum Physics, Cosmic Memory

I. Introduction

Cosmic consciousness is argued for in this paper. By consciousness we mean: any macroscopic or microscopic system that operates through the use of both memory and choice. Cosmic consciousness means, any cosmological system that requires both memory and choice to operate it. Of course the word 'choice' is a deeply philosophical concept and will be left for future discussions, such as those made by Robert Kane (1996) and his idea of 'ultimate responsibility'. For now, we will simply accept Maxwell's approach of an intelligent, or conscious, creature that attempts to violate the Second Law of Thermodynamics, as now discussed.

The validity of the assertion for the existence of cosmic consciousness (sometimes referred to as cosmic intelligence, intelligent creature, or cosmic creature, in this paper), begins with the seminal thought experiment made by the eminent mathematician and physicist James Clerk Maxwell, in a letter to Peter Guthrie Tait (Maxwell 1867). Maxwell envisaged a tiny intelligent being who opens and closes a valve connecting separate chambers; each chamber filled with the same gas under the same conditions. This intelligent creature, referred to as 'Maxwell's demon' in literature (Leff and Rex 2002; Thomson 1897), functions to open a valve to allow the faster gas molecules to flow into one chamber, and slower gas molecules into the other chamber. In effect, the demon creates a temperature difference between the two compartments (It can be shown in classical thermodynamics that the temperature of the gas is strictly dependent on the speed of gas molecules). In this context, intelligent creature is used in the historical sense, as described by Maxwell (1871): "… if we conceive of a creature whose faculties are so sharpened that he can follow every molecule..."And subsequently nicknamed 'intelligent demon' by William Thomson (1874a; 1879): "This process of diffusion could be prevented by an army of Maxwell's 'intelligent demons'… separating the hot from the cold… "

If indeed, such an intelligent demon could perform such a separation task without expending energy (using a frictionless valve, which the creature opens slowly), the result would violate the Second Law of Thermodynamics (Serreli et al 2007). This is so, because such a self-contained system could perpetually perform work without the need for any external energy source.

The thought experiment proposed by Maxwell, has provoked a long-line of arguments in the scientific community as to whether or not the Second Law of Thermodynamics (SLT) can be violated or not. This ongoing battle has produced an extensive, and impressive list of authors who have introduced their own unique approach (as well as related discussions), and includes, in part: Thomson (1874b; 1879;), Poincare (1893), Planck (1922), Szilárd (1929a), Lewis (1930), von Neumann (1932); Gamow (1944); Born, M. (1948); Wiener (1950); Bohm (1951), Jacobson (1951), Brillouin (1951), Saha (1958), Feynman (1963), Asimov (1965), Bell (1968), Penrose, O. (1970; 1979), Beauregard and Tribus (1974), Popper (1982), Davies (1986), Bennett (1987), Hawking (1987), Landauer (1987), Rex (1987), Leff (1990), Christensen (1991), and more recently Penrose, R. (2002).

In this paper, we argue for the preservation of the Second Law of Thermodynamics by relating entropy with memory, as did Einstein's close friend, Leó Szilárd (1929b). Ever since his publication, his idea of relating entropy to memory has had a life of its own with many writers (information is often exchanged for memory in literature). For example Lewis, G. N. (1930) states: "Gain in entropy always means loss of information, and nothing more. … if any essential data added, the entropy becomes less."Others view the entire universe as a dimensional information structure (Lloyd 2002; Davies 2004a).

The approach taken in this paper, rests on three assumptions:

1) SLT is the fundamental principle of space-time. That is, although different universes, or multiverses (soon to be discussed), may each have their own distinct physical parameters, constants and even dynamics, nevertheless they must obey some form of the Second Law of Thermodynamics both globally and locally, or the combination of the two; that any thought experiment, or actual system under consideration, must preserve the Second Law of Thermodynamics, otherwise such thought experiments are to be discounted, and any physical system that violates SLT is to be interpreted is not real, or at least that not all the variables are known.

2) Memory and entropy are deeply related aspects of each other, in much the same way that various forms of energy are related and can be converted into one form or the other without loss.

3) Any system converting entropy into memory, or memory into entropy, which also involves choice (such as opening or closing a gate), thus con-

tributes to running the system, we characterizes as having intelligence or consciousness. If such a system is the universe itself, or multiverses, we say cosmic consciousness is involved the operation of the cosmology.

Since the goal in this article is to argue for the existence of cosmic consciousness, we take the prudent choice of developing a cosmological model following thought experiment to developed by James Clerk Maxwell.

2. Multiverses and Proposed Cosmological Model

Is our universe just one among an ensemble of many? Just an infinitesimal part of an elaborate structure, which consists of numerous universes, possibly an infinite number of universes, sometimes referred to as multiverses? Such questions were recently discussed by Rüdiger Vaas (2010). Vaas explains: "… there are many different multiverse accounts (see, e.g., Carr 2007; Davies 2004b; Deutsch 1997; Linde 2008; Mersini-Houghton 2010; Rees 2001; Smolin 1997a; Tegmark 2004, Vaas 2004b, 2005, 2008b, 2010; Vilenkin 2006), and even some attempts to classify them quantitatively (see, e.g., Deutsch 2001 for manyworlds in quantum physics, and Ellis, Kirchner & Stoeger 2004 for physical cosmology). They flourish especially in the context of cosmic inflation (Aguirre 2010; Linde 2005, 2008, Vaas 2008b), string theory (Chalmers 2007, Douglas 2010) and a combination of both, as well as, in different, quantum gravity scenarios, that seek to resolve the big bang singularity and, thus, explain the origin of our universe."

With so many possible types of multiverses to choose from, which one can we legitimately select to develop a cosmology that parallels the thought experiment of James Clerk Maxwell and his intelligent demon? The answer is, we must rely on various observations to guide us, and when those are not available, to be guided by sound theoretical models.

What is needed then, is, instead of using two chambers, we need a cosmology that incorporates a pair of universes separated by some kind of cosmic gate [for the sake of simplicity, (Occam's razor), a pair of universes is chosen, rather than numerous universes joined by a cosmic gate].

But what kind of gate could not only join two universes together, but be verified by empirical evidence? What is imagined is a black hole dynamically altering spacetime to create a multiverse, and which acts as a doorway between our universe and the newly formed universe There is much empirical evidence to support the existence of black holes and dynamics, in the context of this article: (Rees, 1989; Begleman et al, 1984; Casares, 1992; Remillard, 1992; Reynolds, 2008).

The cosmological model proposed thus far, is somewhat similar, yet distinct from, the model proposed by Lee Smolin (Smolin 1992a; Smolin 1997b): where locally, a collapsing black hole causes the emergence of a new universe, via the dynamical properties of the black hole. Our model allows for, as Smolin suggests, alternative universes to which fundamental constants or parameters such as the speed of light, gravitational constant, and so forth, can be different from our own universe. Multiverses are also referred to as Fecund Universes (Smolin 1992b ; Tegmark 2003). At the quantum level, Deutsch (Deutsch 2002) connected information to such multiverses, as we semi-classically do in this paper. Deutsch states: "The structure of the multiverse can be understood by analyzing the ways in which information can flow in it. We may distinguish between quantum and classical information processing. In any region where the latter occurs-which includes not only classical computation but also all measurements and decoherent processes (mechanism by which quantum systems interact with their environments to exhibit probabilistically additive behavior) the multiverse contains an ensemble of causally autonomous systems, each of which resembles a classical physical system."

The model also resembles string-theory (Susskind 2006) multiple-dimension arguments for why gravity is so weak, that is, it is leaking from one dimension into another (Horava and Witten 1996; Maartens and Koyama 2010); although CMS Collaboration is reporting results on microscopic black holes that makes this scenario less likely (CMS Collaboration 2010). However, the model proposed here has its distinctions; mainly that it places the Second Law of Thermodynamics as the 'prime mover,' by providing revitalized energy and matter back into our universe, as will be soon discussed.

To be clear, the cosmological model proposed in this article, has nothing to do with the "Big Crunch" scenario, in which the expansion of space eventually reverses and the universe collapses, eventually ending as a black hole singularity. Nor, do multiverses forming in various regions when space-time stops expanding, in so doing, form bubbles found in chaotic inflation theory proposed by Alan Guth (2007).

So just how does our universe work in tandem with the multiverse formed by a black hole? To begin with, it is assumed that massive particles trapped within the black hole are reduced to a common fundamental particle; each one being of equal size, spin and mass (Joseph 2010). When this occurs, identical particles pass through the black hole into the multiverse.

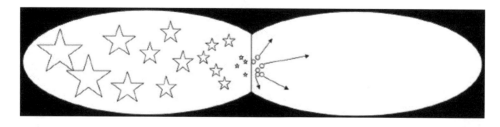

Because these particles are identical, with identical spins, gravity waves will cause the particles to repel each other (Penrose, R. 1960; Stacey, et al 1987;

Economou, 1992; Gasperini 1998). What happens then? As they accelerate away from each other in all directions, they near the speed of light. When this occurs, it is assumed, following Maxwell's thought experiment, that a cosmic creature opens a gate leading to small pathways joining the two universes. Such gates are referred to as a string gates (See Figure 3).

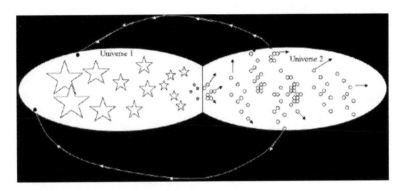

As for massless particles (for example photons), they too pass into the multiverse via the black hole and continue moving at the speed of light c. Again the cosmic creature opens a string gate leading back into our universe. Overall, we have matter and energy leaving our universe, entering a multiverse via a black hole, then returning, via a string-gate (Stefano Ansoldi, Antonio Aurilia and Euro Spallucci, 2002), back into our universe. In effect, this process reseeds our universe with new energy and fast moving particles. Such a cosmology is reminiscent of a publication by Rhawn Joseph (2010). In his article he states: "The infinite, eternal universe continually recycles energy and mass at both the subatomic and macro-atomic level, thereby destroying and then reassembling atoms, molecules, stars, planets and galaxies. Mass, molecules, atoms, protons, electrons, and elementary particles are continually created and destroyed, and matter and energy, including hydrogen atoms, are continually recycled and recreated by super massive black holes and quasars at the center of galaxies, and via infinitely small gravity holes also known as "black holes", "Planck Particles", "Graviton Particles", and "Graviton-holes. ...Passageways may exist within these infinitely small spaces, which may lead to other dimensions (Appelquist and Chodos 1983; Aharony, et al., 2000; Greene 2003; Keeton and Petters 2005; Randall and Sundrum 1999) and thus to "other worlds" or another space-time (Hawking, 1988). A singularity may exist on both sides of a hole in space-time, simultaneously, thereby creating duality from singularity. An infinite number of holes would yield an infinite number of possibilities and would enable a singularity to have not just duality, but multiplicity via multiple dimensions existing on the other side of an infinity of holes." However, 'Joseph cosmology' violates SLT because it creates a perpetual machine, and should viewed as an incomplete model. Whereas the cosmology presented in this paper, which also is a perpetual machine, preserves SLT by having:

1) A cosmic creature (analogous to Maxwell's demon) make choices to open a string-gate to allow fast moving particles to pass back into our universe; or if these particles are not moving near the speed of light, to keep the string-gate closed.

2) Memory and entropy be different aspects of each other, suggested by Szilárd (1929c).

With these two points taken together (that an intelligent creature appears to create a perpetual machine-like-universe that violates SLT, but upon further consideration, the information gained by the creature restores SLT), the cosmology argued for in this article preserves SLT, and so meets the assumptive criteria to be considered a valid cosmology.

3. Cosmic Creature

By what mechanism does this cosmic creature observe massive and massless particles moving near or at the speed of light respectively? The mechanism is somewhat similar to what occurs when particles collide and scatter off of each other, as described by the Standard Model (Glashow 1961; Weinberg 1967). In particle physics, during a collision, virtual particles are emitted. These virtual particles can mediate force and transfer momentum to other particles (Eisberg and Resnick, 1985; O'Reilly, 2002), causing the colliding particles to alter their trajectory (Guralnik 1964). The mechanism is also similar to when a Goldstone boson (Goldstone 1961) is absorbed by bosons making them become massive.

Likewise, for the cosmological model presented here, we assume is that virtual particles are emitted by the accelerating massive particles traversing the multiverse. When a massive particle has a speed much less than that of light, its de Broglie wavelength will be longer, to that of a particle accelerating close to the speed of light, consequently the mediating virtual particle will also have a longer wavelength. This characterizes the massive particle and informs the cosmic creature to either choose to keep the string-gate open or closed. If the particle has a wavelength associated with moving at the speed of light, the cosmic creature opens the string gate --that is the creature absorbs the virtual particle and opens the string-gate so that the string absorbs the massive particle; in so doing the massive fundament particle passes from the multiverse, back into our universe (Joseph 2010; Gubser, 1997) (See Figure 3). As for any massless particles entering the multiverse via the black hole, they too return to our universe, via cosmic string-gate.

Thus our universes and the multiverse creates a kind of perpetual machine, but one that does not violate the Second Law of Thermodynamics because the decrease in entropy is balanced by the increase of memory, just as Leo Szilárd (1929d) did, in his thought experiment, where he considered particles having different chemical parameters inside a box with functioning pistons. Szilárd writes (1929e): "A perpetual motion machine therefore is possible if— according to the general method of physics— we view the experimenting man as a sort of deus ex machina, one who is continuously and exactly informed of the existing state of nature and who is able to start, or interrupt, the microscopic course of nature at any moment without expenditure of work."

Instead of an experimenting deus ex machina, we propose both memory and choice compensates for the decrease of entropy in the proposed cosmological model, hence cosmological consciousness allows for perpetual operating universes thereby preserving the Second Law of Thermodynamics. Leo Szilárd further writes: "We shall realize that the Second Law is not threatened as much by this entropy decrease as one would think, as soon as we see that the entropy decrease resulting from the intervention is compensated completely in any event, if the execution of such a measurement were, for instance, always accompanied by production of k ln2 units of entropy. In that case it will be possible to find a more general entropy law, which applies universally to all measurements." In regards to this paper, it is assumed since the task of opening the string gate involves selection, and that the reduction of entropy is compensated for by the intelligent creature gaining a equivalent amount of entropy (k ln2), no violation of the Second Law of Thermodynamics occurs. This the minimal condition necessary to claim, in such coupled universes, the existence of cosmological consciousness exists. In particular, we see the production of k ln2 units of entropy (equivalent to conscious memory and selection) are required to compensate for the perpetual workings, of the proposed cosmological universe, if and only if the Second Law of Thermodynamics is to be preserved.

Stated another way, over time, all matter in our gravitationally attractive universe must [according to various cosmological scenarios, such as the heat death of the universe first proposed by Lord Kelvin (Crosbie et al 1989)] eventually die out because of maximum entropy increase. However the view here is that of a cosmological consciousness that gains information so that our universe can be reseeded energetic matter and revitalized energy.

4. Other Considerations

Before going any further we must consider Brillouin's argument. Basically he argues that such an intelligent being opening the valve to let in fast gas molecules, needs light in order to see the fast gas molecules (Brillouin 1951). Brillouin states: "In an enclosure at constant temperature, the radiation is that of a "blackbody", and the demon cannot see the molecules. Hence, he cannot operate the trap door and is unable to violate the second principle (law). If we introduce a light source, the demon can see the molecules, but the over-all balance of entropy is positive." However, the thought experiment proposed in this paper, the system of coupled universes is not in thermal equilibrium and the intelligent creature cannot "see" via radiation. Instead the creature absorbs a virtual particle as previously discussed, which provides k ln2 information related to positive entropy, which is necessary to balance the decrease the entropy of the system when energetic particles are permitted to pass from the multiverse back into our universe. Thus no photons are required to detect fast moving elementary particles, only virtual particles.

Let us consider more closely the characteristics of the virtual particle under discussion. First of all, we realize that since all particulate matter is discrete, it is countable; that is there is a one-to-one correspondence between the natural numbers N, to the individual particles in the multiverse. Whether there are ten particles, ten billion, or even an arbitrarily large number of such particles residing in the multiverses, they are nevertheless countable. Secondly, if we assume the virtual particles are in one-to-one correspondence with the set of real numbers R, rather than the restricted set of natural numbers N, then the intelligent creature will never run out of information to keep the perpetual cosmic machine running.

To understand this, consider the closed interval between the numbers one and two [1, 2]. In this interval there exactly two natural counting numbers, i.e. number one and two. Whereas in the same interval there is uncountable real numbers such as the square root of two and $\pi/^2$. What this implies is that, even though the set of natural numbers {N} is infinite, the set of real numbers {R} is a larger infinite set that cannot be counted. Hence in set notation: {N} < {R}.

By assuming the virtual particles are associated with the set of real numbers R, they are uncountable. Consequently, even if the same particle cycles through our universe, into the multiverse and back to our universe an infinite amount of times, and each time the intelligent creature selects to absorb one of these virtual particles associated with the recycled countable particles, there will still be an uncountable many virtual particles left over to operate our universe with new energy; that is the cyclic process goes on forever, and the paired universes form a perpetual machine powered on memory and selection by a cosmic creature. Furthermore, if the cosmological model presented actually existed, then we are led to the conclusion that, either the Second Law of Thermodynamics is deeply flawed (or very limited domain), which contradicts all empirical observations, or indeed our universe powered by cosmological consciousness.

5. Conclusion

Though the cosmological model presented here was developed in part from a thought experiment argued by James Clerk Maxwell, and that the cosmology presented was purely hypothetical, should our current understanding of the universe prove to be incorrect, for example no big bang occurred, the question arises what powers a forever and infinite universe? Or put another way, if the universe has been here forever, why has not all the useable energy been consumed long ago? Such a scenario was first considered by William Thomson (Lord Kelvin), who argued mechanical energy loss in nature from the point of view of the 1st and 2nd Laws of thermodynamics. One solution might be the universe, coupled to a multiverse via a black hole, is powered by cosmological consciousness--an intelligent creature must make a choice to separate fast moving particles from slow moving particles. That is, these coupled universes form a perpetual machine. It is memory and choice that preserves the Second Law of Thermodynamics and make valid the proposed cosmological model. Both selection and memory imply cosmic awareness.

In conclusion, since it has been well established by empirical observation, that the Second Law of Thermodynamics is not observed to be violated, we argued, using a cosmological model relating memory, choice, entropy, that coupled universes are necessarily conscious.

Acknowledgement: I wish to thank Harvey Leff for his generous input into this paper. When I first entered University, he soon became one of my great professors and mentors; then a few years later my department chairperson, and finally a dear friend. I also wish to thank John Jewett who helped me with several early publications; I will always treasure our discussions about mathematics and quantum physics. John Fang, of course, is my very, very good friend. Over the years, with many side conversations, John taught me far more than a particle approach to gravity; he allowed a genuine friendship to develop while visiting both him and his gracious wife Stephanie, at their home, sometimes listening to such great composers as Brahms and Debussy. Also Kai Lam's wonderful conversations will always be remembered and cherished. Through the years I realize Kai is one of the most honest and sincere persons I have ever met. I would also like to commend with deep gratitude and friendship, Steven McCauley, who stood strong and noble on my behalf during a brief period of difficult times I had. Finally I wish to thank Kip Thorne for taking the time to converse with me at the 22nd Gravitational meeting in Santa Barbara, and then to subsequently correspond with kind intelligent honesty. And to Edward Witten who answered my questions by leaving the answers up to me.

References

Aguirre, A. (2010). Eternal Inflation: Past and Future. In: Vaas, R. (ed.). Beyond the Big Bang. Springer, Heidelberg.

Aharony, et al., (2000) Large N Field Theories, String Theory and Gravity. Phys.Rept.323, pp.183-386.

Ansoldi, Aurilia and Spallucci. (2002) Fuzzy Dimensions and Planck's Uncertainty Principle from p-brane Theory. Classical and Quantum Gravity 19, 3207.

Appelquist and Chodos (1983) Quantum effects in Kaluza-Klein theories. Phys. Rev. Lett. 50, pp. 141–145.

Asimov, I. (1965). Life and Energy. Bantam Books, New York, pp. 65-76.

Beauregard, O. and Tribus, M. (1974). Information theory and thermodynamics. Helv. Phys. Acta 47, pp. 238-247.

Begelman, M. C., Blandford, R. D. & Rees, M. (1984) Theory of extragalactic radio sources. J. Rev. mod. Phys. 56, pp. 255–351.

Bell, D. A. (1968). Information Theory. Sir Isaac Pitman & Sons, pp. 20-21; 212-219.

Bennett, C. H. (1987) Demons, engines and the second law. Sci. Am. 257, pp. 108-116.

Bohm, D. (1951); Quantum Theory. Prentice-Hall, Inc., Englewood Cliffs, New Jersey, pp. 608-609.

Born, M. (1948) Die Quantenmchanik und der Zweite Hauptsatz der Thermodynamik. Annalen der Physik 3, 107.

Brillouin, L. (1951). Maxwell's Demon Cannot Operate: Information and Entropy I. J. Appl. Phys. 22, pp. 334-337.

Carr, B. (2007). The Anthropic Principle Revisited. In: Carr, B. (ed.). The Universe or Multiverse? Cambridge University Press, Cambridge, pp. 77–89.

Casares, J., Charles, P. A., Naylor, T. (1992) Nature 355, 614–617.

Christensen, W. J. (1991) A Gedanken-experiment for the Plausible Existence of Cosmological Intelligence. JOPIPA Cal Poly Pomona University. Vol. 8 Number 1 Editor John Jewett.

CMS Collaboration (Dec 2010). Search for Microscopic Black Hole Signatures at the Large Hadron Collider, CERN-PH-EP/2010-073 CMS-EXO- 10-017; arXiv:1012.3375.

Davies, P. C. W. (1986). The ghost in the atom. Cambridge University Press pp. 20-22.

Davies, P. C. W. (2004). Multiverse cosmological models. Mod. Phys. Lett. A19, 727–744; arXiv:astro-ph/0403047.

Deutsch, D. (1997). The Fabric of Reality. Allen Lane, London, New York.

Deutsch, D. (2001). The Structure of the Multiverse; arXiv:quant-ph/0104033.

Deutsch D. (2002). The Structure of the Multiverse. Proc. R. Soc. Lond. A 458 pp. 2911-2923.

Douglas, M. (2010). The String Landscape: Exploring the Multiverse. In: Vaas, R. (ed.). Beyond the Big Bang. Springer, Heidelberg.

Economou, A. (1992). Gravitational effects of traveling waves along global cosmic strings, Phys. Rev. D 45, pp. 433–440.

Eisberg, R., and Resnick. R. (1985). Quantum Physics of Atoms, Molecules, Solids, Nuclei, and Particles. Wily, Fouqué, P.; Solanes, J. M.; Sanchis, T.; Balkowski, C. (2001). Structure, mass and distance of the Virgo cluster from a Tolman-Bondi model . Astronomy and Astrophysics 375: 770–780.

Ellis, G.F.R., Kirchner, U., Stoeger, W.R. (2004). Multiverses and physical cosmology. M.N.R.A.S. 347, 921–936; arXiv:astro-ph/0305292.

Feynman, R. P. Leighton, R. B. And Sands, M., (1963); Feynman Lectures on Physics. Addison-Wesley, Reading, Massachusetts. Vol 1, pp 46.1-46.9.

Gamow, G. (1944). Mr. Tompkins in Paperback. Cambridge University Press, (1971), pp. 95-111. Originally published in Mr. Tompkins Explores the Atom, 1944.

Gasperini, M. (1998) Repulsive Gravity in the very early Universe. General Relativity and Gravitation, Volume 30, Number 12, 1703-

1709, DOI: 10.1023/A:1026606925857.

Glashow, S. L (1961) Partial-symmetries of weak interactions Nuclear Physics Volume 22, Issue 4, February 1961, pp. 579-588. doi:10.1016/0029- 5582(61)90469-2.

Goldstone, J (1961). Field Theories with Superconductor Solutions, Nuovo Cimento 19: pp. 154-164, doi:10.1007/BF02812722

Greene, B. (2003) The Elegant Universe: Superstrings, Hidden Dimensions, and the Quest for the Ultimate Theory. W.W. Norton & Co.

Gubser, S. et al, (1997). String theory and classical absorption by three-branes, Nuclear Physics B Volume 499, Issues 1-2, pp. 217-240.

Guralnik, G. S. et al (1964) Global Conservation Laws and Massless Particles, Physical Review Letters 13: pp. 585–587. doi:10.1103/PhysRevLett.13.585

Guth, A. (2007) Eternal inflation and its implications. J. Phys. A 40 pp. 6811- 6826, arXiv:hep-th/0702178v1.

Hawking, S. W., (1987). The direction of time. New Sci. 115, No. 1568, pp. 46-49.

Hawking, S. W., (1988) Wormholes in spacetime. Phys. Rev. D 37, 904–910.

Horava, P. and Witten, E. (1996). Nucl. Phys. B 460 506 [arXiv:hepth/ 9510209]; Nucl. Phys. B 475 94 [arXiv:hep-th/9603142].

Jacobson, H. (1951). The role of information theory in the inactivation of Maxwell's demon. Trans. N. Y. Acad. Sci. 14, pp. 6-10.

Joseph, K. (2010). The Quantum Cosmos and Micro-Universe: Black Holes, Gravity, Elementary Particles, and the Destruction and Creation of Matter Journal of Cosmology, Vol 6.

Kane, R. (1996) The Significance of Free Will. Oxford University Press, ISBN 0-19-512656-4.

Keeton, C. R. and Petters, A. O (2005) Formalism for testing theories of gravity using lensing by compact objects. III. Braneworld gravity. Physical Review D 73:104032.

Landauer, R. (1987). Computation: A fundamental physical view. Phys. Scr. 35, pp. 88-95.

Leff H. (1990). Maxwell's demon, power, and time. Am. J. Phys. 58, pp. 135- 142.

Leff H. & Rex, A. F. (2002) Maxwell's Demon 2: Entropy, Classical and Quantum Information, Computing, (Taylor & Francis).

Lewis, G. N. (1930). The symmetry of time in physics. Science 71, pp. 569- 577.

Linde, A. (2005). Particle Physics and Inflationary Cosmology. Contemp. Concepts Phys. 5 1–362; arXiv:hep-th/0503203.

Linde, A. (2008). Inflationary Cosmology. Lect. Notes Phys. 738, 1–54; arXiv:0705.0164.

Lloyd, S. (2002) Computational Capacity of the Universe . Physics Review Letters; American Physical Society 88 (23): 237901.

Maartens, R. and Koyama, K. (2010). BraneWorld Gravity. Living Rev. Relativity 13, (2010), 5 Living Rev. Relativity 13, (2010), 5 arXiv.org:hepth/ 1004.3962. p 26.

Maxwell J. C. (1871) Theory of Heat. Longmans, Green, and Co;, London. Ch 12.

Maxwell J. C. Letter to P. G. Tait, (1867). Quoted in C. G. Knott, Life and Scientific Work of Peter Guthrie Tait, Cambridge University Press, London, 1911, pp. 213-214; and reproduced in The Scientific Letters and Papers of James Clerk Maxwell Vol. II 1862-1873 (Ed.: P. M. Harman, 1871), Cambridge University Press, Cambridge University Press, Cambridge, 1995, pp. 331-332. Also J. C. Maxwell, Theory of Heat, Longmans, Green and Co., London, Chapter

Mersini-Houghton, L. (2010). Selection of Initial Conditions: The Origin of Our Universe from the Multiverse. In: Vaas, R. (ed.). Beyond the Big Bang. Springer, Heidelberg.

Neumann, J. von (1955). Mathematical Foundations of Quantum Mechanics. Princeton University Press, Princeton. (published originally in German 1932) Ch V.

Penrose, O. (1970). Foundations of statistical mechanics. Pergamon Press, Oxford, pp 221-238.

Penrose, O. (1979). Foundations of statistical mechanics. Rep. Prog. Phys. 42, pp. 1937-2006.

Penrose, R. (1960). A spinor approach to gravity, Ann. Phys. 10, p. 171.

Penrose, R. (2002) The emperor's new mind: concerning computers, minds, and the laws of physics, Oxford university press, [ISBN 0-19-286198-0]

Planck, M. (1922). Treatise on Thermodynamics. Dover Publications, Inc., New York, (translated from the 7th German Edition)) pp. 203-207.

Poincare, J. (1893). Mechanism and experience. Revue de Metaphysique et de Morale 1, pp 534-537. Reprinted in S. G. Brush (1966). Kinetic Theory . Vol. 2. –Irreversible Processes, pp. 203-207.

Popper, K. R. (1982). Quantum Theory and the Schism in Physics. Roman & Littlefield, p. 114.

Randall, L., and Sundrum, R. (1999). Large Mass Hierarchy from a Small Extra Dimension. Phys. Rev. Lett. 83, 3370–3373.

Rees, M. J. A. (1989) Rev. Astr. Astrophys. 22, pp. 471–506.

Rees, M. (2001). Our Cosmic Habitat. Princeton University Press, Princeton.

Remillard, R. A., McClintock, J. E., Bailyn, C. D. (1992). Astrophys. J. 399, L145–L149.

Reynolds, C. S. (2008). Bringing Black Holes into Focus. Nature 455, 39-40 (4 September 2008).

Saha, M. N. and Srivastava, B. N. (1958). A Treatise on Heat. The Indian Press, Calcutta, p 320.

Serreli V. et al (2007), Exercising Demons: A molecular information ratchet. Nature 445, 523-527.

Smith, C. and Wise, N. M. (1989). Energy and Empire: A Biographical Study of Lord Kelvin. (pg. 500). Cambridge University Press.

Smolin, L. (1992). Did the universe evolve? Class. Quant. Grav. 9, pp. 173–191.

Smolin, L. (1997ab). The Life of the Cosmos. Oxford University Press, Oxford. Stacey, F. D. Tuck, G. J. and Moore, G. I. (1987). Quantum gravity: Observational constraints on a pair of Yukawa terms, Phys. Rev. D 36, pp. 2374–2380.

Stefano Ansoldi, Antonio Aurilia and Euro Spallucci, (2002). Fuzzy dimensions and Planck's Uncertainty, Principle for p-branes, Classical and Quantum Gravity 19, 3207.

Susskind, L. (2006) The Cosmic Landscape, Back Bay Books; Little Brown (253 –292).

Szilárd, L. (1929a). On the extension of entropy in a thermodynamic system by the intervention of intelligent beings. Z. F. Physik 53, pp. 840-856. English Translations: Behavioral Science 9, pp. 301-310 (1964); B. T. Feld and G. Weiss Szilárd, The Collected Works of Leo Szilárd: Scientific Papers, (MIT Press, Cambridge, (1972), pp. 103-129; and J. A. Wheeler and W. H. Zurek, Quantum Theory and Measurement (Princeton University Press) pp. 539-548.

Szilárd, L. (1929a-f). On the Decrease of Entropy in a Thermodynamic System By the intervention of Intelligent Beings, pp. 124 - 131 In H. Leff and A. Rex (1990 Book Maxwell's Demon . Princeton University Press. Princeton, New Jersey.

Tegmark, M. (2003) Parallel Universes. Scientific American.

Tegmark, M. (2004). Parallel Universes. In: Barrow, J., Davies, P.C.W., Harper jr., C.L. (eds.). Science and Ultimate Reality. Cambridge University Press, Cambridge, pp. 459–491; arXiv:astro-ph/0302131

Thomson, W. (1874). The Kinetic Theory of the Dissipation of Energy. Nature 9, pp. 441-444.

Thomson, W. (1879). The Sorting Demon of Maxwell. R. Inst. Proc. IX, 113. Reprinted in Lord Kelvin's, Mathematical and Physical Papers. Vol 5 (Cambridge, U. P. London 1911) pp. 21-23.

Vaas, R. (2004a). Time before Time. Classifications of universes in contemporary cosmology, and how to avoid the antinomy of the beginning and eternity of the world. arXiv:physics/0408111.

Vaas, R. (2004b). Ein Universum nach Maß? Kritische Überlegungen zum Anthropischen Prinzip in der Kosmologie, Naturphilosophie und Theologie. In: Hübner, J., Stamatescu, I.- O., Weber, D. (eds.) (2004). Theologie und Kosmologie. Mohr Siebeck, Tübingen, pp. 375–498.

Vaas, R. (2005). Tunnel durch Raum und Zeit. Kosmos, Stuttgart.

Vaas, R. (2008a). Phantastische Physik: Sind Wurmlöcher und Paralleluniversen ein Gegenstand der Wissenschaft? In: Mamczak, S., Jeschke, W. (eds.). Das Science Fiction Jahr 2008. Heyne, München, pp. 661–743.

Vaas, R. (2008b). Hawkings neues Universum. Kosmos, Stuttgart.

Vaas, R. (2010). Beyond the Big Bang, Springer, Heidelberg.

Vaas R. (2010). Multiverse Scenarios in Cosmology: Classification, Cause, Challenge, Controversy, and Criticism. Journal of Cosmology, Vol 4, pp. 664- 673.

Vilenkin, A. 2006. ManyWorlds in One. Hill and Wang, New York.

Weinberg, S. (1967). A model for the Leptons. Phys. Rev. Lett. 19, 1264–1266.

Wiener, N. (1950). Entropy and information. Proc. Symp. Appl. Math. Amer. Math. Soc. 2 89.

4. The Quantum Hologram And the Nature of Consciousness

Edgar D. Mitchell, Sc.D.[1], and Robert Staretz, M.S.

[1]Apollo Astronaut, 6th Man to Walk on the Moon

Abstract

We present a new model of information processing in nature called the Quantum Hologram which we believe is supported by strong evidence. This evidence suggests that QH is also a model that describes the basis for consciousness. It explains how living organisms know and use whatever information they know and utilize. It elevates the role of information in nature to the same fundamental status as that of matter and energy. We speculate that QH seems to be nature's built-in vast information storage and retrieval mechanism and one that has been used since the beginning of time. This would promote QH as a theory which is a basis for explaining how the whole of creation learns, self-corrects and evolves as a self-organizing, interconnected holistic system.

KEY WORDS: Awareness, Consciousness, Entanglement, Information, Intention, Intuition, Mind, Non-Locality, Perception, Phase Conjugate Adaptive Resonance, Quantum Hologram, Resonance, Zero Point Field

1. Definition of Consciousness

One common dictionary definition of consciousness is "the ability to be aware of and to be able to perceive the relationship between oneself and one's environment". The most basic definition, however, is simply "awareness". Another definition suitable for more complex organizations of matter such as animals with a brain includes a description which contains some of the following ideas: "thoughts, sensations, perceptions, moods, emotions, dreams, and awareness of self". Just like life itself, consciousness is one of those things that is easy to recognize but very difficult to define. It has been debated by philosophers in the West since the time of ancient Greek civilization over twenty five hundred years ago.

Eastern traditions have been wrestling with the concept of consciousness for millennia and seem to have a much better handle on it although still not nearly complete. In the West, explanations of consciousness have been mostly ignored or left to our religious traditions. This is certainly true since the time of Descartes and the philosophy of Cartesian duality. It has only been in very recent times that a serious effort to understand mind or consciousness has been undertaken by the scientific community. Much of the effort now underway is based on the assumption of epiphenomenalism, that consciousness, or mind if you prefer, is a byproduct of the functioning of underlying physical structures of the brain and that mind is confined entirely within the brain's processes. However, there is a considerable amount of accumulating experimental and anecdotal evidence suggesting that this interpretation is not correct (Chalmers 1996; Penrose 1994).

At a basic level consciousness seems to be associated with a sense of separation and awareness of the surrounding environment from the conscious entity. It also seems to be associated with the ability to process, store and / or act on information gathered from that external environment. But is consciousness restricted to a functioning brain? Are microscopic organisms such as viruses, amoeba, and algae conscious in some primitive sense? Clearly they do not have brains let alone a nervous system or even neurons. And yet they demonstrate purposeful behavior and are aware of their environment. Amoeba, for example, search for food by moving on pseudo pods toward prey that they eventually surround, engulf and digest. Several types of algae are so versatile that they change the process how they obtain food based on available sunlight. When light is plentiful, they gravitate towards it, which they sense through a photoreceptor at one end of the cell. If the light is too bright, they will swim away toward more suitable lighting conditions.

At a more primitive level viruses are considered by many scientists as non-living because they do not meet all the criteria commonly used in the definition of life. They do, however exhibit some aspects of consciousness or at least some rudimentary form of an awareness of their surroundings. Unlike most organisms, viruses are not made of complete cells. They reproduce by invading and taking over the machinery of their target host cell. When a virus comes into contact with a potential host, it inserts itself into the genetic material of the host's cell. The infected cell is then instructed to produce more viral protein and genetic material instead of performing its normal functions. Is that purposeful behavior or intentionality by the invading virus?

It would seem that based on our first definition, even simple living entities are conscious to some degree, since they display a level of awareness and intentionality to, in some way, manipulate their environment. And, it's not just restricted to living entities. We find certain properties all the way down to the subatomic level, particles in some sense aware of their environment. How is this possible? At the molecular, atomic and subatomic levels it is through the quantum phenomenon of entanglement and non-locality that particles act and react to other particles with which they have become entangled.

2. The Roots of Consciousness

Could it be that at the most fundamental level consciousness begins with these ubiquitous quantum events? We believe that is, in fact, the case. Furthermore recent evidence suggests that certain quantum phenomena (Schempp 1998, 2007, 2008; Mitchell 1995, 2003, 2008) operate at the macro level as well as the micro level and are responsible for many phenomena that living entities experience that cannot be otherwise explained. This would explain how twins or mother and child seem to communicate telepathically when at least one of them is under extreme duress as we have seen is so often reported anecdotally in the literature. In fact, as we shall soon see, several of these so called quantum group effects including a whole class of so-called psychic phenomena have been documented throughout recorded history. As we shall show shortly, some of these phenomena have been either demonstrated or suggested in recent laboratory experiments.

Just like everything else in nature, moving up the evolutionary chain of increasing complexity in organisms is built upon the foundation of what has come before. For consciousness, we propose an evolutionary scaffolding as illustrated in Figure 1. At the lowest level resides the most basic aspects of undifferentiated awareness built upon the quantum principles of entanglement, non-locality and coherent emission / absorption of photons. These phenomena are ubiquitous throughout the world of matter. At this most elementary level all matter seems to be interconnected with all other matter and this interconnection even transcends space and time. We postulate that this is the basis for the most fundamental aspect of consciousness which we describe as undifferentiated awareness and this mechanism of basic perception extends up the entire evolutionary chain of increasing complexity of living organisms. The differences in consciousness being in degree and not in kind as one moves from left to right up the consciousness ladder.

Moving beyond the simplest level of the consciousness towards mentality (e.g. higher functions of consciousness / mind), the next level pertains to the consciousness of simple life composed primarily of single celled organisms. Here we have the beginnings of a crude capability of awareness through the use of molecular structures that are sensitive to their environment utilizing either chemical and / or electro-magnetic means. In the latter case this is especially prevalent at those frequencies in the EM spectrum corresponding to visible light, infrared and ultra-violet waves. Sensing the external environment by these means have been considered the primary mechanisms of perception that have been the focus of classical science for quite some time now. For simple organisms, like plants, amoeba, viruses, etc. clearly there are no brain structures to facilitate perception. Marcer (1997) has applied a theory called the Quantum Hologram (see below) to propose that life at the most basic level, including such things as prokaryote cells and neurons in higher organisms, exchange information with their environment by utilizing the quantum property of non-locality. The implication here is that all organisms from the simplest to the most complex are interconnected at a very fundamental level using information obtained by nonlocal quantum coherence (Ho 1997). Furthermore they are even interconnected with their external environment by their coherent quantum emissions via the mechanism of the Quantum Hologram as we shall soon demonstrate (Marcer et al. 1997).

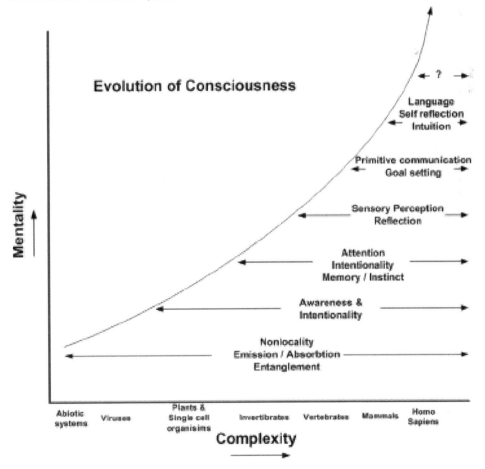

Figure 1 - Evolution of Consciousness

Nature always seems to evolve into mechanisms and structures that enhance an organism's survivability in its environment. Other higher levels of perception and awareness are necessary to locate objects in space-time in addition to including the non-local quantum simple awareness effects as we have just described. So, as we move further up the evolutionary ladder, we continue to enhance mentality as shown in Figure 1. At each level, the organism has access to the perceptual mechanisms of the levels below. Clearly, at each level the organism is utilizing information (e.g. patterns of energy and matter) obtained from its environment. This implies that there is a process (e.g. consciousness) that uses and assigns meaning to this information. Note that "meaning" is also information that places the perceived information into context for use by the organism.

Penrose and Hamereoff (1998) have proposed that microtubules in brain cells might be responsible for more fundamental forms of perception. They also postulate that these microtubules provide the foundation for the emergence of higher orders of consciousness in species with a brain. Microtubules are hollow cylindrical polymers of the protein tubulin which organize cellular activities. These protein lattices exist in the cell's cytoskeleton found within the brain's neurons. Penrose and Hamerehoff claim that tubulin states are governed by quantum mechanical effects within each tubulin interior and these effects function as a quantum computer using "quantum bits" that interact non-locally with other tubulins and with the Quantum Hologram. When enough tubulins are entangled long enough to reach a certain threshold a "conscious event" occurs. Each event results in a state which regulates classical neural activities such as triggering neural firings that ultimately affect perception, learning and / or memory.

At first glance, quantum states in biological systems seem difficult to maintain in brains because these quantum states generally require extreme cold (e.g. close to 0°K) to eliminate thermal noise produced by the environment. Some researchers argue that this is necessary to prevent decoherence of these quantum states. However Penrose and Hameroff claim that decoherence may be prevented by the hollow microtubules themselves which act as shields to the surrounding "noisy" environment.

3. Holographic Processing

It has been suggested that the brain processes and stores information holographically as a massively parallel processing and associative computer system. Pribram (1999) and others have studied this extensively and demonstrated it in both the laboratory with animals and in operating theaters on humans. In the latter case the brain has been exposed and stimulated with low voltage electrical signals while the patient was conscious to describe the resulting experience. These subjects have recalled extremely detailed and vivid memories as if they were actually reliving the experiences being recalled. Animals that have had portions of their brains damaged or removed have been able to recall memories (ex. optimum ways to run a maze) even when the damage has been extensive. These experiments and several others provide evidence that suggest that brains store information holographically (e.g. stored as images contained within interference patterns). Marcer has further extended this to postulate that not only information is stored in this manner but that information is processed holographically in the brain as well. He has also attributed this processing to, in effect, creating a detailed three dimensional movie generating the stream of consciousness that the mind experiences.

Holographic processing is accomplished with the brain acting as a phase conjugating device (e.g. a phase gate) which is type of logic circuit where the inputs are sensitive to the phase of the input signal. The result is a "virtual" signal which is a mirror image of the quantum emissions (e.g. photons of light) actually emitted from the object being perceived. The brain acts as an information receptor utilizing adaptive resonance with a specific range of EM frequencies (e.g. wavelengths) in its input path. The input signals received are a representation of the external object resonating with similar virtual signals generated (output) by the brain. This sets up a resonance condition which may be interpreted as a standing wave between the object and the brain. The input signal is really the quantum emission spectrum of the object being perceived.

Like all holographic processing, the associative pattern that is created facilitates retrieval of information in a resonant loop utilizing the overlapping reference signals of quantum emissions from the external object. It enables the perceiving organism's brain structures to perform pattern classification and recognition of the resonating signals. This resonance process is called phase conjugate adaptive resonance (PCAR). We believe that PCAR is the basis for the most fundamental level of perception in all living organisms in the evolutionary tree of life (Mitchell 2001). As an example think of bats, dolphins, whales that use sonar to send out signals and receive reflections back to locate targets. PCAR is the brain analog of that process.

One of the most important aspects of a laser hologram is that it exhibits the distributive property. This means that even a small part of an entire holographic record contains the entire record of the recorded image but with less resolution (e.g. definition) when reconstructed. Figure 2 is an actual quantum hologram or wave interference pattern of a patient's brain that would appear when exposed on a photographic plate. The left column labeled "A" represents the resulting interference pattern and the right column labeled "B" shows the corresponding 3 dimensional brain image. These images were produced by a typical Magnetic Resonance Imaging (MRI) machine similar to the ones used in medical diagnosis. In the left column (labeled "A") in the middle, and bottom row of pictures, the outside and inside respectively of the entire interference pattern shown in A (top row) have been removed to show the reduced resolution of B, compared to B (top) to illustrate A's holographic nature.

Quantum holography operates similarly in that quantum emissions from complex matter, for example, bio-matter, carry information about the entire organism. Stem cell research supports this concept. The fact that living cells in any organism evolve and grow from more simple stem cells, implies quantum entanglement throughout the organism and its composite parts, with an associated instantaneous exchange of information through PCAR. Thus some information about the entire organism is carried in the quantum emissions from its parts.

For those readers who are familiar with recent developments in ground based astronomical telescopes, a problem that has plagued astronomers since the invention of the telescope has been dealing with the aberrations in the telescopic images caused by the shimmering from the earth's fluid atmosphere. Recent developments in laser technology and high speed computing have allowed astronomers to eliminate these aberrations. A coherent laser beam is targeted to follow the same path that the telescope is focused on. As the laser beam is reflected off the shimmering atmosphere back to a receiver, the phase delay in the returning signal is processed and compared in real time against the reference beam transmitted. This comparison enables the computer to correct for aberrations with the telescope's optical imaging system caused by the atmospheric distortions. The result is that we are now able to receive clear images on earth based telescopes just like we can do with the Hubble Space Telescope which, of course, is outside the earth's atmosphere. This concept of self correcting optical imaging telescopes is not unlike the PCAR process described above.

PCAR is necessary for the brain to perceive objects as they really exist in three dimensional space. If the brain had to rely solely on the visible light spectrum that was reflected off the external object and onto the retina of the eyes, the object would appear two dimensional just as it would be if a picture of the object was recorded photographically with a camera. Contrary to the popular opinion that we see objects in three dimensions entirely because of binocular vision, just close one eye and observe an external object with the remaining open eye. The object appears "out there" and not as an image "in the brain" because of PCAR. This clearly presents a survival advantage to an organism allowing it to accurately see and locate objects (especially predators and food) in three dimensional space.

Holographic processing is not restricted to processing sensory information in the visible light portion of the electromagnetic spectrum but it applies to enhancing all of the five normal senses. Consider snapping your fingers. The sound seems to originate from the location of the fingers in 3-D space and not at a point within the brain. As before, this experience results from the fact that the signal carrying the sound to the brain is resonating with the conjugate virtual signal created in the brain.

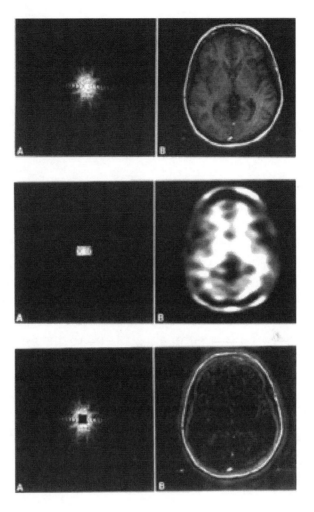

Figure 2. Quantum Hologram. Illustration retrieved from: http://www.bcs.org.uk/siggroup/cyber/quantumholography.htm with permission

Figure 3 is an illustration describing the PCAR process. Emissions from the object of attention (e.g. the apple) are received (e.g. input) by the brain. The brain in turn creates phase conjugate (mirror image) "virtual" waves to identify the object. The standing wave that results allows the brain to locate and associate the object in space. The standing waves are created by interference of the two waves traveling in opposite directions. Recall that standing waves are waves that do not appear to propagate but are fixed in position and just move in the vertical direction about the zero point on the reference line. This standing wave creates the resonant condition that allows the brain to process the information so as to locate the object in 3 dimensional space.

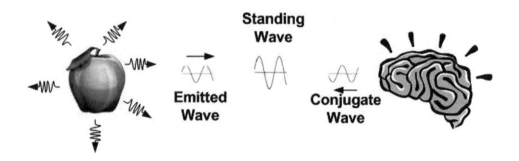

Figure 3. - The PCAR Process

Max Planck, considered by most as the founder of Quantum theory solved the problem of the so-called ultraviolet catastrophe in the late 1890's. He postulated a theory, now known as the Planck postulate, that electromagnetic energy could be emitted or absorbed only in discrete quanta which formed the basis of the description of black body radiation. This was ultimately extended and used to describe how all matter absorbs and reemits photons (quanta of energy) from and into the quantum foam of the zero point field (ZPF) that pervades all matter and even the vacuum of space (Haisch et al. 1997). Normally these emissions are random exchanges of energy between particles and the ZPF. However, the emissions from complex matter (e.g. living organisms) have been shown to exhibit quantum coherence and also carry information non-locally. Recall that the quantum phenomena of non-locality implies instantaneous transmission of information across space and time (Darling 2005).

4. Quantum Emissions and Non-Locality

As we shall demonstrate, non-locality also applies to macro scale objects and is referred to in this paper as a "group" phenomenon (Schempp 1998 and 2008). As a simple example, consider visiting a sacred place of worship such as the Notre Dame Cathedral in Paris. As one enters the cathedral, it is hard not to feel a sense of hush, awe and reverence. Over the centuries, countless people have entered this majestic cathedral with these very feelings. And these feelings were literally absorbed by the structure over the years through the process of quantum emission from the people and resonance with the very atoms and molecules of the cathedral structures. The longer the exposure to this resonance, the more coherence has been achieved with the molecules and atoms in the structure. This coherence re-manifests as emissions back into the environment which then resonate with the visiting people (via the cellular mechanism proposed above) entering the structure. The result is the subjective feelings of hush, awe and reverence which is exactly what a visitor experiences when entering the cathedral. This is another example of PCAR.

In Figure 3, we showed the emissions from the apple resonating with and being absorbed by the brain. However, the opposite is also true. The associated biomass of the entire body is also emitting coherent quantum fluctuations that are being absorbed by the apple and is therefore having some effect on the apple as well. Now add in everything else in the environment surrounding the apple and the human, which would lead one to conclude that every object, in some sense, has an effect on every other object. Like it or not, we appear to live in a participatory universe - there is no such thing as pure objective reality and we influence everything that we interact with. Perhaps this is the same mechanism why we "feel" positive energy from some people while others seem to emit negative energy?

Quantum Holography (QH), which we have alluded to several times above, is a recently discovered attribute of all physical matter and has been validated by experimental work with functional magnetic resonating imaging (fMRI). In his work with MRI tomography, Schempp (1999) used a mathematical formalism to expand quantum information theory. He validated his approach by significantly improving the definition and specificity of MRI and, in the process, discovered the inherent information content of the emitter-absorber model of quantum mechanics. This work provides a model to understand quantum level information processing of biological systems, specifically how the reception and processing of information leads to the functions of memory, awareness, attention and intention. We extend Schempp's work and postulate that all the cells of any biological entity and all its other organ systems, including the brain, have evolved as a massively parallel, learning, computing system. And the key ingredient of understanding this computing system and its processes is the quantum hologram which we will now describe.

As we have previously described, quantum emissions from any material entity carry information non-locally about the event history (e.g. an evolving record of everything that has happened) of the quantum states of the emitting matter. Recall that these quantum emissions are in the form of EM waves of many different wavelengths (or frequencies if you prefer) and that the information associated with these emissions is contained in both the amplitude and the phase relationships of the emitted waves as interference patterns. This is similar to the way that information is stored in the interference pattern on a holographic plate as described previously. These interference patterns can carry an incredible amount of information including the entire space-time history of living organisms.

5. The Quantum Hologram

Mounting evidence seems to indicate that every physical object (both living and nonliving) has its own unique resonant holographic memory and this holographic image is stored in the Zero Point Field (Marcer et al. 1997). Information in the ZPF is stored non-locally and cannot be attenuated. Furthermore this information can be picked up via the mechanism of resonance as we described above. This information, its storage and its access is collectively called the Quantum Hologram (QH).

We can think of an organism's QH as its nonlocal information store in the ZPF that is created from all the quantum emissions of every atom, molecule and cell in the organism. Every objective or physical experience, along with every subjective experience is stored in our own personal hologram and we are in constant resonance with it. Each of us has our own unique resonant frequencies or our unique QH which acts as a "fingerprint" to identify our non local information stored in the zero point field. Since the event history of all matter is continually broadcast non-locally and stored in the QH, the QH can be viewed as a three dimensional vista / movie evolving in time which fully describes everything about the states of the object that created it. Not only do we each have our own unique QH, but it is also possible for others to tap into parts of it through resonance. We shall develop this idea more fully later in this paper.

To illustrate how resonance with an object's or living organism's QH might work, consider the following example. Take two identical guitars, tune the corresponding strings on each guitar to the same frequencies and place them on the opposite sides of a room. Now pluck a string on one of the guitars and notice what happens to the corresponding string on the guitar at the other side of the room. It will begin to vibrate in resonance with the first guitar. This is not unlike the example with singer Ella Fitzgerald and the shattered champagne glass that has been popularized on TV commercials several years ago. It turns out that the vibrating strings on each guitar will produce a standing wave. The vibrations cause the wave to travel down the string (incident wave) to the point where the string is attached to the guitar. The wave will reflect off that point (producing a reflected wave) and travel back in the opposite direction as a mirror image of the incident wave. As the waves meet they will interfere with each other and produce our standing wave which is then propagated through the air.

Now add a third identical guitar with the corresponding strings also tuned to the identical frequencies as the other two. If the corresponding strings on each of any two of the guitars are plucked simultaneously at the same point on the string, the corresponding string on the third guitar will again begin to vibrate as before but with one slight difference. Since both plucked strings were struck at the same point and at the same time, the sound waves produced from each will constructively interfere and reinforce each other in the air resulting in sound waves of greater amplitude. So as before, the string on the third guitar will begin to resonate again but this time with greater amplitude. This assumes that the guitars are place at appropriate locations such that the sound waves from the other two guitars arrive so that the waves arrive in phase. The result will be that they will constructively interfere with each other. If the distances are such that the signals arrive out of phase and destructively interfere, they may cancel out at the third guitar.

Now, repeat this process adding a 4th guitar with the proper placement and so on. In each case the amplitude of the resonating standing wave will continue to increase (unless the responding guitar string should happen to break from the large amplitude of the wave it is receiving). This analogy of the strengthening of the resonating wave shall be of particular importance when accessing information stored in the ZPF. The simple reason is the larger the amplitude of the standing wave, the easier it is for another object to resonate with it.

The analogy of the guitars is similar to how information is stored in the Quantum Hologram with the ZPF. Since the brain operates as a massively parallel quantum computer. The brain does this by setting up a resonant condition with microtubules scatted throughout the brain tuned to the same frequency as the standing waves of the same frequency located in the ZPF. As we mentioned before every macro-scale physical object in nature has its own unique quantum hologram. It exists in 4D space time reality and is a non-local information structure that never attenuates. And, most importantly, it carries the entire event history of the physical object it was created from or, in the case of living organisms, the entire subjective and objective reality experienced by that organism. All this information about the entity is carried in the amplitude, frequencies (e.g. wavelengths) and in the phase relationships of those waves from the emitted entity. Perhaps, most important of all, is the fact that the information stored in the QH is recoverable through the process of resonance not only with the individual organism that created it but with other organisms as well if they are "tuned" to it. This seems to happen most commonly in humans when strong emotional connections exist between the two.

Note that we are not suggesting that we are all virtual beings living in a "literal" holographic reality as interference patterns on nature's holographic plate. We are all real beings living in a very real material existence consisting of matter, energy and information just as we experience it. Also the quantum hologram is not about the discovery of some new kind of subtle energies such as "élan vital", qi or prana suggested by many throughout history. Neither is it about multidimensional theories nor living in other planes of existence other than our normal 4 dimensional space time reality. But this does mean that we now have a mechanism to describe how mind can manipulate matter. We will have considerably more to say about this later in this paper.

The Quantum Hologram is a model of how reality works. Like all models it enables us to make predictions and create interpretations about how nature operates. We can test those predictions to validate and refine the model and perhaps someday even design, build and utilize technologies that implement various aspects of the model's predictions. But the map is not the territory and, like all models, it must be refined as more information becomes available and our understanding improves. And, most importantly, models are subject to interpretation based on our prior knowledge and experiences.

Our QH model seems to explain many effects including aspects of mind, memory, stream of consciousness, various factors affecting health, psychic events, Jung's collective unconscious, the Akashic record and other phenomena that arise out of the resonance with the QH residing in the zero point field. Before the discovery of quantum holography we had no mechanism to model or account for these phenomena let alone for information transfers between objects that these effects imply.

We believe that QH is supported by both experimental and anecdotal evidence suggesting it is also a model that describes the basis for consciousness. It explains how living organisms know and use whatever information they know and utilize. It elevates the role of information in nature to the same fundamental status as that of matter and energy. In fact the QH seems to be nature's built-in vast information storage and retrieval mechanism and one that has been used since the beginning of time. This would promote QH as a theory which is basis for explaining how the whole of creation learns, self-corrects and evolves as a self-organizing, interconnected holistic system. Since the laws of nature appear to be the same throughout the universe there is no reason that it should not also apply to extraterrestrial consciousness as well.

6. Applications & Implications of QH

We will now look at many previously unexplained phenomena and describe how they can all be explained with the quantum holographic model. Then we will attempt to describe some of the profound implications and ramifications to which this theory leads. We will discuss anecdotal evidence, actual experiments, their implications and potential further applications of the Quantum Holographic (QH) model. Before we begin, let us briefly summarize what was earlier in this paper. We described how QH describes a real phenomenon of nature that has been validated in the laboratory. We postulated that it is a description of reality that is based on a mathematical formalism (e.g. a theoretical model) of how nature implements and utilizes information, memory, perception, attention and intention. Furthermore we suggest that QH explains many phenomena in nature where no adequate mechanisms were previously known to describe them. This is particularly true in accounting for the transfer of information between material objects or between objects and their environment.

QH offers a hypothesis and convincing evidence that explains how living terrestrial organisms know and how they utilize information. In doing so, it elevates information to the same fundamental status throughout the universe as matter and energy. Furthermore when energy, matter and information are utilized in processes, QH leads us to the very basis of consciousness itself. So, perhaps the most profound implication of all is that the QH model provides a basis for explaining how the whole of creation learns, self corrects, evolves by being, in some sense, conscious of itself. In other words, QH describes the universe as a self-organizing inter-connected conscious holistic system.

We postulate that the storage mechanism for the Quantum Hologram resides in the zero point field (ZPF). This field is ubiquitous, nonlocal, cannot be attenuated, lasts indefinitely (e.g. never loses coherence), can store unlimited quantities of information and any portion of it encodes the whole just as a hologram does. It can be thought of not only as nature's information storage mechanism but also as nature's information transfer mechanism. QH information is contained in the amplitude, frequencies and the phase relationships of the underlying interference patterns from the emitted quanta. This information is emitted and absorbed by all objects and exists in four dimensional space / time reality. QH applies to all scale sizes from the smallest subatomic particle to the largest structures in the cosmos and takes place at all temperatures (even down to absolute zero). It exists simultaneously beneath the classical descriptions of how information is exchanged between non living objects and below the normal five senses for living organisms.

For living organisms QH applies to intra and inter communications between cells, organs and organ systems, and finally between organisms as well as with the larger environment as suggested by Lipton (2005), Sheldrake (1981) and several others. It applies to all living organisms on earth as well as to all biological entities that exist throughout the cosmos. Whether abiotic or biotic, the entire event history of all matter anywhere in the universe from the micro scale to the macro scale is being continuously broadcast non-locally by coherent quantum emissions. This history is also reabsorbed by (e.g. received) and interacts with all other matter and the ZPF through the exchange of quantum information.

The mechanism of QH applies to all of the cells of the human body (approximately 50-100 trillion) and answers the question that is often posed dealing with how all these cells cooperate and work together to make the whole human. While they are actively cooperating, thousands of cells are dying continuously every second; so many, in fact, that over the course of one week the body will have billions of new cells and yet you remain you with the same memories and the same functionality and the same distinct features. How does that happen?

Every one of our cells contains the same genetic blueprint of DNA and clearly that DNA and environmental influences exert major influences over the development and functioning of our cells. But cells are not only subordinate to DNA; they also function and maintain homeostasis by communicating and cooperating simultaneously with many other cells of the body and by the information they receive from the environment. Much of this inter-cellular signaling is electrochemical in nature but biologists still struggle with some aspects of the mechanisms utilized in this information transfer. With up to 100 trillion cells it is hard to imagine how that many cells can remain in harmony by the slow process of electrochemical signaling especially under times of great distress when survival of the entire organism is at stake and / or requires extremely rapid and coordinated responses.

The primacy of DNA as the master blueprint for an organism has been the central dogma for biology for a long time. There is now very convincing evidence that all organisms on earth from plants to mammals acquire characteristics through the interaction with their environment and can then pass these characteristics on to their offspring (Lipton 2005). This process is called "epigenetic inheritance" and has spawned a new field in biology called epigenetics which is the study of the mechanisms by which the environment influences cells and their offspring without changing genetic codes. This is forcing scientists to rethink evolutionary theory and harks back to the days of Lamarckian evolution. Lamarck's theory, developed 50 years before Darwin, hypothesized that evolution was based on cooperative interaction between organisms and their environment. This interaction enabled these organisms to pass on adaptations necessary for survival as the environment changed.

Lipton states that "results from the Human Genome Project are forcing biologists to the recognition that they no longer can just use genetics to explain why humans are at the top of the evolutionary ladder on earth. From this effort, it turned out that there is not much difference in the total number of genes found in humans and those found in primitive organisms". So where does the information come from that defines who we are? Lipton further goes on to state that "cellular constituents are woven into a complex web of crosstalk, feedback and feedforward communication loops and that thousands of scientific studies over the years have consistently revealed that EM signaling affects every aspect of biological functioning".

How does this mechanism work? Could QH offer an explanation? In addition to DNA and environmental influences in all earth based living organisms, intercellular communication is especially critical in embryonic development. From a single fertilized egg, the embryo divides thousands of times and each time producing identical offspring cells called stem cells. Then at some critical point when the embryo has reached a certain size, something truly miraculous happens. Cells begin to differentiate and form groups of like cells that will eventually become all the highly specialized tissues and organs that make up the human body. Out of the entire mass of undifferentiated cells making up the embryo, how does a particular stem cell suddenly know that it is to transform into a heart cell, liver cell, neuron, etc? Clearly, some of the differentiation results from electrochemical signaling with the immediate surrounding cells. This exchange is certainly necessary and provides information about how a cell must change to express itself correctly to become the right type of cell at the right place and at the right time. But is it possible that this signaling, by itself is not sufficient to explain the full development of the embryo into a complete organism?

Sheldrake (1997) has studied this problem and has proposed a theory called the Hypothesis of Formative Causation. It describes an alternative explanation for how the structure and form (morphology) of an organism develops. In his model, developing organisms are shaped by fields which exist within and around them and these fields contain the form and shape of the organism. He proposes that each species has its own information field, and within each organism there are fields nested within fields. All of these fields contain information derived from previous expressions of the same kind of organisms. He further states:

> That a field's structure has a cumulative memory, based on what has happened to the species in the past. This idea applies not only to living organisms but also to protein molecules, crystals, even to atoms. In the realm of crystals, for example, the theory would say that the form a crystal takes depends on its characteristic morphic field. Further, the morphic field is a broader term which includes the fields of both form and behavior;

Sheldrake's view is that nature forms habits (e.g. memories) and over time these habits strengthen and influence following generations. Similarly, other habits atrophy over time from lack of continued use. In fact Sheldrake is not alone in proposing such a mechanism. The great psychologist Carl Jung has proposed "the collective unconscious" which represents a vast information store containing the entire religious, spiritual and mythological experiences of the human species. According to Jung, these archetypes have existed since ancient times and are inherited where they exist deep with the human psyche and heavily influence the thinking mind. In a similar manner, Teilhard de Chardin proposed the concept of the "noosphere" which represents the collective consciousness of the human species that emerges from the interaction of human minds. De Chardin asserted that as individuals and the global society evolve into more complex networks, the noosphere evolves along with it.

Finally there is the Akashic record which was developed in the Sanskrit and ancient Indian culture. It is described as an all pervasive foundation that contains not only all knowledge of the human experience but also the entire history of the universe. Our normal five senses cannot access this information but it can be accessed through spiritual practices such as meditation. In the last few years, Laszlo (2004) has also been promoting a theory he has named the A-field which contains many aspects that are also very similar to the concepts described in this paper. All these concepts imply a mechanism very similar to our description of QH.

7. Resonance and the Quantum Hologram

Whatever name this mechanism is called, we postulate that the primary means for accessing transcendent information is via the process of resonance. Remember, that since the laws of nature appear to operate the same everywhere in the universe these is no reason to conclude that QH would not also apply to biological entities anywhere in the universe. In higher organisms with brains, the massively parallel processing capabilities of the brain structures are capable of simultaneously resonating with QH information at an incredible range of frequencies. This is shown in Figure 4 where the effects resulting from varying degrees of resonance with the QH is depicted.

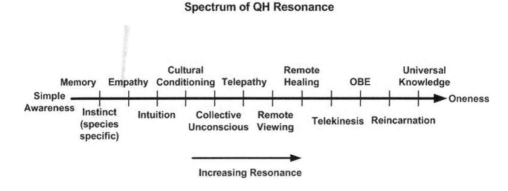

Figure 4. - RESONANCE SPECTRUM

The simplest form of resonance (e.g. entanglement) is shown on the left side of the graph. Moving to the right we show phenomena that manifests with increasing degrees of resonance and frequencies. We have included several phenomena on this graph such as Out-of-Body-Experiences (OBEs) and reincarnation but, as we shall soon see, our explanations of them are based on QH theory and require a different interpretation than those commonly found in popular literature. Finally, we shall describe how the degrees of resonance can occur along with techniques to facilitate them.

We have indicated earlier that information is nothing more than what is contained in patterns of matter or energy. The meaning that is derived from these patterns is developed in the mind of the percipient based on prior experience (e.g. knowledge and memory). In other words, all events are subject to interpretation and that in turn is based on the prior experiences and beliefs of the percipient. The more knowledge and experience we gain the more likely we will interpret an event closer to actual reality.

Changing beliefs especially when they are not based on knowledge but instead are based on faith, however, are another matter entirely. For example, in colonial America, most people attributed thunder and lightning to evil spirits. At that time the obvious solution to the problem was to ward off the evil spirits by ringing church bells. Needless to say that was not very effective. More than a few well intentioned souls were electrocuted while in the bell tower ringing those bells. We now know that the premise of evil spirits is totally incorrect and that thunder and lightning are merely the results of electrostatic discharges between the atmosphere and the ground. How did we get to this understanding? We evolved our understanding and beliefs by investigating and learning about how nature really works through trials, by observation and by experimentation (remember the Ben Franklin experiments with kites we learned about in grade school).

8. Experimental Evidence

We shall now describe some of the experiments testing the concept of non-locality and their findings. There is considerable experimental and anecdotal evidence, although some of it controversial, to suggest that simple organisms perceive and respond to nonlocal information. In the area of human experimentation, results have likewise been mixed for much of the last 75 years. However meta-analysis by Radin (1997) and independently by Utts (1991) across a large and appropriate spectrum of experiments demonstrates compelling statistics that the perception of non-local information exists and is real. Meta-analyses is a new tool which has become essential in many of the soft sciences including ecology, psychology, sociology and medicine. The essence of meta-analysis is that outcomes of collections of previous experiments are analyzed by statistical methods which combine results from different existing studies that address a set of related research hypotheses.

Perhaps if there were a larger body of experimental evidence for simple life forms, similar results of meta-analysis would emerge. Failure to replicate results in well constructed experiments does not, in the case of subtle consciousness phenomena, prove that the phenomenon is missing but rather that a hidden mechanism below the threshold of classical measurement may be operating. For example, the most telling experimental evidence to explain the sometimes inconsistent results relates to direct nonlocal and / or experimenter effects. These effects are unintentional biasing effects on the results of an experiment caused by the expectations, beliefs or preconceptions on the part of the experimenter.

Schmiedler (1972) isolated the "sheep / goat" ("sheep" is a label for believers and "goat" is a label for non-believers in psychic experiments) effect in human experiments decades ago. Experimenters and /or participants in human telepathy (or similar nonlocal) experiments exhibited results statistically above or below chance results depending on their subjective bias towards the experiment. In other words 100% wrong answers would be as statistically significant as 100% correct answers in such tests, and in addition would betray the mind set or intention of the subject whereas only chance results would be inconclusive. More recently, a series of experiments by Schlitz (1997) investigating intentionality clearly demonstrated that experimenter bias (intentionality) affected the outcome even in double blind experiments. Thus, in the subtle realms of mind and consciousness studies, bias, belief and intention clearly have an effect.

The lack of an existing theoretical structure in classical science to support any type of perception of non-local information, much less to support bias, belief or intention as having a nonlocal effect, is quite sufficient to account for anomalous results in many scientific experiments. The prevailing dogma of classical science against any type of non-local action at the level of macro scale reality has not prevented experiments from successfully being conducted. It has sometimes caused positive results to be dismissed as anomalous, of faulty design or outright fraud when in most of these cases the results were defensible had proper nonlocal theory been available.

Radin (2006) describes a series of experiments conducted in the latter half of the 20th century on a whole range of psychic phenomena that suggests that experimenters are obtaining far more correlations than can be expected by chance. This was done by performing meta-analyses for random number generators (RNG) studies subjected to psycho-kinesis (PK) intention which resulted with the odds against chance of 35 trillion to one over the entire database. This analysis followed a decade long series of experiments by Dunne and Jahn (1988) at Princeton University provided overwhelming evidence that human subjects could produce statistically skewed results in mechanical processes normally considered to be driven by random processes. A similar study with Ganzfeld meta-analysis by Radin demonstrated results with odds against chance of 29×10^{18} to one (e.g. one chance in 29,000,000,000,000,000,000). Radin goes on to describe several other studies showing similar results.

Radin has also discovered that audiences watching stage performances would skew the output of random number generators during periods of high emotional content in the performance. In a wide-ranging audience participation experiment, he recorded the output of computer random number generators during the television broadcasts of the O.J. Simpson murder trial. Most television news programs covered this event live for weeks on end with millions of viewers. Again, the results of random number generators set up to monitor this event were skewed corresponding to emotional peaks during the trial drama and corresponding to the number of people watching. A similar effect was noted on 9/11/2001 at the time of the World Trade Center disaster in New York City.

The thesis in the Princeton experiments was that participant intentionality created non random effects to bias the skewed distribution. In the Radin experiments, the results were not the result of intentionality because the participants were unaware of the experiment, but his hypothesis was that rapt attention drove the system away from randomness and toward greater order. These results suggest that attention and intention provide closely correlated outcomes and further, that randomness may not be a property of nature but what may be perceived as random noise in a system may just be awareness that is not in resonance at that moment with the particular perceptual system.

Many types of mind-to-mind or mind-to-object experiments have been rigorously and routinely conducted for decades with statistical significance but they are often dismissed or ignored by mainstream science because the implications of non-local action are so foreign to the mainstream view of objectivism and the possibility of mind-matter interactions. However, if we consider the condition of resonance is necessary (specifically PCAR as described earlier), then we must also consider the perceived object (e.g. the target) and the percipient's perceptual system as entrained in a phase locked resonant feedback loop. The incoming wave from the target carrying the emitted information may be labeled as "perception" from the view point of the percipient, and the return path may be labeled as attention (or intention) depending on what the percipient is trying to achieve. Note however that this is a two way street, the act of perceiving also affects the target object being perceived! Therefore we do live in a participatory universe. There is no such things as pure objectivity.

In the case of non-local effects at a distance, outside the body, simple correlation of entangled particles is the most basic form of perception. And these correlations between entangled particles are reciprocal. Action on one particle creates an effect on other entangled particles instantaneously and even across large distances. This phenomenon is no less important for macro scale objects.

Sheldrake (1999) has conducted experiments with dogs whereby the animals correctly anticipated their owner's departure from a remote location to return home. He has also conducted other successful experiments on previously unexplained behaviors of animals. In one example rats that were learning to traverse a new maze benefited non-locally from the experience of others that had previously learned the maze in the total absence of classical space time information. Other examples include distant (e.g. nonlocal) awareness of deaths and accidents, animals that heal humans and those sensitive to forebodings of natural disasters.

It is not surprising then, that humans exhibit an even wider range of reactions to non-local information. The evidence suggests that humans can perceive, recognize and give meaning to nonlocal information across a broad range of complexity, from inanimate objects, simple organisms, animals and other humans (refer to Figure 4). The existence of QH provides an adequate informational structure to permit a theory for the observed results. This is a classical example, where results are repeatedly observed over time that fall outside the prevailing paradigm, and must await new developments in science before the phenomenon can be adequately explained. Perhaps this explains psychic abilities.

In humans, it is a well established meditation principle that prolonged focused attention on an object of meditation causes the percipient and the target object to appear to merge so that a much deeper level of understanding about the object is obtained. This includes information such as its history or internal functioning that would not be available through classical space-time information. The quantum holographic theory describes how this phenomenon might take place. Further, it is accepted that the mind and associated brain with its 100 billion neurons function together as a massively parallel pattern matching (e.g. information) processor, capable of performing many tasks simultaneously. Most of this processing is done subconsciously or in the right hemisphere which is attributed to the intuitive part of the mind.

Conscious focused attention is a unique and singular task that takes place sequentially mostly in the left hemisphere in the cognitive part of the brain. The condition of attention deficit disorder (ADD) is precisely the problem of a percipient being unable to maintain a singular focus for a sufficient time to complete a desired task or observation. Thus the action of focusing attention by a percipient may be construed as a necessary condition for resonance (PCAR) to be established with the perceived object. Even for people with such a handicap, reducing stress, eliminating distractions, and quieting the mind via meditation may also improve one's ability to focus thereby improving the resonance condition.

Healers typically report such a focusing to create a resonance with the object of their healing activities. Once in resonance, they often report sensing in their mind some sort of picture which appears as a type of 3-D holographic image. They maintain that diseased or damaged tissues in the target often appear as fuzzy or appear somehow different from the normal tissue surrounding it. Sometimes they describe it as sensing energy blockages. They claim to be able to focus energy or somehow manipulate (e.g. intentionality) the diseased tissue which over time causes the image to change and take on the same characteristics of the healthy tissue surrounding it. Could this be the result of the act of intention of the healer resonating with the quantum emissions and subsequent absorptions by the diseased tissues?

Healers and other psychically sensitive individuals often enter into resonance with the object of their focused attention (or intention) by using an icon (e.g. a representation of the object of interest). Similarly people praying for others (not in a religious sense of supplication to a higher being) are suggestive of initiating a non-local resonance process with a target object. Healing prayer has existed in all cultures for millennia. If prayer did not produce some positive results, it is likely that religion would have abandoned it centuries ago. For most of its history healing prayer was attributed to supernatural agency rather than resonance with the target's QH. This is simply another example of phenomenology waiting while science catches up as in our colonial lightning example above.

In recent times Dossey (1993) and many others have attempted to document the efficacy of prayer, particularly healing prayer. Some claim the results establish the case for healing prayer. However the difficulties of controlling all the variables, the experimenter affect, etc. in such clinical studies leave many avenues for valid criticism. The fact that Radin's many studies demonstrated that attention alone produced non-local results in REGs (random event generators) and other machines in reducing randomness (e.g. increasing order) confirms that information has a nonlocal effect and may be correctly formulated as negative entropy. These results apply to healing prayer as well. In these cases, icons are often used to facilitate this resonant process. Icons can be an image, picture, representation or an article associated with the target object of the intention. What each of these modalities has in common is that they appear to provide a mechanism for the intender to "tune in" or resonate with the target. Touching an icon seems to satisfy the resonant (PCAR) requirement and probably allows the intender access to the information about the target not available from normal space-time information. Police agencies often use this modality with psychics who then focus their attention to gain information about a crime scene often with considerable success.

Healers and people praying may also utilize icons but in this case with focused intentionality to resonate with the person to be targeted by similar means. The use of icons to retrieve nonlocal information also suggests an explanation of water memory and homeopathy. Molecules of toxic substances from an original solution are removed by serial dilution. Could some of the water molecules resonate with the emitted photons from the original toxic substances and later resonate with the human immune system when absorbed by it?

If, as required in the theory of the Quantum Hologram, the icon has been in the presence of the individual or contains the signature of the person about whom information or healing is desired, the event history of the icon and that of the individual intersect. The phase relationships of the quantum emissions of the icon contains a record of the target object's journey in three dimensional space and time, as well as the quantum states through which it has passed on this journey. The sensitive individual, with a honed talent, often seems to be able to decode the information coded in these phase relationships of the photons emitted from the icon about the individual or object sought. It may also be the case with the bloodhound that additional non-local information has been gained about the subject, even though the classical explanation is that the animal is operating only with heightened olfactory sensing.

Although perception in the three dimensional world requires and utilizes resonance (PCAR), most humans do not routinely bring to conscious awareness non-local information when operating in ordinary three dimensional reality. We perceive objects as presented by space-time information, that is, shape, color, function (tree, chair, table, etc) but are not usually aware of the additional non-local information (location in space, threats, etc) unless there is strong emotional connection. Consider the case of an infant separated from its parents during time of war or unprecedented disaster. Years later, by a chance reunion, the now unfamiliar child and / or birth mother sense a strong connection while others sense nothing. Could this be because of the resonance between mother and child during pregnancy and through the birth process?

It usually takes training as provided by many esoteric traditions and / or certain naturally sensitive individuals to routinely perceive the non-local holographic information associated with a particular target object. There is considerable evidence to suggest that the brain / mind has these latter capabilities at birth. The development of language, suppression of these capabilities by cultural conditioning and subsequent lack of practice all contribute to the atrophy of natural ability of conscious, intuitive perceptions. Perhaps cultural conditioning is one of the reasons why so called reincarnation experiences are so common in children in eastern cultures while virtually unheard of in the west. The late Dr. Ian Stevenson (2001) of the department of Psychiatric Medicine at the University of Virginia traveled around the world and investigated children usually from the ages between 2 and 5 who claim to have lived previous lives.

> "At the same time they have often displayed behaviors or phobia that were either unusual in their family or not explained by any current life events. In many cases of this type the child's statements have been shown to correspond accurately to facts in the life and death of a deceased person; in many of these cases the families concerned have had no contact before the case developed."

Our view is that although the reincarnation event is a real non-local event experienced by the child, the interpretation of the event is not correct. We believe that the person is in a high state of resonance with the quantum hologram of the deceased and is able to retrieve QH information about the deceased from that resonance condition. As the child ages, rational left brain processing begins to dominate and the child is no longer able to resonate with the QH of the deceased unless the child has been trained to maintain that state of altered consciousness. We would attribute a similar effect with someone who experiences an out-of-body experience (OBE). Again, this most likely represents a high state of resonance with the remote location and the experiencer is retrieving and processing the QH of the objects at the remote location being visited non-locally.

In cases like the ones just described, meditators, mystic adepts and natural psychics routinely demonstrate that non-local information is perceptible from physical objects and icons by focusing attention, quieting the left brain and allowing intuitive perceptions to enter conscious awareness. Those most practiced in meditation experience an altered sense of space-time, the dissolution of self, have access to universal knowledge and sometimes feel a unified sense of oneness with all of existence. Along with this sense of oneness comes a feeling of immense bliss and a great clarity of mind. We postulate that they have entered into a state of high resonance with the QH and have access to all the information that is implied by such unification. This seems to describe the epiphany that I (Edgar Mitchell) experienced on my return flight from the moon.

Particularly in western tradition, academic interest has been on left brain or rational processing rather than right brain intuitive functions. It is the left brain cognitive ability in humans that provides acceptable labeling of the intuitive, creative and artistic processes taking place in the right brain. Given the fact that with training and practice, all individuals can reestablish and deepen their cognitive access to intuitive, non-local information demonstrates that learning recall is taking place within the whole brain itself and involves enhanced coherence and coordination between the hemispheres and with the QH. This process is different and distinct from the left brain function of extending and extrapolating factual data and forming conclusions based on logical deduction to leap to an "intuitive" conclusion, while omitting the immediate steps leading to that conclusion.

When an object or person of interest is not in the immediate vicinity of the percipient so that space-time information obtained by normal senses is unavailable for receiving and interpreting nonlocal information, the method is somewhat different in obtaining resonance with the target. The case in point is the subject of Remote Viewing (RV) which is another latent ability we all have to some degree. RV allows us to describe and experience activities and events that are normally precluded from us with ordinary perception from our normal five senses.

Remote viewing has been researched extensively by Putoff (1996) and Putoff and Targ (1976) at Stanford Research Institute since the mid 1970's. Their work attracted the attention and funding from the U.S. Central Intelligence Agency and was conducted in secret for almost 20 years. Some of the work involved exploring the limits of what remote viewing could do and also in improving the quality and consistency of the result. Much of the remainder of the effort was in training operatives to collect intelligence information against foreign adversaries. The government funding of the effort ended after the collapse of the Soviet Union in the 1990's.

For the purposes of our discussion with RV, the questions we are interested in pertain to the "reference signal" used to decode the quantum holographic information in the absence of any classical space time signals and also how the condition of resonance (PCAR) is established by the percipient. Experimental protocols from RV normally provide clues to the location of the target object such as a description, a picture or location by latitude and longitude or an icon representing the target. These clues seem to be sufficient for the percipient to establish resonance with the target. Space-time information (as perceived by the normal five senses) about the target is not perceived by the percipient, nor does the object usually appear at its physical location in space time like a photograph or map in the mind. Rather the information is perceived and presented as internal information and the percipient must associate the perceptions with his / her internal data base of experience in order to recognize and describe the target's perceived attributes.

In the case of complex objects being remotely viewed, the perceived information is seldom so unambiguous as to be instantly recognizable as correct. Sketches, metaphors and analogies are usually employed to recognize and communicate the nonlocal information. A considerable amount of training, teamwork and experience are necessary to reliably and correctly extract complex nonlocal information from a distant location. The information appears to the percipient as sketchy, often dream-like and wispy, subtle impressions of the remote reality. Very skilled individuals may report the internal information as frequently vivid, clear and unambiguous. The remote viewing information received in this case is strictly non-local and, based on the hypothesis of QH, the received information is missing the normal space time component information from any of the five normal senses about the object necessary to completely identify and specify it via resonance.

It has been demonstrated that this intuitive mode of perception can be enhanced by training in most individuals. Perhaps additional training and greater acceptance of this capability will allow percipients to develop greater detail, accuracy and reliability in their skill. In principle, training will not only enhance the remote viewing skill and its accuracy, but should also cause the associated neural circuitry to become more robust as well.

In the absence of normal perceptual sensory signals such as light or sound to establish the resonance condition to provide a basis for decoding the target object's quantum hologram, an icon representing the object seems to be sufficient to allow the mind to focus on the target and to establish the resonant (PCAR) condition as we have described earlier. However a reference signal is also required to provide decoding of the encoded holographic phase dependant information. It has been suggested by Marcer (1998) by that any waves reverberating through the universe remain coherent with the waves at the source, and are thus sufficient to serve as the reference signal to decode the holographic information from any object's quantum hologram emanating from a remote location.

We conclude our discussion of potential QH applications with the experiments conducted by George De La Warr in the 1940's and 1950's. De La Warr was a British engineer who became interested in understanding the mechanisms associated with remote diagnosis and healing. His work was documented by Day (1966) in the 1960's. De La Warr began experiments with his wife, an accomplished psychic healer, to detect the radiation emitted in such processes. At first he thought this mechanism was related to some form of EM radiation but later realized that it was associated with resonance. He eventually built a diagnostic device which acted as a resonant cavity. Perhaps the strangest aspect of the discovery was that when the device was operated by his wife, she could focus her attention on a living target object and was able to produce a resonant condition between the target and the measuring device. She was also able to "project" this resonance condition and expose a blank photographic plate. She was eventually able to pick up resonances from plants, trees, humans and even diseased tissues.

Over time the De La Warrs built up a library of several hundred such photographic plates. Many years later Benford (2008) came across this library and had some of the photographs analyzed by modern 3-D CAD/CAM software (Bryce® 4). The analysis showed that the images were spatially encoded with a 3-D effect similar to those produced by fMRI machines but with much higher resolution. (Earlier we described the discovery made by Shempp that fMRI machines encode quantum information holographically). Recall that fMRI machines were not in existence until many years after these photographic plates were exposed by Mrs. De La Warr. These experiments along with recent discoveries associated with fMRI machines seem to provide compelling evidence that macro-scale quantum holography is a real phenomenon and is produced by conscious attention and intention by a percipient on objects of interest.

9. How Nature Learns

We end this section with our model of QH summarizing how nature (and all living entities) perceives, learns, adapts and evolves in its environment. This model is shown in figure 5. In this model, we show how establishing resonance (PCAR) between a percipient and a target object, the phase conjugate (mirror image) signaling paths connecting the two, can be labeled "perception" on the input side and either "attention" or "intention" on the output side. In the case where the object is a simple physical object (like an apple), our interest is on the non-local information perceived by the percipient about the apple. However from the point of view of the apple, information about the percipient is also available to the apple. The resonant condition between the two is a reciprocal relationship.

Nature's Learning Mechanism

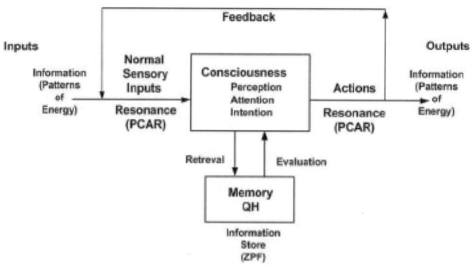

Figure 5. - How Nature Learns

The Quantum Holographic model predicts that the history of events of the target object (apple in this case) is carried in the apple's QH which implies that the "attention" or "intention" focused on the apple by the percipient causes that event to be recorded in the apple's QH. Clearly we cannot query the apple to inquire about its experience but none-the-less the interaction will create a phase shift in the apple's QH (interference pattern) which should be detectable. Although we are using anthropic labeling as we are discussing human perception with the apple, this phenomenon is rooted in natural (and primitive) nonlocal physical processes which are fundamental to the interaction between all objects whether living or not. The evolved complexities of perception, cognition, etc., associated with the brain, as yet have no obvious analogous label other than "non-locality and entanglement" to describe the interactive experience with the environment for simple objects like apples.

Once the resonance condition is established, the percipient can evaluate the results (via the feedback mechanism shown in Figure 5) and can then change its mind state with regard to the object being perceived. The perceived information can then be processed by brain functions so that cognition occurs with respect to the perceived information and thus allowing meaning to be assigned to it. Cognition and meaning require finding a relationship between the perceived information and the information residing in the percipient's memory and this information will be interpreted based on the percipient's beliefs and prior experience stored in its memory. The percipient can then form intent with respect to the object. In such cases the output labeled "action" changes from "attention" (passive state) to "intention" (pro-active state).

In self aware animals (e.g. those with a brain) cognition, meaning and intent with respect to an external object can often be described in simple terms, for example: enemy; fight or flight; food, eat; greet, etc. The nonlocal component of information, although present and creating effect, is operating below the level of conscious perception in humans and results in "instinctual" subconscious behaviors in animals. Classical modeling of this autonomous activity describes it in terms of classical information and energy flow in the central nervous system and the brain. However, as QH suggests, non-locality is operating at all levels of activity, certainly there are resonances involving this non-local information operating throughout all the cells of an organism in parallel with classical space-time functions as described earlier in this paper.

The results for intentional effects of non-locality should be no more difficult to accept than the results for perception -- normal perception using the five senses. The resonant condition (PCAR) implies a symmetry whereby information flows in both directions between the object and the percipient such that each is both target object and percipient to the other. Only the complexity of the more ordered normal sensory mechanisms suggests a non-symmetrical relationship. In general, humans seem to have great difficulty accepting that thoughts, specifically intentionality, can cause action at a distance (remember Einstein's "spooky action at a distance"). Yet, it has been observed for centuries and only in recent decades has it been subjected to scientific scrutiny.

The case of resonance conditions via PCAR to create remote effects by transfer of non-local information between equally complex percipients like humans is not difficult to understand. Indeed, hundreds of successful experiments have established the case. In all these cases no energy transfer is required, only nonlocal information, as each percipient / target object has access to its own energy source. The case for intentionality creating remote effects in inanimate objects is more puzzling. Teleportation of quantum states has been successfully accomplished for particles as described by Darling (2005) and now has practical applications in quantum computing. Numerous studies by Radin (1997), and earlier by Dunne and Jahn (1988) show that macro-scale objects can also be changed or moved, but the energy transfer mechanism by which the classical states of a remote object are affected remains elusive but perhaps is related to utilizing energy directly from the zero point field.

10. Summary and Implications

Someone recently requested the authors describe quantum holography and its implications in two pages, a very difficult task indeed. It has taken us considerably more than that to get here. Nature is extremely complex and does not give up her secrets willingly. Humankind's efforts at understanding her rests on the shoulders of countless dedicated men and women who have come before and are yet to come. Clearly we have a long way to go before we understand it all. Perhaps what is truly most amazing about nature is that it appears to be knowable at all. Our investigations into the nature of consciousness leads us to believe that the best way to survive and sustain humankind as a civilization and to thrive as well is dependent upon the emergence of a new world view, one that understands our proper place in the larger scheme of nature. This includes a worldview that properly addresses, in verifiable scientific terms, our collective relationship to each other, to the biosphere, to the environment, and to the entire cosmos. Towards that end, the evidence that we have presented suggests that we live in a universe that operates according to the following principles. It is:

Self-organizing - All non living and living matter seems to be the result of the emergent complexity adapting and evolving in response to changes in the environment.

Intelligent - The universe utilizes information, processes it and assigns meaning to it. It seems to evaluate new experiences against stored information and "chooses" actions based on that evaluation based on feedback mechanisms.

Creative - All matter in the universe appears to be interconnected and communicates with itself to continually form more complex systems. These systems seem to regulate and organize themselves in ways that are flexible, adaptable and exhibit some form of purposeful behavior.

Trial and error - The habits of nature, its laws, and its operating principles, seem to adapt and evolve by trial and error. The more successful an adaptation is the more it is reinforced. The less successful it is, the more likely it would be to atrophy and eventually die out or fade away into disuse.

Interactive - All matter continually interacts with all other matter. There is no such thing as independent action. Everything is defined in relationship to everything else.

Learning - Experience is retained in nature's memory, the Quantum Hologram. Once information is created it is always available and never forgotten.

Participatory -The role of intention in conscious matter has demonstrable effect.

Evolving - Since its beginnings, nature has been developing into ever increasing levels of complexity in response to environmental changes or pressures resulting from natural processes.

Non-locally connected - All things in nature are interconnected in a very fundamental way beyond time and space. The exchange of information between any two objects occurs instantaneously no matter their space time separation and these interconnections cannot be shielded or attenuated.

Based on Quantum Principles - From the micro scale of subatomic particles to the largest objects in the cosmos and everything in between, all matter displays the quantum characteristics of entanglement, coherence, correlation and resonance.

This universe seems, in some sense, to be a living, evolving, adapting universe that utilizes information to organize itself and to create ever increasing levels of complexity. We are a part of it and cannot be separated from it and are interconnected with it all. Furthermore it appears to be a self referencing system (see Figure 6). As nature learns, habits form and those that lead to useful outcomes solidify and effectively become "hard coded". Even then these "habits of nature" (including us) adapt and evolve by trial and error as change occurs. It appears that nature has bootstrapped itself not only into existence but has evolved itself into the current state of complexity that we now observe all around us. Most astounding of all is that humankind has evolved to the point that we can ask questions and have begun to gain understanding fundamental to nature's very existence. Perhaps, then, we and all sentient beings really are one of nature's way of knowing about and experiencing itself. Not only that, in some sense, we seem to be able to influence its very evolution.

Our hypothesis of interconnectedness and oneness suggested by quantum attributes and processes have been espoused by ancient sages, avatars, mystics, spiritual leaders and shamans throughout all times and by all cultures. Just as modern man has evolved from our ape-like ancestors, so too must we evolve to the next level of sophistication and refinement, and by inference our civilization as well. Change, adaptation to that change and evolution seem to be nature's intrinsic mandate built in to the very fabric of reality. All creation must either perish or constantly evolve. Nature has demonstrated this principle throughout its entire history and has seen to it that there are no alternatives. The arrow of time flows in one direction only.

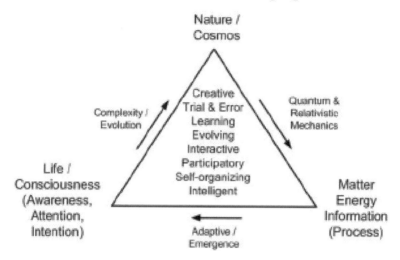

Figure 6 - Nature's Bootstrap Process

We have presented our hypothesis as a map of reality that appears to match observations and experimental evidence fairly well. It seems to account for many phenomena in nature that here-to-for had no explanations to account for them. However, as we have said throughout this manuscript, the map is not the territory. Instead it is nothing more than a model of that territory that makes predictions about how the territory will behave under certain circumstances. We believe that the cornerstones of our theory are built upon known and verified properties and processes of nature and perhaps some yet to be discovered. However, like all theories in science, all that we have proposed is testable. Those parts of it that are not validated will have to be modified, revised or discarded and replaced. Such is the nature of scientific inquiry. At the very least we hope that we will encourage discussion and research to further enhance humankind's understanding of nature.

It has been said that democracy requires an informed electorate to thrive and prosper. It would seem that that is excellent advice in most areas of human endeavor. Sound bites, personal biases, self serving interests have no place if we are to adapt and evolve in our understanding. We must remain open, be willing and desirous to be informed and, most of all, willingly engage in learning and discovering new knowledge about this world in which we live and our true place within it. The issues we face are too important to ignore either by willful neglect or lack of understanding. Our very survival and the survival of all life on earth depend upon it.

We leave you this ancient Sanskrit proverb:

God sleeps in the minerals,
Awakens in plants,
Walks in animals and,
Thinks in man.

References

Benford, M. (2000). Empirical Evidence Supporting Macro-Scale Quantum Holograph in Non-Local Effects, Journal of Theoretics, retrieved from archives at www.journaloftheoretics.com/Articles/2-5/Benford.htm.

Chalmers, D. (1996). The Conscious Mind, Oxford University Press, New York.

Church, D. (2007). The Genie in your Genes, Elite Books, Santa Rosa, CA.

Darling, D. (2005). Teleportation - The Impossible Leap, John Wiley & Sons. Hoboken, NJ.

Davies, P. (2008). The Goldilocks Enigma, Houghton Mifflin, Boston MA.

Day, L., De La Warr, G. (1966). Matter in the Making. Vincent Stuart LTD, London, UK.

Dossey, L. (1993). Healing Words- The Power of Prayer and the Practice of Medicine, Harper, San Francisco, CA.

Dunne,B., Jahn, R.G. (1988). Operator Related Anomalies in a Random Mechanical Cascade, JSE 2:pp 155-180.

Haisch, B. (2006). The God Theory, Red Wheel/Weiser LLC, San Francisco, CA.

Haisch, B., Reuda, A., Putoff, H. (1997). Physics of the Zero Point Field: Implications for Inertia, Gravity and Mass, Speculations in Science and Technology, Vol. 20, pp. 99-114.

Hameroff, S. (1994). Quantum Coherence in Microtubules: A Neural Basis for Emergent Consciousenss?, Journal of Consciousness Studies, Vol. 1, pp 91-118.

Hameroff, S. (1998). Quantum Computation in Brain Microtubules? The Penrose-Hameroff "Orch Or" Model of Consciousness. Philosophical Transactions: Mathematical, Physical and Engineering Sciences, Royal Society of London, Aug 25, 1998; 356(1743): 1869-1896.

Hameroff, S., Woolf, N. (2001), A Quantum Approach to Visual Consciousness, Trends in Cognitive Sciences, Vol. 5, Issue 11, pp. 472-478.

Harman W., Clark, J. (1994). New Metaphysical Foundations of Modern Science, Institute of Noetic Sciences, Sausalito, CA.

Ho, M. (1997). Quantum Coherence and Conscious Experience, Institute of Science in Society, Kybernetes 26, pp. 265-276.

Laszlo, E. (2004). Science and the Akashic Field, Inner Traditions, Rochester, VT.

Lipton, B. (2005). The Biology of Belief, Elite Books, Santa Rosa, CA, p 64.

Marcer, P. J. (1998). The Jigsaw, the Elephant and the Lighthouse, Proceedings of ANPA 20.

Marcer P., Schempp W. (1997). The Model of the Prokaryote Cell as an Anticipatory System Working by Quantum Holography, In: Dubois D. Proceedings of the First International Conference on Computing Anticipatory Systems, Liege, Belgium, August 11-15:307-313.

Marcer, P., Schempp, W. (1997). Model of the Neuron Working by Quantum Holography, Informatica Vol. 21, pp. 519-534.

Marcer, P.J. (1998). A Quantum Mechanical Model of the Evolution of Consciousness.

Mitchell, E. and Williams, D. (2008), The Way of the Explorer, New Page Books, Franklin Lakes, NJ.

Mitchell, E. (1995). A Dyadic Model of Consciousness, World Futures, Vol. 46. pp. 69-78.

Mitchell, E. (2001). Natures Mind: The Quantum Hologram, Institute of Noetic Sciences, Petalma, CA.

Mitchell, E., (2003). Quantum Holography: A Basis for the Interface Between Mind and Matter, Bioelectromagnetic Medicine, Marcel Dekker, New York.

Penrose, R. (1994). Shadows of Mind, Oxford University Press, Oxford.

Pribram, K. (1999). Brain and the Composition of Conscious Experience, Journal of Consciousness Studies, No. 5, pp. 19-42.

Pribram, K. (1999) Quantum Holography: Is it relevant to Brain Function?, Information Sciences Vol. 115, pp. 97-102.

Puthoff, H.E. (1996). CIA Initiated Remote Viewing Program at Stanford Research Institute, JSE 10:63-76.

Puthoff, H.E. and Targ, R. (1976). A Perceptual Channel for Information Transfer over Kilometer Distances: Historical Perspective and Recent Research, proceedings of the IEEE 64:329-354.

Radin, D. (1997). The Conscious Universe, Harper, San Francisco, CA.

Radin, D. (2006). Entangled Minds, Paraview, NY.

Schempp, W. (1998). Magnetic Resonance Imaging: Mathematical Foundations and Applications, Wiley-Liss, New York.

Schempp W. (1999). Sub-Reimannian Geometry and Clinical Magnetic Resonance Tomography, Math Meth Appl Sci 22:867-922.

Schempp, W. and Mueller, K. (2007), Bandpass filter processing strategies in non-invasive sympectic spinor response imaging, Inf. Computation. Sci. 4, pp. 211-232.

Schempp, W. (2008), The Fourier holographic encoding strategy of symplectic spinor visualization, New Directions in Holography, pp. 479-522.

Schmeidler, G.R., Craig, J.G. (1972). Moods and ESP Scores in Group testing, Journal of ASPR, Vol. 66, no. 3, pp 280-287.

Schlitz, M., Wiserman, R. (1997). Experimenter Effects and the Remote Detection of Staring, Journal of Parapsychology, Vol. 61, September.

Sheldrake, R. (1981). A New Science of Life: The Hypothesis of Formative Causation, Tarcher, Los Angeles.

Sheldrake, R. (1997). Mind, Memory, and Archetype: Morphic Resonance and the Collective Unconscious, Psychological Perspectives.

Sheldrake, R. (1999). Dogs That know When Their Owners Are Coming Home, Random House, NY, NY.

Stevenson, I. (2001). Children Who Remember Previous Lives: A Question of Reincarnation, revised ed., Jefferson, NC, McFarland & Company.

Utts, J.M. (1991). Replication and Meta-Analysis in Parapsychology, Statistical Science, 6: 363-382.

5. How Consciousness Becomes the Physical Universe

Menas Kafatos, Ph.D.[1], Rudolph E. Tanzi, Ph.D.[2], and Deepak Chopra, M.D.[3]

[1]Fletcher Jones Endowed Professor in Computational Physics,
Schmid College of Science, Chapman University,
One University Dr., Orange, California, 92866, U.S.A.

[2]Joseph P. and Rose F. Kennedy Professor of Neurology,
Harvard Medical School Genetics and Aging Research Unit
Massachusetts General Hospital/Harvard Medical School
114 16th Street Charlestown, MA 02129

[3]The Chopra Center for Wellbeing,
2013 Costa del Mar Rd. Carlsbad, CA 92009

Abstract

Issues related to consciousness in general and human mental processes in particular, remain the most difficult problem in science. Progress has been made through the development of quantum theory, which, unlike classical physics, assigns a fundamental role to the act of observation. To arrive at the most critical aspects of consciousness, such as its characteristics and whether it plays an active role in the universe, requires us to follow hopeful developments in the intersection of quantum theory, biology, neuroscience and the philosophy of mind. Developments in quantum theory aiming to unify all physical processes have opened the door to a profoundly new vision of the cosmos, where observer, observed, and the act of observation are interlocked. This hints at a science of wholeness, going beyond the purely physical emphasis of current science. Studying the universe as a mechanical conglomerate of parts will not solve the problem of consciousness, because in the quantum view, the parts cease to be measureable distinct entities. The interconnectedness of everything is particularly evident in the non-local interactions of the quantum universe. As such, the very large and the very small are also interconnected.

Consciousness and matter are not fundamentally distinct but rather are two complementary aspects of one reality, embracing the micro and macro worlds. This approach of starting from wholeness reveals a practical blueprint for addressing consciousness in more scientific terms.

KEY WORDS: Quantum Universe, perennial philosophy, consciousness, wholeness, Reality

1. Introduction

We realize that the title of our paper is provocative. It is aimed at providing a theory of how the physical universe and conscious observers can be integrated. We will argue that the current state of affairs in addressing the multifaceted issue of consciousness requires such a theory if science is to evolve and encompass the phenomenon of consciousness. Traditionally, the underlying problem of consciousness has been excluded from science, on one of two grounds. Either it is taken as a given that it has no effect on experimental data, or, if consciousness must be addressed, it is considered subjective and therefore unreliable as part of the scientific method. Therefore, our challenge is to include consciousness while still remaining within the methods of science.

Our starting point is physics, which recognizes three broad approaches to studying the physical universe: classical, relativistic, and quantum. Classical Newtonian physics is suitable for most everyday applications, yet its epistemology (method of acquiring knowledge) is limited -- it does not apply at the microscopic level and cannot be used for many cosmic processes. Between them, general relativity applies at the large scale of the universe and quantum theory at the microcosmic level. Despite all the attempts to unify general relativity with quantum theory, the goal is still unreached. Of the three broad approaches, quantum theory has clearly opened the door to the issue of consciousness in the measurement process, while relativity admits that observations from different moving frames would yield different values of quantities. Many of the early founders of quantum mechanics held the view that the participatory role of observation is fundamental and the underlying "stuff" of the cosmos is processes rather than the construct of some constant, underlying material substance.

However, quantum theory does not say anything specific about the nature of consciousness -- the whole issue is clouded by basic uncertainty over even how to define consciousness. A firm grasp of human mental processes still remains very elusive. We believe that this indicates a deeper problem which scientists in general, are reluctant to address: objective science is based on the dichotomy between subject and object; it rests on the implicit assumption that Nature can be studied ad infinitum as an external objective reality. The role of the observer is, at best, secondary, if not entirely irrelevant.

2. Consciousness and Quantum Theory

In our view, it may well be that the subject-object dichotomy is false to begin with and that consciousness is primary in the cosmos, not just an epiphenomenon of physical processes in a nervous system. Accepting this assumption would turn an exceedingly difficult problem into a very simple one. We will sidestep any precise definition of consciousness, limiting ourselves for now to willful actions on the part of the observer. These actions, of course, are the outcome of specific choices in the mind of the observer. Although some mental actions could be automated, at some point the will of the conscious observer(s) sets the whole mechanical aspects of observation in motion.

The issue of observation in QM is central, in the sense that objective reality cannot be disentangled from the act of observation, as the Copenhagen Interpretation (CI) clearly states (cf. Kafatos & Nadeau 2000; Kafatos 2009; Nadeau and Kafatos, 1999; Stapp 1979; Stapp 2004; Stapp 2007). In the words of John A. Wheeler (1981), "we live in an observer-participatory universe." The vast majority of today's practicing physicists follow CI's practical prescriptions for quantum phenomena, while still clinging to classical beliefs in observer-independent local, external reality (Kafatos and Nadeau 2000). There is a critical gap between practice and underlying theory. In his Nobel Prize speech of 1932, Werner Heisenberg concluded that the atom "has no immediate and direct physical properties at all." If the universe's basic building block isn't physical, then the same must hold true in some way for the whole. The universe was doing a vanishing act in Heisenberg's day, and it certainly hasn't become more solid since.

This discrepancy between practice and theory must be confronted, because the consequences for the nature of reality are far-reaching (Kafatos and Nadeau, 2000). An impressive body of evidence has been building to suggest that reality is non-local and undivided. Non-locality is already a basic fact of nature, first implied by the Einstein-Podolsky-Rosen thought experiment (EPR, 1935), despite the original intent to refute it, and later explicitly formulated in Bell's Theorem (Bell, 1964) and its relationship to EPR – for further developments, see also experiments which favor QM over local realism, e.g. Aspect, Grangier, and Roger, 1982; Tittel, Brendel, Zbinden & Gisin, 1998. One can also cite the Aharonov-Bohm (1959) effect, and numerous other quantum phenomena.

Moreover, this is a reality where the mindful acts of observation play a crucial role at every level. Heisenberg again: "The atoms or elementary particles themselves ... form a world of potentialities or possibilities rather than one of things or facts." He was led to a radical conclusion that underlies our own view in this paper: "What we observe is not nature itself, but nature exposed to our method of questioning." Reality, it seems, shifts according to the observer's conscious intent. There is no doubt that the original CI was subjective (Stapp, 2007). However, as Bohr (1934), Heisenberg (1958) as well as the other developers of CI stated on many occasions, the view that emerged can be summarized as, "the purpose is not to disclose the real essence of phenomena but only to track down... relations between the multifold aspects of our experience." Stapp (2007) restates this view as "quantum theory is basically about relationships among conscious human experiences." Einstein fought against what he considered the positivistic attitude of CI, which he took as equivalent to Berkeley's dictum to be is to be perceived (Einstein 1951), but he nevertheless admitted that QM is the only successful theory we have that describes our experiences of phenomena in the microcosm.

Quantum theory is not about the nature of reality, even though quantum physicists act as if that is the case. To escape philosophical complications, the original CI was pragmatic: it concerned itself with the epistemology of quantum world (how we experience quantum phenomena), leaving aside ontological questions about the ultimate nature of reality (Kafatos and Nadeau, 2000). The practical bent of CI should be kept in mind, particularly as there is a tendency on the part of many good physicists to slip back into issues that cannot be tested and therefore run counter to the basic tenets of scientific methodology.

To put specifics into the revised or extended CI, Stapp (2007) discusses John von Neumann's different types of processes. The quantum formalism eloquently formalized by von Neumann requires first the acquisition of knowledge about a quantum system (or probing action) as well as a mathematical formalism to describe the evolution of the system to a later time (usually the Schrödinger equation). There are two more processes that Stapp describes: the first, according to statistical choices prescribed by QM, yields a specific outcome (or an intervention, a "choice on the part of nature" in Dirac's words); the second, which is primary, preceding even the acquisition of knowledge, involves a "free choice" on the part of the observer. This selection process is not and cannot be described by QM, or for that matter, from any "physically described part of reality" (Stapp, 2007).

These extensions (or clarifications) of the original orthodox CI yield a profoundly different way of looking at the physical universe and our role in it (Kafatos and Nadeau, 2000). Quantum theory today encompasses the interplay of the observer's free choices and nature's "choices" as to what constitutes actual outcomes. This dance between the observer and nature gives practical meaning to the concept of the participatory role of the observer. (Henceforth, we won't distinguish between the original CI and as it was extended by von Neumann—referring to both as orthodox quantum theory.) As Bohr (1958) emphasized, "freedom of experimentation" opens the floodgates of free will on the part of the observer. Nature responds in the statistical ways described by quantum formalism.

Kafatos and Nadeau (2000) and Nadeau and Kafatos (1999) give extended arguments about these metaphysically-based views of nature. CI points to the limits of physical theories, including itself. If any capriciousness is to be found, it should not be assigned to nature, rather to our mindset about how nature ought to work. As we shall see, there are credible ways to build on quantum formalism and what it suggests about the role of consciousness.

3. Quantum Mechanics and the Brain

It is essential that we avoid the mistake of rooting a physical universe in the physical brain, for both are equally rooted in the non-physical. For practical purposes, this means that the brain must acquire quantum status, just as the atoms that make it up have. The standard assumption in neuroscience is that consciousness is a byproduct of the operation of the human brain. The multitude of processes occurring in the brain covers a vast range of spatio-temporal domains, from the nanoscale to the everyday human scale (e.g. Bernroider and Roy, 2004). Even though they differ on certain issues, a number of scientists accept the applicability of QM at some scales in the brain (cf. Kafatos 2009).

For example, Penrose (1989, 1994) and Hameroff and Penrose (1996) postulate collapses occurring in microtubules induced by quantum gravity. In their view, quantum coherence operates across the entire brain. Stapp (2007) prefers a set of different classical brains that evolve according to the rules of QM, in accordance with the uncertainty principle. He contends that bringing in (the still not developed) quantum gravity needlessly complicates the picture.

In order for an integrative theory to emerge, the next step is to connect the quantum level of activity with higher levels. As a specific example of applying quantum-like processes at mesoscale levels, Roy and Kafatos (1999b) have examined the response and percept domains in the cerebellum. They have built a case that complementarity or quantum-like effects may be operating in brain processes. As is well known, complementarity is a cornerstone of orthodox quantum theory, primarily developed by Niels Bohr. Roy and Kafatos imagine a structuresmeasurement process with a device that selects only one of the eigenstates of the observable A and rejects all others. This is what is meant by selective measurement in quantum mechanics. It is also called filtration because only one of the eigenstates filters through the process. In attempting to describe both motor function and cognitive activities, Roy and Kafatos (1999a) use statistical distance in setting up a formal Hilbert-space description in the brain, which illustrates our view that quantum formalism may be introduced for brain dynamics.

It is conceivable that the overall biological structures of the brain may require global relationships, which come down processes to global complementarity—every single process is subordinated to the whole. Not just single neurons, but massive clusters and networks communicate all but instantaneously. One must also account for the extreme efficiency with which biological organisms operate in a holistic manner, which may only be possible by the use of quantum mechanical formalisms at biological, and neurophysiological relevant scales (cf. Frohlich, 1983; Roy and Kafatos, 2004; Bernroider and Roy, 2005; Davies, 2004, 2005; Stapp, 2004; Hameroff et. al., 2002; Hagan et. al., 2002; Hammeroff and Tuszynski, 2003; Rosa and Faber, 2004; Mesquita et. al., 2005; Hunter, 2006; Ceballos et al., 2007).

Stepping into the quantum world doesn't produce easy agreement, naturally. The issue of decoherence (whereby the collapse of the wave function brings a quantum system into relationship with the macro world of large-scale objects and events) is often brought up in arguing against relevant quantum processes in the brain. However, neuronal decoherence processes have only been calculated while assuming that ions, such as K+, are undergoing quantum Brownian motion (e.g. Tegmark, 2000). As such, arguments about decoherence (Tegmark, 2000) assume that the system in question is in thermal equilibrium with its environment, which is not typically the case for bio-molecular dynamics (e.g. Frohlich, 1986; Pokony and Wu, 1998; Mesquita et. al., 2005).

In fact, quantum states can be pumped like a laser, as Frohlich originally proposed for biomolecules (applicable to membrane proteins, and tubulins in microtubules, see also work by Anirban, present volume). Also, experiments and theoretical work indicate that the ions themselves do not move freely within the ion-channel filter, but rather their states are pre-selected, leading to possible protection of quantum coherence within the ion channel for a time scale on the order of 10-3 seconds at 300K, ~ time scale of ion-channel opening and closing (e.g. Bernroider and Roy, 2005). Similar timescales apply to microtubular structures, as pointed out by Hameroff and his co-workers. Moreover, progress in the last several years in high-resolution atomic X-ray spectroscopy from

MacKinnon's group (Jang et al. 2003) and molecular dynamics simulations (cf. Monroe 2002) have shown that the molecular organization in ion channels allows for "pre-organized" correlations, or ion trappings within the selectivity filter of K+ channels. This occurs with five sets of four carbonyl oxygens acting as filters with the K+ ion, bound by eight oxygens, coordinated electrostatic interactions (Bernroider and Roy 2005). Therefore, quantum entangled states of between two subsystems of the channel filter result.

Beyond the brain, evidence has mounted for quantum coherence in biological systems at high temperatures, whereas in the past coherence was thought to apply to systems near absolute zero. For proteins supporting photosynthesis (Engel, et.al., 2007), solar photons on plant cells are converted to quantum electron states which propagate or travel through the relevant protein by all possible quantum paths, in reaching the part of the cell needed for conversion of energy to chemical energy. As such, new quantum ideas and laboratory evidence applicable to the fields of molecular cell biology and biophysics will have a profound impact in modeling and understanding the process of coherence within neuro-molecular systems.

4. Bridging the Gap: A Consciousness Model

Our purpose here is not to settle these technical issues – or the many others that have arisen as theorists attempt to link quantum processes to the field of biology – but to propose that technical considerations are secondary. What is primary is to have a reliable model against which experiments can offer challenges. Such a model isn't available as long as we fail to account for the disappearance of the material universe implied by quantum theory. This disappearance is real. There is, at bottom, no strictly mechanistic, physical foundation for the cosmos. The situation is far more radical than most practicing scientists suppose. Whatever is the fundamental source of creation, it itself must be uncreated. Otherwise, there is a hidden creator lying in the background, and then we must ask who or what created that.

What does it mean to be uncreated? The source of reality must be self-sufficient, capable of engendering complex systems on the micro and macro scale, self-regulating, and holistic. Nothing can exist outside its influence. Ultimately, the uncreated source must also turn into the physical universe, not simply oversee it as God or the gods do in conventional religion. We feel that only consciousness fits the bill, for as a prima facie truth, no experience takes place outside consciousness, which means that if there is a reality existing beyond our awareness (counting mathematics and the laws of physics as 1 part of our conscious experience), we will never be able to know it. The fact that consciousness is inseparable from cognition, perception, observation and measurement, is undeniable; therefore, this is the starting point for new insights into the nature of reality.

What is the nature of consciousness in our model? We take it as a field phenomenon, analogous to but preceding the quantum field. This field is characterized by generalized principles already described by quantum physics: complementarity, non-locality, scale-invariance and undivided wholeness. But there is a radical difference between this field and all others: we cannot define it from the outside. To extend Wheeler's reasoning, consciousness includes us human observers. We are part of a feedback loop that links our conscious acts to the conscious response of the field. In keeping with Heisenberg's implication, the universe presents the face that the observer is looking for, and when she looks for a different face, the universe changes its mask.

Consciousness includes human mental processes, but it is not just a human attribute. Existing outside space and time, it was "there" "before" those two words had any meaning. In essence, space and time are conceptual artifacts that sprang from primordial consciousness. The reason that the human mind meshes with nature, mathematics, and the fundamental forces described by physics, is no accident: we mesh because we are a product of the same conceptual expansion by which primordial consciousness turned into the physical world. The difficulty with using basic terms like "concept" and "physical" is that we are accustomed to setting mind apart from matter; therefore, thinking about an atom isn't the same as an atom. Ideas are not substances. But if elementary particles and all matter made of them aren't substances either, the playing field has been leveled. Quantum theory gives us a model that applies everywhere, not just at the micro level. The real question, then, isn't how to salvage our everyday perception of a solid, tangible world but how to explore the mysterious edge where micro processes are transformed into macro processes, in other words, how Nature gets from microcosm to macrocosm. There, where consciousness acquires the nature of a substance, we must learn how to unify two apparent realities into one. We can begin to tear down walls, integrating objects, events, perceptions, thoughts, and mathematics under the same tent: all can be traced back to the same source.

Physics can serve a pivotal role in transitioning to this new model, because the entire biosphere operates under the same generalized principles we described from the quantum perspective, as does the universe itself. This simple unifying approach must be taken, we realize, as a basic ontological assumption, since it cannot be proven in an objective sense. We cannot extract consciousness from the physical universe, despite the fervent hope of materialists and reductionists. They are forced into a logical paradox, in fact, for either the molecules that make up the brain are inherently conscious (a conclusion to be abhorred in materialism), or a process must be located and described by which those molecules invent consciousness -such a process has not and never will be specified. It amounts to saying that table salt, once it enters the body, finds a way to dissolve in the blood, enter the brain, and in so doing, learns to think, feel, and reason.

Our approach, positing consciousness as more fundamental than anything physical, is the most reasonable alternative: Trying to account for mind as arising from physical systems in the end leads (at best) to a claim that mathematics is the underlying "stuff" of the universe (or many universes, if you are of that persuasion). No one from any quarter is proposing a workable material substratum to the universe; therefore, it seems untenable to mount a rearguard defense for materialism itself. As we foresee it, the future development of science will still retain the objectivity of present-day science in a more sophisticated and evolved form. An evolved theory of the role of the observer will be generalized to include physical, biological, and most importantly, awareness aspects of existence. In that sense, we believe the ontology of science will be undivided wholeness at every level. Rather than addressing consciousness from the outside and trying to devise a theory of everything on that basis, a successful Theory Of Everything (TOE) will emerge by taking wholeness as the starting point and fitting the parts into it rather than vice versa. Obviously, any TOE must include consciousness as an aspect of "everything," but just as obviously current attempts at a TOE ignore this and have inevitably fallen into ontological traps.

The time has come to escape those traps. An integrated approach will one day prevail. When it does, science will become much stronger and develop to the next levels of understanding Nature, to everyone's lasting benefit.

References

Aharonov, Y., and Bohm, D. (1959) Significance of Electromagnetic Potentials in the Quantum Theory, Phys. Rev., 115, 485-491.

Aspect, A., Grangier, P. and Roger, G. (1982) Experimental Realization of Einstein-Podolsky-Rosen-Bohm Gedankenexperiment: A New Violation of Bell's Inequalities, Phys. Rev. Lett., 49, 91-94.

Bell, J.S. (1964) On the Einstein-Podolsky-Rosen paradox, Physics, 1, 195.

Bernroider, G. and Roy, S. (2004) Quantum-classical correspondence in the brain: Scaling, action distances and predictability behind neural signals. FORMA, 19, 55–68.

Bernroider, G. and Roy, S. (2005) Quantum entanglement of K+ ions, multiple channel states, and the role of noise in the brain. In: Fluctuations and Noise in Biological, Biophysical, and Biomedical Systems III, Stocks, Nigel G.; Abbott, Derek; Morse, Robert P. (Eds.), Proceedings of the SPIE, Volume 5841-29, pp. 205-214.

Bohr, N. (1934) Atomic Theory and the Description of Nature, Cambridge, Cambridge University Press. Bohr, N. (1958) Atomic Physics and Human Knowledge, New York: Wiley.

Ceballos, R., Kafatos, M., Roy, S., and Yang, S., (2007) Quantum mechanical implications for the mind-body issues, Quantum Mind 2007, G. Bemoider (ed) Univ. Salzburg, July, 2007.

Davies, P. (2004) Does Quantum Mechanics play a non-trivial role in Life? BioSystems, 78, 69–79.

Davies, P. (2005) A Quantum Recipe for Life, Nature, 437, 819.

Einstein, A. (1951) In: Albert Einstein: Philosopher-Physicist, P.A. Schilpp (Ed.) New York, Tudor.

Einstein, A., Podolsky, B., and Rosen, N. (1935) Can Quantum-Mechanical Description of Physical Reality Be Considered Complete?, Phys. Rev., 47, 777-780.

Engel, G.S., Calhoun, T.R., Read, E.L., Ahn, T.K., Mancal, T., Cheng, Y.C., Blankenship, R.E., and Fleming, G.R. (2007) Evidence for wavelike energy transfer through quantum coherence in photosynthetic systems, Nature, 446, 782-786.

Fröhlich, H. (1983) Coherence in Biology, In: Coherent Excitations in Biological Systems, Fröhlich, H. and Kremer, F. (Eds.), Berlin, Springer-Verlag, pp. 1-5.

Fröhlich, H. (1986) In: Modern Bioelectrochemistry, F. Gutman and H. Keyzer (Eds.) Springer-Verlag, New York.

Hagan, S., Hameroff, SR., and Tuszynski, JA. (2002) Quantum computation in brain microtubules: Decoherence and biological feasibility, Physical Review E., 65(6), Art. No. 061901 Part 1 June.

Hameroff, S. and Penrose, R. (1996) Conscious Events as Orchestrated Space-Time Selections, Journal of Consciousness Studies, Vol 3, No. 1, 36-53.

Hameroff, S. and Tuszynski, J. (2003) Search for quantum and classical modes of information processing in microtubules: Implications for the living state, In: Bioenergetic organization in living systems, Eds. Franco Mucumeci, Mae-Wan Ho, World Scientific, Singapore.

Hameroff, S., Nip, A., Porter, M. and Tuszynski, J. (2002) Conduction pathways in microtubules, biological quantum computation, and consciousness, Biosystems, 64(1-3), 149-168.

Heisenberg, W. (1958) Physics and Philosophy, New York, Harper.

Hunter, P. (2006) A quantum leap in biology. One inscrutable field helps another, as quantum physics unravels consciousness, EMBO Rep., October, 7(10), 971–974.

Jiang, Y.A., Lee, A., Chen, J., Ruta, V., Cadene, M., Chait, B.T., and MacKinnon, R. (2003) X-ray structure of a voltage-dependent K+ channel, Nature, 423, 33-41.

Kafatos, M. and Nadeau, R. (1990; 2000). The Conscious Universe: Parts and Wholes in Physical Reality, New York: Springer-Verlag.

Kafatos, M. (2009) Cosmos and Quantum: Frontiers for the Future, Journal of Cosmology, 3, 511-528.

Mesquita, M.V., Vasconcellos, A.R., Luzzi, R., and Mascarenhas, S. (2005) Large-scale Quantum Effects in Biological Systems, Int. Journal of Quantum Chemistry, 102, 1116–1130.

Monroe, C. (2002) Quantum information processing with atoms and photons, Nature, 416, 238-246.

Nadeau, R., and Kafatos, M. (1999) The Non-local Universe: The New Physics and Matters of the Mind, Oxford, Oxford University Press.

Penrose, R. (1989) The Emperor's New Mind, Oxford University Press, Oxford, England.

Penrose, R. (1994) Shadows of the Mind,

Oxford University Press, Oxford, England.

Pokorny, J., and Wu, T.M. (1998) Biophysical Aspects of Coherence and Biological Order, Springer, New York.

Rosa, L.P., and Faber, J. (2004) Quantum models of the mind: Are they compatible with environment decoherence? Phys. Rev. E, 70, 031902.

Roy S., and Kafatos, M. (1999a) Complemetarity Principle and Cognition Process, Physics Essays, 12, 662-668.

Roy, S., and Kafatos, M., (1999b) Bell-type Correlations and Large Scale Structure of the Universe, In: Instantaneous Action at a Distance in Modern Physics: Pro and Contra, A. E. Chubykalo, V. Pope, & R. Smirnov-Rueda (Eds.), New York: Nova Science Publishers.

Roy, S., and Kafatos, M. (2004) Quantum processes and functional geometry: new perspectives in brain dynamics. FORMA, 19, 69.

Stapp, H.P. (1979) Whiteheadian Approach to Quantum Theory and the Generalized Bell's Theorem, Found. of Physics, 9, 1-25.

Stapp, H. P. (2004) Mind, Matter and Quantum Mechanics (2nd edition), Heidelberg: Springer-Verlag.

Stapp, H.P. (2007) The Mindful Universe: Quantum Mechanics and the Participating Observer, Heidelberg: Springer-Verlag.

Tegmark, M. (2000) Importance of quantum decoherence in brain processes, Phys Rev E Stat Phys Plasmas Fluids Relat Interdiscip Topics, 2000 Apr, 61(4 Pt B), 4194-206.

Tittel, W., Brendel, J., Zbinden, H. and Gisin, N. (1998) Violation of Bell Inequalities by Photons More Than 10km Apart, Phys. Rev. Lett., 81, 3563-3566.

Wheeler, J.A. (1981) Beyond the Black Hole, In: Some Strangeness in the Proportion, H. Woolf (Ed.), Reading, Addison-Wesley Publishing Co.

6. Conscious States Are a Crosstalk Mechanism for Only a Subset of Brain Processes

Ezequiel Morsella, Ph.D.[1,2], and Tiffany Jantz,

[1]Department of Psychology, San Francisco State University, San Francisco, California [2] Department of Neurology, University of California, San Francisco

Abstract

There is a consensus that conscious states are associated with only a subset of the many sophisticated processes that have been identified in the human nervous system. In this review, we identify both the cognitive processes and brain activities that can occur unconsciously and, through this process of elimination, isolate the cognitive mechanisms (and underlying neural mechanisms) that are intimately associated with conscious states. The approach reveals that, consistent with the integration consensus (i.e., that conscious states permit for otherwise independent information processing in the brain to be integrated for adaptive action), conscious states establish a form of intra-brain communication for only a subset of the many kinds of crosstalk within the brain. The form of crosstalk associated with these elusive states seems to be intimately related to the control of voluntary action and the skeletal muscle output system.

KEY WORDS: Consciousness, dissociation, alien hand, cognition.

1. Introduction

There is a consensus that conscious states are associated with only a subset of the many sophisticated processes that have been identified in the human nervous system (Baars, 2002; Crick & Koch, 2003; Dehaene & Naccache, 2001; Gray, 2004; Merker, 2007; Morsella, Krieger, & Bargh, 2010). By 'conscious state' we are referring to the most basic form of consciousness, the kind of consciousness that has fallen under the rubrics of 'basic awareness,' 'sentience,' and 'phenomenal state.' This most basic form of consciousness has been defined best by the philosopher Nagel (1974), who claimed that an organism has phenomenal states if there is something it is like to be that organism--something it is like, for example, to be human and experience pain, love, breathlessness, or yellow afterimages.

The conscious state is 'everything' to us, because it encompasses the totality of our human experience. However, knowledge of nervous function reveals that, normally unbeknownst to us, many of the complicated functions in the nervous system are carried out beneath the horizon of basic consciousness. For example, unconscious processes include (a) low-level perceptual processing, such as the putting together of perceptual features both within and across sensory modalities (e.g., vision and touch), and (b) motor control, as in the control of the impulses that contract some muscle fibers but not others when carrying out an action. Moreover, sophisticated processes such as those constituting that syntax and the parsing of sentences are largely unconscious. Appreciating all that can be achieved unconsciously in the brain leads one to the question, What do conscious states contribute to nervous function? To begin to answer this question, it is helpful as a first step to isolate the cognitive/brain processes that seem to be most intimately associated with conscious states.

We will first present the results of investigations seeking to isolate the cognitive processes that are most associated with consciousness; then we will review data isolating consciousness to a subset of brain processes.

2. The Subset of Basic Cognitive Processes Associated with Conscious States

Most cognitive operations occur unconsciously. For example, there is substantial evidence for the unconscious nature of low-level perceptual analysis and of the integration of sensory information within a sensory modality, such as the 'binding' of shape and color of an orange (Zeki & Bartels, 1999), and across modalities, as in the countless audiovisual integrations responsible for illusions such as the ventriloquism effect and the classic and dramatic McGurk effect (McGurk & MacDonald, 1976). (In this effect, an observer views a speaker mouthing 'ga' while presented with the sound 'ba.' Surprisingly, the observer is unaware of any intersensory interaction, perceiving only 'da.') Regarding semantics (e.g., the meaning of words), evidence suggests that many of its workings, too, are unconscious. For example, a speaker does not know that, when naming a cat, the meaning of the concept DOG was activated, at least to some extent (Levelt, 1989).

Regarding action, there is evidence that action plans can be activated, selected, and even expressed unconsciously (e.g., during neurological disorders; see review in Morsella & Bargh, 2011). This is obvious in actions such as the pupillary reflex, reflexive pain withdrawal, and in behavioral responses to stimuli that have been rendered subliminal through techniques such as 'visual masking' (Hallet, 2007). Convergent evidence for the existence of unconscious action is found in neurological cases where, following brain injury in which a general awareness is spared, actions can are decoupled from consciousness, as in blindsight (Weiskrantz, 1997), in which patients report to be blind but still exhibit visually guided behaviors. Similarly, in alien hand syndrome (Bryon & Jedynak, 1972), anarchic hand syndrome (Marchetti & Della Sala, 1998), and utilization behavior syndrome (Lhermitte, 1983), brain damage causes hands and arms to function autonomously. These actions include relatively complex goal-directed behavior (e.g., the manipulation of objects; Yamadori, 1997) that are maladaptive and, in some cases, can be at odds with a patient's reported intentions (Marchetti & Della Sala, 1998). In addition, Goodale and Milner (2004) report neurological cases in which there is a dissociation between action and conscious perception. Patient D.F., suffering from visual form agnosia, was incapable of reporting the orientation of a tilted slot, but could nonetheless negotiate the slot accurately when inserting an object into it. Complex integrations that are wholly unconscious occur regularly in motor control (Grossberg, 1999; Jeannerod, 2006; Rosenbaum, 2002) and in the control of smooth muscle (e.g., the pupillary reflex; Bartley, 1942; Morsella, Gray, Krieger, & Bargh, 2009).

Which processes tend to always be associated with conscious states? According to the integration consensus (Morsella, 2005), conscious states appear to furnish the nervous system with a form of internal communication that integrates neural activities and information-processing structures that would otherwise be independent, permitting diverse kinds of information to be gathered in some sort of global workspace, thus allowing adaptive action to emerge (cf., Baars, 2002; Dehaene & Naccache, 2001; Merker, 2007; Morsella, 2005). The kinds of unconsciously-mediated actions described above lack this form of integration and thus appear irrational and impulsive, meaning that the actions are not constrained by information that, for adaptive action, should be influencing them. But exactly which kind of integration requires the conscious state? As mentioned above, many integrational processes (e.g., intersensory illusions) occur in the nervous system unconsciously.

To address this issue, Supramodular Interaction Theory (SIT; Morsella, 2005) proposes that conscious states are necessary to integrate diverse sources of information/processes, but only certain kinds of information/processes. For example, conscious states are unnecessary to integrations between perceptual processes, as in afference binding (e.g., intra- or inter-sensory interactions; Morsella & Bargh, 2011). Similarly, basic stimulus-response (S -> R) associations (efference binding; Haggard, Aschersleben, Gehrke, & Prinz, 2002), such as inhaling reflexively or pressing a button in response to a subliminal stimulus (Fehrer & Biederman, 1962; Fehrer & Raab, 1962; Taylor & McCloskey, 1990, 1996) can occur unconsciously. (See review of correct motor responses to subliminal stimuli in Hallett, 2007.) Figure 1 reviews all the many unconscious forms of interaction (or 'crosstalk') in the brain.

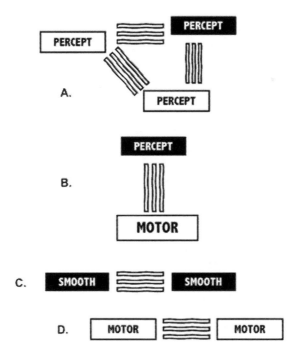

Figure 1. The many forms of interaction (or 'crosstalk') that can occur unconsciously in the nervous system. A. Unconscious crosstalk between perceptual systems within a sensory modality (signified by the boxes sharing the same hue) and between different sensory modalities (signified by the boxes bearing distinct hues). B. Unconscious interactions between perceptual and motor processes, as in the case of unconscious efference binding. C. Unconscious interactions in the control of smooth muscle effectors, as in the case of the consensual processes in the pupillary reflex. D. Unconscious

interactions in motor control.

According to SIT, the difference between conscious and unconscious interactions is not simply about the complexity, controllability, or feedback properties of the processes involved, nor is it about the degree to which the processes require memory, semantics, action-related mechanisms, or 'top-down' processes (see extensive treatment of this in Morsella, 2005). Rather, what most reliably distinguishes conscious from unconscious interactions pertains to the nature of the effectors involved (Morsella, 2005).

For example, research reveals that conscious conflicts (e.g., holding one's breath or suppressing a pain-withdrawal reflex) are special in that they involve conflicting tendencies toward the skeletal muscle output system. Specifically, consciousness is reliably perturbed when there is the simultaneous activation of two conflicting streams of efference binding toward the skeletal muscle output system (e.g., signaling inhale and do not inhale when holding one's breath underwater; see evidence in Morsella, Berger, & Krieger, in press). Such efference-efference binding results in integrated actions (Morsella & Bargh, 2011) such as holding one's breath, breathing faster for a reward, carrying a hot dish of food, or suppressing socially-inappropriate behavior (Figure 2).

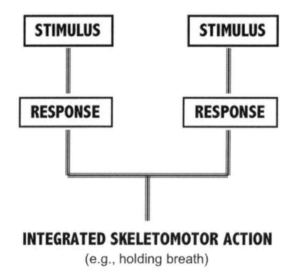

Figure 2. Efference-efference binding, a form of interaction (or 'crosstalk') that requires the instantiation of the conscious state. In this case, two Sensory-Response streams influence action simultaneously, leading to integrated actions such as holding one's breath.

Conscious states are not necessary for intersensory conflicts, smooth muscle actions, or skeletomotor actions that are unintegrated (Morsella & Bargh, 2011), that is, those driven by a single stimulus-response (S -> R) stream, such as withdrawing one's hand from a hot stove or responding to a subliminal stimulus. SIT is unique in its ability to explain the subjective effects of conflicts from action conflicts (e.g., holding one's breath) and the lack of subjective effect from intersensory conflicts or smooth muscle conflicts. From this standpoint, the skeletal muscle system is a multi-determined effector controlled by many brain systems, each potentially having distinct phylogenetic origins and operating principles. This form of multi-determination reveals that action selection at the organismic level suffers from a 'degrees of freedom' problem (Rosenbaum, 2002): There are simply too many things that one can decide to do next. For instance, during the trial of a laboratory experiment about action conflict, a subcortical brain region may want the eyes to glance leftwards (because a bright flash occurred there), and a cortical region may want the eyes to look rightwards (because of the experimenter's instruction). The challenge of multidetermination in action selection is met, not by unconscious motor algorithms (as in the case of motor control; Rosenbaum, 2002), but by the ability of conscious states to constrain what the organism does by having the inclinations of multiple systems co-exist in the conscious field and thereby constrain skeletomotor output.

Figuratively speaking, the skeletal muscle system is like a big steering wheel that different parts of the brain are trying to control at the same time (Morsella, Krieger, & Bargh, 2009). When these different systems are in conflict, as when one voluntarily holds one breath--and part of us wants to inhale and another part of us does not want to inhale--we have a strong conscious experience. Consciousness is the way the different regions communicate with each other. Without consciousness, there would be no crosstalk between the different systems and one would inhale reflexively. One does not need consciousness to withdraw from a painful stimulus, but one does need it in order to keep touching it. Another way to think about it is as follows. There are many quasiindependent computers in the brain, and each can do complicated things and influence overt action, which we can think of as influencing the actions of a printer. Each computer can influence the printer, but in order for two computers to interact and then influence the printer, you need a wifi system. Consistent with the integration consensus (stating that consciousness integrates information/processes that would be independent otherwise), in this way consciousness functions as a wifi system to integrate the different processes in the brain and yield adaptive action. One can certainly imagine integration among systems occurring without anything like a conscious state, but in this descriptive approach, and for the reasons that only the tinkering and happenstance process of evolution could explain, it was these states that were selected to solve this particular integration problem in the brain (Morsella, 2005). SIT is unique in also explaining why skeletal muscle is voluntary muscle: Skeletomotor actions are at times 'consciously mediated' because they are directed by multiple, encapsulated systems that, when in conflict, require consciousness to yield adaptive action.

Through the process of elimination (i.e., identifying all the cognitive operations that the nervous system can carry out unconsciously), we have attempted to identify what conscious states are for by isolating the few basic processes that seem to be intimately related with these states and that seem incapable of occurring unconsciously. Can a similar eliminative approach isolate the neural substrates of conscious states?

3. The Subset of Brain Areas/Processes Associated with Conscious States

When homing in on the conscious state neuroanatomically, the evidence not as straightforward as the evidence for isolating the function of conscious states in processing, partly because consciousness may not so much be associated with a specific brain region as with a particular mode of processing among regions. It seems that the mode of interaction among regions is as important as the nature and loci of the regions (Buzsáki, 2006). For example, the presence or lack of interregional synchrony leads to different cognitive and behavioral outcomes (Hummel & Gerloff, 2005; see review of neuronal communication through 'coherence' in Fries, 2005). That the mode of interaction between areas is important for conscious states is evident in binocular rivalry. In binocular rivalry (Logothetis & Schall, 1989), an observer is presented with different visual stimuli to each eye (e.g., an image of a house in one eye and of a face in the other).

It might seem reasonable that, faced with such stimuli, one would perceive an image combining both objects--a house overlapping a face. Surprisingly, however, an observer experiences seeing only one object at time (a house and then a face), even though both images are always present. At any moment, the observer is unaware of the computational processes leading to this outcome; the conflict and its resolution are unconscious. Consistent with the view that the mode of interaction between areas is important for consciousness, during binocular rivalry, it is only the conscious percept that, neurally, is coupled to both perceptual brain activity and motor-related processes in frontal cortex (Doesburg, Green, McDonald, & Ward, 2009)

Unconscious processes involve smaller networks of brain areas than their conscious counterparts (Gaillard et al., 2009; Sergent & Dehaene, 2004; See review in Morsella, Krieger, & Bargh, 2010.) For example, unconsciously-mediated actions (e.g., reflexive pharyngeal swallowing) involve substantially fewer brain regions than their voluntary counterparts (e.g., volitional swallowing; Kern, Jaradeh, Arndorfer, & Shaker, 2001; Ortinski & Meador, 2004). Regarding gross neuroanatomy, consciousness has been linked to the 'ventral processing stream' of the brain, which is not necessary for action execution but for knowledge-based action selection (Goodale & Milner, 2004). It has been proposed that conscious states require a form of thalamocortical interaction (or resonance) between thalamic 'relay' neurons and cortical neurons (Coenen, 1998;

Edelman, Baars, & Seth 2005; Llinás & Ribrary, 2001; Ojemann, 1986), but this is inconsistent with the fact that we consciously experience aspects of olfaction even though the afferents from the olfactory sensory system bypass the thalamus and directly target regions of the ipsilateral cortex (Morsella, Krieger, & Bargh, 2010; Shepherd & Greer, 1998). This is not to imply that conscious olfaction does not require the thalamus: in post-cortical stages of processing, the thalamus receives inputs from cortical regions that are involved in olfactory processing (Haberly, 1998). Buck (2000) proposes that conscious aspects of odor discrimination depend primarily on the activities of the frontal and orbitofrontal cortices; Barr and Kiernan (1993) propose that olfactory consciousness depends on the pyriform cortex. These proposals appear inconsistent with subcortical accounts of consciousness (Merker, 2007; Penfield & Jasper, 1954).

Regarding the cerebral cortex, experiments on 'split-brain' patients (Wolford, Miller, & Gazzaniga, 2004), binocular rivalry (Logothetis & Schall, 1989), and splitbrain patients experiencing binocular rivalry (O'Shea & Corballis, 2005) strongly suggest that basic consciousness does not require the nondominant (usually right) cerebral cortex nor the commissures linking the two cortices. Evidence also suggests that, although the absence of the spinal cord or cerebellum leads to sensory, motor, cognitive, and affective deficits, it does not seem to eradicate the basic conscious state (Schmahmann, 1998).

Similarly, although extirpation of the amygdalae or hippocampi leads to anomalies including severe deficits in affective processing (LeDoux, 1996) and episodic memory (Milner, 1966), respectively, it seems that a basic conscious state persists without these structures. Investigations regarding prefrontal lobe syndromes (Gray, 2004), the phenomenology of action (Desmurget et al., 2009; Desmurget & Sirigu, 2010), and the psychophysiology of dream consciousness, which involves prefrontal deactivations (Muzur, Pace-Schott, & Hobson, 2002), suggest that, although the prefrontal lobes are involved in cognitive control (see review in Miller, 2007), they are not essential for the generation of basic consciousness. According to Gray (2004), one is conscious, not of high-level executive processes or motor efference, but only of perceptual-like contents, whether they precede an action, as in the case of anticipated action effects, or whether they follow an action, as in the case of experiencing the actual action effects. This is consistent with recent experiments by Sirigu and colleagues. Research from the Sirigu laboratory (Desmurget et al., 2009; Desmurget & Sirigu, 2010) reveals that direct electrical stimulation of parietal areas of the brain gives subjective urges and that increased activation makes them believe that they actually executed the corresponding action, even though no action was performed. Motor activations (e.g., in premotor areas) can lead to the actual action, but subjects believe that they did not perform any action. Interestingly, this approach is consistent with what ideomotor approaches (Hommel, 2009; Hommel, Müsseler, Aschersleben, & Prinz 2001) to action control have proposed: That awareness of our actions occurs only in a 'sensorium' (a term used by Johannes Müller) with motor processes being unconscious.

In contrast to cortical accounts of consciousness, Penfield and Jasper (1954) proposed that these states are primarily a function of subcortical structures. This 'cortical-subcortical' controversy arose from Penfield and Jasper's (1954) observations of awake patients undergoing brain surgeries involving ablations and direct brain stimulation. Penfield and Jasper concluded that, though the cortex may elaborate the contents of consciousness, it is not the seat of consciousness. Recently, based on clinical observations of anencephaly, Merker (2007) re-introduces this hypothesis in a theoretical framework in which consciousness is primarily a phenomenon associated with mesencephalic areas. It seems reasonable to conclude that consciousness can persist even when great quantities of the cortex are absent (Merker, 2007). The outstanding question is whether an identifiable form of consciousness can exist despite the non-participation of all cortical matter. Cortical areas that seem most associated with the conscious state are the sensory and parietal areas. Future research on the cortical-subcortical controversy may evaluate the necessary role of parietal areas (e.g., those identified by Desmurget and Sirigu, 2010) in the generation of basic consciousness.

4. The Place of Consciousness in the Physical World

In summary, isolating the cognitive underpinnings of conscious states (some kind of integrative process for adaptive skeletomotor function) has proven easier than isolating the neural underpinnings.

We adopted an inductive and descriptive approach, in which nervous function is described as is, and not as it (perhaps) should be--what would be a normative rather than a descriptive approach. Hence, intuitions regarding how the nervous system should work take a back seat to actual data revealing how it actually works, whether optimally or suboptimally. Thus, it is premature to propose that these states are 'epiphenomenal,' that is, serving no function whatsoever. Until one understands the place of a given phenomenon in nature, and how it emerges from nature, one should not make the strong claim that the phenomenon be epiphenomenal. Second, the current approach may identify what these states are for, but it sheds no light on why 'subjectivity' is associated with the tightly circumscribed, integrative function that these states appear subserve.

References

Baars, B. J. (2002). The conscious access hypothesis: Origins and recent evidence. Trends in Cognitive Sciences, 6, 47 - 52.

Barr, M. L., & Kiernan, J. A. (1993). The human nervous system. An anatomical viewpoint, sixth edition. Philadelphia: Lippincott.

Bartley, S. H. (1942). A factor in visual fatigue. Psychosomatic Medicine, 4, 369 - 375.

Bryon, S., & Jedynak, C. P. (1972). Troubles du transfert interhemispherique: A propos de trois observations de tumeurs du corps calleux. Le signe de la main etrangère. Revue Neurologique, 126, 257 - 266.

Buck. L. B. (2000). Smell and taste: The chemical senses (pp. 625-647). In E. R. Kandel, J. H. Schwartz, & T. M. Jessell (Eds.), Principles of neural science, fourth edition. New York: McGraw-Hill.

Buzsáki, G. (2006). Rhythms of the brain. New York: Oxford University Press. Coenen, A. M. L. (1998). Neuronal phenomena associated with vigilance and consciousness: From cellular mechanisms to electroencepalographic patterns. Consciousness and Cognition, 7, 42-53.

Crick, F., & Koch, C. (2003) A framework for consciousness. Nature Neuroscience, 6, 1-8.

Dehaene, S., & Naccache, L. (2001). Towards a cognitive neuroscience of consciousness: Basic evidence and a workspace framework. Cognition, 79, 1 - 37.

Desmurget, M., Reilly, K. T., Richard, N., Szathmari, A., Mottolese, C., & Sirigu, A. (2009). Movement intention after parietal cortex stimulation in humans. Science, May 8th, 324 (5928), 811 - 813.

Desmurget, M., & Sirigu, A. (2010). A parietal-premotor network for movement intention and motor awareness. Trends in cognitive sciences, 13, 411 - 419.

Doesburg, S. M., Green, J. L., McDonald, J. J., & Ward, L. M. (2009). Rhythms of consciousness: Binocular rivalry reveals large-scale oscillatory network dynamics mediating visual perception. PLoS, 4, 1 - 14.

Edelman, D. B., Baars, B. J., & Seth, A. K. (2005). Identifying the hallmarks of consciousness in non-mammalian species. Consciousness and Cognition, 14, 169-187.

Fehrer, E., & Biederman, I. (1962). A comparison of reaction time and verbal report in the detection of masked stimuli. Journal of Experimental Psychology, 64, 126 - 130.

Fehrer, E., & Raab, D. (1962). Reaction time to stimuli masked by metacontrast. Journal of Experimental Psychology, 63, 143 - 147.

Fries, P. (2005). A mechanism for cognitive dynamics: Neuronal communication through neuronal coherence. Trends in Cognitive Sciences, 9, 474-480.

Gaillard, R., Dehaene, S., Adam, C., Clémenceau, S., Hasboun, D., Baulac, M., et al. (2009). Converging intracranial markers of conscious access. PLoS Biology, 7, e1000061. doi:10.1371/journal.pbio.1000061

Goodale, M., & Milner, D. (2004). Sight unseen: An exploration of conscious and unconscious vision. New York: Oxford University Press.

Gray, J. A. (2004). Consciousness: Creeping up on the hard problem. New York: Oxford University Press.

Grossberg, S. (1999). The link between brain learning, attention, and consciousness. Consciousness and Cognition, 8, 1 - 44.

Haberly, L. B. (1998). Olfactory cortex (pp. 377 - 416), in G. M. Shepherd (Ed.), The synaptic organization of the brain, fourth edition. New York: Oxford University Press.

Haggard, P., Aschersleben, G., Gehrke, J., & Prinz, W. (2002). Action, binding and awareness. In W. Prinz & B. Hommel (Eds.), Common mechanisms in perception and action: Attention and performance (Vol. XIX, pp. 266-285). Oxford, UK: Oxford University Press.

Hallett, M. (2007). Volitional control of movement: The physiology of free will. Clinical Neurophysiology, 117, 1179 - 1192.

Hommel, B. (2009). Action control according to TEC (theory of event coding). Psychological Research, 73, 512 - 526.

Hommel, B., Müsseler, J., Aschersleben, G., & Prinz, W. (2001). The theory of event coding: A framework for perception and action planning. Behavioral and Brain Sciences, 24, 849 - 937.

Hummel, F., & Gerloff, C. (2005). Larger interregional synchrony is associated with greater behavioral success in a complex sensory integration task in humans. Cerebral Cortex, 15, 670 - 678.

Jeannerod, M. (2006). Motor cognition: What action tells the self. New York: Oxford University Press.

Kern, M. K., Jaradeh, S., Arndorfer, R. C., & Shaker, R. (2001). Cerebral cortical representation of reflexive and volitional swallowing in humans. American Journal of Physiology: Gastrointestinal and Liver Physiology, 280, G354 - G360.

LeDoux, J. E. (1996). The emotional brain: The mysterious underpinnings of emotional life. New York: Simon and Schuster.

Levelt, W. J. M. (1989). Speaking: From intention to articulation. Cambridge, MA: The MIT Press.

Lhermitte, F. (1983). "Utilization behaviour" and its relation to lesions of the frontal lobe. Brain, 106, 137 - 255.

Llinás, R. R., & Ribary, U. (2001). Consciousness and the brain: The thalamocortical dialogue in health and disease. Annals of the New York Academy of Sciences, 929, 166 - 175.

Logothetis, N. K. & Schall, J. D. (1989). Neuronal correlates of subjective visual perception. Science, 245, 761 - 762.

Marchetti, C., & Della Sala, S. (1998). Disentangling the alien and anarchic hand. Cognitive Neuropsychiatry, 3, 191 - 207.

McGurk, H. & MacDonald, J. (1976). Hearing lips and seeing voices. Nature, 264, 746 - 748.

Merker, B. (2007). Consciousness without a cerebral cortex: A challenge for neuroscience and medicine. Behavioral and Brain Sciences, 30, 63 - 134.

Miller, B. L. (2007). The human frontal lobes: An introduction. In B. L. Miller & J. L. Cummings (Eds.), The human frontal lobes: Functions and disorders, second edition (pp. 3 - 11). New York: Guilford.

Milner, B. (1966). Amnesia following operation on the temporal lobes (pp. 109 - 133). In C. W. M. Whitty & O. L. Zangwill (Eds.), Amnesia. London: Butterworths.

Morsella, E., & Bargh, J. A. (2011). Unconscious action tendencies: Sources of 'unintegrated' action. In J. T. Cacioppo & J. Decety (Eds.), The handbook of social neuroscience (pp. 335 - 347). New York: Oxford University Press.

Morsella, E. (2005). The function of phenomenal states: Supramodular interaction theory. Psychological Review, 112, 1000 - 1021.

Morsella, E., Gray, J. R., Krieger, S. C., & Bargh, J. A. (2009). The essence of conscious conflict: Subjective effects of sustaining incompatible intentions. Emotion, 9, 717 - 728.

Morsella, E., Berger, C. C., & Krieger, S. C. (in press). Cognitive and neural components of the phenomenology of agency. Neurocase.

Morsella, E., Krieger, S. C., & Bargh, J. A. (2009). The function of consciousness: Why skeletal muscles are "voluntary" muscles. In E. Morsella, J. A. Bargh, & P. M. Gollwitzer, Oxford handbook of human action (pp. 625-634). Oxford University Press.

Morsella, E., Krieger, S. C., & Bargh, J. A. (2010). Minimal neuroanatomy for a conscious brain: Homing in on the networks constituting consciousness. Neural Networks, 23, 14 - 15.

Muzur, A., Pace-Schott, E. F., & Hobson, J. A. (2002). The prefrontal cortex in sleep. Trends in Cognitive Sciences, 6, 475 - 481.

Nagel, T. (1974). What is it like to be a bat? Philosophical Review, 83, 435 - 450.

Ojemann, G. (1986). Brain mechanisms for consciousness and conscious experience. Canadian Psychology, 27, 158 - 168.

Ortinski, P., & Meador, K. J. (2004). Neuronal mechanisms of conscious awareness. Archives of Neurology, 61, 1017 - 1020.

O'Shea, R. P., & Corballis, P. M. (2005). Visual grouping on binocular rivalry in a splitbrain observer. Vision Research, 45, 247-261.

Penfield, W. and Jasper, H. H. (1954). Epilepsy and the functional anatomy of the human brain. New York: Little, Brown.

Rosenbaum, D. A. (2002). Motor control. In H. Pashler (Series Ed.) & S. Yantis (Vol. Ed.), Stevens' handbook of experimental psychology: Vol. 1. Sensation and perception (3rd ed., pp. 315 - 339). New York: Wiley.

Schmahmann, J. D. (1998). Dysmetria of thought: Clinical consequences of cerebellar dysfunction on cognition and affect. Trends in Cognitive Sciences, 2, 362-371.

Sergent, C., & Dahaene, S. (2004). Is consciousness a gradual phenomenon? Evidence for an all-or-none bifurcation during the attentional blink. Psychological Science, 15, 720 - 728.

Shepherd, G. M., & Greer, C. A. (1998). Olfactory bulb (pp. 159 - 204) in G. M. Shepherd (Ed.), The synaptic organization of the brain, fourth edition. New York: Oxford University Press.

Taylor, J. L., & McCloskey, D. I. (1990). Triggering of preprogrammed movements as reactions to masked stimuli. Journal of Neurophysiology, 63, 439 - 446.

Taylor, J. L., & McCloskey, D. I. (1996). Selection of motor responses on the basis of unperceived stimuli. Experimental Brain Research, 110, 62-66.

Weiskrantz, L. (1997). Consciousness lost and found: A neuropsychological exploration. New York: Oxford University Press.

Wolford, G., Miller, M. B., & Gazzaniga, M. S. (2004). Split decisions. In M. S. Gazzaniga (Ed.), The cognitive neurosciences III (pp. 1189 - 1199). Cambridge, MA: The MIT Press.

Yamadori, A. (1997). Body awareness and its disorders. In M. Ito, Y. Miyashita, Y., & E. T. Rolls (Eds.), Cognition, computation, and consciousness (pp. 169-176). Washington, DC:, USA: American Psychological Association.

Zeki, S., & Bartels, A. (1999). Toward a theory of visual consciousness. Consciousness and Cognition, 8, 225 - 259.

7. The Neuroanatomy of Free Will: Loss of Will, Against the Will, "Alien Hand"
Rhawn Joseph, Ph.D.
Emeritus, Brain Research Laboratory, California

Abstract

The neuroanatomy of "free will" is described, detailed, and supplemented by case histories of individuals who were compelled to behave against their will, who lost control over their will, and who suffered a complete loss of free will. In all instances the frontal lobe, the medial regions in particular are implicated in the mediation of "free will." The frontal lobes serve as the "Senior Executive" of the brain and personality, acting to process, integrate, inhibit, assimilate, and remember perceptions and impulses received from the limbic system, striatum, temporal parietal and occipital lobes, and neocortical sensory receiving areas. Through the assimilation and fusion of perceptual, volitional, cognitive, and emotional processes, the frontal lobes engages in decision making and goal formation, modulates and shapes character and personality and directs attention, maintains concentration, and participates in information storage and memory retrieval. Further, the frontal lobes, the SMA and medial regions in particular, can direct behavior by controlling movement and the musculature of the body, and in this manner, it serves what is best described as "free will."

KEY WORDS: Consciousness, Free Will, Frontal Lobes, Medial Frontal Lobes, Supplementary Motor Area, Catatonia, Alien Hand, Split-Brain

1. FREE WILL & FUNCTIONAL LOCALIZATION

If the brain and mind are synonymous, is unknown. However, if the brain is damaged, the mind too is effected. Disturbances in brain functioning, be it due to drugs, alcohol, injury, tumor, stroke, emotional trauma, seizures, or electrode stimulation, directly affect consciousness (Joseph 1982, 1986a, 1988a, 1996, 1998, 1999a, 2001, 2003). Often specific aspects of the conscious mind are directly impacted (Joseph 1988a,b) including what has been referred to as "free will" (Joseph 1996, 1999b). This is because specific functions are localized to specific regions of the brain.

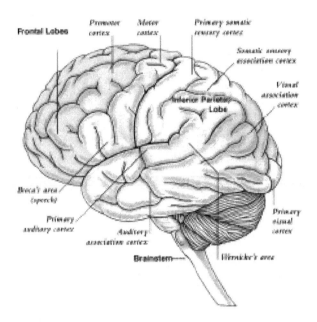

For example, in most humans, severe injury to the left frontal lobe can abolish the ability to speak words or intelligible sentences, a condition classically referred to as Broca's expressive aphasia (Joseph 1982, 1996, 1999b). Although the "Will" to speak remains intact, those afflicted may be capable only of expressing their frustrations by cursing which is mediated by the right frontal lobe, as is the ability to sing (Joseph 1982, 1988a, 1996, 1999b). Hence, patients can curse and may be able to sing words they can't say.

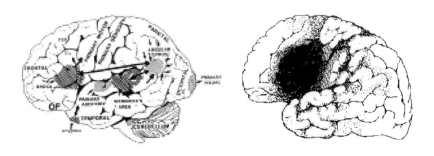

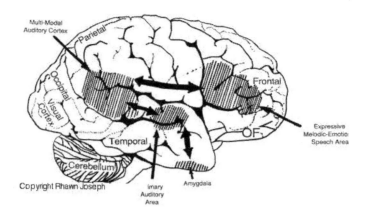

If, however, the damage to the left frontal lobe is widespread and extremely deep, penetrating into the medial (middle) portions of the anterior cerebral hemispheres, not just the "will to speak", but "free will" may be abolished and those afflicted may be forced to act "against their will" (Joseph 1986, 1988a, 1999b). What we call "free will" appears to be localized to the frontal lobes, the medial most portions in particular.

2. DISCONNECTION SYNDROMES

No region of the brain functions in isolation, unless isolated following a very circumscribed lesion thereby disconnecting it from other areas of the cerebrum; a condition classically referred to as "disconnection syndrome" (Geschwind 1981; Joseph 1982, 2009). For example, whereas Broca's area subserves the capacity to speak, the grammatical and denotative aspects of language (the words which will be spoken) are organized and assimilated by a multi-modal association area in the angular gyrus of the inferior parietal lobe (IPL) immediately adjacent to Wernicke's receptive language area in the superior temporal lobe (Joseph 1982, 1986, 1988a, 1996, 2000). The IPL and Wernicke's and Broca's areas are linked together by a rope of nerve fibers, the arcuate fasciculus.

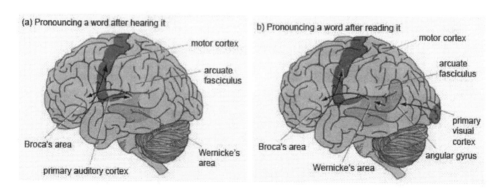

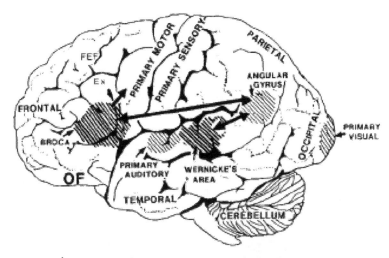
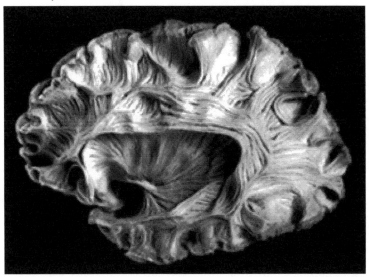
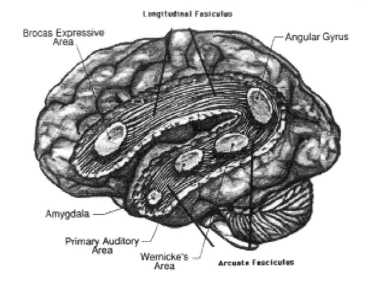

However if the arcuate fasciculus is damaged such as due to stroke, those afflicted will know what they want to say, but will be unable to say it and will suffer severe word finding difficulty (Joseph 1982). Temporary functional disconnections occur even in the normal brain, where the missing word is known but can't be found, and this condition is experienced as "tip of the tongue" phenomenon. Thus one part of the mind is disconnected from another (Geschwind 1981; Joseph 1982, 2009). The "will to speak" remains intact (due to preservation of the frontal language areas), whereas the missing words are locked away in the posterior regions of the forebrain.

One rather severe form of of disconnection is "locked-in syndrome" which is due to destruction of the pyramidal (cortico-spinal) nerve fiber pathways linking the brainstem pontine area with the forebrain (Smith and Delargy 2005). Although completely paralyzed and seemingly comatose or "brain dead", patients are believed to be fully conscious and aware, and maintain the "will" to move and speak, but are unable with the exception of, in some cases, the capacity to blink and thus communicate through the eyes (Allain et al., 1998; Smith and Delargy 2005). Consciousness is maintained and cognitive functions are preserved because the forebrain is intact and completely functional with only mild reductions in cerebral metabolism noted (Allain et al., 1998; Zeman 2003).

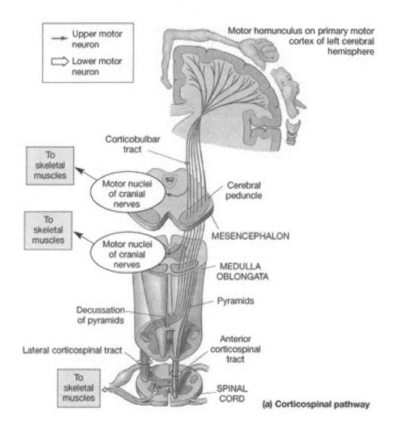

(a) Corticospinal pathway

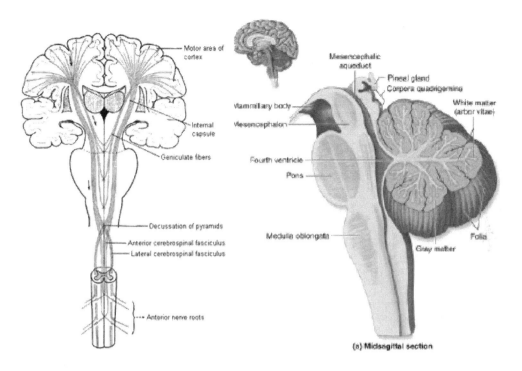

Parisian journalist Jean-Dominique Bauby suffered a stroke in December 1995, which severely damaged the neural pathways leading to and from the frontal lobes and brainstem. He was completely paralyzed and was initially believed to be in a vegetative state (Bauby 1998). However, when he began to cry, and by blinking his left eye, his caretakers realized he was in fact fully conscious. Over the next two years he learned to communicate by blinking a code for different letters of the alphabet, and dictated his memoir, The Diving Bell and the Butterfly (Bauby 1998). He died of pneumonia two days after it was published.

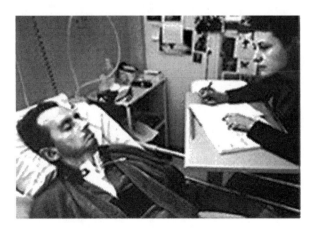

Consciousness is maintained in locked-in syndrome, because the forebrain which includes the frontal lobes, remains intact, and the inferior pons and the medulla of the brainstem which mediates vital life-sustaining functions are undamaged. However, because the motor areas in the forebrain are disconnected from the motor centers in the brainstem, the patient can't move their body and may appear to be in a vegetative state (Smith and Delargy 2005; Zeman 2003). They are no longer able to act according to their free will, which is locked-in.

3. FREE WILL AND THE FRONTAL LOBES

Various aspects of personality, memory, attention, perception, emotion, the body image, and consciousness may be variably compromised with damage to different regions of the forebrain (Joseph 1982; 1986a,b; 1988a, 1992, 1994, 1998, 1999a,b, 2001, 2003, 2009), e.g. amygdala (emotion), hippocampus (memory), temporal lobe (memory, language, personality), parietal lobe (body image, hand-in-space). Certainly damage to these and other brain areas may limit and restrict what we call "free will". However, insofar as "free will" is defined as the ability to make plans, consider alternatives, and chose among and act upon them, if the frontal lobes remain intact and consciousness and movement are preserved, patients can still make choices and act on them, and they do not lose their free will.

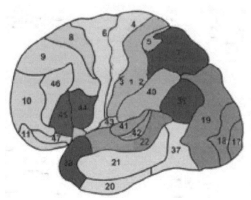
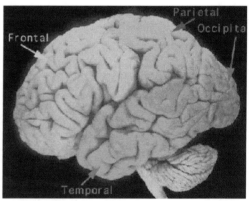

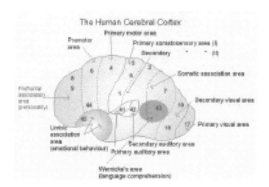

By contrast, if the frontal lobes are damaged, or if the neural pathways between and within different frontal areas are compromised or disconnected, patients may not only lose their "free will" but other brain areas may act against their "will" and engage in behaviors which the patient cannot willfully resist (Joseph 1986a, 1988a, 1996, 1999b). Free will is localized to the frontal lobes, the medial frontal areas in particular.

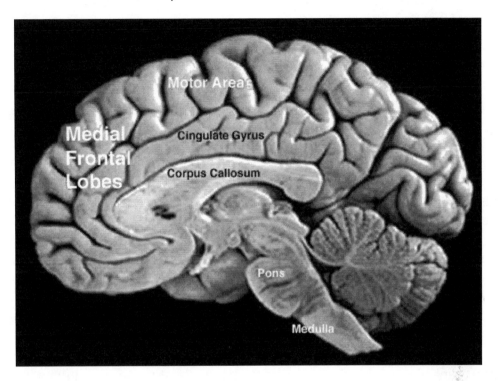

The frontal lobes are not a homologous tissue but serve myriad functions, and are "interlocked" via converging and reciprocal connections, with the brainstem, midbrain, thalamus, limbic system, striatum, anterior cingulate, and throughout the neocortex including the primary and secondary sensory receiving areas (Fuster 1997; Jones and Powell 1970; Miller & Cummings 2006 Pandya & Yeterian 1990; Petrides & Kuypers 1969; Pandya & Vignolo 1969; Risberg & Grafman 2006). Different areas of the frontal lobe are continually informed about activity in other areas of the brain, and in this way can direct attention, control behavior, and act according to their free will. In consequence, different aspects of "free will" can be compromised if neuronal pathways from the frontal lobes to other brain areas have been severed and "disconnected" and depending on which regions of the frontal lobe have been injured (Joseph 1986, 1988a, 1999b). For example, injuries to the right frontal lobe can result in a loss of control over free will.

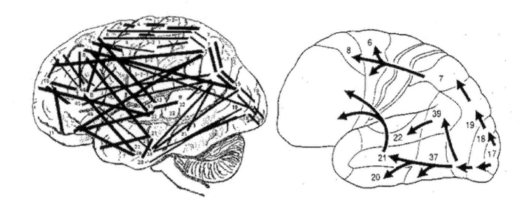

4. THE RIGHT FRONTAL LOBE: FREE WILL UNRESTRAINED

Depending on the degree of damage, individual with right frontal lobe injuries may become unrestrained, overtalkative, and tactless, saying whatever "pops into their head", with little or no concern as to the effect their behavior has on others or what personal consequences may result. They may become inordinately disinhibited and influenced by the immediacy of a situation, buying things they cannot afford, lending money when they themselves are in need, and acting and speaking "without thinking" (Fuster 1997; Joseph 1986a, 1988a, 1999b; Miller & Cummings 2006; Risberg & Grafman 2006). Seeing someone who is overweight or obese they may begin making sounds like a hog ("oink oink") or laughingly call the person a "a fat pig." If they enter a room and detect a faint odor, its: "Hey, who farted?" If they see food they like, they may grab it off another person's plate.

Following severe right frontal lobe injuries there may be periods of gross disinhibition which may consist of loud, boisterous, and grandiose speech, singing, yelling, and beating on walls. The destruction of furniture and the tearing of clothes is not uncommon. Some patients may impulsively strike doctors, nurses, or relatives, or sexually proposition family members or complete strangers, and thus behave in a thoroughly labile, aggressive, callous and irresponsible manner.

One patient, with a tumor involving the right frontal area, threw a fellow patient's radio out the window because he did not like the music. He also loudly sang opera in the halls. Indeed, during the course of his examination he would frequently sing his answers to various questions (Joseph 1986a, 1988a, 1999b).

D.F. tried to commit suicide by sticking a gun in his mouth, but aimed it wrong and blew out his right frontal lobe. After he had been discharged from the hospital, he returned for his doctor's appointment laughing, making ridiculous remarks, and dressed inappropriately with a toothbrush, toothpaste, hair brush, and wash rag sticking out of his shirt pocket (Joseph 1988a).

When asked, pointing at his pocket, "Why the toothbrush?" he replied with a laugh, "That's just in case I want to brush my teeth," and in so saying he quickly drew the toothbrush from his pocket, climbed up on top of the doctor's desk, and began to brush his teeth. Then he began to dance on the desk and wanted to demonstrate how the skin flap which covered the hole in his head (from the craniotomy and bullet wound) could bulge in or out when he held his breath or held his head upside down (Joseph 1988a).

Nevertheless, despite his bizarre and inappropriate behavior, D.F.'s overall IQ was above 130 (98% rank: "Very Superior"). Unfortunately, although he had a high IQ, he could no longer control or employ that intelligence, intelligently. He had lost control over his "free will." During one lucid moment, he stated that "I don't know why I act this way. I just can't control it."

One frontal patient described as formerly very stable, and a happily married family man, became excessively talkative, restless, grossly disinhibited, sexually preoccupied and would approach complete strangers and proposition them for sex. He also extravagantly spent money and recklessly purchased a business which soon went bankrupt.

In another case, a 46-year old woman with a right frontal lobe tumor began walking around outside in just a slip and bra and then stripping naked in front of neighbors. She claimed she was descended from queens, was fabulously wealthy, and that many men wanted to divorce their wives and marry her (Joseph 1996).

A 19-year old man with seizure activity in the right frontal lobe, felt compelled to take his penis out of his pants, in public, and to masturbate. He was subsequently arrested for exposing himself in public. He claimed to have no control over his behavior. When interviewed he suddenly exposed his penis and urinated in the direction of his doctor who by grace of very fast reflexes moved aside just in time to avoid getting drenched (Joseph 1996).

A very conservative, highly reserved, successful, brilliant , happily married engineer with over 20 patents to his name suffered a right frontal injury when he fell from a ladder. He became sexually indiscriminate and began patronizing up to 3 prostitutes a day, whereas before his injury his sexual activity was limited to once weekly with his wife. He spent money lavishly, depleted his considerable savings, suffered delusions of grandeur where he thought he was a senator and a billionaire, and camped out at Disney Land and attempted to convince personnel to fund his ideas for a theme park on top of a mountain (Joseph 1986a, 1996). At night had dreams where President Kennedy and Senator Kennedy would appear and offer him advice --and he was a republican!

Even with "mild" to moderate right frontal lobe injuries patients may initially demonstrate periods of tangentiality, grandiosity, irresponsibility, laziness, hyperexcitability, promiscuity, silliness, childishness, lability, personal untidiness and dirtiness, poor judgment, irritability, fatuous jocularity, and tendencies to spend funds extravagantly. Unconcern about consequences, tactlessness, and changes in sex drive and even hunger and appetite (usually accompanied by weight gain) may occur (Fuster 1997; Joseph 1986a, 1988a, 1999b; Miller & Cummings 2006; Passingham, 1997; Risberg & Grafman 2006).

These individuals, however, are not acting according to their free will, but often against their will. A janitor, who following his frontal lobe injury, suddenly believed he was a multi-millionaire congressman, quit his job, tried to take over the local congressman's office, stood on corners making speeches, and made extravagant purchases of items he could not afford or pay for. Later he explained that "I know its not true, but I believe its true and can't stop myself" (Joseph 1996).

Essentially, with right frontal lobe injuries, that aspect of the brain which serves and controls free will becomes disconnected from those brain areas which act on free will.

5. MOVEMENT AND THE FRONTAL MOTOR AREAS

To exercise one's free will requires access to and control over the motor systems of the brain and the body musculature. Motor control and movement are controlled, at the level of the frontal lobes, by the primary, secondary, and supplementary motor areas. These forebrain neocortical tissues are interconnected and communicate with other motor centers, such as the brainstem via a thick ropes of neurons known as the pyramidal (cortico-spinal) tracts (Fuster 1997; Joseph 1986a, 1988a, 1999a; Miller & Cummings 2006; Passingham, 1997; Risberg & Grafman 2006).

There is a one-to-one correspondence between single neurons in the primary motor area and single muscles such that the musculature of the entire body surface is represented according to their importance (Chouinard & Paus 2006; Dum & Strick 2005; Verstynen, et al. 2011). For example, the hands, fingers, face and mouth, have a greater representation than the musculature of the back. It is the primary motor areas which control fine motor functioning (such as when typing or writing).

The representation of the musculature is more diffuse in the secondary premotor area. The premotor transmits its motor impulses to the primary motor areas and subcortical regions of the brain but receives its marching orders from the supplementary motor area (Joseph 1999b; Nachev et al. 2008).

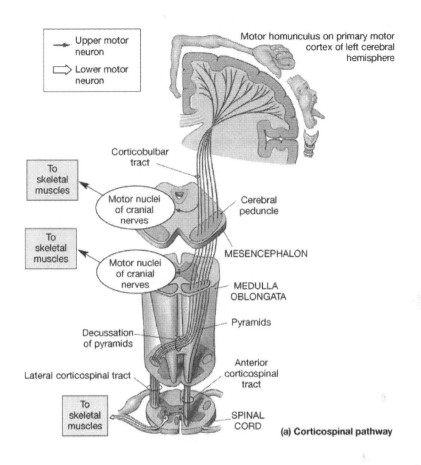

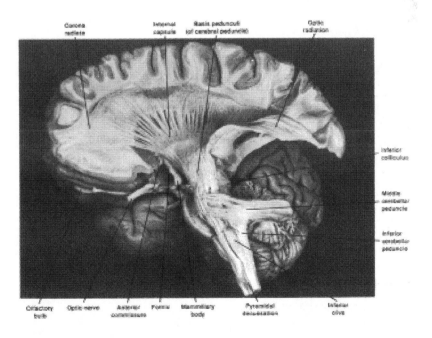

Nerve Fiber Pathways to and From the Neocortex to brainstem

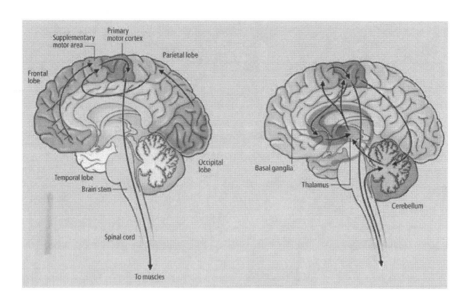

6. FREE WILL AND THE MEDIAL, SUPPLEMENTARY MOTOR AREAS

The SMA is more concerned with the general problem of guiding and co-ordinating the movements of the extremities through space (Andres, et al., 1999; Nachev et al. 2008; Passingham, 1997; Stephan, et al., 1999). However, it is the SMA which programs and exerts executive control over the secondary and primary motor areas, and it is the SMA and medial frontal lobes which are more closely linked with "free will."

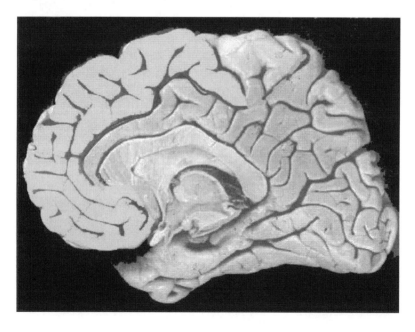

The supplementary motor areas (SMA) originates within the medial walls of the right and left frontal lobe, is extensively interconnected with the limbic system, brainstem, anterior cingulate, and striatum, and extends up and over the medial walls to the lateral walls of the frontal lobes.

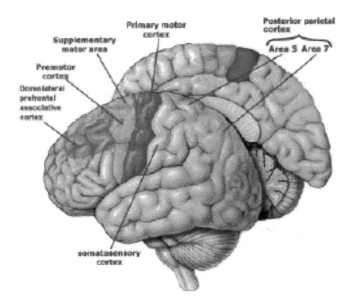

The SMA is especially concerned with preparing the hands, arms, and body to move, and becomes active simultaneously with the intention to move, but before movement and prior to activation of the secondary and primary motor areas. Single cell recordings (Brinkman & Porter, 1979; Tanji & Kurata, 1982) and studies of blood flow studies (Orgogozo & Larsen, 1979; Shibasaki et al. 1993), movement related evoked potentials (Ikeda et al. 1992) and other functional indices (Nachev et al., 2008) indicate increased activity in the SMA prior to moving, and when just thinking about moving the body, hand, arm, or leg. Thus activity begins in the SMA well before movements are initiated and prior to activation within the premotor and primary motor areas.

For example, when anticipating or preparing to make a movement, but prior to the actual movement, neuronal activity will first begin and then dramatically increase in the SMA, followed by activity in the premotor and then the primary motor area, and then the subcortical striatum (caudate, putamen, globus pallidus) and finally the brainstem (Alexander & Crutcher, 1990; Mink & Thach, 1991; Nachev et al., 2008). This indicates that the "will" to move begins in the SMA and medial frontal lobes and exert executive control over the secondary, primary, and subcortical motor areas which then perform these "willed" actions.

The SMA in fact exerts executive control over the arms and hands, and can willfully direct the extremities towards items of interest (e.g. coffee cups, tools, pens, keys, breasts) which are then grasped, manipulated and utilized (Andres

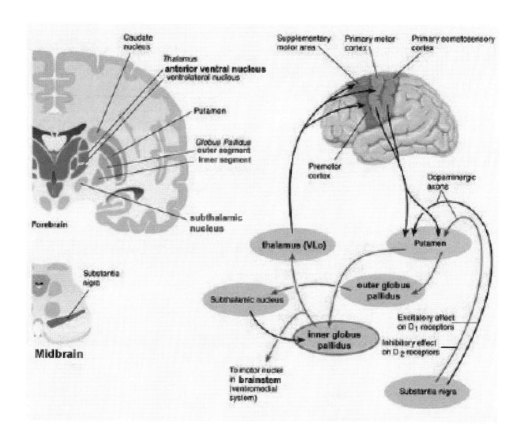

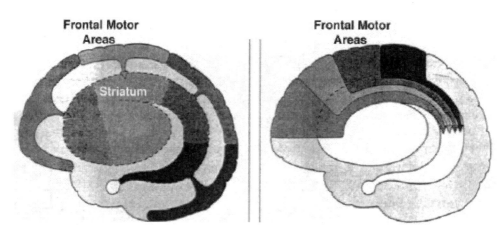

et al., 1999; Passingham, 1997; Stephan et al., 1999). Direct electrical stimulation of the SMA will also trigger complex semipurposeful movements of the hands, arms, legs, and feet (Penfield & Jasper, 1954). The same is not true of the primary or secondary motor areas which, if activated by electrode, may only trigger the twitching of single muscles. Although guiding fine motor movements of the hands, the pre- and primary motor areas are under the control of the SMA and medial frontal lobes.

7. DISCONNECTION: ACTING AGAINST THE WILL

If the neural pathways linking the SMA / medial frontal lobes to the secondary and primary area are severed or grossly injured, the disconnected primary and secondary motor areas may act against the patient's will. Behavior will be controlled, deterministically, by external stimuli (Joseph 1996, 1999b).

Patients will involuntarily engage in complex coordinated movements involving their arms and hands, and may reach out and grasp and manipulate or even use various objects even though they don't want to, and try to willfully resist. Patients will lose control over their arms, hands and legs, and compulsively pick up and use tools, pens, cups, or other objects, such that in the extreme the right or left hand may act completely independently of the "conscious" mind (Denny-Brown, 1958; Gasquoine 1993; Goldberg & Bloom 1990; Lhermitte, 1983). Instead of acting voluntarily according to their "free will", behavior is determined and controlled by their environment.

In fact, the mere visual presence of a cup, pen, hammer, saw or scissors near the hand may trigger groping movements as well as grasping and involuntary use of the object. For example, they may compulsively reach out and take the examiner's pen or swipe their glasses from their face and put them on. One patient put three pair of spectacles on and wore them simultaneously.

If a hammer is placed on the testing table the patient may involuntarily pick it up and begin hammering on the table or walls, even when told not to, and will continue even after told to stop. If a glass of water is placed on the table, they may pick it up and drink from it, although they are not thirsty and despite their efforts to willfully oppose these actions (Gasquoine 1993; Goldberg and Bloom 1990; Lhermitte 1983; McNabb et al. 1988). Denny-Brown (1958) has referred to this condition as "magnetic apraxia" and "complusive exploration".

Sometimes just the presence of a pencil and a piece of paper may cause a patient to pick up the pencil and begin writing an endless letter that might include the mechanical repitition of the same word line after line and even page after page.

Not only do they act against their will, but once they take hold of a pen, cup, or other object, they may be unable to let go and cannot release their grip. In fact, the entire arm and hand may become increasingly stiff and rigid, such that the hand will become frozen and seemingly "stuck" to the pen, paper, cup, or whatever they were holding or touching. While walking they may move more and more slowly until they seem unable to move, as if their feet are stuck to the floor as if glued (Denny-Brown, 1958).

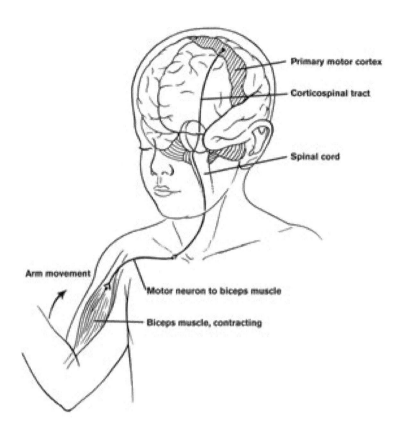

Gegenhalthen (counterpull), i.e. involuntary resistance to movement of the extemities, appears to be exclusively associated with medial frontal lobe/SMA abnormalities (Joseph 1996, 1999b). If the patient raises an affected arm, it will stiffen and become increasingly rigid as the will to move it increases. If the patient tries to resist or to relax the arm, it instead becomes frozen in place.

The same condition also results if a nurse or physician attempts to move the arm; it becomes increasingly rigid and patients are forced, against their will, to maintain their afflicted arm even in uncomfortable postures, frozen in place, for long time periods. However, over time, the affected limb may slowly return to a resting posture--a condition referred to as "waxy flexibility."

Likewise, if they are presented with yet another cup, pen, or tool, the frozen state may slowly disappear and they may reach out and compulsively take hold of the new object.

Naturally, individuals so afflicted become quite upset, frustrated and embarrassed by the misbehavior of their body, and will complain these actions occur against their will.

8. THE ALIEN HAND

With injury or damage to the medial frontal lobes, SMA, and their interconnections with the other motor areas, one or both hands may take on a "life of their own"; a condition referred to as the "alien hand" (Joseph 1986a, 1988a,b, 1996, 1999b). In fact, with damage to or disconnection of these medial frontal tissues, involuntary uncontrolled movements may become so purposeful and complex that they appear to be directed not by the external environment, but by a consciousness with its own free; an alien consciousness that is disconnected from another aspect of consciousness which can speak but which cannot willfully oppose these alien behaviors by an alien hand (Joseph 1986a, 1988a,b, 1996, 1999b).

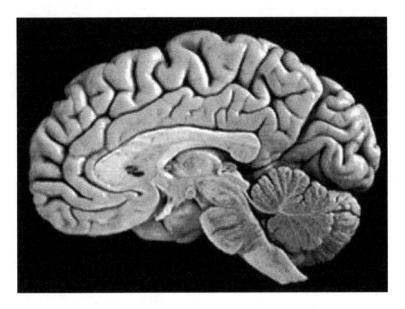

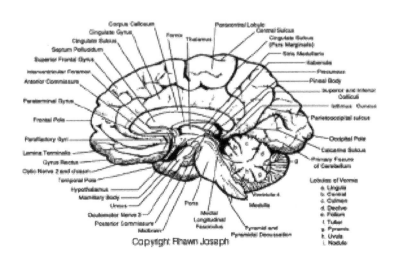

For example, a patient described by Gasquoine (1993) had a propensity to reach out and touch female breasts with his right hand. He couldn't stop touching and fondling pencils, cups, nik naks, doctors, nurses, and persons within in reach. He reported this caused him great embarrassment and that his hand acted against his will. However, as the alien behavior was confined to one hand, the right, he would grab hold of it with his left hand and try to wrestle it under control. He also learned to trick the alien hand. He would give it something to hold and it would be unable to let go, thus preventing it from grabbing breasts, buttocks, or property which was not his own. In addition to his alien hand, he also complained that he would speak his private thoughts out loud, against his will.

McNabb et al. (1988, pp. 219, 221) describe one woman with extensive damage involving the medial left frontal lobe and anterior corpus callosum, who was forced to punish her misbehaving right hand by slapping it with the left. However, sometimes the right hand would fight back and interfere with actions performed by the left. She complained that her right hand showed an uncontrollable tendency to reach out and take hold of objects and then be unable to relese them. When the right hand behaved mischievously she attempted to restrain it by wedging it between her legs or by holding or slapping it with her left hand. She repeatedly express astonishment at these actions by her hand. A second patient frequently experienced similar difficulties. When "attempting to write with her left hand the right hand would reach over and attempt to take the pencil. The left hand would respond by grasping the right hand to restrain it."

Problems of an even more severe nature have plagued patients following complete (surgical) destruction of the corpus callosum and thus the neural pathways linking the two medial motor areas of the frontal lobes (Joseph, 1988ab). These independent "alien" behaviors usually involve the left hand and the half of the body, and were purposeful, intentional, complex and obviously directed by an awareness maintained by the disconnected right hemisphere (which controls the left hand). These alien actions were often completely against the "will" of the consciousness maintained in the left hemisphere.

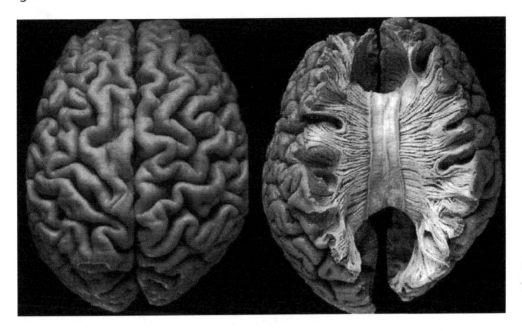

(Left) Superior View of Right & Left Hemisphere. (Right) Partially Dissected Split Brain View Show Corpus Callosum

These "alien" disturbances were so purposeful, and often so well thought out, it was as if these "split-brain" patients had developed two independent "free wills" maintained by independent minds housed in the right and left half of the brain (Joseph, 1986b, 1988a,b); two free wills and two minds which were unable to communicate, and each of which had a "mind of its own."

As originally described by Nobel Lauriate Roger Sperry (1966, p. 299), "Everything we have seen indicates that the surgery has left these people with two separate minds, that is, two separate spheres of consciousness. What is experienced in the right hemisphere seems to lie entirely outside the realm of awareness of the left hemisphere. This mental division has been demonstrated in regard to perception, cognition, volition, learning and memory."

For example, one patient's left hand would not allow him to smoke and would pluck lit cigarettes from his mouth. He reported that he had been trying to quit, unsuccessfully, for years, but it was only after the surgery that he found he couldn't smoke, because the left hand wouldn't let him (Joseph, 1988a).

Each frontal lobe has its own primary, secondary, and supplementary motor areas and medial (and lateral) tissues. Since free will is associated with the SMA and medial frontal lobe, disconnecting the connections between the right and left created two "free wills" one maintained by the right the other by the left frontal lobe.

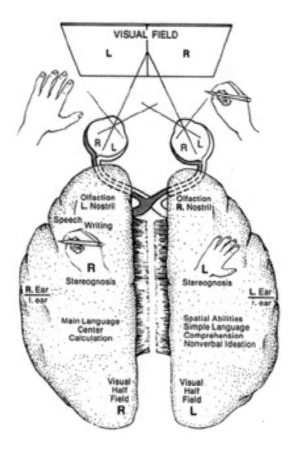

Akelaitis (1945, p. 597) describes two patients with complete corpus callosotomies who experienced extreme difficulties making the two halves of their bodies cooperate. "In tasks requiring bimanual activity the left hand would frequently perform oppositely to what she desired to do with the right hand. For example, she would be putting on clothes with her right and pulling them off with her left, opening a door or drawer with her right hand and simultaneously pushing it shut with the left. These uncontrollable acts made her increasingly irritated and depressed."

Another patient experienced difficulty while shopping, the right hand would place something in the cart and the left hand would put it right back again. Yet another had the same problem when trying to decide what to eat, first picking one item, and the other hand putting it back and picking another. Of course, similar difficulties in "making up one's mind" also plague those who have not undergone split-brain surgery.

A recently divorced split-brain patient reported that on several occasions while walking about town he was forced to go some distance in another direction by the left half of this body. He resisted with the right half of the body and essentially engaged in a battle of the wills. As it turned out, the left half of the body was trying to take him to the home of his ex-wife, but the right half of his body refused to have anything to do with her.

Another split-brain patient who had recently broken up with his girlfriend voiced considerable anger toward her and stated an intention to never see her again. He also admitted that they had broken up several times, but always got back together. This time, however, he was adamant: He never wanted to see her again. When asked to indicate with "thumbs up" or "thumbs down" if he still liked her, the right hand gave a "thumbs down" and the left hand a "thumbs up."

Geschwind (1981) reports a callosal patient who complained that his left hand on several occasions suddenly struck his wife--much to the embarrassment of his left (speaking) hemisphere. In another case, a patient's left hand attempted to choke the patient himself and had to be wrestled away.

Bogen (1979) indicates that almost all of his "complete commissurotomy patients manifested some degree of intermanual conflict in the early postoperative period." One patient, Rocky, experienced situations in which his hands were uncooperative; the right would button up a shirt and the left would follow right behind and undo the buttons. For years, he complained of difficulty getting his left leg to go in the direction he (or rather his left hemisphere) desired. Another patient often referred to the left half of her body as "my little sister" when she was complaining of its peculiar and independent actions.

A split-brain patient described by Dimond (1980, p. 434) reported that once when she had overslept "my left hand slapped me awake." This same patient, in fact, complained of several instances where her left hand had acted violently toward other people, and this caused her considerable distress and embarrassment.

Yet another split brain patient, "2-C" complained of instances in which his left hand tried to strike a relative (Joseph 1988a,b). Once, after he had retrieved something from the refrigerator with his right hand, his left took the food, put it back on the shelf and retrieved a completely different item "Even though that's not what I wanted to eat!" he complained. Once while watching and enjoying a program TV, the left half of his body dragged him from his seat, and then changed the channels and then returned to his seat to watch a different program, even though: "That's not what I wanted to watch!" On several occasions, his left leg refused to continue "going for a walk" and would only allow him to return home.

9. MIRROR NEURONS, ALIEN HANDS, SPLIT-BRAINS & FREE WILL

In 1996 Rizzolatti et al., reported the discovery of "mirror" neurons which were most densely concentrated in the SMA. According to Rizzollati et al. (1998) "mirror neurons appear to form a system that allows individuals to recognize motor actions made by others by matching them with an internal motor copy," thus enabling them to perform and mimic the actions of others (Rizzolatti & Craighero 2004, Rizzollati et al. 2009).

Since the SMA and medial frontal lobes are implicated in the expression of all "alien" actions, is it possible these "alien behaviors" are merely reflexive, and that the "alien" hand is merely mimicking and mirroring the behavior it is observing, and thus doing the opposite?

The answer was provided in 1988 when it was conclusively demonstrated that these "alien" behaviors are purposeful, goal directed, and under the control of a "free will" and a mind separate from the dominant stream of consciousness which has access to language (Joseph 1988ab). In one test, both hands of split-brain patients were provided information and given multiple choices so as to make a choice which precluded mimicry. Specifically, the patient sat before a box with two separate holes (on the right and the left) in which he inserted his hands. Above each hole, in plain sight, were small patches of textured material, one above the other, i.e. sandpaper, wire mesh, smooth metal, velvet. While in the hole and out of sight, both hands were stimulated simultaneously, but with completely different materials, e.g. velvet to the right, sand paper to the left. They were then required to point to the material they experienced. On every test the right and left hands made the correct responses. The "alien" left hand did not mimic the right and did not interfere with its choices.

However, these patients (that is, the language dominant left hemispheres) had not been told they would be experiencing two different fabrics, and then expressed shock and dismay when the alien left hand chose a different material. In one case, the patient's normal right hand, repeatedly reached over and tried to force the left (alien) hand to make a different choice, that is, to chose the same material experienced by the right hand, although the left hand responded correctly! However, in this case, the alien hand resisted and refused to make a different choice, even when the patient (his left hemisphere) vocalized: "Thats wrong!" Repeatedly, he reached over and grabbed his left hand with his right but it refused to point at the wrong item. He became so angry with his left hand that he yelled at it, said "I hate this hand" and then began punching and hitting it. Finally the two hands began to fight!

In this and in other instances, it was demonstrated that the alien hand was not misbehaving, but acting purposefully and making rational choices according to its own free will - and which was opposed by the free will residing in the other, disconnected half of the brain.

10. CATATONIA, THE SMA / MEDIAL FRONTAL LOBES & FREE WILL

Different regions of the brain are specialized for performing specific functions, and interact with yet other regions to coordinate and make possible complex cognitive and behavioral activity. The frontal lobes serve as the senior executive of the brain and personality and the MFL and SMA provide the neuroanatomical substrate which executes what has been traditionally described as "free will."

If other areas of the brain are disconnected from the SMA / MFL the patient may feel compelled to act against their will. However, if the MFL and SMA are destroyed, "free will" is abolished. Ideas and thoughts are no longer generated, and the "will" to speak or to initiate or complete a voluntary movement may become completely attenuated and abolished (Hassler 1980; Laplane et al. 1977; Luria 1980; Penfield and Jasper 1954; Penfield and Welch 1951). Patients may lose even the will to speak and become mute, unresponsive, stiff, frozen, unmoving, motionless, and catatonic (Hassler 1980; Joseph, 1999b; Laplane et al. 1977; Luria 1980; McNabb et al. 1988; Penfield and Jasper 1954; Penfield and Welch 1951).

In one case, a soldier developed gegenhalten, waxy flexibility, mutism, and catatonia after a gunshot wound that passed completely through the frontal lobes, destroying much of the SMA and MFL and disconnecting these regions from other brain tissue. For two months he laid in a catatonic-like stupor, always upon one side with slightly flexed arms and legs, never changing his uncomfortable position. He did not obey commands, was incontinent, made no complaints, gazed steadily forward and showed no interest in anything. However, over the ensuing weeks, he periodically showed signs of lucidity, and could be persuaded to talk, and would answer quite correctly about his personal affairs. When questioned he explained that during his catatonic periods, although he was aware of his surroundings, his mind was empty, devoid of thoughts, and it just did not occur to him to move, speak, or eat. So he did nothing. Incredibly, the patient "was eventually returned to active duty" (Freeman and Watts 1942).

In another case, a 42 year old male with no previous psychiatric history, developed gegenhalten, mutism, and catatonia after a beating and suffering frontal and midline subdural hematomas (which later required the drilling of burr holes for evacuation). He resisted the efforts of others to move him, and would sit motionless and unresponsive for hours in odd and uncomfortable positions. The patient's symptoms seemed to wax and wane such that he demonstrated some periods of seeming normality (Joseph 1996). When he was in one of his "normal" periods, and was asked about his behavior he replied: "Its not that I don't want to move, its that I don't feel the want". During another "normal" period he explained his catatonic state thusly: "I feel nothing. No thoughts. Everything is going on outside my head, but nothing is going on inside my head. Like I am a rock. I am just there." Thus, while in his catatonic state his mind was a blank and there was simply no reason to even move.

It is not uncommon with SMA / MFL injuries for patients to become catatonic, mute, and remain in odd, uncomfortable postures for long time periods

(Joseph 1996, 1999b). Upon recovery many patients may later remark they had completely lost the will to speak, that thoughts did not enter their head, that they were unable to think or generate ideas, and instead experienced a motivational-ideational void, a complete emptiness without feelings, which left them without any reason to move or function (Brutkowski 1965; Hassler 1980; Laplane et al. 1977; Luria 1980; Mishkin 1964). As those afflicted have no interest in eating, drinking, or even moving, essentially they had also lost the will to live.

With massive SMA MFL injuries, free will is abolished, and this is because, free will is localize to the frontal lobes.

11. CONCLUSIONS: THE NEUROANATOMY OF FREE WILL

The frontal lobes serve as the "Senior Executive" of the brain and personality, acting to process, integrate, inhibit, assimilate, and remember perceptions and impulses received from the limbic system, striatum, temporal parietal and occipital lobes, and neocortical sensory receiving areas. Through the assimilation and fusion of perceptual, volitional, cognitive, and emotional processes, the frontal lobes engages in decision making and goal formation, modulates and shapes character and personality and directs attention, maintains concentration, and participates in information storage and memory retrieval. Further, the frontal lobes, the SMA and medial regions in particular, can direct behavior by controlling movement and the musculature of the body, and in this manner, it serves what is best described as "free will."

Because "free will" is localized to the right and left frontal medial motor areas, surgical destruction of the neural pathways linking the right and left SMAs and medial frontal lobes will result in two independent streams of mental activity which act according to their own "free will."

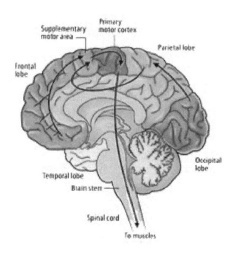

Destruction and disconnection of the pathways leading from the SMA and medial frontal lobes to the secondary and primary motor areas leads to difficulties where the patient's body will act "against their will", and perform actions they are unable to prevent or control. Compulsive utilization, and forced groping and grasping, are obviously compulsive in nature, and reminiscent of obsessive-compulsive disorders, which are also associated with frontal lobe abnormalities (Joseph 1996), especially of the orbital-medial frontal lobes and the striatum /anterior cingulate which are buried within the depths of the frontal lobes.

Patients may experience unwanted, recurrent, perseverative ideas, or compulsions to repetitively perform certain acts, e.g. hand washing. They may also experience intrusive recurring thoughts, feelings, or impulses to perform certain actions against their will. Motorically obsessive compulsions may involve repetitive, stereotyped acts including the perseverative manipulation and touching of objects.

Moreover, abnormal or electrical activation of the medial-orbital frontal lobes has been reported to trigger recurrent and intrusive ideational activity ("forced thinking"), as well as compulsive urges to perform aberrant actions, e.g. shouting or the manifestations of various motor tics (Penfield and Jasper 1954; Ward 1988). Similar disturbance have been reported for those who are afflicted with Tourretts syndrome (Singer 2005; Walkup et al., 2006). Those with Tourretts may involuntarily shout out obscenities, and inappropriate remarks, or engage in spontaneous, but repeptitive and stereotyped movements which are experienced as "unvoluntary" (Tourette Syndrome Classification Study Group, 1993). Although the exact etiology is unknown, abnormalities in deep medial frontal lobe structures have been reported (Walkup et al., 2006).

Damage to the neural pathways to and from the medial frontal lobes often simultaneously disrupt the orbital, lateral, and inferior frontal lobes, and can produce a unique constellation of compulsive, perseverative symptomology, including "forced thinking" and uncontrolled obsessive compulsions. That is, a patient may suffer from a difficulty suppressing or inhibiting previous thoughts or behaviors, which then occur again and again.

However, with massive damage to the medial frontal lobes and supplementary motor areas, instead of acting against their will, they lose the ability to will; free will is abolished, and this is because what has been called "free will" is localized to the frontal lobes, the medial frontal lobes in particular.

References

Alexander, G. E., & Crutcher, M. D. (1990). Preparation for movement. Neural representations of intended direction in three motor areas of monkey. J. Neurophysiology, 64, 164-178.

Allain P, Joseph PA, Isambert JL, Le Gall D, Emile J. Cognitive functions in chronic locked-in syndrome: a report of two cases. Cortex 1998;34: 629-34.

Akelaitis, A. J. (1945). Studies on the corpus callosum. American Journal of Psychiatry, 101, 594-599.

Bauby, J.-D., (1998), The Diving Bell and the Butterfly: A Memoir of Life in Death, Vintage.

Bogen, J. (1979). The other side of the brain. Bulletin of the Los Angeles Neurological Society. 34, 135-162.

Brinkman, C., & Porter, R. (1979). Supplementary motor area. Journal of Neurophysiology, 42, 681-709.

Brutkowski, S. (1965) Functions of the prefrontal cortex. Physiological Review, 45, 721-746.

Chouinard, P. A., Paus, T. (2006). The Primary Motor and Premotor Areas of the Human Cerebral Cortex, Neuroscientist April 2006 vol. 12 no. 2 143-152.

Denny-Brown, D. (1958). The nature of apraxia. Journal of Nervous and Mental Disease, 15-56.

Dimond, S. J (1980). Neuropsychology. Buttersworth.

Dum, R. P., Strick, P. L. (2005). Frontal Lobe Inputs to the Digit Representations of the Motor Areas on the Lateral Surface of the Hemisphere The Journal of Neuroscience,25(6):1375-1386.

Freeman, W., & Watts, J. W. (1942). Psychosurgery. Springfield, IL: Charles C. Thomas.

Fuster, J.M. (1997). The prefrontal cortex. Anatomy, physiology, and neuropsychology of the frontal lobes. New York: Ravens-Lippincott.

Fuster, J. M. (1995). Neuropsychiatry of frontal lobe lesions. In B.S. Fogel & R. B. Schiffer (Eds). Neuropsychiatry. Baltimore, Williams & Wilkins.

Gasquoine, P. G. (1993). Bilateral alien hand signs following destruction of the medial frontal cortices. Neuropsychiatry, Neuropsychology & Behavioral Neurology, 6, 49-53.

Geschwind N (1965) Disconnection syndromes in animals and man. Brain. 882, 237-274, 585-644.

Geschwind, N. (1981). The perverseness of the right hemisphere. Behavioral Brain Research, 4, 106-107.

Goldberg, G., & Bloom, K. K. (1990). The alien hand sign. American Journal of Physical Medicine & Rehabilitation. 69, 228-38.

Hassler, R. (1980). Brain mechanisms of intention and attention with introductory remarks on other volitional processes. Progress in Brain Research, 54, 585-614.

Ikeda, A., Luders, H. O., Burgess, R. C., and Shibasaki, H. (1992) Movement-related potentials recorded from supplementary motor area and primary motor area. Brain, 115, 1017-1043.

Jones, E. G., & Powell, T. P. S. (1970). An antomical study of converging sensory pathways within the cerebral cortex of the monkey. Brain, 93, 793-820.

Joseph, R. (1982). The Neuropsychology of Development. Hemispheric Laterality, Limbic Language, the Origin of Thought. Journal of Clinical Psychology, 44 4-33.

Joseph, R. (1986a). Confabulation and delusional denial: Frontal lobe and lateralized influences. Journal of Clinical Psychology, 42, 845-860.

Joseph, R. (1986b). Reversal of language and emotion in a corpus callosotomy patient. Journal of Neurology, Neurosurgery, & Psychiatry, 49, 628-634.

Joseph, R. (1988a) The Right Cerebral Hemisphere: Emotion, Music, Visual-Spatial Skills, Body Image, Dreams, and Awareness. Journal of Clinical Psychology, 44, 630-673.

Joseph, R. (1988b). Dual mental functioning in a split-brain patient. Journal of Clinical Psychology, 44, 770-779.

Joseph, R. (1992) The Limbic System: Emotion, Laterality, and Unconscious Mind. The Psychoanalytic Review, 79, 405-455.

Joseph, R. (1994) The limbic system and the foundations of emotional experience. In V. S. Ramachandran (Ed). Encyclopedia of Human Behavior. San Diego, Academic Press.

Joseph, R. (1996). Neuropsychiatry, Neuropsychology, Clinical Neuroscience, 2nd Edition. Williams & Wilkins, Baltimore.

Joseph, R. (1998). Traumatic amnesia, repression, and hippocampal injury due to corticosteroid and enkephalin secretion. Child Psychiatry and Human Development. 29, 169-186.

Joseph, R. (1999a). The neurology of traumatic "dissociative" amnesia. Commentary and literature review. Child Abuse & Neglect. 23, 71-80.

Joseph, R. (1999b). Frontal lobe psychopathology: Mania, depression, aphasia, confabulation, catatonia, perseveration, obsessive compulsions, schizophrenia. Psychiatry, 62, 138-172.

Joseph, R. (2000). Limbic language/language axis theory of speech. Behavioral and Brain Sciences. 23, 439-441.

Joseph, R. (2001). The Limbic System and the Soul: Evolution and the Neuroanatomy of Religious Experience. Zygon, the Journal of Religion & Science, 36, 105-136.

Joseph, R. (2003). Emotional Trauma and Childhood Amnesia. journal of Consciousness & Emotion, 4, 151-178.

Joseph, R. (2009). Quantum Physics and the Multiplicity of Mind: Split-Brains, Fragmented Minds, Dissociation, Quantum Consciousness, Journal of Cosmology, 3, 600-640.

Joseph, R., Forrest, N., Fiducia, N., Como, P., & Siegel, J. (1981). Electrophysiological and behavioral correlates of arousal. Physiological Psychology, 1981, 9, 90-95.

Laplane, D., Talairach, J., Meininger, V., Bancaud, J., & Orgogozo, J. M. (1977). Clinical consequences of cortisectomies involving the supplementary motor area in man. Journal of the Neurological Sciences, 34, 301-314.

Lhermitte, F. (1983). "Utilization behaviour" and its relation to lesions of the frontal lobes. Brain, 106, 237-255.

Luria, A. (1980). Higher cortical functions in man. New York: Basic Books.

McNabb, A. W., Caroll, W. M., & Mastaglia, F. L. (1988). "Alien hand" and loss of bimanual coordination after dominant anterior cerebral artery territory infaction. Journal of Neurology, Neurosurgery, and Psychiatry, 51, 218-222.

Miller, B. L., & Cummings, J. L. (2006). The Human Frontal Lobes. Guilford Press.

Mink, J. W., & Thach, W. T. (1991). Basal ganglia motor control. I, II, & III. Journal of Neurophysiology, 65, 273-351.

Mishkin, M. (1964). Perseveration of central sets after frontal lesions in monkeys. In J. M. Warren & K. Akert (Eds.), The frontal granular cortex and behavior. (pp 219-241). New York: McGraw-Hill.

Nachev, P., Kennard, C., & Husain, M. (2008). Functional role of the supplementary and pre-supplementary motor areas Nature Reviews Neuroscience 9, 856-869.

Orgogozo, J. M., & Larsen, B. (1979). Activation of the suplementary motor area during voluntary movement in man suggest it works as a supramodal motor area. Science, 206, 847-850.

Pandya, D. N. & Yeterian, E. H. (1990) Architecture and connectivity of cerebral cortex: Implications for brain evolution and function. In: Neurobiology of higher cognitive function. eds. A. B. Scheibel & A. F. Wechsler. Guilford.

Pandya, D. N., & Kuypers, H. G. J. M. (1969). Corticocortical connections in the rhesus monkey. Brain Research, 13, 13-36.

Pandya, D. N., & Vignolo, L. A. (1969). Corticocortial connections in the rhesus monkey. Brain Research, 13, 13-16.

Passingham, R. (1997). Functional organization of the motor system. In Frackowiak, R. S. J., et al., (Eds.,) Human Brain Function. Academic Press, San Diego.

Penfield, W. (1952) Memory Mechanisms. Archives of Neurology and Psychiatry, 67, 178-191.

Penfield, W., & Jasper, H. (1954). Epilepsy and the functional anatomy of the human brain. Boston: Little-Brown & Co.

Penfield, W., & Perot, P. (1963) The brains record of auditory and visual experience. Brain, 86, 595-695.

Penfield, W., & Welch, K. (1951). Supplementary motor area of cerebral cortex. Clinical and experimental study. Archives of Neurology & Psychiatry, 66, 289-317.

Petrides, M. & Pandya, D. N.. (1988) Association fiber pathways to the frontal cortex from the superior temporal region in rhesus monkey. Journal of Comparative Neurology 273:52-66.

Risberg J., Grafman, J. (2006). The Frontal Lobes: Development, Function and Pathology, Cambridge University Press

Rizzolatti, G., Craighero, L. (2004). THE MIRROR-NEURON SYSTEM, Annu. Rev. Neurosci. 27, 169–92.

Rizzolatti G, Fadiga L, Fogassi L, Gallese V. (1996). Premotor cortex and the recognition of motor actions. Cogn. Brain Res. 3:131– 141.

Rizzolatti G, Luppino G, Matelli M. (1998). The organization of the cortical motor system: new concepts. Electroencephalogr. Clin. Neurophysiol. 106:283–296.

Rizzolatti, G., Fabbri-Destro, M., Cattaneo, L. (2009). Mirror neurons and their clinical relevance. Nat Clin Pract Neurol 5 (1): 24–34.

Smith, E., and Delargy, M. (2005). Locked-in syndrome, BMJ. 330, 406-408.

Shibasaki, H., Sadato, N., Lyshkow, H., et al. (1993). Both primary motor cortex and supplementary motor area play an important role in complex finger movement. Brain, 116, 1387-1298.

Singer, H.S. (2005). Tourette's syndrome: from behaviour to biology". Lancet, 4, 149–59.

Sperry, R. (1966). Brain bisection and the neurology of consciousness. In F. O. Schmitt and F. G. Worden (eds). The Neurosciences. MIT press.

Stephan, K. M., Binkofski, F., Halsband, U., et al., (1999). The role of ventral medial wall motor areas in bimanual coordination. Brain, 122, 351-368.

Tanji, J., & Kurata, K. (1982). Comparison of movement-related neurons in two cortical motor areas of primates. Journal of Neurophysiology, 40, 644-653.

Tourette Syndrome Classification Study Group (1993). "Definitions and classification of tic disorders". Arch Neurol. 50, 1013–16.

Verstynen, T., et al. (2011). In Vivo Mapping of Microstructural Somatotopies in the Human Corticospinal Pathways. Journal of Neurophysiology, 105, 336-346

Walkup JT, Mink JW, Hollenback PJ, (2006). Advances in Neurology, Vol. 99, Tourette Syndrome. Lippincott, Williams & Wilkins. .

Ward, C.D. (1988). Transient feelings of compulsion caused by hemispheric lesions: Three cases. Journal of Neurology, Neurosurgery, and Psychiatry, 51, 266-268.

Zeman A. (2003). What is consciousness and what does it mean for the persistent vegetative state? Adv Clin Neurosci Rehabil, 3, 12-4.

8. Brain, Consciousness, and Causality

Andrea Nani[1], Andrea E. Cavanna[2,3]

[1]School of Psychology, University of Turin, Italy. [2]Department of Neuropsychiatry, BSMHFT and University of Birmingham, UK. [3]Department of Neuropsychiatry, Institute of Neurology and University College London, UK

Abstract

Consciousness seems to be a fundamental ingredient of human life: our common sense tells us that without it we would not behave in the same way. However since the end of the XIX century among some philosophers and scientists have become increasingly familiar with a counterintuitive position on the place of consciousness in nature, known as epiphenomenalism. Epiphenomenalism excludes from scientific accounts of human behavior any appeal to conscious processes occurring in the brain. Its main claim is that conscious experience is an epiphenomenon of brain activity, without causal powers in terms of volition and action. This paper examines the issue whether consciousness can be regarded as a mere epiphenomenon from both the theoretical and empirical perspective. The epiphenomenalist theory is analyzed with reference to the work of leading neuroscientist Gerald Edelman and neurological syndromes defined by key alterations in conscious domains. It is argued that conscious states are likely to play essential causal roles in the scientific account of how the brain brings about the voluntary actions that contribute to form our deepest personal identities.

KEY WORDS: Causality, Consciousness, Edelman, Epiphenomenalism, Neurology

1. Introduction. The Temptation of Epiphenomenalism

One of the most enduring and intriguing questions for both philosophical and scientific researchers is whether we have conscious minds capable to control and produce the motivations for all our actions. In the light of our common experience, an affirmative answer to that question would be but a platitude. Indeed, brains seem to be capable to create a great variety of mental events. Love, hate, sadness, joy, sorrow, pleasure, shame, grief, delight, and resentment are only a few of the many different psychological states composing our rich mental lives. However, on more accurate reflection the solution to the problem of the nature of consciousness would not appear as evident as it may seem at first sight.

There are, in fact, philosophers and neuroscientists who firmly believe that mental properties, particularly the conscious ones, are wholly epiphenomenal with respect to brain processes (Edelman 2004, 2007; Fuster 2003). According to epiphenomenalism, mental properties are superfluous by-products of the function of our cerebral mechanism, just like the images which are reflected by mirrors are not made by glass, and the shadows which objects cast on the ground are not parts of those objects. This view is not new: in a famous conference held at the British Association for the Advancement of Science, Thomas Henry Huxley compared consciousness to the steam whistle of a locomotive (Huxley 1884). In contrast with Descartes, Huxley did not consider animals as unconscious machines, but was very perplexed with regard to the exact function of consciousness and hypothesized that conscious states played no role in behavioral mechanisms (Huxley 1874). Just as the steam whistle of a locomotive did not influence the work of the locomotive's motor, he thought, so animal consciousness could neither cause nor modify animal behavior. Being also a strenuous advocate of Darwin's theory of natural selection, Huxley assumed for reasons of biological continuity that there are no differences between animal and human consciousness (Huxley 1884).

We can affirm that the modern shape of the problem of epiphenomenalism was set with Huxley, even though he never used this word in his writings. In effect, the Modern Age (post-Cartesian) thinkers did not tend to contrast sharply the concepts of mind and body. Ancient Greeks had a much broader idea of mind, closely linked with bodily functions (Bremmer 1983). Mind (i.e. the soul) was considered the principle of life capable to animate the body in order for it to perform its basic biological processes, such as breathing, digestion, procreation, growth, motion, and, for humans, also other sophisticated processes of life, such as thinking, perceiving, imagining, and reasoning. Of course every school of ancient Greek philosophy had its own concept of "soul". For instance, Aristotle thought of the soul as the body's system of active abilities to accomplish the vital functions that organisms naturally perform, e.g. nutrition, movement or thought (Nussbaum and Rorty 1992). On the other hand, Stoic philosophy of mind conceived the soul itself as a corporeal entity (Inwood 2003). This position was similar to that of Epicurus, who taught the soul to be a kind of body composed of atomic particles (Kerferd 1971). Perhaps Plato and the Pythagoreans held the closest concept of soul to the Cartesian view. They maintained it to be as something incorporeal and able to exist independently of the body (Lorenz 2008; Huffman 2009). Nevertheless, it is only after Descartes' philosophy that the debate of mental causation was to be set down in its modern form. An important echo of this debate is to be found in the discussion upon the automatism of behavior raised by Huxley in the second half of the nineteenth century.

Positions similar to those embraced by Huxley are still held by some contemporary philosophers and neuroscientists. Epiphenomenalists would be willing to explain the origin of consciousness in the same way as we can physically explain how mirrors produce reflected images or bodies cast their shades on the ground. According to this explanatory model, all our psychological states should, in theory, be accounted for entirely in terms of scientific vocabularies which contain no mental concepts.

A similar approach, at least in relation to its practical consequences for psychological research, was maintained by Burrhus Skinner with his theory of "radical behaviorism". Skinner supported the view that mental terms could be completely paraphrased in behavioral terms, or eliminated from explanatory discourse altogether (Skinner 1974). Therefore, all accounts of human behavior would have to be given in neutral and objective terms, such as stimulus, response, conditioning, reinforcement, and so on. This position has some analogies with epiphenomenalism in that it considers consciousness a nonphysical entity which has nothing to do with behavior.

Such epiphenomenalist lines of reasoning imply that a rigorous discourse on human actions should deny the reality of consciousness, and thereby of all mental states correlated with this phenomenon. In fact, when human beings express propositions about conscious states, they actually intend to speak of other things, specifically of certain brain physical states which are to be the unique causes of all their bodily dispositions. In William James' words, "consciousness ... would appear to be related to the mechanism of [the] body simply as a collateral product of its working, and to be completely without any power of modifying that working" (James 1890). In a sense, arguing that conscious mental phenomena are causally real would be like arguing that the black spots which are on the tails of some fishes are real eyes capable of seeing and not just evolutionary tricks for misleading predators.

The solution offered by epiphenomenalism to the problem of the nature of consciousness can be roughly summarized in the assertion that such a problem does not exist because we have no consciousness, but only the illusion of having it. Although epiphenomenalism could be somehow attractive, it does not present a satisfactory solution to the problem of what consciousness really is. The present paper will show that the epiphenomenalist perspective does not offer a consistent account of certain neuropsychological phenomena which seem to be intrinsically subjective.

We will start by examining the arguments that drive some philosophers and neuroscientists to regard consciousness as an empty concept involving outdated thinking.

2. Causal Links

Epiphenomenalism upholds that the conscious mind is not part of the physical world. This implies that, given the physical causal closure of the universe, conscious mental events cannot interact with the physical reality in any way. If we think of ourselves as consciously acting agents, the epiphenomenalist claim sounds counterintuitive: is it plausible that our conscious will cannot influence the physical world?

In order to answer this question we first need a better understanding of exactly what it means to claim that conscious mental events cannot influence physical events in any way. This doctrine implies that the only possible type of causation we have to deal with is the so-called bottom-up causality, i.e. the causation that goes from the physical level to the conscious one. According to this concept of causality, the irresistible desire for an apple pie cannot actually cause the act of eating a slice of pie, since the account of our behavior is to be determined only at the physical level, where physical entities move other physical entities. However, if all causation processes belong to the physical level, how could our ontological catalogue list phenomena which are not included in causal accounts of behavioral expressions given in physical terms only? A reasonable philosophical precept (epitomized by Occam's razor) warns us that ontology should not be expanded without necessity. Moreover, there is the problem of defining the nature of those phenomena. If these entities were completely different from physical processes, then conscious mental events would necessarily belong to a distinct ontological domain, but it is easy to conclude that in such a case epiphenomenalism would be just like a spurious kind of dualism. On the other hand, if conscious mental events were physical phenomena of a very special nature, how could we distinguish the physical events capable to cause other physical events from the ones which have no causal power at all?

We cannot actually make a distinction of this kind by means of the third person vocabulary that scientists generally use to depict their objective vision of the world. Therefore, those who trust epiphenomenalism would have to include in their ontological catalogue states, events, and processes susceptible to be described exclusively in terms of the first person perspective, and to identify these states, events, and processes with non-causal physical phenomena. In addition, this sort of phenomena would have to be put together with all the other states, events, and processes liable both to be described in terms of the third person perspective and to be identified with causal physical phenomena. However, it is by no means clear why some processes – whose nature is basically physical – would have to be described exclusively in terms of the first person perspective rather than third person perspective.

In addition to these conceptual difficulties, a further grave quandary is whether we maintain the division between causal and non-causal physical events. In other words, taken for granted such a division, what in this case would the principle of the physical causal closure precisely mean? The principle of the physical causal closure states that if a physical event has a cause that occurs at t, it has a physical cause that occurs at t. Jaegwon Kim (2005) correctly observes that the "physical causal closure does not by itself exclude non-physical causes, or causal explanations, of physical events." For instance, there could be a nonphysical causal explanation of a physical event being the first ring of a chain of other numerous physical events. According to Kim (2005), in order to rule out this kind of explanation, we need an exclusion principle such as the following: if an event e has a sufficient cause c at t, no event at t distinct from c can be a cause of e.

Following the principle of exclusion, the sufficient cause c of e at time t may be either physical or mental. However, this instance is ruled out if the principle of exclusion is linked to the principle of the physical causal closure and the event e is identified with an event whose nature is purely physical. As a result, the two principles joined together hold that only physical events can cause other physical events.

It is important to highlight that these two principles are not stricto sensu in contrast with the folk psychology view that there is a conscious mind in every human being. Moreover, these assumptions are consistent with the hypothesis of psycho-physical parallelism, according to which there could be a distinct domain of specific conscious mental events coming to occur whenever other particular physical events come to occur. Still it is unconditionally denied that conscious mental phenomena can causally interact with physical phenomena. In fact, in order to be closed, any causal explanation is to be expressed as a chain of purely physical events. This leads us back to our previous question: if we accept that both epiphenomenal events and non-causal physical entities can by no means be part of scientific accounts given in causal physical terms, then why should we expand our ontology without necessity by including allegedly redundant conscious phenomena?

3. An Argument from Quantum Physics?

An interesting argument against the plausibility of epiphenomenalism can be derived from a specific interpretation of the theory of quantum mechanics. It is well-known that the act of measurement is a crucial aspect from the perspective of quantum theory. The implications and the account of this process has been the subject of controversy for more than seven decades and the debate does not seem to be closed yet.

The fundamental point of the argument from quantum physics holds that the observer's consciousness plays a key role in the collapse of the wave function to a certain state, described by a second-order differential equation by Erwin Schrödinger. The root of this idea can be found in the so-called Copenhagen interpretation of quantum mechanics. Although the Copenhagen interpretation is not a homogenous view (Howard 2004), Heisenberg (1955) appears to be the one who first coined the term and developed the underlying philosophy as a unitary interpretation (Heisenberg 1958). On the other hand, Niels Bohr – the Danish physicist commonly regarded as the father of the Copenhagen interpretation – never seems to have emphasized or privileged the role of the observer in the wave packet collapse (Howard 1994). Bohr argued for an interpretation of complementarity with regard to the wave-particle duality which is incompatible with Heisenberg's interpretation of wave function collapse (Gomatam 2007). Bohr's view regarding the wave function was more moderate than Heisenberg's and based on epistemological concerns, rather than ontological commitments. When Bohr referred to the subjective character of quantum phenomena he was not referring to the conscious intervention of the observer in the process of measurement, but to the context-dependent status of all physical observations (Murdoch 1987; Faye 1991). However the drastic theoretical move – outlined in Heisenberg's writings – that quantum measurement is to be understood by involving the observer's act in addition to a physical process, has become the core of the so-called Copenhagen interpretation.

This idea was further developed by other physicists. Von Neumann (1932) postulated an ad hoc intervention of an observing system in order to account for the collapse or reduction of the wave function through measurement. Quite cautiously, he never claimed that the observing system had to be conscious (von Neumann 1932). In contrast to von Neumann's view, London and Bauer (1932) attributed to the oberver's consciousness only the key role in understanding the process of quantum measurement. Such a proposal was later expanded on by the physicist Eugene Wigner, according to whom "it was not possible to formulate the laws (of quantum theory) in a fully consistent way without reference to consciousness" (Wigner 1967).

The proposal to consider consciousness as causally involved in physical state reductions was further developed following Heisenberg, von Neumann, and Wigner (Stapp 1993, 1999, 2006; Schwartz et al. 2005). Undoubtedly, this approach challenges the epiphenomenal position. In fact, how is it possible for consciousness to be non-causal if it can bring about the collapse of the wave packet? If those who champion the Copenhagen interpretation are right, then epiphenomenalism should be completely refuted. On the one hand, this approach gives a fundamental causal power to consciousness in understanding the universe; on the other hand, it has the serious shortcoming of putting consciousness outside the physical world. In fact, if consciousness really causes the collapse of the wave function, then it must be a process that is not be describable by Schrödinger equation, because otherwise it would be caught in an infinite regress. Based on these arguments, consciousness should be a nonphysical entity. Therefore, if both the principle of the physical causal closure and the principle of exclusion discussed in the previous section are true, the so-called Copenhagen interpretation of quantum mechanics does not provide a strong argument for refuting epiphenomenalism.

4. Consciousness and Causality

In view of the foregoing reflections, we would have to be very reluctant to claim the distinction between causal and non-causal physical events or between causally efficacious physical and causally efficacious nonphysical events, although these distinction are of course logically possible. Accordingly, both philosophers and neuroscientists well-disposed to epiphenomenalism have to maintain that the nature of conscious mental processes cannot be physical, since, by definition, a state/event/process exists in physical terms only if it has the property of influencing and exerting causal effects over other physical entities.

An interesting type of neuroscientific approach is exemplified by the epiphenomenalist position held by Gerald Edelman, a leading neuroscientist who has given remarkable contributions to the study of mind and consciousness and their place in nature (Edelman 1989, 1993, 2007). According to Edelman's approach ("Neural Darwinism"), different configurations or patterns of neurons compete with each other to gain constancy and stability within the brain. The Neural Darwinism approach holds that groups of neurons and the neural patterns and configurations which nerve cells form ("neural networks") are subject to natural selection, just like biological species are evolutionarily selected by the environment. Specifically, Edelman's theory of "neuronal group selection" postulates that anatomical connectivity in the brain occurs via selective mechanochemical events that take place epigenetically during development. This process creates a structurally diverse primary repertoire by differential reproduction. A second selective process occurs during postnatal behavioral experience through epigenetic modifications in the strength of synaptic connections between neuronal groups, thus creating a diverse secondary repertoire by differential amplification.

In Edelman's view, human consciousness depends on and arises from the uniquely complex physiology of the human brain. He advanced a theory of how the brain generates different levels of consciousness through multiple parallel re-entrant connections between individual cells and between larger neuronal groups, in which he endorsed an epiphenomenalist position with regard to consciousness, which is central to the scope of this paper (Edelman 2004, 2007).

It has been argued that the question whether consciousness can have a causal role in determining behavior and other mental states should find an answer supported by both conceptual and empirical considerations. (Flanagan 1992; Heil and Mele 1993; Searle 2004). Contrary to this view, Edelman seems to give pre-eminence to theoretical arguments over empirical results. In fact his thesis – that consciousness is not causal – is almost exclusively based upon the following theoretical argument:

> This account [which is that conscious processes arise from enormous numbers of re-entrant interactions between different areas of the brain] implies that the fundamental neural activity of the reentrant dynamic core converts the signals from the world and the brain into a "phenomenal transform" – into what it is like to be that conscious animal, to have its qualia. The existence of such a transform (our experience of qualia) reflects the ability to make high-order distinctions or discriminations that would not be possible without the neural activity of the core. Our thesis has been that the phenomenal transform, the set of discriminations, is entailed by that neural activity. It is not caused by that activity but it is, rather, a simultaneous property of that activity. (Edelman, 2004).

Edelman's idea is that some cerebral processes entail certain phenomenal transforms, which are the contents of our conscious mental states. Following him, we can call the cerebral processes C' and the phenomenal transforms C. We can put both C' and C in a row and index them to indicate their successive states in time: C'_0-C_0; C'_1-C_1; C'_2-C_2; C'_3-C_3; and so forth. It is crucially important to highlight how in that view only the underlying cerebral processes are endowed with causal powers, whereas the phenomenal processes entailed by those brain states are not. However the relationship between the cerebral processes and the phenomenal transforms is considered by Edelman as necessary. This necessary correlation appears to be of a metaphysical kind, i.e. a correlation that holds in every possible world. Therefore, Edelman's position does not appear to be consistent with the philosophical "zombie argument", which assumes the existence of an individual capable to behave just in the same way as conscious human beings do, but in the absence of any subjective conscious experience (in Edelman's words, an individual having C' but not C). In fact, Edelman claims that "The argument we are making here implies, however, that if C' did not entail C, it could not have identical effects" (Edelman 2004; the emphasis is not ours). Consequently, an individual lacking C (phenomenal consciousness) cannot show the same behavior of an individual who has C. Indeed, specific activities of the nervous system necessarily give rise to particular conscious sensations, which in turn cannot exist without a specific underlying activity of the brain. As a result, the zombie hypothesis should be utterly inconsistent (Jackson 1982).

Edelman's theory is central to the discourse on epiphenomenalism advanced so far. In fact, the phenomenal properties which he refers to are necessarily implied by the underlying neural activity of the brain. Strictly speaking, those phenomenal properties are not redundant but absolutely non-causal, even though they have a sort of physical nature. In addition, those properties have to be seen as by-products, since they cannot play any specific role in our scientific account of natural phenomena. Therefore, neither the principle of physical causal closure nor the principle of physical exclusion seem to be violated by Edelman's perspective.

In our view, the arguments put forward by Edelman in order to demonstrate the epiphenomenalist nature of consciousness raise a number of issues. What is most unclear is the very nature of the necessary relationship between the causal physical events (i.e., the cerebral processes or C') and the non-causal physical events (i.e., the mental processes or C). If we accept that this relationship is necessary, there is no reason to assume an ontology in which conscious mental processes and physical processes within the nervous system are distinct. If a certain property is necessarily implied by certain physical processes (in such a way that the latter could not bring about the same effect without the former, as Edelman claims), then either that very property and those physical processes are different aspects of the same entity, or that very property is part of the co-occurring physical processes. From the logical point of view, an effect cannot find its cause in one event which is the result of the sum of a causal physical state and a non-causal physical state, since the non-causal physical process cannot play any causal role at all (Heil and Mele 1993). In fact, what Edelman believes to be non-causal, "the phenomenal transform", must be provided with causal powers.

In this sense, Edelman's theory with regard to the epiphenomenalist nature of consciousness appears to be anomalous. A "true" epiphenomenalist would, in fact, plausibly think of the causal relationship between physical and conscious states as contingent rather than necessary. Thus, epiphenomenalists should be willing to accept the zombie argument, since it is logically possible to accept that, if conscious mental processes are contingent, there could be possible worlds in which they do not bring about any behavioral effect.

In addition to these theoretical considerations, empirical data can raise other reservations with regard to epiphenomenalism. For example, if we agree on depriving mental entities of their causal role, we would encounter difficulties accounting for a host of well described neurological conditions. These include, but are not limited to, blindsight (cortical blindness with preserved ability to locate objects), unilateral neglect syndrome (loss of ability to detect information coming from the left side of the body), allochiria (experience of a sensory stimulation at the contralateral side to the applied stimulus), anosognosia (denial of gross neurological deficit), prosopognosia (inability to recognize familiar faces), and somatoparaphrenia (a condition in which patients deny ownership of a limb or an entire side of their body). Arguably, these neurological disorders can be explained at least to some extent in terms of a dysfunction in the causal role played by consciousness in dealing with perceptive or proprioceptive information.

The understanding of somatoparaphrenia is an exemplar case, based on the concept of verbal manifestations commonly referred to as propositional attitudes in the tradition of analytic philosophy (Bisiach and Geminiani 1991). Propositional attitudes are all the expressions whose contents consist of subjective beliefs, desires, intentions, fears, etc. In case of a patient showing somatoparaphrenic symptoms, it is plausible to suppose that a dysfunction in the conscious processing of proprioceptive information about the patient's limb (for instance, the left leg), results in the patient holding the belief that the leg does not belong to his body. Patients with somatoparaphrenia will therefore verbally deny that they own that leg, and some of them have in fact been reported trying to reject the limb that they perceive as alien (Critchley 1974).

Somatoparaphrenia has been described, with a few exceptions, in patients suffered from right parietal (or parieto-occipital) lobe injury – and almost invariably concerns the left side of the body. This condition is usually associated with motor and somatosensory deficits, and with the syndrome of unilateral spatial neglect. In a study on 79 acute stroke right-brain-damaged patients (Baier and Karnath 2008), 12 patients showed anosognosia for hemiplegia. Eleven out of these 12 patients exhibited somatoparaphrenic symptoms, and 6 among them displayed the strong belief that their limbs belonged to another person. In other cases, body parts can be just felt by the patient as separated from the body (Starkstein et al. 1990). More complex symptoms have been described: for instance, a patient can refer to the affected limb as "a make-believe leg" (Levine et al. 1991), or as "a baby in bed" (Richardson 1992).

The spectrum of somatoparaphrenic symptoms is wide, however a distinction can be drawn between misidentifications that can be corrected by patients when the error is pointed out by the examiner, and delusions that stubbornly resist to the examiner's demonstration (Feinberg et al. 2005). It is referred to fully-fledged somatoparaphrenia only in the second category of symptoms.

In sum, somatoparaphrenic phenomena do not imply a mental illness and can be characterized as follows (Vallar and Ronchi 2009):

- the feeling of estrangeness and/or separation of the affected body parts;
- delusional beliefs of disownership of the affected body parts;
- delusional beliefs that the affected body parts belong to another person;
- complex delusional misidentifications of the affected body parts;
- associated disorders, such as supernumerary limbs, personification, and misoplegia (hatred for the affected limbs).

Overall, available data suggest that patients showing somatoparaphrenic symptoms suffer from impairments in the higher-level processes concerned with body awareness and ownership. Therefore it seems reasonable to hypothesize that if certain behaviors do not occur without specific conscious sensations accompanied by the beliefs which refer to them (i.e. mental representations of the body), then those specific conscious sensations and their consequent beliefs must play an important causal role in the process that produces this kind of behavior.

5. Conclusion

In specific scientific contexts, it seems mandatory to apply concepts which carry a commitment for a causal role for consciousness. For the sake of the unity of science, it seems justified to take the physical causal closure for granted; on the other hand, it does not seem as well justified to take for granted the clear-cut distinction traced by epiphenomenalism between the physical world and the conscious mind. However, neither the common version of epiphenomenalism (in which conscious states are contingent), nor its variant proposed by Edelman (in which consciousness and the physical world are necessarily intertwined), appear to be a valid theory to explain the nature of consciousness. The theoretical and empirical arguments advanced in this article show that it is likely for consciousness to play fundamental roles in the genesis of behavior. Undoubtedly much more is to be done, especially on the side of the empirical research, in order to unravel the actual brain mechanisms of conscious causal processes.

References

Baier B., Karnath H.O. (2008), Tight link between our sense of limb ownership and self-awareness of actions. Stroke: 486-488.

Bisiach E., Geminiani G. (1991), Anosognosia related to hemiplegia and hemianopia. Prigatano G.P., Schacter D.L. (eds.), Awareness of deficit after brain injury, Oxford: Oxford University Press.

Bremmer, J. (1983), The Early Greek Concept of the Soul, Princeton: Princeton University Press.

Critchley M. (1974), Misoplegia or hatred of hemiplegia. Mt.Sinai Journal of Medicine 41: 82-87.

Edelman, G.M. (1989), The Remembered Present: A Biological Theory of Consciousness, New York: Basic Books.

Edelman, G. M. (1993), Neural Darwinism: The Theory of Neuronal Group Selection. Neuron 10: 115-125.

Edelman, G.M. (2004), Wider than the Sky. The Phenomenal Gift of Consciousness, Yale University Press.

Edelman, G.M. (2007), Second Nature, Brain Science and Human Knowledge, Yale University Press.

Faye, J. (1991), Niels Bohr: His Heritage and Legacy. An Antirealist View of Quantum Mechanics, Dordrecht: Kluwer Academic Publisher.

Feinberg, T.E., DeLuca, J., Giacinto, J.T., Roane, D.M, Solms, M. (2005), Right-hemisphere pathology and the self. Feinberg, T.E., Keenan, J.P. (eds.) The lost self. Pathologies of the brain and identity, Oxford: Oxford University Press, pp. 10-130.

Flanagan, O. (1992), Consciousness reconsidered, MIT Press, Cambridge Mass.

Fuster, J. M. (2003), Cortex and Mind, Oxford University Press.

Gomatam, R. (2007), Niels Bohr's Interpretation and the Copenhagen Interpretation — Are the two incompatible? Philosophy of Science, 74, December issue.

Heil, J. and Mele, A. eds. (1993), Mental Causation, Oxford University Press.

Heisenberg, W. (1955), The Development of the Interpretation of the Quantum Theory. In W. Pauli (ed.), Niels Bohr and the Development of Physics, 35, London: Pergamon.

Heisenberg, W. (1958), Physics and Philosophy: The Revolution in Modern Science, London: Goerge Allen & Unwin.

Howard, D. (1994), What Makes a Classical Concept Classical? Toward a Reconstruction of Niels Bohr's Philosophy of Physics. In Faye, J., and Folse, H., eds., Niels Bohr and Contemporary Philosophy. Series: Boston Studies in the Philosophy of Science, vol. 158. Dordrecht: Kluwer Academic Publisher.

Howard, D. (2004), Who Invented the "Copenhagen Interpretation?" A Study in Mythology. Philosophy of Science 71, pp. 669-682.

Huffman, C. A. (2009), The Pythagorean Conception of the Soul from Pythagoras to Philolaus. Body and Soul in Ancient Philosophy, D. Frede and B. Reis (eds.), Berlin: Walter de Gruyter.

Huxley, T.H. (1874), On the Hypothesis That Animals Are Automata, and Its Hystory. Fortnightly Review 95: 555-580. Reprinted in Collected Essays, London: Macmillan, 1893.

Huxley, T.H. (1884), Animal Automatism, and Other Essays, Humboldt Library of Popular Science Literature, New York: I. Fitzgerald.

Inwood, B. (2003), The Cambridge Companion to the Stoics, Cambridge: Cambridge University Press.

Jackson, F. (1982), Epiphenomenal qualia. Philosophical Quarterly 32: 127-136.

James, W. (1890), The Principles of Psychology, reprint. New York: Dover, 1950.

Kerferd, G., 1971, Epicurus' doctrine of the soul. Phronesis 16: 80–96.

Kim, J. (2005), Physicalism, or Something Near Enough, Princeton University Press, Princeton.

Levine, D.N., Calvanio, R., Rinn, W.E. (1991), The pathogenesis of anosognosia for hemiplegia. Neurology 41: 1770-1781.

London, F., and Bauer, E. (1939), La théorie de l'observation en mécanique quantique. Hermann, Paris. English translation: The theory of observation in quantum mechanics. In Wheeler J.A., and Zurek, W.H., eds., (1983), Quantum Theory and Measurement, Princeton University Press, Princeton.

Lorenz, H. (2008), Plato on the Soul. The Oxford Handbook of Plato, G. Fine (ed.), Oxford: Oxford University Press.

Murdoch, D. (1987), Niels Bohr's Philosophy of Physics, Cambridge: Cambridge University Press.

Neumann, J. von (1932), Die mathematischen-Grundlagen der Quantenmechanik, Springer, Berlin. Reprinted in English (1955), Mathematical Foundations of Quantum Mechanics. Princeton University Press, Princeton.

Nussbaum, M. C. & Rorty, A. O., eds., (1992) Essays on Aristotle's De Anima, Oxford: Clarendon Press.

Richardson, J.K. (1992), Psychotic behavior after right hemispheric cerebrovascular accident: a case report. Arch. Phys. Med. Rehabil. 73: 381-384.

Robinson, H. M. (1978), Mind and Body in Aristotle. The Classical Quarterly 28: 105-124.

Schwartz, J.M., Stapp, H.P., and Beauregard, M. (2005), Quantum theory in neuroscience and psychology: a neurophysical model of mind/brain interaction. Philosophical Transactions of the Royal Society B 360, 1309-1327.

Searle, J. R. (2004), Mind, a brief introduction, Oxford University Press.

Skinner, B. (1974), About Behaviorism, New York: Vintage.

Stapp, H.P. (1993), A quantum theory of the mind-brain interface. Mind, Matter, and Quantum Mechanics, Springer, Berlin, pp. 145-172.

Stapp, H.P. (1999), Attention, intention, and will in quantum physics. Journal of Consciousness Studies 6(8/9), pp. 143-164.

Stapp, H.P. (2006), Clarifications and Specifications In Conversation with Harald Atmanspacher. Journal of Consciousness Studies 13(9), pp. 67-85.

Starkstein, S.E., Berthier, M.L., Fedoroff, P., Price, T.R., Robinson, R.G. (1990), Anosognosia and major depression in 2 patients with cerebrovascular lesions. Neurology 40: 1380-1382.

Vallar, G., Ronchi, R. (2009), Somatoparaphrenia: a body delusion. A review of the neuropsychological literature. Experimental Brain Research 192: 533-551.

Wigner, E.P. (1967), Symmetries and Reflections, Indiana University Press, Bloomington.

9. The Dissipative Brain and Non-Equilibrium Thermodynamics
Walter J. Freeman, Ph.D.[1], and Giuseppe Vitiello, Ph.D.[2]

[1]Department of Molecular & Cell Biology, Division of Neurobiology, University of California at Berkeley, Berkeley CA. [2]Facoltà di Scienze Matematiche, Fisiche e Naturali e Istituto Nazionale di Fisica Nucleare Università di Salerno, Italia

Abstract

Cognitive neurodynamics describes the process by which brains direct the body into the world and learn by assimilation from the sensory consequences of the brain directed actions. Repetition of the process comprises the action-perception cycle by which knowledge is accumulated in small increments. The global memory store is based in a rich hierarchy of attractor landscapes comprising a library of increasingly abstract generalizations. In this paper we briefly summarize the dissipative many-body model of brain which provides the theoretical scheme aimed to describe the basic dynamics underlying the neurological activity described above. Energy dissipation as heat manifests itself in the disappearance and emergence of the ground state coherence. The emergence of classicality out of the microscopic dynamics is a central feature of the dissipative many-body model.

KEY WORDS: Brain modeling, Many-body physics, Coherence, Quantum Dissipation, Consciousness,

Observations and data analysis carried on in the past decades (Freeman, 1975-2006) have shown that the brains of animal and human subjects engaged with their environments exhibit coordinated oscillations of populations of neurons, changing rapidly with the evolution of the relationships between the subject and its environment, established and maintained by the action-perception cycle (Freeman, 2004a-2006; Vitiello, 2001; Freeman & Vitiello, 2006-2010). Our analysis of electroencephalographic (EEG) and electrocorticographic (ECoG) activity has shown that cortical activity during each perceptual action creates multiple spatial patterns in sequences that resemble cinematographic frames on multiple screens (Freeman, Burke and Holmes, 2003; Freeman, 2004a,b). In this paper we will briefly review some of the features of the dissipative model of brain which has been formulated in recent years (Vitiello, 1995, 2001; 2004; Freeman & Vitiello, 2006- 2010).

The sources of these patterns are identified with large areas of the neocortical neuropil (the dense felt-work of axons, dendrites, cell bodies, glia and capillaries forming a continuous sheet 1 to 3 mm in thickness over the entire extent of each cerebral hemisphere in mammals). The carrier waves of these patterns are identified with narrow band oscillations (±3-5 Hz) in the beta (12-30 Hz) and gamma (30-80 Hz) ranges (Freeman, 2005, 2006, 2009). The change in the dynamical state of the brain with each new frame resembles a collective neuronal process of phase transition (Freeman & Vitiello, 2006-2009) requiring rapid, long-distance communication among neurons for almost instantaneous re-synchronization of vast numbers of neurons (106 to 108). Several mechanisms such as dendritic loop currents, propagated action potential, and diffusion of chemical transmitters have been proposed to explain the observed temporal precision and fineness of spatial texture of synchronized cortical activity. The documented rapid changes in synchronization over distances of mm to cm (Freeman, 2005) are incompatible with the mechanisms of long-range diffusion and the extracellular dendritic currents of the ECoG, which are much too weak. The length of most axons in cortex is a small fraction of the observed distances of long-range correlation, which cannot easily be explained even by the presence of relatively few very long axons creating small world effects (Barabási, 2002).

The occurrence of such collective neuronal processes with their observed properties has suggested to us to use the formalism of many-body field theory to model the brain functional activity. In such an approach the brain appears to be a macroscopic quantum system (Umezawa, 1993; Vitiello, 1995, 2001, 2004; Freeman & Vitiello, 2006-2010), namely a system whose macroscopic behaviour cannot be explained without recourse to the microscopic dynamics of its elementary components. Here it has to be specified that neurons, glia cells and other microscopic organelles are considered to be classical elements in the dissipative many-body model. The quantum degrees of freedom are the quanta of the electrical dipole fields of the biomolecules and water molecules, the matrix in which all the biological cells are embedded. The existence of macroscopic quantum systems in other physical domains, such as crystals, ferromagnets, superconductors, etc., shows indeed that the domain of validity of quantum field theory (QFT) is not restricted to the microscopic physics (Umezawa, 1993; Blasone, Jizba and Vitiello, 2011).

The use of the QFT formalism in the study of the brain does not mean that the traditional classical tools of biochemistry and neurophysiology might be abandoned. Rather, these classical tools might receive further boost from the understanding of the underlying microscopic dynamics. It was in such a line of thoughts that Ricciardi and Umezawa (Ricciardi & Umezawa, 1967; Stuart, Takahashi and Umezawa, 1978; 1979) formulated the many-body model of brain. In the '40s, motivated by his experimental observations, Karl Lashley wrote: "Here is the dilemma. Nerve impulses are transmitted from cell to cell through definite intercellular connections. Yet all behaviour seems to be determined by masses of excitation. ... What sort of nervous organization might be capable of responding to a pattern of excitation without limited specialized paths of-conduction? The problem is almost universal in the activities of the nervous system" (Lashley, 1942, p. 306). The observations by Lashley were confirmed by other neuroscientists, such as Karl Pribram who proposed (Pribram, 1971) a holographic model to explain psychological field data. The understanding of such data in the frame of the available theory of condensed matter systems was the aim of the many-body model. The crucial mechanism on which the model is based is the one of the spontaneous breakdown of symmetry (SBS).

QFT is based on a dual level of description: the dynamical level, where the dynamical field equations and their symmetry properties are postulated, and the physical level of the fields in terms of which the observables are described. The whole QFT computational machinery consists in solving the dynamical field equations (the dynamical level) in terms of physical fields acting on the space of the physical states (the physical level). The point is that there are many "non-equivalent copies" of spaces of physical states. In other words, there are (infinitely) many possibilities in which the same basic dynamics may be realized in terms of physical observables: there are many possible "physically different worlds" in which the same basic dynamics may manifest itself. "Different" (or, in mathematical language, "non-unitarily equivalent") spaces of physical states means that the physical observables acquire different values depending on which one is the space of physical states (the world) we choose (or we are forced by some specific boundary conditions) to work with. Due to such a peculiar property of QFT (it is this property that makes QFT fundamentally different from Quantum Mechanics!!), it may happen that the symmetry properties of the space of the physical states are not the same as the ones of the basic dynamical equations: the basic symmetry gets broken in the process of mapping the dynamical level to the physical level of description. Since, we have access to (we "live" in) this last level, it is the dynamically rearranged symmetry the one that we observe, not the one of the basic dynamical field equations. In particular, in the process of symmetry breakdown an observable variable emerges, called the order parameter, which characterizes the macroscopic behaviour of the physical system, as a whole. The order parameter expresses in a highly non-linear way the microscopic behaviour of the myriads of elementary constituents of the system. The order parameter thus emerges as a classical field and marks the transition from the microscopic scale to the macroscopic scale. It is a measure of the complexity of the basic dynamics ruling the system, which cannot be reduced to or derived from the sum of the behaviours of the elementary components (Umezawa, 1993; Blasone, Jizba and Vitiello, 2011). In our model we conceive the order parameter as the density of the synaptic interactions at every point in the cortical neuropil, and we interpret the ECoG recorded at each point as an experimentally observable correlate of the neural order parameter.

One further point, turning out to be a very important one in our brain modelling, is that the symmetry breakdown is spontaneous: this means that, under given boundary conditions (e.g. at given temperature), the specific form into which symmetry gets rearranged is chosen by the dynamics of the system, i.e., by its inner dynamical evolution. SBS is thus a dynamical process, different from the explicit breakdown obtained by introducing at the dynamical level constraints explicitly violating the symmetries of the basic field equations. In neurobiological terms, these constraints are stimuli, typically an impulse in the form of a click, flash, or electric shock, able to reduce the functional activity of the brain into slavery. In the SBS, instead, the external stimulus acts only as a trigger of the inner evolution.

One central theorem in QFT states that SBS implies the existence of particles, called Nambu- Goldstone (NG) modes or fields, that are massless and are bosons, i.e., they can be collected or condensed in the same physical state without any restriction on their number and, since they are massless, they can span the whole system volume and are therefore responsible for the occurrence of long-range correlations, namely of the ordering which thus is established in the system: Order appears as a result of the symmetry breakdown; order is lack of symmetry. The lowest energy state, called the vacuum or the ground state, thus appears as a condensate of such NG modes.

The order parameter provides a measure of the condensation density of the NG modes in the vacuum state and therefore a measure of the long-range correlation. On the other hand, the condensation process is described by the transformation $B \rightarrow B + \alpha$, where B denotes the NG field and α is complex number, $\alpha = |\alpha| \exp(i\theta)$, which may also depend on space-time. In such a last case, we have space-time dependent condensation; otherwise we have homogeneous condensation. It is well known that the transformation $B \rightarrow B + \alpha$ generates a coherent state. The number of the condensed NG field indeed is given by $|\alpha|^2$ and thus we see that the vacuum is characterized by an unique phase θ: the NG modes share the same phase, which is characteristic of coherent states. In the Ricciardi and Umezawa (RU) brain model, memory is described by SBS triggered by an external stimulus; long-range correlations are then generated by the inner dynamics of the brain and NG modes are condensed in the vacuum; the memory code is taken to be the condensation density $|\alpha|^2$. Note that the memory thus associated to a specific triggering stimulus is not a representation of that stimulus (Freeman & Vitiello, 2006-2010).

Fhrölich (1968) and Del Giudice et al (1985; 1986) have proposed that the electrical polarization density arising from the water matrix and the other biomolecules might be considered to be the order parameter in the study of biological matter, also with reference to the formation and the dynamical properties of microtubules in the cell, thus characterizing the living phase of the matter.

Jibu and Yasue (1995) and Jibu, Yasue and Pribram (1996) then proposed that the symmetry breakdown in the RU model was the one of the rotational symmetry of the electrical dipoles. We propose a further refinement: the order parameter accounts for the density of dipole moment exerted by neuron populations at each point in the neuropil through synaptic interactions. Moreover, the model has been extended to include the dissipative dynamics describing the fact that the brain is permanently open on the external world (Vitiello, 1995; 2001). The starting observation on which the dissipative model is based is indeed that there is no question that brains are open thermodynamic systems operating far from equilibrium. Brains burn glucose to store energy in glycogen ("animal starch") and high-energy adenosinetriphosphate (ATP), and in transmembrane ionic gradients; they dissipate free energy in proportion to the square of the ionic current densities that are manifested in epiphenomenal electric and magnetic fields, and that mediate the action-perception cycle (Freeman & Vitiello, 2006). Brain imaging techniques such as fMRI are indirect measures of metabolic dissipation of free energy, relying on secondary increases in blood flow and oxygen depletion. The dendrites dissipate 95% of the metabolic energy in summed excitatory and inhibitory ionic currents, the axons only 5% in action potentials that carry the summed output of dendrites by analog pulse frequency modulation. One of the main tasks of the dissipative model is thus the one of formulating the thermodynamic features involved in the action-perception cycle. The model indeed shows how the process of energy dissipation as heat manifests itself in the disappearance and emergence of coherence.

On the other hand, dissipation enables brains to form an indefinite variety of different ground states, which is prerequisite for high memory capacity (Vitiello, 1995: 2001). Indeed, introducing dissipation solves the memory capacity problem plaguing the RU model, where any subsequent stimulus would trigger a new NG condensation erasing the previous condensate (memory overprinting). In the dissipative model, under the influence of an external stimulus, the brain inner dynamics selects one of the possible (inequivalent) ground states, each of them thus being associated to a different memory. Infinitely many memories may thus be stored and, due to the unitarily inequivalence of the (vacuum) states, they are protected from reciprocal interference. In the dissipative model we regard the NG condensate as an expression of a transiently retrieved memory (thought, percept, recollection) that has been accessed by a phase transition.

The possibility to exploit the whole variety of unitarily inequivalent vacua arises as a consequence of the mathematical necessity in quantum dissipation to "double" the system degrees of freedom so as to include the environment in which the brain is embedded. That reflective fraction of the environment is thus described as the Double of the system, which turns out to be the system time- reversed copy. The entanglement between the brain and its environment is thus described as a permanent coupling, or dynamic dialog between the two, which may be related to consciousness mechanisms. Consciousness thus appears as a highly dynamic process rooted in the dissipative character of the brain dynamics, which, ultimately, is grounded into the non-equilibrium thermodynamics of its metabolic activity.

In recent years, the dissipative model has been developed also considering the available experimental observations and data analysis (Freeman & Vitiello, 2006-2010). The reader can find in the quoted literature a list of properties and predictions of the model, as compared to observations, which here for brevity we do not report. The data analysis shows that one can depict the brain non-linear dynamics in terms of attractor landscapes. Each attractor is based in a nerve cell assembly of cortical neurons that have been pair-wise co-activated in prior Hebbian association and sculpted by habituation and normalization (Kozma & Freeman, 2001). Its basin of attraction is determined by the total subset of receptors that has been accessed during learning. Convergence in the basin to the attractor gives the process of abstraction and generalization to the category of the stimulus. The memory store is based in a rich hierarchy of landscapes of increasingly abstract generalizations (Freeman, 2005; 2006). The continually expanding knowledge base is expressed in attractor landscapes in each of the cortices.

In conclusion, the dissipative many-body model of brain provides the theoretical scheme aimed to describe the basic dynamics underlying the neurological activity. It illustrates the observed formation and properties of imploding and exploding conical phase gradients and the occurrence of null spikes that have been identified in multichannel records of ECoG signals. Energy dissipation is shown to incorporate the observed feature of null spikes, which are transient extreme reduction in macroscopic energy, in which order disappears and symmetry is momentarily re-established. The extreme localization in space, time and spectrum (Freeman, 2009) indicate that the null spikes are observable manifestations of a singularity by which the symmetry is broken. Classical Maxwell equations and current fields are derived from the quantum dynamics (Freeman & Vitiello, 2010), thus confirming that functional aspects of brain dynamics are derived as macroscopic manifestations of the underlying many-body dynamics: the emergence of classicality out of the microscopic dynamics thus appears to be a central feature of the dissipative many-body model. The model also describes the size, number and time dependence of the transient non-homogeneous patterns of percepts appearing during non-instantaneous phase transitions, such as those observed in brain. Further developments of the dissipative model considering other features of non-equilibrium neuronal thermodynamics are under study.

References

Barabási, A-L. (2002). Linked. The New Science of Networks. Perseus: Cambridge MA.

Blasone, M., Jizba P. and Vitiello, G. (2011). Quantum Field Theory and Its Macroscopic Manifestations: Boson Condensation, Ordered Patterns and Topological Defects. Imperial College Press, London.

Del Giudice, E., Doglia, S., Milani, M. and Vitiello, G. (1985). A quantum field theoretical approach to the collective behaviour of biological systems. Nucl. Phys. B 251 [FS 13], 375-400.

Del Giudice, E., Doglia, S., Milani, M. and Vitiello, G. (1986). Electromagnetic field and spontaneous symmetry breakdown in biological matter. Nucl. Phys. B 275 [FS 17], 185-199.

Freeman, W. J. (1975). Mass Action in the Nervous System. Academic, New York.

Freeman, W. J. (2000). Neurodynamics. An Exploration of Mesoscopic Brain Dynamics. Springer, Berlin.

Freeman, W. J. (2004a). Origin, structure, and role of background EEG activity. Part 1. Analytic amplitude. Clin. Neurophysiol. 115, 2077-2088; (2004b) Part 2. Analytic phase. Clin. Neurophysiol. 115, 2089-2107; (2005) Part 3. Neural frame classification. Clin. Neurophysiol. 116 (5), 1118-1129.

Freeman, W. J. (2006). Definitions of state variables and state space for brain-computer interface. Part 1. Multiple hierarchical levels of brain function. Cognitive Neurodynamics 1(1), 13-14.

Freeman, W. J. (2009). Deep analysis of perception through dynamic structures that emerge in cortical activity from self-regulated noise. Cognitive Neurodynamics 3(1), 105-116.

Freeman, W. J. and Vitiello, G. (2006). Nonlinear brain dynamics as macroscopic manifestation of underlying many-body field dynamics. Physics Life Rev. 3, 93-118.

Freeman, W. J. and Vitiello, G. (2008). Brain dynamics, dissipation and spontaneous breakdown of symmetry. J. Phys. A: Math. Theor. 41, 304042.

Freeman, W. J. and Vitiello, G. (2009). Dissipative neurodynamics in perception forms cortical patterns that are stabilized by vortices. J. Physics: Conf. Series 174, 012011.

Freeman, W. J. and Vitiello, G. (2010). Vortices in brain waves. Int. J. Mod. Phys. B 24, 3269–3295.

Freeman, W. J., Burke B. C. and Holmes, M. D. (2003). Aperiodic phase re-setting in scalp EEG of beta-gamma oscillations by state transitions at alpha-theta rates. Human Brain Mapping 19 (4), 248-272.

Frhölich, H. (1968). Long range coherence and energy storage in biological systems. Int. J. Quantum Chemistry 2, 641-649.

Jibu, M. and Yasue, K. (1995). Quantum Brain Dynamics and Consciousness. John Benjamins, Amsterdam.

Jibu, M., Pribram, K. H. and Yasue, K. (1996). From conscious experience to memory storage and retrieval: the role of quantum brain dynamics and boson condensation of evanescent photons. Int. J. Mod. Phys. B 10, 1735-1754.

Kozma, R. and Freeman, W. J. (2001). Chaotic resonance: Methods and applications for robust classification of noisy and variable patterns. Int. J. Bifurc. Chaos 10, 2307-2322.

Lashley, K. S. (1942). The problem of cerebral organization in vision. In: Cattell J. (Ed.), Biological Symposia VII: 301, p. 306.

Pribram, K. H. (1971). Languages of the Brain. Prentice-Hall, Englewood Cliffs, NJ.

Ricciardi L. M. and Umezawa, H. (1967). Brain physics and many-body problems. Kibernetik 4, 44-48; reprinted in: In: Globus, G. G., Pribram, K. H., Vitiello, G. (Eds.), Brain and Being, John Benjamins Publishing Co., Amsterdam, pp. 255-266.

Stuart, C. I. J., Takahashi, Y. and Umezawa, H. (1978). On the stability and non-local properties of memory. J.Theor. Biol. 71, 605-618.

Stuart, C. I. J., Takahashi, Y. and Umezawa, H. (1979). Mixed system brain dynamics: neural memory as a macroscopic ordered state. Found. Phys. 9, 301-327.

Umezawa, H. (1993). Advanced Field Theory: Micro, Macro and Thermal Concepts. AIP, New York.

Vitiello, G. (1995). Dissipation and memory capacity in the quantum brain model. Int. J. Mod. Phys. B 9, 973-989.

Vitiello, G. (2001). My Double Unveiled. John Benjamins Publishing Co., Amsterdam.

Vitiello, G. (2004). The Dissipative Brain. In: Globus, G. G., Pribram, K. H., Vitiello, G. (Eds.), Brain and Being, John Benjamins Publishing Co., Amsterdam, pp. 315-334.

10. Electromagnetic Bases of the Universality of the Characteristics of Consciousness: Quantitative Support
Michael A. Persinger, Ph.D.

Laurentian University, Sudbury, Ontario, Canada

Abstract

The similar magnitudes of magnetic fields involved with the cerebral operations correlated with consciousness and interstellar and galactic measures may be preconditions for a more universal existence of the conditions that facilitate consciousness phenomena. Quantitative estimates from induced magnetic moments, averaged densities of energies in the universe, electromagnetic interactions, and the temporal solutions derived from these measures suggest that the presence of consciousness-like phenomena within large extracerebral spaces may be more probable than anticipated.

KEY WORDS: picoTesla magnetic fields; cerebral fields; extragalactic magnetic fields; Bohr magneton; probabilistic determinism; consciousness

1. Introduction: The Electromagnetic Bases of Consciousness

There is converging evidence that consciousness associated with the cerebral volume of the human brain may be identical with or strongly correlated with complex configurations of magnetic fields whose intensities are within the picoTesla (10^{-12} T) range (Anninos et al, 1991; Persinger and Lavallee, 2010). In comparison, magnetic field strengths in the order of 10^{-10} T within galaxies and galaxy clusters (Opher et al, 2009) and (unamplified) values with upper limits in the 10^{-13} T range (Neronov and Vovk, 2010) for extragalactic fields have been measured or inferred. They might be considered superimposed upon cerebral values. In this paper, convergence of quantitative estimates and theoretical arguments suggest that the conditions for consciousness may be more universal than anticipated.

There are multiple approaches that suggest the importance of the picoTesla range for the operational magnetic field strength of the cerebral volume that is congruent with the occurrence of consciousness. This approach can assume that either the chemical reactions, neuroelectromagnetic fields, and synaptic spaces within the human cerebral volume create consciousness or the specific neuronal configurations within brain space allow the congruence with electromagnetic conditions that do not originate from neuronal processes, per se, that manifest consciousness. In other words consciousness can be assumed to emerge from the chemical interactions occurring between the approximately 10^{13} synapses within the cerebral cortices or as a field that is superimposed within the accompanying vast extracellular and perisynaptic space of this complex volume. The focus here is not to test either hypothesis directly but to discern the quantitative likelihood that the properties of consciousness may exhibit both non-local and local similarities.

Obviously the observation that "similar plus similar equals the same" is not necessarily valid. In addition, scalar identities do not easily reveal the complexity and information density (McFadden, 2002) that must exist as structure within the range of shared magnitudes of magnetic fields. Such spatial-temporal complexity (Tononi and Edelman, 1998) has been argued to be a fundamental characteristic of consciousness (Pribram and Meade, 1999). However a convergence of quantitative solutions could give perspective by which previously disparate levels of discourse might be integrated.

Some quantitative estimates from classical and nonclassical approaches are illustrative. The magnetic field associated with the typical current for an ion channel within the plasma membrane of the neuron is about 10^{-12} A. Because the intrinsic energetic characteristics of the ion channel converge with multiple spatial properties that contribute to the resting membrane potential and those of the cerebral volume (Persinger and Lavallee, 2010) this value could be considered fundamental. Assuming an axon is a wire the resulting magnetic field $B=(\mu i)/2\pi r$ where μ is permeability, i is the current (pA) and r=the distance across the membrane (10 nm) is within the 10 to 100 pT range.

The change in frequency, or a derivative of it, with which consciousness is often associated (Persinger et al, 2010) in non-angular systems from the application of a magnetic field with a strength of 10^{-11} T would be, according to standard Zeeman solutions, within the 1 to 100 Hz range, which involves most of the band-width for the major cerebral correlates of consciousness. In this case, the change in frequency is the product of the strength of the magnetic field and a unit charge (1.6×10^{-19} A s) divided by a constant times the mass of an electron. More specifically, a 40pT field in a non-angular system would produce a change in angular frequency of about 7 Hz. This frequency, and more specifically the theta (4 Hz to 7 Hz) band, is the pivotal frequency that relates the "40Hz" consciousness pattern (Llinas and Pare, 1991) of the cerebral cortical activity with the theta range of the hippocampal formation (Holtz et al, 2010; Lisman and Idiart, 1995; Persinger and Lavallee, 2010), the gateway to memory. From this context it may not be spurious (although unorthodox) that the mass (kg) of an electron divided by a unit charge (A s) multiplied by 7 Hz (1/s) yields a magnetic field strength of 40×10^{-12} T.

2. Effects of Galactic-level Intensities Upon the Magnetic Moment of Cerebral Matter

In previous papers and experimental research we have been pursuing the concept that minute changes in the intrinsic motions of electrons associated with weakly applied magnetic fields (B) may facilitate understanding of entanglement (Hu and Wu, 2008; Persinger et al, 2008; Persinger and Koren, 2007). The induced magnetic moment corresponding to the change in angular velocity (opposite to B) can be described as:

$$\Delta m = -[e^2 r^2 / 4 m_e] B$$

where "e" is the unit charge, r=the Bohr radius and m_e is the mass of the electron. Classically one applies strong fields. However suppose the intrinsic cerebral magnetic field was applied, i.e., ~10 pT? If this is pursued and the appropriate values are multiplied (*) or divided then :{[(1.6 x 10^{-19}) $A^2 s^2$ * (5.1 x 10^{-11}) m2]/4 * 9.1 x 10^{-31} kg}*10^{-11} T results in 1.8 x 10^{-40} Am^2 or J/T. This quantity is a very small value.

However, if one assumes that a second extracerebral field such as that intrinsic within galaxies is superimposed upon a comparable consciousness field of ~10 pT, then the energy associated with this would be 1.8 x 10^{-40} J/T * 10^{-11} T or 1.8 x 10^{-51} J. The significance of this value becomes apparent because it approaches the rest mass of a photon when m^2/s^2 approaches zero, i.e, between 10^{-52} kg (Tu et al, 2005) and 10^{-51} kg. What occurs at these boundaries is still a matter of speculation although experimental procedures to these implications have begun (Dotta et al, 2011).

What is theoretically more enlightening from the perspective of the considerations in this paper is that even this small amount of energy would have a period or a frequency which can be determined by the dividing Planck's constant by this increment of energy. Hence, 6.624 x 10^{-34} J s/ 1.8 x 10^{-51} J is equal to 3.67 x 10^{17} s. In larger temporal aggregates this is 11.7 billion years, which is within the order of magnitude and variability of the coefficient of the age of the universe. There are several implications of this quantitative association. We have noted in neural systems that narrow bands of energy or ionic concentrations are interconnected and functionally related. For example the "neuroquantal" increment of $\sim 10^{-20}$ J (Persinger, 2010), the energy associated with a single action potential, the presumed corporeal bases of consciousness and thinking, ($\Delta v = 1.2 \times 10^{-1}$ V * 1.6×10^{-19} As) also emerges: 1) at the distance between the charges on the neuronal membrane that create the resting membrane potential, 2) during the intersynaptic transformation of an action potential, 3) during phosphorylation, 4) as the intrinsic energy for the electromagnetic wavelength equivalent of a cell (~ 10 μm), and, 5) from the energy differential between the solution for the discrepancy between the classical and Compton radius for the electron (Persinger et al, 2008). This congruence between the generalized background of magnetic fields within and between galaxies and the magnetic field coupled to cerebral function would suggest a potential continuity that would not necessary require but also not exclude a neurophysics derived from the philosophies of Spinoza (see Grene, 1973) or Teilhard de Chardin (1955).

One possible model is that the intrinsic magnetic field of the galaxy constitutes the first field that affects the Δm but the application of the second, from consciousness, is required to produce the energy whose temporal solution is the age of the universe. If the latter is valid then changing the intensity of the secondary applied field associated with consciousness would by solution potentially change the age of the universe. That the act of measurement or awareness can affect the measurement is an intrinsic feature of Heisenberg's uncertainty. It may be relevant that if we assume complete certainty of the location of an electron with a classical radius of 2.82×10^{-15} m then the uncertainty of momentum is 6.624×10^{-34} J s/2.82×10^{-15} m or 2.35×10^{-19} kg m/s (Aczel, 2002). At the average bulk velocity of the transcerebral magnetic fields associated with consciousness, ~ 4.5 m/s (Nunez, 1995; Persinger and Lavallee, 2010), the energy would be 10^{-20} J, the essential quantal unit of thinking and by extension consciousness.

3. The Implication of One Neuron Affecting Cerebral Function

The implications of the recent replications that the firing of only one neuron is sufficient to alter the microstate of the entire cerebral manifold (Li et al, 2009) or determine the 0,1 condition for a rat (Houwelling and Brecht, 2008) to respond or not respond overtly (which requires hundreds of thousands of coordinated neurons) have relevance to any model of consciousness which assumes a distribution over immense numbers of synapses. Given the typical power density of 10^{-13} W/m^2 of cosmic ray (proton) incidence from potentially distant galaxies upon the earth's surface, this means that the energy within the area of a single cortical column (with a width of about 0.5 mm or $.25 \times 10^{-6}$ m2), which is often statistically determines (controlled) by a single neuron, is about 10^{-20} J/s, the energy associated with a single action potential.

We (Koren and Persinger, 2010) have previously estimated that the energy density of the entire universe which can be discerned by the energy equivalence of its mass ($\sim 10^{52}$ kg) is 10^{69} J. When divided by a likely volume of 10^{79} m^3, the average value is about 10^{-1} J/m^3. The energy density within the human brain associated with 10^6 neurons each firing at 10 Hz with an action potential energy of 10^{-20} J is 10^{-13} J and when divided by the volume of the human brain (about 10^{-3} m^3) would be 10^{-10} J/m^3. Consequently if we assume that conditions with shared vector solutions can potentially interact or at least be resonant, the energy within brain space would be congruent with this average of the whole volume of the universe and would satisfy one of the conditions for a hologram.

4. "Free Will" versus Cosmic Statistical Determinism

In light of the recent interest in radio stars and the measurements of 10^{-11} Tesla averaged intergalactic and intragalactic magnetic fields, which overlap with the operating intensity of cerebral functions, one obvious question is would there be congruent solutions with the energy of these stars? The energy flux from radio stars is $\sim 10^{-16}$ [W/m^2]/Hz. If we assume the intrinsic brain resonance frequency is 7 Hz (the theta range), then the energy is 6.624×10^{-34} J s * 7 Hz or 46.37×10^{-34} J per second (watts). The equivalent within cross-sectional space would be that value divided by 10^{-22} W/m^2 or effectively an area with a diameter of about 10 μm (7 μm to be more exact) which is the within the range of the width of an average cell soma, including the neuron. As mentioned, the electromagnetic energy equivalence of 9 to 10 μm, is 10^{-20} J (Persinger, 2010). The differences in coefficients would be within measurement error and the normal distributions around a central tendency.

In other words the energy equivalence in a 1 Hz band from radio stars would match the energy of a 7 Hz photon applied to cell-level space. Does this suggest that under specific conditions the change in output superimposed upon or reflected as alterations in "noise", that is cosmic ray incidence, could influence the theta activity range of the human brain and subsequently affect perception, memory, consciousness, and even the perception of "choice" or free will? It may be relevant that Charles Fort (Persinger and Lafreniere, 1977) frequently noted the "coincidence" of mass human events and environmental oddities at the time that "new" stars were first observed or "old" stars "disappeared". These "excess correlations" were dismissed by Fort's contemporaries. However, if only one neuronal quantum can change the behavior of a rat or alter the microstate of the entire cerebral manifold, the question must at least be considered: at what distance can we exclude a potential source?

5. Energy Densities of the Human Brain and the Biosphere

The issue of similarities, perhaps manifested as "scale invariance", between macroscopic and cerebral electromagnetic phenomena is at present difficult to answer for galactic magnetic fields because of the limited resolution. However, the marked similarities of electromagnetic patterns within the earth's surface ionosphere shell is not only revealing philosophically but potentially important for understanding human cerebral function. The energy density from 10^{10} neurons within the entire cerebral human volume would be about 10^{-7} J/m^3. There are about 70 to 100 lightning flashes per second world wide, most of which occur within a narrow shell of about 2 km within the biosphere. Assuming a typical 10 Coulomb (C) flow of electrons across a potential difference of 10^8 V, the energy would be 10^9 J per flash for a total of 10^{11} J/s world wide. The volume of the shell within which the electromagnetic fields associated with lightning patterns occurs is about 10^{18} m^3 which yields a density of 10^{-7} J/m^3.

This value is convergent with the average energy density within the human cerebrum associated with thinking and neuronal activity. If about 5 C is distributed within a lightning channel with an average current of 100 A, and is contained within a radius of 1 cm, the resulting cross-sectional current density is about 10^5 A/m^2. The area of the annulus around a 1 µm width axon of average length is about 10^{-14} m^2. Given the average current of 10^{-9} A from the approximately 10^3 ion channels each with 1 pA capacities, the cross-sectional current density would be equivalent to $\sim 10^5$ A/m^2. Such identities of quantitative solutions are not incidental. For example both lightning and the action potential of the axon share the correlates of nitric oxide production, shared salutatory and pulse patterns, similar mass-charge velocity ratios, and interface times between ground-stroke and axon-dendritic back-propagations. The average magnitude of the magnetic field of the primarily 7 Hz to 40 Hz oscillations within the earth-ionosphere cavity is in the order of 1 pT to 10 pT.

6. Conclusions and Implications

Quantitative similarities and calculated solutions for the intensities of magnetic fields associated with cerebral function and those that exist within intra- and extragalactic space suggest that the energetic conditions associated with consciousness and its many variants may be more universal than anticipated. If scale invariance is operative, then the volumes within which the conditions associated with human varieties of consciousness emerge may be substantially larger but would display comparable characteristics. The marked similarities between action potentials and the lightning strikes that generate the earth's own 1 to 10 pT, 7 Hz to 40Hz "manifold" also strongly indicate that if consciousness is simply a property of specific electromagnetic patterns and intensities, a variety of these conditions may exist within extracerebral systems.

References

Aczel, A. D. (2002). Entanglement: the greatest mystery in physics. Raincoast Books: Vancouver.

Anninos, P.A., Tsagas, N., Sandyk, R., Derpapas, K. (1991). Magnetic stimulation in the treatment of partial seizures. International Journal of Neuroscience, 60, 141-71.

Dotta, B. T., Buckner, C. A., Lafrenie, R. M., Persinger, M. A. (2011). Photon emissions from human brain and cell culture exposed to distally rotating magnetic fields shared by separate light-stimulated brains and cells. Brain Research, in press.

Grene, M. (1973). Spinoza. Anchor Books: N.Y. Holtz, E.M., Glennon, M., Pendergast, K., Sauseng, P. (2010). Theta-gamma phase synchronization during memory matching in visual working memory. NeuroImage, 52, 326-335.

Houwelling, A. R., Brecht, M. (2008). Behavioural report of a single neuron in somatosensory cortex. Nature, 451, 65- 68.

Hu, H., Wu, M. (2006). Thinking outside of the box: the essence and implications of quantum entanglement. NeuroQuantology, 4, 5-16.

Koren, S. A., Persinger, M. A. (2010). The Casimir force along the universal boundary: quantitative solutions and implications. Journal of Physics, Astrophysics and Physical Cosmology, 4, 1-4.

Nunez, P. L. (1995) Neocortical dynamics and human EEG rhythms. Oxford University Press: London.

Li, C.Y., Poo,M.M., Dan, Y. (2009). Burst-firing of a single cortical neuron modifies global brain state. Science, 324, 643-646.

Lisman, J.E., Idiart, M. A. (1995). Storage of 7 +/- 2 short-term memories in oscillatory subcycles. Science, 267, 1512-1515.

Llinas, R., Pare, D. (1991). Of dreaming and wakefulness. Neuroscience, 44, 521-535.

McFadden, J. (2002). The consciousness electromagnetic information (Cemi) field theory: the hard problem made easy? Journal of Consciousness Studies, 9, 45-55.

Neronov, A., Vovk, I. (2010). Evidence of strong extragalactic magnetic fields from Fermi observations of TeV blazars. Science, 328, 73-75.

Opher, M., Bibi, F. A., Toth, G., Richardson, J.D., Izmodenov, V. V., Gombosi, T. I. (2009). A strong, highlytilted magnetic field near the Solar System. Nature, 462, 1036-138.

Persinger, M. A. (2010). 10^{-20} Joules as a neuromolecular quantum in medicinal chemistry: an alternative approach to myriad molecular pathways? Current Medicinal Chemistry, 17, 3094-3098.

Persinger, M. A., Koren, S. A. (2007). A theory of neurophysics and quantum neuroscience: implications for brain function and the limits of consciousness. International Journal of Neuroscience, 117, 157-175.

Persinger, M. A., Koren, S. A., Lafreniere, G. F. (2008). A neuroquantologic approach to how human thought might affect the universe. NeuroQuantology, 2008, 262-271.

Persinger, M. A., Lafreniere, G. F. (1977). Space-time transients and unusual events. Nelson-Hall: Chicago.

Persinger, M. A., Lavallee, C. F. (2010). Theoretical and experimental evidence of macroscopic entanglement between human brain activity and photon emission: implications for quantum consciousness and future application. Journal of Consciousness Exploration & Research, 7, 785-807.

Persinger, M. A., Saroka, K. S., Koren, S. A., St-Pierre, . S. (2010). The electromagnetic induction of mystical and altered states within the laboratory. Journal of Consciousness Exploration and Research, 7, 808-830.

Persinger, M. A., Tsang, E. W., Booth, J. N., Koren, S. A. (2008). Enhanced power within a predicted narrow band of theta activity during stimulation of another by circumcerebral weak magnetic fields after weekly spatial proximity: evidence of macroscopic entanglement? NeuroQuantology, 6, 7-21.

Pribram, K.H., Meade, S. M. (1999). Consciousness awareness: processing in the synaptodendritic web. New Ideas in Psychology, 17, 205-209.

Teilhard de Chardin, P. (1955). The phenomenon of man. William Collins Sons: London.

Tononi, G., Edelman, G. M. (1998). Consciousness and complexity. Science, 282, 1846-1851.

Tu, L. C., Luo, J., Gilles, G. T. (2005). The mass of the photon. Reports on Progress in Physics, 68, pp. 77.

11. Does 'Consciousness' Exist?

William James, Ph.D.

Harvard University, Cambridge, MA

First published in Journal of Philosophy, Psychology, and Scientific Methods, 1, 477-491, 1904

'Thoughts' and 'things' are names for two sorts of object, which common sense will always find contrasted and will always practically oppose to each other. Philosophy, reflecting on the contrast, has varied in the past in her explanations of it, and may be expected to vary in the future. At first, 'spirit and matter,' 'soul and body,' stood for a pair of equipollent substances quite on a par in weight and interest. But one day Kant undermined the soul and brought in the transcendental ego, and ever since then the bipolar relation has been very much off its balance. The transcendental ego seems nowadays in rationalist quarters to stand for everything, in empiricist quarters for almost nothing. In the hands of such writers as Schuppe, Rehmke, Natorp, Munsterberg -- at any rate in his earlier writings, Schubert-Soldern and others, the spiritual principle attenuates itself to a thoroughly ghostly condition, being only a name for the fact that the 'content' of experience is known. It loses personal form and activity - these passing over to the content -- and becomes a bare Bewusstheit or Bewusstsein überhaupt of which in its own right absolutely nothing can be said.

I believe that 'consciousness,' when once it has evaporated to this estate of pure diaphaneity, is on the point of disappearing altogether. It is the name of a nonentity, and has no right to a place among first principles. Those who still cling to it are clinging to a mere echo, the faint rumor left behind by the disappearing 'soul' upon the air of philosophy. During the past year, I have read a number of articles whose authors seemed just on the point of abandoning the notion of consciousness,[1] and substituting for it that of an absolute experience not due to two factors. But they were not quite radical enough, not quite daring enough in their negations. For twenty years past I have mistrusted 'consciousness' as an entity; for seven or eight years past I have suggested its non-existence to my students, and tried to give them its pragmatic equivalent in realities of experience. It seems to me that the hour is ripe for it to be openly and universally discarded.

To deny plumply that 'consciousness' exists seems so absurd on the face of it -- for undeniably 'thoughts' do exist -- that I fear some readers will follow me no farther. Let me then immediately explain that I mean only to deny that the word stands for an entity, but to insist most emphatically that it does stand for a function. There is, I mean, no aboriginal stuff or quality of being, contrasted with that of which material objects are made, out of which our thoughts of them are made; but there is a function in experience which thoughts perform, and for the performance of which this quality of being is invoked. That function is knowing. 'Consciousness' is supposed necessary to explain the fact that things not only are, but get reported, are known. Whoever blots out the notion of consciousness from his list of first principles must still provide in some way for that function's being carried on.

I

My thesis is that if we start with the supposition that there is only one primal stuff or material in the world, a stuff of which everything is composed, and if we call that stuff 'pure experience,' the knowing can easily be explained as a particular sort of relation towards one another into which portions of pure experience may enter. The relation itself is a part of pure experience; one if its 'terms' becomes the subject or bearer of the knowledge, the knower,[2] the other becomes the object known. This will need much explanation before it can be understood. The best way to get it understood is to contrast it with the alternative view; and for that we may take the recentest alternative, that in which the evaporation of the definite soul-substance has proceeded as far as it can go without being yet complete. If neo-Kantism has expelled earlier forms of dualism, we shall have expelled all forms if we are able to expel neo-Kantism in its turn.

For the thinkers I call neo-Kantian, the word consciousness to-day does no more than signalize the fact that experience is indefeasibly dualistic in structure. It means that not subject, not object, but object-plus-subject is the minimum that can actually be. The subject-object distinction meanwhile is entirely different from that between mind and matter, from that between body and soul. Souls were detachable, had separate destinies; things could happen to them. To consciousness as such nothing can happen, for, timeless itself, it is only a witness of happenings in time, in which it plays no part. It is, in a word, but the logical correlative of 'content' in an Experience of which the peculiarity is that fact comes to light in it, that awareness of content takes place. Consciousness as such is entirely impersonal -- 'self' and its activities belong to the content. To say that I am self-conscious, or conscious of putting forth volition, means only that certain contents, for which 'self' and 'effort of will' are the names, are not without witness as they occur.

Thus, for these belated drinkers at the Kantian spring, we should have to admit consciousness as an 'epistemological' necessity, even if we had no direct evidence of its being there.

But in addition to this, we are supposed by almost every one to have an immediate consciousness of consciousness itself. When the world of outer fact ceases to be materially present, and we merely recall it in memory, or fancy it, the consciousness is believed to stand out and to be felt as a kind of impalpable inner flowing, which, once known in this sort of experience, may equally be detected in presentations of the outer world. "The moment we try to fix out attention upon consciousness and to see what, distinctly, it is," says a recent writer, "it seems to vanish. It seems as if we had before us a mere emptiness. When we try to introspect the sensation of blue, all we can see is the blue; the other element is as if it were diaphanous. Yet it can be distinguished, if we look attentively enough, and know that there is something to look for."[3] "Consciousness" (Bewusstheit), says another philosopher, "is inexplicable and hardly describable, yet all conscious experiences have this in common that what we call their content has a peculiar reference to a centre for which 'self' is the name, in virtue of which reference alone the content is subjectively given, or appears.... While in this way consciousness, or reference to a self, is the only thing which distinguishes a conscious content from any sort of being that might be there with no one conscious of it, yet this only ground of the distinction defies all closer explanations. The existence of consciousness, although it is the fundamental fact of psychology, can indeed be laid down as certain, can be brought out by analysis, but can neither be defined nor deduced from anything but itself."[4]

'Can be brought out by analysis,' this author says. This supposes that the consciousness is one element, moment, factor -- call it what you like -- of an experience of essentially dualistic inner constitution, from which, if you abstract the content, the consciousness will remain revealed to its own eye. Experience, at this rate, would be much like a paint of which the world pictures were made. Paint has a dual constitution, involving, as it does, a menstruum[5] (oil, size or what not) and a mass of content in the form of pigment suspended therein. We can get the pure menstruum by letting the pigment settle, and the pure pigment by pouring off the size or oil. We operate here by physical subtraction; and the usual view is, that by mental subtraction we can separate the two factors of experience in an analogous way -- not isolating them entirely, but distinguishing them enough to know that they are two.

II

Now my contention is exactly the reverse of this. Experience, I believe, has no such inner duplicity; and the separation of it into consciousness and content comes, not by way of subtraction, but by way of addition -- the addition, to a given concrete piece of it, other sets of experiences, in connection with which severally its use or function may be of two different kinds. The paint will also serve here as an illustration. In a pot in a paint-shop, along with other paints, it serves in its entirety as so much saleable matter. Spread on a canvas, with other paints around it, it represents, on the contrary, a feature in a picture and performs a spiritual function. Just so, I maintain, does a given undivided portion of experience, taken in one context of associates, play the part of a knower, of a state of mind, of 'consciousness'; while in a different context the same undivided bit of experience plays the part of a thing known, of an objective 'content.' In a word, in one group it figures as a thought, in another group as a thing. And, since it can figure in both groups simultaneously we have every right to speak of it as subjective and objective, both at once. The dualism connoted by such double-barrelled terms as 'experience,' 'phenomenon,' 'datum,' 'Vorfindung' -- terms which, in philosophy at any rate, tend more and more to replace the single-barrelled terms of 'thought' and 'thing' -- that dualism, I say, is still preserved in this account, but reinterpreted, so that, instead of being mysterious and elusive, it becomes verifiable and concrete. It is an affair of relations, it falls outside, not inside, the single experience considered, and can always be particularized and defined.

The entering wedge for this more concrete way of understanding the dualism was fashioned by Locke when he made the word 'idea' stand indifferently for thing and thought, and by Berkeley when he said that what common sense means by realities is exactly what the philosopher means by ideas. Neither Locke nor Berkeley thought his truth out into perfect clearness, but it seems to me that the conception I am defending does little more than consistently carry out the 'pragmatic' method which they were the first to use.

If the reader will take his own experiences, he will see what I mean. Let him begin with a perceptual experience, the 'presentation,' so called, of a physical object, his actual field of vision, the room he sits in, with the book he is reading as its centre; and let him for the present treat this complex object in the commonsense way as being 'really' what it seems to be, namely, a collection of physical things cut out from an environing world of other physical things with which these physical things have actual or potential relations. Now at the same time it is just those self-same things which his mind, as we say, perceives; and the whole philosophy of perception from Democritus's time downwards has just been one long wrangle over the paradox that what is evidently one reality should be in two places at once, both in outer space and in a person's mind. 'Representative' theories of perception avoid the logical paradox, but on the other hand they violate the reader's sense of life, which knows no intervening mental image but seems to see the room and the book immediately just as they physically exist.

The puzzle of how the one identical room can be in two places is at bottom just the puzzle of how one identical point can be on two lines. It can, if it be situated at their intersection; and similarly, if the 'pure experience' of the room were a place of intersection of two processes, which connected it with different groups of associates respectively, it could be counted twice over, as belonging to either group, and spoken of loosely as existing in two places, although it would remain all the time a numerically single thing.

Well, the experience is a member of diverse processes that can be followed away from it along entirely different lines. The one self-identical thing has so many relations to the rest of experience that you can take it in disparate systems of association, and treat it as belonging with opposite contexts. In one of these contexts it is your 'field of consciousness'; in another it is 'the room in which you sit,' and it enters both contexts in its wholeness, giving no pretext for being said to attach itself to consciousness by one of its parts or aspects, and to out reality by another. What are the two processes, now, into which the room-experience simultaneously enters in this way?

One of them is the reader's personal biography, the other is the history of the house of which the room is part. The presentation, the experience, the that in short (for until we have decided what it is it must be a mere that) is the last term in a train of sensations, emotions, decisions, movements, classifications, expectations, etc., ending in the present, and the first term in a series of 'inner' operations extending into the future, on the reader's part. On the other hand, the very same that is the terminus ad quem of a lot of previous physical operations, carpentering, papering, furnishing, warming, etc., and the terminus a quo of a lot of future ones, in which it will be concerned when undergoing the destiny of a physical room. The physical and the mental operations form curiously incompatible groups. As a room, the experience has occupied that spot and had that environment for thirty years. As your field of consciousness it may never have existed until now. As a room, attention will go on to discover endless new details in it. As your mental state merely, few new ones will emerge under attention's eye. As a room, it will take an earthquake, or a gang of men, and in any case a certain amount of time, to destroy it. As your subjective state, the closing of your eyes, or any instantaneous play of your fancy will suffice. In the real world, fire will consume it. In your mind, you can let fire play over it without effect. As an outer object, you must pay so much a month to inhabit it. As an inner content, you may occupy it for any length of time rent-free.

If, in short, you follow it in the mental direction, taking it along with events of personal biography solely, all sorts of things are true of it which are false, and false of it which are true if you treat it as a real thing experienced, follow it in the physical direction, and relate it to associates in the outer world.

III

So far, all seems plain sailing, but my thesis will probably grow less plausible to the reader when I pass from percepts to concepts, or from the case of things presented to that of things remote. I believe, nevertheless, that here also the same law holds good. If we take conceptual manifolds, or memories, or fancies, they also are in their first intention mere bits of pure experience, and, as such, are single thats which act in one context as objects, and in another context figure as mental states. By taking them in their first intention, I mean ignoring their relation to possible perceptual experiences with which they may be connected, which they may lead to and terminate in, and which then they may be supposed to 'represent.' Taking them in this way first, we confine the problem to a world merely 'thought of' and not directly felt or seen. This world, just like the world of percepts, comes to us at first as a chaos of experiences, but lines of order soon get traced. We find that any bit of it which we may cut out as an example is connected with distinct groups of associates, just as our perceptual experiences are, that these associates link themselves with it by different relations,[6] and that one forms the inner history of a person, while the other acts as an impersonal 'objective' world, either spatial and temporal, or else merely logical or mathematical, or otherwise 'ideal.'

The first obstacle on the part of the reader to seeing that these non-perceptual experiences have objectivity as well as subjectivity will probably be due to the intrusion into his mind of percepts, that third group of associates with which the non-perceptual experiences have relations, and which, as a whole, they 'represent,' standing to them as thoughts to things. This important function of non-perceptual experiences complicates the question and confuses it; for, so used are we to treat percepts as the sole genuine realities that, unless we keep them out of the discussion, we tend altogether to overlook the objectivity that lies in non-perceptual experiences by themselves. We treat them, 'knowing' percepts as they do, as through and through subjective, and say that they are wholly constituted of the stuff called consciousness, using this term now for a kind of entity, after the fashion which I am seeking to refute.[7]

Abstracting, then, from percepts altogether, what I maintain is, that any single non-perceptual experience tends to get counted twice over, just as a perceptual experience does, figuring in one context as an object or field of objects, in another as a state of mind: and all this without the least internal self-diremption on its own part into consciousness and content. It is all consciousness in one taking; and, in the other, all content.

I find this objectivity of non-perceptual experiences, this complete parallelism in point of reality between the presently felt and the remotely thought, so well set forth in a page of Münsterberg's Grundzuge, that I will quote it as it stands.

"I may only think of my objects," says Professor Munsterberg; "yet, in my living thought they stand before me exactly as perceived objects would do, no matter how different the two ways of apprehending them may be in their genesis. The book here lying on the table before me, and the book in the next room of which I think and which I mean to get, are both in the same sense given realities for me, realities which I acknowledge and of which I take account. If you agree that the perceptual object is not an idea within me, but that percept and thing, as indistinguishably one, are really experienced there, outside, you ought not to believe that the merely thought-of object is hid away inside of the thinking subject. The object of which I think, and of whose existence I take cognizance without letting it now work upon my senses, occupies its definite place in the outer world as much as does the object which I directly see."

"What is true of the here and the there, is also true of the now and the then. I know of the thing which is present and perceived, but I know also of the thing which yesterday was but is no more, and which I only remember. Both can determine my present conduct, both are parts of the reality of which I keep account. It is true that of much of the past I am uncertain, just as I am uncertain of much of what is present if it be but dimly perceived. But the interval of time does not in principle alter my relation to the object, does not transform it from an object known into a mental state.... The things in the room here which I survey, and those in my distant home of which I think, the things of this minute and those of my long vanished boyhood, influence and decide me alike, with a reality which my experience of them directly feels. They both make up my real world, they make it directly, they do not have first to be introduced to me and mediated by ideas which now and here arise within me.... This not-me character of my recollections and expectations does not imply that the external objects of which I am aware in those experiences should necessarily be there also for others. The objects of dreamers and hallucinated persons are wholly without general validity. But even were they centaurs and golden mountains, they still would be 'off there,' in fairy land, and not 'inside' of ourselves."[8]

This certainly is the immediate, primary, naïf, or practical way of taking our thought-of world. Were there no perceptual world to serve as its 'reductive,' in Taine's sense, by being 'stronger' and more genuinely 'outer' (so that the whole merely thought-of world seems weak and inner in comparison), our world of thought would be the only world, and would enjoy complete reality in our belief. This actually happens in our dreams, and in our day-dreams so long as percepts do not interrupt them.

And yet, just as the seen room (to go back to our late example) is also a field of consciousness, so the conceived or recollected room is also a state of mind; and the doubling-up of the experience has in both cases similar grounds.

The room thought-of, namely, has many thought-of couplings with many thought-of things. Some of these couplings are inconstant, others are stable. In the reader's personal history the room occupies a single date -- he saw it only once perhaps, a year ago. Of the house's history, on the other hand, it forms a permanent ingredient. Some couplings have the curious stubbornness, to borrow Royce's term, of fact; others show the fluidity of fancy -- we let them come and go as we please. Grouped with the rest of its house, with the name of its town, of its owner, builder, value, decorative plan, the room maintains a definite foothold, to which, if we try to loosen it, it tends to return and to reassert itself with force.[9] With these associates, in a word, it coheres, while to other houses, other towns, other owners, etc., it shows no tendency to cohere at all. The two collections, first of its cohesive, and, second, of its loose associates, inevitably come to be contrasted. We call the first collection the system of external realities, in the midst of which the room, as 'real,' exists; the other we call the stream of internal thinking, in which, as a 'mental image,' it for a moment floats.[10] The room thus again gets counted twice over. It plays two different rôles, being Gedanke and Gedachtes, the thought-of-an-object, and the object-thought-of, both in one; and all this without paradox or mystery, just as the same material thing may be both low and high, or small and great, or bad and good, because of its relations to opposite parts of an environing world.

As 'subjective' we say that the experience represents; as 'objective' it is represented. What represents and what is represented is here numerically the same; but we must remember that no dualism of being represented and representing resides in the experience per se. In its pure state, or when isolated, there is no self-splitting of it into consciousness and what the consciousness is 'of.' Its subjectivity and objectivity are functional attributes solely, , realized only when the experience is 'take,' i.e., talked-of, twice, considered along with its two differing contexts respectively, by a new retrospective experience, of which that whole past complication now forms the fresh content. The instant field of the present is at all times what I call the 'pure' experience. It is only virtually or potentially either object or subject as yet. For the time being, it is plain, unqualified actuality, or existence, a simple that. In this naïf immediacy it is of course valid; it is there, we act upon it; and the doubling of it in retrospection into a state of mind and a reality intended thereby, is just one of the acts. The 'state of mind,' first treated explicitly as such in retrospection, will stand corrected or confirmed, and the retrospective experience in its turn will get a similar treatment; but the immediate experience in its passing is [11]'truth,' practical truth, something to act on, at its own movement. If the world were then and there to go out like a candle, it would remain truth absolute and objective, for it would be 'the last word,' would have no critic, and no one would ever oppose the thought in it to the reality intended.[12] I think I may now claim to have made my thesis clear. Consciousness connotes a kind of external relation, and does not denote a special stuff or way of being. The peculiarity of our experiences, that they not only are, but are known, which their 'conscious' quality is invoked to explain, is better explained by their relations -- these relations themselves being experiences -- to oneanother.

IV

Were I now to go on to treat of the knowing of perceptual by conceptual experiences, it would again prove to be an affair of external relations. One experience would be the knower, the other the reality known; and I could perfectly well define, without the notion of 'consciousness,' what the knowing actually and practically amounts to -- leading-towards, namely, and terminating-in percepts, through a series of transitional experiences which the world supplies. But I will not treat of this, space being insufficient.[13] I will rather consider a few objections that are sure to be urged against the entire theory as it stands.

V

First of all, this will be asked: "If experience has not 'conscious' existence, if it be not partly made of 'consciousness,' of what then is it made? Matter we know, and thought we know, and conscious content we know, but neutral and simple 'pure experience' is something we know not at all. Say what it consists of -- for it must consist of something -- or be willing to give it up!"

To this challenge the reply is easy. Although for fluency's sake I myself spoke early in this article of a stuff of pure experience, I have now to say that there is no general stuff of which experience at large is made. There are as many stuffs as there are 'natures' in the things experienced. If you ask what any one bit of pure experience is made of, the answer is always the same: "It is made of that, of just what appears, of space, of intensity, of flatness, brownness, heaviness, or what not." Shadworth Hodgson's analysis here leaves nothing to be desired. (1) Experience is only a collective name for all these sensible natures, and save for time and space (and, if you like, for 'being') there appears no universal element of which all things are made.

VI

The next objection is more formidable, in fact it sounds quite crushing when one hears it first.

"If it be the self-same piece of pure experience, taken twice over, that serves now as thought and now as thing" -- so the objection runs - "how comes it that its attributes should differ so fundamentally in the two takings. As thing, the experience is extended; as thought, it occupies no space or place. As thing, it is red, hard, heavy; but who ever heard of a red, hard or heavy thought? Yet even now you said that an experience is made of just what appears, and what appears is just such adjectives. How can the one experience in its thing-function be made of them, consist of them, carry them as its own attributes, while in its thought-function it disowns them and attributes them elsewhere. There is a self-contradiction here from which the radical dualism of thought and thing is the only truth that can save us. Only if the thought is one kind of being can the adjectives exist in it 'intentionally' (to use the scholastic term); only if the thing is another kind, can they exist in it constitutively and energetically. No simple subject can take the same adjectives and at one time be qualified by it, and at another time be merely 'of' it, as of something only meant or known."

The solution insisted on by this objector, like many other common-sense solutions, grows the less satisfactory the more one turns it in one's mind. To begin with, are thought and thing as heterogeneous as is commonly said?

No one denies that they have some categories in common. Their relations to time are identical. Both, moreover, may have parts (for psychologists in general treat thoughts as having them); and both may be complex or simple. Both are of kinds, can be compared, added and subtracted and arranged in serial orders. All sorts of adjectives qualify our thoughts which appear incompatible with consciousness, being as such a bare diaphaneity. For instance, they are natural and easy, or laborious. They are beautiful, happy, intense, interesting, wise, idiotic, focal, marginal, insipid, confused, vague, precise, rational, causal, general, particular, and many things besides. Moreover, the chapters on 'Perception' in the psychology books are full of facts that make for the essential homogeneity of thought with thing. How, if 'subject' and 'object' were separated 'by the whole diameter of being,' and had no attributes and common, could it be so hard to tell, in a presented and recognized material object, what part comes in through the sense organs and what part comes 'out of one's own head'? Sensations and apperceptive ideas fuse here so intimately that you can no more tell where one begins and the other ends, than you can tell, in those cunning circular panoramas that have lately been exhibited, where the real foreground and the painted canvas [14].

Descartes for the first time defined thought as the absolutely unextended, and later philosophers have accepted the description as correct. But what possible meaning has it to say that, when we think of a foot-rule or a square yard, extension is not attributable to our thought? Of every extended object the adequate mental picture must have all the extension of the object itself. The difference between objective and subjective extension is one of relation to a context solely. In the mind the various extents maintain no necessarily stubborn order relatively to each other, while in the physical world they bound each other stably, and, added together, make the great enveloping Unit which we believe in and call real Space. As 'outer,' they carry themselves adversely, so to speak, to one another, exclude one another and maintain their distances; while, as 'inner,' their order is loose, and they form a durcheinander in which unity is lost.(1) But to argue from this that inner experience is absolutely inextensive seems to me little short of absurd. The two worlds differ, not by the presence or absence of extension, but by the relations of the extensions which in both worlds exist.

Does not this case of extension now put us on the track of truth in the case of other qualities? It does; and I am surprised that the facts should not have been noticed long ago. Why, for example, do we call a fire hot, and water wet, and yet refuse to say that our mental state, when it is 'of' these objects, is either wet or hot? 'Intentionally,' at any rate, and when the mental state is a vivid image, hotness and wetness are in it just as much as they are in the physical experience. The reason is this, that, as the general chaos of all our experiences gets sifted, we find that there are some fires that will always burn sticks and always warm our bodies, and that there are some waters that will always put out fires; while there are other fires and waters that will not act at all. The general group of experiences that act, that do not only possess their natures intrinsically, but wear them adjectively and energetically, turning them against one another, comes inevitably to be contrasted with the group whose members, having identically the same natures, fail to manifest them in the 'energetic' way. I make for myself now an experience of blazing fire; I place it near my body; but it does not warm me in the least. I lay a stick upon it, and the stick either burns or remains green, as I please. I call up water, and pour it on the fire, and absolutely no difference ensues. I account for all such facts by calling this whole train of experiences unreal, a mental train. Mental fire is what won't burn real sticks; mental water is what won't necessarily (though of course it may) put out even a mental fire. Mental knives may be sharp, but they won't cut real wood. Mental triangles are pointed, but their points won't wound. With 'real' objects, on the contrary, consequences always accrue; and thus the real experiences get sifted from the mental ones, the things from out thoughts of them, fanciful or true, and precipitated together as the stable part of the whole experience-chaos, under the name of the physical world. Of this our perceptual experiences are the nucleus, they being the originally strong experiences. We add a lot of conceptual experiences to them, making these strong also in imagination, and building out the remoter parts of the physical world by their means; and around this core of reality the world of laxly connected fancies and mere rhapsodical objects floats like a bank of clouds. In the clouds, all sorts of rules are violated which in the core are kept. Extensions there can be indefinitely located; motion there obeys no Newton's laws.

VII

There is a peculiar class of experience to which, whether we take them as subjective or as objective, we assign their several natures as attributes, because in both contexts they affect their associates actively, though in neither quite as 'strongly' or as sharply as things affect one another by their physical energies. I refer here to appreciations, which form an ambiguous sphere of being, belonging with emotion on the one hand, and having objective 'value' on the other, yet seeming not quite inner nor quite outer, as if a diremption had begun but had not made itself complete.

Experiences of painful objects, for example, are usually also painful experiences; perceptions of loveliness, of ugliness, tend to pass muster as lovely or as ugly perceptions; intuitions of the morally lofty are lofty intuitions. Sometimes the adjective wanders as if uncertain where to fix itself. Shall we speak of seductive visions or of visions of seductive things? Of healthy thoughts or of thoughts of healthy objects? Of good impulses, or of impulses towards the good? Of feelings of anger, or of angry feelings? Both in the mind and in the thing, these natures modify their context, exclude certain associates and determine others, have their mates and incompatibles. Yet not as stubbornly as in the case of physical qualities, for beauty and ugliness, love and hatred, pleasant and painful can, in certain complex experiences, coexist.

If one were to make an evolutionary construction of how a lot of originally chaotic pure experience became gradually differentiated into an orderly inner and outer world, the whole theory would turn upon one's success in explaining how or why the quality of an experience, once active, could become less so, and, from being an energetic attribute in some cases, elsewhere lapse into the status of an inert or merely internal 'nature.' This would be the 'evolution' of the psychical from the bosom of the physical, in which the esthetic, moral and otherwise emotional experiences would represent a halfway stage.

VIII

But a last cry of non possumus will probably go up from many readers. "All very pretty as a piece of ingenuity," they will say, "but our consciousness itself intuitively contradicts you. We, for our part, know that we are conscious. We feel our thought, flowing as a life within us, in absolute contrast with the objects which it so unremittingly escorts. We can not be faithless to this immediate intuition. The dualism is a fundamental datum: Let no man join what God has put asunder."

My reply to this is my last word, and I greatly grieve that to many it will sound materialistic. I can not help that, however, for I, too, have my intuitions and I must obey them. Let the case be what it may in others, I am as confident as I am of anything that, in myself, the stream of thinking (which I recognize emphatically as a phenomenon) is only a careless name for what, when scrutinized, reveals itself to consist chiefly of the stream of my breathing. The 'I think' which Kant said must be able to accompany all my objects, is the 'I breath' which actually does accompany them. There are other internal facts besides breathing (intracephalic muscular adjustments, etc., of which I have said a word in my larger Psychology), and these increase the assets of 'consciousness,' so far as the latter is subject to immediate perception; but breath, which was ever the original of 'spirit,' breath moving outwards, between the glottis and the nostrils, is, I am persuaded, the essence out of which philosophers have constructed the entity known to them as consciousness. That entity is fictitious, while thoughts in the concrete are fully real. But thoughts in the concrete are made of the same stuff as things are.

I wish I might believe myself to have made that plausible in this article. IN another article I shall try to make the general notion of a world composed of pure experiences still more clear.

References & Footnotes

1. Articles by Bawden, King, Alexander, and others. Dr. Perry is frankly over the border

2. In my Psychology I have tried to show that we need no knower other than the "passing thought." [Principles of Psychology, vol. I, pp. 338 ff.]

3. G.E. Moore: Mind, vol. XII, N.S., [1903], p.450

4. Paul Natorp: EinleitungindiePsychologie, 1888, pp. 14, 112.

5. "Figuratively speaking, consciousness may be said to be the one universal solvent, or menstruum, in which the different concrete kinds of psychic acts and facts are contained, whether in concealed or in obvious form." G.T.Ladd: Psychology, Descriptiveand Explanatory, 1894, p.30.

6. Here as elsewhere the relations are of course experienced relations, members of the same originally chaotic manifold of nonperceptual experience of which the related terms

themselves are parts.

7. Of the representative functions of non-perceptual experience as a whole, I will say a word in a subsequent article; it leads too far into the general theory of knowledge for much to be said about it in a short paper like this.

8. Munsterberg: Grundzugeder Psychologie, vol. I, p. 48.

9. Cf. A.L. Hodder: The Adversaries of the Sceptic, pp.94-99.

10. For simplicity's sake I confine my exposition to "external" reality. But there is also the system of ideal reality in which the room plays its part. Relations of comparison, of classification, serial order, value, also are stubborn, assign a definite place to the room, unlike the incoherence of its places in the mere rhapsody of our successive thoughts.

11. Note the ambiguity of this term, which is taken sometimes objectively and sometimes subjectively.

12. In the Psychological Review for July [1904], Dr. R.B. Perry has published a view of Consciousness which comes nearer to mine than any other with which I am acquainted. At present, Dr. Perry thinks, every field of experience is so much "fact." It becomes "opinion" or 'thought" only in retrospection, when a fresh experience, thinking the same object, alters and corrects it. But the corrective experience becomes itself in turn corrected, and thus the experience as a whole is a process in which what is objective originally forever turns subjective, turns into our apprehension of the object. I strongly recommend Dr. Perry's admirable article to my readers.

13. I have given a partial account of the matter in Mind, vol. X, p. 27, 1885, and in the Psychological Review, vol. II, p. 105, 1895. See also C.A. Strong's article in the Journal of Philosophy, Psychology and Scientific Methods, vol I, p. 253, May 12, 1904. I hope myself very soon to recur to the matter.

14. Spencer's proof of his 'Transfigured Realism' (his doctrine that there is an absolutely non-mental reality) comes to mind as a splendid instance of the impossibility of establishing radical heterogeneity between thought and thing. All his painfully accumulated points of difference run gradually into their opposites, and are full of exceptions.

12. Consciousness -- What Is It?

L. Dossey[1], B. Greyson[2], P.A. Sturrock[3], and J. B. Tucker[4]

[1]Explore, Santa Fe, NM 87501 [2]University of Virginia Health System, Division of Perceptual Studies, Charlottesville, VA 22902 [3]Center for Space Science and Astrophysics, Stanford University, Stanford, CA 94305 [4]University of Virginia Health System, Division of Perceptual Studies, Charlottesville, VA 22902

Abstract

Conventionally, there is a tendency to view consciousness as simply a property or activity of the brain. One can explain a lot about consciousness in this way – but not everything. In this article, we draw attention to certain aspects of consciousness that resist the conventional interpretation including, in particular, out-of-body experiences, past-life memories, the apparent linked consciousnesses of twins, and healing at a distance.

KEY WORDS: consciousness, shared consciousness, twins, dissociation, near death experiences

1. INTRODUCTION

Oscar Wilde began a mini-essay with the words "Ah! Meredith! Who can define him?" (Wilde, 1889) Perhaps we should begin this essay with "Ah! Consciousness! Who can define it?" However, before discussing consciousness, it may be helpful to distinguish two different types of science.

What we might call "Type I" science is a "forward" or "deductive" process, in which we begin with agreed concepts and an established base of knowledge such as physical laws, and explore the consequences. The bulk of "everyday science" seems to fall into this category.

What we might call "Type II" science is a "reverse" or "inductive" process in which, beginning with a phenomenon that is not understood and handicapped by the absence of useful concepts, we attempt to establish appropriate concepts and then derive the principles governing the phenomenon. Major paradigm shifts such as quantum mechanics belong in this category.

In discussing a complex problem such as consciousness, it makes a huge difference which scientific approach one adopts. Following the pattern of Type I science, it is natural to begin with the brain and to examine the possibility that consciousness can be fully understood as a brain activity. This approach is by no means straightforward, and raises challenging problems. If you had terminal health problems, a future surgeon might be able to remove your brain and implant it in the healthy body of some young person who had just been killed. Then we would suppose that, when you wake up from the operation, you recognize your own mind with its memories and peculiarities, and are happy to see that you now have a handsome, strong, and healthy young body.

But now let us fast-forward one hundred thousand years, when computers can read everything in your brain, and use that information to program another brain to have precisely the same informational structure and content as your original brain. Now, when you wake up from that operation, where are you? If you are in one body, which is it? Or can you be in both bodies at the same time? We see that there may be conceptual problems with the consciousness-is-brain-activity hypothesis.

However, to return to our main theme—the nature of consciousness, it is important to note that we do not test a hypothesis by looking for more and more facts that are in agreement with that hypothesis. On the contrary, we test a hypothesis by carefully searching for facts that are incompatible with that hypothesis. If no such facts come to light, the case for the hypothesis is greatly strengthened. If such facts do come to light, the hypothesis must be abandoned or at least modified.

The purpose of this article is to highlight four phenomena that appear to be incompatible with the consciousness-is-brain-activity hypothesis. One of these comprises "out-of-body" experiences (OBEs), in which a person reports being separated from his or her body and acquiring information that it would have been impossible for the person to obtain by normal means. This topic is discussed in Section 2. A second topic is reincarnation, indicated by evidence that a child remembers a previous life, when careful investigation finds correspondences between the child's memories and facts concerning the "previous personality." This topic is discussed in Section 3. We discuss evidence for interactions between consciousnesses in Section 4, healing at a distance in Section 5, and offer some concluding thoughts in Section 6.

2. OUT-OF-BODY EXPERIENCES

In out-of-body experiences (OBEs), a person's consciousness is experienced as having separated from the body. A tentative estimate is that at least 10% of the 5/26 general population have experienced one or more OBEs (Alvarado, 2000, pp. 184– 186). Models of consciousness that link it inextricably with the brain have included neural mechanisms to account for the experience of being out of the body and perceiving events as if from a different location. Sometimes these proposed mechanisms have been taken to imply that out-of-body experiences are nothing more than hallucinations or illusions produced by altered brain physiology (Churchland, 1986; Crick, 1994; Pinker, 1997).

For example, Joseph (1999, 2001, 2009) has written extensively about the role of hyperactivation of the amygdala, hippocampus, and inferior temporal lobe in splitting consciousness from the body under traumatic circumstances. However, Joseph prudently acknowledged that these neuroanatomical data are ambiguous with regard to whether altered brain physiology causes out-of-body sensations that are merely hallucinatory or, alternatively, whether it enables the mind truly to separate from the body, permitting accurate out-of-body perceptions that accord with external reality (Joseph, 2001, p. 132). If out-of-body perceptions are merely hallucinatory, then the neurophysiological findings elaborated by Joseph and others may provide sufficient explanation. Even if out-of-body perceptions are not hallucinatory, but are in fact veridical representations of external reality beyond the reach of the senses, these neurophysiological models may still contribute significantly to our understand of the phenomenon. But in the latter case, they do not provide sufficient explanation without some further explication of the mind- body relationship (Kelly et al., 2007).

Many out-of-body perceptions are entirely subjective, providing no evidence that the person actually separated from the body, rather than simply imagined separating. However, in other cases experiencers report that, while out of the body, they became aware of events either occurring at a distance or that in some other way would have been beyond the reach of their ordinary senses. Some of these accurate perceptions included unexpected or unlikely details, such as a woman in childbirth who reported being out of her body and seeing her mother, a non-smoker, smoking a cigarette in the waiting room (Cook, Greyson, & Stevenson, 1998, p. 391). Notably, Ring and Cooper (1999) reported 31 cases of blind individuals (nearly half of them blind from birth) who experienced during their OBEs quasi-visual and sometimes veridical perceptions of objects and events.

A frequent criticism of these reports of perceptions of events at a distance from the body is that they often depend on the experiencer's testimony alone. The paucity of corroborating testimony in many cases has encouraged commentators to dismiss such reports cases as anecdotal. However, some cases have been corroborated by others (e.g., Clark, 1984; Hart, 1954; Ring & Lawrence, 1993). Van Lommel et al. (2001, p. 2041), for example, reported a case in which a cardiac arrest victim was brought into the hospital comatose and cyanotic, and even after restoration of his circulation he remained in a coma and on artificial respiration in the intensive care unit for more than a week. When he regained consciousness and was transferred back to the cardiac care unit, he immediately recognized one of the nurses, saying that this was the person who had removed his dentures during the resuscitation procedures. He said further that he had watched from above the attempts of hospital staff to resuscitate him in the emergency room, and he described "correctly and in detail" the room and the people working on him, including the cart in which the nurse had put his dentures. The nurse corroborated and verified his account. Cook, Greyson, & Stevenson (1998, pp. 399–400) reported a case of this type in which a patient undergoing open-heart surgery described leaving his body and watching the cardiac surgeon "flapping his arms as if trying to fly." The surgeon verified this detail by explaining that after "scrubbing in", and to keep his hands from possibly becoming contaminated, he had flattened his hands against his chest, while rapidly giving instructions to the surgical interns by pointing with his elbows.

A dramatic OBE from the 19th century involved Mr. Wilmot and his sister Eliza who were traveling by ship from Liverpool, England, to New York in a severe storm (Sidgwick, 1891, pp. 41-46). More than a week after the storm began, Mr. Wilmot's wife in Connecticut, worried about the safety of her husband, had an experience while she was awake during the middle of the night, in which she seemed to go to her husband's stateroom on the ship, where she saw him asleep in the lower berth and noticed another man in the upper berth looking at her. She hesitated, kissed her husband, and left. The next morning, Mr. Wilmot's roommate asked him somewhat indignantly about the woman who had come into their room during the night. Eliza Wilmot corroborated this story, saying that the next morning, before she had seen her brother, his roommate asked her if she had been in to see Mr. Wilmot during the night, and when she replied no, he said that he had seen a woman come into their room in the middle of the night and kiss Mr. Wilmot.

Of course, not all OBEs are veridical in nature, and most provide no evidence of anything more than a subjective experience. Nonetheless, some OBE reports are corroborated by independent observers; the Wilmot case is not unique (see, e.g, Cook, Greyson, & Stevenson, 1998; Kelly, Greyson, & Stevenson, 1999–2000). Hart (1954) analyzed 288 published OBE cases in which persons reported perceiving events that they could not have perceived in the ordinary way. In 99 of these cases the events perceived were verified as having occurred, and the experience had been reported to someone else before that verification occurred. A type of OBE that particularly strains models that link consciousness inextricably to the brain involves "reciprocal apparitions," again exemplified by the Wilmot case. In such cases, while one person is having an OBE, or having a dream in which he or she seems to go to a distant location, a person at that location, unaware of the first person's experience, sees an apparition of that person. Hart (1954) summarized 30 such cases that had been published up to that time (see also Hart & Hart, 1933). In one unpublished case from the University of Virginia collection, a nurse became friends with a quadriplegic man who required several hospitalizations for pneumonia and other complications. During one of these hospitalizations, the nurse, feeling guilty that she had not recently visited this patient, had a dream in which she seemed to go to him in the hospital, stood at the end of his bed, and told him to keep fighting.

Shortly afterward, the patient's sister told this nurse that he had reported seeing her standing at the foot of his bed, telling him to keep fighting.

A few individuals may have OBEs repeatedly or voluntarily, making them potentially amenable to observation under controlled conditions. Tart (1968) studied a woman who was able, while monitored by EEG in the laboratory, to have an OBE in which she read a five-digit number that was randomly selected and placed as a target on a shelf out of range of her normal sight. In another experiment, a person who could induce OBEs at will attempted during randomly selected periods to go to a specified location during an OBE and influence a variety of detectors located there, including his pet kitten, which showed significantly less movement and less vocalizing during the OBE periods than during the control periods (Morris, et al., 1978). Osis and McCormick (1980) tasked another person who claimed to induce OBEs at will to view a randomly-generated target that appeared as an illusion visible only from one particular point in space. Unbeknownst to the subject, a strain gauge sensor was situated at that location. The strain gauge activation was significantly higher during hits than during misses. These experiments, as well as the spontaneous reciprocal apparitions described above, suggest that veridical OBE perception may be objectively real, implying that some aspect of consciousness can under certain circumstances separate from the physical body.

3. REINCARNATION - CHILDREN'S REPORTS OF PAST-LIFE MEMORIES

Children's reports of memories of previous lives have been the subject of systematic study for the last fifty years (Stevenson, 2001). Beginning with Stevenson, a number of researchers have now collected over 2,500 cases. Though easiest to find in cultures with a general belief in reincarnation, cases have been identified all over the world, including in the U.S. (Tucker, 2005) and in Western Europe (Stevenson, 2003). The strongest cases have included statements that have been verified to 9/26 accurately describe the life of one particular deceased individual. The following is one example.

The Case of Kumkum Verma. Stevenson (1975) reported the case of Kumkum Verma, a girl in India who began talking about a previous life when she was three years old. She said she had lived in a place called Darbhanga, a city of 200,000 people 25 miles from her village. She described a life as a woman there and named the section of the city where she said she had lived, a commercial district of artisans and craftsmen. She gave numerous details, and her aunt made notes of some of Kumkum's statements six months before any attempt was made to verify them. Though some of the notes were lost, Stevenson obtained a partial list and had it translated into English. It revealed 18 statements from Kumkum, all of which matched the life of a blacksmith's wife who died in Darbhanga five years before Kumkum was born. These included the name of the city section, her son's name and the fact that he worked with a hammer, her grandson's name, the name of the town where her father lived, the location of his home near mango orchards, and the presence of a pond at her house. Also included were personal details that were accurate for the deceased woman, with Kumkum saying she had an iron safe at her house, a sword hanging near her cot, and a snake near the safe to which she fed milk.

Kumkum's family noted that she used some unusual expressions and spoke with an accent, both of which they associated with the lower classes of Darbhanga. Kumkum appeared to have had no access to the information about the woman through normal means. By all accounts, the two families involved had been completely unknown to each other and were separated not only by distance but also by social class, as Kumkum's father was an educated landowner.

The children in these cases generally begin talking about a previous life at an early age, with the average being 35 months. They usually stop by the age of six or seven. They typically start their past-life talk spontaneously with no urging from their parents. Indeed, parents often try to get the children to stop talking about a past life, even in places with a general belief in reincarnation. The children usually describe recent, ordinary lives, with the one exceptional aspect of the life frequently being the death, as many of the previous individuals died at an early age and 70% died by unnatural means. Some children say they were deceased family members, but others report being strangers at other locations as Kumkum did. Almost all describe a previous life in the country in which they live, though some report being soldiers of another nationality who were killed in the child's home country during a war. This includes 24 Burmese children who said they had been Japanese soldiers killed in Burma during World War II, a claim parents would have been most unlikely to encourage given how despised the Japanese Army was there (Stevenson & Keil, 2005).

Along with their statements, many of the children show behaviors that appear associated with their apparent memories. They often show strong emotions about the previous life and also emotions appropriate for the relationship the previous person had with different individuals, being deferential toward the previous husband, for instance, but bossy toward the previous person's younger siblings. Other behaviors include phobias, with the children showing intense fears related to the mode of death in 35% of the unnatural death cases (Tucker, 2005). Many of the children also show themes in their play that appear connected to their apparent memories. This most often involves acting out the occupation of the previous person, with some children engaging compulsively in such play for hours on end.

It is significant to note that a number of the children had birthmarks or birth defects that matched wounds—usually the fatal wounds—on the bodies of the previous individuals. Stevenson (1997) published a collection of 225 such cases that included a variety of dramatic or unusual defects. These included stubs for the fingers of only one hand—when the previous person had lost the fingers of one hand in a fodder chopping machine—and an underdeveloped side of the face of one child with only an accompanying stump for an ear—when the previous person had been shot in the side of the head at close range. Stevenson listed 18 cases in which children had double birthmarks, corresponding to both the entrance and exit wounds suffered by gunshot victims.

Such marks are reminiscent of work in other areas demonstrating that mental images can produce specific somatic effects in at least some individuals, such as hypnotized subjects who develop blisters after being told they are being burned by a hot object (Gauld, 1992). The birthmarks and birth defects are consistent with a process in which the consciousness of the previous individual, containing the final mental images from the previous life, affects the development of the fetus and produces defects similar to the wounds that individual had suffered. In sum, the children appear to possess memories, emotions, and mental images that previously belonged to a deceased individual. Though the mechanism that might enable their transfer to a new body remains to be understood, the cases provide evidence for the persistence of consciousness after death.

4. LINKED CONSCIOUSNESS IN TWINS

Another area that provides evidence that consciousness is not confined to the brain involves cases of twins who appear to share a non-physical connection. This phenomenon is most thoroughly described in Playfair (2008). Some examples follow.

A student at Stony Brook University awoke out of a deep sleep at 6 a.m. and cried out that her twin sister was in trouble. She told her roommate and soon called her mother. She learned that at the time she woke up, a bomb had exploded outside her sister's apartment in Arizona, shattering her window and leading her and her husband to rush out of the building (Playfair, 2008, p. 60).

A young girl was with her mother in the kitchen of their house when she suddenly said, "Hurry, Elizabeth has fallen off Jack's bicycle and hurt her knee!" Her mother followed her as she ran out of the house and down the road, where they found her twin sister still lying on the ground where she had fallen (Gaddis & Gaddis, 1972, pp. 99-100).

A set of twins, a physician and a London banker, reported that as teenagers, one of them was walking down a road when she felt threatened by a car that kept turning around and approaching her. The girl became panicked and started to run. She imagined her sister and thought, "Alison, if there's anything you can do, tell Dad to come quick!" Alison, at home studying in the room the two girls shared, suddenly felt as if her sister was there. She reported experiencing "a feeling of real panic— like 'Get Dad! Get Dad!' I suddenly knew there was something wrong with Aily" (Playfair, 2008, p. 53).

The connections at times appear to manifest somatically rather than mentally. The girls in the last example seemed to share pain at times. One tripped in a pothole during a run and sprained her ankle. At that moment, her sister experienced a sudden burning sensation that started in her feet and then spread over her body. Another time, one of them was in bed when she experienced pain in her nose so severe she got up and took a painkiller. At the same time sixty miles away, her sister was in a pool when another swimmer shot up from the bottom and hit her in the face, breaking her nose (Playfair, 2008, pp. 53-54).

In another case, a woman who had moved to Japan phoned her twin and asked her to send some bras because her breasts had become tender and swollen. Later that day, her symptoms led her to wonder if her sister was pregnant, and by the time they spoke again later that week, her sister could confirm that indeed she was (Playfair, 2008, pp. 51-52).

Such reports are not rare. In a survey in which 600 twins or parents of twins completed questionnaires, 183 reported either experiences that might be explained 13/26 by reading each other's minds or instances of being surprised by having the same illness or pain simultaneously (Rosambeau, 1987). Such connections can appear evident at a very early age. One example involved 3 day-old twins, where one began shrieking and shaking as his brother was face down in pillows and turning blue, the child's screams saving his brother's life (Playfair, 2008, pp. 44-45).

It may not be surprising then that apparent connections can occur in twins separated in infancy. The Minnesota Study of Twins Reared Apart involved 135 pairs, including the "Jim twins" (Segal, 1999, pp. 116-118). Reared by different adoptive families in Ohio and reunited at age 39, they were both named Jim and had been married twice, first to a woman named Linda and then to a woman named Betty. They had sons with the same name (though different spellings): James Alan and James Allan. As children, they each had a dog named Toy. They had taken family vacations to the same three-block strip of Florida beach (without ever meeting), both arriving in light blue Chevrolets. Both worked part-time as sheriffs, and they consumed the same brand of cigarettes and the same brand of beer.

Since such similarities can hardly be ascribed to genetics, one might be inclined to blame simple coincidence. That explanation becomes more strained when considering the American twins who, reunited at age 25 after being raised in dissimilar environments, discovered they both used the same rare Swedish toothpaste (Segal, 1999, p. 119). Such cases suggest a persistent non-physical linkage in some twins, even those who are reared apart from each other.

5. CONNECTIONS BETWEEN CONSCIOUSNESSES, AND HEALING AT A DISTANCE

In Section 2 and 3, we have presented evidence suggesting that consciousness has an existence independent of the body—and therefore independent of the brain. But what properties could one or should one assign to that consciousness other than the fact that it is in some way related to, or perhaps comprises, someone's "personality" or "essential identity"? In particular, can there be a linkage between the consciousnesses of two different individuals?

This leads one to the consideration of evidence concerning "ESP" or "extrasensory perception," for which there is voluminous experimental evidence. [See, for instance, Bem (2011), Jahn and Dunne (1987), Radin (1997).] It also leads to another important and relevant area of current research—that of anomalous healing, such as healing brought about by the unknown influence of a healer on a healee who may be in a remote location. According to the conventional view that human consciousness is simply an activity or property of the brain, such healing is impossible. Yet there is abundant evidence that it does in fact occur. In recent years, many researchers have undertaken clinical and laboratory studies designed to answer two fundamental questions: (1) Do the compassionate healing intentions of humans affect biological functions in remote individuals who may be unaware of these efforts? And (2) can these effects be demonstrated in nonhuman processes, such as microbial growth, specific biochemical reactions, or the function of inanimate objects?

The first question is extraordinarily difficult to study (Schwartz and Dossey, 2010). There are studies which demonstrate significant effects of distant healing in cardiopulmonary (Byrd, 1988) and AIDS (Sicher, et.al., 1998) patients, for example. But the methodological and ethical challenges involved in studying healing effects on humans at a distance are formidable. For instance, in distant healing prayer studies, can it be assumed that the "not prayed for group" really did not receive any prayer or healing thoughts from either themselves or their loved ones?

The second question, whether healing can affect non-human processes, is far easier to address. The pioneering work of Bernard Grad at McGill University set the standards for systematic laboratory work on healing. Most notably, Grad studied the effects of healers on wound healing in mice and the growth rates of "shocked" plants (Grad, 1965). In both areas Grad found that wounded mice healed 15/26 significantly faster after having been treated by a healer, and shocked plants similarly had higher germination and faster growth rates. In more recent times, Bengston and Krinsley (2000) have found that inexperienced skeptical volunteers acting as healers can produce full cures in mice infected with a normally fatal dosage of mammary adenocarcinoma. Subsequent mouse studies using the same mammary adenocarcinoma model also indicated a curious "resonant bonding" between experimental and control mice, so that healing intention directed towards the treated experimental animals somehow also affects the untreated control animals (Bengston and Moga, 2007). It is interesting that these experiments on laboratory mice seem to produce patterns that mimic placebo responses in human studies.

What has been accomplished? In 2003, Jonas and Crawford (2003) found "over 2,200 published reports…and other writings on spiritual healing, energy medicine, and mental intention effects. This included 122 laboratory studies, 80 randomized controlled trials, 128 summaries or reviews, 95 reports of observational studies and nonrandomized trials, [and] 271 descriptive studies, case reports, and surveys…."

How significant are these clinical and laboratory studies? Using the strict CONSORT (Consolidated Standards of Reporting Trials; CONSORT 2010) criteria, Jonas and Crawford gave an A grade to studies involving the effects of intentions on inanimate objects such as sophisticated random number generators. They also gave a high grade (B) to intercessory prayer studies involving humans, and to similar laboratory experiments involving nonhumans such as plants, cells, and animals. In order to relate this phenomenon to brain activity, one would perhaps need to hypothesize some form of radiation that emanates from the brain and somehow influences the healee. However, if this approach proves not to be fruitful, one could regard this as another phenomenon that, along with OBE's and reincarnation, needs a fundamentally different type of explanation. Recently, for example, Hendricks, Bengston, and Gunkelman (2010) demonstrated interpersonal EEG coupling between healer and subject pairs. The healer's EEG data showed harmonic frequency coupling across the spectrum, followed first by between-individual EEG frequency entrainment effects, and then by instantaneous EEG phase locking. The healer produced a pattern of harmonics consistent with Schumann's resonances, with an entrainment of the subject's EEG by the healer's resonance standing waves, and with eventual phase coupling between the healer and healee. The authors speculate that healing may involve a Schumann-resonance-type standing electric field as a connectivity mechanism (Hendricks, Bengston, & Gunkelman, 2010).

6. DISCUSSION

If consciousness is not simply a brain activity, what might it be? One approach is to modify the "brain" concept, and to hypothesize that consciousness is the activity of some other entity. For this purpose, we could reactivate (one might say "resuscitate") the dated and little-used term "soul." We might then suppose that it is the soul that leaves the body and rises in the operating room to view the body from above. We might also suppose that the soul leaves the body of someone who dies, to later reenter a new person who is about to be born. But these may prove to be overly simplistic concepts: just as computing may be carried out either in a desktop computer or in the "cloud", so it may prove that the consciousness associated with one person is intrinsically inseparable from consciousnesses related to additional—perhaps many other—persons.

It is important to keep in mind the extraordinary tentativeness of almost anything that can be said about the nature of consciousness. According to John Maddox (then editor of Nature), "What consciousness consists of...is...a puzzle. Despite the marvelous successes of neuroscience in the past century..., we seem as far from understanding cognitive process as we were a century ago." (Maddox, 1999.) The philosopher Jerry A. Fodor expressed a similar opinion, saying, "Nobody has the slightest idea how anything material could be conscious. Nobody even knows what it would be like to have the slightest idea about how anything material could be conscious. So much for the philosophy of consciousness."(Fodor, 1992) In a similar vein, Stuart Kauffman, the theoretical biologist and complex systems theorist, wrote "Nobody has the faintest idea what consciousness is…. I don't have any idea. Nor does anybody else, including the philosophers of mind." (Kauffman, 2011.)

The need for humility in approaching the subject of consciousness has long been emphasized by theoretical physicists: physics, as currently understood, may not be up to the task of deciphering the nature of the mind. Wigner (1983) expressed the view that "It [physics] will have to be replaced by new laws, based on new concepts, if organisms with consciousness are to be described.... [I]n order to deal with the phenomenon of life, the laws of physics will have to be changed, not only reinterpreted." Penrose (2003) has stated "My position [on consciousness] demands a major revolution in physics.... [T]here is something very fundamental missing from current science. Our understanding at this time is not adequate and we're going to have to move to new regions of science...." Nick Herbert, a physicist, has expressed his thoughts more colorfully: "Science's biggest mystery is the nature of consciousness. It is not that we possess bad or imperfect theories of human awareness; we simply have no such theories at all. About all we know about consciousness is that it has something to do with the head, rather than the foot." (Herbert, 1987.)

Experimental results such as those discussed in preceding sections may point to an unknown mechanism of linkage between consciousnesses. Schrödinger, one of the fathers of quantum mechanics, coined the term "entanglement" (Schrödinger, 1935) and later proposed that the consciousnesses of all individuals are united (Schrödinger, 1969, 1983). "Entanglement" is a property of a quantum-mechanical system containing two or more components that have once been in contact. Even though they may later be separated, they remain linked in such a way that the quantum state of any one of them cannot be adequately described without full consideration of the others (Schrödinger, 1935). Though resisted by Einstein as 18/26 "spooky action at a distance," quantum entanglement has been demonstrated experimentally, including over kilometer distances (Tittel et al. 1998; Nadeau & Kafatos, 1999, pp. 65-82).

Although physicists originally believed entangled states between distant particles were of no practical consequence, evidence now suggests that the effects of quantum entanglement may "scale up" into our macroscopic world, such as linking separated human neurons in vitro. (See, for instance, Pizzi, et al., 2004.) If separated neurons can be entangled in vitro, might whole brains be entangled at a distance? Several experiments using fMRI and EEG-based protocols suggest that this is the case. In these experiments, the stimulation of one individual's brain appears to be registered simultaneously in a distant individual's brain by fMRI or EEG (Standish et al., 2003, 2004; Wackerman et al., 2003). These experiments suggest that the idea of united, linked minds may be more than philosophical speculation.

For Schrödinger (1935), entanglement was the key insight dividing classical from modern physics. He said, "I would not call ... [entanglement] one but rather the characteristic trait of quantum mechanics, the one that enforces its entire departure from classical lines of thought." He further wrote (Schrödinger, 1983): "To divide or multiply consciousness is something meaningless. In all the world, there is no kind of framework within which we can find consciousness in the plural; this is simply something we construct because of the spatiotemporal plurality of individuals, but it is a false construction…. The category of number, of whole and of parts are then simply not applicable to it; the most adequate…expression of the situation is this: the self-consciousness of the individual members are numerically identical with [one an]other and with that Self which they may be said to form at a higher level." He also remarked (Schrödinger 1969): "Mind is by its very nature a singulare tantum. I should say: the overall number of minds is just one."

To paraphrase Schrödinger's statement, one might say that consciousnesses are inextricably entangled. This concept plays an important role in modern psi 19/26 research. According to Radin, "There are theoretical descriptions showing how tasks can be accomplished by entangled groups without the members of the group communicating with each other in any conventional way. Some scientists suggest that the remarkable degree of coherence displayed in living systems might depend in some fundamental way on quantum effects like entanglement. Others suggest that conscious awareness is caused or related in some important way to entangled particles in the brain. Some even propose that the entire universe is a single, self- entangled object." (Radin, 2006, p.1.)

But we must be cautious: Invoking "entanglement" may simply substitute one mystery for another. While it is true that distant individuals appear to be linked in some sense (for instance in the correlated behaviors of identical twins raised apart, or in apparent healing at a distance), there is as yet no definitive evidence that "human entanglement" is a manifestation of "quantum entanglement." We may be dealing with correspondences in terminology and nothing more. It is important to bear in mind that in studies of apparent remote healing, for instance, factors such as compassion, love, and empathy seem to play a key role (Achterberg et al., 2005), but these factors are not to be found in the equations of quantum physics. Moreover, physicists agree that the nonlocal connections between entangled particles cannot be used to transfer information. (Nadeau & Kafatos, 1999, pp. 80-81.) In contrast, it appears that information can be transferred between distant, entangled humans (as in remote healing; in identical twins raised apart); in correlated fMRI or EEG patterns between distant human beings; or in ostensible telepathic exchanges. (See, for instance, Radin 1997, 2006.) It therefore seems unlikely that "entangled particles" can fully account for the entangled actions and emotions of human beings. For all that, "entanglement" is a useful metaphor for distant correlated human experiences. The fact that entanglement is now recognized to exist at the subatomic quantum level should at the very least encourage us to explore similar (but probably different) phenomena at the human level.

The idea that humans may be linked collectively through space and time is ancient, and is one of the underlying philosophies of several Eastern wisdom traditions. In the West it emerged in the philosophy of Plato, Plotinus, and Swedenborg. It formed the basis of Emerson's view of the Over-Soul. Swiss psychologist Carl G. Jung invoked this view in his concept of the collective unconscious that unites all minds — past, present, and future. However, research concerning consciousness is clearly research of the "Type II" variety. Rather than attempt to develop a theory based on information now in hand, we may be better advised to regard current research on OBEs, reincarnation, and anomalous healing as the beginning of a major long-term program of developing more powerful and more fruitful concepts that can elucidate the nature of consciousness. We have recently pointed out that phenomena such as those discussed in this article, together with other anomalous phenomena, may require a revision of our current "Model of Reality," and that our revised model may involve the concept of hyperspace (Sturrock, 2009).

ACKNOWLEDGEMENT We wish to acknowledge a valuable contribution from Professor William Bengston, of the Department of Sociology at St. Joseph's College, and helpful suggestions from two referees.

References

Achterberg J, Cooke K, Richards T, Standish L, Kozak L, Lake J., 2005, Evidence for correlations between distant intentionality and brain function in recipients: a functional magnetic resonance imaging analysis. J Alternative and Complementary Medicine, 11(6), 965-971.

Alvarado, C. S., 2000. Out-of-body experiences. In E. Cardeña, S. J. Lynn, & S. Krippner, Eds., Varieties of anomalous experience: Examining the scientific evidence, pp. 183-218, American Psychological Association, Washington, DC.

Bengston, W., & Krinsley, D., 2000. The Effect of the Laying On of Hands on Transplanted Breast Cancer in Mice. Journal of Scientific Exploration, 14, 353-364.

Bengston, William, & Moga, Margaret. 2007. Resonance, Placebo Effects, and Type II Errors: Some Implications from Healing Research for Patients. Experimental Methods. Journal of Alternative and Complementary Medicine, 13, 317-327.

Bem, D. J., 2011. Feeling the Future: Experimental Evidence for Anomalous Retroactive Influences on Cognition and Affect, Journal of Personality and Social Psychology, 100, 407-425.

Byrd, Randolph. 1988. Positive Therapeutic Effects of Intercessory Prayer in a Coronary Care Population. Southern Medical Journal, 81, 826-829.

Churchland, P. S., 1986. Neurophilosophy:

Toward A Unified Science Of The Mind/Brain. MIT Press, Cambridge, MA. Clark, K., 1984. Clinical interventions with near-death experiencers. In B. Greyson & C. P. Flynn, Eds., The Near-Death Experience: Problems, Prospects, Perspectives, pp. 242-255. Charles C. Thomas, Springfield, IL.

CONSORT, 2010. Transparent Reporting of Trials. 2010, http://www.consort-statement.org/. Accessed December 9, 2010.

Cook, E. W., Greyson, B., & Stevenson, I., 1998. Do any near-death experiences provide evidence for the survival of human personality after death? Relevant features and illustrative case reports. Journal of Scientific Exploration, 12, 377-406.

Crick, F. (1994). The Astonishing Hypothesis: The Scientific Search For The Soul. Simon & Schuster, London.

Fodor J., 1992, The big idea: Can there be a science of mind? Times Literary Supplement. July 3, 1992, 5-7.]

Gaddis, V., & Gaddis, M., 1972. The Curious World of Twins. Hawthorn, New York, NY.

Gauld, A., 1992. A History of Hypnotism. Cambridge University Press, Cambridge.

Grad, B. 1965. Some biological effects of the "Laying-On of Hands": A review of experiments with animals and plants. Journal of the American Society for Psychical Research, 59: pp. 95-127.

Hart, H., 1954. ESP projection: Spontaneous cases and the experimental method. Journal of the American Society for Psychical Research, 48, 121-146.

Hart, H., & Hart. E. B., 1933. Visions and apparitions collectively and reciprocally perceived. Proceedings of the Society for Psychical Research, 41, 205-249.

Hendricks, L., Bengston, W., & Gunkelman, J., 2010. The healing connection: EEG harmonics, entrainment, and Schumann's resonances. Journal of Scientific Exploration, 24, 655-666.

Herbert, N., 1987, Quantum Reality, Anchor/Doubleday, Garden City, NY, 249.

Jahn, R. F., & Dunne, B.J., 1987. Margins of Reality: The Role of Consciousness in the Physical World. Houghton Mifflin Harcourt, New York, NY.

Jonas, W.B., & Crawford, C.C. (eds.), 2003, Healing, Intention and Energy Medicine. Churchill Livingstone, New York, NY, pp. xv-xix.

Joseph, R. 1999. The neurology of traumatic "dissociative" amnesia: Commentary and literature review. Child Abuse & Neglect, 23, 715-727.

oseph, R., 2001. The limbic system and the soul: Evolution and the neuroanatomy of 23/26 religious experience. Zygon, 36, 105-136.

Joseph, R., 2009. Quantum physics and the multiplicity of mind: Split-brains, fragmented minds, dissociation, quantum consciousness. J. Cosmology, 3, 600-640.

Kauffman, S., 2011, Quoted in: God enough. Interview of Stuart Kauffman by Steve Paulson. www.salon.com/env/atoms_eden/w008/11/19/stuart_kauffman/index1.html. Accessed February 15, 2011.

Kelly, E. W., Greyson, B., & Stevenson, I., 1999–2000. Can experiences near death furnish evidence of life after death? Omega, 40, 513-519.

Kelly, E. F., Kelly, E. W., Crabtree, A., Gauld, A., Grosso, M., Greyson, B., 2007. Irreducible Mind: Toward a Psychology for the 21st Century. Rowman and Littlefield, Lanham, MD.

Maddox, J., 1999, The unexpected science to come. Scientific American. Dec. 1999;

281(6):62-7.

McFarland, Jefferson, NC. Stevenson, I., & Keil, J., 2005. Children of Myanmar who behave like Japanese soldiers: a possible third element in personality. Journal of Scientific Exploration, 19, 171-183.

Morris, R. L., Harary, S. B., Janis, J., Hartwell, J., & Roll, W. G., 1978. Studies of communication during out-of-body experiences. Journal of the American Society for Psychical Research, 71, 1-21.

Nadeau, R., Kafatos, M., 1999. Over a distance and in "no time": Bell's Theorem and the Aspect and Gisin experiments, in The Non-Local Universe, Oxford University Press, New York, NY.

Newman, H. H., 1940. Multiple Human Births. Doubleday, Duran, New York.

Osis, K., & McCormick, D., 1980. Kinetic effects at the ostensible location of an out-of-body projection during perceptual testing. Journal of the American Society for Psychical Research, 74, 319-329.

Penrose, R., 2003. Quoted in: Karl Giberson. The man who fell to Earth. Interview with Sir Roger Penrose. Science & Spirit. March/April; 2003: 34-41.]

Pinker, S. (1997). How The Mind Works. New York: Norton. Pizzi ,R, Fantasia, A, Gelain, F, Rossetti, D, & Vescovi, A. , 2004. Non-local correlation between separated human neural networks. In: Donkor, E., Pirick, A.R. & Brandt, H.E. (eds.) Quantum Information and Computation II. Proceedings of SPIE. 107-117. 24/26

Playfair, G. L., 2008. Twin Telepathy. (2nd ed.). History Press, Stroud, Gloucestershire, UK. Radin, D. 1997. The Conscious Universe: The Scientific Truth of Psychic Phenomena. HarperEdge, San Francisco, CA.

Radin, D., 2006. Entangled Minds: Extrasensory Experiences in a Quantum Reality, Paraview Pocket Books, New York, NY.

Ring, K., & Cooper, S., 1999. Mindsight: Near-Death And Out-Of-Body Experiences In The Blind. William James Center, Institute of Transpersonal Psychology, Palo Alto, CA

Ring, K., & Lawrence, M., 1993. Further evidence for veridical perception during near-death experiences. Journal of Near-Death Studies, 11, 223-229.

Rosambeau, M., 1987. How Twins Grow Up. The Bodley Head, London.

Schrödinger E., 1935, Discussion of probability relations between separated systems. Mathematical Proceedings of the Cambridge Philosophical Society, 31, 555-563.

Schrödinger E., 1969. What is Life? and Mind and Matter. Cambridge University Press, Cambridge, 145.

Schrödinger E., 1983. My View of the World. Ox Bow Press, Woodbridge, CT, 31-34.

Schwartz, S.A., & Dossey, L., 2010. Nonlocality, intention, and observer effects in healing studies: laying a foundation for the future, Explore (NY), 6, 295-307.

Segal, N. L., 1999. Entwined Lives: Twins and What They Tell Us About Human Behavior. Dutton, New York.

Sicher, F., Targ, E., Moore, D., & Smith, H. 1998. A Randomized, Double Blind Study of Distant Healing in a Population with Advanced AIDS. Western Journal of Medicine, 169(6), 356-363.

Sidgwick, E. M., 1891. On the evidence for clairvoyance. Proceedings of the Society for Psychical Research, 7, 30-99.

Standish, L., Johnson, L.C., Richards, T., & Kozak, L., 2003. Evidence of correlated functional MRI signals between distant human brains. Alternative Therapies in Health and Medicine, 9, 122-128.

Standish, L., Kozak, L., Johnson, C., Richards, T. 2004. Electroencephalographic 25/26 evidence of correlated event-related signals between the brains of spatially and sensory isolated human subjects. J. Alternative and Complementary Medicine 10, 307-314.

Stevenson, I., 1975. Cases of the Reincarnation Type. Volume 1, Ten Cases in India. Charlottesville, VA: University Press of Virginia.

Stevenson, I., 1997. Reincarnation and Biology: A Contribution to the Etiology of Birthmarks and Birth Defects. Praeger, Westport, CT.

Stevenson, I., 2001. Children Who Remember Previous Lives: A Question of Reincarnation, rev. ed. McFarland, Jefferson, NC. Stevenson, I., 2003. European Cases Of The Reincarnation Type.

Sturrock, P.A., 2009. A Tale of Two Sciences, Exoscience, Palo Alto, CA.

Tart, C. T., 1968. A psychophysiological study of out-of-body experiences in a selected subject. Journal of the American Society for Psychical Research, 62, 3-27.

Tittel, W., Brendel, J., Zbinden, H., & Gisin, N., 1998. Violation of the Bell inequalities by photons more than 10km apart. Physical Review Letters, 81, 3563-3566.

Tucker, J. B., 2005. Life Before Life: A Scientific Investigation of Children's Memories of Previous Lives. St. Martin's Press, New York, NY.

Van Lommel, P., van Wees, R., Meyers, V., & Elfferich, I., 2001. Near-death experiences in survivors of cardiac arrest: A prospective study in the Netherlands. Lancet, 358, 2039-2045.

Wackerman, J, Seiter, C, Keibel, Walach, H., 2003. Correlations between brain electrical activities of two spatially separated human subjects. Neuroscience Letters, 336, 60-64.

Wigner, E. P., 1983. Remarks on the mind-body problem. In: Quantum Theory And Measurement (John Wheeler and Zurek Wojciech, eds.) Princeton University Press, Princeton, NJ, p. 99.

Wilde, O., 1889. The Decay of Lying.

13. Consciousness: Solvable and Unsolvable Problems

Prof. Dr. Etienne Vermeersch

Emeritus Professor, Ghent University, 9000 Ghent, Belgium

Abstract

In this article, I propose three theses concerning consciousness:
(1) "The hard problem", as Chalmers (1996) called it, "the question of how physical processes in the brain give rise to subjective experience" is not a 'mystery'. Instead, its formulation makes a straightforward answer impossible in principle.
(2) Although consciousness has some unique characteristics in the normal adult human being, there is 'pre-consciousness' (PC) present throughout the animal kingdom (PC1, PC2, …PCn), where PC10 would be its highest evolution (in chimpanzees, dolphins…?) before the full-fledged human consciousness (HC)
(3) Neither PC nor HC are 'epiphenomena': they have a real function as PC or HC. In other words, an animal 'zombie' (Searle, 1992, Chalmers, 1996) could never have evolved further than a primitive stage, let alone to a 'zimbo' (Dennett, 1991).

KEY WORDS: Consciousness, *Forms, Emotion, I-system, E-system, information

1. Contents of Consciousness

Contents of consciousness, such as pain, pleasure, anxiety or 'self-awareness', can be linked to neurophysiological processes but the latter can only be 'observed' and described in an 'observer oriented language'. The former, however, are experiences of the internal states of an organism by the organism itself. Colour can be experienced by many organisms to some extent, while guesses can be made about the 'qualities' of the private experiences of others. But the experiences themselves (not their observable aspects) can only be expressed in an 'ego oriented' language. This is obviously the case when these contents of consciousness are exclusive experiences –through internal detectors – of an organism about its internal states. It is contradictory to say that the experience by an organism of its internal states could be had by another organism. How can my typical heart 'angor' be experienced by another person? I can say to myself, "Yesterday, at 10.30 AM, I had this type of feeling". But even the best cardiologist cannot imagine how exactly I felt, let alone feeling that himself. It is obvious that this type of experiences can never be formulated in an 'observer language' characteristic of the natural and social sciences.

1.1. Lack of insight into this inherent property of these types of experiences inclines some people to think that the 'observer approach' is about common 'material things', whereas the 'ego approach' refers to entities of another type. But there is no reason to think that the processes within an organism experienced through its internal detectors are less material than those accessible to external observation. Many discussions about the 'mind-body-problem' suffer from the absence of this elementary distinction between these two types of languages and their mutual untranslatability. Some think that 'subjective' manipulations cannot be realized in material systems; how do they know this? This basic 'insight' seems to depend on the 'principle of non-distinction of the distinct' viz. the distinction between 'observer language' and 'ego language' and the impossibility of translation between them, whether the systems are material or not.

1.2. Even if a complete human-like robot could be constructed, whose humanity could be proved by an extended Turing test, all we could say is that it has emotions and consciousness like us. But that would not enable us to know how it 'feels', any more than we can know that of another person. I repeat: the inner experiences of a person, whether biological or man-made, are experiences accessible only to herself. Of course, we believe that only human beings and some animals appear capable of experiencing internal states and we cannot prove in detail that they are strictly material systems. We have to wait until completely material man-made robots tell us that they have such experiences. Denying that they are full-fledged persons would lead us to negate the existence of their 'other minds' and that amounts to extreme solipsism or anti-machine racism. In this sense, the solution to Chalmers' problem sounds somewhat like the answer to: "how does life function in a dead organism?"

2. A Novel Theory

Before taking up the other two theses, I propose a rather novel theory that might make further discussions more convincing.

2.1.1. First, some basics about my world view. (a) In the whole universe, next to space/time, there is only E/M: things with mass or energy as referred to in Einstein's equation ($E = Mc^2$). (b) E/M can be measured variously by a variety of instruments. The differences lie in the different 'states' of E/M substrates. (c) Human knowledge cannot describe all particular states; it refers mostly to sets (classes) of states, sets of sets, etc. Without such a hierarchy of sets, our world would be a chaos and useful actions would be impossible. (d) To introduce organisation into this chaos, we use what I call I-systems (information processing systems). The sim- plest type has an entry which can detect different states (the input) of an E/M-substrate (e.g. different sound waves) and an exit which can exhibit different states (the output) of an EM-substrate (different light waves). The essential property of a simple I-system is that it systematically realizes a one-to-one correspondence between sets of input states and sets of output states.

2.1.2. We can generalize this elementary idea. Every system that systematically realizes an isomorphism between subsets of states of the input and subsets of states of the output is an I-system. One can also say that such a system identifies ('considers' as equivalent) these subsets of states of the input and discriminates them from other subsets of states.

2.1.3. The basic definition of this article is the following: a set of states of an E/M substrate that is identified and discriminated from other sets of states is a *form. Generally speaking, (to be qualified later) such discrimination and identification is realized by an I-system.

Where this definition is meant, a star * is added to the term 'form'. Further: (i) When there is isomorphism between the *forms of input and output, some E/M manipulation ('transformation') on the substrates is implied in most cases. (*Forms of sound waves may be mapped onto *forms of electric current etc.) (ii) *Forms can travel from one I-system to another, since output *forms of the first can be input *forms for the second. etc. (iii) I-systems can contain internal I-systems, whose existence can be deduced from its overt behaviour. (iv) Using different identification and discrimination criteria, the input of an I-system can identify different types of *forms, (v) Subsystems of a total system are mostly interlinked in different ways.

2.1.4. Here is the basic postulate of this '*form theory': notions like 'form', 'shape', 'figure', 'configuration', 'type', 'structure', 'design', 'melody', 'fragrance', 'meme' etc. can be defined as a '*form. More importantly, whatever is called 'a signal', 'a sign', 'a message', ' a stimulus' and especially a unit of 'information', an item of 'perception' or a piece of 'knowledge', can be rigorously defined as a *form.

2.1.5. The case of 'information' is rather special, since this is the first real, physically de- scribed and consistently applicable definition of the term. (Compare in The Stanford Encyclopedia of Philosophy for instance: "information": 767 entries, "information AND definition": 456 entries, "information AND concept": 541 entries.)

(i) Shannon's theory, for instance, is not about "information"; it provides a measure of the 'amount of information' of a *form (the way ph is a measure of acidity of a substance). At the source, this measure depends on the probability of a particular transmitted *form. Regarding the receiver, it refers to the uncertainty about the incoming *form.

(ii) There is no such thing as the 'content' of information'. Of course, *forms may have dif- ferent connotations and functions, but these must be studied by analysing the relevant I- systems.

(iii) Since *forms are sets of states of an EM substrate identified by I-systems, sets of *forms identified as such are also *forms. Hence, one can introduce an indefinite number of *forms at different levels.

2.1.6. It follows that there are no *forms without EM-substrates and I-systems to identify sets of states of them. There is one important exception, though.

2.2.1. One of the most remarkable, perhaps the most remarkable event in the history of the universe, is the fact that *forms detected by I-systems in the 20th century did in fact emerge more than three billion years ago as sets of states of an E/M substrate that could generate other sets of states of an E/M substrate with which they were isomorphic. These, in turn, produce equally isomorphic sets of states. In other words, *forms become capable of replicating themselves without the interference of an I-system or any other external system.

They are not only *forms for us but also *forms in their own right. This event, of course was the origin of life. But since DNA chains are rather well known, I will focus on the special role of the different types of phenotypic *forms in this development.

2.2.2. Suppose that, somewhere in the Cambrian sea, there is a worm-like animal - a tube with two openings through which water with nutrients can flow in and out. Suppose further that it can sexually reproduce. When the distribution of the density of nutrients is not homogeneous, Variation and Selection (VS) will favour the development of locomotion to increase the probability of contact with nutrients. But a random movement uses a lot of energy. VS will thus also favour the development of detection systems to locate nutrients. Say that these are characterized by degrees of acidity. If a chemical detector of a ph-*form initiates and directs its locomotion system to the nutrients, our tube has an I-system. Other types of sensors could also emerge: tactile, sound, light, etc. Their usefulness depends on the existence of a non-homogeneous distribution of M/E substrates: a subset (a *form) of these is typical for the nutrient of our tube. Since the detection of *forms saves energy, VS will cause our tube to develop a wide range of complex I-systems to detect *forms of smell, taste, touch, sound, light etc. Those familiar with ethology will recognize that its 'releasers' are *forms of input and that its "innate motor patterns" are *forms of output. The "innate releasing mechanism" is the E/M transformation mechanism which realises the isomorphism. Many such I-systems studied by ethologists are innate, but it is clear that VS will tend to develop systems capable of learning: activities essential for survival are made adaptable to different situations.

Throughout the animal world, the basic needs and activities related to survival have to do with a search for nutrients, avoidance of danger (predators or poison) and mating. In the 'struggle for life' concerning these needs, there can be no question about the advantage of the existence of a wide variety of efficient I-systems. When environments are variable and complex, this is even more the case. At some levels, the innate input-output links might no longer be efficient. Then, through feedback loops, VS will tend to introduce and link a variety of input *forms to more variable output *forms. This could happen through the linking of sets of neural complexes (sets of internal *forms) to different input *forms. Even the whole set of *forms of visual inputs could be isomorphic with the set of 'visual' neural *forms. That would bring about a kind of internal neural model of the visual world of the animal. Similar subsystems could be developed for auditory, olfactory, tactile inputs.

The advantage of such internal sets of *forms is obvious when the input *forms become unreliable because of darkness, noise, etc. It is self-evident that, in primitive stages, these internal 'models' are not detailed 'representations'; the *forms detect only aspects of the en- vironment relevant to the basic needs and activities mentioned above. These internal models should also be linked to the diverse output *forms (directing movements; sending warning signals). An even greater advantage is achieved when these models have feedback loops; the auditory *form of a roar could evoke a response in the neural (visual) *form of the lion. I cannot expatiate here on the possible extensions of the identification capacities of an ever growing inter-linked I-system. A more detailed and coherent general picture of the Umwelt might emerge. Equally, the organism as a whole would be tightly unified, whose adaptive value is evident: a coordination of activities that guarantee the fulfilment of basic needs (food, mating, danger) is essential for survival.

2.2.3. I must emphasize that the theoretical distinction between the different types of I- systems, and their external and internal inputs and outputs, including the links between them, does not imply that they can always be observed within the nervous system as clearly distinct components. These 'functional units' may result from the interaction of neuron complexes in different parts of the brain. A visual or auditory I-system refer to *forms as I define them. Thus, the E/M basis remains essential and they explain how a chaotic whole can control our actions.

2.2.4. To my astonishment, however, in many general textbooks on biology, one extremely important subsystem is rarely mentioned. A link between, say, the visual system and the motor system is useless, if there is no system capable of mobilizing, intensifying, inhibiting these and other systems and, when required, switching from one system to another.

When an animal is saturated with food, its I-systems should be used for mating rather than eating. If it detects a *form linked to danger while eating or mating, surely, the escape (or aggression) behaviour should be compulsorily mobilized. There is no point to eating or mating if you die before the aim is attained.

To regulate the interaction of the different needs (or drives), a special I-system is needed. That is the E-system (emotional) which manages the hierarchy of types of behaviour. It coordinates all important subsystems to produce the type of required behaviour in that situation. When the hunger drive is at rest, mating can play a central role; but when a danger *form I is detected, the animal as a whole has to produce the appropriate response. Complex animals can survive only when they are totally engaged in the most appropriate activity at any given moment.

2.2.5. As all other *forms, these coordinated *forms are also identified sets of states of E/M substrates. But it is remarkable that what human beings call feelings or emotions (hunger, thirst, eating and sexual pleasure, disgust, pain, fear, anger…) occur in behaviour situations connected to the basic needs of animals in general. They not only occur in the same situations but, in many cases, they also have the same effect. The burning of the skin, which is a nociceptive stimulus for an animal and a pain stimulus for us, leads to the same reaction. The *form of a tiger causes escape behaviour in animals and human beings; sometimes this is accompanied with what we interpret as signs of 'anxiety'.

Of course, the degree of complexity of the subsystems of different animals varies. So does the degree of integration of their various subsystems. But I submit that the E-system contributes to the increasing unification and coordination of the activities of the whole system. My thesis is that the more the whole organism reacts as a unit ('pleasure' of eating, 'lust' of 'mating', 'anxiety' of fleeing), the more does this unity bring about a specific state, which is the most primitive form what we call 'emotions'. What I would call PC1…PC3…PCn types of consciousness are in reality degrees of coherence and coordination in some or in many situations. To what degree this unification is also experienced by the animal as a state of itself is difficult to know. Anyhow, next to the visual, auditory, etc. models of animals, human beings also have a model of *forms where our linguistic capabilities function. It is complex, but it consists nevertheless of sets of sets of *forms linked to the other systems including the E-system. Within that linguistic model, thanks to the properties of our languages, metareflections are possible. Hence there is something special to human consciousness. The awareness of the unity of the organism is considerably enhanced, including its past and future. This may lead to a more intense form of emotions and consciousness. But the very fact that we can refer to the development and intensification of a number of linked characteristics in the course of evolution, seems to make the hypothesis of pre-human states of pre-consciousness, PC1, PC2… PC9 rather plausible. The idea that a special soul is needed to become human beings is obsolete. Why would such an unintelligible leap be necessary in the case of emotions and consciousness?

3. Emotions and I-Awareness

Full-fledged emotions and accompanying general moods are typical for human beings and, as I explained above, we will never be able to have contact with their inner direct I- awareness. The same problem exists in an even more intense way as far as animals are con- cerned. But, since there can be no doubt about the causal survival value of the E-system in animals, there is no reason to consider its function in human beings as an epiphenomenon.

References

Chalmers, D., (1996), The Conscious Mind. Oxford: Oxford University Press.

Dennett, D., (1991), Consciousness Explained. Harmondsworth: Allen Lane.

Searle, J., (1992), The Rediscovery of Mind. Massachusetts: The MIT Press.

Vermeersch, E., (1977), "An analysis of the concept of culture." In Bernardi, Bernardo, ed., The Concept and Dynamics of Culture, World Anthropology, The Hague: Mouton, , pp. 9- 73.

14. Decoding the Chalmers Hard Problem of Consciousness: Qualia of the Molecular Biology of Creativity and Thought

Ernest Lawrence Rossi[1] & Kathryn Lane Rossi[2]

Professors of Neuroscience[1] and Psychotherapy[2], La Nuova Scuola Di Neuroscienze Ipnosi Terapeutica (The New School of Therapeutic Hypnosis) and l'Istituto Mente-Corpo (The Mind-Body Institute), San Lorenzo Maggiore, Italy. [1,2] Directors of The Milton H. Erickson Institute of California Central Coast, 125 Howard Ave., Los Osos, CA, 93402, USA.

Abstract

This update and decoding of the Chalmers hard problem of consciousness integrates philosophy, psychology and the humanities with current evolutionary concepts of molecular biology. We summarize research that suggests how the natural variation and selection of molecular processes in primordial RNA world evolved into RNA/DNA transcription/translational dialogues with the qualia of human experience in rhythms of performance, stress and healing today. Pilot research on a new creative psychosocial genomic healing protocol, which activates the bioinformatics of qualia-dependent gene expression associated with stem cell healing and a reduction in chronic inflammation and cellular oxidation stress, is now a priority for translational research on human health and well being.

KEY WORDS: Chalmers, consciousness, DNA microarrays, dreams, evolution, experience-dependent gene expression, psychotherapy, qualia, RNA world.

1. Introduction

This paper reviews Chalmers (1996) "Hard Problem of Consciousness" that was originally formulated as follows:

"The really hard problem is the problem of experience. When we think and perceive, there is a whir of information processing, but there is also a subjective aspect... Why should physical processing give rise to a rich inner life at all? ... Here, the topic is clearly the hard problem – the problem of experience... What makes the hard problem hard and almost unique is that it goes beyond problems about the performance of functions... Why is the performance of these functions accompanied by experience?... There is an explanatory gap between functions and experience, and we need an explanatory bridge to cross it. ..A full theory of consciousness must build an explanatory bridge... To account for conscious experience, we need an extra ingredient in the explanation ... What is your extra ingredient, and why should that account for conscious experience?" (pp. 6-13).

Throughout this paper we will note the "extra ingredients" we are contributing to bridge the explanatory gap between biological functions and the qualia of psychological experience for a more satisfactory theory of consciousness to guide further research.

2. Darwin's Daily & Hourly Work of Evolution

We introduce a new perspective on bridging the explanatory gap between the qualia of consciousness and their biological functions by quoting what Darwin wrote about the daily and hourly operation of evolution in chapter four of The Origin of Species.

> "It may be said that natural selection is a daily and hourly scrutinizing, throughout the world, every variation, even the slightest; rejecting that which is bad, preserving and adding up all that is good; silently and insensibly working, whenever and wherever opportunity offers, at the improvement of each organic being in relation to its organic and inorganic conditions of life. We see nothing of these slow changes in progress, until the hand of time has marked the long lapses of ages, and then so imperfect is our view into long past geological ages, that we only see that the forms of life are now different from what they formerly were." (Italics added here)

We take this apparently casual, intuitive and highly speculative comment by Darwin very seriously because it helped us catch a glimpse of the natural time parameters of the molecular-genomic dynamics underpinning the qualia of consciousness, healing, and problem solving that we witness in our daily and hourly work with our patients in psychotherapy (Lloyd & Rossi, 1992, 2008;

Rossi & Nimmons 1992). Figure 1 illustrates how an evolutionary co-creative complex adaptive system that includes culture, qualia, eRNA, genes and the brain can move molecules to facilitate new conscious experiences.

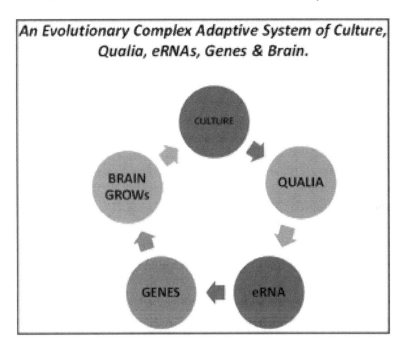

Figure 1. The evolutionary complex adaptive system of culture, qualia, eRNA, genes & brain growth.

In brief, current epigenetic research explores why and how qualia of experience "dialogues" with the DNA of our experience-dependent gene expression via the signaling functions of RNA molecules (Mattick, 2010; Yarus, 2010). We propose that novel qualia of human experience contribute to the adaptive updating of consciousness on the molecular-genomic RNA/DNA level by modulating brain plasticity via natural variation and selection daily and hourly (Rossi, 1972, 1985, 2000, 2002, 2004, 2007, 2011). We support this proposal in the following sections by reviewing research on this adaptive function of qualia in the daily and hourly cycles of human experience from the quantum to the cognitive-behavioral levels.

3. Evolutionary Complex Adaptive Systems of Consciousness via Quantum Entanglement

Consciousness and non-locality in quantum physics and psychology have generated profound quandary seeking resolution (Rossi, 1988 a, b, c). Rapparini (2010), for example, outlined the evolution of complex adaptive systems of the qualia of consciousness via quantum entanglement as follows:

"In the course of an evolution process begun with inorganic chemical compounds and proceeding in steps of ever increasing complexity, life, thought and consciousness have emerged as a bonus that cannot be explained on a purely computational basis. Because in quantum physics we no longer have observables representing the ontology of the world, but only observations representing epistemic knowledge, it implies a psychological dimension which so far has been neglected. It is therefore necessary, in order to complete the picture, to introduce the psychological dimension best represented in the western tradition by Jung and by Buddhist practices in the East. Contrary to Buddhist tradition however, the brain not the mind, is considered here a sensory organ on a par with the other five. In fact, because of quantum entanglement between observer and the observed, it is possible to state that the observer (the brain) is part of the physical world (the observed) whose representations (qualia), as generated by the brain includes the representation of the brain itself. This leads to an innovative definition of qualia and their role in the emergence of consciousness: the mind is the qualia of the brain's neural mechanisms, this is how the perceiver perceives himself, from within. This is how consciousness emerges (Rapparini, 2010, p. 169). This is what qualia are for: the objectivity of cognition as obtained from the subjectivity of feelings of what happens... The closing of the explanatory gap makes it possible for the mind to change the brain by self-directed neuroplasticity in agreement with the Buddhist belief on the mental power of meditation through bare attention. Finally, the concept of entanglement has been instrumental in reaching a definition of consciousness based on the innovative role of qualia." (Rapparini, 2010, p. 173).

This leads to the co-evolution of complex adaptive systems of life and the qualia of consciousness from the primordial RNA world over 4 billion years ago (Atkins, Gesteland, & Cech, 2011).

4. Qualia of Complex Adaptive Molecular Systems of Consciousness via RNA World

How and why do the complex adaptive systems (Mitchell, 2009) of culture, qualia, eRNAs, genes and brain plasticity co-evolve? Kim et al. (2010) recently described the identification of a new class of 12,000 enhancer Ribonucleic Acid molecules (eRNAs) that are involved in regulating gene expression during neuronal activity in mouse cortical brain tissue (Ren, 2010). The significance of this research as an "extra ingredient" for bridging the explanatory gap between the novel qualia of consciousness and their molecular-genomic infrastructure became apparent in related research on the genome of the zebra finch described as follows by Warren, Clayton et al. (2010. p. 758).

"The zebra finch is an important model organism in several fields with unique relevance to human neuroscience. Like other songbirds, the zebra finch communicates through learned vocalizations, an ability otherwise documented only in humans and a few other animals . . . We show that song behavior engages gene regulatory networks in the zebra finch brain, altering the expression of long non-coding RNAs, microRNAs, transcription factors and their targets. We also show evidence for rapid molecular evolution in the songbird lineage of genes that are regulated during song experience. These results indicate an active involvement of the genome in neural processes underlying vocal communication and identify potential genetic substrates for the evolution and regulation of this behavior."

Figure 2. Micro eRNAs respond to thought by modulating transcription/translation via qualia-dependent gene expression.

Clayton, one of the co-authors made the salient comment, "this is the first time a microRNA has been shown to respond to a particular thought process" (Saey, 2010). We now propose eRNAs are an "extra ingredient" mediating between the novel qualia of activity or experience-dependent gene expression that underpins brain plasticity and consciousness (Kempermann, 2006; Van Pragg et al., 1999). Consciousness is a novelty-seeking modality that evolved as a sensitive detector or qualia to facilitate rapid and creative adaptation to environments manifesting constant change with natural variation and selection (Rossi, 2002, p.135). Molecular biologists believe that "Cells are masters of regulating genes in response to environmental cues." (Liu & Arkin, 2010, p. 1185). Culler et al. (2010, p. 1251) recently reported a profound proof-of-principle experiment wherein they showed how RNAs function as "sensing-actuation devices" transmitting the qualia of information from the environment to modulate the gene expression of DNA within cells as follows.

"Cellular decisions, such as differentiation, response to stress, disease progression, and apoptosis, depend upon regulatory networks that control enzymatic activities, protein translocation, and genetic responses. Central to the genetic programming of biological systems is the ability to process information within cellular networks and link this information to new cellular behaviors, in essence rewiring network topologies... RNA is a promising substrate for platforms to interface with cellular networks because of the versatile sensing and actuation functions that RNA can exhibit and the ease with which RNA structures can be designed. RNA-based sensing-actuation devices have been engineered that respond predominantly to externally [environmental] applied small-molecule and nucleic acid inputs and control gene expression through diverse mechanisms." (p. 1251, italics added here).

It appears that sensitive and fragile "RNA-based sensing-actuation devices" were the primordial molecular qualia signaling epigenetic information from the environment of early earth to the more stable DNA molecule. One possibility is that DNA may have been a mutation in RNA world that began to function as memory molecule which eventually made natural evolution possible from one generation to the next possible. We propose that this adaptive coordination between the sensing (qualia), signaling, and catalytic self-replicating properties of RNA (with A-U, G-C base pairings) interacting with the more stable memory properties of DNA (due to A-T, G-C base pairing) was the original bridge over the explanatory gap between the molecular-genomic qualia and functions of life and consciousness. Wang et al. (2011, pp. 279 & 289) expressed it in this way:

> "A major surprise arising from genome-wide analyses has been the observation that the majority of the genome is transcribed, generating non-coding RNAs (ncRNAs). It is still an open question whether some or all of these ncRNAs constitute functional networks regulating gene transcription programs. However, in the light of recent discoveries and given the diversity and flexibility of long ncRNAs ... it becomes likely that many or most ncRNAs act as sensors and integrators of a wide variety of regulated transcriptional responses and probably epigenetic events... Together, the ncRNA sensor code appears to be a robust and critical strategy underlying a wide variety of gene regulatory programs."

If we are willing to take a philosophical, linguistic and quantum leap from "ncRNA sensor code" to "ncRNA qualia code" such research could be another "extra ingredient" in our evolving theory of the origin of life and the qualia consciousness via the dynamics of RNA/DNA coordination during transcription and translation. Current research implies that the more fragile but versatile molecular RNA signaling software of RNA world became integrated with the more stable DNA memory hardware to initiate the evolution of life as we know it. It is interesting to note that when Gilbert (1986) first introduced the concept of "RNA world" in a bottoms-up approach to the evolution of life and mind, Rossi (1986a, 1986b) began exploring such research from a top-down approach. A comprehensive theory, of course, requires an integration of both approaches to fill in the qualia gaps that remain between them. In the following sections we review the incredibly wide range of qualia-dependent molecular-genomic processes that underpin the deep psychobiological rhythms of human consciousness and experience.

5. Qualia of Consciousness during the Basic Rest-Activity Cycle (BRAC)

The top of figure 3 illustrates the 90-120 minute Basic Rest-Activity Cycle (BRAC) of human performance that is the chronobiological foundation of many of the ultradian rhythms (less than the 24 hour circadian cycle) on many levels (Lloyd & Rossi, 1992, 2008). The lower part of Figure 3 illustrates the selection-amplification SELEX (SElection of Ligands by EXponential amplification) cycle in modern experiments of the possible evolution of RNA world (Yarus, 2010). We propose the SELEX experiments as molecular-genomic analogues of the evolution of the BRAC. The qualia of the BRAC are the phenotypes of the RNA/DNA genotypes of transcription and translation.

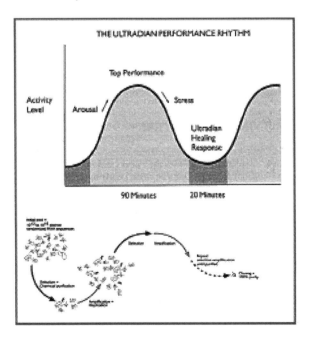

Figure 3. The Basic Rest-Activity Cycle (BRAC) and ultradian rhythms of human performance & stress (top part) analogous to the SELEX RNA dynamics of selection and amplification (bottom part).

The contrast between the qualia of the BRAC as typically experienced during the Ultradian Healing Response, when people take appropriate rest-breaks every 90-120 minutes or so throughout the day and the qualia of the Ultradian Stress Response; when people chronically attempt to forego taking appropriate rest periods is presented in table 1 (Rossi, 1996; Rossi & Nimmons, 1991).

Table 1. The contrast between the qualia experienced during the ultradian healing response, when people take appropriate breaks throughout the day, versus the ultradian stress response leading to behavioral, cognitive and emotional problems or psychosomatic symptoms when they do not take appropriate healing breaks throughout the day.

THE ULTRADIAN HEALING RESPONSE	THE ULTRADIAN STRESS SYNDROME
1. Recognition Signals: An acceptance of nature's call for your need to rest and recover your strength and well-being leads you into an experience of comfort and thankfulness.	1. Take-a-Break Signals: A rejection of nature's call for your need to rest and recover your strength and well-being leads you into an experience of stress and fatigue.
2. Accessing the Deeper Breath: A spontaneous deeper breath comes all by itself after a few moments of rest as a signal that you are slipping into a deeper state of relaxation and healing. Explore the deepening feeling of comfort that comes spontaneously. Wonder about the possibilities of mind-gene communication and healing with an attitude of "dispassionate compassion."	2. High on your Hormones: Continuing effort in the face of fatigue leads to the release of stress hormones that short-circuits the need for ultradian rest. Performance goes up briefly at the expense of hidden wear and tear so that you fall into further stress and a need for artificial stimulants (caffeine, nicotine, alcohol, cocaine, etc.).
3. Mind-Body Healing: Spontaneous fantasy, memory, feeling-toned complexes, active imagination, and numinous states of being are orchestrated for healing and life reframing.	3. Malfunction Junction: Many mistakes creep into your performance, memory, and learning; emotional problems become manifest. You may become depressed or irritable and abusive to yourself and others.
4. Rejuvenation and Awakening: A natural awakening with feelings of serenity, clarity, and healing together with a sense of how you will enhance your performance and well-being in the world.	4. The Rebellious Body: Classical psychodynamic symptoms now intrude so that you finally have to stop and rest. You are left with a nagging sense of failure, depression and illness.

It appears as if the art and quality of life is determined by how well we negotiate the qualia-dependent molecular-genomic processes of these alternations between performance, stress and healing in the arts, humanities and sciences as well as the work and play of everyday life.

6. Qualia of Consciousness during the 4-Stage Creative Cycle

A cartoon of the qualia of consciousness as experienced during the 4-stage creative cycle in the arts, sciences and everyday life is illustrated in figure 4.

Quantum Physics, Evolution, Brain, Mind

Figure 4. A cartoon of the Qualia of the 4-Stage Creative Process (Tomlin, 2005, with permission).

A profile of qualia of the 4-stage creative process: initiation, incubation, aha!, and verification during the 90-120 minute BRAC experienced 12 times a day while awake, sleeping and dreaming is illustrated in figure 5.

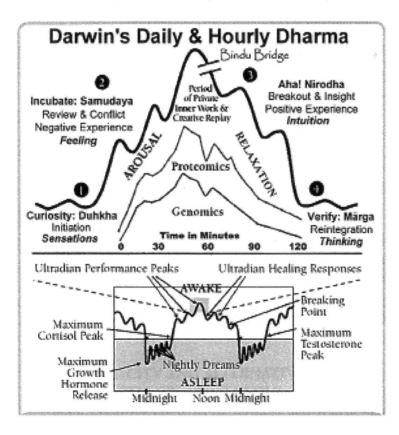

Figure 5. Qualia of Darwin's daily and hourly work of the 4-Stage Creative Cycle in one 90-120 minute basic rest-activity cycle (top) of the ~24 hour circadian cycle (bottom). The Sanskrit terms represent

the qualia of Buddha's Four Nobel Truths. The proteomics (protein) profile in middle curve depicts the energy landscape for protein folding within neurons of the brain (Cheung et al., 2004). The functional concordance of co-expressed genes are illustrated by the genomics profile below it (Levsky et al., 2002). The lower diagram illustrates how the qualia of human experienced are typically experienced as the BRAC within the circadian cycle of waking and sleeping.

Table 2 outlines creative experiences that turn on qualia-dependent gene expression and brain plasticity, which we call, "the Novelty-Numinosum-Neurogenesis Effect," from two complementary perspectives: the bottoms-up approach of neuroscience (Rossi, 2002, 2007, 2011) and the top-down approach of heightened states of purported spiritual experience that are called, "the numinosum" (Otto, 1923; Jung, 1958).

Table 2. Three qualia associated with experience-dependent gene expression and brain plasticity by neuroscience research and the corresponding three qualia experienced during the heightened states of consciousness associated with purported spiritual experience.

Qualia of Human Experience that Activate The Novelty-Numinosum-Neurogenesis Effect	
Neuroscience	Numinosum
• Activity • Novelty • Enrichment	• Fascination • Mysterious • Tremendous
Kempermann, 2006; Ribeiro et al., 2008	Otto, 1923; Jung, 1958

Recent research reveals many possibilities for exploring the molecular-genomic evolution of the qualia of the 4-stage creative cycle. Ramakrishana (2011), for example, presents an overview of the translation pathway of the eukaryote cell in 4 stages: initiation, elongation, release, and recycling that looks remarkably similar to the dynamics of 4-stage creative cycle. Further, within the process of transcription/translation, Breaker (2011) describes how riboswitches regulate experience-dependent gene expression to form Boolean logical gates, which could function as the molecular infrastructure for the qualia of creativity and thought.

"The term riboswitch was established to define RNAs that control gene expression by binding metabolites without the need for protein factors. More recently, the name has begun to be used for riboswitch-like RNAs that respond to temperature... Riboswitches need to form molecular architectures with sufficient complexity to carry out two main functions:

molecular recognition and conformational switching. Simple riboswitches each carry one aptamer that senses a single ligand and one expression platform that usually controls gene expression via a single mechanism. . . Although simple riboswitches only respond to one ligand type, this restriction in signaling complexity can be overcome by stacking tandem riboswitches from different ligand-binding classes such that gene expression is responsive to more than one chemical signal. Indeed, a natural example of such two-input Boolean logic gate has been observed . . . the tandem arrangement functions as Boolean NOR gate . . . integration of multiple aptamers . . .functions as a Boolean AND gate" (p. 64 -75).

Orchestrating RNA/DNA transcription and translation via riboswitches forming "Boolean logic gates" is another extra ingredient" spanning the explanatory gap for decoding the Chalmers hard problem of consciousness. Riboswitch Boolean logic gates could be the molecular-genomic basis of mind-body computation, creativity and thought that takes place in every cell of the brain and body. Further research could now assess how the binding kinetics of simple and complex riboswitches (Breaker, 2011, p. 72) could account for the scalloped fractal dynamics of the basic rest-activity cycle illustrated in figure 5.

7. Qualia of the Molecular Biology of Memory & Learning during Sleep and Dreaming

Our psychosocial genomic model of creative psychotherapy that emulates the natural process whereby novel qualia experienced in our waking hours induce mind-brain-gene dialogues during slow wave sleep and REM state dreaming in Ribeiro's evolutionary theory of sleep and dreaming illustrated in figure 6.

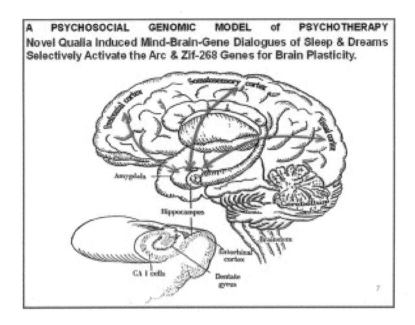

Figure 6. A functional psychosocial genomic approach to creative psychotherapy modeled on the natural dynamics of qualia-dependent gene expression during REM dreaming.

The unexpected role of the qualia in the molecular biology of sleep and dreaming was outlined as follows (Rossi, Erickson-Klein, Rossi, 2008, pp. 344-345):

"The central hypothesis of Ribeiro's (2004) evolutionary theory of sleep and dreaming is that dreams are probabilistic simulations of past events and future expectations. The adaptive function of such simulations is to construct and explore novel behaviors for future survival. A salient function of dreams is to utilize memories processed during the circadian cycle of waking, sleeping, and dreaming for the creation, selection and generalization of adaptive scenarios about the world. Ribeiro et al. (1999, 2002, 2004, 2008) provide extensive details about what they call the "cognitive role" of focusing activity-dependent gene expression and brain plasticity for adaptive behavior during the two major phases of sleep. This theory proposes that the first phase of slow-wave (SW) sleep evolved from rest in early reptiles as a quiescent, "offline state" suitable for the consolidation of new memory and learning. Consistent with much current neuroscience research, these researchers believe that this cognitive role takes place through the reverberation of novel waking patterns of neuronal activity during SW-sleep. The second major phase of sleep, rapid-eye-movement (REM) dreaming, which is characterized by heightened cerebral activity, first evolved in early birds and mammals as a post SW-sleep state that was capable of facilitating memory consolidation by activating gene expression to make the proteins needed for generating the activity-dependent synaptic plasticity of neurons, that became the neural correlates of adaptive behavior. Mammals then evolved extended REM states of dreaming to prolong neuronal reverberation in novel ways that could promote memory reconstruction in a behaviorally adaptive manner rather than mere rote record of past events. In brief, sleep and dreaming became an inner stage for integrating past events with current novel experiences to simulate and creatively replay the present as a rehearsal for future adaptive behavior."

The intensity of important life turning points (adolescence, marriage, divorce, trauma, war etc.) are associated with vivid dreaming that activates qualia-dependent gene expression and brain plasticity illustrated in figure 7.

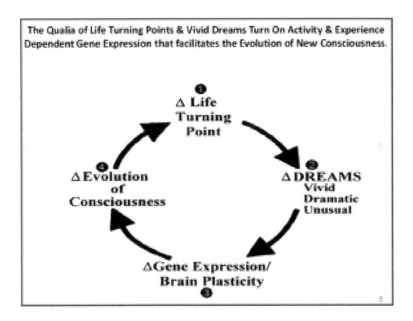

Figure 7. Significant life turning points that activate vivid, dramatic and unusual dreams associated with qualia-dependent gene expression and brain plasticity.

These insights into the role of qualia during memory, learning, sleep and dreaming motivated us to construct a new model of evidence-based psychotherapy, which we call "The Creative Psychosocial Genomic Healing Experience."

8. Qualia of the Creative Psychosocial Genomics Healing Experience (CPGHE)

An overview of our new protocol for facilitating the "Creative Psychosocial

Genomics Healing Experience" in psychotherapy is illustrated in figure 8.

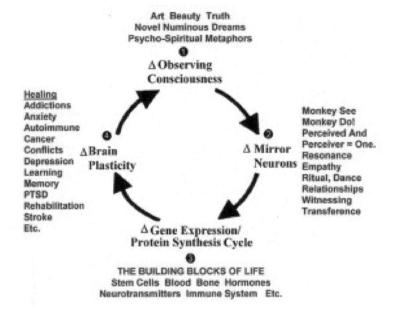

Figure 8. An overview of the Creative Psychosocial Genomic Healing Experience (CPGHE).

We used DNA microarrays in a pilot study to explore the molecular-genomic underpinning of the qualia of experience associated with mind-body healing and problem solving (Rossi et al., 2008; Atkinson et al., 2010). We hypothesized that a top-down creatively oriented experience could modulate qualia-dependent gene expression associated with memory and learning as well as wide range of brain plasticity and psychoneuroimmune effects such as the healing placebo.

A DNA microarray analysis of the white blood cells of human subjects was performed immediately before, within one hour after, and 24 hours after being administered the Creative Psychosocial Genomic Healing Experience (CPGHE). The rationale, administration, and scoring of this new protocol now is freely available at "http://www.ernestrossi.com/ernestrossi/Neuroscienceresearchgroup.html. We found that the qualia of the CPGHE are associated with (1) a molecular-genomic signature for the up-regulation (heightened activity) of genes characteristic of stem cell growth, (2) a reduction in cellular oxidative stress, and (3) a reduction in chronic inflammation as illustrated in figures 9 and 10.

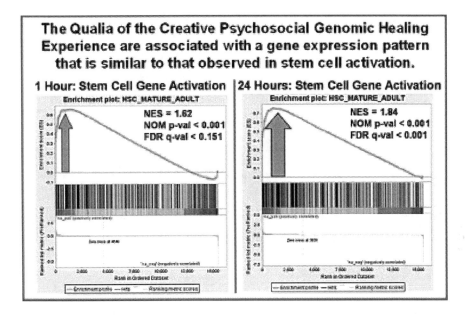

Figure 9. The Creative Psychosocial Genomic Healing Experience associated with qualia-dependent gene expression similar to stem cell healing. What appears to be a bar code are genes that are expressed during DNA microarray experiments. The up-regulation of gene expression generated by the CPGHE is emphasized by the large upward pointing arrow.

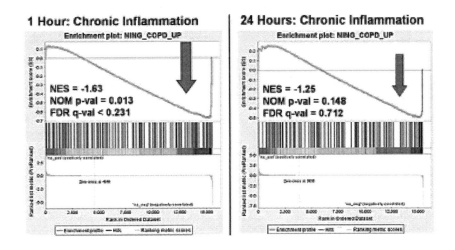

Figure 10. The CPGHE associated with the qualia-dependent gene expression that is the opposite of chronic inflammation and cellular oxidative stress is emphasized by the large downward pointing arrow.

These three empirical associations are an initial beta version of the molecular-genomic signature of the qualia of the CPGHE that requires further replication. These results have been partially confirmed in related research using DNA microarrays to compare qualia-dependent gene expression in beginning and advanced students of meditation. It was found that the "relaxation response" associated with meditation also reduces chronic inflammation and oxidative cellular stress (Dusek et al., 2010).

9. Summary

This update and decoding of the Chalmers Hard Problem of Consciousness has skirted the precipices of metaphysics in seeking what is currently known about qualia of the molecular biology of life and mind that are ready for further experimental evaluation.

We reviewed research on the evolutionary continuum of RNA world from quantum entanglement to the qualia of consciousness during the 90-120 minute basic rest-activity cycle of the 4-stage creative process in the chronobiology of performance and stress.

The epigenetic and informational dynamics of riboswitches that form "Boolean logic gates" for mediating adaptive changes in behavior and consciousness via RNA/DNA transcription/translation suggests how decoding the Chalmers Hard Problem of Consciousness could generate the foundations for a new science of the qualia of the molecular biology of creativity and thought.

There is experimental evidence for the role of novel qualia-dependent gene expression and brain plasticity encoding adaptive memory and learning while awake and during REM dreaming. This motivated our pilot study of the Creative Psychosocial Genomic Healing Experience as a new model for evidence-based molecular-genomic research on psychotherapy.

Replicating research on how the Creative Psychosocial Genomic Healing Experience facilitates qualia-dependent gene expression associated with the up-regulation of (1) stem cell activation, (2) the down-regulation of chronic inflammation and (3) oxidative stress is now a priority for translational research on human health and well being.

References

Atkins, J., Gesteland, R., & Cech. (2011). RNA Worlds: From Life's Origins to Diversity in Gene Regulation. Cold Spring Harbor Laboratory Press, Cold Spring Harbor, N.Y.

Atkinson, D., Rossi, E. et al. (2010). A New Bioinformatics Paradigm for the Theory, Research, and Practice of Therapeutic Hypnosis. American Journal of Clinical Hypnosis, 53 (1).

Breaker, R. (2011). Riboswitches and the RNA World. In Atkins, J., Gesteland, R., & Cech (Eds.). RNA Worlds. Cold Spring Harbor Laboratory Press, Cold Spring Harbor, N.Y., 63-77.

Chalmers, D. (1996). Facing Up to the Problem of Consciousness. In Hameroff, S., Kaszniak, A., Scott, A. (Eds.) Toward a Science of Consciousness: The First Tucson Discussions and Debates. The MIT Press, Cambridge, 7-28.

Cheung, M., Chavez, L., and Onuchic, J. (2004). The energy landscape for protein folding and possible connections to function. Polymer, 45, 547–555.

Culler, S., Hoff, K., & Smolke, C. (2010). Reprogramming Cellular Behavior with RNA Controllers Responsive to Endogenous Proteins. Science, 330, 1251-1255.

Dusek, J., Otu, H., Wohlhueter, A., Bhasin, M., Zerbini, L., Joseph, M., Benson, H. & Libermann, T. (2010). Genomic counter-stress changes induced by the relaxation response. PLoS ONE, 3(7), e2576. doi:10.1371/journal.pone.0002576.

Erickson, M. (1964/2008). The Burden of Responsibility in Effective Psychotherapy. In Rossi, E., Erickson-Klein, R. & Rossi, K. (eds.) Volume 3: Opening the Mind: Innovative Psychotherapy, The Collected Works of MHE, pp. 67-71. MHE Foundation Press, Phoenix, Arizona.

Gilbert, W. (1986). The RNA World. Nature, 319, 618.

Jung, G. (1958). Psychology and Religion: West and East. Pantheon Books: N. Y.

Kempermann, G. (2006). Adult Neurogenesis: Stem Cells and Neuronal Development in the Adult Brain. Oxford University Press, N.Y.

Kim, T., Hemberg, M., et al. (2010). Widespread transcription at neuronal activity-regulated enhancers. Nature, 465, 182-187.

Levsky, J., Shenoy, S., Pezo, C., and Singer, R. (2002). Single-cell gene expression profiling. Science, 297, 836–840.

Liu, C. and Arkin, A. (2010). The Case for RNA. Science, 330, 1185-1186.

Lloyd, D. & Rossi, E. (Eds.) (1992). Ultradian Rhythms in Life Processes: An Inquiry into Fundamental Principles of Chronobiology and Psychobiology. Springer-Verlag: N.Y.

Lloyd, D. & Rossi, E. (Eds.) (2008). Ultradian Rhythms from Molecules to Mind: A New Vision of Life. Springer, NY.

Mattick, J. (2010). RNA as the Substrate for Epigenome-Environment Interactions. Bioessays, 32: 548-552.

Mitchell, M. (2009). Complexity: A Guided Tour. Oxford University Press, N.Y., US.

Otto, R. (1923). The Idea of the Holy. Oxford University Press, London.

Ramakrishnan, V. (2011). The Ribosome: Some Hard Facts about its Structure and Hot Air about its Evolution. In Atkins, J., Gesteland, R., & Cech (Eds.). RNA Worlds. Cold Spring Harbor Laboratory Press, Cold Spring Harbor, N.Y., 155-164.

Rapparini, R. (2010). Complex Systems of Consciousness. In: Rana, U., Srinivas, K., Aery, N., Purohit, A. (Eds.), The Philosophy of Evolution. Yash Publishing House, 1 E-14, Pawan Puri, Bikaner-334003, Camp: Jodhpur (Rajasthan), India, 169-174.

Ren, B. (2010). Transcription: Enhancers make non-coding RNA. Nature, 465, 173–174.

Ribeiro, S. (2004). Towards an evolutionary theory of sleep and dreams. A MultiCiência: Mente Humana, 3, 1-20.

Ribeiro, S., Goyal, V., Mello, C. & Pavlides, C. (1999). Brain gene expression during REM sleep depends on prior waking experience. Learning & Memory, 6: 500-508.

Ribeiro, S., Mello, C., Velho, T., Gardner, T., Jarvis, E., & Pavlides, C. (2002). Induction of hippocampal long-term potentation during waking leads to increased extra hippocampal zif-268 expression during ensuing rapid-eye-movement sleep. Journalof Neuroscience, 22(24), 10914-10923.

Ribeiro, S., Gervasoni, D., Soares, E., Zhou, Y., Lin, S., Pantoja, J., Lavine, M., & Nicolelis, M. (2004). Long-lasting novelty-induced neuronal reverberation during slow-wave sleep in multiple forebrain areas. Public Library of Science, Biology. (PLoS), 2 (1), 126-137.

Riberio, S., Simões, C. & Nicolelis, M. (2008). Genes, Sleep and Dreams. In Lloyd & Rossi (Eds.) Ultradian rhythms from molecule to mind. Springer. N.Y., 413-430.

Rossi, E. (1986a). Altered States of Consciousness in Everyday Life: The Ultradian Rhythms. In B. Wolman (Ed.), Handbook of Altered States of Consciousness. Van Nostrand Reinhold. N. Y., 97 – 132.

Rossi, E. (1986b). The Indirect Trance Assessment Scale (ITAS): A Preliminary Outline and Learning Tool. In Yapko, M. (Ed.), Hypnotic and Strategic Interventions: Principles and Practice. Irvington, N.Y.

Rossi, E. (1988a). Nonlocality in Physics and Psychology: An Interview with John Stewart Bell. Psychological Perspectives, 19(2), 294-319.

Rossi, E. (1988b). Beyond relativity and quantum theory: An interview with David Bohm. Psychological Perspectives, 19, 25-43.

Rossi, E. (1988c). Perspectives: Consciousness and the New Quantum Psychologies. Psychological Perspectives, 19(1), 4-13.

Rossi, E. (1990). The new yoga of the west: Natural rhythms of mind-body healing. Psychological Perspectives, 22, 146-161.

Rossi, E. (1996). The Symptom Path to Enlightenment: The New Dynamics of Hypnotherapeutic Work. Zeig, Tucker, Theisen: N.Y.

Rossi, E. (2002). The Psychobiology of Gene Expression: Neuroscience and Neurogenesis in Hypnosis and the Healing arts. W. W. Norton, N.Y.

Rossi, E. (2004). A Discourse with Our Genes: The Psychosocial and Cultural Genomics of Therapeutic Hypnosis and Psychotherapy. Available in English and Italian. Editris s.a.s., San Lorenzo Maggiore, Italy. Zeig, Tucker & Theisen Phoenix, Arizona, US.

Rossi, E. (2007). The Breakout Heuristic: The New Neuroscience of Mirror Neurons, Consciousness and Creativity in Human Relationships: Selected Papers of Ernest Lawrence Rossi Vol. 1. The M.H.E. Foundation Press, Phoenix.

Rossi, E. (2011). Creating Consciousness: How Psychotherapists Can Facilitate Wonder, Wisdom, Beauty & Truth. Vol. 2, Selected Papers of Ernest Lawrence Rossi Vol. 2. The M.H.E. Foundation Press, Phoenix.

Rossi, E, Erickson-Klein, R, and Rossi, K (2008). The future orientation of constructive memory: An evolutionary perspective on therapeutic hypnosis and brief psychotherapy. American Journal of Clinical Hypnosis, 50:4, 343-350.

Rossi, E. & Lippincott, B. (1992). The wave nature of being: Ultradian rhythms and mind-body communication. In Lloyd, D. & Rossi, E. (Eds.) Ultradian Rhythms in Life Processes. Springer-Verlag: N.Y. 371-402.

Rossi, E. & Nimmons, D. (1991). The Twenty-Minute Break: The Ultradian Healing Response. Jeremy Tarcher, Los Angeles.

Rossi, E. & Rossi, K. (2008). Open Questions on Mind, Genes, Consciousness, and Behavior:

The Circadian and Ultradian Rhythms of Art, Beauty, and Truth in Creativity. In Lloyd & Rossi (Eds.) Ultradian rhythms from molecule to mind. Springer. N.Y. 391-412.

Rossi, E. et al. (2010). What Makes Us Human: A Neuroscience Prolegomenon for the Philosophy of Evolution and Consciousness. In: Rana, U., Srinivas, K., Aery, N., Purohit, A. (Eds.), The Philosophy of Evolution. Yash Publishing House, 1 E-14, Pawan Puri, Bikaner-334003, Camp: Jodhpur (Rajasthan), India, pp. 15-38.

Tomlin, S. (2005). Dramatizing maths: What's the plot? Nature, 436, 622-623.

Van Praag, H., Kempermann, G., and Gage, F. (1999). Running increases cell proliferation and neurogenesis in the adult mouse dentate gyrus. Nature Neuroscience, 2, 266-270.

Wang, X., Song, X., Glass, C., & Rosenfeld, M. (2011). The long arm of long noncoding RNAs: Roles as sensors regulating gene transcription programs. In Atkins, J., Gesteland, R., & Cech (Eds.). RNA Worlds. Cold Spring Harbor Laboratory Press, Cold Spring Harbor, N.Y., pp. 279-292).

Warren, W., Clayton, D. et al. (2010). The genome of a songbird. Nature, 464, 757-762.

Yarus, M., (2010). Life from an RNA World: The Ancestor Within. Harvard University Press, Cambridge, MA, US.

15. Protophenomena and their Physical Correlates

Bruce J. MacLennan, Ph.D.[1]

[1]Department of Electrical Engineering and Computer Science,
University of Tennessee, Knoxville, TN 37996

Abstract

Philosophers distinguish several senses of "consciousness," but the most problematic is phenomenal consciousness, which refers to one's subjective awareness of the external world and of one's own mental state, to the experience of being someone. Understanding how phenomenal consciousness is related to physical processes has been dubbed by David Chalmers the Hard Problem of consciousness, because it addresses, without evasion, the relation of mind and matter. This paper presents a method that addresses the Hard Problem in terms of protophenomena, elementary units of embodied subjectivity, and permits the formulation of detailed, experimentally verifiable hypotheses about protophenomena and their physical correlates. Moreover, these techniques reveal theoretical preconditions for phenomenal consciousness in non-biological systems, such as robots, and allow formulation of empirically verifiable hypotheses relating to non-biological consciousness.

KEY WORDS: Hard Problem, neurophenomenology, phenomenal consciousness, protophenomena, qualia, robot consciousness, zombie problem

1. THE HARD PROBLEM

Philosophers distinguish functional consciousness, which refers to the cognitive and behavioral functions of consciousness in an organism, from phenomenal consciousness, which refers to the subjective experience of awareness (e.g., Block, 1995). Functional consciousness presents many challenging scientific problems, including the integration of multimodal sensory information, memory, emotion, motivation, planning, and action in order to promote the survival of the organism and its species, and how these functions are implemented in a nervous system. Although these problems are challenging, they present no fundamental epistemological problem, because they can addressed by the usual methods of empirical science. Unfortunately, these methods are unsuitable for addressing the principal question of phenomenal consciousness, which David Chalmers (1995, 1996) has called the Hard Problem: the relation of subjective experience to physical processes in the nervous system (see also Strawson, 1994, 2006; MacLennan, 1995, 1996a). The problem is that current physical theory says nothing about subjective experience, and so there seems to be no inconsistency in the possibility of these neurological processes taking place exactly as they do, but without accompanying subjective experience — the so-called "zombie problem" (Campbell, 1970; Kirk, 1974; Kripke, 1980). Closely related is the problem of robot consciousness, which is a good test case for any proposed theory of consciousness. Could a robot be conscious, and if so, under what conditions?

Because of a fundamental difference in kind between the shared experience of third-person observation, on which most science is based, and the private experience of first-person awareness, it is an epistemological error to attempt a reduction of first-person phenomena to third-person phenomena (Strawson, 1994, 2006; Chalmers, 1996, Pt. 2; MacLennan, 1995, 1996a). Therefore a different empirical approach is required to solve the Hard Problem.

2. PROTOPHENOMENAL ANALYSIS

Protophenomenal analysis is based on the observation that the conscious state has parts that stand in relation to each other (including relations in time). Therefore, although a reduction of phenomenal consciousness to physical processes is impossible, it is possible to reduce conscious phenomena to simpler conscious phenomena. As a simple example, visual experience, which is extended in perceptual space, can be divided into elementary patches of brightness, color, texture, and so forth (although even vision is more complicated than this example suggests; e.g., Pribram, 2004, p. 10; MacLennan, 2003,

2010). In many cases these simpler phenomena can be correlated with activity in specific brain regions. Therefore, phenomenal consciousness can be investigated through neurophenomenological reduction, that is, parallel reductions in the realms of subjective experience and neurobiology. Obviously the latter depends on neuroscientific investigation, but the former requires skill in experimental phenomenology (Ihde, 1986; McCall, 1983), which elucidates the structure of conscious experience by means of careful and publicly validated internal observation and experiment. The parallel reductive processes inform each other, suggesting hypotheses and experiments on each side.

Reduction must stop somewhere, and on the neurological side one usually stops at the single neuron as the unit of neural activity. (The unit of analysis in not uncontested, since various researchers have defended both larger units, such as the microcolumn, and smaller units, such as the dendritic spine.) Likewise, while it is conceivable that the conscious state is a continuum, and infinitely divisible, a good working hypothesis is that there are smallest units of subjectivity, which have been called protophenomena (Chalmers, 1996, pp. 126–7, 298–9; Cook, 2000, 2002a, 2002b, chs. 6–7, 2008; MacLennan, 1996a; cf. proto-qualia in Llinas, 1988; phenomenisca in MacLennan, 1995). William James referred to them as "aboriginal atoms of consciousness" and "mental atoms" (1890/1955, vol. I, ch. 6, p. 149). Since, apparently, one's conscious state is correlated with the activity of masses of neurons, to a first approximation a protophenomenon can be understood as the contribution to the conscious state made by a single neuron. For example, neurons in primary sensory cortices have receptive fields, which are simple ranges of stimuli to which they respond (e.g., an oriented edge at a particular location in the visual field), and the corresponding protophenomenon would be something like the visual experience of that edge.

These simple visual examples are easy to understand, but apt to mislead, suggesting that protophenomena are "raw sense data" (such as a localized "red-here-now"). Therefore, we need to emphasize that protophenomena are taken to be the elements of the entire conscious state, including perceptions and their interpretations, but also recalled memories, discursive thoughts, imagination, intuitions, intentions, emotions, moods, motor actions, and so on — all that constitutes conscious mental life.

Phenomenology studies the structure of phenomena, which in this context are things that appear (Greek, phainetai) in conscious experience. Protophenomena are the elementary constituents of phenomena, but we might not be conscious of them in isolation, for they are very small. For example, if we suppose that there is one protophenomenon for each neuron, there may be 30 to 100 billion protophenomena in the conscious state. It sounds paradoxical to say that we might not be conscious of the elementary constituents of the conscious state, but the claim can be understood by analogy, for protophenomena are analogous, in the subjective realm, to the individually imperceivable atoms of which macroscopic physical objects are made. Solid, visible objects are made from atoms, but in isolation atoms are neither visible nor solid, though visibility and solidity are consequences of their properties and relations when combined in large numbers. So also conscious phenomena are a consequence of the coherent activity of vast numbers of individual protophenomena ("atoms of consciousness"), whose individual contributions are generally below our level of awareness. (To put it more operationally, the addition or deletion of a protophenomenon to a phenomenon is no more likely to affect one's behavior than is the addition or deletion of an atom to a macroscopic object.)

The atomic analogy is also helpful in understanding protophenomenal interdependencies, for macroscopic objects owe their existence and properties to interatomic and intermolecular forces that bind the individual atoms into coherently behaving wholes. So also there are dependencies among protophenomena that parallel, in phenomenal space, the connections among neurons in physical space.

As a neuron's activity is a function of its own past activity and of the activity of the neurons from which it gets its input, so the intensity of a protophenomenon in the conscious state is a function of the protophenomena on which it depends and its own past intensity. (MacLennan, 1996b, presents a first approximation to a mathematical description of protophenomenal interdependence.) Since there does not seem to be any essential difference between the neurons in different sensory cortices, indeed throughout cerebral cortex, it is likely that sensory qualities are a consequence of the interconnections among neurons rather than of properties inherent to the neurons. Protophenomenal analysis thus implies a structural theory of qualia; that is, the qualitative characters of protophenomena are not inherent but arise by virtue of their interconnections. This is supported by a variety of observations, including the well-known phenomenon of referred pain, in which neurons in sensory cortex reassign themselves from an amputated limb to other parts of the body (e.g., Karl, Birbaumer, Lutzenberger, Cohen & Flor, 2001), and recent experiments by Sur (2004) demonstrating that neurons in auditory cortex could be "rewired" to support visual phenomena. Thus the protophenomenal quality of a neuron is not inherent to the neuron, but depends on the neuron's interconnections with other neurons, and likewise on the dependence of the corresponding protophenomenon on the protophenomena corresponding to these other neurons.

A related issue is the fact that much of what our brains do is unconscious, and so protophenomenal analysis must account for the fact that some neurophysiological processes have associated conscious phenomena, but others do not. Possible solutions are presented elsewhere (MacLennan 1996a, 2008). The importance of protophenomena for the Hard Problem is that they correspond to small-scale neurophysiological processes, which we call the correlative activity of the protophenomena, which take place at certain activity sites in the brain. These processes have not been identified definitively, but there are several likely candidates, including the binding of neurotransmitters to their receptors, somatic membrane potential, and the opening of ion channels when a nerve impulse is generated. Their identity is an empirical matter, but experimental evidence is currently lacking.

Therefore, the possibility of robot consciousness depends on what sort of physical processes have accompanying protophenomena, and in particular on whether protophenomena accompany nonbiological (or even non-neural) processes. Hence, identifying what physical processes support protophenomena is fundamental to determining whether robots could have phenomenal consciousness (MacLennan, 2009).

3. PHYSICAL SUBSTRATES FOR PHENOMENAL CONSCIOUSNESS

The simplest hypothesis is that the intensity of a protophenomenon in consciousness is correlated to some activity in a corresponding neuron, for example, its impulse rate. But if we try to be more precise, we see that there many physical processes that could be the correlative activity. A few of the possibilities are the impulse rate at the axon hillock, the ion flux across the somatic membrane, and the somatic membrane potential. However, some neuroscientists have argued that conscious experience is more likely associated with graded electrical fields and currents in the dendritic tree, rather than in the all-or-nothing firing of action potentials (Pribram, 1971, pp. 104–5, 1991, pp. 7–8, 2004). This suggests that the relevant physical processes might be neurotransmitter flux across the synapse, binding of neurotransmitter molecules to receptors, or the membrane potential of a dendritic spine. We do not yet have the required experimental evidence, but identifying the correlative activities will be relevant to whether physical processes in a robot might have protophenomena.

Seeing the variety of possible correlative activities in a neuron raises the question of why some physical processes should have protophenomena but not others. Norman D. Cook has suggested one answer, arguing that a protophenomenon corresponds to the opening of a nerve cell to its intercellular environment when the impulse is generated (Cook, 2000, 2002a, 2002b, chs. 6–7, 2008). He explains that the momentary opening of the cell membrane at the time of the action potential is the single-cell protophenomenon ... underlying "subjectivity" – literally, the opening up of the cell to the surrounding biochemical solution and a brief, controlled breakdown of the barrier between cellular "self" and the external world (Cook, 2002a).

This opening of a cell to its environment increases the correlation between its internal and external states, increasing the mutual information between the interior and the exterior and decreasing the entropy of the entire system. This is a fundamental information process, an elementary act of protocognition, and so it is an attractive as the possible physical correlates of protophenomena.

Addressing the question of what sorts of physical processes support consciousness, Chalmers (1996, ch. 8) suggested that any physical information space has associated protophenomena, which would permit non-biological consciousness, but his idea is based on a vague notion of "differences that make a difference." The central issue is that information, in a cognitive sense, depends on distinctions that have a function or purpose, or that are relevant. In a biological sense, relevance can be grounded, ultimately, in the survival of the individual or its group, that is, in inclusive fitness (Burghardt, 1970). In our experiments demonstrating the evolution of communication in a population of machines, we have shown how this idea can be transferred to nonbiological nonequilibrium systems (MacLennan, 1992, 2006; MacLennan & Burghardt, 1993). Information is relevant to the persistence of the population in its nonequilibrium state, to its "survival."

Hence our approach to the Hard Problem combines Cook's and Chalmers' insights with thermodynamics, since living things must act to maintain internal organization, to keep their entropy low. From the pioneering work of Schrödinger (What is Life? 1945/1967) to recent work by Eric Schneider (2005), it is apparent that living systems maintain themselves in a far-from-equilibrium state by using free energy and matter to create and sustain internal order, that is, to decrease internal entropy. Furthermore, even the simplest life forms use information, which is mathematically equivalent to negative entropy, to improve their ability to survive. When a cell senses its environment, it decreases the independence of the internal and external states, thus decreasing entropy (increasing information). Although information theory has a problematic relation to thermodynamics through the shared concept of entropy, it nevertheless provides a basis for addressing the issue of relevance in information. Thus entropy, with ties to organization and survival on one side and to information and intelligence on the other, is fundamental to understanding "differences that make a difference," and to explicating the physical information spaces that correspond to protophenomena.

Therefore we have been developing an empirically verifiable theory of phenomenal consciousness that relates the intensities of protophenomena and their mutual interdependence to the fundamental thermodynamics of order and information. Such a theory will provide a basis for determining the possibility or actuality of phenomenal consciousness in robots and in the non-human universe at large.

There are several significant research questions: Can Chalmers' notions of physical information spaces and "differences that make a difference" be explained in terms of the thermodynamics of order and information? What does this theory suggest are the most plausible physical activity sites associated with protophenomena? Which psychophysical experiments could discriminate among hypotheses concerning the physical processes that could be the correlative activity of protophenomena?

4. THEORETICAL INVESTIGATIONS

In the remainder of the paper, we will present our ongoing investigations of protophenomena. A principal task is to consolidate the preceding hypotheses into a theory integrating protophenomenal analysis, information theory, and thermodynamics in order to account for phenomenal consciousness and its physical correlates. The approach is primarily theoretical and uses entropy as the principal organizing concept. We are developing the theory mathematically and in a sufficiently general way so that we can understand its applicability to non-biological self-sustaining non-equilibrium systems (i.e., artificial life forms). The goal is a sufficiently precise theory of phenomenal consciousness (both biological and non-) to permit experimental investigation. Eventually the theory will have to be validated in the context of human phenomenal consciousness (which is a given), and the physical processes associated with protophenomena and their relationships need to be identified more precisely. More specifically, the mathematical analysis is grounded in thermodynamics, information theory, cellular biochemistry, and neuroscience. We are formalizing the idea that information processes in neurons serve functions that can be explained in terms of thermodynamics (in particular the Maximum Entropy Production Principle). To the extent that this analysis is independent of the specifics of neural cell biology, it will suggest (but not establish) physical preconditions for non-biological phenomenal consciousness.

Therefore, we are analyzing the information processing of a single neuron in these terms. Aside from the obvious connection between the brain and consciousness, the neuron is the primary agent making "decisions that make a difference." Our current focus is the pyramidal cell, but we do not expect the results to depend on the cell type. The challenge is that a neuron does not survive as an individual, independent organism, so the connection to thermodynamics is very indirect (via the survival of the organism and, more accurately, the survival of its species).

This analysis can be carried to a deeper level, focusing on those processes that mediate the connection of a cell (and in particular a neuron) to its environment: receptor binding and gated ion channels. In addition to their role in opening the cell to its environment and increasing the mutual information between "self" and "other," these receptors and channels can be considered elementary decisionmaking units, which contribute to the function of the neuron.

We can focus deeper yet, on the allosteric proteins that constitute these receptors and gates, and that implement the information processing networks within the cell (Bray, 1995). The binding of regulator molecules to an allosteric protein changes its function in the cell, which is a very direct embodiment of decisions that make a difference, and so it approaches the hypothesized fundamental requirements for an activity site and brings the information-theoretical and thermodynamical notions of entropy nearly to identity.

Finally, we must close the loop between these very low level physical information processes and the entropy-decreasing self-organization of entire organisms, and indeed of species and ecosystems. The goal is an overarching framework for a theory of physical activity sites and their corresponding protophenomena.

5. EMPIRICAL INVESTIGATIONS

Another objective of our research is to define experimental protocols to identify physical correlates of protophenomena, which will be based on existing neuroscience research and adapted to artificial systems. Eventual conduct of these experiments would serve to confirm or disconfirm the possibility of phenomenal consciousness in robots and other physical systems.

The first task is to identify, from among the various potential activity sites, those that are most plausible on the basis of the theoretical analysis.

The second task is to outline one or more experimental protocols capable of confirming or refuting hypotheses that particular physical processes are the correlative activity of protophenomena. The basic idea can be explained easily, but the actual experiments could be formidable. The approach depends on identifying at least one activity site whose protophenomenon is salient and relatively isolable in consciousness so that a human subject can report the experiences resulting from physical interventions in its neural locus. The complications are the "smallness" (in experiential terms) of individual protophenomena and their typically dense interdependence with other protophenomena, which corresponds to the dense interconnectivity of most neurons. Roughly speaking, the goal is to identify one or more neurons whose individual activity can be determined reliably through conscious experience. Therefore, we are surveying the neuropsychological literature to identify promising protophenomena with known neural loci.

Within a protophenomenon's neural locus there are several candidates for correlative activity (e.g., neurotransmitter flux across the synapse, neurotransmitter receptor activity, dendritic spine potential, somatic potential, ion flux at the axon hillock). For each possible activity site the hypothesis to be tested is that a certain kind of correlative physical activity is necessary and sufficient for the corresponding protophenomenon's presence in the conscious state. Its sufficiency is established by creating the physical activity artificially in the absence of the normal stimulus and with more proximal connections suppressed, and showing that this intensifies the protophenomenon. Necessity is established by replacing the activity site by another physical system that is functionally equivalent but makes use of different physical processes. Persistence in consciousness would show that the natural process is not necessary, and would demonstrate at least one artificial process sufficient to maintain the protophenomenon's presence. Conversely, its absence from consciousness would tend to disconfirm these conclusions, suggesting that the correlative activity depends on the natural process. By systematically investigating each possible correlative activity in this way, we can determine what sort of neural process corresponds to protophenomenal intensity and — more importantly for robot consciousness — we may discover other physical processes that have associated protophenomena. If none of the individual possibilities show clear evidence of being activity sites, then it might be necessary to investigate combinations of two or more possibilities by similar methods. It is also likely that conscious state, content, and process (Pribram, 2004) have different kinds of correlative activity, would be another topic for investigation.

Obviously, experimental interventions of this sort present enormous practical challenges, and ethical ones as well. Nevertheless, by defining them clearly we plan to demonstrate that hypotheses about protophenomena have empirical content, and are thus scientific, even if the experiments cannot be conducted at this time.

Therefore we are reviewing current and proposed experimental techniques in neuropsychology to determine which experiments may be feasible now or in the near future. For example, Losonczy, Makara, and Magee (2008) have developed techniques for delivering individual neurotransmitter molecules to individual dendritic spines with a spatial resolution of 1 micron and a time resolution of 1 millisecond. Although these procedures were not applied in vivo, their work shows that the sorts of experiments we have in mind are not impossible. The recently developed optogenetics technique uses light-sensitive proteins to control channels and enzymes with neuron-level millisecond-precision in intact animals (Zemelman, Lee, Ng & Miesenböck, 2002; Zhang, Wang, Brauner, et al., 2007). More precise control can be achieved by two-photon glutamate uncaging, which the molecule to be activated can be controlled by the coincidence of two lights (Pettit, Wang, Gee & Augustine 1997). By means such as these we may develop a series of "crucial experiments" to determine the nature of activity sites in humans and to identify preconditions for artificial activity sites (if any). These experiments would constitute an agenda for solving the Hard Problem scientifically and applying it to artificial phenomenal consciousness.

References

Block, N. (1995). On a confusion about a function of consciousness. Behavioral and Brain Sciences, 18, 265–66.

Bray, D. (1995). Protein molecules as computational elements in living cells. Nature, 376, 307–12.

Burghardt, G. M. (1970). Defining 'communication'. In: Johnston, J. W., Jr., Moulton, D. G., Turk, A. (Eds.), Communication by Chemical Signals, Appleton-Century-Crofts, New York, pp. 5–18.

Campbell, K. K. (1970). Body and mind. Doubleday, New York.

Chalmers, D. J. (1995). Facing up to the problem of consciousness. Journal of Consciousness Studies, 2, 200–219.

Chalmers, D. J. (1996). The conscious mind. Oxford University Press, New York.

Chalmers, D. J. (2002). Consciousness and its place in nature. In: Chalmers, D. J. (Ed.), Philosophy of Mind: Classical and Contemporary Readings, Oxford, Oxford.

Cook, N. D. (2000). On defining awareness and consciousness: The importance of the neuronal membrane. In: Proceeding of the Tokyo-99 Conference on Consciousness, World Scientific, Singapore.

Cook, N. D. (2002a). Bihemispheric language: How the two hemispheres collaborate in the processing of language. In: Crow, T. (Ed.), The Speciation of Modern Homo Sapiens, Proceedings of the British Academy, London.

Cook, N. D. (2002b). Tone of voice and mind: The connections between intonation, emotion, cognition and consciousness. John Benjamins, Amsterdam.

Cook, N. D. (2008). The neuron-level phenomena underlying cognition and consciousness: Synaptic activity and the action potential. Neuroscience, 153(3), 556–70.

Freeman, A. (Ed.) (2006). Consciousness and its place in nature: Does physicalism entail panpsychism? Imprint Academic, Charlottesville, VA. Revision of Journal of Consciousness Studies, 13 (2006), nos. 10–11.

Ihde, D. (1986). Experimental phenomenology: An introduction. State University of New York Press, Albany.

James, W. (1890/1955). The Principles of Psychology, Authorized Ed. Dover edition of Henry Holt edition, New York.

Karl, A., Birbaumer, N., Lutzenberger, W., Cohen, L.G., Flor, H. (2001). Reorganization of motor and somatosensory cortex in upper extremity amputees with phantom limb pain. The Journal of Neuroscience, 21, 3609–18.

Kirk, R. (1974). Zombies versus materialists. Aristotelian Society, 48 (suppl.), 135–52.

Kripke, S. A. (1980). Naming and necessity. Harvard University Press, Cambridge, MA.

Llinas, R. R. (1988). The intrinsic electrophysiological properties of mammalian neurons. Science, 242, 1654–64.

Losonczy, A., Makara, J. K., Magee, J. C. (2008). Compartmentalized dendritic plasticity and input feature storage in neurons. Nature, 452, 436–40.

MacLennan, B. J. (1992) Synthetic ethology: An approach to the study of communication. In: Langton, C. G., Taylor, C. Farmer, J. D., Rasmussen, S. (Eds.), Artificial Life II. The Second Workshop on the Synthesis and Simulation of Living Systems, MIT Press, Redwood City, pp. 631–658.

MacLennan, B. J. (1995). The investigation of consciousness through phenomenology and neuroscience. In King, J., Pribram, K. H. (Eds.), Scale in conscious experience: Is the brain too important to be left to specialists to study? Lawrence Erlbaum, Hillsdale, NJ, pp. 25–43.

MacLennan, B. J. (1996a). The elements of consciousness and their neurodynamical correlates. Journal of Consciousness Studies, 3 (5/6), 409–24. Reprinted in: Shear, J. (Ed.), Explaining consciousness: The hard problem. MIT, Cambridge, MA, 1997, pp. 249–66.

MacLennan, B. J. (1996b). Protophenomena and their neurodynamical correlates (Technical Report UT-CS-96-331). Knoxville, TN: University of Tennessee, Knoxville, Department of Computer Science. Available at .

MacLennan, B. J. (1999a). Neurophenomenological constraints and pushing back the subjectivity barrier. Behavioral and Brain Sciences, 22, 961–63.

MacLennan, B. J. (1999b) The protophenomenal structure of consciousness with especial application to the experience of color: Extended version (Technical Report UT-CS-99-418). Knoxville, TN: University of Tennessee, Knoxville, Department of Computer Science. Available at .

MacLennan, B. J. (2003). Color as a material, not an optical, property. Behavioral and Brain Sciences, 26, 37–8.

MacLennan, B. J. (2006) Making meaning in computers: Synthetic ethology revisited. In: Loula, A., Gudwin, R., Queiroz, J. (Eds.), Artificial Cognition Systems, IGI Global, Hershey, ch. 9 (pp. 252–83).

MacLennan, B. J. (2008). Consciousness: Natural and artificial. Synthesis Philosophica, 22(2), 401–33.

MacLennan, B. J. (2009). Robots react but can they feel? A protophenomenological analysis. In: Vallverdú, J., Casacuberta, D. (Eds.), Handbook of Research on Synthetic Emotions and Sociable Robotics: New Applications in Affective Computing and Artificial Intelligence, IGI Global, Hershey, NJ, pp. 133–53.

MacLennan, B. J. (2010). Protophenomena: The elements of consciousness and their relation to the brain. In: Batthyány, A. Elitzur, A., Constant, D. (Eds.), Irreducibly Conscious: Selected Papers on Consciousness, Universitätsverlag Winter, Heidelberg & New York, pp. 189–214.

MacLennan, B. J., Burghardt, G. M. (1993). Synthetic ethology and the evolution of cooperative communication. Adaptive Behavior, 2, 161–188.

McCall, R. J. (1983). Phenomenological psychology: An introduction. With a glossary of some key Heideggerian terms. University of Wisconsin Press, Madison.

Pettit, D. L., Wang, S. S., Gee, K. R., & Augustine, G. J. (1997). "Chemical two-photon uncaging: a novel approach to mapping glutamate receptors." Neuron, 19(3), 465–71.

Pribram, K. H. (1971). Languages of the brain: Experimental paradoxes and principles in neuropsychology. Prentice-Hall, Englewood Cliffs.

Pribram, K. H. (1991). Brain and perception. Holonomy and structure in figural processing. Lawrence Erlbaum, Hillsdale.

Pribram, K. H. (2004). Consciousness reassessed. Mind and Matter, 2(1), 7–35.

Schneider, E. D. (2005). Into the cool: Energy flow, thermodynamics, and life. University of Chicago Pr., Chicago.

Schrödinger, E. (1967). What is life? The physical aspect of the living cell & mind and matter. Cambridge: Cambridge University Pr.

Strawson, G. (1994). Mental reality. MIT Pr., Cambridge.

Strawson, G. (2006). Realistic monism. In Freeman (2006), pp. 3–31.

Sur, M. (2004). Rewiring cortex: Cross-modal plasticity and its implications for cortical development and function. In: Calvert, G. A., Spence, C., Stein, B. E (Eds.), Handbook of multisensory processing, MIT Press, Cambridge, MA, pp. 681–94.

Zemelman, B. V., Lee, G. A., Ng, M., Miesenböck, G. (2002). "Selective photostimulation of genetically chARGed neurons." Neuron, 33(1): 15–22. doi:10.1016/S0896-6273(01)00574-8.

Zhang, F., Wang, L. P., Brauner, M., et al. (2007). "Multimodal fast optical interrogation of neural circuitry." Nature, 446 (7136): 633–9. doi:10.1038/nature05744.

16. The Macro-Objectification Problem and Conscious Perceptions

GianCarlo Ghirardi

Emeritus, Department of Physics, the University of Trieste, the Abdus Salam International Centre for Theoretical Physics, Trieste, and Istituto Nazionale di Fisica Nucleare, Sezione di Trieste, Italy.

ABSTRACT

We reconsider the problem of the compatibility of our definite perceptions with the linear nature of quantum theory. We review some proposed solutions to the puzzling situation implied by the possible occurrence of superpositions of different perceptions and we argue that almost all are not satisfactory. We then discuss the way out which makes explicit reference to consciousness and we underline its pros and cons. In the second part of the paper we reconsider this problem in the light of the recently proposed collapse models, which overcome the difficulties of the standard theory by adding nonlinear and stochastic terms to the evolution equation and, on the basis of a unique dynamical principle, account both for the wavy behaviour of microsystems as well as for definite macroscopic events. By taking into account that different microscopic situations can trigger different displacements of an enormous number of particles in our brains which, in turn, lead to different and definite perceptions, we make plausible that such models do not assign a peculiar role to the conscious observer. Simply, the characteristic amplification mechanism leading to the collapse implies the suppression of all but one of the nervous stimuli corresponding to different perceptions. Thus, collapse models, at the nonrelativistic level, qualify themselves as theories which can consistently account for all natural processes, among them our definite perceptions.

Part I

1 A Quantum description of physical processes

Since the galilean revolution, the natural language of any scientific theory has been mathematics. In particular, different physical situations characterizing an individual physical system are usually described by appropriate mathematical entities which uniquely specify the "state" of the system under consideration. A paradigmatic example is given by Newtonian mechanics: the state of a system is uniquely specified by the assignment of the positions and velocities of all its constituents. Besides the states, a crucial role is played by the physically observable quantities, such as the momentum, the energy and so on. In classical mechanics these quantities are simply functions of the positions and the velocities of the constituents and, as such, they always possess precisely definite values. Finally, any theory must have a predictive character. This is usually embodied in an evolution equation for the "state" of the system which uniquely assigns a state at any time t once the initial state (at time 0) is specified. Obviously the formal scheme must also contain the prescription which, once the state is known, allows one to infer the value he will get when subjecting the system to a "measurement" of the physical observable he is interested in.

To supply, in a quite elementary way, the reader who is not familiar with quantum theory with the formal elements which are necessary to understand what follows, we consider it appropriate to summarize the "rules of the game" in the case of quantum mechanics. In reading this section it may be useful to give a look at Fig.1 which puts in evidence all formal aspects we are going to describe, by making reference to the oversimplified case of a linear vector space of dimension 2.

- The states of the system are associated to normalized (i.e. of length 1) vectors (which, for this reason are also called state vectors) of a linear vector space. Let us denote, following Schrödinger, as $|\psi)$, $|\varphi)$ two such states. The linear nature of the space of the states implies that if the two just mentioned states are possible states for a system, then also any normalized combination of them with complex coefficients $\alpha|\psi) + \beta|\varphi)$, $|\alpha|2 + |\beta|2 = 1$ is also a possible state of the system.

- The physical observables of the system are associated to appropriate operators acting on the state space. Here, an extremely important and innovative aspect of the formalism consists in the emergence of the phenomenon of quantization: the physical observables (in general) cannot assume any value in appropriate continuos intervals(as it happens in classical physics) but they can take only some discrete values, called eigenvalues, which are the values for which the eigenvalue equation, Eq.(1) below, admits a solution. If we denote as Ω the operator corresponding, within quantum mechanics, to an appropriate classical observable(like the angular momentum and similar) then we must look for values ωi and states $|\varphi i)$ satisfying:

$$\Omega |\varphi i) = \omega i|\varphi i). \qquad (1)$$

The eigenvalues ω_i (due to some precise formal requests on the operators representing observables) turn out to be real, they are usually discrete, the eigenvectors turn out to be pairwise ortogonal and, as a set of states, they are "complete", a technical expression to stress that any vector of the space can be expressed as a linear combination of them. Accordingly, the eigenstates of an observable yield an orthogonal system of axes for the space itself.

- The evolution equation is a deterministic differential equation, the celebrated Schrödinger's equation. Its most relevant feature, for what interests us here is that it is linear, i.e., if $|\psi, 0\rangle$ and $|\varphi, 0\rangle$ are two initial states and $|\psi, t\rangle$ and $|\varphi, t\rangle$ are their evolved at time t, then the evolved of the initial state $\alpha|\psi, 0\rangle + \beta|\varphi, 0\rangle$ is $\alpha|\psi, t\rangle + \beta|\varphi, t\rangle$.
- The predictions of the theory are fundamentally probabilistic and are embodied in the following rule. If the system is described by the state $|\Psi\rangle$ and if one is interested in the predictions concerning the outcome of a measurement of an observable Ω, one must express the state as a linear combination of the eigenstates of the observable itself(something we know that is always possible):

$$|\Psi\rangle = \sum_i c_i |\varphi_i\rangle. \qquad (2)$$

Then the theory makes precise probabilistic predictions concerning any outcome, let us say ω_k, which one can get in a measurement of Ω (and the same procedure allows to determine the probabilities of the outcomes for the other observables). Actually, the probability $P(\Omega = \omega_k || \Psi\rangle)$ of such an outcome for such an observable when the system is in the indicated state is simply given by the modulus square of the corresponding coefficient c_k of Eq.(2) (Note that since the vector has length one the sum of the squares of all its components equals one, i.e., one of the possible outcomes is obtained with certainty). From this rule one sees that when one and only one of the c_i's is different from zero (and therefore it equals 1) we can predict with certainty the outcome itself. In such a case we say that the observable Ω, possesses with certainty the value ω_k.

- It has to be stressed that, in general, different operators do not commute $\Omega\Gamma \neq \Gamma\Omega$. The most relevant implication of this fact is that, in general, the eigenvectors of a pair of such operators are not aligned. This means that if one prepares the system in a state that corresponds to a definite value of an observable (and therefore it is an eigenstates of the observable itself) the corresponding statevector will have non zero projections on at least two (in general many) eigenstates of the other noncommuting observable. As a consequence there are nonzero probabilities of get-

Quantum Physics, Evolution, Brain, Mind

ting one among various outcomes for it: the considered variable does not have a precise value. Accordingly, making sharp the value of an observable makes indefinite the value of other observables.

This is the uncertainity principle which holds, in particular, between the position and momentum variables: the more precise we make one of the two observables, the less precise we make the other.

It has to be stressed that if we consider a set of observables which commute among themselves (a set having this property is called an abelian set) then a theorem ensures that they admit precisely the same eigenstates. There obviously follows that, when such a common eigenstate describes the state of the system, all the considered observables possess precisely definite values.

- The theory specifies also the effect of the measurement process: if we make a system in a state like the one of Eq.(2)(with various c_i different from zero), to interact with the measuring apparatus, and we obtain the outcome, let us say ωj, the state of the system changes instantaneously from $|\Psi\rangle$ to the eigenstate corresponding to the eigenvalue of the measured observable: $|\Psi\rangle \rightarrow |\varphi j\rangle$. Note that, contrary to the standard evolution, which is linear and deterministic, the change induced by the measurement process is nonlinear(since the probabilities are given by the squares of the moduli of the coefficients) and stochastic (since all outcomes corresponding to non zero coefficients may occur with the specified probabilities).

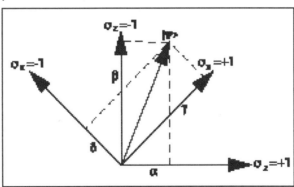

Figure 1: The formal structure of quantum mechanics in a pictorial form.

As already anticipated, we have chosen to depict, in Fig.l, the situation we have just described by making reference to the simplest vector space which occurs in the theory, the one related to the spin degree of freedom of a spin 1/2 particle. In such a case the vector space is two dimensional. Moreover, the observables corresponding to the projections of the spin along any given direction can take only the values ±1, in units of $h/2\pi$, h being Planck's constant.

They correspond to the observables σz, the projection of the spin along the z-axis (in the indicated units), and σx, the projection of the spin along the x-axis, respectively. From the figure one can grasp all relevant points of the formalism: the modulus square of the components α (β)of the state vector along the horizontal (vertical) axis give the probability of getting the outcome +1(−1) when the spin is measured along the z-axis. Similarly, the modulus square of γ and δ yield the probabilities of getting the two above mentioned outcomes in a measurement of the spin component along the x-axis.

As one clearly sees, making precise and equal to +1, e.g., the value of the spin component along the x-axis, which means aligning the statevector |Ψ⟩ with the line at 45°, implies, since this unit vector has components $1/\sqrt{2}$ along the horizontal and vertical axes that, for such a state, there is a probability 1/2 of getting the outcome +1 or -1 in the measurement of the z-spin component. Making absolutely precise one observable, i.e. σx, renders thus maximally indeterminate the other one since equal probabilities are attached to its two possible outcomes. We also recall that, within the standard theory, a measurement changes instantaneously the state of the system. In our case, if we measure the x-component of the spin and we get the result +1, the statevector |Ψ⟩ is transformed into the unit vector at 45°.

2 The position representation

To allow the reader to follow the discussion in the second part of the paper we are compelled to make an important specification: not all observable quantities are quantized, some of them can take any value within a continuous interval. This implies some formal mathematical refinements which we will not discuss in detail. Typical examples of this situation are the position and the momentum observables; both of them can assume, just as in classical mechanics, all values lying between $-\infty$ and $+\infty$. Let us consider the analogous, in the continuous case, of the discrete case discussed above. For the moment, let us assume that we are dealing with a one-dimensional problem, i.e. a particle moving along a line, and let's denote as X its position variable. The eigenvalue equation (1) will be replaced by:

$$X|x\rangle = x|x\rangle, \qquad (3)$$

x being the value of the position occupied by the particle, while Eq.(2) will be replaced by:

$$|\Psi\rangle = \int_{-\infty}^{+\infty} dx\, \phi(x)|x\rangle. \qquad (4)$$

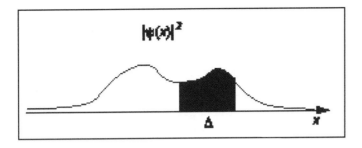

Figure 2: The probability of finding the particle within the interval Δ is given by the black area in the figure.

It is important to clarify the physical meaning of the coefficient (the wavefunction) ψ(x) appearing in this equation. First of all, the normalization condition (the fact that |Ψ⟩ has length 1) now reads:

$$\int_{-\infty}^{+\infty} dx |\psi(x)|^2 = 1. \qquad (5)$$

Secondly, just as the squares of the moduli of the coefficients Ci give the probabilities of getting the outcome Wi in a measurement of the observable D, now the modulus square of the wavefunction |ψ(x)|2 yields the probability density of finding the particle at x in a position measurement. Physically this means that the area subtended by the modulus square of this quantity in a given interval Δ of the x-axis, gives the probability of finding the particle in the indicated interval when subjected to a position measurement. We have depicted the situation in Fig.2.

3 The quantum measurement problem

With the above premises, we can now formulate in a precise way the quantum measurement problem. The idea is quite simple. Suppose we have a microscopic system in an eigenstate of a specific observable, and we want to ascertain its value, which, as we know, coincides with certainty with the associated eigenvalue. If we denote as usual as |φi⟩ the eigenstate of the microobservable Ω we are interested in, and we assume that we can get knowledge of its definite value ω_i, we must consider an apparatus A characterized by a "ready state" |A$_0$⟩ and a micro-macro interaction leading to the evolution:

$$|\varphi_i\rangle \otimes |A_0\rangle \rightarrow |\varphi_i\rangle \otimes |A_i\rangle, \qquad (6)$$

where the macrostates |A$_i$⟩ are macroscopically and perceptually distinguishable. The standard example is the one of a macroscopic pointer that, after the measurement, will "point" at the obtained value i.

Equation(6) deserves some comments. In it the product of two states, referring to two different systems appears(the use of the circled cross makes reference to the formal fact that the space of the composite system is the direct product of the spaces of the constituents). However, from a physical point of view, the states appearing in Eq.(6) have a precise meaning: at the l.h.s. we have a microsystem in the state which corresponds to its having the precise value ω_i for an appropriate observable, while the macroscopic apparatus is in a "ready" state in which its pointer points at 0. Concerning the state at the r.h.s. the microsystem is in the same state and still has the property $\Omega = \omega_i$, while the macroscopic pointer is also in a definite, different from the previous one, state: its pointer points at i on the scale. We have ascertained the possessed value of the observable Ω.

Here comes the problem. In fact we can easily prepare the microsystem in a state which, instead of being an eigenstate of the observable we are interested in, is a linear superposition of eigenstates belonging to different eigenvalues (the reader should remember that the eigenstate of the σ_x component of the spin associated to the eigenvalue +1 is a linear superposition, with equal coefficients of the two eigenstates of σ_z belonging to the eigenvalues +1 and −1). Moreover, we can put the microsystem in such a superposition into interaction with the macroapparatus designed to "measure" Ω. At this point we are in troubles since we can argue as follows. The unfolding of the process which implies the micro-macro (system-apparatus) interaction must be governed by our fundamental theory, quantum mechanics. In fact, what reason whatsoever there could be for all micro-constituents of the universe to be governed by the quantum laws, while a macrosystem, which is nothing more than an assembly of nuclei, atoms, molecules and so on, should be ruled by different laws? However, as we have repeatedly stressed, due to the linear character of the evolution, if the microsystem-macroapparatus interaction is described by Eq.(6) when the triggering state is an eigenstate of the observable Ω, then, in the present case, one has:

$$\sum_i c_i |\varphi_i\rangle \otimes |A_0\rangle \equiv \sum_i c_i [|\varphi_i\rangle \otimes |A_0\rangle] \rightarrow \sum_i c_i |\varphi_i\rangle \otimes |A_i\rangle. \quad (7)$$

The final state is not a product (factorized) state of the system and the apparatus, but a linear superposition of such states. Technically it is referred as an entangled system-apparatus state. The puzzle arises from the fact that, as the theory tells us, such a state does not describe neither a microsystem with a definite property for Ω(and this is not particularly problematic) nor a macroscopic object whose pointer points at a precise position. Actually the theory implies that the pointer "does not have a precise location" and only if we decide to "measure its position" we will get one of the potentially possible outcomes: the pointer will be found to point at, let us say, k. Which meaning can be attached to a state in which a superposition of macroscopically and perceptually different states appears? How to interpret this state of affairs?

As we have already anticipated, the "orthodox" way out is to claim that the final situation is described, with probability $|c_i|^2$, by one of the terms of the superposition, all of which correspond to the pointer "pointing at a precise position". So, standard quantum mechanics contains two evolution laws, the linear one typical of microsystems and the nonlinear and stochastic one corresponding to Wave Packet Reduction(WPR), and accounting for the measurement processes. The big problem derives from the fact that the theory does not contain any formal element which identifies when one or the other type of evolution occur. Actually, macroscopic systems which require a genuinely quantum treatment exist. The problem has been emphasized with admirable lucidity by the late J. Bell:

> Nobody knows what quantum mechanics says exactly about any situation, for nobody knows where the boundary really is between wavy quantum systems and the world of particular events.

4 The von Neumann chain

John von Neumann has been the first to dig deeply into the measurement problem. He made an important remark. If one takes into account Eq.(7), one cannot consider it as exhaustively describing the measurement process. In fact, as it is obvious, the final state leads to the natural question: what actually is the outcome of the measurement? (i.e. where does the pointer point?). The answer is obvious: we must consider a further process aimed to ascertain which one of the eigenstates $|A_i\rangle$ actually occurred. In this way the so-called von Neumann's chain finds its origin. The r.h.s of Eq.(7) must be enriched by taking into account the further measuring apparatus (let us call it B) devised to identify in which of the macrostates the pointer A is. Accordingly, we must consider successive measurement procedures aiming to identify what the actual state of affairs is:

$$\sum_i c_i|\varphi_i\rangle \otimes |A_0\rangle \otimes |B_0\rangle \otimes \ldots \to \sum_i c_i|\varphi_i\rangle \otimes |A_i\rangle \otimes |B_i\rangle \otimes \ldots \quad (8)$$

and so on. The chain never ends but it exhibits a quite interesting feature. No matter at which point one chooses to break it, if the linear superposition is replaced by one of its terms (let us say the ith) one gets a consistent set of outcomes: the particle is found in state $|\varphi_i\rangle$), the macroapparatus has its macroscopic pointer pointing at the value i of the scale, the apparatus B reveals that the macroapparatus A actually points at i and so on. In brief, there is a full "final" consistency provided one breaks (at a certain level) the chain.

Some remarks are at order:

- This approach leaves open the stage at which the breaking must be assumed to occur, provided it occurs at some point. In this sense, it is not surprising that von Neumann and Wigner have chosen (see below) to break it at the level in which consciousness enters into play.
- von Neumann's has tacitly made some quite drastic assumptions concerning the unfolding of the process (typically that the measurement has 100% efficiency, that the final states are strictly ortogonal etc., a set of assumptions which are very difficult to be verified in actual experiments) and this is why one usually refers to the just described scheme as "The Ideal von Neumann's Measurement Process".

5 The unavoidability of the problem

Some authors (Primas, 1990) have suggested that the previous argument arises from having described the measurement process by a too idealized scheme. That this is not the case has been proved in absolute generality in a recent paper (Bassi & Ghirardi, 2000), in which it has been shown that, if quantum mechanics has unrestricted validity, the occurrence of the embarrassing superpositions of macroscopically and perceptually different states of macrosystems cannot be avoided.

The original paper by Bassi and Ghirardi has given rise to a stimulating debate with B. d'Espagnat: (d'Espagnat, 2001). The relevant conclusion of this author, for what concerns us here, is:

> What Bassi & Ghirardi has proved is that we must either accept the break [i.e. to abandon the superposition principle] or grant that man-independent reality - to the extent that this concept is meaningful - is something more remote from anything ordinary human experience has access to than most scientists were up to now prepared to believe.

In the appendix the author has felt the necessity to state, as Bassi and Ghirardi have argued, that there are only two ways out of this impasse: either one accepts the break or one must assume that it is consciousness that leads to WPR.

6 Some attempts to overcome the difficulties

The measurement problem has seen an alive and never ending debate as well as many attempts for a consistent solution in the last 90 years. The actors of this "drama" have been scientists of the level of Bohr, Einstein, de Broglie, Schrödinger, Born, Jordan, von Neumann, Wigner and many others.

6.1 Superselection rules

An interesting proposal (Daneri et al. 1962), strongly supported by Rosenfeld, assumes that in principle the set of all conceivable observables of a macroscopic object is an abelian set, so that, when macrostates enter into play, the state (2) cannot be distinguished from the statistical mixture ensuing to WPR i.e., to the situation in which one has an ensemble of systems the fraction $|ck|^2$ of which is in the state $|\varphi k\rangle \otimes |Ak\rangle$. In fact, if all observable quantities commute, they have a common set of eigenstates so that the implications of the assumption that one has an ensemble of systems in these eigenstates distributed according to the probabilistic law $|ck|^2$, makes the ensuing situation indistinguishable from the one associated to the state(4). This has been made more precise by Jauch (Jauch, 1964). The proposal meets serious difficulties since, if the assumption is correct, also the hamiltonian should commute with all observables, and, as a consequence, it could not drive a pointer state from one to a macroscopically different state. However, this certainly occurs, because the state $|A_O\rangle$, corresponding to the ready state of the apparatus, is transformed by the measurement process in the state corresponding to a position of the pointer which differs from 0. This fact in turn implies that the energy of the whole system is not an observable, a quite peculiar and nonsensical fact.

6.2 Many Universes and Many Minds

Important proposals are "The Many Universes Interpretations" of Everett III (Everett, 1957) and DeWitt (de Witt, 1971) and "The Many Minds Interpretation" of Albert and Lower (Albert & Loewer, 1988). The first proposal suggests that all potentialities of the state, referring to different macroscopic situations, become actual in different universes. There is a continuous multifurcation of the Universe associated to superpositions of macroscopically different states. So, when a state like the one of Eq.(7) is dynamically brought into play by the measurement interaction, one must think that there are infinitely many universes, each of them corresponding to one and only one of the perfectly meaningful terms of the superposition.

The Many Minds Interpretation assumes that in place of all potentialities becoming actual in different universes, all possible perceptions occur in appropriately correlated different "sheets" of our brains.

I will not discuss these proposals here. I will limit myself to stress that all of them are affected by an unsatisfactory vagueness. In fact it is not clear when the superselection rules of (Daneri et al. 1962) become effective, as well as when a splitting of the universe or of the brain should occur.

6.3 Bohmian Mechanics

The basic idea (Bohm, 1952a,b) of this approach is that quantum mechanics is an incomplete theory and that, in order to characterize the state of a system, further variables (hidden and inaccessible) are necessary. The hidden variables that supplement the wave-function are the initial positions of the particles of the system. The general scheme goes then as follows: one starts with a given wavefunction $\psi(r_1, ... r_2, ..., 0)$ at time t =0, one solves the Schrödinger equation and derives the wavefunction at time t. In terms of the known wavefunction at the various times one introduces a velocity field which drives the various particles. The fundamental feature of the theory is that it is constructed in such a way that, if the initial density distribution of the particles agrees with $|\psi(r_1, ... r_2, ...;0)|^2$, then, at any subsequent time the quantum density distribution of the particles which propagate deterministically from their precise initial positions in agreement with the velocity field, is given by $|\psi(r_1, ... r_2, ...;t)|^2$, and, as such, it coincides with the one implied by standard quantum mechanics. The theory is completely deterministic, it fully agrees with the quantum predictions concerning the probability density distributions of the positions and it claims that what the theory is about are exclusively the positions of all particles of the universe. The scheme overcomes the difficulties related to WPR and it is mathematically precise. For our purposes it is important to stress that this approach does not meet any difficulty with the psycho-physical correspondence if one accepts that our perceptions are fully determined by the locations of the particles in our brain.

6.4 Decoherence

A lot of attention (Zurek, 1981, 1982, 1991; Griffiths, 1984, 1996; Gell-Mann & Hartle, 1990) has been paid to approaches making a precise reference to the decoherence induced on any quantum system, and in particular on macroscopic ones, by the unavoidable interaction with the environment. The idea is rather simple, and can be depicted by enriching the von Neumann chain by taking into account that the different macrostates $|A_i\rangle$ become correlated with ortogonal states $|E_i\rangle$ of the environment. One should then replace Eq.(7) by the more physically appropriate equation:

$$\sum_i c_i|\varphi_i\rangle \otimes |A_O\rangle \otimes |E_O\rangle \rightarrow \sum_i c_i|\varphi_i\rangle \otimes |A_i\rangle \otimes |E_i\rangle. \qquad (9)$$

Since the experimenter has no control of the degrees of freedom of the environment, i.e. of the states $|E_i\rangle$, he must disregard them. Doing this one ends up dealing, as in the case of Daneri et al. and of Jauch, with a statistical mixture, precisely the one implied by WPR.

This argument characterizes Zurek's approach. Griffiths, Gell-Mann and Hartle "Decoherent Histories" scheme is a more refined approach based on analogous considerations. Some remarks are appropriate.

- These approaches make clear that the unavoidable interactions with the environment and, subsequently, with the whole universe, make extremely difficult to put into evidence the superpositions of different macroscopic states. In this sense they show that the standard theory with the inconsistent WPR postulate works well FAPP, For All Practical Purposes.
- It does not seem acceptable to claim that linear superposition of states corresponding to different perceptions occur but that this fact is irrelevant because each of them is associated, e.g., to a different location of a molecule of the environment (the argument actually rests entirely and simply on the orthogonality of the states $|E_i\rangle$ of the environment).
- The proposal deals essentially with ensembles. But in practice one mostly deals with individuals physical systems.
- Contrary to the classical case, in quantum mechanics, different statistical ensembles may give rise to precisely the same physics. In particular the ensemble containing an equal number of states "|Pointer Here⟩" and "|PointerThere⟩" is physically indistinguishable from the one containing an equal number of states "|Pointer Here⟩ + |PointerThere⟩" and "|Pointer Here⟩ - |PointerThere⟩". Both attach the same probabilities to all conceivable outcomes of prospective measurements (Obviously, these probabilities differ from those implied by the quantum formalism when the state is the final one of Eq.(7)). This being the situation, what makes legitimate to make the first choice and to ignore the possibility of a mixture of states involving the superposition of macroscopically and perceptively different states?

Even the more convinced supporters of decoherence have been compelled to face this problem and have been lead to recognize (Joos & Zeh, 1985) that:

> Of course, no unitary treatment of the time dependence can explain while only one of these dynamically independent components is experienced.

6.5 von Neumann and Wigner

von Neumann himself and, subsequently, Wigner, have taken a clear cut attitude: in nature "physical processes" and "conscious acts of perception" occur and they obey different laws. In brief, they accept that quantum mechanics governs only the first set of processes. von Neumann stresses that, in his chain, the final state always refers to the act of perception of a conscious observer, a process which, in his view, is not governed by quantum mechanics but by WPR. The idea is fascinating and it has a certain consistency. Its limitation derives from the fact that it does not make precise the borderline between the two levels. As J. Bell has put it:

> What is conscious? The first living cell or a Ph.D student?

This concludes the first part of our analysis and leads us to consider modifications of the standard theory.

Part II

7 Collapse or GRW models

Quite recently a new approach to the problem of interest has been proposed. It is based on the idea that (Bell, 1990):

> Schrodinger's equation is not always true.

One adds to the standard evolution equation nonlinear and stochastic terms which strive to induce WPR at the appropriate level, leading to states which correspond to definite macroscopic outcomes. The theory, usually referred as the GRW theory (Ghirardi et al. 1986), is a rival theory of quantum mechanics and is experimentally testable against it. Its main merits are that it qualifies itself as a unified theory governing all natural processes, in full agreement with quantum predictions for microscopic processes and inducing the desired objectification of the properties of macroscopic systems. Let us be more precise.

- A first problem concerns the choice of the preferred basis: if one wants to objectify some properties, which ones have to be privileged? The natural choice is the one of the position basis, as suggested by Einstein (Einstein, 1926):

    > A macro body must always have a quasi-sharply defined position in the objective description of reality

- A second problem, and the more difficult, is to embody in the scheme a triggering mechanism implying that the modifications to the standard theory be absolutely negligible for microsystems while having a remarkable effect at the macroscopic level.

The theory is based on the following assumptions:

- Let us consider a system of N particles and let us denote as $\psi(r_1, ..., r_N)$ their configuration space wavefunction. The particles, besides obeying the standard evolution, are subjected, at random times with a mean frequency λ, to random and spontaneous localization processes around appropriate positions. If a localization affects the i-th particle at point x, the wavefunction is multiplied by a Gaussian function $G_i(x) = (\frac{\alpha}{\pi})^{3/4} \exp[-\frac{\alpha}{2}(r_i - x)^2]$.

- The probability density of a localization for particle i at point x is given by the length(< 1) of the wavefunction immediately after the hitting: $G_i(x)\psi(r_1, ..., r_N)$. This implies that localizations occur with higher probability where, in the standard theory, there is a larger probability of finding the particle.

Obviously, after the localization has occurred the wavefunction has to be normalized.

It is immediate to realize that a localization, when it occurs, suppresses the linear superposition of states in which a particle is well localized at different positions separated by a distance larger than $1/\sqrt{\alpha}$. The situation is depicted in Fig.3.

The most important feature of the model stays in the trigger mechanism. To understand its role, let us consider the superposition of two macroscopic pointer states $|H\rangle$ and $|T\rangle$, corresponding to two macroscopically different locations of the pointer. Since the pointer is "almost rigid" and contains an Avogadro's number of microscopic constituents one immediately realizes that a localization of anyone of them suppresses the other term of the superposition: the pointer, after the localization of one of its constituents, is definitely either Here or There (see Fig.4 for an intuitive understanding of the process).

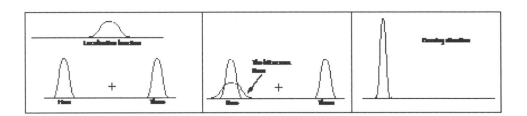

Figure 3: The localization affecting a particle in the superposition of two far-away position states.

With these premises we can choose the values of the parameters of the theory (which Bell has considered as new constants of nature): the mean frequency of the localizations A and their accuracy $1/\sqrt{\alpha}$. In the original proposal these values have been chosen (with reference to the processes suffered by nucleons, since the frequency is proportional to the mass of the particles) to be:

$$\lambda = 10^{-15} \text{sec}^{-1}, \quad \frac{1}{\sqrt{\alpha}} = 10^{-5} \text{cm} \tag{10}$$

So, a microscopic system suffers a localization, on average, every hundred millions years. This is why the theory agrees to an extremely high level of accuracy with quantum mechanics for microsystems. On the other hand, due to the trigger mechanism, one of the constituents of a macroscopic system, and, correspondingly, the whole system, undergoes a localization every 10^{-7} seconds.

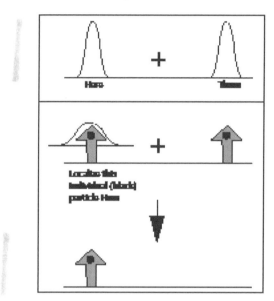

Figure 4: The trigger mechanism: in the case of a macroscopic body, localizing one of its constituents amounts to localizing the system itself.

Few comments are at order:

- The theory allows to locate the ambiguous split between micro and macro, reversible and irreversible, quantum and classical. The transition is governed by the number of particles which are well localized at positions further apart than 10^{-5} cm in the two states whose coherence is going to be dynamically suppressed.

- The theory is testable against quantum mechanics. Various proposals have been put forward (Rae, 1990; Rimini, 1995; Adler, 2002; Marshall et al. 2003). The tests are difficult to be performed with the present technology. The theory identifies appropriate sets of mesoscopic processes which might reveal the failure of the superposition principle.

- Most of the physics does not depend on both parameters, but on their product OOA. A change of one or two orders of magnitude of this value will already conflict with experimentally established facts. So, in spite of its appearing "ad hoc", if one chooses to make objective the positions (and no other alternatives are practicable), almost no arbitrariness remains.

An interesting feature deserves a comment. Let us make reference to a discretized version of the model. Suppose we are dealing with many particles and, accordingly, we can disregard Schrodinger's evolution of the system because the dominant effect is the collapse. Suppose that we divide the universe in elementary cells of volume $10^{-15} cm^3$, the volume related to the localization accuracy. Denote as |n1, n2, ...) a state in which there are ni particles in the i-th cell and let us consider the superposition of two states |n1, n2, ...) and |m1, m2, ...) which differ in the occupation number of the various cells. It is then quite easy to prove that the rate of the dynamical suppression of one of the two terms is governed by the quantity:

$$\exp\left\{-\lambda t \sum_i (n_i - m_i)^2\right\}, \tag{11}$$

the sum running over all cells of the universe.

It is interesting to remark that, being $\lambda = 10^{-16}\,sec^{-1}$, if one is interested in time intervals of the order of perceptual times (i.e. about 10^{-2} sec) this expression implies that the universal dynamics characterizing the theory does not allow the persistence for perceptual times of a superposition of two states which differ for the fact that 10^{18} nucleons (a Planck's mass) are differently located in the whole universe. This remark suggests a relation with the idea of Penrose who, to solve the measurement problem by following the quantum gravity line, has repeatedly claimed that it is the Planck mass which defines the boundary between the wavy quantum universe and the world of our definite perceptions.

8 Collapse theories and definite perceptions

An interesting objection to GRW has been presented (Albert & Vaidman, 1989; Albert, 1992). By taking advantage of the extreme sensitivity of our perceptual apparatuses, the authors have remarked that one can devise situations leading to definite perceptions which do not involve the displacement of a sufficient number of particles (such as those of a pointer) in order that the GRW mechanism enters into play.

Albert and Vaidman have considered a spin 1/2 particle with the spin along the x-axis which goes through a Stern-Gerlach apparatus devised to test the value of the z-spin component. The particle ends up in the superposition of moving along a trajectory pointing upwards and one pointing downwards (see Fig.5). After the apparatus there is a fluorescent screen such that when it is hit by the particle at a given point, about 10 atoms with an extremely short life-time are excited and decay immediately. In this way, we produce the linear superposition of two rays originating from two points A and B, differently located on the screen.

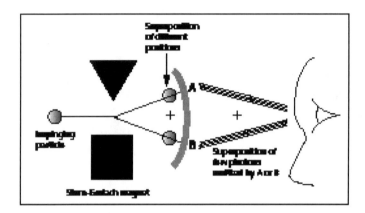

Figure 5: The set up suggested by Albert and Vaidman which should prove that also the GRW theory must attach a specific role to conscious perceptions.

The argument goes on: due to the fact that the process is triggered by a single particle and that it involves few photons, there is no way for the GRW dynamics to induce reduction on one of the two states. On the other hand we know that, due to the sensitivity of our visual apparatus, an observer looking at the screen will have a definite perception concerning the fact that the luminous signal comes from A or from B. Accordingly, in Albert and Vaidman's opinion, collapse theories must accept reduction by consciousness in order to account for definite perceptions by the observer.

Note that since microscopically different situations usually trigger macroscopic changes in the material world, only in exceptional cases one should resort to consciousness. Apart from this fact (which has not been appropriately recognized by the above authors) we have taken the challenge represented by the smart suggestion of Albert and Vaidman. Our counterargument goes as follows:

- We agree that the superposition of the considered microstates persists for the time which has to elapse before the eye is stimulated (actually it must persist because in place of observing the rays we could make them interfere, proving that the superposition is there).
- However, to deal with the process in the spirit of the GRW theory, one has to consider all systems which enter into play (triggering particle, screen, photons and brain). A quite prudential estimate of the number

of ions (Na and K ions going through the Ranvier nodes and trasmitting the electrical signal to the lateral geniculate body and to the higher visual cortex) which are involved, makes perfectly reasonable (we refer the reader to (Aicardi et al. 1991) for details) that a sufficient number of particles are displaced by a sufficient spatial amount (recall that the myelin sheet is precisely 10^{-5} cm thick) to satisfy the conditions under which, according to the GRW theory, the suppression of the superposition of the two signals will take place within the perception time scale.

- We are not attaching any particular role to the conscious observer. The observer's brain is the only system which is present in which a superposition of two states involving different locations of a large number of particles occur. As such, it is the only place where the reduction can and actually must occur according to the theory. If in place of the eye of a conscious observer one puts in front of the photon beams a spark chamber or a device leading to the displacement of a macroscopic pointer, reduction will equally take place. In the example, the human nervous system is simply a physical system, a specific assembly of particles, which performs the same function as one of these devices, if no other device interacts with the photons before the human observer does.

Before concluding we feel the need of a specification. The above analysis could be taken as indicating a naive and oversimplified attitude towards the deep problem of the mindbrain correspondence. There is no claim and no presumption that the GRW theory allows a physicalistic explanation of conscious perceptions. We only point out that, for what we know about the purely physical aspects of the process, before the nervous pulse reaches the higher visual cortex, the conditions guaranteeing, according to collapse models, the suppression of one of the two electric pulses are verified. In brief, a consistent use of the dynamical reduction mechanism accounts for the definiteness of the conscious perception, even in the peculiar situation devised by Albert and Vaidman.

References

Aicardi, F., Borsellino, A., Ghirardi, G.C., Grassi, R. (1991). Dynamical Models for State Reduction: do they Ensure that Measurements have Outcomes? Found. Phys. Lett., 4, 109-128.

Adler, S. (2002). Environmental Influence on the Measurement Process in Stochastic Reduction Models. J. Phys., A 35, 841-858.

Albert, D.Z. (1992). Quantum Mechanics and Experience. Harvard University Press, Cambridge, Mass.

Albert, D.Z., Loewer, B. (1988). Interpreting the Many Worlds Interpretation. Synthese, 77, 195-213.

Albert, D.Z., Vaidman, L. (1989). On a Proposed Postulate of State-Reduction. Phys. Lett., A 139, 1-4.

Bassi, A., Ghirardi, G.C. (2000). A General Argument Against the Universal Validity of the Superposition Principle. Phys. Lett., A 275, 373-381.

Bell, J.S., (1990). Are there quantum jumps? in: Schroedinger - centenary celebration of a polymath. Kilmister C.W. (Ed.), Cambridge University Press, Cambridge. 41-52.

Bohm, D. (1952a). A Suggested Interpretation of Quantum Theory in Terms of 'Hidden Variables' I. Phys. Rev., 85, 166-179.

Bohm, D. (1952b). A Suggested Interpretation of Quantum Theory in Terms of 'Hidden Variables' II. Phys. Rev., 85, 180-193.

Daneri, A., Loinger, A., Prosperi, G.M. (1962). Quantum Theory of Measurement and Ergodicity Conditions. Nuclear Phys., 33, 297-319.

d'Espagnat, B. (2001). A Note on Measurement. Phys. Lett., A 282, 133-137.

DeWitt, B.D. (1971). The Many Universes Interpretation of Quantum Mechanics. in: Foundations of Quantum Mechanics. d'Espagnat, B. (Ed.), Academic Press, N.Y.

Einstein, A. (1926), in: Pauli, W., Born, M. (1971). The Born-Einstein letters. Walter and Co., N.Y. 90-91.

Everett, H. (1957). 'Relative State' Formulation of Quantum Mechanics. Rev. Mod. Phys., 29, 454-462.

Gell-Mann, M., Hartle, J.B. (1990). Quantum Mechanics in the Light of Quantum Cosmology. in: Proceedings of the 3rd International Symposium on the Foundations of Quantum Mechanics in the Light of New Technology. Kobayashi, S. et al. (Eds.) Physical Society of Japan, Tokio, 1990. 321-343.

Ghirardi, G.C., Rimini, A., Weber, T. (1986). Unified Dynamics for Microscopic and Macroscopic Systems. Phys. Rev., D 34, 470-491.

Griffiths, RB. (1984). Consistent Histories and the Interpretation of Quantum Mechanics. J. Stat. Phys., 36, 219-272.

Griffiths, RB. (1996). Consistent Histories and Quantum Reasoning. Phys. Rev., A 54, 2759-2774.

Jauch, J.M. (1964). The problem of Measurement in Quantum Mechanics. Helv. Phys. Acta, 37, 293-316.

Joos E., Zeh, H.D. (1985). The Emergence of Classical Properties through the Interaction with the Environment. Zeit. Phys., B 59, 223-243.

Marshall, W., Simon, C., Penrose R., Bouwmeester, D. (2003). Towards Quantum Superpositions of a Mirror. Phys. Rev. Lett., 91, 130401-130404.

Primas, H. (1990). The Problem of Measurement in Quantum Mechanics. Panel Discussion II. in: Symposium on the Foundations of Modern Physics 1990, Discussion Sections. Laurikainen K.V., Viiri, J. (eds.), Joensuu, Finland. 42-50.

Rae, A.I.M. (1990). Can GRW Theory be tested by Experiments on SQUIDS? J. Phys., A 23, L57-L60.

Rimini, A. (1995). Spontaneous Localization and Superconductivity. in: Advances in Quantum Phenomena. Beltrametti E. et al (Eds.), Plenum Press, N.Y. US. 321-333.

Zurek, W. (1981). Pointer Basis of Quantum Apparatus: into what Mixture does the Wave Packet Collapse? Phys. Rev., D 24, 1516-1525.

Zurek, W. (1982). Environment Induced Superselection Rules. Phys. Rev., D 26, 18621880.

Zurek, W. (1991). Decoherence and the Transition from Quantum to Classical. Physics Today, 44, v. 10, 36-44

17. A Mindful Alternative to the Mind/Body Problem
Ellen Langer, Ph.D.
Harvard University, Cambridge, MA

The age-old "mind/body" problem (i.e. How can something non-material, such as a thought, affect the material body?) continues to challenge philosophers and scientists alike. The implicit assumption, however – that mind and body are separate entities – may be the problem that needs to be addressed. From Plotinus to Nagarjuna to Spinoza, a long line of thinkers through the ages have proposed that mind and body are but two sides of the same coin. That many such thinkers were often dwelling over concerns of philosophy or religion when they developed this idea may unfortunately have caused this insight to be met with suspicion, even outright derision, by the modern scientific academe. Current findings from fields as diverse as social psychology, neurobiology, and cognitive science, however, indicate that the tides of popular sentiment may again be turning.

As early as 1976/77, Judith Rodin and I gave nursing home elders mindful choices to make in a controlled randomized trial and found an increase in longevity (Langer and Rodin, 1976; Rodin and Langer, 1977). We replicated the effect of a mindfulness intervention on longevity in other experimental investigations (Alexander, et al, 1989; Langer, et al 1984) as well. Our research indicated that merely changing the pattern of one's thinking could indeed generate significant effects in the body and that mind and body were not as divorced from one another as the dominant scientific paradigm at that time had theretofore assumed. Today it is more or less taken for granted that mind effects body although the pathways are still unknown but mind-body monism is not accepted by the modern scientific world because pathways are still unknown.

In this essay, I am proposing a reworking of our understanding of the relationship between mind and body where the search for pathways from one to the other may be misguided, and do so from the perspective of mindfulness theory. To better understand the context of this proposal, it will perhaps be helpful to first review just what I mean by mindfulness and its counterpart, mindlessness.

Mindfulness is defined as an active state of mind characterized by novel distinction–drawing that results in being 1) situated in the present; 2) sensitive to context and perspective and 3) guided (but not governed) by rules and routines. The phenomenological experience of mindfulness is the felt experience of engagement. The process of noticing or creating novelty reveals inherent uncertainty. When we recognize that we don't know the person, object, or situation as well as we thought we did, our attention naturally goes to the target.

Mindlessness, by contrast, is defined as an inactive state of mind characterized by reliance on distinctions and categories drawn in the past. Here 1) the past over-determines the present; 2) we are trapped in a single perspective but oblivious to that entrapment; 3) we are insensitive to context; and 3) rules and routines govern rather than guide our behavior. Moreover, mindlessness typically comes about by default rather than by design. When we accept information as if unconditionally true, we become trapped by the substantive implications of the information. Even if it is to our advantage in the future to question the information, if we mindlessly processed it, it will not occur to us to do so (Chanowitz and Langer, 1981). The same rigid relationship results from mindless repetition (Langer and Imber, 1979).

Mindfulness allows for doubt and that allows for choice and thus free will. Being in a mindful state, removed from rigid routines, introduces possibilities from which one can make alternative choices and thus exercise their free will. When mindless, by contrast, our behavior is predetermined by the past, closing us off to choice and new possibilities. We live in a world governed by the principles of science. The precision with which we can now measure the world in and around us is, however, only as useful as the degree of mindfulness we employ to analyze it. Science becomes mindless when we automatically begin to conflate precision with certainty. Certainties lead to mindlessness; when we think we know, there is no reason to find out. Too often scientists observe a phenomenon, create a theory to explain it and then collect data to prove their theory. Not surprisingly, confirmation is found. Theory is supposed to be understood as possibility but at least in the social sciences, most often is taken as absolute fact leaving little experienced difference between laws and theories. These theories build upon each other with the result of a series of concatenated probabilities making it harder and harder to question the basic assumptions of the original proposition. Scientific evidence can only yield probabilities but science in use takes these probabilities and converts them into absolutes. This practice makes it hard to question basic assumptions.

Take medicine for example. Many diseases are labeled chronic. Chronic is understood as uncontrollable. If something is understood to be uncontrollable we would be foolish to try and control it. Yet no science can prove uncontrollability. All science can prove is that something is possible or it is indeterminate. Indeterminate is very different from uncontrollable. Moreover, by generalizing the findings to the general population because of methodological considerations like random assignment without due regard to the subject population used (i.e. all of those people who self heal are missing from the medical experiment) we are discouraged from trying to self heal. In any experiment the researcher has to make many hidden decisions regarding the parameters of the study (i.e. who the subjects are, the time and circumstances in which they'll be tested, the amount of the independent variable to administer, etc). With these dimensions out of mind, findings seem more certain than they might otherwise seem. Couple this with the mistaken tendency of people to seek certainty and confuse the stability of their mindsets with the stability of the underlying phenomena, and we end up with an illusion of knowing and unnecessary limits to what we might otherwise find out.

This illusory sense of knowing is pervasive, extending even to the point where we misconstrue the nature of our own mental processes. What are we actually doing when we hold a certain concept in our mind's eye? Picture a car, for example. Now, start taking away individual elements that seem essential to the "car-ness" of it all, and ask yourself if you'd still know it's a car. A car without wheels? Still a car. Minus a steering wheel, or a bumper or an engine? Still seen as a car (albeit perhaps not one you'd want as yours.). A Jeep and a station wagon and a Smartcar all somehow fit into this same category of "car" despite their clear diversity in features and appearance. Wittgenstein (1953) famously performed a similar dissection of conceptual categories, effectively demonstrating (in his case, with the concept of "game") the inherent illusion that our mental categories for things are actually based upon some identifiable set of core features. So what is it that makes a car a car? Not much, as it turns out.

Recent findings in the field of cognitive neuropsychology have begun to indicate that this assertion – that conceptual categories lack inherent unifying features – is backed by more than just sound logic. Barsalou (2009) and Wilson-Mendelhall et al. (2010) have established that the brain doesn't actually use a set of core concepts to define mental categories of objects and phenomena – rather, our thought processes remain in a perpetual state of collection, assessment, and reaction to incoming information. It is only at the point of higher-level cognitive processes that we begin to grow lazy and assume that all examples of cars have some inherent "car-ness" about them (Or, for that matter, that all instances of fear, or anger, or pride, must necessarily be connected by some unifying element). In reality, the idea of "car" (or "fear", or any other concept) is actually represented in our brains as a loose amalgam of instances (this morning on the way to work in traffic, on a showroom floor, in a junkyard), specific examples (a smartcar, a station wagon, a jeep), functions (creating momentum, providing shelter, controlling climate), and other characteristics of certain objects that we learn at some point to clump together. In short, there's no core element that makes a car a car every time, all the time. Mindfulness requires that we engage the world with this same degree of dynamism and flexibility.

No matter what we are doing, we are doing it mindlessly or mindfully and the consequences of being in one state or the other are enormous. Research described in over 150 research papers and four books on the topic of mindfulness reveals that the simple processes of creating and noticing novelty are literally and figuratively enlivening. We've found increases in well being, health, competence, relationship satisfaction, effective leadership and creativity to name a few of the many findings. Perhaps the most startling findings are the most recent. In one study we instructed a group of symphony musicians to play a familiar piece of music and "make it new in very subtle ways that only they would know." Another group of musicians was told to "recall a performance of the music that they were very pleased with and replicate it." We taped the performances and played them for audiences, blind to our instruction, and they overwhelmingly preferred the mindfully played piece. The musicians showed a similar preference (Langer, Russell, and Eisenkraft, 2009). An interesting aspect to this work is that rather than cacophony, when everyone "did it their own way," superior coordinated performance resulted. In other work we also showed that mindfulness seems to leave its imprint in the products of our labor.

More important to the present discussion is recent work that follows up on research originally conducted in 1981. The idea was and is deceptively simple. Mind and body are just words, concepts to which we rigidly adhere. Consider artificial boundaries like North vs South Korea or old (>65) vs. young (<65), where the concepts may have been mindfully generated initially but then took on a life of their own.

What would happen, we asked, if we got rid of the distinction between mind and body? If we put the mind and body back together so to speak, then wherever the mind is, so too would be the body. Within this understanding there is no reason to search for mediating mechanisms. Whatever is going on at the level of the brain is happening simultaneously with the thought and is just another level of analysis. With this view in mind we conducted a series of investigations where we put minds in healthy places and took physical measurements.

In the first of these studies, elderly men were taken to a timeless retreat retrofitted to 20 years earlier. To firmly anchor their minds in that earlier time they would speak in the present tense about the past for the full week they spent there. A comparison group of men lived the week at the retreat reminiscing about the past. For them, their minds were firmly in the present. The results were notable, especially considering that the study was conducted back in 1981 before there was much mind-body research and before 80 became the new 60. Despite how enfeebled these men in their 80's were at the start of the study, both groups improved significantly from where they started. Hearing, vision, memory and grip strength were significantly different after the week. The experimental group showed greater improvement differing significantly from the comparison group with respect to manual dexterity, IQ: 63 % > (only 44% of the control group); height, gait, posture, joint flexibility, diminished symptoms of arthritis. We photographed everyone before and after the week and found that all the experimental participants looked noticeably younger at the end of the study.

In my view, it was the change in mindset, much the same way a placebo works, that accounted for the difference between the two groups. By priming a time when they were vital, their mindsets of old age as a time of debilitation became irrelevant. Of course over the week many things could have varied that we couldn't possible control in such an ambitious undertaking. We were able to use tighter controls, however, in more recent investigations. Two things should be addressed regardless of the explanation for the findings one may choose. The first is the widespread belief that elders are not supposed to improve their hearing and vision—or indeed improve on any of the measures we took. I'll return to this in a later discussion of science. The second issue to consider is that the idea of mind-body unity led to these findings and thus, at the least, the theory serves a heuristic purpose.

Ali Crum and I (Crum and Langer, 2007) tested this mind-body hypothesis in a very different setting. We ran the study with chambermaids. We started by inquiring about how much exercise they thought they got in a typical week. Surprisingly, they thought they didn't get exercise, despite the fact that their work is exercise. Exercise, they thought was what one did after work. If exercise is good for our health and they get more than the surgeon general recommends, then we should expect that they would be healthier than socioeconomically-equivalent others who do not exercise as much or as consistently. Interestingly, they were less healthy. While noteworthy, this was not the focus of the study. We randomly divided the participants into two groups and taught one group to change their mindset and to regard their work as exercise. We took as many measures as we could think of regarding food eaten during the course of the month between tests, exercise intensity at work and exercise outside of work. There were no differences between the two groups on any of these measures. One group continued to see their work as exercise and one group did not. We found significant differences on measures of waist to hip ratio, weight loss, body mass index and blood pressure. We attribute these improvements for the experimental group to the change in mindset.

We tested this mind-body hypothesis in another series of experiments. Here we focused on vision. The standard eye chart has letters that get progressively smaller as one reads down the chart. Implicitly this creates the expectation that soon we will not be able to see. We reversed the eye chart so that the letters get progressively larger, thereby creating the mindset that soon we will be able to see. With the change in mindset, participants tested were able to see what they couldn't see before. There is also an expectation on the eye chart that we will start to have difficultly around two-thirds of the way down the chart. Accordingly, we created a chart that began a third of the way the standard chart. Again, participants could see what they couldn't see before.

In another study we took advantage of the assumption that pilots have excellent vision. We had men don the clothes of air force pilots and fly a flight simulator. Control participants simulated flying the simulator. Vision improved for those embodying the mindset of pilot (Langer, Djikic, Pirson, Madenci, & Donohue 2010).

Finally, we wanted to see if we could condition improved vision. Participants read a chapter of one of my books where the font of either the letter "a" or the letter "e" was imperceptibly small or they read the chapter in normal font size. Over time, participants would of course come to know what the "dot" represented. Consider seeing "c.n, t.ke, and m.ny" for example. We found: When participants were exposed to the small a and then later tested for the letter a: 19.3% of the people in the experimental group saw better compared to those in the control group ($p=0.036$). When exposed to the small a and tested for the letter e: 50.7% of individuals in the experimental group saw better than compared to those in the control group ($p<0.001$). When exposed to the small a and tested for the letter e : 44% of participants in the experimental group saw better compared to the control group ($p <0.001$). When exposed to the small e and tested for the letter a: 26.8% of participants in the experimental group saw better compared to the control group ($p<0.01$). When exposed to the manipulation of both letter a and e and then tested for all letters in Snellen eye chart the experimental group missed 5.44 of the small letters a and the control group missed 8.74 letters. In the Snellen eye chart test, the experimental group missed 5.88 of the small letters e and the control group missed 9.37 letters. ($p<.05$) (Pirson and Langer, 2010)

Our accepted theories and mindsets tell us that vision is not supposed to improve. But from where do these mindsets come? We accept negative mindsets, e.g. vision will necessarily worsen over time, and we create theories of the eye to show why this must be. The expectation becomes self-fulfilling, further validating the original supposition. Yet with this simple understanding that our own minds may create our seeming limitations, we may come to see and function beyond where alternative mind-body views currently enable.

References

Alexander, C., Langer, E., Newman, R., Chandler, H. & Davies, J. (1989). Aging, mindfulness and meditation. Journal of Personality & Social Psychology, 57, 950-964.

Barsalou, L.W. (2009). Simulation, situated conceptualization, and prediction. Philosophical Transactions of the Royal Society of London: Biological Sciences, 364, 1281-1289.

Chanowitz, B. & Langer, E. (1981). Premature cognitive commitment. Journal of Personality and Social Psychology, 41, 1051-1063.

Crum, A. & Langer, E. (2007) Mindset Matters: Exercise as a Placebo. Psychological Science, 18, 2, 165-171.

Langer, E. & Rodin, J. (1976). The effects of enhanced personal responsibility for the aged: A field experiment in and institutional setting. Journal of Personality and Social Psychology, 34, 191-198.

Langer, E. & Imber, L. (1979). When practice makes imperfect: the debilitating effects of overlearning. Journal of Personality and Social Psychology, 37, 2014-2025.

Langer, E., Beck, P., Janoff-Bulman, R. & Timko, C. (1984). The relationship between cognitive deprivation and longevity in senile and non-senile elderly populations. Academic Psychology Bulletin, 6, 211-226.

Langer, E., Russell, T., & Eisenkraft, N. (2009) Orchestral performance and the footprint of mindfulness. Psychology of Music.

Langer, E., Madenci, A., Djikic, M., Pirson, M. and Donahue, R., (2010) Believing is seeing: reversing vision inhibiting mindsets. Psychological Science, In press.

Pirson, M., Langer, EJ. and Ie, Amanda, (2010) Seeing What We Know, Knowing What We See: Challenging the Limits of Visual Acuity. Fordham University Schools of Business Research Paper No. 2010-023.

Rodin, J. & Langer, E. (1977). Long-term effects of a control-relevant intervention among the institutionalized aged. Journal of Personality and Social Psychology, 35, 897-902.

Wilson-Mendenhall, C.D., Barrett, L.F., Simmons, W.K., & Barsalou, L.W. (2010). Grounding emotion in situated conceptualization. Neuropsychologia. IN press.

Wittgenstein, Ludwig (1953). Philosophical Investigations. Oxford: Blackwell Publishing.

18. Consciousness: The Fifth Influence
Michel Cabanac[1], Rémi Cabanac[2], the late Harold T. Hammel[3]

[1]Département de psychiatrie & neurosciences, Faculté de médecine Université Laval, Québec, Canada G1K 7P4

[2]Institut de Recherche en Astrophysique et Planétologie, Université de Toulouse, Univ. Paul Sabatier, Centre national de la recherche scientifique, Tarbes, France

[3]Department of Physiology and Biophysics, Indiana University Bloomington, IN, U. S. A.

Abstract

This article is a theoretical consideration on the role of sensory pleasure and mental joy as optimizers of behaviour. It ends with an axiomatic proposal. When they compare the human body to its environment, Philosophers recognise the cosmos as the Large Infinite, and the atomic particles as the Small Infinite. The human brain reaches such a degree of complexity that it may be considered as a third infinite in the Universe, a Complex Infinite. Thus any force capable of moving that infinite deserves a place among the forces of the Universe. Physicists have recognized four forces, the gravitational, the electromagnetic, the weak, and the strong nuclear forces. Forces are defined in four dimensions (reversible or not in time) and it is postulated that these forces are valid and applicable everywhere. Pleasure and displeasure, the affective axis of consciousness, can move the infinitely complex into action and no human brain can avoid the trend to maximize its pleasure. Therefore, we suggest, axiomatically, that the affective capability of consciousness operates in a way similar to the four forces of Physics, i e. influences the behaviour of conscious agents in a way similar to the way the four forces influence masses and particles. However, since a mental phenomenon is dimensionless we propose to call the affective capability of consciousness the fifth influence rather than the fifth force.

KEY WORDS: consciousness evolution optimisation force pleasure/joy

1. INTRODUCTION

In the Seventeenth Century the Mathematician-Philosopher Blaise Pascal (1670-1672), considering the world of his time, recognized anxiously that the human was balanced between two abysses: the infinitely small abyss of atomic particles, raccourci d'atome, and the infinitely large abyss of the cosmos, l'univers. In our times, the Pascalian anguish would have worsened with the knowledge accumulated since the Seventeenth Century. Both the small and the large infinites have gained several orders of magnitude since Pascal's time. However, in the mid Twentieth Century the Palaeontologist-Philosopher Pierre Teilhard de Chardin (1965), in his attempt to reconcile Biology with Physics and Astronomy (It is often considered that Teilhard's main thrust was an attempt to reconcile science and faith, however in this precise book his aim was indeed what we state here), removed the Pascalian anguish. His reconciliation was the result of his reckoning that the human phenomenon represents also an infinite. He considered that matter is organised in living beings and, as a rough estimate of this organisation, he counted the number of atoms co-ordinated in autonomous entities such as the human body. In this organism the number reached 10^{25}, i.e. a number similar to the magnitude of the positive and negative exponents for the size of the Cosmos and the atom measured in centimeters (Fig. 1).

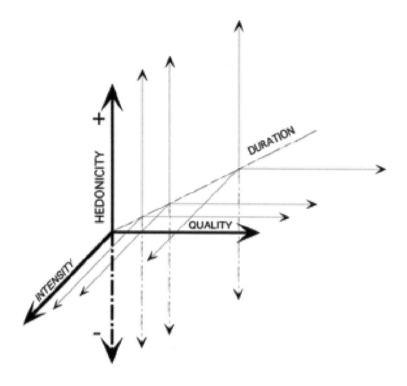

Figure 1. Teilhard de Chardin's natural curve of complexity. In ordinates the length of identifiable entity-objects measured in centimetres (long. en cm.). Homme, human; Terre (rayon), Earth (radius); Voie lactée, Galaxy; Univers, Universe. In abscissa complexity (Complexité, en n. d'atomes) as estimated in number of atoms. Cellule, Cell; Lemna, water lentil (duckweed);Homme (cerveau), Human (brain). a, appearance of Life; b, appearance of Homo (from Teilhard de Chardin, 1965, with permission).

2. THE BRAIN AND THE INFINITE COMPLEXITY

Our purpose is not to confirm or support Teilhard de Chardin's philosophical point of view. Moreover, it may seem irrelevant to compare numbers with different dimensions. However, we acknowledge Teilhard de Chardin's philosophical breakthrough when he recognized infinite complexity as a relevant feature of Nature. Common language use the word 'infinite' more liberally than Mathematicians or Physicists. Infinity is the acknowledgement of an immensity in comparison with the usual human environment. For the physicist, infinite is nearer to the mathematical definition, i.e. • $\int 1/0$. The Infinite of the physicist may be a starting paradigm or framework (e.g. Joseph, 2010) but when physicists find an infinite emerging from their equations, they mean they reached the end of their model. In this article, we use 'infinite' in its common rather than its mathematical acceptation.

The size of the observable universe is the distance of the farthest object the light of which reaches us i.e. the causal horizon (sphere of light since the beginning of the universe); at greater distance no information can be gathered, the universe is unknown. This distance is now estimated of the order of 3000 Megaparsec. Expressed in metres, the unit of the Systeme International closest to the size of the human body, the size of the observable universe turns out to be of the order of 10^{26} m. Elementary particles are defined by their energy equivalent rather than their radius; yet, the radius of the atom of hydrogen is about 1 Å and that of the proton constituent of this atom's nucleus is about 10^{-15} m; the electron is estimated to be less than 10^{-18} m. The three quarks, constituents of a proton, have the same radius as the proton. Expressed in meters the sizes of the smallest identified particles turn out to be of the order of 10^{-18} m. These figures are the present state of our knowledge; they do not change Teilhard de Chardin's message.

Gell-Mann (1944), has underlined the difficulty of defining and, furthermore, quantifying complexity. The quantitative comparing of biological complexity to the dimensions of universe and quarks makes little sense in terms of numbers because complexity has no dimension and therefore cannot be compared to units of length. However, it is of interest to obtain numbers to illustrate the astronomic complexity reached in living beings. The number of atoms in the human body, used by Teilhard de Chardin as a gross estimate of complexity can no longer be accepted; the number of atoms organised in a Sequoia or a whale is much larger than in a human body and nothing indicates that these entities are more complex than humans. As suggested by Teilhard de Chardin himself, the number of atoms is not the best index of complexity. Among other candidate indexes of complexity, such as the genetic code, the central nervous system is more likely to be the locus of the highest and most organised complexity in nature. The human central nervous system is generally accepted as containing about 10^{11} to 10^{12} neurones organised in a single whole entity. Each neurone is capable to communicate synaptically with other neurones. Rather than the raw number of cells contained in the brain, synapses are a better indicator of the brain complexity. It is possible to make a rough estimate of the number of synapses accumulated in the human central nervous system (Kandel & Schwartz, 1985). The number of synapses per neurone is extremely variable from 2 in the highly specialised bipolar retinal cells, to 150 000 in Purkinje cells. A conservative figure may be estimated ca. 10 000 as the average number of synapses per neurone. The resulting number of synapses would thus reach ca. 10^{15}. This is a conservative estimate since 10^{-14}-10^{15} was proposed as the number of synapses in the human cortex only Changeux (1983). Such a number defies the imagination, but can be compared to other phenomena to illustrate its magnitude: if the universe is 10-20 billion years old, the number of synapses in a human brain is the same as the number of minutes since the big bang. Yet, this is not the ultimate of brain complexity. Although the synaptic transmission of an action potential is an all or nothing phenomenon, a synapse is capable of modulating information in a far more complex way than this simple Boolean computation. The mechanism involves neurotransmitter vesicles that operate as quanta of synaptic transmission. The number of vesicles in a given synaptic button is variable but may be estimated conservatively as 10^2. In turn, the index we have selected to estimate complexity reaches now 10^{17}. This figure is given to illustrate the brain complexity. Other criteria would provide numbers of similar or higher magnitude e.g. Penrose (1994) estimated that the brain is able to perform 10^{24} operations per second. Thus, Teilhard de Chardin's solution to Pascal's anguish may be transposed from the human body to the single human brain, the complexity of which may be considered as a third infinite, the infinite complex.

A somewhat similar line was developed in what is commonly designated as the Anthropic principle (Barrow & Tipler, 1986; Bertola & Curi, 1993; Breuer, 1991). However, the argument defended in the present article is independent of the anthropic principle. In this article we address the question of the 'how' agents are brought into motion, not the question of the 'why' they are motioned. We may now consider what can move this infinite complex, what force can induce the brain, taken as an individual whole entity, to behave.

Before going further we must acknowledge that the brain is the locus of the second emergence of Evolution. After emergence of life from matter, thinking and consciousness emerged from complex nervous systems. We should not be arrogant enough to shun the possibility that other forms of thinking complexity may exist in the cosmos but, in the present state of our knowledge, the human brain is the most advanced locus of thinking. In the following we shall accept that cognition/consciousness takes place in the brain, and will not be concerned by the (necessary) structures underlying consciousness and pleasure (see, Shizgal 1997, Shizgal & Arvanitogiannis 2003; Berridge 1996 ; Berridge & Winkielman, 2003, Berridge & Kringelbach 2008) but will restrict our approach to the infinite complexity of the thinking brain on its emerging properties and, thus, will concentrate on, sensation and consciousness.

PLEASURE AND THE PHILOSOPHERS Moralists have always recognized that the seeking of pleasure is the motor of action. Here is what but a sample of a few thinkers wrote: "mankind have nothing better under the sun than to eat and drink and rejoice" (Ecclesiastes 900 B.C.); "one may also think that, if all humans seek pleasure, that is because they desire to live" Aristotle (Aristote, 4th B.C.); "nature gives us organs to inform us through pleasure about what we should seek and through pain what we should avoid" Condillac (1754); "pleasure is the spring of action" Bentham (1742-1832, in Bowring, 1962); "pleasure is always useful" (Dostoievski: Alexis Ivanovitch, The Player); "the Creator[...]

gave us the inducement of appetite, the encouragement of taste, and the reward of pleasure" (Brillat-Savarin 1825); "pleasures and pains represent the sole genuine basis for understanding human motives" Jevons (1871); "from every point of view the affective process must be regarded as motivational in nature" Young (1959).

Finally, the Fathers of the American Constitution, also, acknowledged that the pursuit of happiness is the main aim of the human life. In the following, it will be argued that the seeking of pleasure is the motor of human behaviour. In "the principles of psychology" William James (1890) claimed to be putting psychology on a firm foundation as a natural science in the positivist sense and considered pleasure as a natural motivation. Thus, he clearly considered a mental cognition as deserving a scientific approach an attitude that contrasted with that of his follower Sigmund Freud (1920, 1922). We do not claim that the seeking of pleasure and the avoidance of displeasure explain brain function, but that they explain behaviour as foreseen by Philosophers. Yet, this knowledge now will be based, not on intuition, like the Moralists of the past, but on scientific evidence. We acknowledge that proximal explanation of behaviour may seem unnecessary when examined from the point of view of natural selection. However, proximal explanations of behaviour must be provided. In addition, the emergence of thinking, consciousness, and awareness must, and can be viewed from an evolutionary perspective (Cosmides et al. 1992; Cabanac et al. 2009). The problem is not that we can understand evolution phenomenologically without consciousness, the problem is that consciousness is there, we must deal with it, and understand its role in nature.

3. PHYSIOLOGICAL ROLE OF PLEASURE

The starting point of the chain of reasoning is the observation that sensory pleasure tags stimuli that are useful for optimal physiological functioning and survival and, obversely, that sensory displeasure indicates useless or noxious stimuli (Cabanac, 1971, 1979).

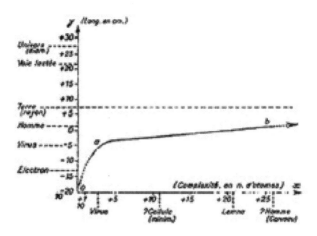

Figure 2. Sensation seen as a multidimensional event in response to a stimulus. Quality, describes the nature of the stimulus; quantity, describes the intensity of the stimulus; affectivity (pleasure or displeasure), describes the physiological usefulness (survival value) of the stimulus. Duration is added to the three other fundamental axes of sensation. It is argued, here, that this model describes equally sensation and consciousness (from Cabanac, 1996, with permission).

Sensation is the mental representation of a stimulus that reaches a sensory organ. Whenever a stimulus excites a sensory ending, the sensation aroused is a four-dimensional phenomenon (Fig. 2). The first dimension, the quality of the sensation indicates the nature of the stimulus. The second dimension, the intensity of the sensation, indicates the magnitude of the stimulus. The third dimension, the affectivity of the sensation (pleasure/displeasure), indicates the potential usefulness of the stimulus. The fourth dimension, the duration of the sensation, indicates the duration of the stimulus. Let us examine the third dimension. The word alliesthesia was coined to describe the fact that a same given stimulus can feel pleasant or unpleasant according to the internal state of the subject and that the resulting sensation can move up and down the affectivity axis. This is especially obvious in the determinism of thermo-regulatory behaviour (Attia, 1984), and of the control of food intake (Fantino 1984, 1995; McBride, 1997). When a subject is hypothermic, a warm skin feels pleasant and a cool one unpleasant. When the subject is hyperthermic a warm skin feels unpleasant and a cool one pleasant. The efficacy of this mechanism is almost limitless: already, the simplest behaviour, such as the seeking of pleasure by a hyperthermic subject who immerses only his hand in stirred 20°C water, can extract as much as 70 W from the hand, i.e. as much as the subject's basal heat production (Cabanac et al., 1972).

The adaptation of sensory pleasure and displeasure to physiological need is such that very similar equations describe thermal preference (Cabanac et al., 1972; Bleichert et al. 1973), and the autonomic regulatory responses, evaporative heat loss (Stolwijk et al. 1968), or chemical thermogenesis (Nadel et al. 1970). In addition, thermal pleasure and displeasure is adapted to the defense not only of core temperature but to the difference Tset-Tcore, the error signal of temperature regulation. The case of alimentary pleasure is similar, with pleasure occurring from food stimuli when applied to hungry subjects and displeasure when applied to satiated subjects (Fantino, 1984, 1995). The similarity with temperature regulation was recognized when alliesthesia to sweet stimuli disappeared in subjects whose body weight was reduced with dieting, and recovered when body weight returned to control value. A last example of the functional role of pleasure may be found in postural adjustments: the most precise gestures are reached with the help of the affective capability of musculo-articular sensation (Rossetti et al. 1994).

Sensory pleasure is thus the axis of sensation that allows optimisation of behaviour, as seen from the physiologist point of view, that of immediate improvement of physiological function, and survival of the subject.

4. THE COMMON CURRENCY

One of the basic postulates of Ethology is that behaviour is the product of natural selection and thus is optimal (Baerends, 1956; Tinbergen, 1950). However, Ethologists seldom question the proximate cause of behaviour and optimisation of behaviour. McFarland and Sibly (1975) have theorised that since behaviour is a final common path there must exist a common currency within the brain-mind to allow trade-offs between motivations and thus rank them by order of priority (McFarland & Sibly, 1975; McFarland, 1985; McNamara & Houston, 1986). Over the following decades a series of experiments led to the conclusion that pleasure is this common currency (Cabanac, 1995). More generally, the common currency is the affective axis of consciousness. The rationale was as follows:

Since there is need for a common currency, if sensory pleasure/displeasure is the mental signal that allows us to produce useful behaviours adapted to their physiological goals, then pleasure/displeasure could be this common currency. This was confirmed in several experimental conflicts of motivations involving physiological functions such as temperature regulation, fatigue aroused by muscular exercise, and taste. Useful behaviours were selected by the subjects who simply maximized the algebraic sum of pleasures/displeasures aroused simultaneously in both sensory modalities involved. The same mechanism works with purely mental motivations. When subjects were placed in situations where a motivation involving a sensory modality (e.g. thermal discomfort, or hunger) was confronted with a purely mental motivation (e.g. the pleasure of playing a video-game, or the displeasure of losing money) they behaved in the same way. They maximized the algebraic sum of pleasures/

displeasures aroused simultaneously in both sensory and mental modalities involved. The principle of their decisions can be described by Fig. 3. Finally, it was demonstrated in rats that motivations for acquiring rewards as different from one another as intra-cranial electrical stimulation and the sweet taste of sugar, activated the same area of the brain (Shizgal & Conover, 1996). Thus, pleasure was also a common currency in animals.

	Resulting hedonic experience	Action
behaviour 1	a -----> A	yes
behaviour 2	B -----> b	no
behaviour 1 + behaviour 2	a + B -----> A + b	yes

with, a < A, and B > b
and with a + B < A + b

Figure 3. Mechanism by which a behaviour (behaviour 2) that produces displeasure (B to b) can be chosen by a subject if another behaviour (behaviour l) that produces pleasure (a to A) is simultaneously chosen. The necessary and sufficient condition for the behaviour 2 to occur (action) is that the algebraic sum of affective experience (pleasure) of the yoked behaviours is positive (a+B<="" dir="">

5. CONSCIOUSNESS

The paramount role of pleasure and displeasure can be recognized not only in sensation and physiology, but in the whole realm of consciousness (McKenna, 1996; Warburton, 1996). An evolutionary psychology perspective (Cosmides et al. 1992; Cosmides & Tooby, 1995; Cabanac et al. 2009) is useful here to extrapolate from sensation to consciousness, and leads to a postulate. It may be postulated that consciousness evolved from sensation (Cabanac, 1996). A corollary of this postulate is that when consciousness evolved from sensation, consciousness inherited the properties of sensation, i.e. figure 2 describes not only sensation but also the four-dimensional structure of all conscious events.

We may accept the above conclusions as a result of mere introspection: I can analyse any thought I have and recognise a quality, an intensity, an affectivity, and a duration. However, the postulate entails its corollary according to the following steps:

1) The affective axis of sensation, pleasure and displeasure, is the sign of the physiological usefulness of a stimulus. Pleasure is both the tag of a

useful stimulus and the force that orients behaviour to approach and eventually consume this stimulus.

2) Since behaviour is a final common path, the brain needs a common currency to rank the motivations for access to behaviour in a time-sharing pattern (McFarland & Sibly, 1975; McNamara & Houston, 1986).

3) Since maximisation of sensory pleasure is the motivation for behaviours adapted to physiological goals and since motivations with physiological goals compete with other motivations such as playing, aesthetic, and social ones for access to the behavioural final common path, pleasure is this common currency (Cabanac, 1992). In the same way as sensory pleasure tags the usefulness of a stimulus, joy tags the usefulness of any other conscious experience This was recently confirmed in a series of experiments where pleasure/joy tagged efficacious mentaloperation in decisions involving playing and poetry (Cabanac et al. 1997), grammar (Balaskó et al. 1998), mathematics (Cabanac et al. 2002), aggressiveness (Ramírez et al. 2005), rationality (Cabanac & Bonniot-Cabanac, 2007), politics and gambling, (Bonniot-Cabanac & Cabanac, 2009), social behaviour (Bonniot-Cabanac & Cabanac, 2010), curiosity (Perlovsky et al. 2010), music (Hudson, 2011), mood and sleep (Roberts, 2004). Thus, pleasure/joy seems to be the key to all decision making processes (McKenna, 1996).

4) The affective axis is thus the motivational capability of consciousness.

5) If consciousness possesses the affective axis, then the other axes of sensation also were inherited by consciousness.

The reader will have recognized in these five steps the same rationale as used earlier. The behavioural final common path, and the common currency, lead to the challenging conclusion that all conscious events bear the structure sketched in Fig. 2. It follows that affectivity might be the general motivation for any behaviour and that all the properties of sensory pleasure also would belong to the affective axis of consciousness. Joy is, in consciousness, congruent with pleasure, in sensation. Sensory pleasure possesses several characteristics: Pleasure is contingent, pleasure is the sign of a useful stimulus, pleasure is transient, pleasure motivates behaviour (Cabanac, 1971, 1979). Therefore, it follows from the same premises that joy keeps with global consciousness the same properties as pleasure with sensation. Joy possesses the same characteristics as sensory pleasure: joy is contingent, joy is the sign of a useful thought, joy is transient, joy motivates behaviour (Cabanac, 1986, 1996; Warburton, 1996).

It follows that pleasure/joy is the ineluctable motor of action of the human mind and bears the properties of an impelling force, the application of which will actuate the human being. This conclusion does not exclude, as mentionned above, that animals are susceptible to obey also the law of maximization of pleasure; interested readers will find elements of discussion in Griffin (1992) and Dawkins (1993). The evolutionary advantage of sensory pleasure, for the animals that first had it, is that it saved the nervous system from storing astronomical numbers of potentially useful or noxious stimulus-response reflexes. Sensory pleasure added flexibility to the behavioural pattern of those animals that possessed it and passed it to their offspring. Similarly joy saves the brain from storing an infinite number of rules of thumb, and provides a formidable advantage both in amount of information stored and in flexibility of the behavioural responses. "The important role of affect in guiding evaluation and choice is apparent.... » (Finucane et al. 2003); now we understand the fundamental rôle of affective factors and their interaction with cognition. That mechanism improved not only physiology, as initially demonstrated, but also mental.

The major advantages of affective decision process as compared to rational (or reflexive) pre affective decision process (in living or inanimate agents) is to facilitate the ranking of priorities and to add flexibility to this process. The affective dimension of consciousness may be what makes a conscious agent different from an algorithmic Türing machine, as pointed out by Penrose (1994).

6. CONCLUSION AND PROPOSAL : THE FIFTH INFLUENCE

Historically, the Physicists have defined as a Newtonian force an influence operating on a body so as to produce an alteration, or tendency to alteration. Later, they have recognized four influences defined in a different framework but with similar conceptual results, the gravitational force, the electromagnetic force, the weak, and the strong nuclear forces (Table I). The gravitational force – formally a curvature of space-time in General Relativity – is the weakest of all interactions but exerts its influence most noticeably in the cosmos. Its range is infinite and all particles are submitted to its influence. The electromagnetic force is much stronger than the former, it also exerts its influence over infinite distances but on charged particles only. It is accepted that the universe is neutral and only hot and dense regions in galaxies (stars, planets, plasma), and high-temperature clouds of gas in clusters of galaxies are submitted to the electromagnetic force. Photons and electromagnetic waves, so conspicuous in our everyday life, result from the electromagnetic force. The weak and the strong nuclear forces exert influences at atomic and sub-atomic levels. They cause radioactivity and nuclear reactions in star cores. Unlike the other forces, the strong force increases with separation of the quarks. All forces have in common the property to actuate the bodies and particles on which they are exerted.

Table 1. The four fources recognized by the Physicists and the fifth influence proposed in the present paper

INFLUENCE	FIELD	RANGE	MACROSCOPIC MODEL
gravitational	cosmos	infinite	planetary systems
electromagnetic	charged particles	infinite	earth magnetosphere
weak	subatomic	atom	beta decay, neutrinos
strong nuclear	subatomic	atom	nuclear reactions
fifth influence	brain	infinite (behavior)	human

6.1 Proposal The Physicists have hypothesized that matter was under the influence of a unified force in the beginning of the universe. As the energy density went down with time, the unified force separated, through symetry breaking, into gravitation and great unified force (GUT), then GUT separated into strong and electro-weak forces, then electro-weak forces separated into electro-magnetic and weak forces. It may be of interest to speculate on the emergence of the fifth influence from matter as the last symetry breaking of the forces; thus, eventually, in biological times, pleasure emerged from complexity as the last 'symetry breaking', resulting from a critical level of complexity reached by animal brains.

Having recalled that complexity is a relevant feature of the world, and that the brain is the locus of the highest known organised complexity; having recalled that the affective capability of consciousness is the ineluctable motivation for behaviour; and having finally recalled what a force is, we can now reach a conclusion with a proposal:

We suggest that the affective axis of consciousness, pleasure/displeasure and joy/distress, operates in conscious animals in a way similar to the way the four forces of Physics influence masses and particles, i.e. influences their behaviour.

Since pleasure can move the infinitely complex into action, and since no thinking brain can escape the trend to maximize pleasure/joy, pleasure deserves to be ranked as the fifth 'force' of the universe. However, since a mental phenomenon is dimensionless we propose to consider the affective capability of consciousness, as the fifth influence rather than the fifth force. The above proposal is an axiom, and as such should not require demonstration. Yet, the experiments on which it is based were as many attempts to falsify it. This article is not the place to discuss when in phylogeny, presumably among higher Vertebrates, consciousness and the fifth influence emerged. Yet, we may accept that its early action was to actuate behaviours oriented toward the survival of individuals. With the advent of consciousness and humankind, the fifth influence led individuals to experience that not only biological behaviours but social ones are rewarding as well. Therefore, the fifth influence may be considered as just starting to drive the world towards increasing complexity by organising the more and more complex system of links that humankind weaves between individuals, the network that Teilhard de Chardin (1955) called "Noosphère."

Acknowledgements: We wish to thank our colleagues C. Barrette (Dept. Biology) and J. R. Roy (Dept. Physics) for their kind improving earlier versions of the present article.

References

Aristote 4th century BC. De anima (Fr. Transl.). Librairie philosophique J. Vrin, Paris.

Attia M. (1984) Thermal pleasantness and temperature regulation in man. Neuroscience and Biobehavioral Reviews 8, 335-343.

Baerends G. P. (1956) Aufbau des tierischen Verhaltens. In Handbuch der Zoologie. (Edited by Kükenthal W and Krumbach T) De Gruyter & Co, Berlin, p. Teil 10 (Lfg 7).

Balaskó M., Cabanac M. (1998) Grammatical choice and affective experience in a second-language test, Neuropsychobiology, 37, 205-210.

Barrow J. D., Tipler F. T. (1986) The Anthropic Cosmological Principle. Oxford University Press, Oxford, UK

Berridge K. C. (1996), Food reward: brain substrates of wanting and liking. Neuroscience and Biobehavioral Reviews, 20, 1-25.

Berridge K. C., Kringelbach M. L. (2008), Affective neuroscience of pleasure: reward in humans and animals. Psychopharmacology, 199, 1432-2072.

Berridge K. C., Winkielman P. (2003), What is an unconscious emotion? (The case for unconscious "liking"). Cognition and Emotion, 17, 181-211.

Bertola F, Curi U (1993) The Anthropic Principle: Proceedings of the Second Venice Conference on Cosmology and Philosophy. Cambridge University Press, Cambridge, UK.

Bleichert A, Behling K, Scarperi, M, Scarperi S (1973) Thermoregulatory behaviour of man during rest and exercise. Pflüggers Archiv, 338, 303-312.

Bonniot-Cabanac M.-C., Cabanac M. (2009) Pleasure in decision making situations: Politics and gambling. Journal of Risk Research, 12, 619-645.

Bonniot-Cabanac M.-C., Cabanac M. (2010) Do government officials decide more rationally than the rest of us? A study using participants from the legislature, the executive, and the judiciary. Social Behavior and Personality, 38, 1147-1152.

Bowring J. (1962) The works of Jeremy Bentham published under the superintendance of his executor John Bowring 1833-1844. Russell & Russell inc.,

New York, USA.

Breuer R. (1991) The Anthropic Principle. Man as the Focal Point of Nature. Birkhäuser, Boston, USA.

Brillat Savarin A. (1828) Physiologie du Gout. Sautelet & Cie., Paris, France.

Cabanac M. (1971) Physiological role of pleasure. Science 173, 1103-1107.

Cabanac M. (1979) Sensory pleasure. Quarterly Reviews of Biology, 54, 1-29.

Cabanac M. (1986) Du confort au bonheur. Psychiatris Française 17, 9-15.

Cabanac M. (1992) Pleasure: the common currency. Journal of Theoretical Biology, 155, 173-200.

Cabanac M. (1995) La quète du plaisir. Liber, Montréal, Canada.

Cabanac M. (1996) On the origin of consciousness, a postulate and its corrollary. Neuroscience and Biobehavioral Reviews, 20, 33-40.

Cabanac M., Bonniot-Cabanac M.-C., (2007) Decision making: Rational or hedonic? Behavioral and Brain Functions, 3, 1-45.

Cabanac M., Cabanac A. J., Parent A. (2007) When in phylogeny did consciousness emerge? Behavioural and Brain Research, 198, 267-272.

Cabanac M., Guillaume J., Balaskó M., Fleury A. (2002) Pleasure in decision-making situations, BiomedCentral, http://www.biomedcentral.com/imedia/9977974521128571_ARTICLE.PDF.

Cabanac M., Massonnet B., Belaiche R. (1972) Preferred hand temperature as a function of internal and mean skin temperatures. Journal of Applied Physiology, 33, 699-703.

Cabanac M., Pouliot C., Everett J. (1997) Pleasure as a sign of efficacy of mental activity. European Psychologist, 2, 226-234.

Changeux J. P. (1983) L'homme neuronal. Fayard, Paris, France.

Condillac E. B-de- (1754) Traité des sensations. (2nd ed.) Arthème Fayard, Paris, France.

Cosmides L., Tooby J. (1995) From evolution to adaptations of behaviour. In Biological Perspectives on Motivated Activities. (Edited by Wong R) Ablex Publ. Co., pp 11-74. Northwood, New Jersey, USA.

Cosmides L., Tooby J., Barkow J. H. (1992) Introduction: evolutionary psychology and conceptual integration. In The Adapted Mind. (Edited by Barkow JH, Cosmides L, Tooby J.), pp 3-15. Oxford University Press, New York, USA.

Dawkins M. S. (1993) Through our eyes only? W. H. Freeman & Co, Oxford, UK.

Fantino M. (1984) Role of sensory input in the control of food intake. Journal of the Autonomic Nervous System, 10, 326-347.

Fantino M. (1995) Nutriments et alliesthésie alimentaire. Cahiers de Nutrition et de Diététique, 30, 14-18.

Finucane M. L., Peters E., Slovic P. (2003) Judgment and decision making: The dance of affect and reason, in Schneider S. L. & Shanteau J., Emerging Perspectives on Decision Research, Cambridge University Press, Cambridge, 327-364.

Freud S. (1920) As Des Lustprinzips, Essay.

Freud S. (1922) Beyond the pleasure principle Hubback, C. J. M. (Transl.) International psychoanalytical library, pp. 1-83, The International Psycho-Analytical Press.

Gell-Mann M. (1994) The Quark and the Jaguar. W. H. Freeman & Co, New York, USA.

Griffin D. R. (1992) Animal Minds. University

of Chicago Press, Chicago, USA.

Hudson N. J. (2011) Musical beauty and information compression: Complex to the ear but simple to the mind? BMC Research Notes, 4, http://www.biomedcentral.com/1756-0500/4/9.

James W. (1890) The Principles of Psychology, Dover Publications Inc.

Jevons W. S. (1871) The theory of political economy. MacMillan, London, UK.

Joseph R. (2010) The Quantum Cosmos and Micro-Universe: Black Holes, Gravity, Elementary Particles, and the Destruction and Creation of Matter. Journal of Cosmology, 4, 780-800.

Kandel E. R., Schwartz J. H. (1985) Principles of Neural Science. Elsevier Science Publ Co, Amsterdam, Nederland.

McBride R. L. (1998) On the relativity of everyday pleasures. In The value of pleasures and the question of guilt (Edited by Warburton DM, Sherwood N) John Wiley & Sons Ltd., Chichester, UK.

McFarland D. J. (1985) Animal Behaviour. Pitman, London, UK.

McFarland D. J., Sibly R. M. (1975) The behavioural final common path. Philosophical Transactions of the Royal Society. London, 270, 265-293.

McKenna F. (1996) Understanding risky choices. In Pleasure and Quality of Life. (Edited by Warburton DM, Sherwood N) pp 157-170. John Wiley & Sons Ltd., Chichester, UK.

McNamara J. M., Houston A. (1986) The common currency for behavioural decisions. American Naturalist, 127, 358-378.

Nadel E. R., Horvath S. M., Dawson C. A., Tucker A. (1970) Sensitivity to central and peripheral thermal stimulation in man. Journal of Applied Physiology, 29, 603-609.

Pascal B. (1670) Pensées (réedit. 1972) Librairie Générale Française, Paris, France.

Penrose R. (1994) Shadows of the Mind. Oxford University Press, Oxford, UK.

Perlovsky L. I., Bonniot-Cabanac M.-C., Cabanac M. (2010) Curiosity and pleasure. WebmedCentral Psychology, 1, WMC001275.

J. M. Ramírez J. M., Bonniot-Cabanac M.-C., Cabanac M. (2005) Can Aggression Provide Pleasure? European Psychologist, 10, 136-145.

Roberts S. (2004) Self-experimentation as a source of new ideas: Ten examples about sleep, mood, health, and weight. Behavioral Brain Sciences, 27, 227–288.

Rossetti Y., Meckler C., Prablanc C .(1994) Is there an optimal arm posture? Deterioration of finger localization precision and comfort sensation in extreme arm-joint postures. Experimental Brain Research, 99, 131-136.

Shizgal P. (1997) Neural basis of utility estimation. Current Opinion in Neurobiology, 7, 198-208.

Shizgal P., Arvanitogiannis A. (2003) Gambling on dopamine. Science, 299, 1856-1858.

Shizgal P., Conover K. (1996). On the neural computation of utility. Current Directions in Psychological Science , 5, 37-43.

Stolwijk J. A. J., Saltin B., Gagge A. P. (1968). Physiological factorsassociated with sweating during exercise. Aerospace Medicine , 39:1101-1105.

TeilharddeChardin P. (1955) Le phénomène humain. Éditions du Seuil, Paris.

TeilharddeChardin P. (1965). La place de l'homme dans la nature. Éditions du Seuil,

Paris.

Tinbergen N. (1950). The hierarchical organization of mechanisms underlying instinctive behaviour. Symposia of the Society for Experimental Biology, 4:305-312.

Warburton D. M. (1996). The functions of pleasure. In Pleasure and Quality of Life. (Edited by Warburton D. M., Sherwood N.) John Wiley & Sons, Chichester, p. 1-10.

Young P. T. (1959). The role of affective processes in learning and motivation. Psychological Reviews, 66:104-123.

19. The Spread Mind: Seven Steps to Situated Consciousness
Riccardo Manzotti

Institute of Communication and Behavior, "G. Fabris", IULM University, Via Carlo Bo, 8, 20143 Milano, Italy

Abstract

This paper outlines a radical version of phenomenal vehicle externalism dubbed "The Spread Mind" which suggests that both the content and the vehicles of phenomenal experience are identical to a process beginning in the environment and ending in the cortex. In seven conceptual steps, the Spread Mind outlines a counterintuitive yet logically possible hypothesis – namely that the physical underpinnings of consciousness may comprehend a part of the environment and thus may extend in space and time beyond the skin. If this view had any merit, consciousness would be situated in a strong sense.

KEY WORDS: Consciousness, Phenomenal experience, Externalism, Time, Ontology, Causation, Representation

1 Where to Look for the Physical Basis of Phenomenal Experience

The quest for the physical underpinnings of consciousness is still an unresolved one (Koch 2007; Tallis 2010; van Boxtel and de Regy 2010). When one perceives a red patch, what is the necessary and sufficient physical basis of such a phenomenal experience? Indeed what is the physical phenomenon that is one's phenomenal experience of a red patch? Such questions single out the hard-problem (Chalmers 1996): "From all the low-level facts about physical configurations and causation, we can in principle derive all sorts of high-level facts about macroscopic systems, their organization, and the causation among them. One could determine all the facts about biological function, and about human behavior and the brain mechanisms by which it is caused. But nothing in this vast causal story would lead one who had not experienced it directly to believe that there should be any consciousness. The very idea would be unreasonable; almost mystical, perhaps" (Chalmers 1996, p. 102). More recently, Christof Koch wrote that "How brain processes translate to consciousness is one of the greatest unsolved questions in science. The scientific method [...] has utterly failed to satisfactorily explain how subjective experience is created" (Koch 2007).

Such a lack of a physical explanation might be a consequence of one or more ill chosen assumptions as to the nature of the physical world (Strawson 2006; Skrbina 2009; Strawson 2011). Of course, there is plenty of evidence showing that neural activity is correlated and indeed necessary to consciousness. In the last twenty years, several researchers presented outstanding and remarkable results as to the ways in which conscious experience is related with brain activity (Logothetis 1998; Zeki 2001; Andrews, Schluppeck et al. 2002; Crick and Koch 2003; Changeux 2004; Buzsáki 2007; Hohwy 2009; Laureys and Tononi 2009; van Boxtel and de Regy 2010). Yet what is still missing is a theory outlining a conceptual and causal connection between neural activity and phenomenal experience. Thus it may be worth to consider other hypotheses however counterintuitive they may seem to be. After all, "If no theories seem to be capable of accounting for conscious experience, this probably means that there is something inherent in our assumptions that divorce theories from conscious experience" (Rockwell 2005, p. 49).

An assumption that is sometimes taken for granted is that the physical underpinnings of consciousness have to be internal to the nervous system. Neuroscientists like Christof Koch, Atti Revonsuo, Giulio Tononi, Semir Zeki have explicitly made this assumption. For instance, Atti Revonsuo stated that "sensory input and motor output are not necessary for producing a fully realized phenomenal level of organization. The dreaming brain creates the phenomenal level in an isolated form, and in that sense provides us with insights into the processes that are sufficient for producing the phenomenal level" (Revonsuo 2000, p. 58). Along the same line, Christof Koch believes that "The goal [of the scientific study of consciousness] is to discover the minimal set of neuronal events and mechanisms jointly sufficient for a specific conscious percept" (Koch 2004, p. 16). My opinion is that these authors assume that, although the environment has a key role in shaping neural networks during development, once the brain is developed and running, there is a set of neural events sufficient for specific conscious percepts [Crick and Koch 1990]. Such a hypothesis has become widely accepted not only in neuroscience but also in philosophy of mind. Consider a philosopher like Jaegwon Kim stating that "if you are a physicalist of any stripe, as most of us are, you would likely believe in the local supervenience of qualia – that is, qualia are supervenient on the internal physical/biological states of the subject" (Kim 1995, p. 160). Yet, is such an assumption unavoidable? After all, if you are a physicalist, you ought to look for physical phenomena – any kind of physical phenomena – and not just for an "internal physical/biological state of the subject". What occurs outside the body is physical too. Neural processes are only a subset of a much larger domain of feasible physical processes.

Here, I will consider a rather counterintuitive hypothesis – dubbed the spread mind – that might shed a new light on the issue of the physical underpinnings of phenomenal experience (Manzotti 2006c). This hypothesis is a somewhat more radical version of other related views (Varela, Thompson et al. 1991; O'Regan and Noë 2001; Rockwell 2005; Honderich 2006; Thompson 2007; Noë 2009; Velmans 2009). In a nutshell, I will consider whether it makes any sense to suppose that the physical underpinnings of consciousness are temporally and spatially larger than the subject body. Is consciousness situated in the environment?

2 Beyond the Skin

Instead of focusing on the brain and the nervous system, a few authors considered whether the relation between the mind and the environment is closer than traditionally assumed (Varela, Thompson et al. 1991; Clark and Chalmers 1998; O'Regan and Noë 2001; Rockwell 2005; Honderich 2006; Thompson 2007; Noë 2009; Velmans 2009). This insight gave rise to various views both in cognitive science and in philosophy of mind (Rowlands 2003; Robbins and Aydede 2009a; Hurley 2010): the embodied mind, the embedded mind, the extended mind, and the larger stance dubbed situated cognition. The insight has unfolded by degrees. For instance, for some authors, the environment provides the right place for the development of cognitive functions (Anderson 2003; Gallagher 2005) while, for other scholars, the environment allows off-loading cognitive work (Clark 1989; Wilson 2004). Eventually, David Chalmers and Andy Clark suggested that, somehow, the cognitive mind leaks into the world (Clark and Chalmers 1998; Clark 2008). During the last twenty years, many scholars gave rise to a heated debate whether and to what extent the mind may extend into environment (Varela, Thompson et al. 1991; Thompson and Varela 2001; Chrisley and Ziemke 2002; Anderson 2003; Pfeifer, Lungarella et al. 2007; Chemero 2009; Robbins and Aydede 2009a; Robbins and Aydede 2009b; Rupert 2009; Shanahan 2010). Although most of these authors limited their proposal to the cognitive mind (Clark and Chalmers 1998; Wilson 2004; Clark 2008) nevertheless they collected their share of criticism (Rupert 2004; Adams and Aizawa 2008; Rupert 2009). Since they focus on cognitive skills rather than on phenomenal experience, these views may be grouped under the label of situated cognition (Anderson 2003; Robbins and Aydede 2009b). Only a handful of authors ventured to suggest that although the mechanisms of the conscious mind remain safely inside the brain, phenomenal content may literally either depend on or be constituted by the outside world (Dretske 1996; Velmans 2000; Lycan 2001; O'Regan and Noë 2001; Honderich 2006; Noë 2009). Yet, even these phenomenal externalists have continued to distinguish between internal representations and the external world. For instance, Fred Dretske stated that "sensory experience gives primary representation to the properties of distant objects and not to the properties of those more proximal events on which it (causally) depends" (Dretske 1981, p. 165). Three decades later he still remarks that "The experiences themselves are in the head […] but nothing in the head needs have the qualities that distinguish these experiences" (Dretske 1996, p. 144-145). As a result they are often labeled an supporters of phenomenal content externalism – namely the view that although the content of experience depends on states of affairs external to the body, nevertheless the vehicles of experience remains inside the body.

Here, I will venture one step further. I will openly consider a perilous hypothesis: are the vehicles of phenomenal experience spread in time and space beyond the boundaries of the skin? Is phenomenal experience itself extended in time and space? Is consciousness situated in a strong sense? So far, many authors stepped back from this counterintuitive view. For instance, David Chalmers, in the foreword to Andy Clark's book on the extended mind, wrote that "[the extended mind does] not rule out the supervenience of consciousness on the internal" (Chalmers 2008, p. 6). As we have seen, Kim's dictum explicitly rules out such a possibility. Likewise, many objections have been raised against the hypothesis that the processes underpinning phenomenal experience might be totally or partially external to the body (Kim 1995; Clark and Chalmers 1998; Wilson 2004; Velmans 2007; Adams and Aizawa 2008; Clark 2008).

And yet, what are the strong arguments against strong situated consciousness – namely the hypothesis that the physical processes constituting consciousness are larger than the nervous system? Here I will venture to consider and put under scrutiny such a hypothesis. I will consider whether our phenomenal experience might indeed be extended in time and space beyond the limits of the nervous system.

3 Towards Situated Consciousness: The Case of the Rainbow

At the onset, I will make use of an example of a physical process in which the traditional separation between subject and object seems to collapse and vanish. Consider a physical phenomenon like the rainbow (Manzotti 2006a), which is not a perceptual phenomenon but a physical coloured shape taking place in its own right. Not only the rainbow is a physical phenomenon, but it cannot be defined autonomously without an observer. In fact, a rainbow is a process that requires a further physical system in order to take place. Where is the rainbow? Where is the experience of the rainbow? Is there a rainbow without an observer? Is there a rainbow-observer without a rainbow in the cloud? As we will see, these questions may have a surprising answer that blurs the traditional boundaries between experience and physical world. I will try to show that the rainbow is a physical phenomenon and yet that it shares many properties of phenomenal experience.

For one thing, although the rainbow is not a hallucination, each observer has her own private rainbow. It is a private physical phenomenon and to be private is usually considered to be a prerogative of phenomenal experience. Let me explain. Call A the cloud at the horizon. Call B the body of an observer. Call C the process that starts from the surface of each reflecting and refracting drop and ends in B. If B were not there staring at the cloud, the process C would not occur. Yet, the rain drops would still float in the air; the sun rays would still meet the drops and bounce back at an angle proportional to their frequency. However, the reflected and refracted rays would not meet anywhere. They would simply scatter around and eventually disperse. The rainbow would not occur. But if B is in the right place at the right time, what happens? A subset of lightrays (those whose origin corresponds to a semicircular distribution of rain drops) enters into the retina of the subject and proceeds, by means of neural signals, up to the cortex where it finally produces a joint effect because of its shape and colours (Manzotti 2006b; Manzotti 2006d; Manzotti 2006c; Manzotti 2009). The rainbow is a process that would not occur if it weren't for B. This process C is the rainbow. The rainbow is neither A nor B alone. Where is C? C is neither inside the observer B nor in the cloud A. C is spread in time and space.

The example of the rainbow may suggest an idealist or constructivist standpoint. Yet, this is not the case, since I propose neither that the rainbow is a mental creation of the beholder nor that is anything like a sense datum. On the contrary, I draw attention to the physical nature of the rainbow and I stress that it is a physical process taking place in the environment. The hypothesis under scrutiny here remains safely inside the physicalist domain: it couldn't be farer from any idealistic perspective. I suggest that phenomenal experience may be a distributed phenomenon and thus that consciousness may be situated in a very strong sense. Nevertheless, it may seem that the above example is concocted to make the reader believe that the rainbow does not exist if no one is looking at it. It is a fair bet. In fact, I suggest that when no suitable physical system (like a human subject) is interacting with the cloud, there is no rainbow. Of course, such a conclusion does not imply that the rainbow is a mental phenomenon. Rather it implies that the rainbow is a physical process that requires the proper physical conditions – among which I would list a cloud, the sun, and a suitable structure where the light rays may meet and produce a joint effect (the last structure is the body of a perceiving observer).

To convince yourself that the rainbow is a physical phenomenon that cannot occur in isolation, it is sufficient to set aside any tentative observer and ask to a physicist where is the rainbow. She would have no way to tell. The fact is that there is no way to single out a rainbow without assuming that the cloud is indeed seen from a certain view point. Without an observer, the rainbow is everywhere in the cloud (or nowhere). There is no way to pinpoint where a rainbow is without including the position and the causal role of the observer. Similarly, there is no way to tell whether a rainbows exists or not, without a proper observer. In short, it is impossible, from a physical perspective, to disentangle the rainbow from its observer. There are many aspects that the rainbow, as a physical phenomenon, nonetheless shares with phenomenal experience. First, the rainbow is a private phenomenon – a property that has always been traditionally attributed to mental states. The same rainbow cannot be seen by two different observers, since each of them will single out a different rainbow due to the fact that they occupy two different positions in space. Processes are necessarily private and yet physical. Secondarily, the rainbow is something that takes place. It is not a static entity. The rainbow takes place and it is extended in time and in space. Third, the rainbow singles out a part of reality bringing into existence a form (the coloured arch) that was not there before. Is it a mental form or the physical occurrence of a form?

In the next seven sections, I will try to show that conscious experience may correspond to a physical process akin to the rainbow. Further, I will suggest that if the physical world is conceived as made of processes, it has the resources to host consciousness once it is accepted that the physical processes which is phenomenal experience, may indeed be larger than those taking place in the nervous system.

4 First Step: Extension in Space

The example of the rainbow outlines a process that is physically extended in space. How much? Well, roughly, from the surface of the object to the neural activity in the brain. Should this extension upset us? I don't see why. After all, as far as we know, all physical phenomena are extended in space.

For one, neural activity, which is the traditional candidate to underpin phenomenal experience, is extremely extended in space. Consider what happens inside the brain when a signal is received. The nervous signal travels at a speed of 100-120 m/s. The completion of a complex perceptual process such as recognizing a face or a complex scene takes 200-300 msec (Quiroga, Mukamel et al. 2008). This means that signals travels across the brain for roughly 20-30 meters before completing their job. Put simply, a neural implementation of an average perceptual process requires 20-30 meters to complete. The total length required by such a process is much longer than that of the agent's brain or body. This is not an opinion but a fact that derives out of empirical evidence in neurophysiology (Biederman and Kalocsai 1997; Farah, Wilson et al. 1998; Andrews, Schluppeck et al. 2002). Therefore, in neuroscience a physical process spanning dozens of meters is commonly taken to be capable of producing phenomenal experience. Hence there is no reason to reject other putative physical processes only because they are extended on relatively large spatial distances like the rainbow. The Spread Mind suggests considering a process that is spatially extended to such an extent that it can contain the external object of perception. Put simply, if one perceives a rainbow on a cloud a few kilometers far away, according to the spread mind, the physical underpinnings of her phenomenal experience begins on the cloud and eventually ends in her brain? In principle, what's wrong with a physical process of a few kilometers instead of dozens of meters?

5 Second Step: Extension in Time

It may seem odd to suggest that phenomenal experience may be identical with a physical process beginning before the first triggered neural firing. Could consciousness be constituted (at least to a certain extent) also by events that occur before any causal influence is exerted on the nervous receptors? To answer to such a question, I will make use of the same strategy adopted in the first step. I will argue that the same kind of temporal extension required for the rainbow is required for neural processes too. Thus I will argue that there is no reason to reject the hypothesis of temporally extended processes since neural processes are extended too. Consider the evidence provided by Benjamin Libet that a continuous and uninterrupted neural activity of 0,5 sec is required for consciousness: "What the on-time experiments have demonstrated is that threshold awareness pops in rather suddenly when the activities persist for the full 500 msec requirement!" (Libet 2004, p. 112) This evidence has been quantitatively refined by further studies (Libet 1966; Intraub 1980; Quiroga, Mukamel et al. 2008) but it is commonly accepted that there is a minimum temporal threshold during which a sustained neural activity has to occur. This minimum time span entails that the physical underpinning of consciousness takes some time to complete. Whether this time is 30 msec, 500 msec or a different amount of time, it is not going to make any difference as to the issue at stake here. The point is that consciousness is spread in time. Even considering only neural processes, the physical underpinnings of consciousness seem to require a finite time span.

By and large, all physical phenomena require some time to complete. There are no instantaneous phenomena. Most macroscopical phenomena require a lot of time to complete: a gesture, a process of oxidation, the recognition of a face, and the like. There is no reason to fix by fiat a maximum temporal span for "legitimate" process. As far as we know, a 30 msec process is as feasible to foster consciousness as a 30 sec (or longer) process. This may seem a rather bold claim, but the burden of the proof ought to weight on the shoulder of those who deny it. Is there is any know fact or theory that explicitly link phenomenal experience to the duration of the underlying physical process? At the best of my knowledge there isn't. Of course, there is some evidence as to the temporal length of neural phenomena involved with phenomenal experience, but this is an entirely different matter (Varela 1999; Le Poidevin 2007; Droege 2009; Slagter, Johnstone et al. 2010).

6 Third Step: No Objects But Processes

The rainbow, presented here as a paradigmatic example of coupling between the external world and the neural activity, is a process. It is neither an object in the commonsensical view of a static and autonomous entity nor a mental construct inside the brain. It is not a hallucination concocted by the mind. From a physical point of view, the rainbow is the process taking place from the cloud and ending in the observer's cortex after passing through her eyes, her peripheral nerves and a few intermediate neural nuclei. Hence, the rainbow is spatially and temporally extended. But is the rainbow a unique and indeed exceptional case – perhaps like reflections, mirror images, and other optical phenomenon – or is it one example of a more general condition?

To a certain extent, isn't all we are conscious of – namely fields of view, sounds, uttered words, chairs, faces, tactile patterns, tastes, smells, colors, dots of light – like the rainbow? If we consider the alleged external targets of conscious perception, they are all observer-dependent entities: they are wholes whose unity is the result of the interaction with a given observer. For instance, visual patterns require sighted subjects to exert any effect; sound patterns require a certain auditory competence; tactile structures require skills and actual contact. It is a very well known fact that has been pointed out by many scholars (von Uexküll 1909; Gibson 1966; Bateson 1979). Here, I dare to go a little further. Each target of experience is a whole made of a collection of simpler physical phenomena. The occurrence of the whole is made possible by the interaction with the observer's body and brain.

Consider a face. Does a face exist without someone capable of recognizing it? I suggest a negative answer to this question. If all human subjects in the world were affected by prosopoagnosia, there would be no faces insofar as the facial features on the front side of each head would not be able to produce any effect as a whole. Of course, facial features would remain. Yet, facial features are not faces (van Inwagen 1990; Merricks 2001). The confidence in the existence of faces – and objects – independently of being involved in causal interaction is related with the belief in a world of static and autonomous entities. However, the physical world is not static. It is made of processes. Even stones and static objects are processes. We do not perceive them as dynamic process because of the temporal limitations of our perceptual apparatus (Holcombe 2009). And yet stones and static objects keep changing.

There are also less obvious perceptual targets. For instance, a color is not an absolute perception either of a certain wavelength or of a certain spectral reflectance. Rather a color is the integration of many separate aspects such as the average spectral composition of the background, the received wavelength, certain complex textural features, and so on (Zeki and Marini 1998; Byrne and Hilbert 2003; Hardin 2008). Notwithstanding what the layman believes, a color is not a simple property of the world but rather a whole made of a complex collection of otherwise separate facts and processes coming into existence thanks to a proper physical system.

It might be argued that everything that is perceived is a whole constituted by a process (Manzotti 2009) – a whole that is in the external world in the same sense in which an object is in the external world; yet a whole whose existence is constituted by the completion of the process ending in the beholder's brain.

Again, let me stress that this is not idealism. It is a very physical framework. I clam neither that the subject creates its objects nor that objects are mental entities. The rainbow is not a mental image but a real physical process made of rain drops, sunrays, an arch shaped collection of drops over a cloud, neural firings, chemical reactions, and so forth. Instead of considering only a portion of it, I suggest to take it as a whole. Rather I point out that the object – in the sense that it is understood by most people – is a conceptual simplification of the actual physical process taking place. Such a process is occurring also because there is the subject's body. I am trying to stress the continuity between the "external" and the "internal". No preexisting subject is assumed as a necessary condition for the occurrence of reality. Subject and object occur together as the two sides of a coin.

The Spread Mind is totally different from Berkeley's idealism. His famous claim as to the tree falling in the forest was based on a dualistic ontology and on a British empiricist epistemology. For Berkeley, experience was made of ideas whose existence was possible because of an omnipotent beholder. On the contrary, the model presented here suggests that experience is a dynamic physical process. The Spread Mind is a radical physicalist view suggesting a physical candidate for phenomenal experience. The fact that such a process is constitutive of the subject entails that the beholder is part of the physical reality. Hence the subject's absence makes a difference or, similarly, a difference in reality makes the subject. To recap, if a tree falls in a forest and no one is around to hear it, does it make a sound? It causes sound waves but sound waves are not "sounds". They are just pressure waves. However, the spread mind does not suggest that sounds are ether "object of senses", or ideas, or sensations, or neural representations. The spread mind suggests that sounds are complex physical processes beginning in the tree crash and ending in someone's brain – on the one hand, they are partially external and physical; on the other hand, they are also constituted by the beholder's brain. Thus the answer to the classic question about the tree falling in the forst is negative like in the case of idealism, but the explanation is entirely different.

At the best of our knowledge, the physical world is made of basic entities such as atoms and energy units. However, they are not what we are aware of. Our experience is not made of atoms and energy waves but rather of macroscopic wholes like sounds, colors, chairs, tree, cars, and human beings. The macroscopic world of our everyday experience requires a suitable ontology. I suggest here that such ontology is grounded on processes rather than on perduring or enduring individuals. As to the bottom level, I won't argue whether such a fundamental level is itself made of smaller processes as the string theory would indeed suggest: I leave the issue to much more skilled physicists than myself. However, whatever the nature of the basic level, physics ought to explain how to step from such a low level to the macroscopic level we experience in our everyday conscious life. The Spread Mind suggests that this issue may be addressed by processes that like the rainbow bring into existence macroscopic wholes. In this regard, the issue of consciousness may help to understand the structure of the physical world rather than being and uninvited nuisance to the physical picture of reality. The physical world may be made of the aforementioned basic stuff (elementary particles and energy packets) and by the processes in which that basic stuff gets involved. In my opinion, processes are perfectly respectable citizens of the physical domain. Yet, processes take place only when there the right conditions are met.

Consider once more the problem of faces. The atoms of the head are the basic stuff. The face is not identical to the atoms of the head; the face is a process in which a subset of the atoms composing the head is the cause of a unified process. The face is akin the sound made by the tree falling in the forest or like the rainbow. Without the right conditions, the process might not take place. Such atoms are not enough. To act as a whole they need something more as it was the case with the rainbow and the tree. A human subject capable of recognizing a face is a suitable candidate. Now call A the subset of atoms of Alex's head that can be one of the necessary conditions for the perception of a face. Call B Beatrix's body in normal working conditions. Call C the process you get when B is in front of A and Beatrix perceives Alex's face. Because of the interaction between B and A, the process C occurs. C is the face and is physical. Yet C would not occur if either only B or only A were there. C is causally akin to the rainbow. The above argument is very important. The world we have an experience of, is made of processes, not of static entities that we observe as such. Rainbows, faces, chairs, and the like are processes that happen in the way they do because of the possibilities that our body offers to them. Let me stress again that this is not, by any means, a profession of idealism. I am not saying that the body (not to say the brain or the mind) creates its world. I am saying that the world we experience is made of physical processes whose occurrence require both the environment and the perceiver's body.

7 Fourth Step: No More Representations

The Spread Mind suggests that the world is made of processes like the rainbow and that some of these processes are one and the same with the experiences of the subject. Instead of having a physical chair, a neural chair and a phenomenal chair, the Spread Mind suggests that there is only a chair-process occurring. In other words, there is an identity between the world and our representation of it. Hence, we don't need any longer to assume a separation between the representation and what it represents. The process is, at the same time, the object of representation, the vehicle of that representation as well as the process constituting the subject. These three aspects are allegedly separate entities according to the most widespread view in philosophy of mind, psychology, and neuroscience (Dennett 1991; Searle 1992; Logothetis 1998; Zeki 2001; Kay, Naselaris et al. 2008; Tye 2009, among the many). For the Spread Mind, they are equally partial and incomplete ways to describe a process that is physically unique. The Spread Mind considers seriously the apparently preposterous hypothesis that the representation and the represented external object might be one and the same.

This fourth step may appear as irredeemably wrong. It seems to run against one of the most entrenched assumptions both of philosophy of mind and cognitive science – namely that the mind "deals with" representations which are numerically and physically separate from what they represent. The claim that the apple one is perceiving is the same as the perception of the apple is rather hard to swallow: one may be tempted to believe that the suggested identity is an elementary slip-up – how could it be that perceiving a X is identical with X? The identity does not make any sense taken literally insofar as it is assumed that X is an entity existing autonomously and that X is different from Y. However, accepting such assumption would beg the question. Once more, consider a face. The Spread Mind suggests 1) reconsidering what conscious experience is and 2) reconsidering what the physical world is (what the face is). The notion of separate representation may be akin to that of center of mass. They are handy notions that, nevertheless, do not correspond to anything real. The center of mass is a useful concept that simplifies many calculations. However, it does not exist in a physical sense. It is just an epistemic shortcut that allows tackling with a more complex state of affairs. The notion of representation may play the same role. It may simplify the way in which describe the causal entanglement between subject and object. And yet, it may not correspond to anything in ontological terms.

If the standpoint endorsed by the Spread Mind is somewhat correct, the issue of representation would be swept away, at least in the case of mental representations, since it would suggest that a representation and the represented thing are not physically different, rather they are conceptually different. My representation of the rainbow (when I look at it) is not something stemming out of the neural activity taking place in my brain. My representation of the rainbow is literally the rainbow. It is the process taking place from the cloud and concluding into my visual cortex.

Consider again A, B, and C, as in the previous section. Beatrix watches Alex and sees a face. It may be misleading to assume that there is a face to be seen and that Beatrix concocts either a neural or a mental representation of the face – a representation that is nonetheless separate from Alex's face. The Spread Mind tells a different story. Neither A nor B are a face. The face is C, the process. C is a process that is at the same time the face and the representation of the face.

Hence the Spread Mind suggests that identity is the solution to the issue of representation. To represent something is to be that something. To represent the world means to be identical to that world. Consciousness is situated in the strongest possible sense – it is literally spread in the environment. The mind spreads into the world since it is made of those processes that made the particular world one has an experience of. After all, it is well known that individuals with different motor and sensor apparatuses live in different worlds (von Uexküll 1909). Faces, chairs, colors, sounds, and the like would not exist as wholes without a body that allows them to take place. Once they take place, they are what the mind is made of.

8 Fifth Step: No More Separation Between

Phenomenal World and Physical Reality

It is sometimes assumed that the physical world does not have the same properties attributed to phenomenal experience – this is indeed one of the assumption of the abovementioned hard problem (Chalmers 1996). Philosophers stress properties that do not seem to match between the physical and the phenomenal domain such as quality, perspectivalness, and unity – among the others. For instance, consider colors. Our phenomenal chromatic space is closed while the electromagnetic spectrum is open. As a result, some have argued that light cannot be literally colored (Beau Lotto and Purves 2002; Cohen 2007; Hardin 2008) – which is true. So far, other proposed physical spaces haven't scored much better. Surprisingly one of the better account of phenomenal experience takes into consideration the sensorimotor contingencies between the properties of the surfaces, the actual reflected light, the ambient light, and the observer's perceptual skills (Bompas and O'Regan 2006) a result that may support the present paper.

Conceiving consciousness as a situated and indeed spread physical phenomenon may offer significant advantages as to the issue of the phenomenal character of experience. The phenomenal experience of something might not entail the emergence either of an unexpected phenomenal world or of a phenomenal character out of a physical world, rather it might mean that certain parts of the physical world act together – this unity might be consciousness. What has been called phenomenal character may be the result of the way in which processes constitutes wholes out of simpler parts.

Consider this example. You look around in a room. You see a flower in a vase, a computer and a bookshelf overflowing with books. Because of you, those things – which are processes according the argument defended above – take place at once. If you were not there, those processes would remain isolated. If you were not conscious of your visual field as a whole, those processes would not intermingle together. At least from a behavioral perspective being unable to react to the room as a whole is a sign of a lack of consciousness of it. I skip other possibilities such as blindsight. Hence there is a macroscopic and physical difference. Your physical presence – your conscious presence – makes the difference. I would dare to say that consciousness is precisely that difference. To be conscious of something is to allow to that something to act as a whole. To be conscious of a room is to allow that room to take place as a whole. This wording unfolds in a different way an insight that I believe to share with recent Ted Honderich's work (Honderich 2006). Of course, this belief of mine does not entail anything on Honderich's behalf. His formulation is that "to be conscious of a room is a way for that room to exist". I fear that the emphasis of existence may underestimate the importance of the dynamic character of reality.

Moreover, the objects of the above examples are not unities in themselves. They are wholes occurring because of your body and your nervous system.

They would not take place without your body in that room. The hypothesis is that their occurrence is coextentive with what you call your conscious experience of the room. Thus, conscious experience might be identical with the process-unity resulting from a collection of scattered events.

Thus, is phenomenal experience of a room different from the room? Yes and no. The room, without being engaged with a perceptual system of a conscious being, is a scattered collection of relatively simpler processes. When you look at it, you allow the occurrence of a new and much larger process that, in turn, permit to a significant subset of those processes to act as a whole. This process would not have happened without your body in that room. The room we have an experience of is that process too. Is there any true difference between your conscious experience of a room and that process? They occur at the same time, in the same place. Furthermore, they put together the same aspects of reality (the books, the computer screen, the shelf, the flower in the vase, and the like). Why should we consider you're your experience and the room you have an experience of are ontologically and numerically different?

9 Sixth Step: World and Body in Time

Take a conscious human subject. Each moment of consciousness corresponds to the occurrence of a world – something that William James called a "specious present" (James 1890). Yet, such a world is not a phenomenal world neither in the traditional phenomenological nor in the reflexive sense suggested by Max Velmans (Velmans 2007). The world considered here is a physical world taking place and having at its center the subject's body. Each cognitive agent defines a world which is not made of qualitative phenomenal ghosts, but of physical processes. We act in the world of our experience and we see things happen. Hence, if we take the body to be the physical portion of reality that is identical with a subject, a conceptual revision is mandatory – there are two notions of the body to consider. On one hand, there is the traditional body singled out by the skin boundaries. On the other hand, there is the Spread Mind that is made, at each moment of time, by all those processes that identify the subject's consciousness.

The Spread Mind is thus at the same time the world, our consciousness, and, in a subtle sense, the subject's instantaneous body. It is a body that includes part of the traditional environment and that it may eschews components of the traditional body. For instance, at a given moment, consciousness might be the same irrespective of organs that, although very useful to preserve the subject's life, do not intermingle with our awareness.

The identity between our experience and the world is consistent with the often mentioned transparency of perception (Harman 1989; Tye 2002) that suggests that there is nothing between the subject and the perceived object. The Spread Mind takes this insight at face value. There is literally nothing since the perceived object (which is the process) is part of the instantaneous body of the subject. If consciousness is literally situated, it follows that the environment we have an experience of, the world of our experience, and the body are one and the same.

10 Seventh Step: Everything Is Perception

Optimistically, so far the model may be worth of some consideration. Yet a potentially fatal blow lies ahead. How does the model explain all those cases where it seems that there is no conceivable continuity with the external world? For instance, consider dreams, mental imagery, hallucinations, after images, phosphenes, and the like. Surprisingly, the Spread Mind tackles easily with these alleged counterexamples. In essence, the proposal is that every phenomenal experience is a case of perception. After all, aren't all cases of mental imagery closely constrained by actual past encounters with the corresponding external entities? Didn't we get personally acquainted with everything we later imagine, dream or hallucinate? Consider dreaming someone who died long ago. Isn't true that we spoke with him a few years ago? Hasn't taken place an uninterrupted causal chain from the time when we've been talking with that person in the past up to the time in which we dream of him? Hasn't a physical process occurred spanning along a few years and ending right now while we sleep and we dream? The Spread Mind suggests that this process, that lasted years instead of milliseconds, is identical to our dream of him. Put simply, dreaming and memory may be cases of postponed and scattered perception. When one's dreaming X, one's perceiving X. Below I will say something as to the possibility that one may dream a conversation never occurred or the traditional flying pink elephant.

I am aware that this final step might seem the most counterintuitive of all. Yet, why should a 30 msec process be scientifically more respectable than a 30 yrs long one? Of course, the shorter the process, the greater its survival value, since a shorter process allows to react to what is taking place to one's temporal and spatial proximity. Yet, as to the contribution to phenomenal experience, there is no known law of nature that gives legitimacy only to relatively short processes. Irrespective of its temporal length, a process may be part of consciousness because of the proper kind of causal entanglement.

In fact, consider dreams. As is well known, far from being that bizarre repertoire of unconstrained imagination that literature and commonsense have always assumed, dream content is extremely conservative. Dream content is creative and productive only insofar as reassembles previous perceptual content. For instance, congenitally born blind subjects do not dream of colors or visual content (Bértolo, Paiva et al. 2003; Lopes da Silva 2003; Kerr and Domhoff 2004).

During everyday perception, causal processes are usually constrained by proximal surroundings. Because of this tight coupling with one's temporal and spatial vicinity, what is perceived is strictly ordered by the structure of the proximal environment. This is not the case when processes, which originated long ago, get to completion in situations in which the causal strength of proximal surroundings is greatly diminished. This diminution of the causal strength of proximal surroundings occurs during sleep or because of some intoxication. Consider sleep. Due to both external and internal reasons, the causal efficacy of proximal surroundings is weakened if not suspended altogether. Yet, this does not mean that what takes place inside the brain, all of a sudden, is autonomous. On the contrary, brain activity is still the result of past experiences and of past moments of acquaintance with the environment. The physical continuity with past events is not lost. The processes taking place inside a brain during a dream are all legitimate physical processes that have began at a certain time and place and that are reaching a conclusion in the subject's brain – not differently from what happens in everyday perception. Consider that everyday normal perception is not instantaneous either. There is a temporal lag between external causes and neural endings in everyday perception, too. It is well known that "What we become aware of has already happened about 0,5 sec earlier. We are not conscious of the actual moment of the present" (Libet 2004, p. 70). So, why should the hypothesis that all mental content is the result of ongoing perceptual processes be so upsetting?

As to the fact that in a dream it is possible to experience a pink elephant, this is the effect of a process that is singling out events that are usually separate in the proximal surrounding. Yet, when the causal ordering effect of surroundings is set aside, past events with no real unity may be unified by new processes. This is something akin to what happens when you look at a landscape. Because of your point of view, buildings that are far away from each other may be perceived as contiguous. Events which occurred in separate instants of time and far away from each other may be perceived as a unity.

11 What is Missing?

The Spread Mind does not require any new unexpected physical phenomenon. The physical world is already equipped with stuff and processes. This is just standard physics. Nothing more than that is required. The Spread Mind suggests that consciousness is situated and physically larger than the body of the subject. Surprisingly, it might turn out that the conscious mind is larger than the cognitive one which is made of the cognitive gears mostly located inside the skull (Adams and Aizawa 2008; Adams and Aizawa 2009).

What is still missing from the sketchy outline presented here? For one, the picture it provides is still incomplete. Many details are still missing. How does this model match with the wealth of evidence provided by neuroscience and psychology? Can the theory fit the available empirical data? Luckily, the

Spread Mind model is not vague as to the emergence of phenomenal experience and their necessary conditions. For instance, the model requires that any phenomenal experience is indeed a process singling out something in the external world.

Every phenomenal content ought to be explained as an actual perception. So, the theory is falsifiable insofar as it makes prediction about the conditions in which consciousness may occur. For instance, a prediction is that no brain in a vat ought to be possible. If any phenomenal content were to take place without a corresponding physical cause, the Spread Mind would be false. Of course, physical events taking place inside the body may still be accounted by the theory insofar as they provide the origin of a suitable process. Henceforth the Spread Mind is amenable to empirical verification.

As a result, it ought to be verified whether there are cases of alleged purely endogenous phenomenal content. Are there any? At the best of my knowledge, there is only a handful of cases that might qualify as valid counterexamples. Among the few, I quote phosphenes due to migraine in congenitally blind subjects, phantom limbs in congenitally limbless patients, phantom penises in female-to-male transsexuals (Saadah and Melzack 1994; Brugger, Kollias et al. 2000; Melzack 2001; Mulleners, Chronicle et al. 2001; Ramachandran and McGeoch 2008). However, these cases are all extremely rare conditions and often rather difficult to interpret. Of course, I don't' want to dismiss them. I am only suggesting a more careful examination of such rather unusual cases. These cases can't be discussed at length here. Yet, just to have an idea how it could be difficult to interpret such remarkable findings, consider that it has hitherto been impossible to settle definitively the debate as to the visual content of congenitally blind subjects (Bértolo, Paiva et al. 2003; Lopes da Silva 2003; Kerr and Domhoff 2004) which is a relatively more common and well-known condition. By the way, on this last issue, the Spread Mind is unambiguous: a congenitally blind subject should not have any visual phenomenal content in her dreams or in other psychological states. Of course she can have experience of space, shapes, and movement due to her historical continuity with such phenomena by means of other senses like proprioception, hearing and touch. Finally I would like to mention another often quoted alleged counterexample to a situated view of consciousness: the direct stimulation of the cortex (for instance, Penfield 1958). Although direct stimulation is considered as an example of autonomous phenomenal experience produced by the brain, I am doubtful. In fact, direct stimulation requires a physical stimulation provided by the experimenter – thus there is an external phenomenon. It is true that the causal path does not pass through the usual sensory nervous paths, yet the cause of the direct stimulation is a physical phenomenon external to the nervous system. Furthermore, the elicited content appears to be derived from past experiences.

A final issue that has not received enough attention is a thorough causal modeling of the processes involved in building up consciousness. Clearly, there are some causes that play a crucial role in cutting the physical world at its joints, so to speak. When and where does a process begin and when and where does a process end? This is not a minor issue and it will need a very careful future examination in order to fulfill all the expectation of a complete theory.

Acknowledgements. My writing has been supported by the bilateral project Italy-Corea ICTCNR and KAIST. Moreover, I would like to thank two anonymous reviewers for their helpful comments and suggestions.

References

Adams, D., Aizawa, K. (2008). The Bounds of Cognition. Blackwell Publishing, Singapore.

Adams, F., Aizawa, K. (2009). Why the Mind is Still in the Head. In: P. Robbins, Aydede, M. (Eds.), The Cambridge Handbook of Situated Cognition, Cambridge University Press, Cambridge, 78-95.

Anderson, M. (2003). Embodied cognition: A field guide. Artificial Intelligence, 149, 91-130.

Andrews, T. J., Schluppeck, D., Homfray, D., Matthews, P., Blakemore, C. (2002). Activity in the Fusiform Gyrus Predicts Conscious Perception of Rubin's Vase-Face Illusion. Neuroimage, 17, 890-901.

Bateson, G. (1979). Mind and Nature: A Necessary Unity. Hampton Press, Cresskill (NJ).

Beau Lotto, R., Purves, D. (2002). The empirical basis of color perception. Consciousness and Cognition, 11, 609-629.

Bértolo, H., Paiva, T., Pessoa, L., Mestre, T., Marques, R., Santos, R. (2003). Visual dream content, graphical representation and EEG alpha activity in congenitally blind subjects. Brain research. Cognitive brain research, 15, 277-84.

Biederman, I., Kalocsai, P. (1997). Neurocomputational Bases of Object and Face Recognition. Philosophical Transactions of the Royal Society of London B, 352, 1203-1219.

Bompas, A., O'Regan, K. J. (2006). Evidence for a role of action in colour perception. Perception, 35, 65-78.

Brugger, P., Kollias, S. S., Müri, R. M., Crelier, G., Hepp-Reymond, M. C., Regard, M. (2000). Beyond re-membering: phantom sensations of congenitally absent limbs. Proceedings of the National Academy of Sciences of the United States of America, 97, 6167-72.

Buzsáki, G. (2007). The structure of consciousness. Nature, 446, 267.

Byrne, A., Hilbert, D. R. (2003). Color realism and color science. Behavioral and Brain Sciences, 26, 3-64.

Chalmers, D. J. (1996). The Conscious Mind: In Search of a Fundamental Theory. Oxford University Press, New York.

Chalmers, D. J. (2008). Foreword. In: Clark, A. (Ed), Supersizing the Mind, Oxford University Press, Oxford, 1-33.

Changeux, J.-P. (2004). Clarifying consciousness. Nature, 428, 603-604.

Chemero, A. (2009). Radical Embodied Cognitive Science MIT Press, Cambridge (Mass).

Chrisley, R., Ziemke, T. (2002). Embodiment. In. Encyclopedia of Cognitive Science, Macmillan, London.

Clark, A. (1989). Microcognition. MIT Press, Cambridge (Mass).

Clark, A. (2008). Supersizing the Mind. Oxford University Press, Oxford.

Clark, A., Chalmers, D. J. (1998). The Extended Mind. Analysis, 58, 10-23.

Cohen, J. (2007). A Relationalist's Guide to Error About Color Perception. Noûs, 41, 335-353.

Crick, F., Koch, C. (1990). Toward a Neurobiological Theory of Consciousness. Seminars in Neuroscience, 2, 263-295.

Crick, F., Koch, C. (2003). A framework for consciousness. Nature Neuroscience, 6, 119-126.

Dennett, D. C. (1991). Consciousness explained. Little Brown and Co., Boston.

Dretske, F. (1981). Knowledge & the flow of information. MIT Press, Cambridge (Mass).

Dretske, F. (1996). Phenomenal Externalism or If Meanings Ain't in the Head, Where Are Qualia? Philosophical Issues, 7, 143-158.

Droege, P. (2009). How or never: How consciousness represents time. Consciousness and Cognition, 18, 78-90.

Farah, M. J., Wilson, K. D., Drain, M., Tanara, J. N. (1998). What is 'special' about face perception? Psychological Review, 105, 482-498.

Gallagher, S. (2005). How the Body Shapes the Mind. Oxford Clarendon Press, Oxford.

Gibson, J. J. (1966). The Senses Considered as perceptual Systems. Houghton Mifflin, Boston.

Hardin, C. L. (2008). Color Qualities and the Physical World. In: Wright, E. (Ed.), The Case for Qualia, MIT Press, Cambridge (Mass), 143-154.

Harman, G. (1989). The Intrinsic Quality of Experience. Philosophical Perspectives, 4, 31-52.

Hohwy, J. (2009). The neural correlates of consciousness: new experimental approaches needed? Consciousness and cognition, 18, 428-438.

Holcombe, A. O. (2009). Seeing slow and seeing fast: two limits on perception. Trends in Cognitive Sciences, 13(5), 216-221.

Honderich, T. (2006). Radical Externalism. Journal of Consciousness Studies, 13, 3-13.

Hurley, S. (2010). The Varieties of Externalism. In: Menary, R. (Ed.), The Extended Mind, MIT Press, Cambridge (Mass), 101-155.

Intraub, H. (1980). Presentation rate and the representation of briefly glimpsed pictures in memory. Journal of Experimental Psychology and Human Learning, 6, 1-12.

James, W. (1890). The Principles of Psychology. Dover, New York.

Kay, K. N., Naselaris, T., Prenger, R. L., Gallant, J. L. (2008). Identifying natural images from human brain activity. Nature, 452, 352-355.

Kerr, N., Domhoff, W. G. (2004). Do the blind literally "see" in their dreams? A critique of a recent claim that they do. Dreaming, 14, 230-

233.

Kim, J. (1995). Dretske's Qualia Externalism. Philosophical Issues, 7, 159-165.

Koch, C. (2004). The Quest for Consciousness: A Neurobiological Approach. Roberts & Company Publishers, Englewood (Col).

Koch, C. (2007). How Does Consciousness Happen? Scientific American, 297, 76-83.

Laureys, S., Tononi, G. (2009). The Neurology of Consciousness. Cognitive Neuroscience and Neuropathology. Elsevier, London.

Le Poidevin, R. (2007). The Images of Time. An essay on temporal representation. Oxford University Press, Oxford.

Libet, B. (1966). Brain Stimulation and the threshold of conscious experience. In: Eccles, J. C. (Ed.), Brain and Conscious Experience, Springer Verlag, New York, 165-181.

Libet, B. (2004). Mind Time. The Temporal Factor in Consciousness. Harward University Press, Cambridge (Mass).

Logothetis, N. K. (1998). Single units and conscious vision. Philosophical Transactions of the Royal Society of London B, 353, 1801-1818.

Lopes da Silva, F. H. (2003). Visual dreams in the congenitally blind? Trends in Cognitive Sciences, 7, 328-330.

Lycan, W. G. (2001). The Case for Phenomenal Externalism. In: Tomberlin, J. E. (Ed.), Philosophical Perspectives, Ridgeview Publishing, Atascadero, 17-36.

Manzotti, R. (2006). An alternative process view of conscious perception. Journal of Consciousness Studies, 13(6), 45-79.

Manzotti, R. (2006). Consciousness and existence as a process. Mind and Matter, 4, 7-43.

Manzotti, R. (2006). A Process Oriented View of Conscious Perception. Journal of Consciousness Studies, 13, 7-41.

Manzotti, R. (2006). A radical externalist approach to consciousness: the enlarged mind. In: Batthyany, A., Elitzur, A. (Eds.), Mind and Its Place in the World. Non-reductionist Approaches to the Ontology of Consciousness, Ontos-Verlag, Frankfurt, 197-224.

Manzotti, R. (2009). No Time, No Wholes: A Temporal and Causal-Oriented Approach to the Ontology of Wholes. Axiomathes, 19, 193-214.

Melzack, R. (2001). Pain and the Neuromatrix in the Brain. Journal of Dental Education, 65, 1378-1382.

Merricks, T. (2001). Objects and Persons. Oxford Clarendon Press, Oxford.

Mulleners, W. M., Chronicle, E. P., Palmer, J. E., Koehler, P. J., Vredeveld, J.-W. (2001). Visual cortex excitability in migraine with and without aura. Headache, 41, 565-72.

Noë, A. (2009). Out of the Head. Why you are not your brain., MIT Press, Cambridge (Mass).

O'Regan, K. J., Noë, A. (2001). A sensorimotor account of vision and visual consciousness. Behavioral and Brain Sciences, 24, 939-73; discussion 973-1031.

Penfield, W. G. (1958). The Excitable Cortex in Conscious Man. Liverpool University Press, Liverpool.

Pfeifer, R., Lungarella, M., Fumiya, L. (2007). Self-Organization, Embodiment, and Biologically Inspired Robotics. Science, 5853, 1088-1093.

Quiroga, R. Q., Mukamel, R., Isham, E. A., Malach, R., Fried, I. (2008). Human single-neuron responses at the threshold of conscious recognition. Proceedings of the National Academy of Sciences of the United States of America, 105(9), 3599-3604.

Ramachandran, V. S., McGeoch, P. D. (2008). Phantom Penises In Transsexuals: Evidence of an Innate Gender-Specific Body Image in the Brain. Journal of Consciousness Studies, 15, 3-26.

Revonsuo, A. (2000). Prospects for a Scientific Research Programme on Consciousness. In: Metzinger, T. (Ed.), Neural Correlates of Consciousness, MIT Press, Cambridge (Mass), 56-75.

Robbins, P., Aydede, M. (Eds.) (2009). The Cambridge Handbook of Situated Cognition, Cambridge University Press, Cambridge.

Robbins, P., Aydede, M. (2009). A Short Primer on Situated Cognition. In: P. Robbins, Aydede, M. (Eds.). The Cambridge Handbook of Situated Cognition, Cambridge University Press, Cambridge, 3-10.

Rockwell, T. (2005). Neither ghost nor brain. MIT Press, Cambridge (Mass).

Rowlands, M. (2003). Externalism. Putting Mind and World Back Together Again. Acumen Publishing Limited, Chesham.

Rupert, R. D. (2004). Challenges to the Hypothesis of Extended Cognition. The Journal of Philosophy, 101, 389-428.

Rupert, R. D. (2009). Cognitive Systems and the Extended Mind. Oxford University Press, New York.

Saadah, E. S., Melzack, R. (1994). Phantom limb experiences in congenital limb-deficient adults. Cortex, 30, 479-485.

Searle, J. R. (1992). The rediscovery of the mind. MIT Press, Cambridge (Mass).

Shanahan, M. (2010). Embodiment and the Inner Life. Cognition and Consciousness in the Space of Possible Minds. Oxford University Press, Oxford.

Skrbina, D. (Ed.) (2009). Mind that abides. Panpsychism in the new millennium, John Benjamins Pub., Amsterdam.

Slagter, H. A., Johnstone, T., Beets, I. A. M., Davidson, R. J. (2010). Neural Competition for Conscious Representation across Time: an fMRI Study. PLos One, 5(5), 1-10.

Strawson, G. (2006). Does physicalism entail panpsychism? Journal of Consciousness Studies, 13, 3-31.

Strawson, G. (2011). Soul Dust. The Observer.

Tallis, R. (2010). Consciousness, not yet explained. New Scientist, 205(2742), 28.

Thompson, E. (2007). Mind in Life. Biology, Phenomenology, and the Sciences of Mind. The Belknap Press of the Harvard University Press, Cambridge (Mass).

Thompson, E., Varela, F. J. (2001). Radical embodiment: neural dynamics and consciousness. Trends in Cognitive Sciences, 5, 418-425.

Tye, M. (2002). Representationalism an the Transparency of Experience. Noûs, 36, 137-151.

Tye, M. (2009). Representational Theories of Consciousness. In: McLaughlin, B. P., Beckermann, A., Walter, S. (Eds.), The Oxford handbook of Philosophy of Mind, Oxford University Pres, New York, 253-268.

van Boxtel, G. J. M., de Regy, H. C. D. G. (2010). Cognitive-neuroscience approaches to the issues of philosophy-of-mind. Consciousness and Cognition, 19, 460-461.

van Inwagen, P. (1990). Material beings. Cornell University Press, New York.

Varela, F. J. (1999). The Specious Present: A Neurophenomenology of Time Consciousness. In: Petitot, J., Varela, F., Pachould, P., Roy, J. M. (Eds.). Naturalizing

Phenomenology: Issues in Contemporary Phenomenology and Cognitive Science, Stanford University Press, Stanford (Cal).

Varela, F. J., Thompson, E., Rosh, E. (1991). The Embodied Mind: Cognitive Science and Human Experience. MIT Press, Cambridge (Mass).

Velmans, M. (2000). Understanding Consciousness. Routledge, London.

Velmans, M. (2007). Reflexive Monism. Journal of Consciousness Studies, 15(2), 5-50.

Velmans, M. (2009). Understanding Consciousness (Second Edition). Routledge, London.

von Uexküll, J. (1909). Umwelt und Innenwelt der Tiere. Springer, Berlin.

Wilson, R. A. (2004). Boundaries of the Mind. The Individual in the Fragile Sciences. Cambridge University Press, Cambridge (Mass).

Zeki, S. (2001). Localization and Globalization in Conscious Vision. Annual Review of Neuroscience, 24, 57-86.

Zeki, S., Marini, L. (1998). Three cortical stages of colour processing in the human brain. Brain, 121, 1669-1685.

20. Consciousness Vectors
Steven Bodovitz

Principal, BioPerspectives,
1624 Fell Street, San Francisco, CA 94117 USA

Abstract

One of the defining characteristics, if not the defining characteristic of consciousness is the experience of continuity. One thought or sensation appears to transition immediately into the next, but this is likely an illusion. I propose that consciousness is broken up into discrete cycles of cognition and that the sense of continuity is the result of determining the magnitude and direction of changes between cycles. These putative consciousness vectors are analogous to motion vectors that enable us to perceive continuous motion even when watching a progression of static images. Detailed characterization of consciousness vectors, assuming they exist, would be a significant advance in the characterization of consciousness.

KEY WORDS: brainstorming, consciousness, conscious vector, consciousness vector, continuity, creativity, delay, DLPFC, dorsolateral prefrontal cortex, motion vector, philosophical zombie, sports psychology

> Time is the substance I am made of. Time is a river which sweeps me along, but I am the river; it is a tiger which destroys me, but I am the tiger; it is a fire which consumes me, but I am the fire. - Jorge Luis Borges (1946).

1. Introduction

To paraphrase the eloquence of Borges, we are made of time and consumed by time. The continuity of experience is one of the defining features, if not the defining feature of consciousness. To be more specific, as first explained by Karl Lashley, each thought or sensation is stable, but each is immediately present after the other, fully formed, with no experience of the underlying processing that led each to become conscious (Lashley, 1956).

Another way to think about the continuity is through a thought experiment of the inverse condition. Start by imagining a lower state of continuity, in which each individual thought is stable, but gaps are apparent. Each. Word. For. Example. In. This. Sentence. Is. Separated. The extra breaks affect your experience of reading the sentence, because you have to think about the flow of the words to get the meaning. Now jump to a complete loss of continuity. Each. Word. Is. Completely. Frozen. In. Time. For. A. Moment. Each. Pops. Into. Cognition. And. Is. Replaced. By. Another. Without temporal integrity, we are repeatedly frozen in time. Frozen memories can inform us where we've been, but not where we are going. We become biological computers without sentience.

The concept of separating information processing from sentience has been proposed in a much more colorful manner by David Chalmers, who describes a philosophical zombie that roughly appears to be human, but otherwise has no awareness (Chalmers, 1996). This is not as abstract as it sounds, because one aspect of this concept has been demonstrated by Hakwan Lau and Richard Passingham (Lau & Passingham, 2006). These researchers used a variant of the well-known paradigm of masking. In a simple version, subjects are briefly shown an image, known as the target, followed quickly by a second brief image, known as the mask, and if the timing falls into well-defined parameters, the target is eliminated from conscious awareness (Koch, 2004a). Lau and Passingham used a more complex version known as a type II metacontrast masking, but the underlying principle is the same, and they tested the accuracy of identifying the target, in this case a square or a diamond, followed by asking the subjects to press keys to indicate whether they actually saw the identity of the target or simply guessed what it was. By using different lengths of time between the presentation of the target and the mask, the researchers were able to identify two conditions in which the accuracy of identifying the target was statistically the same, but the subjective assessments of awareness were significantly different (Lau & Passingham, 2006). Cognition (defined as high-level biological computation) and awareness can be separated, although presumably only under certain circumstances and for brief periods of time.

Taken together, the thought experiment and the actual experiment suggest that the output of cognitive processing is transferred into consciousness in a continuous or seemingly continuous process. The normal limit for information transfer is the speed of light, which is clearly faster than human perception, but not instantaneous. True continuity would presumably require a mechanism based on quantum mechanics. The possibility that microtubules mediate coherent quantum states across large populations of neurons was proposed by Stuart Hameroff and Roger Penrose (1996). This hypothesis has received indirect support in recent years. Physicists at the University of Geneva, for example, demonstrated quantum entanglement by observing two-photon interferences well above the Bell inequality threshold (Salart et al. 2008), but this was with isolated pairs of photons. Moreover, physicists at the University of California at Santa Barbara were able to coax a mechanical resonator into two states at once, which showed for the first time that quantum events could be observed in complex objects, but this required cooling to near absolute zero (Cho, 2010; O'Connell et al. 2010). Notwithstanding this recent progress, whether microtubules, which undergo constant remodeling, can be islands of quantum events in the biochemical and electrical cauldron of the human brain at 37 degrees Celsius remains to be observed. I propose an alternative hypothesis that, rather than true continuity, consciousness is broken up into discrete cycles of cognition and that the sense of continuity is the result of determining the magnitude and direction of changes between cycles.

2. Consciousness Is Likely Discontinuous

Even though the experience of continuity is a defining characteristic of consciousness, it is likely an illusion. The experimental evidence for the discontinuity is largely based on the delay between sensory perception and conscious awareness. Briefly, the pioneering work on the delay was performed by Benjamin Libet, in which he and his colleagues showed an undetectable stimulus could become conscious after approximately 500 msec (Libet et al. 1964; Libet et al. 1967; Libet et al. 1991), but it is not clear whether the delay was due to the processing time to reach consciousness or the time for summation of the stimulus to reach threshold, and others have criticized Libet's conclusions (for example, see Gomes, 1998; Pockett, 2002). A better study was designed by Marc Jeannerod and colleagues, in which subjects were trained to grasp one of three dowels following the appropriate signal and performed the task with a reaction time of 120 msec. When the subjects were asked to verbalize when they first became aware of the signal, the response time was 420 msec, or 300 msec longer (Castiello et al. 1991). Even allowing 50 msec for the required muscle contraction for verbalization, the delay is still a quarter of a second (Koch, 2004b). Thus, according to this experiment, the frequency of cycles of cognition is roughly 4 per second. A larger and arguably more compelling body of evidence for the delay comes from the well-established phenomena of masking, as described above, in which a mask eliminates and replaces the awareness of a target. The elimination is not the result of interfering with the sensory input because if a second mask is presented, the first mask can be eliminated and the awareness of the target can be restored (Dember & Purcell, 1967). The elimination is only possible with a delay between sensory perception and consciousness. The delay, in turn, indicates that consciousness is discontinuous (Koch, 2004c; Libet, 1999).

3. Continuity and Consciousness Vectors

If consciousness is discontinuous, but appears to be continuous, then the problem of understanding consciousness becomes better posed: what creates the continuity? I propose that the sense of continuity is the result of determining the magnitude and direction of changes between cycles (Bodovitz, 2008). These putative consciousness vectors, however, are largely undefined. They presumably track the magnitudes and directions of multiple changes in parallel and/or in aggregate. Moreover, they presumably track changes in inherently qualitative information, such as words and concepts. While these open questions leave the key tenet of this hypothesis unsubstantiated, they create opportunities for breakthroughs by experts in advanced mathematics, physics and/or computer science. At the very least, efforts to model consciousness vectors may provide insights into the value of using the flow of information as feedback for better organizing complex and dynamic data.

The most significant substantiation of consciousness vectors is through analogy to motion vectors, which add motion to a series of otherwise discrete images. The standard speed for movies based on film is 24 frames per second, but rather than strobing, we perceive smooth motion. This is because motion vectors are calculated by the visual system, most likely in visual area V5, also known as visual area MT (middle temporal). Without motion vectors, vision becomes a series of still images, a condition known as akinetopsia or visual motion blindness (Shipp et al. 1994; Zihl et al. 1983; Zihl et al. 1991). The strobe effect makes otherwise simple tasks, such as crossing a street, extremely difficult. The cars are a safe distance away, then bearing down, without any sense of the transition. Even though there is memory of where the cars were, there is no sense of where they are going. Likewise, without consciousness vectors, simple cognitive tasks involving even a limited series of steps would be extremely difficult.

Motion vectors, like consciousness vectors, appear relatively simple at first approximation. In fact, the retina is arguably even a two-dimensional, Euclidian array of detectors, and most objects in motion follow standard trajectories. Yet, the exact neural computations to generate motion vectors have been difficult to determine and are the subject of decades of debate (for review, see Born & Bradley, 2005), although there is consensus that they involve the mapping of retinal activity onto higher-order visual processors such as those in V1 and V5. Thus identifying the calculations for motion vectors may provide the ideal model system for identifying the calculations for the more complex consciousness vectors.

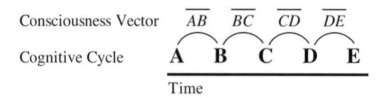

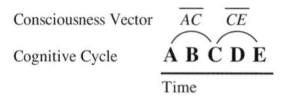

4. Continuity and the Awareness of Change

If consciousness vectors and a sense of continuity are necessary for consciousness, then the corollary is that changes in cognition are necessary for awareness. This corollary is supported by the fact that we only see changes in our visual field. Even though an image may be static, our eyes never are, even during fixation. Our eyes are in constant motion with tremors, drifts and microsaccades. If these fixational eye movements are eliminated, then visual perception fades to a homogenous field (Ditchburn & Ginsborg, 1952; Riggs & Ratliff, 1952; Yarbus, 1967). The significance of these fixational movements for visual processing has long been debated, and no clear consensus has emerged (for review, see Martinez-Conde et al. 2004). The best correlation of neuronal responses to fixational eye movements are specific clusters of long, tight bursts, which might enhance spatial and temporal summation (Martinez-Conde et al. 2004); in addition, a more recent study showed that fixational eye movements improve discrimination of high spatial frequency stimuli (Rucci et al. 2007). But a lack of enhancement or improved discrimination does not explain the complete loss of visual perception. A better explanation is that changes in cognition are necessary for awareness.

5. Discussion

If cognition is broken up into discrete cycles and consciousness vectors create the illusion of continuity, then conscious feedback has pitfalls. It is slow, such that any action that occurs in less than approximately 250 msec will be over before reaching consciousness. Moreover, if events are happening too quickly and/or you are thinking too fast, your conscious feedback will miss changes in cognition, and, to make matters worse, you will have no immediate awareness of any deficiencies and will only be able to deduce the errors afterwards (see figure 1).

In addition, according to this theory, conscious feedback is inherently subtractive. The feedback tracks the magnitude and direction of what has already happened, thereby constraining the introduction of new ideas. All things otherwise being equal, turning down the feedback should inspire more creativity by allocating more energy to new information. Of course, all things otherwise being equal, without feedback, thoughts will be much more disorganized.

Ideally, these practical benefits are only the beginning. If consciousness vectors are real, and if we can begin to understand how they are calculated, we will have a much deeper knowledge of the highest functions of the human brain and possibly be able to apply our insights to artificial intelligence. Unlocking consciousness vectors may unlock human consciousness.

Acknowledgment: I would like to thank Aubrey Gilbert for her insights and careful review. Some of the material in this review was previously published in Bodovitz, 2008. Whereas the previous review included an argument about the possible localization of the brain region responsible for calculating consciousness vectors, this review is focused more on the significance of continuity and the role of putative consciousness vectors.

References

Bodovitz, S. (2008). The neural correlate of consciousness. Journal of Theoretical Biology, 254(3), 594-598.

Borges, J.L. (1946). A New Refutation of Time.

Born, R.T., Bradley, D.C. (2005). Structure and Function of Visual Area MT. Annual Review of Neuroscience, 28, 157-189.

Castiello, U., Paulignan, Y., Jeannerod, M. (1991). Temporal dissociation of motor responses and subjective awareness. A study in normal subjects. Brain, 6, 2639-2655.

Chalmers, D.J. (1996). The conscious mind: in search of a fundamental theory. Oxford University Press, New York, US.

Cho, A. (2010). The first quantum machine. Science, 330(6011), 1604

Dember, W.N., Purcell, D.G. (1967). Recovery of masked visual targets by inhibition of the masking stimulus. Science, 157, 1335-1336.

Ditchburn, R.W., Ginsborg, B.L. (1952). Vision with a stabilized retinal image. Nature, 170, 36-37.

Gomes, G. (1998). The timing of conscious experience: a critical review and reinterpretation of Libet's research. Conscious and Cognition, 7(4), 559-595.

Hameroff, S.R., Penrose, R. (1996). Orchestrated reduction of quantum coherence in brain microtubules: a model for consciusness. In: Hameroff, S.R., Kaszniak, A.W., Scott, A. C. (Eds.) Toward a Science of Consciousness. MIT Press, Cambridge, MA, pp. 507- 540.

Koch, C. (2003). Lecture 15: On Time and Consciousness. In: California Institute of Technology course CNS/Bi 120. http://www.klab.caltech.edu/cns120/videos.php

Koch, C. (2004a). The quest for consciousness: a neurobiological approach. Roberts & Company, Englewood, Colorado, US, pp. 257

Koch, C. (2004b). The quest for consciousness: a neurobiological approach. Roberts & Company, Englewood, Colorado, US, p. 214.

Koch, C. (2004c). The quest for consciousness: a neurobiological approach. Roberts & Company, Englewood, Colorado, pp. 249.

Lashley, K. (1956) Cerebral Organization and Behavior. In: Cobb, S. & Penfield, W. (Eds.) The Brain and Human Behavior. Williams and Wilkins Press, Baltimore, Maryland.

Lau, H.C., Passingham, R.E. (2006). Relative blindsight in normal observers and the neural correlate of visual consciousness. Proceedings of the National Academy of Sciences, 103(49), 18763-18768.

Libet, B. (1999). How does conscious experience arise? The neural time factor. Brain Research Bulletin, 50(5/6), 339-340.

Libet, B., et al. (1967). Responses of human somatosensory cortex to stimuli below threshold for conscious sensation. Science, 158, 2597-1600.

Libet, B., Alberts, W.W., Wright, E.W., Jr., Delattre, L.D., Levin, G., Feinstein, B. (1964). Production of threshold levels of conscious sensation by electrical stimulation of human somatosensory cortex. Journal of Neurophysiology, 27, 546-578.

Libet, B., Pearl, D.K., Morledge, D.A., Gleason, C.A., Hosobuchi, Y., Barbaro, N.M. (1991). Control of the transition from sensory detection to sensory awareness in man by the duration of a thalamic stimulus: the cerebral 'time-on' factor. Brain, 114, 1731-1757.

Martinez-Conde, S., Macknik, S.L. Hubel, D.H. (2004). The role of fixational eye movements in visual perception. Nat Rev Neurosci. 5(3),

O'Connell, A.D., et al. (2010). Quantum ground state and single-phonon control of a mechanical resonator. Nature, 464, 697-703.

Pockett, S. (2002). On subjective back-referral and how long it takes to become conscious of a stimulus: a reinterpretation of Libet's data. Conscious and Cognition, 11(2), 144-161.

Riggs, L.A., Ratliff, F. (1952). The effects of counteracting the normal movements of the eye. J. Opt. Soc. Am. 42, 872-873.

Rucci, M., Iovin, R., Poletti, M., Santini, F. (2007). Miniature eye movements enhance fine spatial detail. Nature, 447, 851-854.

Salart, D., Baas, A., Branciard, C., Gisin, N., Zbinden, H. (2008). Testing the speed of 'spooky action at a distance'. Nature, 454, 861–864.

Shipp, S., de Jong, B.M., Zihl, J., Frackowiak, R.S., Zeki, S. (1994). The brain activity related to residual motion vision in a patient with bilateral lesions of V5. Brain, 117(5), 1023-38.

Yarbus, A.L. (1967). Eye Movements and Vision. New York: Plenum.

Zihl, J., von Cramon D., Mai, N., Schmid, C. (1991). Disturbance of movement vision after bilateral posterior brain damage. Further evidence and follow up observations. Brain, 114(5), 2235-2252.

Zihl, J., von Cramon, D., Mai, N. (1983). Selective disturbance of movement vision after bilateral brain damage. Brain, 106(2), 313-340.

21. Gaia Universalis
Samanta Pino, Ph.D., and Ernesto Di Mauro, Ph.D.

Istituto Pasteur Fondazione Cenci Bolognetti c/o Dipartimento di Genetica e Biologia Molecolare, Università di Roma "Sapienza", Piazzale Aldo Moro, 5, 00185, Roma, Italy.

Abstract

Life relies on two principles: the principle of Continuity and that of Simplicity. Both principles entail its universality. Can we draw the same conclusions applying these principles to consciousness? Summarizing the content of this short assay by an aphorism: consciousness is a property and a manifestation of life, life is universal in principle, consciousness is in principle universal.

KEY WORDS: definition of life, consciousness, simplicity, continuity, complexity.

1. CONSCIOUSNESS AND SELF-CONSCIOUSNESS

Consciousness is the focal point of epistemology. The Cartesian "cogito ergo sum" defines the problem of self-consciousness but not that of our position in the universe, nor the relation of the universe with us. If we try to extend the purport of Cartesian consciousness to the universe as-we-know-it, the only solid starting point is our experience. In so doing it is relevant to keep in mind that the evolution of our brain is the consequence of its adaptation to the evolving environment of planet Earth, necessarily depending upon the Euclidean frame of reference into which the fittest characters were selected for survival. Our brain is well equipped to know where are the right and the left arms, to move quickly the rest of the body in its fourth dimension and, since the appearance of gene mutations that triggered the development of the human cortex (Evans et al. 2005; Mekel-Bobrov et al. 2005), it has also learnt how to answer to problems non solvable in Euclidean terms by developing abstract thought, symbolism, metaphysics. Briefly, self-consciousness is in humans a phenotype determined by, and interacting with, the genotype that encodes it, fruit of local evolution. Is the purport of consciousness extendable beyond nervous systems? And: can the Observer's principle, the cage that apparently compresses all the logic efforts in this quest for a wider horizon, be violated?

2. RHIZOME

Going back in time, each human being is genetically connected with all the other human beings and (there is no need to argue too much on this point) with the rest of living organisms. With the help of symbolic thought, the image that best describes this concept is that of a rhizome, stressing the unity of the living and the ensemble of its general connections. This concept was introduced by Deleuze and Guattari in A Thousand Plateaus (1980). In their original wording the rhizome follows the principles:

1 and 2: Principles of connection and heterogeneity: any point of a rhizome can be connected to any other, and must be.

3: Principle of multiplicity: only when the multiple is effectively treated as a substantive, "multiplicity" that it ceases to have any relation to the One.

4: Principle of asignifying rupture: a rhizome may be broken, but it will start up again on one of its old lines, or on new lines.

5 and 6: Principles of cartography and decalcomania: a rhizome is not amenable to any structural or generative model; it is a "map and not a tracing".

Even though one should be well aware of the dangers intrinsic in what has been defined Post Modern Thought (Sokal & Bricmont, 1997), the evocative power of this metaphor of the living world is high.

In the sense of this metaphor, the living world is a totally connected network in which each single organism is a momentary embodiment of a specific genotype that starts its life (simplifying a little) in the moment of the replication and/or of the recombination (according to the system) of its parental genetic materials, and ends in the moment of the dissolution of its own genome. Each genotype is on-line connected with the genotypes from which it derives and with those that could derive from it. The living network is a unit in the time- and genetic-space extending back to the possibly multi-rooted Last Universal Common Ancestor(s). Before that, individuals were by definition not existing as such, entities immersed in the swamp of combinatorial biochemistry. Briefly, one needs to critically examine the very concept of life.

3. DEFINITION OF LIFE

The accepted definition "life is a self-sustained chemical system capable of undergoing Darwinian evolution" (Joyce, 1994) is rigorous and rigid at the same time. I may recognize myself as being well described and represented by this definition, as could any giraffe or alga or bacterium or even virus I can think of in a Linnaean world. But: a) life is a process and not a system; b) evolution may not be necessarily needed in a omni-comprehensive definition of life. One could imagine an environment in which there are no variations; or in which variations are cyclic and highfrequency, occurring in a time scale that does not commensurably correspond to the time scale of the living entities it harbors and supports. In such environments evolution would not be a categorical property.

Before LUCA, cycles had to establish themselves that could provide the necessary organization of matter, energy control and directional flow of chemical information. Untangling of these biochemical and biophysical hanks has proven difficult and the scenarios in which prebiotic processes are sketched are still biased between metabolism-first, genetic-first or, even, membranes-first. The three "Firsts" should have possibly been three acts of the same comedy, played together on the same stage. The organization of a living (defined and reproducible) entity is difficult if the same Darwinian "warm little pond" (Darwin, 1888) did not allow the simultaneous formation of ins-and-outs by lipid-based membranes, the organization of metabolic cycles as the reductive citric acid cycle (Morowitz, 1970), which is reasonably indicated as the key to ur-metabolism, and a way to store and accurately transmit the necessary coded information. For any sensible and comprehensive theory of life an initial unitary chemical frame from which the three Firsts took place together, or in a parallel and interactive manner, is badly needed.

A compelling example of the proficiency of these interactions was provided (Mansy et al. 2008), showing the promoting effect exerted by nucleic acids on vesicle division. Briefly, life extends beyond the limits of the individual, its lower border being historically indefinable. Nevertheless, a bottom-down approach might indicate the key principles on which it is based.

4. THE STANDARD PRINCIPLES FOR LIFE: 1) CONTINUITY

In terms of physics, the text-book notion underpinning life is a momentary slack of the rush toward entropy, as dictated by the second principle of thermodynamics. In terms of chemistry, life is the concerted unidirectional transfer of chemical energy among different types of molecules following schemes that we call genetic due to their hereditability. The adhesion to genetic instructions may entail highly sophisticated and close to ineludible mechanisms (as in our human case), or may more simply be thought of as an ensemble of molecules elaborating chemical information, partly redirecting the intrinsic energy of their chemical structure, partly using the energy provided by the environment that they were able to harness and retain. Abiotic pre-metabolic cycles (Schwartz & Goverde, 1982; Weber, 2007; Morowitz et al. 2000; Smith & Morowitz, 2004) illustrate this point and provide an idea of the vitality of Darwin's pristine warm little pond. This scenario privileges the "metabolism-first" set of theories in contrast to the "geneticfirst" point of view. These Firsts, we have argued, needed not to be alternative and evolution on this planet was possibly kick-started by the cooperation of the different systems. The involvement of the protein world would require a discussion of its own.

Was it life?

From a point of view of rigorous logics, yes. No interruption in the evolutionary space going between these initial, as yet hypothetical but necessary, genetico-metabolic systems and the more complex (or, so to say, more genetically based) organisms may have taken place, by definition. Any interruption would have lead to a dead-end. Logically, going back in the evolutionary space (and, as far as we know, in time) means going back towards metabolic simplicity, means decreasing the dependence of life on genetic hereditability but does not allow to claim: after this point it was life, before it was not.

In this view we slide straight back into the pristine warm little pond, dipping into the mixture of organics that was allowed to leaven by the happy encounter of combinatorial chance and physical-chemical necessity. The principle of Continuity can be stated with no fear of contradiction for the very reason of our very existence.

5. THE STANDARD PRINCIPLES FOR LIFE: 2) SIMPLICITY

The prerequisite for life is the presence of the four most abundant elements: H, C, O and N (Spaans, 2005). The presence of sulphur S in proteins and in important metabolic nodes of extant organisms on planet Earth is indication of the sophisticated use of chemical alternatives but is metabolically marginal. The information-bearing elements of extant nucleic acids are held together by phosphate bridges. However, genetic chains built on different principles as PNA (Egholm et al. 1992) have provided the proof-of-principle for the existence of possible alternatives. It was suggested that phosphorus P can be replaced in its connecting functions by arsenic As (Wolfe-Simon et al. 2010).

The reactions of the four key elements in space result in a large panel of combinations. The site http://astrochemistry.net provides an idea of their combinatory power even in the harsh conditions they meet in space. If provided with the appropriate environment in terms of temperature, intensity and type of radiations, pressure, relative concentrations, stability of conditions, the catalogue of stable compounds is expected to skyrocket, based on the direct embodiment of the chemical potentialities of the 4 key atoms. The fact that carbon chemistry is particularly fertile is well known. The possibility that the chemistry of HCN has played a pivotal role in the initiation of biogenic reactions was proposed (Oró, 1961), extensively discussed (Orgel, 1998), brought to experimental verification (Saladino et al. 2009).

Simply warming formamide H_2NCOH (the product of the reaction of hydrogen cyanide HCN with water H_2O) in the presence of mineral catalysts of the most common sources (as reviewed in Saladino et al. 2010) results in a large panel of nucleic bases (Saladino et al. 2009), in acyclonucleosides (Saladino et al. 2003), in the phosphorylation of nucleosides (Costanzo et al. 2007). The non-enzymatic polymerization of cyclic nucleotides was reported (Costanzo et al. 2009), showing that the abiotic route from HCN to nucleic polymers is in principle possible in thermodynamically and kinetically sound simple conditions.

6. CONSCIOUSNESS AND THE STANDARD PRINCIPLES OF CONTINUITY AND SIMPLICITY

The interpretation of consciousness was given by R. Penrose as something that enables to perform actions that lie beyond any kind of computational activity (Penrose, 1988). The concept was extended (Penrose, 1994), such as to be applicable to any kind of computational process, implying "a mechanism in brain function whereby a non-computational physical action might indeed underlie our consciously controlled behavior". This mechanism is supposedly based on "subtle and largely unknown physical principles in order to perform the needed non-computational actions". The category of computation (or, more appropriately, its absence) is invoked as the key to understand the essence of consciousness.

Away from the idealized thermal equilibrium state, statistical fluctuations are expected (Penrose, 1994). Consciousness is thus located in the domain of the statistical behavior of matter, which is to say that consciousness is a property of statistical mechanics. This clearcut formulation brings to potentially testable ground a need expressed previously: We must postulate a cosmic order of nature beyond our control in which both the outward material objects and the inward images are subject (Pauli 1948); The same organizing forces that have shaped nature in all her forms are also responsible for the structure of our minds (Heisenberg 1971); Psyche and matter exist in one and the same world, and each partakes of the other (Jung 1970). The 26 years long correspondance between Pauli and Jung (Roscoe 2001) offers numerous examples of these intuitions. If one is allowed to boil down and intellectual correspondence encompassing all these years: the important things are born at the border between order and chaos. Is life computable? Computational models may describe biological systems. Undoubtedly, even though any organism as a whole is more complicated than a field of gravitating bodies or of any problem definable in Laplacean terms (from blood's flow to a human brain), "local" biological phenomena as organisms may be in principle computationally represented.

A different perspective arises if life is considered as a whole. Let us assume, based on what we know from planet Earth biology, that (i) on this planet no interruption of the reproducible flow of the coded complex biochemistry that we dub life has occurred since its emergence, and let us assume, based on what we know from astrochemistry, that (ii) the same potential for dynamic complexity is intrinsic in the reactivity of the components both of interstellar dusts and of their countless aggregation forms, anywhere they might occur. Let us in addition recognize that (iii) the chemistry of hydrogen H, of carbon C, of oxygen O, and of nitrogen N is the same all over the universe. Considered as a whole, the complexity of the potentially biotic chemistry is comparable to the complexity of the universe itself.

Taking into account the formal definitions of complexity, of emergence (and consequently acknowledging life as an emergent phenomenon), and considering the complexity of the reaction vessel involved (the universe itself), life is not computable. And taking also into account, as pointed out by R. Penrose, that "..chemical actions are the result of quantum effects, and [...] one has left the arena of classical physics when considering processes that are dependent upon chemistry" (Penrose, 1994).

Let us draw the logic conclusion from these considerations: gap-free flow of life is present everywhere in the universe in the form of more or less organized, more or less complex chemical reactions. Here we are conservatively limiting the definition of life to the flow of carbon-based reactions and we are implying that the lower limit of life are the processes involving the reactions of the one-carbon molecule HCN, considered as the starting and reactive clump of chemical information.

Can we apply to consciousness the standard principles of Simplicity and Continuity ? If so, we should acknowledge that : being consciousness the attribute of living entities, consciousness is present everywhere there is life in the universe.

Considering consciousness, we spontaneously start from our self-consciousness (which is typical of human beings, putting themselves at the center of every consideration). Like any other phenotype, consciousness is an evolutionary character. Thus the question can be asked about where and when, going back in evolutionary space, did consciousness come into being. Certainly our extant primate relatives are self-conscious, as other mammals have been shown to be. And how could we consider devoid of consciousness the eyes of a crow, or of a lizard? Is the reactivity of amoebas a sign of consciousness? Chapter 7 of (Penrose 1994) points out the implications of the quantic domain in neurophysiology, including the milestone observations in Paramecium. At which point the complex reactions of membranes, of photochemistry, of ion channels observed in protozoans start to differ from those of nervous cells of higher eukaryotes? Should the level of complexity be considered a categorical difference? And so on, backwards. As it is for life, drawing a line separating times before which there was no consciousness from those after which it came into being, is an arbitrary exercise.

Consciousness is considered as the attribute of the complex multicellular structure that we call human brain, and its root behaviour is deductively based on its ability to perturb matter at the quantic level (Penrose, 1988, 1994). Why should we be entitled to assume that simpler living structures do not perturb matter at the same quantic level, and why should we not draw from this the inescapable conclusion? Walking back in evolution, complexity decreases, but the limit of non-life and of non-reactivity is never reached. Drawing borders is a didactic, arbitrary act of our self-consciousness.

As for life, consciousness is an intrinsic property of reacting matter. In this living universe. Two principles are at the basis of contemporary philosophical quest, heavily overlapping with more or less speculative physics: the Observer's and the Anthropic Principles. The first states that physical entities cannot be observed without interfering with them, the second that we are here in this universe because its properties are just fit to allow our existence. A corollary of the Anthropic principle entails that other similar but not identical universes exist forming a sort of possible myriad doppelgangers of the only reality that we can observe. None of the two principles is in contradiction with the vision of life as a possible state of consciousness of the universe.

We may thus try to widen accordingly our parochial Euclidean horizon, extending the Gaia (Lovelock 1979) vision, and accept that life and consciousness are two aspects of the same rhizomatic phenomenon that encompasses the whole universe, and makes it so interesting.

Acknowledgments: we gratefully thank Silvia Lopizzo for helpful contributions.

References

Costanzo, G., Saladino, R., Crestini, C., Ciciriello, F., Di Mauro, E. (2007) Nucleoside phosphorylation by phosphate minerals. J. Biol. Chem. 282, 16729-16735.

Costanzo, G., Pino, S., Ciciriello, F., Di Mauro, E. (2009) Generation of long RNA chains in water J. Biol. Chem. 284, 33206Ð33216.

Darwin, C. (1888) The life and letters of Charles Darwin Vol 3, p.18. Letter to Joseph Hooker. John Murray, London.

Deleuze, G., Guattari, F. (1980) Mille plateaux. Minuit Ed, Paris.

Egholm, M., Buchardt, O., Nielsen, P.E., Berg, R.H. (1992) Peptide Nucleic Acid (PNA). Oligonucleotide analogs with an achiral peptide backbone. J. Am. Chem. Soc. 114, 1895-1897.

Evans, P.D., Gilbert, S.L., Mekel-Bobrov, N., Vallender, E.J., Anderson, J.R., Vaez-Azizi, L. M., Tishkoff, S.A., Hudson, R.R., Lahn, B.T. (2005) Microcephalin, a Gene Regulating Brain Size, Continues to Evolve Adaptively in Humans. Science, 309, 1717-1720.

Heisenberg, W. (1971) Physics and beyond. Cambridge, University Press, p.101

Joyce, J. (1994) in D. W. Deamer and G. R. Fleischaker (eds.), the foreword of ÔOrigins of Life: The Central Concepts', Jones and Bartlett, Boston.

Jung, C. G. (1970). The Structure and Dynamics of the Self. In: Jung, C. G., Collected Works of C.G. Jung, C.V. 9,2, par. 413. Princeton University Press, 1970

Lovelock, J.E. (1979) Gaia, A new look at life on Earth. Oxford University Press, London.

Mansy, S.S., Schrum, J.P., Krishnamurthy, M., Tobé, S., Treco, D.A., Szostak, J.W. (2008) Template-directed synthesis of a genetic polymer in a model protocell. Nature, 454, 122-125.

Mekel-Bobrov, N., Gilbert, S.L., Evans, P.D., Vallender, E.J., Anderson, J.R., Hudson, R.R., Tishkoff, S.A., Lahn, B.T. (2005) Ongoing Adaptive Evolution of ASPM, a Brain Size Determinant in Homo sapiens. Science, 309, 1720-1722.

Morowitz, H.J. (1970) Entropy for Biologists. An Introduction to Thermodynamics, Academic Press. New York-London.

Morowitz, H.J., Kostelnik, J.D., Yang, J., Cody, G.D. (2000) The origin of intermediary metabolism. Proc. Natl. Acad. Sci. USA. 97, 7704-7708.

Orgel, L.E. (1998) The origin of life- a review of facts and speculations. Trends Biochem. Sci. 23, 461-495.

Oró, J. (1961) Mechanism of synthesis of adenine from hydrogen cyanide under possible primitive earth conditions. Nature, 191, 1193-1194.

Pauli, W., (1948) Letter to Marcus Fierz. Quoted in Henry P. Stapp. Mind, Matter and Quantum Mechanics (Berlin: Springer Verlag, 1993) 175.

Penrose, R. (1988) The emperor's new mind : concerning computers, minds, and the laws of physics. Oxford University Press. London.

Penrose, R. (1994) Shadows of the mind. Oxford University Press. London.

Roscoe, D., (trans) (2001) Atom and Archetype: The Pauli/Jung Letters, 1932-1958. C.A. Meyer ed. Princeton University Press.

Saladino, R., Ciambecchini, U., Crestini, C., Costanzo, G., Negri, R., Di Mauro, E. (2003) One-pot TiO2-Catalyzed Synthesis of Nucleic Bases and Acyclonucleosides from Formamide: Implications for the Origin of Life. ChemBioChem 4, 514-521.

Saladino, R., Crestini, C., Ciciriello, F., Pino, S., Costanzo, G., Di Mauro, E. (2009) From formamide to RNA: the roles of formamide and water in the evolution of chemical information. Res. Microbiol. 160, 441-448.

Saladino, R., Neri V., Crestini C., Costanzo, G., Graciotti, M., Di Mauro E.(2010) The role of the Formamide/Zirconia system on the synthesis of nucleobases and biogenic carboxylic acid derivatives" J. Mol. Evol. 71, 100-110.

Schwartz, A.W., Goverde, M. (1982) Acceleration of HCN oligomerization by formaldehyde and related compounds: implications for prebiotic syntheses. J. Mol. Evol. 18, 351-353.

Smith, E., Morowitz, H.J. (2004) Universality in intermediary metabolism. Proc. Natl. Acad. Sci. USA. 101, 13168-13173.

Sokal, A., Bricmont, J. (1997) Impostures Intellectuelles. Odile Jacob Ed, Paris.

Spaans, M. (2005) The Synthesis of the Elements and the Formation of Stars. Astrobiology: Future Perspectives, Astrophysics and Space Science Library (Ehrenfreund, P. et al. Eds.) Vol. 305, 1-16.

Weber, A.L. (2007) The sugar model: autocatalytic activity of the triose-ammonia reaction. Orig. Life. Evol. Biosph. 37,105-111.

Wolfe-Simon, F., Blum, J.S., Kulp, T.R., Gordon, G.W., Hoeft, S.E., Pett-Ridge, J., Stolz, J.F., Webb, S.M., Weber, P.K., Davies, P.C.W., Anbar, A.D., Oremland, R.S. (2010) A bacterium that can grow by using arsenic instead of phosphorus. Science, PMID: 21127214, 1-6.

22. What Consciousness Does: A Quantum Cosmology of Mind

Chris J. S. Clarke, Ph.D.

School of Mathematics, University of Southampton, University Road, Southampton SO17 1BJ, UK

Abstract

This article presents a particular theoretical development related to the conceptualisation of the role of consciousness by Hameroff and Penrose. The first three sections review, respectively: the different senses of "consciousness" and the sense to be used in this article; philosophical conceptions of how consciousness in this sense can be said to do anything; and the historical development of understanding of the role of consciousness in quantum theory. This background is then drawn upon in the last two sections, which present a cosmological perspective in which consciousness and quantum theory are complementary processes governed by different logics.

KEY WORDS: Qualia, Quantum cosmology, Quantum collapse, Epiphenomenalism, Histories interpretation, Quantum logic, Heidegger

1. Consciousness: What Are We Talking About?

"Consciousness" is notoriously difficult to define because it is so fundamental; it is the precondition for our being able to do or know anything. Surveying the voluminous controversies over the meaning of the word suggests, however, that "consciousness" tends to be used in two fairly distinct ways. Broadly considered, just as the word "spirit" has two quite different referential meanings, viz., alcoholic beverage vs religion/parapsychology, so "consciousness" has two meanings: One meaning refers essentially to subjective experience: our moment- by-moment qualitative awareness of what is happening both internally (thoughts, feelings) and externally. It is the "what it is like" of Nagel's seminal paper (Nagel, 1974). The other meaning is that used by Dennett (1991). Taking to heart Wittgenstein's dictum that "whereof one cannot speak, thereof one must be silent", he restricts the topic of consciousness to those aspects of experience that we can report on, verbally, to other people. From this he is led to restrict the concept to that part of our internal experience that is contained in our inner dialogue, the almost constant talking to ourselves whereby we make sense of the world to ourselves in verbal terms. Thus consciousness in Dennett's sense of the word is the process of our forming "drafts" of parts of our internal dialogue. I shall call these senses of "consciousness" as used by Nagel and Dennett qualis-consciousness and quid-consciousness, respectively, from the Latin words for "how" and "what". Qualis is related to quale, quality, and with the idea that qualis-consciousness is comprised of qualities (qualia) associated with perception and thinking.

This difference between these two senses is crucial when we consider what we know about the consciousness of other people or other beings. Whereas we can, and by its definition can only, explore the quid-consciousness of another person by talking to them, we can only know the non-verbal part of another's qualis-consciousness through empathy; that is, through our evolved capacity for mirroring the sensations of others in response to a range of bodily cues and contextual information (Berger, 1987). This means that, as cogently argued by Nagel (1974), in the case of an organism like a bat with which it is difficult to have much empathy, we cannot know explicitly that they have qualis-consciousness, even though we might postulate that this is the case because they are mammals like ourselves. On the other hand when it comes to the verbally based quid-consciousness we know that bats, lacking language, cannot have it. The distinction between the two concepts is thus vital when discussing non-human consciousness. Without deeper analysis, we cannot rule out the occurrence of a "hidden" qualis-consciousness from any organism, or even from physical systems that we may not consider organisms at all - a vital point that will be revisited in section 5.

2. What Does Consciousness Do?

A major strand in the philosophy of consciousness concerns the notion of epiphenomenalism - the idea that consciousness is an add-on that appears upon ("epi") information processing without having any functional role. This is not relevant to quid-consciousness, which is actually a part of information processing rather than something added to it. In the case of qualis-consciousness, on the other hand, "epiphenomenalism" seems meaningful while at the same time seeming odd, because the whole notion of causation, in the physical sense, seems problematic in connection with qualis-consciousness. This consciousness does not do things like digesting food or moving limbs in a purely mechanical sense, but it comprises the whole of our experienced world (McGilchrist, 2009) and thereby establishes the context and preconditions as a result of which doing-events like moving limbs take place. The distinction between the two senses of "consciousness" lies not in whether they either do things or not; it lies in the distinct categories of "doing" and "being" that are involved. Quid-consciousness does things in a causal sense as part of a whole control structure of information processing. Qualis-consciousness constitutes a meaningful world within which doing is possible.

Here it becomes a matter of one's philosophical position, whether or not qualis-consciousness is anything other than a sort of emotional fog generated by processing in the brain. If one adopts a scientific-realist position on which the world is entirely reducible to mechanical processes, then qualis-consciousness is indeed such a fog. The alternative to this is to recognise that there is a whole area of discourse concerning existence, value, meaning and so on which is related to the mechanical properties of the world, but which is not equivalent to the mechanical aspect of the world.

Since it is qualis-consciousness that raises the most significant problems in consciousness studies, I shall from now on restrict attention to this sense of the word and, with this understood, I shall usually drop the "qualis" and just call it "consciousness".

3. The Changing View of the Role of Consciousness in Quantum Mechanics.

The earlier history of this topic falls into four phases:
(a)   The "quantum theory" of Planck and Einstein, based on a conventionally mechanical concept of "quanta".
(b)   The "quantum mechanics" of Bohr and Heisenberg from about 1925. This was based on complementarity and the uncertainty principle, which in-

creasingly involved the idea of the collapse of the quantum state (also known as the wave function). It culminated in von Neumann's picture (von Neumann, 1932) of two quite distinct processes: a smooth deterministic evolution of the state under a dynamics, and a discontinuous transition from one state to another related to observation. He did not, however, suppose that consciousness was peculiarly concerned in this, arguing that it was sufficient to consider the human being as an assemblage of rather sensitive physical detectors.

(c) The views of Wigner and London and Bauer (London & Bauer, 1939, 1983) that consciousness was essential for collapse. According to this, the quantum state of the human brain was, through the process of experimental observation, coupled to the quantum state of a microscopic system; then the consciousness of the human being collapsed the joint state of human, apparatus and microsystem. This role for consciousness was strictly limited. Consciousness was not responsible for determining what particular quantity was being measured, because this was determined by the apparatus (a point that will be revisited in section 5). It could not bias the probabilities for different outcomes, because this would undermine the very laws of physics. All that consciousness could do was, somehow, to demand that some definite outcome did emerge, rather than a mixture or superposition of possibilities.

(d) A focus on the quantum-classical distinction. This began with Daneri, Loinger and Prosperi (1962) suggesting that "collapse" was the transition from a quantum state to a classical state, and that this was located not in the brain of the observer, but in the experimental apparatus. They showed that it was the large size of the apparatus, with a large number of possible quantum states all linked to the state of the microsystem being observed, which averaged out the peculiarly quantum mechanical nature of the microsystem, resulting in an essentially classical, non-quantum state for the apparatus. Subsequently Zeh (1970) included in this averaging-out of quantum states the highly effective role of interaction with the wider universe through the phenomenon of "decoherence". By this time the idea that consciousness had a role in quantum theory came to be regarded as superfluous.

In consequence of this history, it has become clear that we are here dealing with two distinct (though interrelated) physical representations. One is the superposition of states, a peculiarly quantum effect resulting in, for instance, the interference patterns produced in the experiment where particles are fired towards two parallel slits. The other is the statistical mixture of states used to represent mathematically a situation such as the result of a rolling a dice, where there is a range of possible outcomes with different probabilities for each. Considered purely mathematically, decoherence turns a superposition into a mixture. This does not, however, explain why we are actually aware, at the end of the process, of one particular outcome as opposed to a fuzzy blur of possibilities. We may recall that this, and only this, was what consciousness was supposed to achieve on London and Bauer's earlier way of looking at things. Despite much clarification between 1939 and 1970, the possible role for consciousness has remained little changed, and its operational details have until recently remained obscure.

4. The Perspective of Cosmology on the Role of Consciousness

More recent arguments from the surprising direction of cosmology now clarify things a great deal. In particular, quantum cosmology starkly underlines the need for something like consciousness. To take a particular example: the WMAP satellite observations of the universe at an age of some 380,000 years confirm a picture in which the universe has evolved as if it started in a perfectly smooth homogenous state (though strictly speaking there can be no "initial state" since the very earliest stages merge into the as yet unknown timeless conditions of quantum gravity). By the epoch observed by WMAP we see minute fluctuations superimposed on this uniform background, of the same character as the quantum fluctuations that can be detected when a uniform beam of radiation is observed in the laboratory. On conventional theory, these cosmological fluctuations grew under the influence of gravity to produce stars, galaxies and ourselves. Note, however, that in quantum theory it is the act of observation that precipitates quantum fluctuations: without observation (in whatever generalised form we may conceive it) a homogeneous initial state evolving under homogeneous laws must remain homogeneous. So the early fluctuations that eventually give rise to the existence of planets, people and WMAP are caused by observations such as those made by people and WMAP! The problem of quantum observation lies at the heart of modern cosmology.

This cosmological perspective makes it clear that the bare mathematical formalism of quantum theory in insufficient on its own. Without some additional ingredient, the universe would remain homogeneous and sterile. Two ideas from quantum cosmology are needed in order to make sense of this. They will also provide the key to the role of consciousness.

The first was introduced by James Hartle (1991), building on the "histories" interpretation of Griffiths (1984). Instead of considering probabilities for different outcomes to a single quantum observation, Hartle examined the probabilities of sets of outcomes for any collection of observations scattered throughout the universe in space and time. The mathematics was almost the same as it would have been if one had assumed a collapse of the wave function simultaneously across the universe with each observation; but strictly speaking the latter concept cannot be used in cosmology because it is not consistent with the fact that in relativity theory "simultaneous" is an observer-dependent concept. By considering this "super- observation" extended over the whole of space-time there is no need to consider either collapse or issues of causality between future and past events.

Hartle gave no indication as to what was actually meant by an "observation" or "observer". This issue was made explicit through the second key idea, first raised by Matthew Donald. He considered quid-consciousness - i.e. information processing - but this cannot help because it is in no way essentially different from any other purely mechanical process. Then, however, the idea was explored by Don Page who focussed on "sensation", which is close to the qualis-consciousness of this paper. The aim of a cosmological theory, he argued, was to explain the universe as we see it, and this is equivalent to requiring that the quantum state of the universe is compatible with an instance of conscious sensation like ours. This in turn is equivalent to the quantum state assigning a non-zero probability to such an instance. This then gives a new way of thinking about the role of consciousness: consciousness does not alter the quantum state of the universe, but it imposes a filter on the state, selecting a component (if there is one) compatible with our capacity for sensation.

The combined work of Hartle and Page gives a picture of a universe arising from the interplay of a background homogeneous quantum cosmology with possible networks in space and time of instances of consciousness. Self-contradictory networks of awareness are ruled out because quantum mechanics assigns to them a zero probability (Everett, 1957; Clarke, 1974). But in addition the networks of awareness are shaped by their own internal logic, manifested by qualis- consciousness and different from the Aristotelian logic of quid-consciousness (Clarke, 2007). This logic brings in elements such as agency and meaning. Consciousness, on this view, "does something", but by selection rather than modification, and in a way which is compatible with and dependent on the known laws of physics.

5. A Theoretical Understanding of Consciousness and Quantum Theory

One final building block still seems required: a non-arbitrary criterion is needed for what physical systems have the capacity for (qualis-)consciousness. Many recent authors (de Quincey, 2002; Skrbina, 2005) have, however, come to the conclusion that no such criterion exists. In other words, everything might be conscious, a position known as "panpsychism". A problem remains, however: if "everything" is conscious, what is a "thing"? The answer of Heidegger (1967) concerned only a pejorative cultural aspect of the word; the answer of Döring and Isham (2011) invokes an ad hoc external mechanism; instead we need to explore naturally occurring physical criteria for what is a thing. A consciousness-carrying "thing" must have some internal unity rather than being an arbitrary aggregate of objects, which suggests that it has an internal coherence. The simplest definition of this is that its parts are in quantum entanglement (Clarke, 2007) . In addition, it must not be merely an arbitrary subset of a larger "thing", so that it must be maximal with respect to this coherence. In other words, it must be on the boundary between the quantum and the classical, a boundary set by the onset of decoherence. The structures considered by Hameroff and Penrose (1996) are of this sort.

It now becomes clearer what consciousness does. At this quantum-classical boundary the question of what "observation" (or, more formally, what algebra of propositions) is to be expressed is not yet determined by decoherence, and so is open to determination by consciousness (Clarke, 2007). Following Hartle, this happens not in isolation, but within the whole network of "things" throughout the universe. Physical causation operates through quantum state of the universe, while consciousness independently filters this into awareness through its own sort of logic (in the sense of the structure of an algebra of propositions). The large scope this gives for future experimental and theoretical research has been outlined in (Clarke, 2007,8). Several candidates for the logic of consciousness are available, allowing us to understand how consciousness brings creativity alongside rational deduction. This model raises for the first time the possibility of a rigorous theoretical framework for parapsychology (Clarke, 2008) without which that subject remains only a semi-science. It turns out that consciousness can itself, through the "Zeno effect", enlarge the length scale for the onset of decoherence, which then offers hope for understanding how small-scale elements can be "orchestrated", in Hameroff's sense (Hammeroff & Penrose, 1996), into the ego-consciousness known to us. In addition, there will be other candidates for what a "thing" is, opening up alternative theories that can be tested against the theory just outlined.

References

Berger, D. M. (1987). Clinical empathy. Northvale: Jason Aronson, Inc.

Clarke, C. J. S. (1974) Quantum Theory and Cosmology. Philosophy of Science, 41, 317-332.

Clarke, C. J. S. (2007). The role of quantum physics in the theory of subjective consciousness. Mind and Matter 5(1), 45-81.

Clarke C. J. S. (2008). A new quantum theoretical framework for parapsychology. European Journal of Parapsychology, 23(1), 3-30.

Daneri, A., Loinger, A., Prosperi, G. M. (1962). Quantum Theory of Measurement and Ergodicity Conditions., Nuclear Physics 33, 297-319.

de Quincey, C. (2002). Radical Nature: Rediscovering the Soul of Matter. Montpelier VT : Invisible Cities Press.

Dennett, D. C., (1991). Consciousness Explained, Allen Lane.

Donald, M. (1990). Quantum Theory and the Brain, Proceedings of the Royal Society (London) Series A, 427, 43-93.

Döring, A., Isham, C. (2011). "What is a Thing?": Topos Theory in the Foundations of Physics. In B. Coecke (Ed.), New Structures for Physics, Lecture Notes in Physics, Vol. 813 (pp 753-941). Berlin: Springer.

Everett, H., (1957). Relative State Formulation of Quantum Mechanics, Reviews of Modern Physics 29, pp 454-462.

Griffiths, R..B. (1984). Consistent histories and the interpretation of quantum mechanics. J. Stat. Phys. 36, 219-272.

Hameroff, S., Penrose, R. (1996). Conscious events as orchestrated space-time selections. Journal of Consciousness Studies, 3(1), 36-53.

Hartle, J. (1991). The quantum mechanics of cosmology. In Coleman, S., Hartle, P., Piran, T., Weinberg, S., (Eds) Quantum cosmology and baby universes. Singapore: World Scientific.

Heidegger M. (1967). What Is a Thing? (Trans. W. B. Barton Jr., V. Deutsch). Chicago: Henry Regnery Company.

London, F., Bauer, E. (1939). La théorie de l'observation en mécanique quantique. Hermann, Paris.

London, F., Bauer, E. (1983). The theory of observation in quantum mechanics (translation of the above). In J. A. Wheeler, W. H. Zurek (Eds), Quantum Theory and Measurement (pp. 217- 259). Princeton: Princeton University Press.

McGilchrist, I. (2009). The Master and his Emissary: the divided brain and the making of the Western world. New Haven and London: Yale University Press.

Nagel, T. (1974). What Is it Like to Be a Bat? Philosophical Review 83(4), 435-450.

Skrbina, D. (2005). Panpsychism in the West. Cambridge MA: Bradford Books.

von Neumann, J. (1932). Mathematical Foundations of Quantum Mechanics, (Beyer, R. T., trans.), Princeton Univ. Press.

Zeh, H. D., (1970). On the Interpretation of Measurement in Quantum Theory, Foundation of Physics, 1, pp. 69-76.

23. The Stream of Consciousness

William James, Ph.D.

Harvard University, Cambridge, MA

First published in Psychology, Chapter XI, 1892

The order of our study must be analytic. We are now prepared to begin the introspective study of the adult consciousness itself. Most books adopt the so-called synthetic method. Starting with 'simple ideas of sensation,' and regarding these as so many atoms, they proceed to build up the higher states of mind out of their 'association,' 'integration,' or 'fusion,' as houses are built by the agglutination of bricks. This has the didactic advantages which the synthetic method usually has. But it commits one beforehand to the very questionable theory that our higher states of consciousness are compounds of units; and instead of starting with what the reader directly knows, namely his total concrete states of mind, it starts with a set of supposed 'simple ideas' with which he has no immediate acquaintance at all, and concerning whose alleged interactions he is much at the mercy of any plausible phrase. On every ground, then, the method of advancing from the simple to the compound exposes us to illusion. All pedants and abstractionists will naturally hate to abandon it. But a student who loves the fulness [sic] of human nature will prefer to follow the 'analytic' method, and to begin with the most concrete facts, those with which he has a daily acquaintance in his own inner life. The analytic method will discover in due time the elementary parts, if such exist, without danger of precipitate assumption. The reader will bear in mind that our own chapters on sensation have dealt mainly with the physiological conditions thereof. They were put first as a mere matter of convenience, because incoming currents come first. Psychologically they might better have come last. Pure sensations were described on page 12 [of James' Psychology] as processes which in adult life are well-nigh unknown, and nothing was said which could for a moment lead the reader to suppose that they were the elements of composition of the higher states of mind.

The Fundamental Fact. -- The first and foremost concrete fact which every one will affirm to belong to his inner experience is the fact that *consciousness of some sort goes on*. *'States of mind' succeed each other in him*. If we could say in English 'it thinks,' as we say 'it rains' or 'it blows,' we should be stating the fact most simply and with the minimum of assumption. As we cannot, we must simply say that *thought goes on*.

Four Characters in Consciousness. -- How does it go on? We notice immediately four important characters in the process, of which it shall be the duty of the present chapter to treat in a general way :

1) Every 'state' tends to be part of a personal consciousness. 2) Within each personal consciousness states are always changing. 3) Each personal consciousness is sensibly continuous. 4) It is interested in some parts of its object to the exclusion of others, and welcomes or rejects -- *chooses* from among them, in a word -- all the while.

In considering these four points successively, we shall have to plunge *in medias res* as regards our nomenclature and use psychological terms which can only be adequately defined in later chapters of the book. But every one knows what the terms mean in a rough way; and it is only in a rough way that we are now to take them. This chapter is like a painter's first charcoal sketch upon his canvas, in which no niceties appear.

When I say *every 'state' or 'thought' is part of a personal consciousness*, 'personal consciousness' is one of the terms in question. Its meaning we know so long as no one asks us to define it, but to give an accurate account of it is the most difficult of philosophic tasks. This task we must, confront in the next chapter; here a preliminary word will suffice.

In this room -- this lecture-room, say -- there are a multitude of thoughts, yours and mine, some of which cohere mutually, and some not. They are as little each-for-itself and reciprocally independent as they are all-belonging-together. They are neither: no one of them is separate, but each belongs with certain others and with none beside. My thought belongs with *my* other thoughts, and your thought with *your* other thoughts. Whether anywhere in the room there be a mere thought, which is nobody's thought, we have no means of ascertaining, for we have no experience of its like. The only states of consciousness that we naturally deal with are found in personal consciousness, minds, selves, concrete particular I's and you's.

Each of these minds keeps its own thoughts to itself. There is no giving or bartering between them. No thought even comes into direct *sight* of a thought in another personal consciousness than its own. Absolute insulation, irreducible pluralism, is the law. It seems as if the elementary psychic fact were not *thought* or *this thought* or *that thought*, but *my thought*, every thought being *owned*. Neither contemporaneity, nor proximity in space, nor similarity of quality and content are able to fuse thoughts together which are sundered by this barrier of belonging to different personal minds. The breaches between such thoughts are the most absolute breaches in nature. Every one will recognize this to be true, so long as the existence of *something* corresponding to the term 'personal mind' is all that is insisted on, without any particular view of its nature being implied. On these terms the personal self rather than the thought might be treated as the immediate datum in psychology. The universal conscious fact is not 'feelings and thoughts exist,' but 'I think' and 'I feel.' No psychology, at any rate, can question the *existence* of personal selves. Thoughts connected as we feel them to be connected are *what we mean* by personal selves. The worst a psychology can do is so to interpret the nature of these selves as to rob them of their *worth*.

Consciousness is in constant change. I do not mean by this to say that no one state of mind has any duration -- even if true, that would be hard to establish. What I wish to lay stress on is this, that *no state once gone can recur and be identical with what it was before*. Now we are seeing, now hearing; now reasoning, now willing; now recollecting, now expecting; now loving, now hating; and in a hundred other ways we know our minds to be alternately engaged. But all these are complex states, it may be said, produced by combination of simpler ones; -- do not the simpler ones follow a different law? Are not the *sensations* which we get from the same object, for example, always the same? Does not the same piano-key, struck with the same force, make us hear in the same way? Does not the same grass give us the same feeling of green, the same sky the same feeling of blue, and do we not get the same olfactory sensation no matter how many times we put our nose to the same flask of cologne? It seems a piece of metaphysical sophistry to suggest that we do not; and yet a close attention to the matter shows that *there is no proof that an incoming current ever gives us just the same bodily sensation twice*.

What is got twice is the same OBJECT. We hear the same *note* over and over again; we see the same *quality* of green, or smell the same objective perfume, or experience the same *species* of pain. The realities, concrete and abstract, physical and ideal, whose permanent existence we believe in, seem to be constantly coming up again before our thought, and lead us, in our carelessness, to suppose that our 'ideas' of them are the same ideas. When we come, some time later, to the chapter [20] on Perception, we shall see how inveterate is our habit of simply using our sensible impressions as stepping-stones to pass over to the recognition of the realities whose presence they reveal. The grass out of the window now looks to me of the same green in the sun as in the shade, and yet a painter would have to paint one part of it dark brown, another part bright yellow, to give its real sensational effect. We take no heed, as a rule, of the different way in which the same things look and sound and smell at different distances and under different circumstances. The sameness of the *things* is what we are concerned to ascertain; and any sensations that assure us of that will probably be considered in a rough way to be the same with each other. This is what makes off-hand testimony about the subjective identity of different sensations well-nigh worthless as a proof of the fact. The entire history of what is called Sensation is a commentary on our inability to tell whether two sensible qualities received apart are exactly alike. What appeals to our attention far more than the absolute quality of an impression is its *ratio* to whatever other impressions we may have at the same time. When everything is dark a somewhat less dark sensation makes us see an object white. Helmholtz calculates that the white marble painted in a picture representing an architectural view by moonlight is, when seen by daylight, from ten to twenty thousand times brighter than the real moonlit marble would be.

Such a difference as this could never have been *sensibly* learned; it had to be inferred from a series of indirect considerations. These make us believe that our sensibility is altering all the time, so that the same object cannot easily give us the same sensation over again. We feel things differently accordingly as we are sleepy or awake, hungry or full, fresh or tired; differently at night and in the morning, differently in summer and in winter; and above all, differently in childhood, manhood, and old age. And yet we never doubt that our feelings reveal the same world, with the same sensible qualities and the same sensible things occupying it. The difference of the sensibility is shown best by the difference of our emotion about the things from one age to another, or when we are in different organic moods, What was bright and exciting becomes weary, flat, and unprofitable. The bird's song is tedious, the breeze is mournful, the sky is sad.

To these indirect presumptions that our sensations, following the mutations of our capacity for feeling, are always undergoing an essential change, must be added another presumption, based on what must happen in the brain. Every sensation corresponds to some cerebral action. For an identical sensation to recur it would have to occur the second time *in an unmodified brain*. But as this, strictly speaking, is a physiological impossibility, so is an unmodified feeling an impossibility; for to every brain-modification, however small, we suppose that there must correspond a change of equal amount in the consciousness which the brain subserves.

But if the assumption of 'simple sensations' recurring in immutable shape is so easily shown to be baseless, how much more baseless is the assumption of immutability in the larger masses of our thought!

For there it is obvious and palpable that our state of mind is never precisely the same. Every thought we have of a given fact is, strictly speaking, unique, and only bears a resemblance of kind with our other thoughts of the same fact. When the identical fact recurs, we *must* think of it in a fresh manner, see it under a somewhat different angle, apprehend it in different relations from those in which it last appeared. And the thought by which we cognize it is the thought of it-in-those-relations, a thought suffused with the consciousness of all that dim context. Often we are ourselves struck at the strange differences in our successive views of the same thing. We wonder how we ever could have opined as we did last month about a certain matter. We have outgrown the possibility of that state of mind, we know not how. From one year to another we see things in new lights. What was unreal has grown real, and what was exciting is insipid. The friends we used to care the world for are shrunken to shadows; the women once so divine, the stars, the woods, and the waters, how now so dull and common! -- the young girls that brought an aura of infinity, at present hardly distinguishable existences; the pictures so empty; and as for the books, what *was* there to find so mysteriously significant in Goethe, or in John Mill so full of weight? Instead of all this, more zestful than ever is the work, the work; and fuller and deeper the import of common duties and of common goods.

I am sure that this concrete and total manner of regarding the mind's changes is the only true manner, difficult as it may be to carry it out in detail. If anything seems obscure about it, it will grow clearer as we advance. Meanwhile, if it be true, it is certainly also true that no two 'ideas' are ever exactly the same, which is the proposition we started to prove. The proposition is more important theoretically than it at first sight seems. For it makes it already impossible for us to follow obediently in the footprints of either the Lockian or the Herbartian school, schools which have had almost unlimited influence in Germany among ourselves. No doubt it is often *convenient* to formulate the mental facts in an atomistic sort of way, and to treat the higher states of consciousness as if they were all built out of unchanging simple ideas which 'pass and turn again.' It is convenient often to treat curves as if they were composed of small straight lines, and electricity and nerve-force as if they were fluids. But in the one case as in the other we must never forget that we are talking symbolically, and that there is nothing in nature to answer to our words. *A permanently existing 'Idea' which makes its appearance before the footlights of consciousness at periodical intervals is as mythological an entity as the Jack of Spades.*

Within each personal consciousness, thought is sensibly continuous. I can only define 'continuous' as that which is without breach, crack, or division. The only breaches that can well be conceived to occur within the limits of a single mind would either be *interruptions, time-gaps* during which the consciousness went out; or they would be breaks in the content of the thought, so abrupt that what followed had no connection whatever with what went before. The proposition that consciousness feels continuous, means two things:

a. That even where there is a time-gap the consciousness after it feels as if it belonged together with the consciousness before it, as another part of the same self;

b. That the changes from one moment to another in the quality of the consciousness are never absolutely abrupt.

The case of the time-gaps, as the simplest, shall be taken first.

a. When Paul and Peter wake up in the same bed, and recognize that they have been asleep, each one of them mentally reaches back and makes connection with but one of the two streams of thought which were broken by the sleeping hours. As the current of an electrode buried in the ground unerringly finds its way to its own similarly buried mate, across no matter how much intervening earth; so Peter's present instantly finds out Peter's past, and never by mistake knits itself on to that of Paul. Paul's thought in turn is as little liable to go astray. The past thought of Peter is appropriated by the present Peter alone. He may have a knowledge, and a correct one too, of what Paul's last drowsy states of mind were as he sank into sleep, but it is an entirely different sort of knowledge from that which he has of his own last states. He remembers his own states, whilst he only conceives Paul's. Remembrance is like direct feeling; its object is suffused with a warmth and intimacy to which no object of mere conception ever attains. This quality of warmth and intimacy and immediacy is what Peter's present thought also possesses for itself. So sure as this present is me, is mine, it says, so sure is anything else that comes with the same warmth and intimacy and immediacy, me and mine. What the qualities called warmth and intimacy may in themselves be will have to be matter for future consideration. But whatever past states appear with those qualities must be admitted to receive the greeting of the present mental state, to be owned by it, and accepted as belonging together with it in a common self. This community of self is what the time-gap cannot break in twain, and is why a present thought, although not ignorant of the time-gap, can still regard itself as continuous with certain chosen portions of the past.

Consciousness, then, does not appear to itself chopped up in bits. Such words as 'chain' or 'train' do not describe it fitly as it presents itself in the first instance. It is nothing jointed; it flows. A 'river' or a 'stream' are the metaphors by which it is most naturally described. In talking of it hereafter, let us call it the stream of thought, of consciousness, or of subjective life.

b. But now there appears, even within the limits of the same self, and between thoughts all of which alike have this same sense of belonging together, a kind of jointing and separateness among the parts, of which this statement seems to take no account. I refer to the breaks that are produced by sudden contrasts in the quality of the successive segments of the stream of thought. If the words 'chain' and 'train' had no natural fitness in them, how came such words to be used at all? Does not a loud explosion rend the consciousness upon which it abruptly breaks, in twain? No; for even into our awareness of the thunder the awareness of the previous silence creeps and continues; for what we hear when the thunder crashes is not thunder pure, but thunder-breaking-upon-silence-and-contrasting-with-it. Our feeling of the same objective thunder, coming in this way, is quite different from what it would be were the thunder a continuation of previous thunder. The thunder itself we believe to abolish and exclude the silence; but the feeling of the thunder is also a feeling of the silence as just gone; and it would be difficult to find in the actual concrete consciousness or man a feeling so limited to the present as not to have an inkling of anything that went before.

'Substantive' and 'Transitive' States of Mind. -- When we take a general view of the wonderful stream of our consciousness, what strikes us first is the different pace of its parts. Like a bird's life, it seems to be an alternation of flights and perchings. The rhythm of language expresses this, where every thought is expressed in a sentence, and every sentence closed by a period. The resting-places are usually occupied by sensorial imaginations of some sort, whose peculiarity is that they can be held before the mind for an indefinite time, and contemplated without changing; the places of flight are filled with thoughts of relations, static or dynamic, that for the most part obtain between the matters contemplated in the periods of comparative rest.

Let us call the resting-places the 'substantive parts,' and the places of flight the 'transitive parts,' of the stream of thought. It then appears that our thinking tends at all times towards some other substantive part than the one from which it has just been dislodged. And we may say that the main use of the transitive parts is to lead us from one substantive conclusion to another.

Now it is very difficult, introspectively, to see the transitive parts for what they really are. If they are but flights to a conclusion, stopping them to look at them before the conclusion is reached is really annihilating them. Whilst if we wait till the conclusion *be* reached, it so exceeds them in vigor and stability that it quite eclipses and swallows them up in its glare. Let anyone try to cut a thought across in the middle and get a look at its section, and he will see how difficult the introspective observation of the transitive tracts is. The rush of the thought is so headlong that it almost always brings us up at the conclusion before we can rest it. Or if our purpose is nimble enough and we do arrest it, it ceases forthwith to itself. As a snowflake crystal caught in the warm hand is no longer a crystal but a drop, so, instead of catching the feeling of relation moving to its term, we find we have caught some substantive thing, usually the last word we were pronouncing, statically taken, and with its function, tendency, and particular meaning in the sentence quite evaporated. The attempt at introspective analysis in these cases is in fact like seizing a spinning top to catch its motion, or trying to turn up the gas quickly enough to see how the darkness looks. And the challenge to *produce* these transitive states of consciousness, which is sure to be thrown by doubting psychologists at anyone who contends for their existence, is as unfair as Zeno's treatment of the advocates of motion, when, asking them to point out in what place an arrow is when it moves, he argues the falsity of their thesis from their inability to make to so preposterous a question an immediate reply.

The results of this introspective difficulty are baleful. If to hold fast and observe the transitive parts of thought's stream be so hard, then the great blunder to which all schools are liable must be the failure to register them, and the undue emphasizing of the more substantive parts of the stream. Now the blunder has historically worked in two ways. One set of thinkers have been led by it to Sensationalism. Unable to lay their hands on any substantive feelings corresponding to the innumerable relations and forms of connection between the sensible things of the world, finding no named mental states mirroring such relations, they have for the most part denied that any such states exist; and many of them, like Hume, have gone on to deny the reality of most relations out of the mind as well as in it. Simple substantive 'ideas,' sensations and their copies, juxtaposed like dominoes in a game, but really separate, everything else verbal illusion, -- such is the upshot of this view. The Intellectualists, on the other hand, unable to give up the reality of relations extra mentem, but equally unable to point to any distinct substantive feelings in which they were known, have made the same admission that such feelings do not exist. But they have drawn an opposite conclusion. The relations must be known, they say, in something that is no feeling, no mental 'state,' continuous and consubstantial with the subjective tissue out of which sensations and other substantive conditions of consciousness are made. They must be known by something that lies on an entirely different plane, by an actus purus of Thought, Intellect, or Reason, all written with capitals and considered to mean something unutterably superior to any passing perishing fact of sensibility whatever.

But from our point of view both Intellectualists and Sensationalists are wrong. If there be such things as feelings at all, then so surely as relations between objects exist in rerum naturâ [sic], so surely, and more surely, do feelings exist to which these relations are known. There is not a conjunction or a preposition, and hardly an adverbial phrase, syntactic form, or inflection of voice, in human speech, that does not express some shading or other of relation which we at some moment actually feel to exist between the larger objects of our thought. If we speak objectively, it is the real relations that appear revealed; if we speak subjectively, it is the stream of consciousness that matches each of them by an inward coloring of its own. In either case the relations are numberless, and no existing language is capable of doing justice to all their shades.

We ought to say a feeling of and, a feeling of if, a feeling of but, and a feeling of by, quite as readily as we say a feeling of blue or a feeling of cold. Yet we do not: so inveterate has our habit become of recognizing the existence of the substantive parts alone, that language almost refuses to lend itself to any other use. Consider once again the analogy of the brain. We believe the brain to be an organ whose internal equilibrium is always in a state of change -- the change affecting every part. The pulses of change are doubtless more violent in one place than in another, their rhythm more rapid at this time than at that. As in a kaleidoscope revolving at a uniform rate, although the figures are always rearranging themselves, there are instants during which the transformation seems minute and interstitial and almost absent, followed by others when it shoots with magical rapidity, relatively stable forms thus alternating with forms we should not distinguish if seen again; so in the brain the perpetual rearrangement must result in some forms of tension lingering relatively long, whilst others simply come and pass. But if consciousness corresponds to the fact of rearrangement itself, why, if the rearrangement stop not, should the consciousness ever cease? And if a lingering rearrangement brings with it one kind of consciousness, why should not a swift rearrangement bring another kind of consciousness as peculiar as the rearrangement itself?

The object before the mind always has a 'Fringe.' There are other unnamed modifications of consciousness just as important as the transitive states, and just as cognitive as they. Examples will show what I mean.

Suppose three successive persons say to us: 'Wait!' 'Hark!' 'Look!' Our consciousness is thrown into three quite different attitudes of expectancy, although no definite object is before it in any one of the three cases. Probably no one will deny here the existence of a real conscious affection, a sense of the direction from which an impression is about to come, although no positive impression is yet there. Meanwhile we have no names for the psychoses in question but the names hark, look, and wait.

Suppose we try to recall a forgotten name. The state of our consciousness is peculiar. There is a gap therein; but no mere gap. It is a gap that is intensely active. A sort of wraith of the name is in it, beckoning us in a given direction, making us at moments tingle with the sense of our closeness, and then letting us sink back without the longed-for term. If wrong names are proposed to us, this singularly definite gap acts immediately so as to negate them. They do not fit into its mould. And the gap of one word does not feel like the gap of another, all empty of content as both might seem necessarily to be when described as gaps. When I vainly try to recall the name of Spalding, my consciousness is far removed from what it is when I vainly try to recall the name of Bowles. There are innumerable consciousnesses of want, no one of which taken in itself has a name, but all different from each other. Such feeling of want is tota cœlo other than a want of feeling: it is an intense feeling. The rhythm of a lost word may be there without a sound to clothe it; or the evanescent sense of something which is the initial vowel or consonant may mock us fitfully, without growing -more distinct. Every one must know the tantalizing effect of the blank rhythm of some forgotten verse, restlessly dancing in one's mind, striving to be filled out with words.

What is that first instantaneous glimpse of some one's meaning which we have, when in vulgar phrase we say we 'twig' it? Surely an altogether specific affection of our mind. And has the reader never asked himself what kind of a mental fact is his intention of saying a thing before he has said it? It is an entirely definite intention, distinct from all other intentions, an absolutely distinct state of consciousness, therefore; and yet how much of it consists of definite sensorial images, either of words or of things? Hardly anything! Linger, and the words and things come into the mind; the anticipatory intention, the divination is there no more. But as the words that replace it arrive, it welcomes them successively and calls them right if they agree with it, it rejects them and calls them wrong if they do not. The intention to-say-so-and-so is the only name it can receive. One may admit that a good third of our psychic life consists in these rapid premonitory perspective views of schemes of thought not yet articulate. How comes it about that a man reading something aloud for the first time is able immediately to emphasize all his words aright, unless from the very first he have a sense of at least the form of the sentence yet to come, which sense is fused with his consciousness of the present word, and modifies its emphasis in his mind so as to make him give it the proper accent as he utters it? Emphasis of this kind almost altogether depends on grammatical construction. If we read 'no more' we expect presently a 'than'; if we read 'however,' it is a 'yet,['] a 'still,' or a 'nevertheless,' that we expect. And this foreboding of the coming verbal and grammatical scheme is so practically accurate that a reader incapable of understanding four ideas of the book he is reading aloud can nevertheless read it with the most delicately modulated expression of intelligence.

It is, the reader will see, the reinstatement of the vague and inarticulate to its proper place in our mental life which I am so anxious to press on the attention. Mr. Galton and Prof. Huxley have, as we shall see in the chapter [19] on Imagination, made one step in advance in exploding the ridiculous theory of Hume and Berkeley that we can have no images but of perfectly definite things. Another is made if we overthrow the equally ridiculous notion that, whilst simple objective qualities are revealed to our knowledge in 'states of consciousness,' relations are not. But these reforms are not half sweeping and radical enough. What must be admitted is that the definite images of traditional psychology form but the very smallest part of our minds as they actually live. The traditional psychology talks like one who should say a river consists of nothing but pailsful, spoonsful, quartpotsful, barrelsful, and other moulded forms of water. Even were the pails and the pots all actually standing in the stream, still between them the free water would continue to flow. It is just this free water of consciousness that psychologists resolutely overlook. Every definite image in the mind is steeped and dyed in the free water that flows round it. With it goes the sense of its relations, near and remote, the dying echo of whence it came to us, the dawning sense of whither it is to lead. The significance, the value, of the image is all in this halo or penumbra that surrounds and escorts it, -- or rather that is fused into one with it and has become bone of its bone and flesh of its flesh; leaving it, it is true, an image of the same thing it was before, but making it an image of that thing newly taken and freshly understood.

Let us call the consciousness of this halo of relations around the image by the name of 'psychic overtone' or 'fringe."

Cerebral Conditions of the 'Fringe.' -- Nothing is easier than to symbolize these facts in terms of brain-action. just as the echo of the whence, the sense of the starting point of our thought, is probably due to the dying excitement of processes but a moment since vividly aroused: so the sense of the whither, the foretaste of the terminus, must be due to the waxing excitement of tracts or processes whose psychical correlative will a moment hence be the vividly present feature of our thought. Represented by a curve, the neurosis underlying consciousness must at any moment be like this:

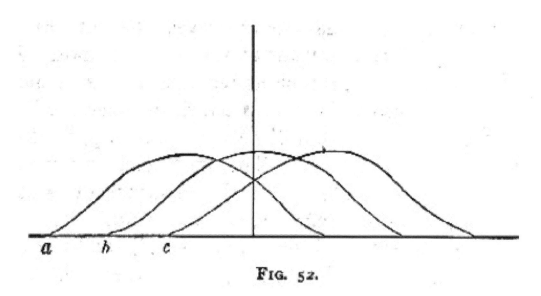

FIG. 52.

Let the horizontal in Fig. 52 be the line of time, and let the three curves beginning at a, b, and c respectively stand for the neural processes correlated with the thoughts of those three letters. Each process occupies a certain time during which its intensity waxes, culminates, and wanes The process for a has not yet died out, the process for c has already begun, when that for b is culminating. At the time-instant represented by the vertical line all three processes are present, in the intensities shown by the curve. Those before c's apex were more intense a moment ago; those after it will be more intense a moment hence. If I recite a, b, c, then, at the moment of uttering b, neither a nor c is out of my consciousness altogether, but both, after their respective fashions, 'mix their dim lights' with the stronger b, because their processes are both awake in some degree.

It is just like 'overtones' in music: they are not separately heard by the ear; they blend with the fundamental note, and suffuse it, and alter it; and even so do the waxing and waning brain-processes at every moment blend with and suffuse and alter the psychic effect of the processes which are at their culminating point.

The 'Topic' of the Thought. -- If we then consider the cognitive function of different states of mind, we may feel assured that the difference between those that are mere 'acquaintance' and those that are 'knowleges-about' is reducible almost entirely to the absence or presence of psychic fringes or overtones. Knowledge about a thing is knowledge of its relations. Acquaintance with it is limitation to the bare impression which it makes. Of most of its relations we are only aware in the penumbral nascent way of a 'fringe' of unarticulated affinities about it. And, before passing to the next topic in order, I must say a little of this sense of affinity, as itself one of the most interesting features of the subjective stream.

Thought may be equally rational in any sort of terms. In all our voluntary thinking there is some TOPIC or SUBJECT about which all the members of the thought revolve. Relation to this topic or interest is constantly felt in the fringe, and particularly the relation of harmony and discord, of furtherance or hindrance of the topic. Any thought the quality of whose fringe lets us feel ourselves 'all right,' may be considered a thought that furthers the topic. Provided we only feel its object to have a place in the scheme of relations in which the topic also lies, that is sufficient to make of it a relevant and appropriate portion of our train of ideas.

Now we may think about our topic mainly in words, or we may think about it mainly in. visual or other images, but this need make no difference as regards the furtherance of our knowledge of the topic. If we only feel in the terms, whatever they be, a fringe of affinity with each other and with the topic, and if we are conscious of approaching a conclusion, we feel that our thought is rational and right. The words in every language have contracted by long association fringes of mutual repugnance or affinity with each other and with the conclusion, which run exactly parallel with like fringes in the visual, tactile, and other ideas. The most important element of these fringes is, I repeat, the mere feeling of harmony or discord, of a right or wrong direction in the thought.

If we know English and French and begin a sentence in French, all the later words that come are French; we hardly ever drop into English. And this affinity of the French words for each other is not something merely, operating mechanically as a brain-law, it is something we feel at the time. Our understanding of a French sentence heard never falls to so low an ebb that we are not aware that the words linguistically belong together. Our attention can hardly so wander that if an English word be suddenly introduced we shall not start at the change. Such a vague sense as this of the words belonging together is the very minimum of fringe that can accompany them, if 'thought' at all. Usually the vague perception that all the words we hear belong to the same language and to the same special vocabulary in that language, and that the grammatical sequence is familiar, is practically equivalent to an admission that what we hear is sense. But if an unusual foreign word be introduced, if the grammar trip, or if a term from an incongruous vocabulary suddenly appear, such as 'rat-trap' or 'plumber's bill' in a philosophical discourse, the sentence detonates as it were, we receive a shock from the incongruity, and the drowsy assent is gone. The feeling of rationality in these cases seems rather a negative than a positive thing, being the mere absence of shock, or sense of discord, between the terms of thought.

Conversely, if words do belong to the same vocabulary, and if the grammatical structure is correct, sentences with absolutely no meaning may be uttered in good faith and pass unchallenged. Discourses at prayer-meetings, reshuffling the same collection of cant phrases, and the whole genus of penny-a-line-isms and newspaper-reporter's flourishes give illustrations of this. "The birds filled the tree-tops with their morning song, making the air moist, cool, and pleasant," is a sentence I remember reading once in a report of some athletic exercises in Jerome Park. It was probably written unconsciously by the hurried reporter, and read uncritically by many readers.

We see, then, that it makes little or no difference in what sort of mind-stuff, in what quality of imagery, our thinking goes on. The only images intrinsically important are the halting-places, the substantive conclusions, provisional or final, of the thought. Throughout all the rest of the stream, the feelings of relation are everything, and the terms related almost naught. These feelings of relation, these psychic overtones, halos, suffusions, or fringes about the terms, may be the same in very different systems of imagery. A diagram may help to accentuate this indifference of the mental means where the end is the same. Let A be some ex-experience [sic] from which a number of thinkers start. Let Z be the practical conclusion rationally inferrible [sic] from it. One gets to this conclusion by one line, another by another; one follows a course of English, another of German, verbal imagery. With one, visual images predominate; with another, tactile. Some trains are tinged with emotions, others not; some are very abridged, synthetic and rapid; others, hesitating and broken into many steps. But when the penultimate terms of all the trains, however differing inter se, finally shoot into the same conclusion, we say, and rightly say, that all the thinkers have had substantially the same thought. It would probably astound each of them beyond measure to be let into his neighbor's mind and to find how different the scenery there was from that in his own.

The last peculiarity to which attention is to be drawn in this first rough description of thought's stream is that --

Consciousness is always interested more in one part of its object than in another, and welcomes and rejects, or chooses, all the while it thinks.

The phenomena of selective attention and of deliberative will are of course patent examples of this choosing activity. But few of us are aware how incessantly it is at work in operations not ordinarily called by these names. Accentuation and Emphasis are present in every perception we have. We find it quite impossible to disperse our attention impartially over a number of impressions. A monotonous succession of sonorous strokes is broken up into rhythms, now of one sort, now of another, by the different accent which we place on different strokes. The simplest of these rhythms is the double one, tick-tóck, tick-tóck, tick-tóck. Dots dispersed on a surface are perceived in rows and groups. Lines separate into diverse figures. The ubiquity of the distinctions, this and that, here and there, now and then, in our minds is the result of our laying the same selective emphasis on parts of place and time

But we do far more than emphasize things, and unite some, and keep others apart. We actually ignore most of the things before us. Let me briefly show how this goes on.

To begin at the bottom what are our very senses themselves, as we saw on pp.10-12 [of James' Psychology], but organs of selection? Out of the infinite chaos of movements, of which physics teaches us that the outer world consists, each sense-organ picks out those which fall within certain limits of velocity. To these it responds, but ignores the rest as completely as if they did not exist. Out of what is in itself an undistinguishable [sic], swarming continuum, devoid of distinction or emphasis, our senses make for us, by attending to this motion and ignoring that, a world full of contrasts, of sharp accents, of abrupt changes, of picturesque light and shade.

If the sensations we receive from a given organ have their causes thus picked out for us by the conformation of the organ's termination, Attention, on the other hand, out of all the sensations yielded, picks out certain ones as worthy of notice and suppresses all the rest. We notice only those sensations which are signs to us of things which happen practically or aesthetically to interest us, to which we therefore give substantive names, and which we exalt to this exclusive status of independence and dignity. But in itself, apart from my interest, a particular dust-wreath on a windy day is just as much of an individual thing, and just as much or as little deserves an individual name, as my own body does.

And then, among the sensations we get from each separate thing, what happens? The mind selects again. It chooses certain of the sensations to represent the thing most truly, and considers the rest as its appearances, modified by the conditions of the moment. Thus my table-top is named square, after but one of an infinite number of retinal sensations which it yields, the rest of them being sensations of two acute and two obtuse angles; but I call the latter perspective views, and the four right angles the true form of the table, and erect the attribute squareness into the table's essence, for æsthetic reasons of my own. In like manner the real form of the circle is deemed to be the sensation it gives when the line of vision is perpendicular to its centre -- all its other sensations are signs of this sensation. The real sound of the cannon is the sensation it makes when the ear is close by. The real color of the brick is the sensation it gives when the eye looks squarely at it from a near point, out of the sunshine and yet not in the gloom; under other circumstances it gives us other color-sensations which are but signs of this -- we then see it looks pinker or bluer than it really is. The reader knows no object which he does not represent to himself by preference as in some typical attitude, of some normal size, at some characteristic distance, of some standard tint, etc., etc. But all these essential characteristics, which together form for us the genuine objectivity of the thing and are contrasted with what we call the subjective sensations it may yield us at a given moment, are mere sensations like the latter. The mind chooses to suit itself, and decides what particular sensation shall be held more real and valid than all the rest.

Next, in a world of objects thus individualized by our mind's selective industry, what is called our 'experience' is almost entirely determined by our habits of attention. A thing may be present to a man a hundred times, but if he persistently fails to notice it, it cannot be said to enter into his experience. We are all seeing flies, moths, and beetles by the thousand, but to whom, save an entomologist, do they say anything distinct? On the other hand, a thing met only once in a lifetime may leave an indelible experience in the memory. Let four men make a tour in Europe. One will bring home only picturesque impressions -- costumes and colors, parks and views and works of architecture, pictures and statues. To another all this will be non-existent; and distances and prices, populations and drainage-arrangements, door- and window-fastenings, and other useful statistics will take their place. A third will give a rich account of the theatres, restaurants, and public halls, and naught besides; whilst the fourth will perhaps have been so wrapped in his own subjective broodings as to be able to tell little more than a few names of places through which he passed. Each has selected, out of the same mass of presented objects, those which suited his private interest and has made his experience thereby.

If now, leaving the empirical combination of objects, we ask how the mind proceeds rationally to connect them, we find selection again to be omnipotent. In a future chapter [22] we shall see that all Reasoning depends on the ability of the mind to break up the totality of the phenomenon reasoned about, into parts, and to pick out from among these the particular one which, in the given emergency, may lead to the proper conclusion. The man of genius is he who will always stick in his bill at the right point, and bring it out with the right element -- 'reason' if the emergency be theoretical, 'means' if it be practical -- transfixed upon it.

If now we pass to the æsthetic department, our law is still more obvious. The artist notoriously selects his items, rejecting all tones, colors, shapes, which do not harmonize with each other and with the main purpose of his work. That unity, harmony, 'convergence of characters,' as M. Taine calls it, which gives to works of art their superiority over works of nature, is wholly due to elimination. Any natural subject will do, if the artist has wit enough to pounce upon some one feature of it as characteristic, and suppress all merely accidental items which do not harmonize with this.

Ascending still higher, we reach the plane of Ethics, where choice reigns notoriously supreme. An act has no ethical quality whatever unless it be chosen out of several all equally possible. To sustain the arguments for the good course and keep them ever before us, to stifle our longing for more flowery ways, to keep the foot unflinchingly on the arduous path, these are characteristic ethical energies. But more than these; for these but deal with the means of compassing interests already felt by the man to be supreme. The ethical energy par excellence has to go farther and choose which interest out of several, equally coercive, shall become supreme. The issue here is of the utmost pregnancy, for it decides a man's entire career. When he debates, Shall I commit this crime? choose that profession? accept that office, or marry this fortune? -- his choice really lies between one of several equally possible future Characters. What he shall become is fixed by the conduct of this moment. Schopenhauer, who enforces his determinism by the argument that with a given fixed character only one reaction is possible under given circumstances, forgets that, in these critical ethical moments, what consciously seems to be in question is the complexion of the character itself. The problem with the man is less what act he shall now resolve to do than what being he shall now choose to become.

Taking human experience in a general way, the choosings of different men are to a great extent the same. The race as a whole largely agrees as to what it shall notice and name; and among the noticed parts we select in much the same way for accentuation and preference, or subordination and dislike. There is, however, one entirely extraordinary case in which no two men ever are known to choose alike. One great splitting of the whole universe into two halves is made by each of us; and for each of us almost all of the interest attaches to one of the halves; but we all draw the line of division between them in a different place. When I say that we all call the two halves by the same names, and that those names are 'me' and 'not-me' respectively, it will at once be seen what I mean. The altogether unique kind of interest which each human mind feels in those parts of creation which it can call me or mine may be a moral riddle, but it is a fundamental psychological fact. No mind can take the same interest in his neighbor's me as in his own. The neighbor's me falls together with all the rest of things in one foreign mass against which his own me stands cut in startling relief. Even the trodden worm, as Lotze somewhere says, contrasts his own suffering self with the whole remaining universe, though he have no clear conception either of himself or of what the universe may be. He is for me a mere part of the world; for him it is I who am the mere part. Each of us dichotomizes the Kosmos in a different place.

24. The Quest for Animal Consciousness

Andrea Nani[1], Clare M. Eddy[2], Andrea E. Cavanna, MD, Ph.D.[2,3]

[1] School of Psychology, University of Turin, Italy
[2] Department of Neuropsychiatry, BSMHFT and University of Birmingham, UK
[3] Department of Neuropsychiatry, Institute of Neurology and University College London, UK

Abstract

The philosophical debate as to whether or not animals possess consciousness has transcended many thousands of years and yet has a very brief history in science. The answer to this question will depend on the selection of criteria used to assess both objective evidence based on structure and function of the central nervous system and more subjective indications of consciousness dependent on observable behavior. Studies investigating the similarity of EEG patterns, neural function and markers of neuronal activity often indicate a degree of overlap amongst humans and other animals, while observational studies have documented a range of behaviors demonstrated by many species which cannot be ascribed to basic stimulus-response associations and are more likely to reflect conscious processing. The available behavioral evidence suggests a possible hierarchical organization of consciousness within the animal kingdom, based on level of sensory awareness, complexity of communicative abilities and social interaction, and the presence of some higher order abilities such as self recognition in a few species.

KEY WORDS: Consciousness; Animal behavior; Criteria; Mammals; Birds; Cephalopods.

1. Introduction

The debate as to whether or not animals possess consciousness has transcended many thousands of years. The Ancient Greeks thought that animals were organisms with a very different soul from humans (Hurley and Nudds 2006). In the Nicomachean Ethics, Aristotle teaches that animals lack the rational part of the soul, which distinguishes and characterizes man. He proposed that they have only the vegetative and sensitive aspects of the soul, thus experiencing sensations relating only to primary personal needs (i.e. thirst, hunger, sleep, sex, etc.), and lacking the capacity to think in rational terms about such needs and to reflect upon their sensations. Consequently, in Aristotle's view animal behavior is assumed to be generated and driven by current impulses. In addition, animals do not have the concept of themselves as beings that exist through time. In other words, they neither extend their lives with the power of imagination into the future, nor dwell on the past.

The picture of animal life illustrated by the French Philosopher René Descartes was even further devoid of conscious experience. Descartes believed that only man is endowed with a soul and can thereby exist within a spiritual dimension, whereas animals live without awareness in a purely physical universe (Descartes 1637 [1984]). In a word, they are simply automata: they just act and react in mechanical way. Unlike man, who derives from the interaction of two kinds of substances (res cogitans and res extensa) and is, by virtue of this fact, an autonomous agent, animals are mere products of the organization of one substance only (res extensa). For this reason, Descartes claimed in his Discours de la Méthode (1637), their behavior is strictly determined by the laws of Nature. They therefore lead a life which is completely "unconscious".

Although the question of the existence of animal consciousness is longstanding for philosophers, it has a very brief history in science. Charles Darwin (1871) was the first thinker to set the problem within a scientific frame. Before Darwin, the idea that there was a huge and unbridgeable gap between animals and men was common. On the contrary, in Darwin's view, since both animals and men are to be the products of natural selection, the difference between them is primarily a question of degree (Darwin 1859). Roughly speaking, there are no properties which men have and animals do not have. They all share a common background (the natural world), and the different capacities expressed in the behavior of men and animals are largely determined by the requirements and constraints of the environments within which they develop (Darwin 1859). Darwin's contribution was to set the problem on the right track, even though we have to recognize that the difference between humans and animals is great, particularly in relation to the astonishing ability shown by human brains in terms of language and the use of symbols.

After Darwin, the issue regarding the nature of animal minds was generally neglected until the end of the last century. Both the development of new techniques for studying the nervous system in order to detect and visualize the brain areas involved in performing certain tasks (i.e. in vivo functional neuroimaging), and the increase of knowledge in the fields of neurophysiology and neuroanatomy, have made it possible to identify the neural correlates of consciousness and better understand the similarities and dissimilarities between humans and animals in terms of conscious experience (Cavanna et al. 2011).

2. Criteria for Defining Consciousness

What level of organization of the nervous system is necessary in order for an animal to possess consciousness? To answer this question, we need a set of criteria which allows us to hypothesize the likelihood that a particular animal is conscious. A useful list of criteria for defining consciousness (Seth, Baars and Edelman 2005) consists of a number of basic brain facts and well-established properties of consciousness.

2.1 Some Basic Brain Facts One approach to determine the presence of consciousness is to compare EEG signatures among species. EEGs of conscious human subjects look very different to those exhibited by subjects who are in unconscious states, such as deep sleep, general anesthesia, epileptic absence seizures, and vegetative states resulting from brain damage (Cavanna and Monaco 2009; Cavanna et al. 2010). We know that consciousness is associated with low-level and irregular activity in the EEG, ranging from 20 Hz to 70 Hz (Berger 1929). Contrastingly, unconscious states are characterized by more regular, slower and high-amplitude waves at less than 4 Hz (Baars, Ramsoy and Laureys 2003). This distinction has been found in all mammalian species studied so far. For example, Cephalopoda can show EEG patterns similar to those of conscious vertebrates (Bullock and Budelmann 1991). Strikingly, similar data have been shown in another invertebrate, the tiny fruit fly (Van Swinderen and Greenspan 2003). Birds which are awake also exhibit EEGs comparable to mammals, although these markedly diverge during sleep (Ayala-Guerrero 1989; Ayala-Guerrero and Vasconcelos-Duenas 1988; Karmanova and Churnosov 1973).

As a second criterion for consciousness, we can assess the activity of the thalamocortical system. Conscious information processing among mammals is associated with specific patterns of interaction between the thalamus and the cortex (Baars, Banks, and Newman 2003). In particular, the lower part of the brain (i.e. the brainstem) is crucial in maintaining arousal, while cortical areas are implicated in the process of creating and manipulating conscious thought (Cavanna et al. 2011). The comparison of anatomical structures among different species implies that the thalamocortical system arose with mammals or mammal-like reptiles more than 100 million years ago; accordingly, consciousness itself might have appeared twice after this divergence, in two reptilian lines which led to birds and mammals (Kardong 1995). This would explain why homologous anatomical structures, which functionally resemble mammalian thalamocortical activity, can also be found in birds (Luo, Ding, and Perkel 2001).

The final criterion relates to the evidence that consciousness implies widespread activity of the brain (Srinivasan, Russell, Edelman, and Tononi 1999). As a result, functional similarities can still be detected between human and animal brains, despite the topological differences in the nervous systems of species whose divergence occurred long ago (Wada, Hagiwara, and Jarvis 2001). For instance, analogies in cerebral functional activity can be inferred by identifying specific molecular markers, such as neurotransmitters, neuropeptides, and neuronal receptors that are present in certain regions of the mammalian brain (Naftolin, Horvath, and Balthazart 2001). Researchers have identified the major neurotransmitters thought to be specific to mammalian brains in the cephalopod nervous system (Hanlon and Messenger 2002).

2.2. The Properties of Consciousness Empirical data suggest that consciousness is characterized by a set of specific properties (Seth, Baars and Edelman 2005). These properties are:

- a wide range of content;
- the high informativeness of conscious representations;
- the fleetingness of conscious scenes;
- an internal coherence of the conscious field such that when two inconsistent stimuli are presented simultaneously only one can be conscious;
- the seriality of the conscious process;
- the binding of perceived object properties (color, shape, smell, etc.);
- the self-attribution of conscious experience;
- accurate reportability of conscious contents;
- sense of subjectivity or privacy of the conscious experience;
- a focus-fringe structure (e.g. awareness of fringe events, like vague feelings of familiarity, the tip-of-the-tongue experience etc.);
- the facilitation of learning conscious information;
- the stability of conscious scenes;
- the allocentric character (externalness) of conscious representation with respect to the observer;
- an internal and external knowing function capable of orienting the de-

cision-making process.

- All these properties are hallmarks of human consciousness, but they can be used as criteria in order to determine the existence of conscious states in animals. For example, even the properties that seem to be particular to the human race - such as the reportability of conscious events and the self-attribution of conscious experience - can plausibly be attributed to animals. Of course, animals cannot make accurate verbal reports, however they can give fairly reliable non-verbal indications of their conscious experience by means of indirect methods. Alternative strategies such as the "commentary key" can be used in experiments to allow monkeys to make behavioral comments about what they consciously see or unconsciously guess to see. For instance, monkeys with lesions to half of the visual cortex are able to make discriminations in their blind field by pointing to the correct location of a target or by choosing between colors, but they cannot distinguish between a stimulus in the blind field and a blank display in the intact part of the visual field (Cowey and Stoerig 1995; Stoerig and Cowey 1997). This condition appears to be very similar to human blindsight, in which patients with cortical blindness demonstrate a preserved ability to locate objects, even though they strongly deny that they consciously perceive them. This is further demonstrated by experiments in which rhesus macaque monkeys were trained to press a lever in order to report perceived stimuli in a "binocular rivalry" condition (Logothetis 1998). Binocular rivalry occurs when two different images are presented simultaneously to the eyes, in such a way that one image is seen only by the left eye while the other image is seen only by the right eye. Rather than perceiving both the images superimposed on one another, subjects report consciously seeing one image first and then the other, in an alternating succession.

The properties listed above make consciousness crucial in order to make accurate discriminations between sensory inputs and, consequently, in decision processes. Studies have shown that only conscious sensory information, and not unconscious, activates cerebral regions subserving executive functions (Frackowiak 2004). It therefore appears that consciousness, at least in its basic level of waking state and perceptual awareness, has a key biological function in evolution. The association between consciousness on the one hand, and reproductive behavior and goal-directed survival on the other, is considered by some authors to be indisputable (Baars 2005). This suggests that consciousness can play a very important role in the continuity of animal life.

3. Consciousness in Animals

Demonstrations of consciousness in animal species may require behaviors which satisfy the criteria listed above and are unlikely to reflect simple conditioned responses.

Many animals discriminate between sensory inputs and flexibly alter behavior. William James (1910) suggested that the recognition and integration of the "sense of sameness is the very keel and backbone of consciousness" (p. 240). Pigeons appear to possess some sense of sameness, and therefore maybe consciousness, given that they can discriminate between visual stimuli. Cook and colleagues report that pigeons can identify the odd area or item across a wide range of visual stimuli. For example, Cook, Kelly, and Katz (2003) showed pigeons either a sequence of same (aaaaaa) or alternating (ababab) photographic stimuli. Pigeons showed that they could discriminate these two types of sequences by exhibiting reward-reinforced differences in pecking rate depending on whether they were shown identical or nonidentical pictures. Following acquisition, the pigeons showed successful transfer of this behaviour to novel stimuli. Pigeons can therefore make a same–different discrimination based on only two different items, and can detect stimulus change quite quickly.

Other good examples of behavioural adaption based on sensory input are tool use and observational learning (Masson and McCarthy, 1999; Shanor and Kanwal, 2009). Tool use has been demonstrated in elephants (Chevalier-Skolnikoff and Liska, 1993) and some types of birds, which include carrion crows in Japan utilizing the movements of cars at traffic lights to crack nuts. Primates are also well known tool users. Orangutans use leaves as gloves for holding prickly fruit and pitcher plants as cups, while chimpanzees have been observed carrying sharp sticks as weapons. Primates in captivity including orangutans readily imitate human behaviors accurately, and octopi can learn skills through observation (Edelman, Baars and Seth, 2005).

Animals may also demonstrate conscious thought through the ability to formulate plans to solve problems, implying they have an understanding of their place within the world and within time. Simple examples of such behaviors include food caching. However, more complex planning abilities were shown by a chimpanzee in a zoo in Sweden, who planned ahead to throw rocks at annoying visitors (Osvath, 2009). The complexity of such behavior seems to imply conscious intention, rather than actions which depend on simple stimulus-response associations. Furthermore, while many animals, including octopus and cuttlefish, can exhibit evidence of short and long term memory, is it possible that this primate's behavior indicates the use of prospective memory, or mental simulation of future events?

An abstract representation of future events may also be exemplified by the behavior of the parrot Alex (Pepperberg, 2008). After witnessing a nut being obscured by a cover and then a food pellet being revealed, rather than simply eating the food pellet, Alex exhibited aggressive behaviors, implying annoyance at the object switch. Alex remembered that the nut was previously in that place, even despite it being invisible to him for a period of time. Furthermore, Alex's reaction seems to imply that he had formed expectations that the nut would still be present when the cover was removed, i.e. an understanding of object permanence. Studies currently debate the age at which this ability develops in human infancy.

Although animals cannot use language to communicate with humans, they may demonstrate a degree of language use or understanding which could reflect conscious thought. Studies of prairie dogs (e.g. Slobodchikoff et al., 1991; Slobodchikoff, 1998) indicate these animals use specific calls to describe different predators. These calls therefore possess true meaning and have clear adaptive survival value. Closer study of prairie dog language indicates the use of modifiers: sounds which reflect differences in attributes such as colour and size. Such language demonstrates evidence of awareness of changing environmental conditions and perhaps even a desire to communicate such changes. In relation to understanding human language, one border collie called Rico has a vocabulary of more than 200 words. Experiments showed that Rico correctly retrieved 37 of 40 randomly chosen toys by name from his collection of 200 toys. But more striking is this dog's ability to learn the names of new toys. When researchers placed a new toy amongst his familiar toys and asked him to get the toy using a word he had never heard before, Rico retrieved the new toy 70% of the time. This could provide evidence of conscious thought and even deductive reasoning, whereby Rico realised that the word must relate to the unknown item as it is was not the same sound associated with the other items he knows (Stein, 2004). There was also evidence that Rico added these new toy names to his vocabulary without further instruction, learning new words in a 'fast-mapping' fashion (Kaminski et al., 2004), which is similar to the way young children aquire new words.

Some animals such as parrots and primates including chimpanzees, gorillas and bonobos, have been taught to communicate in human language using picture boards or sign language. Studies of these animals illustrate key evidence of conscious awareness, as reflected in the animal's ability to describe objects in their environment. Critically, these animals can combine words or sounds they have previously been taught to create a new word to describe a novel object. For example, the parrot Alex (Pepperberg, 2008) described an apple as a ban-erry (combining part of the words banana and cherry) and cake as 'yummy bread'. On the first occasion that Kanzi the bonobo ate kale, she described it as 'slow-lettuce', presumably because it took longer to chew and swallow than the normal lettuce she was familiar with (Savage-Rumbaugh and Lewin,1994). These combinations of new sounds and words can't reflect simple stimulus-response associations and therefore provide relatively strong evidence of accurate report.

An aspect of human consciousness is a personal or internal awareness of the self and reflection on one's own conscious experience, involving elements such as metacognition. Mirror-self recognition (Edelman and Seth, 2009; Shanor and Kanwal, 2009) has thought to have been demonstrated in chimpanzees (Povinelli et al., 2003), orangutans (Suarez and Gallup, 1981), bonobos (Hyatt and Hopkins 1994; Walraven et al. 1995), possibly gorillas (Patterson and Cohn, 1994), rhesus monkeys (Rajala et al., 2010), elephants (Plotnik et al. 2006), dolphins (Reiss and Marino 2001), magpies (Prior et al. 2008) and, more controversially, pigeons (Epstein et al., 1981). For instance, experiments have verified with quantitative and videographic evidence that rhesus monkeys with surgical implants in their heads use mirrors to inspect their implants, as well as other parts of their bodies which they are not able to see (Rajala et al. 2010). Fewer experiments have attempted to investigate metacognition in non-human species, but reports suggest that parrots may show evidence of insight into changes in their own knowledge by discriminating a discrimination (being aware of distinguishing changes: Pepperberg, 2008).

Likewise, dolphins have been shown to be capable of complex intellectual and cognitive activities that some believe are equal or superior to non-human primates (Marino, 2002; Marino et al. 2007). Further, they appear to have a sense of self-image including an ownership of individual body parts (Herman et al. 2001). Dolphins can be trained in the use of gestures which they can learn to associate with specific body parts, and can move, shake or indicate specific body parts when asked, such as by using a specific region of the body to touch a specific object. They will also repeat specific behavior when asked to do so. Moreover, they can mimic not just other dolphins, but humans, and can mirror human actions, such as waving their pectoral fin when a human waves an arm, raising their tails when observing a human raising their leg (Herman 2002). These results suggest that dolphins have self-consciousness and are conscious of their actions as having agency ad ownership (Mercado et al. 1998, 1999; Reiss and Marino 2001).

If animals can understand themselves as individual entities, can they appreciate the existence of other organisms as entities with similar functions? Some researchers argue that certain animals appear to exhibit appreciation of other animals' abstract mental states, or a rudimentary 'Theory of Mind'. Such awareness may be implied by the deceptive behavior of ravens, who re-hide food after conspecifics observe their actions, but not if other birds appear to ignore them (Bugnyar and Heinrich, 2005). When a dog owner points at something, his pet seems to understand he should look not towards the finger, but towards the direction at which it is pointed (Horowitz, 2009). Could this express awareness of his owner's intent? Jackdaws can also use human eye gaze and pointing as cues to behavior suggesting they have some rudimentary understanding of human desire or intent (Shanor and Kanwal, 2009).

Other experiments showed that dominant and subordinate apes competed with each other for food. These findings led researchers to conclude that "at least in some situations chimpanzees know what conspecifics do and do not see and, furthermore, that they use this knowledge to formulate their behavioral strategies in food competition situations" (Hare et al. 2000). It is worth noting here that the theory of mind debate finds its origin in the hypothesis that primate intelligence on the one hand, and human intelligence on the other, are specifically adapted for social cognition (Byrne and Whiten 1988). As a result, it has been argued that evidence for the ability to attribute mental states in a wide range of species might be better sought in natural activities rather than in lab experiments which place the animals in artificial situations (Allen and Bekoff 1997; Hare et al. 2001; Hare and Wrangham 2002).

Perhaps some of the strongest evidence of the likelihood of animals possessing conscious thought are observed behaviors whereby animals act against natural survival responses (Masson and McCarthy, 1999). For example, one report described an elephant in Kenya attempting to rescue a young rhino trapped in mud despite the rhino's mother's aggressive response. In another specific case, when a bull elephant was shot, two younger elephants attempted to rescue and lead him away (Denis, 1963; cf. Carrington, 1958). It has also been observed that dolphins and whales support conspecifics if they need help to breathe (Bearzi and Stanford, 2008), and move injured companions away from attack and danger If we are to accept that such seemingly altruistic responses are unlikely to reflect simple survival instincts, we may have to consider the possibility that they imply some form of conscious will.

4. Conclusions

Investigations of consciousness in animals encompass examination of both the structure and function of the central nervous system and observable behavior. Studies investigating the similarity of EEG patterns, neural function and markers of neuronal activity often indicate a degree of overlap amongst humans and other animals, while observational studies have documented a range of behaviors demonstrated by many species which cannot be ascribed to basic stimulus-response associations and are more likely to reflect conscious processing. Our judgment of whether animals possess consciousness will inevitably depend on the criteria used. However, examination of available behavioral evidence suggests a possible hierarchical organization of consciousness within the animal kingdom (Figure 1).

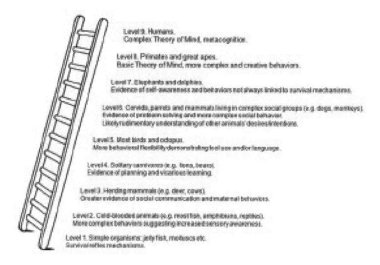

Figure 1.

Criteria which can be used to determine levels of consciousness include level of sensory awareness, degree of behavioral flexibility, complexity of communicative abilities and social interaction, and the presence of higher order abilities such as self recognition, mental reflection and theory of mind (Table 1).

Given the diversity of life in the animal kingdom, it is clear that the question of whether animals possess consciousness is far too simple. Rather, we should ask: at what level does a particular species demonstrate conscious awareness? The challenge for future research will be to conduct investigations which overcome species barriers in order to provide more objective scientific evidence of the conscious abilities easily ascribed to animals based on observed behavior.

Table 1.

Criterion	Explanation
Behavioural flexibility	Ability to remember and learn, plan, and flexibly solve problems (e.g. tool use)
Complexity of social behavior	Maternal behaviors, complexitiy of language and communication, understanding basic intent/desire, Theory of Mind
Sensory awareness	Understanding of abstract concepts, awareness of time, recognition of physical self
Intrinsic self awareness	Self reflection and metacognition

References

Allen, C., and Bekoff, M. (1997), Species of Mind, Cambridge, MA: MIT Press.

Ayala-Guerrero, F. (1989), Sleep patterns in the parakeet Melopsittacus undulates. Physiology and Behavior, 46(5), 787-791.

Ayala-Guerrero, F., and Vasconcelos-Duenas, I. (1988), Sleep in the dove Zenaida asiatica. Behavioral and Neural Biology, 49(2), 133-138.

Baars, B.J. (2005), Subjective experience is probably not limited to humans: The evidence from neurobiology and behavior. Consciousness and Cognition, 14, 7-21.

Baars, B.J., Banks, W.P., and Newman, J. (Eds.) (2003), Essential sources in the scientific study of consciousness, Cambridge, MA: MIT Press.

Baars, B.J., Ramsoy, T., and Laureys, S. (2003), Brain, conscious experience, and the observing self. Trends in Neuroscience, 26(12). 671-675.

Bearzi, M., and Stanford, C.B. (2008), Beautiful minds. The parallel lives of great apes and dolphins. London: Harvard University Press.

Berger, H. (1929), Ueber das Elektroenkephalogramm des Menschen, in Archiv fuer Psychiatrie und Nervenkrankheiten, Berlin, 87, 527-570.

Bugnyar, T. and Heinrich, B. (2005) Ravens, Corvus corax, differentiate between knowledgeable and ignorant competitors. Proceedings in Biological Sciences, 272, 1641–1646.

Bullock, T.H., and Budelmann, T.U. (1991), Sensory evoked potentials in unanesthesized unrestrained cuttlefish: A new preparation for brain physiology in cephalopods. Journal of Comparative Physiology, 168(1), 141-150.

Byrne R. W., and Whiten, A. (1988), Machiavellian Intelligence: social expertise and the evolution of intellect in monkeys, apes and humans, Oxford: Oxford University Press.

Carrington, R. (1958). Elephants, Chatto and Windus, London.

Cavanna, A.E., Cavanna, S., Servo, S., and Monaco, F. The neural correlates of impaired consciousness in coma and unresponsive states. Discovery Medicine 2010;9:431-438.

Cavanna, A.E., and Monaco, F. Brain mechanisms of altered conscious states during epileptic seizures. Nature Reviews Neurology 2009;5:267-276.

Cavanna, A.E., Shah, S., Eddy, C.M., Williams, A., and Rickards, H. (2011) Consciousness: A neurological perspective. Behavioural Neurology in press.

Chevalier-Skolnikoff, S. and Liska, J., 1993, Tool use by wild and captive elephants, Animal Behavior, 46:209–219.

Cook, R. G., Kelly, D. M., and Katz, J. S. (2003).,Successive two-item same–different discrimination and concept learning by pigeons. Behavioural Processes, 62:125–144.

Cowey, A., and Stoerig, P. (1995), Blindsight in monkey, Nature, 373, 247-249.

Darwin, C. (1859 [1964]), The Origin of Species, Cambridge: Harvard University Press.

Darwin, C. (1871 [2004]), The Descent of Man, and Selection in Relation to Sex, Penguin Classics.

Denis, A. (1963). On Safari, Collins, London.

Descartes, R. (1637 [1984]), The Philosophical Writings of Descartes. Translated by J. Cottingham, R. Stoothoff, and D. Murdoch. Cambridge: Cambridge University Press.

Edelman, D.B., and Seth, A.K. (2009), Animal consciousness: A synthetic approach. Trends in Neurosciences 32:476-484.

Edelman, D.B., Baars, B.J., and Seth, A.K. (2005), Identifying hallmarks of consciousness in non-mammalian species. Consciousness and Cognition 14:169-187.

Epstein, Lanza; Skinner, RP; Skinner, BF (1981). ""Self-awareness" in the pigeon". Science 212 (4495): 695–696.

Frackowiak, R. (2004), Functional brain imaging (2nd ed.), London: Elsevier Science.

Hanlon, R.T., and Messenger, J.B., (2002), Cephalopod behavior, Cambridge: Cambridge University Press.

Hare, B., Call, J., Agnetta, B. & Tomasello, M. (2000), Chimpanzees know what conspecifics do and do not see. Animal Behavior, 59: 771–785.

Hare B., Call J., Tomasello M. (2001), Do chimpanzees know what conspecifics know? Animal Behaviour, 63: 139–151.

Hare, B., and Wrangham, R. (2002). Integrating two evolutionary models for the study of social cognition. In Bekoff, Allen, & Burghardt (eds.) (2002), The Cognitive Animal, Cambridge, MA: The MIT Press.

Herman, L. M. (2002). Vocal, social, and self-imitation by bottlenosed dolphins. In: Nehaniv, C., Dautenhahn, K. (Eds.), Imitation in Animals and Artifacts. MIT Press, Cambridge, MA., pp. 63-108.

Herman, L. M., Matus, D. S., Herman, E. Y., Ivancic, M., Pack, A. A. (2001). The bottlenosed dolphin's (Tursiops truncatus) understanding of gestures as symbolic representations of its body parts. Learning & Behavior, 29, 250-264.

Horowitz, A. (2009). Attention to attention in domestic dog (Canis familiaris) dyadic play. Animal Cognition, 12, 107-118.

Hurley, S., and Nudds, M. (eds.), (2006). Rational Animals?, Oxford: Oxford University Press.

Hyatt, C. W., and Hopkins, W. D. (1994). Self-awareness in bonobos and chimpanzees: a comparative approach. In Self-awareness in animals and humans: developmental perspectives, ed. S. T. Parker and R. W. Mitchell and M. L. Boccia, pp. 248-253. New York: Cambridge University Press.

James, W. (1910). Psychology. New York: Holt.

Kaminski, J. et al. (2004). Word learning in a domestic dog: Evidence for "fast-mapping." Science 304, 1682-1683.

Kardong, K.V. (1995), Vertebrates: Comparative anatomy, function, and evolution, Dubuque, Iowa: W.C. Brown.

Logothetis, N.K. (1998), Single units and conscious vision. Philosophical Transaction of the Royal Society of London. Series B: Biological Sciences, 353, 1801-1818.

Luo, M., Ding, L., and Perkel, D.J. (2001), An avian basal ganglia pathway essential for vocal learning forms a close topographic loop. Journal of Neuroscience, 21, 6836-6845.

Marino, L. (2002). Convergence of complex cognitive abilities in cetaceans and primates. Brain, Behavior and Evolution 59, 21-32.

Marino, L., Connor, R. C., Fordyce, R., Herman, L. M., Hof, P. R., Lefebvre, L., Lusseau, D., McCowan, B., Nimchinsky, E. A., Pack, A. A., Rendell, L., Reidenberg, J. S., Reiss, D., Uhen, M. D., Van de Gucht, E., Whitehead, H. (2007). Cetaceans have complex brains for complex cognition. PLoS Biology, 5, 966-972.

Masson, J, and McCarthy, S. (1999), When elephants weep. London: Vintage Press.

Mercado, E. III, Murray, S. O., Uyeyama, R. K., Pack, A. A., Herman, L. M. (1998). Memory for recent actions in the bottlenosed dolphin (Tursiops truncatus): Repetition of arbitrary behaviors using an abstract rule. Animal Learning & Behavior, 26, 210-218.

Mercado, E. III, Uyeyama R. K., Pack, A. A., Herman, L. M. (1999). Memory for action events in the bottlenosed dolphin. Animal Cognition, 2, 17-25.

Naftolin, F., Horvath, T., and Balthazart, J. (2001), Estrogen synthetase (Aromatase) immunohistochemistry reveals concordance between avian and rodent limbic systems and hypothalamus. Experimental Biology and Medicine, 226, 717-725.

Osvath, M. (2009), Spontaneous planning for future stone throwing by a male chimpanzee. Current Biology 19(5):190-191.

Patterson, F. G. P., and Cohn, R. H. (1994). Self-recognition and self-awareness in lowland gorillas. In Self-awareness in animals and humans: developmental perspectives, ed. S. T. Parker and R. W. Mitchell and M. L. Boccia, pp. 273-290. New York: Cambridge University Press.

Pepperberg, I. (2008). Alex and me. London: HarperCollins.

Povinelli, Daniel; de Veer, Monique; Gallup Jr., Gordon; Theall, Laura; van den Bos, Ruud (2003). "An 8-year longitudinal study of mirror self-recognition in chimpanzees (Pan troglodytes)". Neuropsychologia 41 (2): 229–334.

Plotnik, J. M., de Waal, F., and Reiss, D. (2006), Self-Recognition in an Asian Elephant. Proceedings of the National Academy of Sciences, 103: 17053–17057.

Prior, H., Schwarz, A., Güntürkün, O. (2008), Mirror-Induced Behavior in the Magpie (Pica pica): Evidence of Self-Recognition. Public Library of Science/Biology, 6 (8): e202.

Rajala, A.Z., Reininger, K.R., Lancaster, K.M., Populin, L.C., (2010), Rhesus Monkeys (Macaca mulatta) Do Recognize Themselves in the Mirror: Implications for the Evolution of Self-Recognition. PloS ONE, 5(9): e12865. doi:10.1371/journal.pone. 0012865.

Reiss, D., and Marino, L. (2001), Mirror self-recognition in the bottlenose dolphin: A case of cognitive convergence. Proceedings of the National Academy of Science, 98: 5937–5942.

Savage-Rumbaugh, E.S., and Lewin, R. (1994), Kanzi: The ape at the brink of the human mind. New York: John Wiley & Sons.

Seth, A.K., Baars, B.J., and Edelman, D.B. (2005), Criteria for consciousness in humans and other mammals. Consciousness and Cognition, 14, 119-139.

Shanor K., and Kanwal, J. (2009), Bats sing, mice giggle. London: Icon Books Ltd.

Slobodchikoff, C. N., Judith Kiriazis, C. Fischer, and E. Creef. (1991). Semantic information distinguishing individual predators in the alarm calls of Gunnison's prairie dogs. Animal Behaviour, 42: 713-719.

Slobodchikoff, C. N. (1998). The language of prairie dogs. pp. 65-76. in: M. Tobias and K. Solisti-Mattelon, eds. Kinship with the animals. Beyond Words Publishing, Hillsboro, OR.

Srinivasan, R., Russell, D.P., Edelman, G.M., and Tononi, G. (1999), Increased synchronization of magnetic responses during conscious perception. Journal of Neuroscience, 19, 5435-5448.

Stein, R. (2004). 'Common collie or uberpooch?' Washington Post, Friday, June 11, 2004, p. A1.

Stoerig, P., and Cowey, A. (1997), Blindsight in man and monkey, Brain, 120,(Pt 3), 535-559.

Suarez, S.D. and Gallup, G.G. (1981). Self recognition in chimpanzees and orangutans but not gorillas. Journal of Human Evolution, 10(2):175-188.

Van Swinderen, B., and Greenspan, R.J. (2003), Salience modulates 20-30 Hz brain activity in Drosophila. Nature Neuroscience, 6(6), 579-586.

Wada, K., Hagiwara, M., and Jarvis, E.D. (2001), Brain evolution revealed through glutamate receptor expression profiles. Society of Neurosciences Abstracts, Vol. 27, Program No. 538.10.

Walraven, V., van Elsacker, L., and Verheyen, R. (1995). Reactions of a group of pygmy chimpanzees (Pan paniscus) to their mirror images: evidence of self-recognition. Primates 36:145-150.

Wasserman, E., and Zentall, T. (Eds). (2009), Comparative cognition: Experimental explorations of animal intelligence. New York: Oxford University Press.

25. Consciousness and Intelligence in Mammals: Complexity thresholds

David Deamer

Department of Biomolecular Engineering,
University of California, Santa Cruz CA 95064

Abstract

Behavioral responses to sensory input are clearly related to the complexity of animal nervous systems. Here I propose a way to estimate complexity in the mammalian brain using the number of cortical neurons, their synaptic connections and the encephalization quotient. The complexity values correlate reasonably well with expectations based on observation, and suggest that threshold complexities are associated with awareness, self-awareness and consciousness.

KEY WORDS: Consciousness, evolution, neuroscience, nervous system, brain

> "When you measure what you are speaking about, and express it in numbers, you know something about it; but when you cannot measure it, your knowledge is of a meagre and unsatisfactory kind." Lord Kelvin, 1888

Can we measure consciousness? If not, how can we even begin to consider it from a scientific perspective? The main point of this essay is that an experimental approach to understanding consciousness will treat it as the product of an evolutionary process leading to a quantifiable threshold complexity of the nervous system. It follows that consciousness and intelligence are graded phenomena related to increments in complexity. This essay describes a quantitative approach to define the gradation and fulfill Lord Kelvin's challenge in the quotation above.

1. Evolution of Nervous Function

Much of what follows will seem obvious, yet it should be made explicit in order to provide a foundation for later discussion. I will argue that consciousness is best understood in terms of an evolutionary process that began when animal life developed the first differentiated cells, or neurons, associated with communication between different tissues. Fedonkin (2003) provided a detailed and critical review of precambrian animal fossils, and Peterson and Butterfield (2005) used genetic information to calculate that metazoans emerged between 826 and 634 million years ago, in accordance with the fossil record. The earliest animals probably resembled Placozoa, perhaps the simplest form of animal life today. The one known species, Trichoplax adhaerans, has only a few thousand cells of four cell types, and the smallest genome of any animal (Srivastava et al. 2008). There is no defined nervous system present, but one of the cell types is present as a syncytium that may help coordinate movement of the organism. .

During the Cambrian radiation between 580 and 500 million years ago, more complex animals appeared that are now called Bilateria (Shierwater et al. 2009). It is reasonable to assume that these organisms had nervous systems, but it is uncertain whether they represented the single origin of nervous function that later evolved into the nervous systems of today's animals. No doubt it would have been a major selective advantage for the predators and prey of that era to be able to sense their environment and respond with appropriate behavior. The chief characteristic of this level of nervous function is that the response to variable sensory inputs would have been a reflexive sensory-motor response with minimal modulation. This basic function is preserved in higher organisms as well, for instance in the spinal reflex response to a painful stimulus.

The next step came with the ever increasing complexity of animal nervous systems as life evolved into larger aquatic organisms like fish over 500 million years ago, then into terrestrial animal life over four hundred million years ago. As we compare the behavior of fish, reptiles, birds and mammals, it is clear from observations that the vertebrate animals are aware of their environment. Instead of being entirely reflexive, their responses to sensory input can be modulated within certain limitations. Their modulated responses apparently reflect a short term memory measured in seconds, so that intelligent behavior is not possible. With rare exceptions, most birds and mammals are unable to match even the most minimal human intelligence in terms of problem-solving.

The behavior that characterizes self-awareness arose in the increasingly complex nervous systems of primates, and also in other large-brained animals such as elephants and dolphins (Plotnik et al. 2006; Reiss and Marino 2001). A self-aware organism recognizes itself in a mirror, and Homo neanderthalensis

400,000 years ago would likely have had no difficulty passing this test. Self-awareness evolved into modern consciousness 200,000 years ago with the appearance of Homo sapiens in Africa. If a child from that era could somehow be transported forward in time to today's world, it would presumably be indistinguishable from other children in its ability to develop language and adapt to contemporary culture.

The most striking property of a conscious human being is not just self-awareness, but to varying degrees human brains can indefinitely maintain an internal model of sensory input and manipulate the model in order to predict future outcomes. Short term memory is therefore not measured in seconds, but instead can be maintained throughout a problem-solving interval. The word intelligence defines a semi-quantitative measure of the ability of the conscious nervous system to perform such tasks.

2. Three Postulates

I will now present a set of postulates that can be used to clarify the discussion of human consciousness. The postulates, taken together, also suggest experimental and observational tests of hypotheses related to consciousness.

The first postulate is that consciousness will ultimately be understood in terms of ordinary chemical and physical laws. This postulate links consciousness directly to nervous processes in the brain. and arises from a consideration of the principle of parsimony (Occam's Razor). Quantum mechanical involvement, mind-brain duality, and supernatural concepts such as spirit and soul are excluded. The postulate is not simply parsimonious, but is supported by the fact that the conscious state is strongly affected by chemical and physical conditions imposed on the brain. For instance, consciousness is abolished simply by lowering the temperature of the brain by ten degrees, from 310 to 300 degrees Kelvin. When the brain is warmed, consciousness returns. A similar effect is produced by general anesthetics which diffuse from the lungs into the blood, then partition into cell membranes of the brain and interact with protein channels such as GABA and glutamate receptors (Olsen and Li, 2011). Excitability is inhibited, and consciousness disappears. Anesthetics are specific examples of a large number of chemicals that interact with receptors in the cell membranes of cerebral neurons and thereby produce effects ranging from the mild stimulation of nicotine and caffeine to deep anesthesia. If small amounts of such chemicals interacting with neurons can reversibly affect consciousness, it seems inescapable that the mechanisms underlying consciousness most likely involve biochemical and physical processes occurring at the level of cortical neurons and their interactions with one another.

The second postulate is that consciousness is related to the evolution of anatomical complexity in the nervous system. The reason consciousness seems so mysterious at present is that we have not advanced far enough in our knowledge of complex interactions within the brain's neurons. This is analogous to the evolution of computer engineering over the past 70 years. Imagine that somehow a functioning laptop computer could be transported back in time to Los Alamos in 1943, where some of the worlds most brilliant physicists had gathered in wartime to design and test the first nuclear weapon. They would have been astonished by the color screen, the fact that an entire movie could be stored, the WiFi capacity, the internet. No matter how brilliant, their collective genius would be baffled by this seeming miracle. I think that we are like those scientists when today we attempt to understand how the phenomenon of consciousness emerges from nervous function in the brain.

The second postulate suggests that consciousness can emerge only when a certain level of anatomical complexity has evolved in the brain that is directly related to the number of neurons, the number of synaptic connections between neurons, and the anatomical organization of the brain. Again by analogy to the evolution of computers, a certain number of components and interacting connections are required to perform increasingly complex tasks. Consider the evolution of the integrated circuit. The first IC was developed by Kirby and Noyce in the 1950s, and incorporated only a few semiconductor-based transistors. In the late 1960s the number of transistors in an IC had increased to 100s, then to thousands in the mid-1970s. The number increased again to the 100,000 range in the 1980s, to millions in 1990s, and most recently billions. Each of the advances represents a threshold relating the number of transistors to the complexity of computational function.

It is interesting to compare this history to the evolution of the nervous system. The earliest animals were well served by a nervous system having perhaps a few hundred neurons. The different cell types in C. elegans have been counted: there are precisely 302 neurons and a total of 7000 synaptic connections (White et al. 1986). In contrast, the human cerebral cortex is estimated to have 10 - 20 billion neurons and a total of ~10^{15} synapses (see Roth and Dicke, 2005). If in fact consciousness and intelligence are related to complexity of nervous systems, it should be possible to establish a quantitative measure of the complexity, then compare it with our observation of animal behavior.

This brings us to the third postulate, that consciousness, intelligence, self-awareness and awareness are graded, and have a threshold that is related to the complexity of nervous systems. I will now propose a quantitative formula that gives a rough estimate of the complexity of nervous systems. Only two variables are required: the number of units in a nervous system, and the number of connections (interactions) each unit has with other units in the system. The formula is simple: C (complexity) = log(N) * log(Z) where N is the number of units and Z is the average number of synaptic inputs to a single neuron. The idea that complexity arises from interconnecting systems is not a new concept. W. Grey Walter suggested much the same thing in his book The Living Brain published in 1953. A more detailed version of this relationship was previously used to calculate C for the nervous systems of animals ranging from nematodes through insects and frogs and then mammalian brains including humans (Deamer and Evans, 2006). Here I will restrict the list to a set of mammalian species for which I could find estimates of cortical cell number and synaptic junctions per cell. I will present a ranked list calculated from the complexity formula, and then normalize the results to take into account a third variable called encephalization quotient. I will then ask whether threshold levels of complexity can be discerned in the results, and how well they fit our expectations.

3. Measuring Cortical Complexity

The number of neurons (N) increases markedly within the nervous systems of animals ranging from nematodes to humans. The number of synapses per neuron (Z) also varies significantly. Z is difficult to estimate, but has been measured for cortical neurons in the human, rat and mouse brains (DeFilipe et al. 2002). The numbers vary by a factor of 2 within the six cell layers of the neocortex, and again by a factor of 2 when the human, rat and mouse brain are compared in terms of average number of synapses per neuron for all six layers. Each human cortical neuron has approximately 30,000 synapses per cell, each mouse neuron 20,000 synapses per cell and each rat neuron 17,000 synapses per cell. I will assume Z = 30,000 for the brains of primates, dolphins, elephants and monkeys, and Z = 20,000 for the brains of all other animals in the list. For the order of magnitude calculations reported here, these rough estimates of Z are sufficient.

Table I. Mammals ranked by number of cortical neurons. Cortical neuron estimates adapted from Roth and Dicke 2005.

Animal	Brain weight	Cortical neurons (millions)	Log(N)	Log(Z)	C
Human	1350	11500	10.1	4.5	45.5
Elephant	4200	11000	10.0	4.5	45
Chimpanzee	380	6200	9.8	4.5	44.1
Dolphin	1350	5800	9.7	4.5	43.6
Gorilla	480	4300	9.6	4.5	43.2
Horse	510	1200	9.1	4.3	39.1
Dog	64	610	8.8	4.3	37.8
Rhesus	88	480	8.7	4.5	39.1
Cat	25	300	7.6	4.3	32.7
Opossum	7.6	27	7.4	4.3	31.8
Rat	2	15	7.2	4.3	31
Mouse	0.3	4	6.6	4.3	28.4

Table 1 shows the list of mammals ranked according to the number of cortical neurons and log (N). The last two columns show log(Z) and the value of C calculated as log(N)*log(Z). Brain weight is also given for comparison.

The next step is to incorporate the encephalization quotient (EQ) in the analysis. When the amount of brain tissue in a series of animals is plotted against body mass, from mice to elephants, there is a roughly linear relationship (Roth and Dicke, 2005). However, the value for some animals lies significantly above the line, while others are well below the line. A relatively large animal like an elephant needs a greater absolute number of neurons to serve the much larger number of cells in their bodies, but these neurons are not necessarily given over to consciousness or intelligent behavior. Humans, with the highest EQ (7.6) have developed larger brains in relation to body size because our evolutionary pathway happened to select for the nervous activity called intelligence, which presumably requires more brain tissue devoted to that function. Deaner et al. (2007) presented evidence that EQ alone is not the best indicator of intellectual capacity within primates, but that brain mass shows a better correlation. However, here I must use relative EQ to correct for the effect of a much larger range of body size (from mouse to elephant) by normalizing against human EQ. The complexity equation then becomes $C = \log(N \cdot EQ_a/EQ_h) \cdot \log(Z)$, where EQ_a is the animal EQ and EQ_h is the human EQ, taken to be 7.6.

Table II. Mammals ranked by normalized complexity (C). EQ values taken from Roth and Dicke 2005.

Animal	EQ	Normalized C
Human	7.6	45.5
Dolphin	5.3	43.2
Chimpanzee	2.4	41.1
Elephant	1.3	41.1
Gorilla	1.6	40.0
Rhesus	2.1	36.5
Horse	0.9	34.5
Dog	1.2	34.4
Cat	1.0	32.7
Rat	0.4	25.4
Opossum	0.2	24.5
Mouse	0.5	23.2

Table II shows the list of mammals according to EQ, and the values of C normalized with respect to complexity.

4. Discussion and Conclusions

If we asked a hundred thoughtful colleagues to rank this list of mammals according to their experience and observations, I predict that their lists, when averaged to reduce idiosyncratic choices, would closely reflect the calculated ranking. It is interesting that all six animals with normalized complexity values of 40 and above are self-aware according to the mirror test, the rhesus monkey is borderline at 36.5, while the animals with complexity values of 35 and below do not exhibit this behavior. This jump between $C = 36.5$ and 40 appears to reflect a threshold related to self-awareness.

Although mammals with normalized complexity values between 40 and 43.2 are self-aware and are perhaps conscious in a limited capacity, they do not exhibit what we recognize as human intelligence. It seems that a normalized complexity value of 45.5 is required for human consciousness and intelligence, that is, 10 - 20 billion neurons, each on average with 30,000 connections to other neurons, and an EQ of 7.6. Only the human brain has achieved this threshold.

If we take the claim of threshold complexities in nervous systems as a hypothesis, it will only be useful if there are testable predictions. One is that in diseases such as Alzheimer's, the reduced intellectual capacity and lowered state of consciousness begin to occur when the number of active neurons or the number of synaptic connections is reduced below threshold values required for intelligent behavior. In fact, in patients with advanced Alzheimer's disease the number of synapses per neuron was reduced by 25 – 35 percent (see Selkoe 2002 for review). Similarly, when general anesthesia produces an unconscious state, we will find that the threshold is again breached, not in terms of cell number but instead due to a reduced number of functioning synapses and associated membrane receptors caused by the action of the anesthetic compound.

Closely related to the argument presented here is the concept of the connectome, which consists of the white matter connections between different regions of the brain. A study of the human connectome is now underway in several laboratories (Sporns et al. 2005; Thompson and Swanson 2010). Chiang et al. (2009) reported correlations of connectome architecture with human intelligence. When the human connectome can be compared to the brains of other primates, it seems likely we will observe a threshold of anatomical complexity that is related to self-awareness, and a second threshold related to human consciousness.

A third prediction is that because of the limitations of computer electronics, it will be virtually impossible to construct a conscious computer in the foreseeable future. Even though the number of transistors (N) in a microprocessor chip now approaches the number of neurons in a mammalian brain, each chip has a Z of 2, that is, its input-output response is directly connected to just two other transistors. This is in contrast to a mammalian neuron, in which function is modulated by thousands of synaptic inputs and output relayed to hundreds of other neurons. According to the quantitative formula described above, the complexity of the human nervous system is $\log(N) * \log(Z) = 45.5$, while that of a microprocessor with 781 million transistors is $8.9 * .3 = 2.67$, many orders of magnitude less. Of course, what the microprocessor lacks in connectivity can potentially be compensated in part by speed, which in the most powerful computers is measured in terraflops compared with the kilohertz activity of neurons. Interestingly, for the nematode the calculated complexity $C = 3.2$, assuming an average of 20 synapses per neuron, so the functioning nervous system of this simple organism could very well be computationally modeled.

References

Chiang M-C, Barysheva M, Shattuck DW, Lee AD, Madesn SK, Avedissian C, Klunder AD, Toga AW, McMahon KL, de Zubicaray GI, Wright MJ, Srivastava A, Balov N, Thompson PM (2009) Genetics of brain fiber architecture and intellectual performance. J Neuroscience 29:2212-2224.

Deamer DW, Evans J. (2006). Numerical analysis of biocomplexity. In Life As We Know It. J Seckbach, ed. p 201 - 12. New York: Springer.

Deaner RO, Isler K, Burkart J, van Schaik C (2007) Overall brain size, and not encephalization quotient, best predicts cognitive ability across non-human primates. Brain Behav Evol 70:115-124.

DeFelipe J, Alonso-Nanclares L, Arellano JI (2002) Microstructure of the neocortex: Comparative aspect. J Neurocytology 31:299-316.

Fedonkin MA (2003) The origin of the Metazoa in the light of the Proterozoic fossil record. Paleontological Research, 7:9-41.

Olsen RW, Li GD. (2011) GABA(A) receptors as molecular targets of general anesthetics: identification of binding sites provides clues to allosteric modulation. Can J Anaesth 58:206-215.

Peterson KJ, Butterfield NJ. (2005) Origin of the Eumetazoa: testing ecological predictions of molecular clocks against the Proterozoic fossil record. Proc Natl Acad Sci USA 102:9547-52.

Plotnik JM, de Waal FBM, Reiss D (2006) Self-recognition in an Asian elephant. Proc Natl Acad Sci USA 103: 17053-17057.

Reiss D, Marino L (2001) Self-recognition in the bottlenose dolphin: A case of cognitive convergence. Proc Natl Acad Sci USA 98: 5937-5942.

Roth G, Dicke U (2005) Evolution of the brain and intelligence. Trends Cognitive Sciences 9: 250-257.

Schierwater B, Eitel M, Jakob W, Osigus H-J, Hadrys H (2009) Concatenated analysis sheds light on early metazoan evolution and fuels a modern "Urmetazoon" hypothesis. Plos Biol 7(1): e1000020.

Selkoe D (2002) Alzheimer's disease is a synaptic failure. Science 298:789-91.

Sporns O, Tononi, G, Kötter R (2005). The human connectome: A structural analysis of the human brain. Plos Computational Biol 1(4):e42.

Srivastava M, Begovic E, Chapman J, Putnam NH, Hellsten U, Kawashima T, Kuo A, Mitros T (2008). The Trichoplax genome and the nature of placozoans. Nature 454: 955–960.

Thompson RH, Swanson LW. (2010) Hypothesis-driven structural connectivity analysis supports network over hierarchical model of brain architecture. Proc Natl Acad Sci USA. 107:15235-9.

Walter WG (1953) The Living Brain. W.W. Norton and Co. New York.

White JG, Southgate E, Thomson JN, Brenner S (1986) The structure of the nervous system of the nematode Caenorhabditis elegans. Phil Transactions Roy Soc B 314: 1–340.

26. Mirror Self-Recognition on Earth
Thomas Suddendorf, Ph.D.

University of Queensland, School of Psychology,
Brisbane, Australia

Abstract

Adult humans and great apes can recognize themselves in mirrors as measured by a standard task in which a mark is surreptitiously placed above the brow before the subject is presented with a mirror. Lesser apes and monkeys do not search for the mark on their own body upon seeing their reflection. Developmental studies suggest that children who pass the standard task have a rapidly updatable expectation of what they look like from the outside. Analysis of the comparative pattern indicates that the capacity for mirror self-recognition evolved between 18 and 14 million years ago in the common ancestor of extant great apes.

KEY WORDS: self-awareness, phylogenetic reconstruction, ape, child, visual selfrecognition Mirror Self-Recognition on Earth

1. Introduction

Humans spend considerable time in front of mirrors, spawning a multibillion-dollar cosmetic industry. Although a lot of species adjust their posture and appearance to suit different situations (e.g., puffing up to increase the appearance of one's size to predators) there is little evidence to suggest that they are actually aware of what they look like. An exception are our closest living relatives, the great apes, who sometimes show keen interest in their reflections to investigate parts of their bodies that they cannot usually see. When one surreptitiously puts a little paint above the brow of a chimpanzee and then presents a mirror, the ape is likely to examine the mark using its reflection. Members of all great apes have passed this test (Bard, Todd, Bernier, Love, & Leavens, 2006; Gallup, 1970; Lethmate & Dücker, 1973; Posada & Colell, 2007; Povinelli, et al., 1997; Povinelli, Rulf, Landau, & Bierschwale, 1993; Suarez & Gallup, 1981). But there continues to be great disagreement about what this entails about their minds.

2. The Meaning of Self-Recognition

Gordon Gallup, the inventor of this test, advocates a rich interpretation (Gallup, 1970, 1998): Passing the mark test indicates self-awareness - one can become object of one's own attention. However, with perhaps the possible exception of the most shallow of super models, for most of us the term self-awareness entails a lot more than knowing what you look like. It implies knowledge of your inner world, your likes and dislikes, your personality, skills and attitudes. Perhaps most importantly, it implies some knowledge of where you come from and where you are going. So perhaps Gallup's interpretation may over estimate what we can reasonably conclude about great ape minds from their passing the test.

On the other hand, there are lean accounts that appear to severely underrate what the task measures. The comparative psychologist Celia Heyes, for instance, argued that all one needs to pass the task is an ability to distinguish feedback from other types of sensory input (Heyes, 1994, 1998). Effectively, any animal that manages to avoid bumping into things, or that avoids biting itself, has demonstrated such ability. Followers of this view therefore dismiss evidence of mirror self-recognition as not indicating anything particularly interesting – while tacitly ignoring the fact that Heyes' account fails to offer any explanation at all as to why only a few species pass the task and yet many can distinguish feedback from other input in other contexts. For a neutral observer this creates a situation not unlike the climate change debate of recent years: some experts say this, other experts say the opposite, so it is tempting to conclude that we cannot conclude much at all - and it may be best to ignore this all until it is resolved.

As a result scholars and lay people alike may pick whichever interpretation they find more appealing. We can do better than that, and as with understanding climate change, there are some great rewards. One can bring other lines of evidence to bear on the issue. A typical 15-month-old toddler gets very excited when looking in a mirror, but is at a total loss as to where that sticker is that the experimenter put on her fringe. By age two virtually all toddlers pass the test (Amsterdam, 1972; Nielsen & Dissanayake, 2004). My colleagues and I have shown that these children form rapidly updatable expectations of what they look like (Nielsen, Suddendorf, & Slaughter, 2006). In a series of experiments we first demonstrated that young children could equally recognize a mirror image of a mark on their leg as of a mark on their fringe as used in the standard task. In the next experiment, we slipped children into baggy tracksuit pants that we had sewn onto a highchair and then presented them with a mirrored view of their legs. They failed to find stickers on their unfamiliar-looking legs. However, when given 30 seconds exposure to the new pants they were wearing, by removing a tray that blocked the direct view, children passed the task. These results strongly suggest that young children quickly form a mental expectation of what they look like. They rule out theories that place special emphasis on cognition about faces (e.g., Neisser, 1997). The findings do not demand the higher cognitive capacities that Gallup's rich account conjectures. But they clearly indicate more than Heyes' lean account proposes, because children saw the same mirrored feedback of baggy pants in the last two conditions but reacted very differently when they had 30 seconds of prior exposure. Passing the mirror mark test indicates that the subjects have formed a mental expectation of what they look like from the outside.

3. Comparative Conclusions

This strongly suggests that great apes also form such expectations. Chimpanzees develop mirror self-recognition not unlike human children (Bard, et al., 2006). Monkeys, on the other hand, consistently fail the classic task (Anderson & Gallup, 1997; Gallup, Wallnau, & Suarez, 1980; Hauser, Miller, Liu, & Gupta, 2001; Heschel & Burkart, 2006; Roma, et al., 2007). They have also recently failed a leg version of the task (Macellini, Ferrari, Bonini, Fogassi, & Paukner, 2010). High profile cases have been made for success by one elephant (Plotnik, De Waal, & Reiss, 2006), two magpies (Prior, Schwarz, & Gunturkun, 2008) and one dolphin (Reiss & Marino, 2001). However, other studies found negative results (e.g., Povinelli, 1989), so these results clearly require replication. Surprising lessons can be learned from the distribution of the trait among primates.

Emma Collier-Baker and I recently conducted the largest study yet on lesser apes and they all failed the task (Suddendorf & Collier-Baker, 2009). They consistently failed the task in spite of being highly motivated to find the mark. We marked their faces with cake icing to really make sure they want to retrieve the mark when they see it. They greedily scrape it off other body parts they can see directly. We even smeared icing on the mirror itself and the apes would scrape or lick off every last bit of the treat from the mirror and yet ignore the big blob of icing on their own head that was clearly visible in the mirror. These results amount to evidence of absence (rather than the usual common absence of evidence). Great apes and humans do, and lesser apes and monkeys do not recognize themselves, and this has important implication for the evolution of the trait (Suddendorf & Whiten, 2001).

Traits may be shared between species for two profoundly different reasons: analogy and homology (though more complex mechanisms such as horizontal gene transfer also exist). The wings of birds, bats and insects, for example, are independent, analogical solutions to the problem of flight. Homologous traits, on the other hand, are shared because of common descent. To determine whether a trait is shared for homological of analogical reasons, evolutionary biologists compare the number of assumptions each possibility implies about change events that occurred in the past to explain the current distribution of the trait. It is a less parsimonious theory to propose that each bird species independently invented a feathery solution to flying, then to assume that they all inherited the trait by common descent. In our case, if great ape ancestors had evolved a capacity for mirror-self-recognition independently we have to assume that the trait evolved at least on 4 occasions. If great apes have this trait because of common descent, we would have to propose only one such change. It is therefore more parsimonious to conclude that the great ape common ancestor, that in other respects may be quite different from extant apes, acquired this trait before 14 million years ago.

It acquired it before the time that the line leading to orangutans split off and passed it on to all of its descendents. Given that lesser apes fail mirror self-recognition tasks we can now narrow the emergence of this trait down further to a time after the split from lesser apes. The capacity to rapidly form expectations of what one looks like must hence have evolved between 18 and 14 million years ago (Suddendorf & Collier-Baker, 2009). Phylogenetic reconstruction is quite a powerful way to make an inference about extinct minds even without ever having to lay eyes on a fossil of the creature that first evolved it. Indeed, we do not know what this creature looked like, but it probably knew what it looked like.

This comparative analysis not only informs about the evolution of mind, but also offers entirely new practical opportunities. For example, this analysis can narrow down the search for the neurological and genetic basis of these traits. Because great apes share a homologous capacity for self-recognition, the basis of this trait is to be found among the neuronal and genetic characteristics that are shared by humans and these species. Given that the next closely related species, the lesser apes, do not have the capacity for selfrecognition, we can subtract as not sufficient all those characteristics that they share with great apes and humans. Necessary factors underpinning the trait must hence be found among the genetic and neuronal characteristics that are not shared with lesser apes, but are present in all great apes and humans. For example, recent studies have identified one neuronal characteristic that meets these criteria: so-called Von Economo neurons exist in human and great ape brains, but not in those of lesser apes and monkeys (Nimchinsky, et al., 1999). Such systematic comparative approaches have great potential, over and above the inherently interesting contribution they can make to our understanding of the evolution of our minds. We now know that our capacity to recognize ourselves evolved on Earth in our great ape ancestor between 14 and 18 million years ago.

References

Amsterdam, B. K. (1972). Mirror self-image reactions before age two. Developmental Psychobiology, 5, 297-305.

Anderson, J. R., & Gallup, G. G. (1997). Self-recognition in Saguinus? A critical essay. Animal Behavior, 54, 1563-1567.

Bard, K., Todd, B. K., Bernier, C., Love, J., & Leavens, D. A. (2006). Selfawareness in human and chimpanzee infants: What is measured and what is meant by the mark and mirror test. Infancy, 9, 191- 219.

Gallup, G. G. (1970). Chimpanzees: Self recognition. Science, 167, 86-87.

Gallup, G. G. (1998). Self-awareness and the evolution of social intelligence. Behavioural Processes, 42, 239-247.

Gallup, G. G., Wallnau, L. B., & Suarez, S. D. (1980). Failure to find selfrecognition in mother-infant and infant-infant rhesus monkey pairs. Folia Primatologica, 33, 210-219.

Hauser, M. D., Miller, C. T., Liu, K., & Gupta, R. (2001). Cotton-top tamarins (Saguinus oedipus) fail to show mirror-guided selfexploration. American Journal of Primatology, 53(3), 131- 137.

Heschel, A., & Burkart, J. (2006). A new mark test for mirror selfrecognition in non-human primates. [Article]. Primates, 47(3), 187- 198.

Heyes, C. M. (1994). Reflections on self-recognition in primates. Animal Behavior, 47, 909-919.

Heyes, C. M. (1998). Theory of mind in nonhuman primates. Behavioral and Brain Sciences, 21, 101-134.

Lethmate, J., & Dücker, G. (1973). Untersuchungen zum Selbsterkennen im Spiegel bei Orang-Utans und einigen anderen Affenarten [Investigations into self-recogntion in orangutans and some other apes]. Z. Tierpsychol., 33, 248-269.

Macellini, S., Ferrari, P. F., Bonini, L., Fogassi, L., & Paukner, A. (2010). A modified mark test for own-body recognition in pig-tailed macaques (Macaca nemestrina). Animal Cognition, 13(4), 631-639.

Neisser, U. (1997). The roots of self-knowledge. In J. G. Snodgrass & R. L. Thompson (Eds.), The self across psychology (Vol. 818, pp. 19- 33): New York Academy of Sciences.

Nielsen, M., & Dissanayake, C. (2004). Imitation, pretend play and mirror self-recognition: A longitudinal investigation through the second year. Infant Behavior & Development, 27, 342-365.

Nielsen, M., Suddendorf, T., & Slaughter, V.

(2006). Self-recognition beyond the face. Child Development, 77, 176-185.

Nimchinsky, E. A., Gilissen, E., Allman, J. M., Perl, D. P., Erwin, J. M., & Hof, P. R. (1999). A neuronal morphologic type unique to humans and great apes. Proceedings of the National Academy of Sciences of the United States of America, 96(9), 5268-5273.

Plotnik, J. M., De Waal, F. B. M., & Reiss, D. (2006). Self-recognition in an asian elephant. Proceedings of the National Academy of Sciences, 103, 17053-17057.

Posada, S., & Colell, M. (2007). Another gorilla recognizes himself in a mirror. American Journal of Primatology, 69, 576-583.

Povinelli, D. (1989). Failure to find self-recognition in Asian elephants (Elephans maximus) in contrast to their use of mirror cues to discover hidden food. Journal of Comparative Psychology, 103, 122- 131.

Povinelli, D. J., Gallup, G. G., Eddy, T. J., Bierschwale, D. T., Engstrom, M. C., Perilloux, H. K., et al. (1997). Chimpanzees recognize themselves in mirrors. Animal Behaviour, 53, 1083-1088.

Povinelli, D. J., Rulf, A. R., Landau, K. R., & Bierschwale, D. T. (1993). Self-recognition in chimpanzees (Pan troglodytes). Journal of Comparative Psychology, 107, 347-372.

Prior, H., Schwarz, A., & Gunturkun, O. (2008). Mirror-induced behavior in the magpie (pica pica): Evidence of self-recognition. Plos Biology, 6, 1642-1650.

Reiss, D., & Marino, L. (2001). Mirror self-recognition in the bottlenose dolphin: A case of cognitive convergence. [Article]. Proceedings of the National Academy of Sciences of the United States of America, 98(10), 5937-5942.

Roma, P. G., Silberberg, A., Huntsberry, M. E., Christensen, C. J., Ruggier, A. M., & Suomi, S. J. (2007). Mark tests for mirror selfrecognition in Capuchin monkeys (Cebus apella) trained to touch marks. [Article]. American Journal of Primatology, 69(9), 989- 1000.

Suarez, S., & Gallup, G. G. (1981). Self recognition in chimpanzees and orangutans, but not gorillas. Journal of Human Evolution, 10, 175- 188.

Suddendorf, T., & Collier-Baker, E. (2009). The evolution of primate visual self-recognition: evidence of absence in lesser apes. [Article]. Proceedings of the Royal Society B-Biological Sciences, 276(1662), 1671-1677.

Suddendorf, T., & Whiten, A. (2001). Mental evolution and development: evidence for secondary representation in children, great apes and other animals. Psychological Bulletin, 127, 629-650.

Cetaceans and Primates: Convergence in Intelligence and Self-Awareness

Lori Marino

Neuroscience and Behavioral Biology Program, 488 Psychology and Interdisciplinary Sciences Bldg., 36 Eagle Row, Emory University, Atlanta, GA 30322

Abstract

Cetaceans (dolphins, porpoises and whales) have been of great interest to the astrobiology community and to those interested in consciousness and self-awareness in animals. This interest has grown primarily from knowledge of the intelligence, language and large complex brains that many cetaceans possess. The study of cetacean and primate brain evolution and cognition can inform us about the contingencies and parameters associated with the evolution of complex intelligence in general, and, the evolution of consciousness. Striking differences in cortical organization in the brains of cetaceans and primates along with shared cognitive capacities such as self-awareness, culture, and symbolic concept comprehension, tells us that cetaceans represent an alternative evolutionary pathway to complex intelligence on this planet.

KEY WORDS: Consciousness, Cetaceans to consciousness. cetaceans, dolphins, porpoises and whales, self-awareness, astrobiology

1. Why Dolphins and Whales?

Cetaceans (dolphins, porpoises and whales) have been of great interest to the astrobiology community and to those interested in consciousness and self-awareness in animals. Given the likelihood that "water worlds" may be common in the universe (Goertzel and Combs 2010; Tyler, 2010), it is therefore possible that intelligent mammals and other creatures might evolve completely adapted to a life beneath the sea. The study of Cetaceans may provide a window into the evolution of consciousness in extraterrestrials who have evolved in water and not on land.

There is an early historical connection between the study of cetaceans and research on the possibilities for extraterrestrial intelligence beginning with the first scientific meeting on the search for extraterrestrial radio signals at Green Bank, West Virginia in 1961. The meeting participants included Frank Drake, Bernard Oliver and Carl Sagan, as well as renowned dolphin neuroscientist John Lilly (1969, 1987), who regaled the group with accounts of dolphin intelligence which so impressed them that they voted to call themselves the Order of the Dolphin. Despite the fact that some of Lilly's later work was controversial, he brought to light the large complex brain and sophisticated intelligence of dolphins. Moreover, the independent scientific evidence for complex intelligence in dolphins and whales continued to grow. The meeting was the starting point for the Drake Equation (a gross estimate of the number of habitable worlds) and an enduring involvement of cetacean research in the Search for Extraterrestrial intelligence (SETI).

There are, arguably, two, not entirely unrelated, ways in which dolphin research has played a role in astrobiology and SETI. First, the complex and unusual communicative capacities of dolphins, which includes whistles, clicks, echolocation and many other kinds of sounds, is considered by many in the SETI community as a model for probing the factors relevant to communication with an extraterrestrial. To be sure, no one in this domain views dolphins as extraterrestrials. However, they view human-dolphin communication efforts as a way to prepare for the issues that will need to be addressed in deciphering a complex extraterrestrial signal. These kinds of studies range from efforts to communicate with wild and captive dolphins (Herzing, 2010; Reiss and McCowan, 1993) to research using information theory to quantify the complexity within dolphin whistle repertoires, an approach that would be used if an extraterrestrial signal were to be found (McCowan et al., 1999, 2002). These research paths have continued to be productive in illuminating both basic information about dolphin communication and intelligence and its application to SETI.

The second way in which the study of dolphins is highly relevant to astrobiology has more to do with consideration of the evolutionary pathways that lead to highly complex communicative intelligences. The astrobiological paradigm involves using Earth-based data to form hypotheses about what might be possible beyond Earth. Dolphin and whale evolution, in comparison to human evolution, is arguably an alternative way that intelligence and large complex brains have evolved on this planet.

Therefore, studies of dolphin and whale brains and intelligence allow us to probe questions about the evolution of complex intelligence and implications for evolution of awareness. I will focus on this approach in this paper.

2. Cetacean Evolution and Phylogeny

The modern order Cetacea consists of two suborders: Odontoceti (toothed whales, dolphins and porpoises) and Mysticeti (rorqual and baleen whales). Interestingly, the scientific fervor over dolphins and whales is based on knowledge of relatively few species (the bottlenose dolphin, the orca or killer whale, the beluga whale, humpback whale, for instance) among the seventy-seven species known. It is intriguing to consider the vast array of cetacean intelligence that remains unknown to us. The bottlenose dolphin (Tursiops truncatus) has been the focus of the most study and, as such, we know the most about this species. Therefore, in this paper the term "dolphin" will refer to this species unless otherwise specified.

The origin and evolutionary history of cetaceans represents one of the most dramatic transformations in the mammalian fossil record. The earliest cetacean, Pakicetus, was a medium-sized semi-aquatic mammal with carnivorous dentition (Gingerich and Uhen, 1998) is known from 50 million year old fossil evidence. The earliest odontocete specimen is found in the fossil record in the early Oligocene, about 32 million years ago (Fordyce, 2002). The earliest Mysticete is from the late Eocene, approximately 35 million years ago (Mitchell, 1989). Cetacean ancestry is closely tied to that of Artiodactyla, the suborder of even-toed ungulates; molecular and morphological evidence points to a a sister-taxon relationship between modern cetaceans and Hippopotamidae (Geisler and Theodor, 2009; Geisler and Uhen, 2003; Milinkovitch et al., 1998) although the two lineages diverged over 52 million years ago (Gingerich and Uhen, 1998).

3. General Characteristics of Cetaceans

Despite variability in morphology, behavior and ecology across the numerous cetacean species, there are certain features that are characteristic of cetaceans. Here I will focus upon those most relevant to intelligence, that is, ecology, life history, sensory-perceptual capacities, social behavior and communication.

Ecology Dolphins and whales have adapted to a wide range of habitats, including open ocean, coasts, and rivers. Many of the larger odontocetes, such as the sperm whales, dive to depths of over 500 meters. Most odontocetes feed on fish or cephalopods and, in the case of orcas (killer whales) sometimes other marine mammals. Mysticetes are filter feeders and rely on krill and small fish.

Life history Life history has to do with how a species allocates time and resources to growth, reproduction and survival. Although there is variability across species, cetaceans share a life history strategy with other large-brained highly intelligent animals characterized by a long lifespan, long gestation period, and long juvenile period with substantial parental care of a relatively few singly-born offspring over the lifetime (Chivers, 2009). The longer lifespan and long periods of juvenile dependency are related to the importance of learning in the life of the individual; cetaceans, and other large-brained animals, spend a great deal of their time learning about both the physical environment and their social relationships.

Sensory-perceptual systems Audition (hearing) is the most important sensory modality for cetaceans. Auditory structures in odontocete brains are greatly enlarged (Ridgway, 2000) and although less is known about audition in mysticetes, it is clear that they rely heavily on hearing as well. Odontocete hearing is exceptional with a range of 40 to 150 kHz and sensitivity far exceeds that of humans. They have also developed rapid temporal processing of auditory signals which is an integral part of echolocation. Echolocation, the highly specialized sensory-perceptual adaptation involving the emission of sounds and reception of echoes, is superbly refined in odontocetes as they use it for making very fine discriminations in very complex contexts (Au, 2009). Dolphins use echolocation for navigation, prey capture, avoidance of predators, communication and, it has been hypothesized, "communal cognition" (Jerison, 1986). Dolphins are also capable of cross-modal processing. Complexly shaped objects perceived through echolocation alone can be spontaneously recognised through vision alone, and vice-versa (Pack & Herman, 1995; Pack et al., 2002). Mysticetes do not echolocate and rely on lower frequency sounds for communication.

Though they rely heavily on sound to communicate, most cetaceans are able to see fairly well in both water and air. River dolphins are the exception given there is little use for vision in turbid river waters, so their eyes are greatly reduced. Cetaceans also use touch extensively with others in their group and there appear to be particularly sensitive regions on their body, e.g., blowhole area, genital area. A key feature of odontocetes is the complete loss of olfactory structures while mysticetes possess a reduced olfactory sense (Oelschlager and Oelschlager, 2002). Cetacean tongues contain fewer taste receptors than most land mammals and it is not known how well they can use taste, if at all

(Thewissen, 2009).

Social Behavior and Communication Social behavior and communication are intimately connected and, as such, are discussed here together. Cetacean groups vary enormously in size across species and there are often many nested hierarchical levels of social organization that change in composition over time. Perhaps the most relevant aspects of cetacean social behavior for understanding their intelligence and awareness are that they are extremely group-oriented by nature (perhaps at a level not yet understood), maintain highly complex and dynamic social interactions, and possess cultures, i.e., learned traditions passed on from one generation to the next.

Dolphins live in large highly complex societies with dynamic differentiated relationships (Baird, 2000; Connor, 2007; Connor et al., 2000; Lusseau, 2007) that include long-term bonds, higher order nested alliances and cooperative networks (Baird, 2000; Connor, 2007; Connor et al., 2000) that are based upon learning and memory. These complex relationships and social roles are all mediated by an equally-complex system of communication. Cetacean communication includes body postures and movements. However, cetaceans are known for their reliance on acoustic information as the main modality for communication. Odontocetes make three distinct types of sounds: 1) burst pulses or clicks, which can be narrow or broad spectrum, 2) narrow-band frequency-modulated whistles, and 3) percussive sounds produced with the body. Mysticetes do not appear to echolocate and possess a very different repertoire of kinds of sounds, including low-frequency moans, short humps or knocks, chirps and whistles. Humpback whale songs are probably the most recognized and well known form of mysticete vocalization (Dudzinski et al., 2009).

In order to appreciate the complexity of cetacean communication it is important to consider the functional context of these sounds and the role they serve in their social life. A large proportion of vocal variation within cetacean species is likely the result of vocal learning (Rendell & Whitehead, 2001). There is evidence for individual-level variation in the whistle repertoires of dolphins (McCowan & Reiss, 2005). Bottlenose dolphins produce individually distinctive whistles that they apparently use to identify conspecifics (Sayigh et al., 1999; Tyack, 1999) and may also be employed as a cohesion call (Janik & Slater, 1998; McCowan & Reiss, 1995, 2001). Research described earlier, using information theory, has shown that the sequential order of whistles is an important feature of dolphin communication. Data from McCowan et al. (1999, 2002) suggests that at least for two-whistle sequences (second-order entropy) internal structure is present. This suggests there may be a syntactic component to the use of whistles in dolphin communication. Cetacean communication is intimately associated with culture as many cetacean cultural traditions often include acoustic conventions. Multifaceted cultures have been documented from field studies of orcas, sperm whales, bottlenose dolphins, humpback whales, and there is suggestive evidence for other species as well. These impressive cultural traditions involve dialects, foraging sites, feeding strategies, and even tool use. For instance, orcas produce dozens of community, clan and pod-specific call types (Ford, 1991; Yurk et al., 2002). Studies of sperm whales have documented over 33 types of 'coda' vocalisations (rhythmic patterns of clicks) and shown that their use varies among social groups (Rendell & Whitehead, 2003) and there is even matching of codas in sperm whale vocal interactions (Schulz et al., 2008). Marcoux et al. (2007) found evidence for sympatric cultural clans of sperm whales. And cultural transmission of tool use has been documented among bottlenose dolphins, who use sponges to probe into crevices for prey (Krützen et al., 2005) and pass these skills to younger generations (Mann et al., 2008).

In summary, in addition to possessing a general life history pattern associated with strong reliance on learning, cetaceans exhibit complex social and communicative behaviors consistent with a sophisticated level of intelligence and awareness. The experimental work directly addressing intelligence and awareness, described next, supports that conclusion.

4. Cetacean Intelligence and Awareness

There is a large body of data from studies of captive dolphins demonstrating sophisticated cognitive abilities by dolphins and supporting field studies, described above, of apparently complex and sophisticated behavior. In these experimental studies dolphins exhibit prodigious capacities, extraordinary behavioral flexibility, and a profound sense of self.

In the realm of artificial language comprehension, concept formation, and behavioral innovation, dolphins excel. They learn and master not only the semantic features of artificial gestural and acoustic languages, but also the syntactic features (Herman, 1986 for a review; Herman et al., 1993). Human language gains its versatility and communicative power not just through the word, but through the sentence. In human language, syntax allows for the combination of words into an infinite number of possibilities. Herman et al. (1984) showed that dolphins can respond appropriately by situations that involve the use of several different syntactic rules creating thousands of unique sentences constructed from a finite 40-item vocabulary. Among nonhumans, only the great apes, particularly the bonobo (Pan paniscus), have shown this type of ability (e.g., Savage-Rumbaugh et al., 1993). Dolphins also possess an understanding of numerical concepts and can generalize these concepts to novel sets outside of the learned range (Killian et al., 2003). Dolphins demonstrate the ability to learn a variety of types of governing rules for solving abstract problems (Herman et al., 1994). For example, they can reliably classify pairs of objects as "same" or "different" (Mercado et al., 2000).

Dolphins have demonstrated an understanding of the abstract concept of "imitate" (Herman, 2002) and are one of the few species that can imitate both arbitrary sounds and arbitrary behaviors (Richards et al., 1984; Reiss & McCowan, 1993; Herman, 2002). Dolphins can even spontaneously innovate, that is, create new behaviors upon request (Herman, 2006; Mercado et al., 1998).

In addition to showing impressive capacities in the communicative, social, and learning domains, dolphins are among the few species who, thus far, have shown convincing evidence of self-awareness in formal tests. Self-awareness is a sense of personal identity, i.e., what is commonly referred to as the subjective "I". At the bodily level self-awareness is typically labeled self-recognition, the ability to become the object of one's own attention in the physical realm. At a more abstract level self-awareness can take the form of robust psychological continuity over extended time. Dolphins have convincingly demonstrated that they use a mirror to investigate their own bodies, showing that they have a sense of self (Reiss & Marino, 2001). These findings are consistent with further evidence for self-awareness and self-monitoring in dolphins and related cognitive abilities that likely underwrite the complex social patterns observed in many cetacean species. Body awareness has been demonstrated through the dolphin's ability to understand symbolic gestural references to her own body parts and the ability to use those body parts in ways (often novel) specified by the experimenter (Herman et al., 2001). Moreover, McCowan et al. (2000) provided evidence that bottlenose dolphins anticipate, monitor, organize, and modify goal-directed behavior on the basis of contingencies. Finally, awareness of one's own knowledge states has been demonstrated by dolphins reporting when they were "uncertain" about discriminations they were asked to make. They performed comparably to human subjects in the same study showing that they are able to access their mental states about how confident they are of completing a task and acting upon that knowledge appropriately. This is evidence of abstract "meta-knowledge" in dolphins (Smith et al., 1995).

Related to awareness of one's body and knowledge is the capacity to consider the perceptual and mental perspective of others. This capacity, called Theory of Mind, has, again, been evinced in only a few species thus far. And, again, one of them is the dolphin. Dolphins can spontaneously produce pointing (using rostrum and body alignment) to communicate desired objects to a human observer (Xitco et al., 2001), and appear to understand that the human observer must be present and attending to the pointing dolphin for communication to be effective (Xitco et al., 2004).

In summary, consistent with the evidence for considerable behavioral complexity in their natural communication systems and social lives, dolphins exhibit outstanding cognitive capacities in experimental situations and possess a sense of self and possibly a theory of mind. Although the bottlenose dolphin has been the subject of almost all of these formal studies there is every reason to hypothesize that, despite cognitive variability across cetaceans, other species, particularly orcas and other odontocetes, would demonstrate similar capacities. This body of behavioral evidence from both the field and the lab inevitably leads to intriguing questions about the kind of brain that would produce such a complex self-aware intelligence.

5. Cetacean Brain Evolution - An Alternative Pathway to Complex Intelligence

In the past three decades new research has shed light on the complexity of cetacean brains and has begun to lay bare the neurobiological basis for their considerable cognitive abilities. But the story about cetacean brain evolution is one that goes beyond an evolutionary tale about increased brain size. It is also a fascinating example of the way that brain structure-function relationships can follow a complicated pattern of evolutionary divergence and convergence. Studies of cetacean brains has revealed that the human brain is not the only brain that has undergone substantial increases in size and complexity. Cetacean brains have as well but along a different neuroanatomical trajectory, providing an example of an alternative evolutionary route to complex intelligence on earth.

6. The Massive Modern Cetacean Brain

Modern cetacean brains are among the largest of all mammals in both absolute mass and in relation to body size. The largest brain on earth today, that of the sperm whale with an average weight of 8000g for adults (Marino, 2009), is about 60% larger than the elephant brain and six times larger than the human brain. Relative brain size is typically expressed as an Encephalization Quotient or EQ (Jerison, 1973) which is a value that represents how large or small the average brain of a given species is compared with other species of the same average body weight. The EQ for modern humans is 7.0. Our brains are seven times the size one would expect for a species with our body size. Almost all possess above-average encephalization levels compared with other mammals. Numerous odontocete species possess EQs in the range of 4 to 5, that is, they possess brains 4 to 5 times larger than one would expect for their body weights. Many of these values are second only to those of modern humans and significantly higher than any of the nonhuman anthropoid primates (Marino, 1998). Figure 1 illustrates the relative EQ levels of several modern dolphin species and the great apes, including humans.

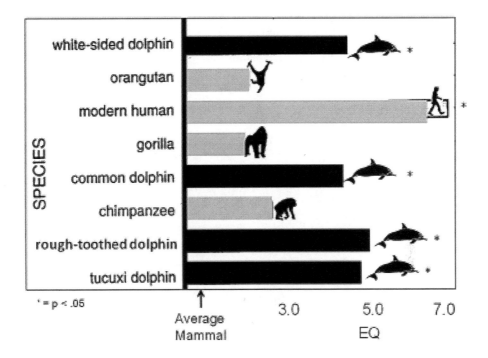

Figure 1. The relative EQ levels of several modern dolphin species and the great apes, including humans. An EQ of 1 (representing mammals with average brain sizes) is labeled for comparison.

EQs of mysticetes are all below 1 (Marino, 2009) because of an uncoupling of brain size and body size in very large aquatic animals. That is, very large animals, such as whales, tend to have smaller brains compared to their weight and this is exacerbated by "aquatic weightlessness". However, mysticete brains are large in absolute size and exhibit high degrees of cortical complexity confirming that mysticete brains have, in addition to odontocete brains, undergone substantial elaboration during the course of their evolution (Oelschlager and Oelschlager, 2009).

7. The Unique Cetacean Brain

The key point about cetacean brains from an astrobiological perspective is not only that they are greatly enlarged and, specifically, highly expanded in the cortical region (the structure involved in high-level processing of information, self-awareness, and generally, abstract intelligence), but represent a strikingly unique combination of cortical characteristics. Whereas subcortical neuroanatomy is shared across mammals (with variations in, mainly, the proportions of various structures), the massive expansion of the cetacean cortex occurred along a very different organizational route than that of other mammal brains (including humans).

Unlike the primate brain, which is organized around four lobes, the cetacean forebrain is arranged around three concentric tiers of tissue and includes an entirely unique region called the paralimbic cortex. The function of the paralimbic cortex in cetaceans is largely unknown and is another signpost of the radical departure of the cetacean brain from the general mammalian pattern. The cetacean forebrain is among the most highly convoluted of all mammals, indicative of a substantial increase in neocortical surface area and volume in cetacean evolution (Ridgway & Brownson, 1984).

The pattern of elaboration of the neocortex in cetaceans has resulted in a highly unusual configuration. The map of sensory projection regions (the cortical regions that receive sensory information) in the cetacean brain stands in striking contrast to that of other large-brained mammals. In primates, for instance, the visual and auditory projection regions are located in the occipital and temporal lobes, respectively. This means that visual information is first processed in the cortex in the back of the brain (occipital region) and auditory information on the side of the brain (temporal region). An expanse of nonprojection or association cortex intervenes between these two regions. Therefore, visual and auditory information must be sent to this intervening cortex from the projection zones if they are to be integrated. In cetaceans, by contrast, the visual and auditory projection zones are located in the parietal region atop the hemispheres, that is, in a very different location than in primates and other large-brained mammals. Moreover, these areas are immediately adjacent to each other (Ladygina et al., 1978, Supin et al., 1978). This arrangement of cortical adjacency is unusual for such a large brain and reveals that not only is the surface map of the cetacean neocortex different from most mammals but the relationship between the visual and auditory processing areas is closer or more highly integrated than in most large-brained mammals. This idiosyncratic pattern of visual-auditory adjacency may allow for the highly developed cross-modal sensory processing abilities in cetaceans discussed earlier. Outside of the borders of the primary sensory regions cetacean brains possess a vast expanse of nonprojection or association cortex for even higher-order cognitive information processing. Figure 2 illustrates the cortical surface configurations of the visual and auditory regions in cetacean brains and human brains.

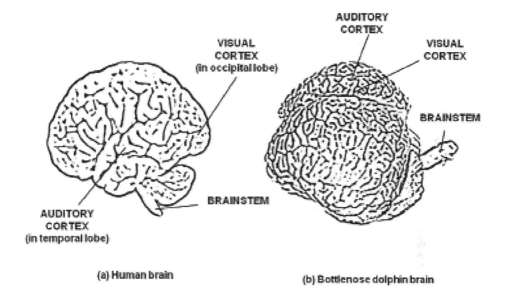

Figure 2. The cortical surface configurations of the visual and auditory regions in cetacean brains and human brains with hindbrain labeled for orientation.

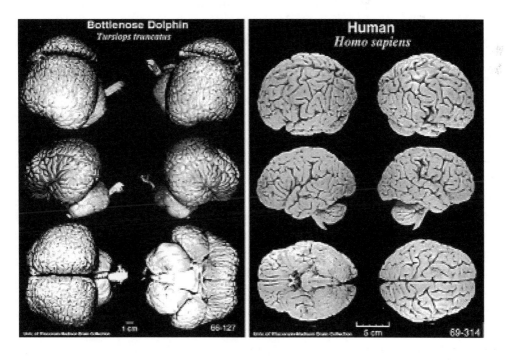

Figure 3. The Dolphin and Human brain.

Recent studies of cetacean neocortical cytoarchitecture reveal extensive neocortical complexity and variability in both odontocetes and mysticetes

(Hof et al, 2005; Hof & Van der Gucht, 2007). The cellular architecture of various regions of the cetacean neocortex is characterized by a wide variety of organizational features, i.e. columns, modules, layers, that are associated with complex brains. Furthermore, there is substantial differentiation across the various neocortical regions.

Whereas there appears to be a high degree of organizational complexity throughout the cetacean neocortex there are specific regions that are especially notable in their apparent degree of elaboration. The cingulate and insular cortices (both situated deeper within the forebrain) in odontocetes and mysticetes are extremely well developed (Hof & Van Der Gucht, 2007; Jacobs et al, 1979) and the expansion of these areas in cetaceans is consistent with high-level cognitive functions such as attention, judgment and social awareness (Allman et al, 2005), Moreover, recent studies show that the anterior cingulate and insular cortices in larger cetaceans contains a type of projection neuron, known as a spindle cell or Von Economo neuron (Hof & Van Der Gucht, 2007). Von Economo neurons are highly specialised projection neurons considered to be involved in neural networks subserving aspects of social cognition (Allman et al, 2005) and have thus far been found in humans and great apes (Allman et al, 2005) and elephants (Hakeem et al, 2008). Spindle cells are thought to play a role in adaptive intelligent behaviour and the presence of these neurons in cetaceans is consistent with their complex cognitive abilities.

Despite similarities in level of complexity and in the presence of specialized neurons, e.g., Von Economo neurons, there is a fundamental distinction in cortical architecture between cetacean and primate brains that relates to a striking difference in connectivity patterns. Specifically, cetacean neocortex is characterized by five layers instead of the six typical of primates and many other mammals. Cetacean neocortex possesses a very thick layer I in combination with the absence of a granular layer IV. In primates, a granular layer IV is the primary input layer for fibers ascending from the midbrain to the cortex and this layer is also the source of important connections within the cortex. However, granular layer IV is absent in cetaceans and this has compelling implications for how information reaches the cetacean cortex and then gets distributed to other areas (Glezer et al, 1988). The current prevailing view is that the thick Layer I is the primary layer receiving incoming fibers. This means that the way information gets to the cortex and distributed within the layers of the cortex in cetacean brains is distinctly different from primates and other mammals.

In summary, cetacean brains are organized around a very different theme than that of primates and other large-brained mammals. There are differences in the arrangement of the cortical project zones, in cytoarchitecture, and in the intra-cortical and input-output relations between the cortex and the rest of the brain. The very different connectivity patterns between primates and cetaceans is a compelling example of distinctly different evolutionary trajectories taken towards neurological complexity.

8. Implications for Astrobiology & Evolution of Extraterrestrial Consciousness

Cetaceans and primates (including humans, who have a typical primate brain) have very large brains that have expanded considerably over their respective evolutionary histories. And they share many mammalian characteristics, including basic neuroanatomy; the brains are similar on the subcortical level as are all mammalian brains. But when one examines the evolutionarily newest structure of mammal brains, the cortex, one finds tremendous differentiation between the cortex of cetaceans and primates. These distinctions are fundamental to the way in which each cortex is contructed, the way it organizes and processes information, and the way it interacts with the rest of the brain. Yet, despite these divergent neuroanatomical evolutionary trajectories cetaceans and primates share a number of cognitive abilities that are found in only a few other animals. Therefore, cetacean-primates comparisons hold tremendous potential for astrobiological questions about the evolution of intelligence.

One of the ongoing debates within the astrobiology community has to do with contingency and convergence, that is, whether, if the "tape of life" were rewound, would complex intelligence evolve again on the earth. I argue that cetacean and primate intelligence is a case of cognitive convergence. Convergence (or homoplasy) involves the appearance of a similar feature in two or more distantly related species whose common ancestor lacked the feature (Nishikawa, 2002; Ridley, 1993). The greater the phylogenetic separation of the two species the stronger the case for convergence, or, in other words, the deeper the convergence. Evolutionary convergence can occur within any domain of biology, from chemistry to morphology to cognition. Cognitive convergence, specifically, is convergence in those processes that comprise the way an organism processes information. These processes include memory, learning mechanisms, sensory and perceptual processing, and levels of awareness. In a general way cognitive convergence refers to convergence in intelligence. Since it is arguably the case that the common ancestor of cetaceans and primates, who lived over 95 million years ago, did not possess many of these shared traits, e.g., self-awareness, symbolic language comprehension, culture, the existence of these traits in these two highly divergent groups of mammals represents a striking case of cognitive convergence. As such, the more we learn about these neuroanatomical differences in the face of shared intelligence the more we learn about the parameters that may shape the evolution of intelligence in general, that is, in an astrobiological context.

References

Allman, J.M., Watson, K.K., Tetrault, N.A., Hakeen, A.Y. (2005). Intuition and autism: a possible role for Von Economo neurons. Trends in Cognitive Science, 9, 367-373.

Au, Whitlow W. L. (2009). Echolocation. In Perrin, W.F., Wursig, B., Thewissen, J.G.M. (Eds), Encyclopedia of Marine Mammals (2nd ed), Academic Press, San Diego, CA., pp. 348-357.

Baird, R. (2000). The killer whale: foraging specializations and group hunting. In Mann, J., Connor, R.C., Tyack, P., Whitehead, H. (Eds), Cetacean societies: Field Studies of Whales and Dolphins,University of Chicago Press, Chicago, pp. 127-153.

Chivers, S.J. (2009). Cetacean life history. . In Perrin, W.F., Wursig, B., Thewissen, J.G.M. (Eds), Encyclopedia of Marine Mammals (2nd ed), Academic Press, San Diego, CA., pp. 215-220.

Connor, R. C. (2007). Complex alliance relationships in bottlenosed dolphins and a consideration of selective environments for extreme brain size evolution in mammals. Philosophical Transactions of the Royal Society: Biological Sciences, 362, 587-602.

Connor, R. C, Wells, R., Mann, J. & Read, A. (2000). The bottlenosed dolphin: social relationships in a fission-fusion society. In Mann, J., Connor, R.C., Tyack, P., Whitehead, H. (Eds). Cetacean Societies: Field Studies of Whales and Dolphins (eds. J. Mann, R. C. Connor, P. Tyack, and H. Whitehead), University of Chicago Press, Chicago, pp. 91-126.

Dudzinski, K. M., Thomas, J. A., Gregg, J. D. (2009). Communication in marine mammals. In Perrin, W.F., Wursig, B., Thewissen, J.G.M. (Eds), Encyclopedia of Marine Mammals (2nd ed), Academic Press, San Diego, CA., pp. 260-269.

Ford, J. K. B. (1991). Vocal traditions among resident killer whales (Orcinus orca) in coastal waters of British Columbia. Canadian Journal of Zoology, 69, 1454-1483.

Fordyce, R.E. (2002). Simocetus rayi (Odontoceti: Simocetidae) (new species, new genus, new family), a bizarre new archaic Oligocene dolphin from the eastern North Pacific. Smithsonian Contributions to Paleobiology,93, 185-222. 22

Geisler, J.H.,Theodor, J.M. (2009). Hippopotamus and whale phylogeny. Nature 458:E1-E4. doi:10.1038/nature07776

Geisler, J.H.,Uhen, M.D. (2003). Morphological support for a close relationship between hippo and whales. Journal of Vertebrate Paleontology 23(4): 991-996.

Gingerich, P.D., Uhen, M.D. (1998). Likelihood estimation of the time of origin of cetacean and the time of divergence of cetacean and Artiodactyla. Paleo-electronica, 2, 1-47.

Glezer, I., Jacobs, M., Morgane, P. (1988). Implications of the "initial brain" concept for brain evolution in Cetacea, Behavioral and Brain Sciences, 11, 75-116.

Goertzel, B., Combs, A. (2010). Water Worlds, Naive Physics, Intelligent Life, and Alien Minds, Journal of Cosmology, 5, 897-904.

Hakeem, A.Y., Sherwood, C. C., Bonar, C.J., Butti, C., Hof, P.R., Allman, J.M. (2009). Von Economo neurons in the elephant brain, The Anatomical Record, 292, 242-248.

Herman, L. M. (2006). Intelligence and rational behaviour in the bottlenosed dolphin. In Hurley, S., Nudds, M. (Eds) Rational animals, Oxford University Press, Oxford, pp. 439-468.

Herman, L. M. (2002). Vocal, social, and self-imitation by bottlenosed dolphins. In . Dautenhahn, K., Nehaniv, C.L. (Eds)

Imitation in animals and artifacts, MIT Press, Cambridge, Mass., pp. 63-108.

Herman, L. M. (1986). Cognition and language competencies of bottlenoseddolphins. In Schusterman, R.J., Thomas, J., Wood, F.G. (Eds), Dolphin Cognition and Behaviour: A Comparative Approach, Lawrence Erlbaum Associates, Hillsdale, New Jersey, pp. 221-251.

Herman, L. M., Kuczaj, S. A. II & Holder, M. D. (1993). Responses to anomalous gestural sequences by a language-trained dolphin: evidence for processing of semantic relations and syntactic information. Journal of Experimental Psychology: General, 122, 184 - 194.

Herman, L.M., Matus, D.S., Herman, E.Y.K., Ivancic, M., Pack, A.A. (2001). The bottlenosed dolphin's (Tursiops truncatus) understanding of gestures as symbolic representations of its body parts. Animal Learning & Behaviour 29, 250-264.

Herman, L. M., Pack, A. A. & Wood, A. M. (1994). Bottlenosed dolphins can generalize rules and develop abstract concepts. Marine Mammal Science, 10, 70-80.

Herman, L, M., Richards, D. G., & Wolz, J. P. (1984). Comprehension of sentences by bottlenosed dolphins. Cognition, 16, 129-219.

Herzing, D.L. (2010). SETI meets a social intelligence: Dolphins as a model for real-time interaction and communication with a sentient species. Acta Astronautica, 67, 1451-1454.

Hof, P., Chanis, R., Marino, L. (2005) Cortical complexity in cetacean brains, The Anatomical Record, 287A, 1142-1152.

Hof, P.R., Van Der Gucht, E. (2007) 'The structure of the cerebral cortex of the humpback whale, Megaptera novaeangliae (Cetacea, Mysticeti, Balaenopteridae)', The Anatomical Record, 290, 1-31.

Jacobs, M. S., McFarland, W. L., Morgane, P. J. (1979) The anatomy of the brain of the bottlenose dolphin (Tursiops truncatus). rhinic lobe (rhinencephalon): the archicortex, Brain Research Bulletin, 4, pp1-108.

Janik, V. M. & Slater, P. J. B. (1997). Vocal learning in mammals. Advances in the Study of Behaviour 26, 59-99.

Jerison, H. J. (1973). Evolution of the Brain and Intelligence. Academic Press, New York.

Jerison, H. J. (1986). The perceptual world of dolphins. In Schusterman, R.J., Thomas, J.A., Wood, F.G. (Eds) Dolphin Cognition and Behaviour: A Comparative Approach, Lawrence Erlbaum, New Jersey, pp. 141-166.

Kilian, A., Yaman, S., Von Fersen, L. & Gunturkun, O. (2003). A bottlenosed dolphin discriminates visual stimuli differing in numerosity. Learning and Behaviour, 31, 133-142.

Krützen, M., Mann, J., Heithaus, M. R., Connor, R. C., Bejder, L. Sherwin, W. B. (2005). Cultural transmission of tool use in bottlenosed dolphins. Proceedings of the National Academy of Sciences USA, 102, 8939-8943.

Ladygina, T. F., Mass, A. M., Supin, A. I. (1978). Multiple sensory projections in the dolphin cerebral cortex. Zh. Vyssh. Nerv. Deiat 28:1-47-1054.

Lilly, J. C. (1969). The Mind of the Dolphin: A Nonhuman Intelligence, Avon, New York.

Lilly, J. C. (1987). Communication Between Man and Dolphin, Julian Press, New York.

Lusseau, D. (2007). Evidence for social role in a dolphin social network. Evolutionary Ecology, 21, 357-366.

Marcoux, M., Rendell, L., Whitehead, H. (2007). Indications of fitness differences among vocal clans of sperm whales. Behavioural Ecology and Sociobiology 61, 1093-1098.

Mann, J., Sargeant, B.L., Watson-Capps, Q.A., Heithaus, M.R., et al. (2008). Why do dolphins carry sponges? PLoS One 3(12), e3868. Doi: 10.1371journal.pone.0003868.

Marino L (2009). Brain size evolution. In Perrin, W.F., Wursig, B., Thewissen, J.G.M. (Eds), Encyclopedia of Marine Mammals (2nd ed), Academic Press, San Diego, CA., pp. 149-152.

Marino, L. (1998). A comparison of encephalization between odontocete cetaceans and anthropoid primates. Brain, Behaviour and Evolution 51, 230-238.

McCowan, B., Doyle, L.R., Hanser, S.F. (2002). Using information theory to assess the diversity, complexity and development of communicative repertoires. Journal of Comparative Psychology, 116 166-172.

McCowan, B., Hanser, S.F., Doyle, L.R. (1999). Quantitative tools for comparing animal communication systems: information theory applied to bottlenose dolphin whistle repertoires. Animal Behaviour, 57, 409-419.

McCowan, B., Marino, L., Vance, E., Walke, L., Reiss, D. (2000). Bubble ring play of bottlenose dolphins: Implications for cognition. Journal of Comparative Psychology, 114, 98-106.

Mercado, III, E. M., Killebrew, D. A., Pack, A. A., Macha, IV, B. & Herman, L. M. (2000). Generalization of same-different classification abilities in bottlenosed dolphins. Behavioural Processes, 50, 79-94.

Mercado III, E., Murray, S.O., Uyeyama, R.K., Pack, A.A. & Herman, L.M. (1998). Memory for recent actions in the bottlenosed dolphin (Tursiops truncatus): repetition of arbitrary behaviours using an abstract rule. Animal Learning and Behaviour, 26, 210-218.

Milinkovitch, M. C. (1992). DNA-DNA hybridizations support ungulate ancestry of Cetacea. Journal of Evolutionary Biology, 5, 149-160.

Mitchell, E.D. (1989). A new cetacean from the late Eocene La Meseta Formation, Seymour Island, Antarctica Peninsula. Canadian Journal of Fisheries and Aquatic Sciences, 46, 2219-2235.

Nishikawa, K. C. (2002). Evolutionary convergence in nervous systems:Insights from comparative phylogenetic studies. Brain, Behavior and Evolution, 59, 240-249.

Oelschlager, H.A. Oelschlager, J.S. (2009). Brains. In Perrin, W.F., Wursig, B., Thewissen, J.G.M. (Eds),Encyclopedia of Marine Mammals (2nd ed), Academic Press, San Diego, CA., pp. 134-149.

Pack, A. A. & Herman, L. M. (1995). Sensory integration in the bottlenosed dolphin: Immediate recognition of complex shapes across the senses of echolocation and vision. Journal of the Acoustical Society of America 98, 722-733.

Pack, A.A., Herman, L.M., Hoffmann-Kuhnt, M. & Branstetter, B.K. (2002). The object behind the echo: Dolphins (Tursiops truncatus) perceive object shape globally through echolocation. Behavioural Processes,58, 1-26.

Reiss, D., Marino, L., (2001). Self-recognition in the bottlenose dolphin: A case of cognitive convergence. Proceedings of the National Academy of Sciences USA, 98 (10), 5937-5942.

Reiss, D., McCowan, B. (1993). Spontaneous vocal mimicry and production by bottlenose dolphins (Tursiops truncatus): Evidence for vocal learning. Journal of Comparative Psychology 107, 301-312.

Rendell, L.E., Whitehead, H. (2003). Vocal clans in sperm whales (Physeter macrocephalus). Proceedings of the Royal Society of London B - Biological Sciences, 270, 225-231.

Rendell, L.E., Whitehead, H. (2001). Culture in whales and dolphins. Behavioural and Brain Sciences, 24, 309-324.

Richards, D., Wolz, J. & Herman, L. M. (1984). Vocal mimicry of computer-generated sounds and vocal labeling of objects by a bottlenosed dolphin, Tursiops truncatus. Journal of Comparative Psychology, 98, 10-28.

Ridgway, S. H. (2000). The auditory central nervous system of dolphins. In Au, W., Popper, A., Fay. R. (Eds) Hearing in Whales and Dolphins, NY: Springer Verlag, p. 273-293.

Ridgway, S. H.,Brownson, R. H. (1984) Relative brain sizes and cortical surface areas of odontocetes, Acta Zoologica Fennica, 172,149-152.

Ridley, M. (1993). Evolution. Blackwell Scientific Publications, Oxford.

Savage-Rumbaugh, E. S., Murphy, J., Sevcik, R. A., Brakke, K. E., Williams, S. L. & Rumbaugh, D. M. (1993). Language comprehension in ape and child. Monographs of the Society for Research and Child Development, serial no. 233, 58, n.3-4, 1-222.

Sayigh, L. S., Tyack, P. L., Wells, R. S., Solows, A. R., Scott, M. D. & Irvine, A. B. (1999). Individual recognition in wild bottlenosed dolphins: a field test using playback experiments. Animal Behaviour, 57, 41-50.

Schulz, T., Whitehead, H., Gero, S., Rendell, L. (2008). Overlapping and matching of codas in vocal interactions between sperm whales: insights into communication function. Animal Behaviour 76, 1977-1988

Supin, A.Y., Mukhametov, L. M., Ladygina, T. F., Popov, V. V., Mass, A. M., Poliakova, I. G. (1978). Electrophysiological studies of the dolphin's brain. Izdatel'ato Nauka, Moscow.

Smith, J.D., Schull, J., Strote, J., McGee, K., Egnor, R. & Erb, L. (1995). The uncertain response in the bottlenosed dolphin (Tursiops truncatus). Journal of Experimental Psychology: General 124, 391-408.

Thewissen, J.G.M. (2009). Sensory biology: overview. In Perrin, W.F., Wursig, B., Thewissen, J.G.M. (Eds),Encyclopedia of Marine Mammals (2nd ed), Academic Press, San Diego, CA., pp. 1003 - 1005).

Tyack, P. L. (1997). Development and social functions of signature whistles in bottlenosed dolphins Tursiops truncatus. Bioacoustics, 8, 21-46.

Tyler, R. (2010). Water Worlds and Oceans May be Common in the Universe Robert Tyler, Ph.D., Applied Physics Laboratory, and Dept. of Earth and Space Sciences, University of Washington, Washington. Journal of Cosmology, 5, 959-970.

Xitco, J. Jr., Gory, J.D., Kuczaj II, S. A. (2001). Spontaneous pointing by bottlenosed dolphins (Tursiops truncatus). Animal Cognition, 4, 115-123.

Xitco, J. Jr., Gory, J.D., Kuczaj II, S. A. (2004). Dolphin pointing is linked to the attentional behaviour of a receiver. Animal Cognition,7, 231-238.

Yurk, H., Barrett-Lennard, L., Ford, J. K. B. & Matkin, C. O. (2002). Cultural transmission within maternal lineages: Vocal clans in resident killer whales in southern Alaska. Animal Behaviour, 63, 1103-1119.

27. Consciousness in Cephalopods?
Jennifer Mather, Ph.D.

Department of Psychology, University of Lethbridge, Lethbridge, AB Canada

Abstract

Behavioral evidence suggest that cephalopods have consciousness, but what might this contain? As they are asocial, they likely do not have awareness of conspecifics. With major neural allocation to peripheral control, especially in the arms of octopuses, they might be only generally conscious of movement and its control. The spectacular skin display system could be mainly open loop, though this has not been well investigated. However, cephalopods monitor and remember their position in space, a parallel with mobile vertebrates and insects. Similarly, their very flexible foraging and food finding, which needs awareness, probably is similar to that of many vertebrates. Despite these abilities, future evolution of this array of competencies may be limited because of the cephalopod asociality.

1. Introduction

Do cephalopod's have consciousness? Recently, Mather (2008) put forward behavioral evidence for a simple form of consciousness in cephalopod mollusks: octopuses, cuttlefish and squid. Similar brain areas to those in 'higher' vertebrates cannot, of course, be found in cephalopods (though see Wells, 1978; Nixon & Young, 2003 on their brains). But behavioral evidence can give us a basis for asking what an alternate evolutionary path to consciousness might be like, and note that Bekoff, Allen and Burghardt (2002) stress the necessity of looking at the adaptive value of intelligence of the particular kind that is ecologically useful for each animal. This paper takes Mather (2008) forward a step to help answer this question, focusing on content and processes of possible cephalopod consciousness.

Why should cephalopods have cognition and even consciousness? Evolutionary pressure might have come in the development of the coleoid cephalopods (most of the present ones, with the exception of the nautiluses) when they lost the protective molluscan shell. Living in the tropical near-shore, one of the most complex environments in the world, may have pressured the drive to develop intelligence (see Richardson, 2010; Godfrey-Smith, 2002; Kashtan, Noor & Alon, 2007). In addition, Packard (1972) suggested that the coleoids evolved at the same time as the bony fishes and in competition with them for ecological niches. Grasso and Basil (2009), on the basis of newly-discovered learning capacity in nautiluses, believe that the early cephalopods might have been pre-adapted for this ability. While structural parallels with vertebrates are impossible, cephalopods do have big brains (Nixon & Young, 2003). Big brains do not automatically denote complex behavior but they allow for this capacity (Jerison, 2002), and Kortscal e al., (1998) point out the correlation of large brain size with complex marine habitats in fishes. Cephalopods have brain areas (vertical and subfrontal lobes) that are associated with visual learning (Wells, 1978) and spatial memory (Alvez, Boal & Dickel, 2008). Brain size and allocation to different functions varies across the sub-class, as octopuses have 3/5 of their neurons in the arms and much brain area allocated to motor control, and squid have large optic lobes (see Grasso & Basil, 2009). Surprisingly, the deep-sea specialized Vampyroteuthis has a very large vertical lobe (Nixon & Young, 2003), and the function of this allocation is unknown.

2. Content

There are several ways in which simple consciousness in cephalopods does not involve the same content as that of vertebrates. They have little awareness about or behavior addressed to conspecifics. Humphrey (1976) historically felt that social cognition, or the picking up of clues about what other members of your species might be doing, could be the foundation of vertebrate consciousness. Recent research on 'theory of mind' in some mammals and birds has reinforced this, as they appear aware of what other individuals are 'thinking' (Whiten & Byrne, 1988). Cephalopod relationships range from apparently asociality in octopuses (though Abdopus is at least a partial exception, see Huffard, et al, 2008, 2010), through vaguely social cuttlefish and to schooling squid (Boal, 2006). Although Fiorito and Scotto (1992) demonstrated octopuses making the same choice as another that they viewed, this observation of conspecific influence has not been extended. A series of studies by Boal (1996) using chemical cues found no evidence of individual recognition in cuttlefish. They maintained dominance hierarchies, which is common in otherwise solitary animals in laboratory situations, but showed no evidence of recognizing other members of the hierarchy. Octopuses' poor performance on Gallup's (1970) mirror test (Mather & Anderson, 2011) supports a lack of visual recognition of individual conspecifics, including oneself.

Another way in which the cephalopods may differ from vertebrates in content of consciousness is in a lack of conscious monitoring of motor output. For example, with 3/5 of their neurons in their arms, octopuses have much local organization and monitoring. This distribution of the nervous system is probably due to their lack of a fixed skeleton. Extensive neural capacity is needed for coordination of the ensuing hydraulic system of flexible allocation of muscles for support and contraction (Kier & Smith, 1985), as well as the major role of the suckers in exploration of the landscape. Manipulation of arm and sucker musculature to perform tasks is very sophisticated (Grasso, 2008), but we do not know how much information is dealt with at the local level and how much is either controlled by the brain or sent as information to it. However, motor output is not somatotopically represented in the motor centres (Zullo, Sumbre, Agnisola, Flash & Hochner, 2009).

If cephalopods are not social and do not routinely monitor what their arms are doing, why might they have evolved intelligence and consciousness? A clue may be in the demanding environment in which they live, that they must make 'decisions' to find prey and avoid predation in an environment that is not only complex but also constantly changing, and see Godfrey-Smith (2002) for the suggestion that environmental complexity might drive the evolution of intelligence.

One unique cephalopod system whose content could be monitored and output planned is the spectacular display complex (Packard, 1995; Messenger, 2001). Octopuses, cuttlefish and squid use pigmented and reflective color, skin texture changes and posture to match any nearby background, with the exception of color mixtures that these color-blind animals cannot distinguish (Kelman, Osorio & Baddeley, 2008). Yet there has been discussion as to whether this system is open loop, and an example is the counter-shading of cuttlefish, which is reflexive and totally dependent on body position (Ferguson, Messenger & Budelmann, 1994). Is this flexible appearance too complex to monitor, and what is it useful for besides camouflage?

Several different authors have found considerable sophistication in the use of the skin display system to foil predators. The cuttlefish camouflage system reveals sophistication in perceptual assessment (Kelman et al. 2008). Cephalopods commonly flee from potential predators (Mather, 2010) but recruit the skin display system if threat is lesser. In the laboratory, Langridge, Broom & Osorio (2009) found that increased threat from a potentially predatory fish caused cuttlefish to escalate the display system, from a general background resemblance to a clear match, then to the appearance of deimatic eyespots which may make the animal appear threatening. But cuttlefish did not use these displays to non-visual predators such as crabs and dogfish. In the field, Mather (2010) saw a similar differentiation of responses by Caribbean reef squid to approaching fish species. Common but herbivorous parrotfish mostly elicited an 'annoyance' striped Zebra display or deimatic eye spots and were allowed to approach closely, whereas potentially predatory bar jack and yellowtail snapper were sometimes given an eyespot but often a jet-propelled escape response and triggered an escape from further away. Squid could produce eyespots on four corners of the dorsal mantle, and spots were differentially 'aimed' at the approaching fish and not at neighboring squid. Such directionality is also true for the eyespots of the cuttlefish, and not the camouflage coloration (Langridge, 2006). Whether it is conscious or not, much calculation must have programmed responses, based on a lifetime of experience with the particular predator species. In squid, an agonistic display can be addressed on one side and a sexual one on the other at the same time to the appropriate partners (Greibel & Mather, 2003), a 'double signalling' that is unique to cephalopods.

In contrast, Hanlon and colleagues (1999) pursued escaping Hawaiian octopuses across the shallow coral reef and watched them assume an apparently random sequence of display patterns, escape jets and emission of ink clouds. Such a sequence might not be programmed, although changes would break a predator's search image (see Treisman, 1988, for research on humans) and foil pursuit nevertheless. Similarly, there is a series of casual observations on octopuses apparently mimicking the appearance of various poisonous fish and sea snakes. In the best-described study (Hanlon et al., 2010), the long-armed octopus appeared to mimic flatfish swimming and behavior. As mimicry is in the eyes of the beholder, such appearances may not have been attempts to consciously look like other species, and Hanlon et al (2010) point out that this could also serve to place the octopus in an advantageous position to hunt prey in the sand (and see foraging behavior).

A display system which evolved to produce camouflage could be used to communicate with conspecifics, and striking sexual skin displays are found in many cephalopod species. Squid of several species (Hanlon, Smale & Sauer, 1994: Moynihan, 1985; Hanlon & Messenger, 1996; Mather, 2004) have a clear repertoire of major displays such as stripes, bars, pale and dark colors, along with peripheral ones like fin stripes, teardrop around the eyes and arm stripes. Giant cuttlefish 'sneaker' males can display a pattern resembling that of a mature female and fool a guarding male, thereby having an opportunity to mate with the female (Norman, Finn & Tregenza, 2001). Juvenile Caribbean reef squid display a deceptive agonistic Zebra when another larger male is courting a female, and the male attends to them and is temporarily prevented from mating (Mather, 2004). The complexity of the Caribbean reef squid display led Moynihan (1985) to suggest that the skin display system was actually a language, with central skin areas taking the role of nouns and verbs, and peripherals as adjective and adverb ones. However, both Hanlon and Messenger (1996) and Mather (2004) state that the recruitment of peripheral information is basically escalation and that the parallel with language is unrealistic.

With a system of such complexity, it is sometimes difficult to see adaptive use of a specific pattern. Many cephalopods make 'passing clouds' on the skin; Packard and Sanders (1971) first described them for common octopuses, and Mather and Mather (2004) investigated them in more detail in Hawaiian octopuses. A dark patch with contrasting pale margins formed on the dorsal mantle, 'moved' forward across the head and down the outstretched arms with web extended between them, 'flowing' off the edge. The apparent target was a crab that froze, as these clouds formed after an octopus had made an unsuccessful web-over capture attempt. In addition displays were somewhat directional, mostly forward but 'aimed' over a 90 degree range and sometimes unilateral. Note the perceptual manipulation, as the octopus is able to move the cloud by apparent movement without moving itself, thereby preventing retinal slip in its eyes and assuring that its vision is not blurred. Both the timing and the aim suggest that this is by no means an automatic response (and see foraging strategies).

3. Processes

Regardless of the content, any animal will use several processes to manipulate information, and cephalopods demonstrate their intelligence in finding prey and avoiding being consumed themselves. It is difficult to define where processing of information would enter consciousness, certainly it is when choices are made about actions and when an event category or Piagetian schema (see Zacks & Tversky, 2001) is constructed, used and modified, then used again (see Figure 1). Octopuses in particular are learning specialists (Mather, 1995; Hochner, Shomrat & Fiorito, 2006), but much of the research on their learning has been carried out in simplistic and highly controlled situations. West-Eberhardt (2003) reminds us that learning follows the sequence exploration-learning-forgetting- learning, and see Shettleworth (2010) on vertebrate exploration. Few explicit studies have been done on exploration by cephalopods but Boal and colleagues (2000) noted that an octopus placed in a new enclosure spent much time the first few hours exploring, as did cuttlefish (Karson, Boal & Hanlon, 2000). Mather and Anderson (1999) found that octopuses given new items first explored them and then some of the animals later showed motor play.

An external event forces an animal to do what Baars (1997) described as 'shine an attentional spotlight' on the situation. At the opening of the top of their tank, octopuses have a variety of reactions--colour changes, head bobs that are thought to generate motion parallax and construct a better three-dimensional image of the environment, moves or shrinking back from the stimulus (Mather & Anderson, 1999). Similarly, cuttlefish exposed to small local water movements change body pattern, move, orient towards the stimulus, swim away or even burrow in the sand (Komak, Boal, Dickel & Budelmann, 2005). When threatened by an overhead visual stimulus, the cuttlefish showed physiological changes, as well as freezing and hyperinflation of the mantle, which would ready them for a jet-propelled escape response (King & Adamo, 2006). Some of these responses allow the animal to gain more information about the situation and others are preparatory for action. Such information retrieval leads the animal to make 'choices' and 'plans' for behavior in situation such as cuttlefish and octopus maze navigation (Alvez et al. 2008). Feedback about a situation leads to modification of a behavior pattern. Boycott (1954) reported that an octopus stung by a sea anemone on the shell of a hermit crab tried different 'cautious' approaches from different angles, apparently to acquire the prey but avoid the anemone.

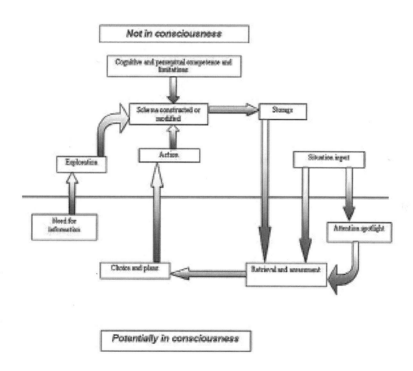

Fig. 1. Steps in processing and using information, likely not in consciousness and potentially in consciousness.

One of the ways in which the content of cephalopod consciousness may be similar to that of animals of other phyla is in calculating and using information about their position in the environment (see Shettleworth, 2010). Octopuses shelter in a protective home, and utilization of unusual shelter such as beer bottles (Anderson, Hughes, Mather & Steele, 1999) and coconut shells (Finn, Tregenza & Norman, 2009) may allow them to live in otherwise unsuitable habitat. While they have size and shape preferences, the ideal shelters are not always available. Instead octopuses explore a prospective 'home', clean out a crevice or dig under a rock, pulling and pushing rubble with the arms and blowing sediment out with jets of water from their funnel (Mather, 1994). The octopus may also bring shells or rocks to block the entrance, and there is a significant correlation between the size of the entrance and the number of rocks the octopus has brought to block it. Both the water jetting and the rock stacking meet Beck's (1980) definition of tool use, although tool use does not automatically denote consciousness. Still, the octopus appears to have a 'mental image' of an appropriate home before it begins to dig.

Like many mobile animals (Shettleworth, 2010), cephalopods have navigational ability and, in some circumstances, spatial memory (Alves et al. 2008). In the wild, octopuses move from their central home on foraging paths across the sea bottom (Mather, 1991a; Forsythe & Hanlon, 1997; Leite, Haimovici & Mather, 2009). Chemical cues are not used in trail-following, as octopuses jet through the water and return from a different direction at the end of a hunting trip. They either retrieve stored information about the location of the home via landmarks (visual memory), or remember the turns of their outward track (path navigation) to choose directions. Over several days, juvenile octopuses in Bermuda foraged in different areas of the nearby sea bottom, thus apparently also having procedural memory of the direction in which they had moved in the last few days (Mather, 1991a) and integrating it into a 'win-switch' foraging strategy. Such information is stored in memory as Mather (1991a) recorded detours when octopuses were out foraging; when displaced from their path, they returned directly home from a different direction.

This spatial memory ability has been found in tests in the laboratory which, by definition, must be simpler than the natural environment. Detour experiments with octopuses by Wells (1964) using visual cues seen through glass, had mixed results as octopuses seemed to need to crawl along the wall of the maze, keeping it fixated with one eye. Animals were successful in visually-guided single-turn mazes constructed by Walker et al., (1970), and Boal et al., (2000). Karson et al (2003) extended this paradigm (escape to remembered deep areas when the water level sank) to cuttlefish, and Hvoreckny et al (2007) showed that both species could solve two different mazes that needed different cues when trials were intermixed, proving their ability to make choices of relevant sensory cues.

Sophisticated assessment has allowed researchers to investigate whether cephalopods use a path navigation (monitoring one's turns) or a visual one (memorizing one's position with respect to a landmark). Mather (1991a) trained octopuses to learn to orient to a visual landmark. Karson, Boal & Hanlon (2003) and Alves, Chichery, Boal & Dickel (2007) found that cuttlefish could use either proximal or distal visual cues for such orientation. Those trained with distal cues relied on path navigation and those using proximal cues used both strategies equally often. These results parallel cue usage in vertebrates and insects (Shettleworth, 2010), and suggest a common utilization of similar information in these totally unrelated animals. Interestingly, lesions of the ventral area of the vertical lobe led to deficits in spatial learning (Graindorge, Alves, Darmaillacq, Chichery, Dickel & Belanger, 2006), suggesting this is a parallel area to the mammalian hippocampus. One area of life in which cephalopods, like other animals, may use their simple consciousness is in hunting for and capturing prey. Cephalopods have a wide array of hunting techniques, described by Hanlon & Messenger (1996) as ambushing, stalking, pursuit, speculative hunting and luring. Luring has been suggested when cephalopods wave their arms before an area that potentially contains prey, though this behavior could also be a distraction. More convincing, though not proven by controlled trials, is the behavior of sepiolid squid buried in the sand who stick out a paled arm tip from the sand and wiggle it (Anderson, Mather & Steele, 2004). This resembles the luring of the angler fish and the actions must have some purpose. It is not just that the cephalopods use several foraging strategies, but that individual species and probably individual animals do.

Hanlon and Messenger (1996) present the case of the Caribbean reef squid, which uses ambush from floating seaweed, stalking small fish in the open water and pursuit of larger ones, speculative hunting by touch on the sand bottom to contact shrimp or other buried prey, and possible luring also near the sea bottom. A particularly good example of speculative hunting was observed by the author, when eels were foraging in the coral rubble and squid followed them, ready to snap up escaped prey (and fish do this to octopuses, see Mather, 1992). This activity must have been both learned with a food reward and planned. Long-term learning plays a major role in cuttlefish hunting. Initially newly hatched cuttlefish have a narrow search image for small mysid crustaceans (Wells, 1962; Messenger, 1977; Dickel, Chichery & Chichery, 2001). After a couple of weeks of life, during which the vertical lobe brain area that stores learned information begins to grow, cuttlefish learn to take alternate prey. But visual exposure to the sight of crabs when the cuttlefish are still in their opaque egg cases changes the preference and they accept crabs immediately after hatching (Darmaillacq, Chichery, Shashar & Dickel, 2006). This is a clear parallel to vertebrate social imprinting, early and relatively permanent learning of one's species (Staddon, 1983).

Learning also affects cuttlefish prey capture behavior. Small prey are captured with the extension of the flexible tentacles and larger ones such as crabs are grabbed by all the arms. Naive cuttlefish often attack crabs frontally and consequently are pinched by the claws; they learn to circle around from behind (Wells, 1962) and avoid this defense. Such trial-and-error learning was not improved by social learning when watching another cuttlefish successfully attack crabs (Boal, Wittenberg & Hanlon, 2000), as control cuttlefish exposed only to the odour of crabs were equally successful. The authors suggest that chemical cues might trigger arousal, in this case focusing Baars (2007) attentional spotlight on the situation.

Prey preference leads to acquisition, and octopuses have been described as generalist predators because they capture a wide variety of molluscan and crustacean prey species (Ambrose, 1984; Mather, 1991b; Leite et al, 2009). This lack of choice suggest an unselective generalist predator that is simply capturing any prey item that it encounters as it moves across the sea floor. But a closer look shows that octopuses move to likely locations, probably visually guided, before beginning intense chemotactile search with the arms, a procedure called saltatory search (O'Brien, Evans & Browman, 1989) and found in foraging birds and fish. Such cooperation of information acquisition by different senses, guided by learned information about likely habitat, must have involved storage and retrieval of information and choice of likely areas as well as memorization of areas previously hunted (see navigation). Further, while a population of octopuses in Bonaire selected a wide variety of prey species, individuals had a much narrower range in a varied habitat (Anderson, Wood & Mather, 2008), probably having learned specific prey locations or mastered chosen penetration techniques. The species was a generalist but individual were specialists. Scheel, Lauster & Vincent (2007) evaluate prey availability and capture, and suggest that giant Pacific octopuses in Alaska may be rate-maximizing foragers. Capture of prey does not mean that consumption can begin immediately, especially if the prey is hard-shelled molluscs. Octopuses have three methods of penetration into the shells of clams: pulling the valves apart, drilling a hole into the shell and injecting a poison that weakens the clam's muscles that hold the valves together, or chipping a small piece off the shell margin to allow the same injection. Drill holes are located at different areas of the shell in different prey species, often near the adductor muscle insertions or over the heart of the clam (Nixon & Macconachie, 1988), and the locations are probably learned. Giant Pacific octopus use different techniques with differential success on several clam prey species. Using a schema of procedures, they first try pulling by trial-and-error; if that is not successful, they drill or chip (Anderson & Mather, 2007). Yet each penetration technique also need a different orientation of the clam within the arms of the octopus, out of sight. Pulling is effective when the umbo of the clam is towards the mouth, chipping when the anterior or posterior margin is placed at the same location and drilling when the side of one valve of the clam is pressed towards the salivary papilla near it. This schema must pair proper orientation of the clam with the penetration procedure used to gain access to it. These studies of process bring us back to content. How much does the brain know about what the arms are doing, and how detailed are the output commands? An autotomized arm of a pygmy octopus, separated from the body, can pick up pieces of food and even 'walk' down an outstretched hand. This may be localized processing, and physics can explain the sequence of actions in some arm movements (Guttfreund et al, 1996). But the arms carry out the penetration preparation mentioned above, and those cannot be autonomous choices. Arms 'walk', and the gait can vary depending on direction of travel as well as whether the octopus needs to be camouflaged as drifting seaweed (Huffard, 2006). Arms reach out for prey, and each individual octopus has a 'favourite' arm to extend into a tunnel to obtain a food reward (Byrne, Kuba, Meisel, Greibel & Mather, 2006). Clearly we need to investigate this very different central-peripheral allocation and cooperation to find out how much information is consciously processed.

Are cephalopods conscious? The evidence presented above suggest that they learn, store schemas, attend, recruit information and actions to fit particular choices and attend to specific aspects of a situation when they are relevant (and see Figure 1). Cephalopods studied so far seldom have Fixed Action Patterns. The exception seems to be digging in sand by sepioids (Mather, 1986; Anderson et al, 2004), using water jets from the funnel to push sand away and throwing sand over their dorsal surface to complete the coverage. Even in this situation, digging is just one choice of several anti-predator actions (Langridge et al, 2009), chosen after chemotactile inspection and with much internal flexibility when investigated in detail (Mather, 1986).

Where might this consciousness develop in thousands of years? Cephalopods are at a metabolic dead end, having made the haemocyanin-based mollusk respiratory system as efficient as possible, with the assistance of three hearts to speed circulation. So they will not likely invade the land. How might their cognitive ability develop? There is no easy passage of information with the overlap of the generations, as in mammals—only octopuses tend young, and only the eggs. Cephalopods are not social, do not cooperate. The skin display system is a magnificent one, its technical capabilities far beyond the sophistication of its use. Squid live in groups and the Caribbean reef squid in small ones, perhaps they will begin to use the display system more nearly to its capacity. Right now the squid have 'nothing important to say' with it. If they become more social, perhaps they will evolve to use it better. Until then, the very solitude of cephalopods blocks their cognitive evolution and further development of consciousness.

References

Alves, C., et al. (2008). Short-distance navigation in cephalopods: a review and synthesis. Cogn Process, 9, 239-247.

Alves, C., et al. (2007). Orientation in the cuttlefish Sepia officinalis: response versus place learning. Anim Cogn, 10, 29-36.

Ambrose, R. F. (1984). Food preferences, prey availability, and the diet of Octopus bimaculatus Verrill. J Exp Mar Biol Ecol, 77, 29-44.

Anderson, R. C., et al. (2004). Determination of the diet of Octopus rubescens Berry 1953 (Cephalopoda, Octopodidae) through examination of its beer bottle dens in Puget Sound. J Moll Stud, 66, 417-419.

Anderson, R. C., Mather, J. A. (2007). The packaging problem: Bivalve mollusc prey selection and prey entry techniques of Enteroctopus dofleini. J Comp Psych, 121, 300-305.

Anderson, R. C., et al. (2004). Burying and associated behaviors of Rossia pacifica (Cephalopoda, Sepiolidae). Vie Mileu, 54, 13

Anderson, R. A., et al. (2008). Octopus vulgaris in the Caribbean is a specializing generalist. Mar Ecol Prog Ser, 371, 199-202.

Baars, B. J. (1999). In the theatre of consciouness: Global workspace theory: A rigorous scientific theory of consciouness. J Consc Stud, 4, 292-309.

Beck, B. (1980). Animal Tool Behavior: The Use and Manufacture of Tools by Animals. Garland, New York, US.

Bekoff, M., et al. (2002) The Cognitive Animal. MIT Press, Cambridge, MA, US.

Boal, J. G. (1996). Absence of social recognition in laboratory reared cuttlefish, Sepia officinalis, L. (Mollusca: Cephalopoda). Anim Behav, 52, 529-537.

Boal, J. G. (2006). Social recognition: a top down view of cephalopod behaviour. Vie Mileu, 56, 69-79.

Boal, J. G., et al. (2000). Experimental evidence for spatial learning in octopuses (Octopus bimaculoides). J Comp Psych, 114, 246-252.

Boal, J. G., et al. (2000). Observational learning does not explain improvement in predation tactics by cuttlefish (Mollusca: Cephalopoda). Behav Proc, 52, 141-153.

Boycott, B. B. (1954). Learning in Octopus vulgaris and other cephalopods. Publ Staz Zool Napoli, 25, 67-93.

Byrne, R. A., et al. (2006). Does Octopus vulgaris have preferred arms? J Comp Psych, 120, 198-204.

Darmaillacq, A-S., et al (2006). Early familiarization overrides innate prey preference in newly-hatched Sepia officinalis cuttlefish. Anim Behav, 71, 511-514.

Dickel, L., et al. (2001). Increase of learning abilities and maturation of the vertical lobe complex during postembryonic development in the cuttlefish, Sepia. Dev Psychobiol, 40, 92-98.

Ferguson, G. P., et al (1991). A countershading reflex in cephalopods. Proc Roy Soc London B, 243, 63-67.

Finn, J. K., Tregenza, T., Norman, M. D. (2009). Defensive tool use in a coconut-carrying octopus. Curr Biol, 19, R1069-R1070.

Forsythe, J. W., Hanlon, R. T. (1997). Foraging and associated behavior by Octopus cyanea Gray, 1849 on a coral atoll, French Polynesia. J Exp Mar Biol Ecol, 209, 15-31.

Gallup, G. G. Jr. (1970). Chimpanzees: Self-recognition. Science, 167, 86-87.

Godfrey-Smith, R. (2002). Environmental complexity and the evolution of cognition. In Sternberg, R. J., Kaufman, J. C. (Eds.), The Evolution of Intelligence, Erlbaum, Mahwah, NJ, US, pp. 223-250.

Graindorge, N., et al. (2006). Effects of dorsal and ventral electrolytic lesions on spatial learning and locomotor activity in Sepia officinalis. Behav Neurosci, 120, 1151-1158.

Grasso, F. W. (2008). Octopus sucker-arm coordination in grasping and manipulation. Am Malacol Bull, 24, 13-23.

Grasso, F. W., Basil, J. A. (2009). The evolution of flexible behavioral repertoires in cephalopod molluscs. Brain Behav Evol, 74, 231-245.

Griebel, U., Mather, J. A. (2003). Double signalling in Sepioteuthis sepioidea. Presented at the CIAC symposium, Phuket, February.

Gutfreund, Y., et al (1996). Organization of octopus arm movements: A model system for studying the control of flexible arms. J Neurosci, 16, 7297-7307.

Hanlon, R. T., Messenger, J. B. (1996). Cephalopod Behaviour. Cambridge University Press, Cambridge, UK.

Hanlon, R. T., et al. (2010). A "mimic octopus" in the Atlantic: Flatfish mimicry and camouflage by Macrotritopus defilippi. Biol Bull, 218, 15-24.

Hochner, B., Shomrat, T., Fiorito, G. (2006). The octopus: a model for comparative analysis of the evolution of learning and memory. Biol Bull, 210, 308-317.

Huffard, C. L. (2006). Locomotion by Abdopus aculeatus (Cephalopods: Octopodidae): Walking the line between primary and secondary defenses. J Exp Biol, 209, 3697-3707.

Huffard, C. L., et al. (2008). Mating behavior of Abdopus aculeatus (d'Orbigny, 1834) (Cephalopoda: Octopodidae) in the wild. Mar Biol, 154, 353-363.

Humphrey, N. K. (1976). The social function of intellect. In: Bateson, P., Hinde, R. A. (Eds.), Growing Points in Ethology, Cambridge University Press, Cambridge, UK, pp. 303-317.

Hvoreckny, L. M., et al. (2007). Octopuses (Octopus bimaculoides) and cuttlefish (Sepia pharaonis, S. officinalis) can conditionally discriminate. Anim Cogn, 10, 449-459.

Jerison, H. J. (2002). On theory in comparative psychology. In: Sternberg, R. J., Kaufman, J. C. (Eds.). The Evolution of Intelligence, Erlbaum, Mahwah, NJ, US, pp. 251- 288.

Karson, M. A., et al. (2003). Experimental evidence for spatial learning in cuttlefish (Sepia officinalis). J Comp Psych, 117, 149-155.

Kashtan, N., Noor, E., Alon, U. (2007). Varying environments can speed up evolution. Proc Nat Acad Sci, 104, 13711-13716.

Kelman, E. J., et al. (2008). A review of cuttlefish camouflage and object recognition and evidence for depth perception. J Exp Biol, 211, 1757-1763.

Kier, W. M., Smith, K. K. (1985). Tongues, tentacles and trunks: The biomechanics of movement in muscular-hydrostats. Zool J Linn Soc, 83, 307-324.

King, A. J., Adamo, S. A. (2006). The ventilatory, cardiac and behavioral responses of resting cuttlefish (Sepia officinalis L.) to sudden visual stimuli. J Exp Biol, 209, 1101-1111.

Komack, S., et al. (2005). Behavioral responses of juvenile cuttlefish (Sepia officinalis) to local water movements. Mar Freshw Behav Physiol, 38, 117-125.

Kotrschal, K., et al. (1998). Fish brains: Evolution and environmental relationships. Rev Fish Biol Fish, 8, 373-408.

Langridge, K. V., Broom, M., Osorio, D. (2009). Selective signalling by cuttlefish to predators. Curr Biol, 17, R1044-R1045.

Leite, T. S., Haimovici, M., Mather, J. A. (2009). Octopus insularis (Octopodidae), evidences of a specialized predator and a time-minimizing forager. Mar Biol, 156, 2355- 2367.

Mather, J. A. (1986). Sand-digging in Sepia officinalis: Assessment of a cephalopod mollusc's "fixed" behavior pattern. J Comp Psych, 100, 315-320.

Mather, J. A. (1991a). Navigation by spatial

memory and use of visual landmarks in octopuses. J Comp Physiol A, 168, 491-497.

Mather, J. A. (1991b). Foraging, feeding and prey remains in middens of juvenile Octopus vulgaris (Mollusca: Cephalopoda). J. Zool, Lond, 224, 27-39.

Mather, J. A. (1992). Interactions of juvenile Octopus vulgaris with scavenging and territorial fishes. Mar Behav Physiol, 19, 175-182.

Mather, J. A. (1995). Cognition in cephalopods. Advances Stud Behav, 24, 316-353.

Mather, J. A. (2008). Cephalopod consciousness: Behavioral evidence. Consc Cogn, 17, 37-48.

Mather, J. A., Anderson, R. C. (1999). Exploration, play and habituation in Octopus dofleini. J Comp Psych, 113, 333-338.

Mather, J. A., Anderson, R. C. (2011) What do octopuses make of their image in a mirror? Ms in sub.

Messenger, J. B. (1977). Prey-capture and learning in the cuttlefish Sepia. Symp Zool Soc Lond, 38, 347-376.

Messenger, J. B.. (2001). Cephalopod chromatophores: Neurobiology and natural history. Biol Rev, 76, 473-528.

Moynihan, M. (1985). Communication and Noncommunication in Cephalopods. Indiana University Press, Bloomington, IN, US

Nixon, M., Young, J. Z. (2003). The Brains and Lives of Cephalopods. Oxford University Press, Oxford, UK.

O' Brien, W. J., Evans, B. I., Browman, H. I. (1989). Flexible search tactics and efficient foraging in saltatory searching animals. Oecologia, 80, 100-110.

Packard, A. (1972). Cephalopods and fish: The limits of convergence. Biol Rev, 47, 241- 307.

Packard, A. (1995). Organization of cephalopod chromatophore systems: A neuromuscular image generator. In: Abbott, N. J., Wiiliamson R., Maddock, L. (Eds.), Cephalopod Neurobiology, Oxford University Press, Oxford, UK, pp. 331-367.

Packard, A., Sanders, G. D. (1971). Body patterns of Octopus vulgaris and the maturation of the response to disturbance. Anim Behav, 19, 780-790.

Richardson, K. (2010). The Evolution of Intelligent Systems: How Molecules Became Minds. Macmillan, Houndsmill, UK.

Shettleworth, S. J. (2010). Cognition, Evolution, and Behavior. Oxford University Press, New York, US.

Staddon, J. E. R. (1983). Adaptive Behavior and Learning. Cambridge University Press, Cambridge, UK.

Treisman, A. M. (1988). Features and objects: The fourteenth Bartlett Memorial lecture. Quart J Exp Psychol, 40B, 201-223.

Walker, J. J., Longo, N., Bitterman, M. E. (1970). The octopus in the laboratory: Handling, maintenance and training. Beh Res Method Instrumen, 2, 15-18.

Wells, M. J. (1962). Early learning in Sepia. Symp Zool Soc Lond, 8, 149-169.

Wells, M. J. (1964). Detour experiments with Octopus. J Exp Biol , 41, 621-642.

West-Eberhardt, M. J. (2003). Developmental Plasticity and Evolution. Oxford University Press, Oxford, UK.

Whiten, A., Byrne, R. W. (1988). Tactical deception in primates. Beh Brain Sci, 11, 233-273.

Zullo, L., Sumbre, G., Agnisola, C., Flash, T., Hochner, B. (2009). Nonsomatoptopic organization of the higher motor centers in Octopus. Curr Biol, 19, 1632-1636.

28. Origins of Thought: Consciousness, Language, Egocentric Speech and the Multiplicity of Mind

Rhawn Joseph, Ph.D.

Emeritus, Brain Research Laboratory, California

Abstract

Consciousness is not a singularity, but a multiplicity. It is this multiplicity which makes self-consciousness (consciousness of consciousness) possible, and which provides the foundation for the development of thought which originates outside of consciousness. Thinking serves as a form of deduction and self-explanation, where one aspect of the mind explains its thoughts to another realm of mind. Thinking can be visual, imaginal, tactile, musical, or take the form of words strung together as a train-of-thought. Insofar as thoughts are verbal, this indicates that one region of the brain is organizing and explaining these verbal thoughts to another region of the mind which comprehends these verbal thoughts. Verbal thinking utilizes the same neural pathways and structures as spoken language; and Broca's expressive and Wernicke's receptive speech areas participate in the expression and comprehension of verbal thoughts. Because these neural pathways and language structures are immature for the first several years of life, and are limited in their ability to communicate within the brain, children initially think out-loud, using a form of language referred to as egocentric speech. As the brain matures, egocentric speech eventually becomes internalized as thought, such that by ages 5 to 6, children have completely internalized egocentric thought production, and think their thoughts in the privacy of their head. However, because the mind is a multiplicity with different tissues of the mind processing different forms of information, the dominant streams of consciousness associated with vision and language, often do not have access to information which might explain the motives for their actions, or how they arrived at certain conclusions or judgments. Because the mind is a multiplicity, in the final analysis, we knowers, remain unknown to ourselves.

KEY WORDS: Consciousness, thinking, egocentric speech, origin of thought, Development of thought, language, mind

1. Consciousness and the Multiplicity of Mind

Consciousness must be conscious of something. Consciousness, to be conscious, requires something which is separate from consciousness, and which becomes an object or focus of consciousness, such as a lamp, dog, cloud, car, singing birds, or the smile of a willing lover. Consciousness in order to be conscious requires something to be conscious of. Even if consciousness is only conscious of being conscious, i.e. self-consciousness, consciousness must be separate from itself to be conscious of itself, thereby becoming an object of consciousness, mirroring and reflecting itself as a duality. The same can be said of the train-of-thought which appears before or within consciousness but is never identical with consciousness (Joseph 1982).

Consciousness is not a singularity, but a multiplicity which, in humans, is often dominated by visual impressions and language (Joseph 1982, 2009). It is this multiplicity which makes consciousness of consciousness possible.

Insofar as consciousness is associated with the brain, then it could be said that the tissues of the mind consists of semi-independent mental realms which are maintained by the brainstem, thalamus, limbic system, the right and left hemisphere, and the occipital, temporal, parietal, and frontal lobes (Joseph 1982, 1986a,b, 1988a,b, 1992, 1999, 2009), each of which speaks and comprehends their own unique language, e.g. visual, tactile, olfactory, auditory. For example, the primary visual cortex perceives, processes, and becomes conscious of visual input but is largely deaf and blind to complex auditory signals, whereas the primary auditory receiving areas are blind and cannot see complex shapes or forms (Joseph 1996).

Because specific regions of the brain are specialized to perform specific functions, information transmitted between these tissues of the mind must undergo a transformation and become translated into a language the other can process and understand (Joseph 1982, 1986a,b, 1988a; Joseph et al., 1984). When the brain talks to itself, it could be said to be thinking; and these thoughts may be visual, verbal, tactile, emotional, or consist of myriad abstractions and symbols which then become objects of the consciousness maintained by the various realms of the multiplicity of mind.

Even if we associate consciousness with the activity of the brain, consciousness is never identical with what it is conscious of. When conscious of the flickering light of a lamp and the wavering shadows on the wall, consciousness and the shadows and flickering lamp light are not one and the same, but are separate and distinguishable. Consciousness is not the lamp, and the lamp is not consciousness; though it could be argued the lamp is in consciousness, just as a lamp may reside in a room next to the bed but remains distinguishable and separate from its surroundings.

Even if the brain is electrically stimulated by a neurosurgeon, any sensations, hallucinations, or memories evoked are experienced as appearing before or in an observing consciousness (Gloor, 1997; Halgren 1992; Penfield 1952; Penfield and Perot 1963). Likewise, when a subject is placed in a sensory reduced environment and deprived of external stimulation, although the brain will produce its own stimulation and generate complex hallucinations, these hallucinatory phenomenon are experienced as detached from the mind (Mason & Brady 2010; Riesen 1975; Zubek 1969).

Consciousness is relational and the same applies to conscious thought which is experienced as separate from consciousness, albeit as taking place inside the head. We may be "in our thoughts" but only insofar as we are thinking the thoughts, and thinking about ourselves as an object of thought which we are conscious of.

2. Thinking and Listening

Producing thoughts and experiencing and becoming conscious of thoughts as they are being produced, are indications of duality in the brain and mind (Joseph 1982, 2009).

In some respects, consciousness could be likened to a witness or an observer, and the same is true when observing or listening to an internally generated train of thought which often takes the form of an internal dialogue or picture show which is being experienced and even heard or seen within one's head (Joseph 1982). Thought, be it thinking in words, musical notes, math symbols, geometric patterns, or picture-images, is not synonymous with consciousness and often originates outside of consciousness and only appears to consciousness after the thoughts are organized and assembled (Ghiselin, 1952; Mandler, 1975; Neisser, 2006; Neisser & Fivush 2008; Nisbet & Wilson, 1977; Wilson, 2004).

Thoughts are the "actors on the stage" and consciousness is the audience. However, until the thoughts emerge, they are hidden from consciousness.

Take for example, "tip of the tongue" word-finding difficulty, in which the missing word is known yet not known. We are conscious that the word can't be found, but are not conscious of the identity of the missing word, though we know it's there. Nevertheless, although aware of the missing word's existence, we are unable to identify it or name it until it appears before consciousness. Awareness vs consciousness and "tip-of-the-tongue" are evidence of duality, and the same is true when considering the nature of consciousness and thought (Joseph 1982).

This duality is most evident in considering the developmental origins of verbal-thoughts. Thinking-in-words initially takes place not inside-the-head, but externally: children first speak their thoughts out-loud, and only gradually, as they grow older, do they begin to think their thoughts inside the privacy of their own head (Piaget, 1974; Vygotsky, 1956). The child speaks their thoughts, and listens to them as they are spoken, and this indicates duality. It is these developmental origins which provide one of the keys to understanding the nature of thought, the purpose of which is to serve the multiplicity of mind.

3. Language, Duality And The Train-Of-Thought

Although an individual may utilize visual, emotional, olfactory, musical, or tactile "imagery" when they think, thinking often takes the form of "words" which might be "heard" or rather, experienced, within one's own head. In fact, one need only listen to one's own thoughts in order to realize that thinking often consists of an internal linguistic monologue, a series of words heard inside the mind.

Thinking-in-words could be considered a form of internal perception, where strings of words, ideas, sentences, are produced by those tissues of the mind which speak the sounds of language, and which are perceived by that aspect of the conscious mind and brain which understands the sounds of language (Joseph 1982, 1986a,b, 1988ab). Thus, the train-of-thought which passes before consciousness, always has an origin which is outside of that aspect of the conscious mind which is listening to the train as it passes.

Thinking-in-words often serves consciousness and the brain as a means of explanation, commentary, or aimless internal chatter (Joseph 1982). Verbal thinking generally consists of an organized temporal-sequential hierarchy of associations, symbols, and labels which appear before an observor, or which are heard by the thinker. It is also a temporal progression, an associative advance and an elaboration which often appears with an initial or leading idea that is followed by a series of related verbal ideations. In the process of thinking in-words, one often acts to organized information which is "not thought out" and not clearly understood, so it may become thought out and thus comprehended in a logical, temporal sequential verbal format. To "think things out", "give it a lot of thought", or "think about it", serves an explanatory or deductive function (Mandler, 1975; Neisser, 2006; Wilson, 2004).

Yet the need to explain things to oneself seems paradoxical. It might be asked, "who is explaining what to whom?" Apparently the "I" that I am thinks (explains) these "things" to the "I" that I am.

We are presented with a curious duality in the nature of consciousness, the purpose of thought, and in the functioning of the brain. Insofar as the train of thought originates in me, in my brain and in my mind then I should know its aim and content prior to (not after) symbolizing the substance of the subject into the temporal-sequential linear organization that the verbal-thinking process generates. However, often we do know; there is an awareness; but the thoughts, idea, memories, and so on, remain hidden from consciousness, as again exemplified by "tip-of-the-tongue" world finding difficulty (Joseph 1982). We become conscious of this information and achieve explicit knowledge only after the information is transferred or made available to that aspect of the conscious mind which is dependent on language for understanding.

Limiting our discussion to word-thoughts, we must conclude, therefore, that the thoughts which will be expressed are not in consciousness before they are expressed, and are formulated and organized by a part of the brain which relies on language for expression, whereas they are comprehended by those regions of the mind-brain which require language for comprehension.

These thoughts only become an object of consciousness after they are organized into a train-of-thought by an aspect of mind which is separate from yet linked with that region of consciousness which experiences the train as it goes by. In other words, one realm of the brain and mind is clearly providing information and often explaining feelings, actions, observations, intentions, or conclusions, to another realm of the brain mind, and this is accomplished when thoughts become language; one region of the mind producing the verbal thoughts, the other listening (Joseph 1982, 1986a, 1988a). And, because these particular forms of thought are structured and perceived as words heard within one's head, then not surprisingly, they come to rely on the same neural pathways and brain structures which subserve the production and perception of language (Friederici, 2002; Kaan & Swaab, 2002; Newman, Just, & Carpenter, 2002), i.e. the inferior parietal lobule/angular gyrus, Broca's expressive speech area in the left frontal lobe, and Wernicke's receptive speech area in the superior temporal lobe (Joseph 1982, 1986a, 1988a; 1999, 2000a). In fact, when engaged in verbal thought, these language areas typically becomes activated as indicated by functional imaging (Kaan & Swaab, 2002; Keller et al., 2001; Paulesu, et al., 1993; Petersen et al., 1988).

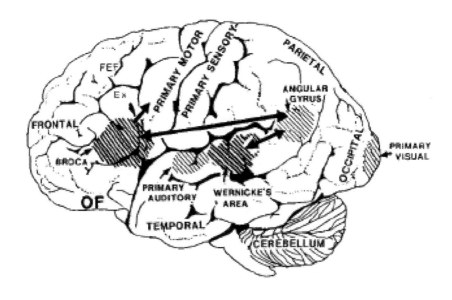

4. Brain and Language

Specifically, it is the lateral surface of the right and left frontal lobe which control vocalization and verbal-though production, the left frontal (Broca's area) producing the words, the right frontal the melody of language (Joseph 1982, 1986a, 1988a, 1996, 1999). The frontal lobes are interlocked with the association areas in the posterior regions of the cerebrum (Petrides & Pandya. 1999, 2001), including Wernicke's area and the angular gyrus - inferior parietal lobe (IPL), as well as the memory centers in the temporal lobe (Joseph 1986a, 1996, 1999). The frontal lobes are therefore continually informed about and have continual access to information processed in these areas of the brain.

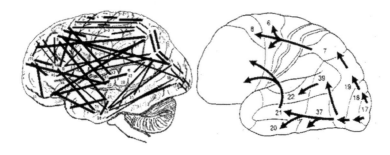

The frontal lobes serve as the senior executive of the brain and personality (Joseph 1986a, 1999, 2010) and play a significant role in searching for and assimilating the information which will be thought about (Christoff and Gabrieli, 2000; Christoff et al., 2001; Joseph 1986, 1999; Newman et al., 2002, 2003; Paulesu et al. 2010). The frontal lobes are responsible for organizing the thoughts which are to be explained and then comprehended by the auditory areas..

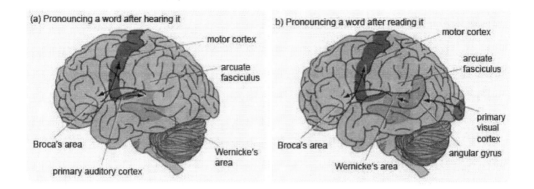

The primary auditory receiving areas are located in the superior temporal lobe. Once received and processed, these auditory signals are transferred to the immediately adjacent Wernicke's area which associates these sounds and comprehends the words of language; whereas the the auditory association area in the right temporal lobe comprehends environmental sounds and melodic and emotional vocalizations (Joseph 1982, 1988a, 2000). Via a massive neural pathways (the corpus callosum) the right and left auditory receiving areas work together when presented with complexity and paralinguistic features which require analysis (Just et al., 1996; Michael et al. 2001; Schlosser et al., 1998).

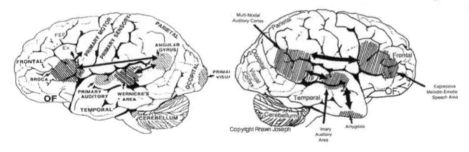

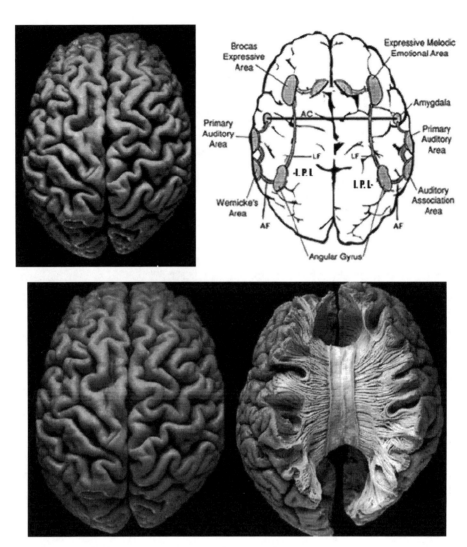

In addition, Wernicke's also plays a major role in the comprehension of verbal-thought, and, in conjunction with the right temporal lobe, the generation of spontaneous thought (Christoff et al., 2004). It is Wernicke's area (in conjunction with the IPL) which provides the words of language to the frontal lobes, and then listens to the train of thought as it passes by.

These temporal lobe auditory areas are linked to the frontal vocalization areas by a rope of nerve fibers, the arcuate fasciculus which passes through the angular gyrus/IPL--an area of the brain which assimilates associations and provides auditory-verbal labels to sensory stimuli. In fact the IPL and frontal lobes become active across a variety of language and non-language problem-solving and thinking tasks (Ben-Shahar et al., 2003; Dapretto & Bookheimer, 1999; Lehmann et al. 2009; Szaflarski et al., 2006; Tyler & Marslen-Wilson 2008; Vigneau et al. 2006).

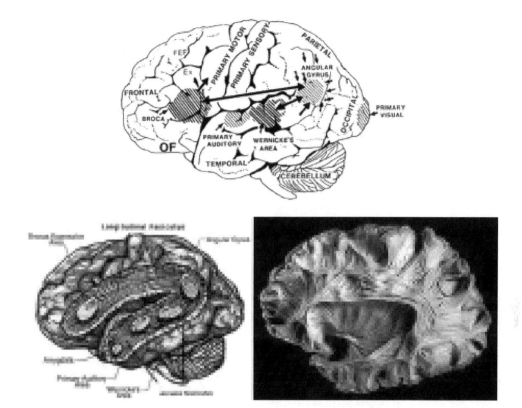

Therefore, whereas Broca's area (in conjunction with the IPL) organizes and expresses the sounds of language and verbal thoughts, Wernicke's area (in conjunction with the IPL) is responsible for comprehending thoughts, ideas, feeling, and so on, after they are put into words.

Thus, it is Broca's area which does the explaining, and it is Wernicke's area which comprehends the train of thought as it goes by. However, when it comes to comprehending internally generated thoughts, Wernicke's area is the last to know.

5. The Development of Language & Thought

Thinking in words is clearly related to language; and human language is a function of the human brain. It is the brain which makes it possible to speak and comprehend language, and to think thoughts, and these thoughts enable one region of the brain to communicate information to another brain area which is dependent on language for comprehension. Thinking in words often serves as a means of organizing, interpreting, and explaining information or impulses so that the language dependent regions of the brain and mind may achieve understanding (Gallagher & Joseph, 1982; Joseph 1982; Joseph & Gallagher, 1985, Joseph et al., 1984).

Initially, and over the course of child development, the thinking of thoughts, that is, the thinking of thoughts as strings of words is externalized and verbally expressed out-loud and is then comprehended as an explanatory commentary, after-the-fact (Joseph, 1982). Children think out-loud, and not in the privacy of their head (Piaget, 1952, 1962, 1974; Vygotsky, 1962).

The brain (or mind) of the child talks to itself by speaking thoughts (instead of thinking the thoughts) which are then heard, by the child, as the thoughts are spoken out-loud. In fact, initially, it appears that children are incapable of thinking inside-their-head but can only understand their thoughts, if they are spoken (Piaget, 1952, 1962, 1974; Vygotsky, 1962).

This indicates that one region of the child's brain must talk out-loud to communicate with another region of the brain which can only receive and comprehend these thoughts if they are spoken. The duality of actor/orator vs audience/listener, is present from the very beginning and it takes place on a stage located not inside the child's head, but outside in the world.

Certainly infants and young children are capable of internalized thinking, but these thoughts are visual and emotional, not verbal (Joseph 1982, 1992; 2003; Piaget, 1974). This is a function of their limited vocabulary and is exemplified by the fact that the first long term memories are emotional and visual and not verbal (Joseph 2003). Thus, the child must acquire language before thinking in words or forming verbal memories. However, the pathways between those areas of the brain which produce thoughts vs those which comprehend thoughts, must also also mature, before word-thoughts can be produced and comprehended internally.

6. Three Linguistic Stages

Broadly considered, there are three maturational stages of verbal development that correspond to the acquisition and development of language and which leads to the thinking of thoughts (Joseph 1982, 1992, 1996). Initially, linguistic expression is reflexive and/or indicative of generalized and diffuse emotions and feelings states. Vocalizations are largely emotional-prosodic in quality, and mediated by limbic and brainstem nuclei. It is only over the course of the first few months, around 1-3 months of age, that these prosodic-melodic utterances become associated with specific moods and emotions (Joseph 1982, 1992). It is at this time that "early babbling" makes it appearance and the infant begins to "coo," "goo," in a repetitive fashion, i.e. "ma ma ma"; and this is referred to as "early babbling."

"Early" babbling is produced by the brainstem and limbic system (Joseph 1982, 1992), and is replaced by "late" babbling which has its onset around 4 months of age. Late babbling is sometimes described as "repetitive babbling" in which the same consonant is repeated, such as "da da da." These transitions are directly related to the maturation of the neocortical speech areas as they gain hierarchical control over the subcortical speech centers (Joseph 1996).

Around one year of age, and once the neocortical speech areas begin to mature and establish hierarchical control over the limbic system and brainstem so as to program the oral-laryngeal motor areas, a new form of vocalization emerges and the infant begins to produce "jargon babbles" and to speak their first words. Syllabication is imposed on the intonational contours of the child's emotional speech by the still immature neocortex of left frontal lobe and motor areas, such that the melodic features of generalized vocal expression come to be punctuated, sequenced, and segmented, and vowel and consonantal elements begin to be produced (Joseph, 1982, 1992). Left hemisphere speech comes to be superimposed over limbic (and right hemisphere) melodic language output, and the infant begins saying actual words. However, due to the immaturity of the neocortex (Churgani, et al. 1987; Blinkov & Glezer, 1968; Conel, 1939; Lecours, 1975), most of the speech produced is "jargon" and resembles the "jargon aphasia" associated with injuries to Wernicke's area (Goodglass & Kaplan, 2000; Joseph 1996). However, rather than due to brain damage, jargon babbling reflects the extreme immaturity of the neocortical speech areas.

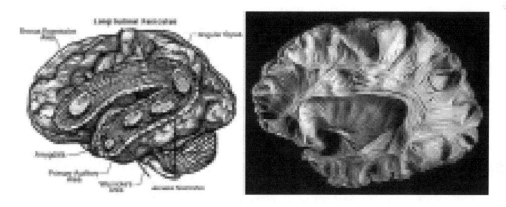

The development of jargon babbling appears to correspond to maturational events taking place in the motor areas of the neocortex which begins to rapidly mature around the first postnatal year (Chi, Dooling, & Gilles, 1977; Gilles et al. 1983; Scheibel, 1991, 1993). Jargon babbling coincides with the production of the first words which are spoken around 11-12 months on average (Nelson, 1981; Oller & Lynch, 1992). In fact, jargon babbling resembles actual speech, and at a distance it may sound as if the infant is conversing and speaking real words, though in fact they are babbling prosodically sophisticated neologistic jargon. Hence, the emergence of the jargon babbling stage signifies an obvious shift in sound production from the limbic system and brainstem to the still immature neocortex.

Jargon babbling not only resembles normal fluent speech but is often produced as the infant is gazing at or making eye-to-eye contact with the listener. Jargon vocalizations are both social and self-directed as the infant also jargon babbles while alone and at play, or while gazing at or exploring some object. When meant for the child's ears alone, these babbles could be described as "egocentric babbling."

Over the ensuing years, social speech emerges from jargon (conversational) babbling, whereas egocentric speech emerges from egocentric babbling. It is egocentric speech which gives birth to thinking-in-words.

Egocentric speech is essentially speech for oneself, and is the first evidence of thinking in words. The child is thinking out loud (Joseph, 1982; Piaget, 1962; Vygotsky, 1962). Egocentric speech is slowly internalized between the ages of 3 and 5, and eventually becomes completely covert; at which point, the child has not only learned to speak in words, but to silently think in words as well (Joseph 1982, 1996; Vygotsky, 1962).

7. Egocentric Speech and the Origins of Thought

At around age 3 the child produces two types of speech: social and egocentric. Part of the time the child engages in social conversational speech which is directed toward others, whereas the remainder of speech activities are egocentric and directed for the sole benefit of the child who listens to their vocalizations as they play (Joseph, 1982, 1996; Piaget, 1962; Vygotsky, 1962).

Egocentric speech is self-directed speech that consists of an explanatory monologue in which children comment on or explain their play and other actions, to themselves. Initially egocentric speech is produced only after the action has occurred. That is, the child essentially talks to themselves, but in an explanatory fashion: They tell themselves what they are doing and what they have done. Egocentric speech is a self-directed form of communication which heralds the first attempts at self-explanation via thinking-out-loud.

Egocentric speech makes its appearance at approximately 3 years of age, and at its peak comprises almost 40-50% of the child's language (Piaget (1952, 1962, 1974). In contrast to conversational social speech in which the child is engaging in a back-and-forth dialogue and is attuned to the listening needs of others, when engaged in egocentric speech the child is oblivious to his/her audience simply because the words spoken are meant for his/her ears alone (Piaget, 1952, 1962, 1974; Vygotsky, 1962). The child is essentially thinking out loud in an explanatory fashion, commenting on and describing his or her actions and is both orator and audience (Joseph, 1982).

When engaged in an egocentric monologue, there is no interest in influencing or explaining to others what in fact is being explained. The child will keep up a running verbal accompaniment to their actions, commenting on their behavior in an explanatory fashion even while alone. Moreover, while engaged in this self-directed external monlogue the child appears oblivious to the responses of others to their statements (Piaget, 1952, 1962, 1974; Vygotsky, 1956). If a playmate were to reply to a statement made during an egocentric monologue, it would not be heard by the child producing the egocentric monologue. It is as if the child has no awareness that others hear him when producing egocentric speech. In fact, many a child has been shocked when, later, his mother (or a friend) repeats or comments upon something he assumed no one else could hear; as if mom can read their thoughts! Thus, when engaged in egocentric speech, the child may not be conscious of the fact they are thinking out-loud.

8. Thinking Out Loud: The Stages of Thought Internalization

Egocentric speech is a dialogue where the child is both the speaker and the listener, the actor and the audience. It is an explanatory monologue, serving as a commentary that is initially produced only after an action has been completed by the child. That is, the child explains to him/herself what they have done after they have observed themselves complete some action, and then they comment on and/or explain what has taken place (Piaget 1974; Vygotsky, 1962). For example the 3-4 year old child will paint a picture and then explain it after it is completed: "This the sun shining on mommy." Or they will crash their toy truck into another toy truck and then remark on and explain what happened: "Trucks crashed."

As the child grows older, instead of explaining after the fact, they will explain what they are doing, as they are doing it. For example the 4-5 year old child will paint a picture and explain it while she is painting, or state the toy trucks are crashing into each other as they smash together. The egocentric monologue accompanies the action.

By time the child reaches age 5-6, they will announced what they are going to do and then do it. For example, the child may state she is going to paint a picture of "mommy and daddy at the beach and they are happy", and then paints it, or he will take his two toy trucks and state "now they are going to crash and everyone will die" and then he will smash the trucks together.

Hence, as the child grows older their comments and explanations occur earlier in the sequence of expression, until finally the child begins to explain his/her actions before they are performed instead of after they have occurred (Piaget, 1952, 1962, 1974; Vygotsky, 1962).

Egocentric speech presents us with an obvious duality: initially, the part of the brain and mind that relies on language, does not know what another part of the brain and mind intends to do, until after the fact, at which point the language regions comment on and explains, to themselves, what they have done and why. This indicates that the aspect of the brain and mind which is talking and listening, does not know until after the actions are performed, and only achieves understanding when language is used to explain their actions to themselves. Presumably, this is due, in part, to functional disconnections between brain areas, secondary to the immaturity of these tissues and their nerve fiber interconnections (Joseph 1982, 1996; Joseph et al., 1984). That is, because different brain areas and their neuronal interconnections can take years to decades to mature (Blinkov & Glezer, 1968; Lecours, 1975; Szaflarski et al., 2006), their ability to communicate is limited. One area of the brain and mind may initiate a behavior, which is witnessed or experienced by other (disconnected) brain areas, only as it occurs outside the brain and body.

Thus, for the first several years of life, it appears that the child's brain as a whole does not know what or why they are engaged in certain actions, or even that they intend to perform certain actions, until after the act has been completed; at which point they explain what they did and why, to those areas of the brain-mind which are dependent on language. This is because one region of the brain initiates the behavior, and another witnesses and then explains to itself what just happened. The child is both actor and witness, explainer and explained to. Clearly, when engaged in an egocentric monologue and thinking-out-loud, the child (the left frontal-temporal-parietal axis) explains their actions to themself.

However, around age 4, and as the child's neural pathways mature, the frontal lobes and language-dependent regions of the brain and mind receive information about their intentions as they engage in these acts, and thus explain, to themselves, what they are doing, while they are doing it. Neuronal interconnections begin to mature and to increasingly communicate and share information.

Around age 5, this information become available before rather than during or after they act. Thus they begin to explain, to themselves, what they are planning to do before they do it. This advanced warning parallels the increasing maturity of the nerve fiber pathways between brain areas (Szaflarski et al. 2006), and the ability of these tissues of the mind to successfully transfer and receive complex information from other regions of the cerebrum (Gallagher & Joseph, 1982; Joseph & Gallagher, 1985; Joseph et al., 1984).

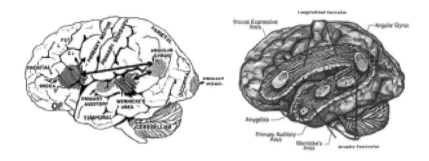

By the time the child reaches age 5-6, egocentric speech has become increasingly internalized (Piaget 1974; Vygotsky, 1962). The pathways between different brain areas have matured (Szaflarski et al., 2006) such as the arcuate fasciculus linking Wernicke's and Broca's area, and the child no longer needs to vocalize their thoughts, but can instead think and comprehend these thoughts internally (Joseph, 1982, 1996). The neural pathways linking these tissues of the mind have sufficiently matured so that information can be readily transferred between brain areas. However, even after thinking has become completely internalized it remains self-directed. Thinking continues to serve an explanatory function.

9. The Internalization of Egocentric Speech

Egocentric speech is never directed at others. Groups of children may be playing together, and each child may be engaged in an egocentric monologue. They are not talking to the other children. They are talking to themselves. They are thinking-out-loud.

Initially egocentric speech is completely external and after the fact (Piaget, 1974; Vygotsky, 1956). Since they are utilizing words, it is thus apparent that they are engaging those areas of the left hemisphere which are dominant for expressing and comprehending language: the frontal-temporal lobes (Kaan & Swaab, 2002; Michael et al., 2001; Schlosser et al. 1998; Tyler & Marslen-Wilson 2008). However, because of the immaturity of the child's brain, vast regions are partially disconnected. Therefore, specific areas of the brain and mind associated with language, are also partially disconnected, essentially creating two brains and two minds in a single head (Gallagher & Joseph, 1982; Joseph 1982; Joseph & Gallagher, 1985; Joseph et al., 1984). In fact, the pattern of neurological activity during the performance of language tasks, does not begin to resemble the adult pattern until the onset of puberty (Holcomb et al., 1992).

Therefore it appears that children must speak their thoughts out loud due to the immaturity of the interconnections between Broca's area and the temporal-parietal area (Szaflarski et al., 2006), and the immaturity of the corpus callosum neural pathways which link the right and left hemisphere (Gallagher

& Joseph, 1982; Joseph 1982, Joseph & Gallagher, 1985; Joseph et al., 1984). Thus, the right half of the brain or limbic system may initiate certain behaviors without the knowledge of the language dependent aspects of consciousness associated with the left hemisphere; which then explains, to itself, what it observes. However, because of the immaturity of the neural pathways within the language areas of the left hemisphere, the frontal lobes must speak the thoughts so they may be heard by Wernicke's area.

Hence, improvements in neocortical transmission in the language areas, and between the right and left hemisphere, parallel the stages of egocentric speech and its internalization as silent thought.

10. Thinking and the Evolution and Development of the Multiplicity of Mind

The mind is a multiplicity, and different regions of the brain and mind speak different languages as they are specialized to perceive or process specific types of stimuli and information. Moreover, although certain areas of the brain are richly interconnected, the neural pathways between yet other tissues of the mind are sparse or non-existant as there is no need for them to communicate, except indirectly and through intermediary tissues such as the angular gyrus of the inferior parietal lobule. For example, the angular gyrus of the inferior parietal lobe is situated at the junction of the association areas for vision, hearing, and somesthesis, and, in conjunction with the frontal lobes and Wernicke's area (Lehmann et al. 2009; Szaflarski et al., 2006), assimilates associations and diverse information variables and then organizes and categorizes them into words and multi-modal linguistic concepts which can be translated into language and the train of thought (Joseph 1982, 1986a). It is the IPL/angular gyrus which enables a person to see, for example, a "cup" and to associate the word "cup" with the visual image, or to imagine a variety of cups of varying size, colors, or utilities, ranging from a "world cup" to the "cup size of a bra." The IPL essentially assimilates diverse associations thereby making it possible to form multi-modal concepts which may be differentially comprehended by different aspects of the mind.

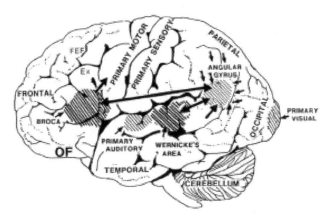

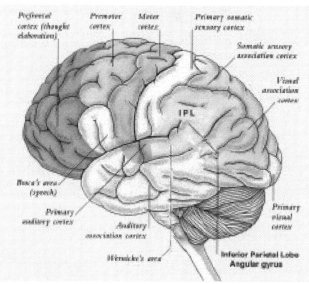

However, not all sensations or information variables have auditory equivalents and cannot be adequately described using language. Words completely fail to describe how it feels to ride a horse, parachute from a falling plane, swim beneath the sea, dilate the pupil of an eye, or experience an orgasm. Moreover, there are myriad behaviors which do not require and which occur independently of the language-dependent aspects of the mind. For example, there are nerve centers in the ancient brainstem which control heart rate, breathing, and pupil dilation; and although influenced by "higher cortical" activity in the forebrain, for the most part these activities take place without the assistance or participation of the conscious mind and the more recently evolved necortical surface layers of the cerebrum. There are also limbic system structures such as the hypothalamus and amygdala which mediate various aspects of emotion and which may hijack the rational, logical, and language dependent aspects of the brain and mind (Joseph 1992); which, like the egocentric child, may act without thinking, and then later may exclaim: "I don't know what came over me" and then search for an answer.

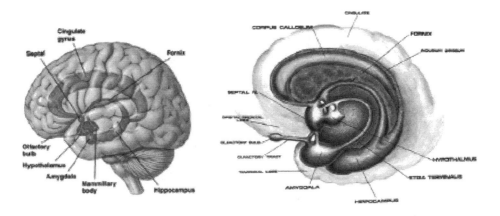

From an evolutionary perspective, it must be recognized that the brain has evolved over the course of the last 600 million years, and it is only around 100 million ago that the six layered neocortical mantle, began to evolve and to slowly cover and envelop the old brain which includes the extremely ancient limbic system and brainstem. Further, it is only within the last 100,000 years that the language-dependent aspects of consciousness began to evolve (Joseph, 1996, 2000b).

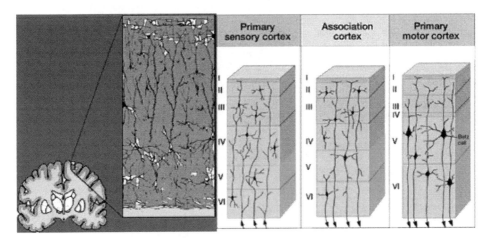

However, be it the brain of a modern human, or that of a reptile, much of behavior is under the control of these more ancient tissues of the mind which do not require language or any type of logical or analytical thinking to perform their functions. Thus, these regions of the brain usually function completely independently of human consciousness which is essentially not-conscious and has no conscious access to the workings of the more ancient regions of the mind.

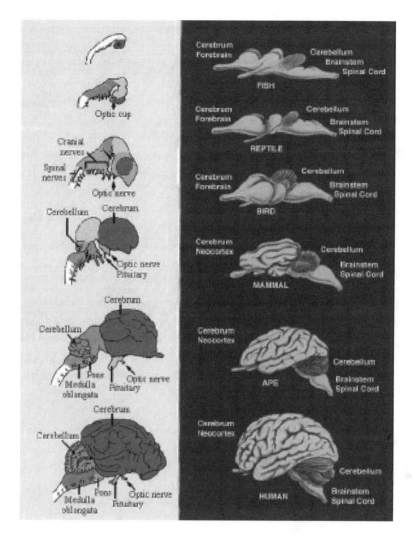

(Left) Human Brain Development 3-Weeks to 9-Months.
(Right) Brain Evolution 500 Million Years to 10,000 Years Ago.

Ontogeny does not always replicate phylogeny. However, the development and maturation of the human brain does in fact parallel the evolution of the brain (Joseph 1996) such that the brainstem structures are fully functional at birth (Joseph 2000c), followed by the limbic system (Joseph 1992, 1996), whereas the more recently evolved neocortex is the last to develop, with some areas taking up to 20 years to fully mature (Blinkov & Glezer, 1968; Conel, 1939; Lecours, 1975).

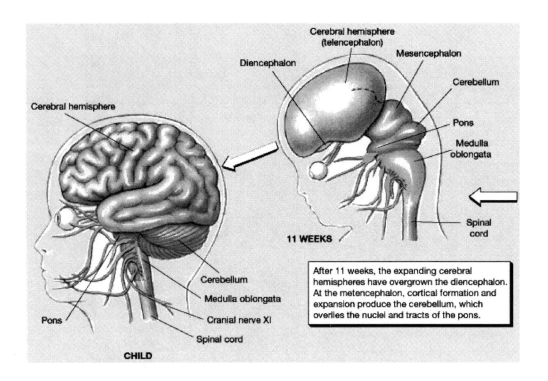

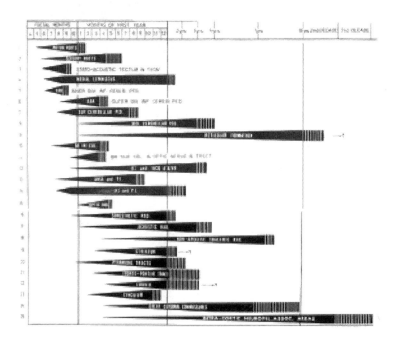

The Myelination of the Human Brain. Myelin is the outer protective coating of the axon and promotes and stabilizes nerve transmission. From Yakovlev & Lecours (1967)

Likewise, emotional, visual, and motor functioning mature well in advance of language. Therefore, in many respects, language, and the language-dependent aspects of consciousness, can be considered an after-thought phyologenetically and ontogenetically, and as such, these areas of the brain and mind often have no access to those tissues controlling behavior, and are therefore often the last to know.

11. The Multiplicity of Mind and the Conscious Unconscious

Seeking knowledge and information, the language dependent aspects of consciousness produce thoughts, to explain to itself why. Thinking, especially thinking-in-words, often serves an explanatory function; and often its purpose is to obtain information or explain behaviors which have a source in the non-linguistic regions of the brain and mind.

Essentially, egocentric speech, and then internalized verbal thoughts, are largely a function of the left hemisphere's attempt to organize, interpret, and make sense of behavior initiated by brain structures not concerned with the denotative, grammatical aspects of human language. Initially, egocentric speech externalized thought production is due to the immaturity of neural pathways which hinder information transfer not just within the left hemisphere, but from the right to left half of the brain (Gallagher & Joseph, 1982; Joseph, 1982; Joseph & Gallgher 1985; Joseph et al., 1984); a function of the immaturity of the corpus callosal fibers connections between the hemispheres (Yakovlev & Lecours, 1967).

However, even in the "normal" intact adult, information transmission within and between the right and left halves of the brain is often incomplete (Joseph 1982, 1988a). Because different brain tissues are functionally specialized to perform different functions, they speak different language and even information which is shared must be interpreted and translated--and much may be misinterpreted or lost in translation. As such, those regions of the mind in the left hemisphere dependent on language sometimes observe (and participates) in behaviors which it did not initiate, and which it does not understand, and will then think up an explanation. However, because it does not have access to this information, the explanations which are thought up, frequently amount to little more than self-deception.

Consider a famous experiment by Nisbet and Wilson (1977) in which they invited shoppers to participate in a consumer survey and to evaluate and indicate their preference for one of four identical pairs of nylon stockings which were laid out in front of them from left to right. Subjects were asked to handle and test the stockings, and to indicate which they thought was the best. Although the stockings were in fact identical, 75% of the subjects believed the stocking on their right was superior to the rest. When asked "why?" subjects confabulated a variety of explanations, claiming differences in color, texture, softness, durability, and so on, even though the stockings were identical! Most (right handed) people show a right-sided response bias (which is why "brand X" is always placed to the left). However, when asked if the position of the article effected their judgment, "virtually all subjects denied it, usually with a worried glance at the interviewer suggesting that they felt either that they had misunderstood the question or were dealing with a madman."

The fact is, based on over a century of psychological studies (Ghiselin, 1952; Mandler, 1975; Neisser, 2006; Neisser & Fivush 2008; Nisbet & Wilson, 1977; Wilson, 2004), it can be concluded that most people have no knowledge as to how they arrived at solving certain problems, made certain judgments, or why they engaged in various behaviors. Even the processes involved in intellectual and creative pursuits are outside the reach of consciousness. In a famous study of creative geniuses, Ghiselin (1952) concluded that "production by a process of purely conscious calculation seems never to occur." Instead, these geniuses often described themselves as observers, as bystanders who witness the fruits of the problem-solving or creative process only as or after it occurs.

The mind is a multiplicity, and those aspects of these mental realms collectively referred to as consciousness, have at best, only limited access to functional knowledge sources. Moreover, the language dependent aspects of consciousness has a limited capacity and can access and accurately report only a small bit of the information amenable to linguistic coding. Not all impulses, feelings, desires, fears, cravings, knowledge, etc., have a label and are not readily translatable into the codes that give rise to thought and language (Joseph, 1982, 1988a). Emotions in particular are easily subject to misinterpretation, misidentification, and misunderstanding, and the same is true of most non-emotional cognitive processes. Most individuals have no understanding of what is involved and on what basis evaluations, judgments, problem-solving strategies, and reasons for initiating certain behaviors were arrived at. This is largely due, however, to the dependence on language. The sequential, organizational linguistic processes which subserve consciousness in the form of verbal thought have no direct access to the basis of these 'processes' in part because they result from them in an attempt to explain them.

12. The Multiplicity of Mind: We Knowers Are Unknown To Ourselves

The brain/mind is a multiplicity (Joseph 2009), and for good reason. These conditions protect the brain and linguistic consciousness from becoming overwhelmed. Different regions of the brain, and thus the mind, are functionally specialized, thereby creating multiple minds which can engage in multiple tasks simultaneously. However, in consequence, we knowers are unknown to ourselves.

Thinking is a product of the multiplicity of mind. However, so too is consciousness. Consciousness, to be conscious, requires something which is separate from consciousness, and which becomes an object or focus of consciousness. Because the mind is a multiplicity, consciousness can be conscious of being conscious. Different regions of the brain may be conscious or at least aware of the mental activity occurring in yet other regions of the cerebrum; and this makes self-consciousness possible. In other words, consciousness does not separate from itself when attempting to know and think about itself, and this is because different forms of consciousness, a multiplicity of minds, dwell within the same head, thereby enabling the mind to be both observer and observed. Thus the distinguishing characteristic of consciousness is that it does not coincide with itself. Consciousness is not a duality—a reflection which is its own reflecting--but a multiplicity.

It is the languages spoken by these multiple minds, and the non-conscious origins of so much of what constitutes human behavior, which are responsible for the development and origin of thought, the primary function of which, is self-explanation: The I that I am, explains these thoughts to the I that I am.

References

Ben-Shahar, M., et al. (2003). The neural reality of syntactic transformations: evidence from fMRI. Psychological Science 14.433–40.

Blakemore, S-J., et al., (2007). Adolescent development of the neural circuitry for thinking about intentions, Soc Cogn Affect Neurosci. doi: 10.1093.

Blinkov, S. M., & Glezer, I. I. (1968). The human brain in figures and tables. Plenum, New York.

Booth, J. R., et al. (2007). The role of the basal ganglia and cerebellum in language processing. Brain Research 1133.136–44.

Bischoff-Grethe, A., et al. (2000). Conscious and unconscious processing of nonverbal predictability in Wernicke's area. Journal of Neuroscience 20.1975–1981.

Chi, J.G., Dooling, E. C., & Gilles, F. H. (1977). Gyral development of the human brain. Ann. Neurology, 1, 86-93.

Christoff, K., and Gabriele, J.D.E. (2000). The frontopolar cortex and human cognition: Evidence for a rostrocaudal hierarchical organization within the human prefrontal cortex. Psychobiology, 28: 168-186.

Christoff, K., et al., (2001), Rostrolateral prefrontal cortex involvement in relational integration during reasoning. NeuroImage, 14: 1136-1149.

Christoff, K., Ream, J.M., and Gabrieli, J. D. E. (2004). Neural basis of spontaneous thought processes. Cortex, (2004) 40, 623-630.

Churgani, H.T., et al. (1987). Positron emission tomography of human brain functional development. Ann Neurology, 22, 487-497.

Conel, J. L.(1939). The postnatal development of the human cerebral cortex. Harvard

University Press. Cambridge.

Dapretto, M., & Bookheimer, S. Y., (1999). Form and content: dissociating syntax and semantics in sentence comprehension. Neuron 24.427–32.

Dikker, S., H. et al. (2009). Sensitivity to syntax in visual cortex. Cognition 110.293–321.

Dehaene. S., et al. (1999). Sources of Mathematical Thinking: Behavioral and Brain-Imaging Evidence Science, 284, 970-974.

Friederici, A. D. (2002). Towards a neural basis of auditory sentence comprehension. Trends in Cognitive Sciences, 6, 78-84.

Gallagher, R. E., & Joseph, R. (1982). Non-linguistic knowledge, hemispheric laterality, and the conservation of inequivalance. Journal of General Psychology, 107, 31-40.

Ghiselin, B. (1952). The Creative Process. New York, Mentor.

Gilles, F.H., Leviton, A., & Dooling, E.C., (1983). The developing human brain. Boston: John Wright.

Gloor, P. (1997). The Temporal Lobes and Limbic System. Oxford University Press. New York.

Goodglass, H., & Kaplan, E. (2000). Boston diagnostic aphasia examination. New York, Lange.

Halgren, E. (1992). Emotional neurophysiology of the amygdala within the context of human cognition. In J. P. Aggleton (Ed.). The Amygdala. New York, Wiley-Liss.

Holcomb, P,. J., Coffey, S. A., & Neville, H. J. (1992). Visual and auditory sentence processing. Developmental Neuropsychology, 8, 203-241.

Joseph, R. (1980). Joseph, R. (1982). The Neuropsychology of Development. Hemispheric Laterality, Limbic Language, the Origin of Thought. Journal of Clinical Psychology, 44 4-33.

Joseph, R. (1986a). Confabulation and delusional denial: Frontal lobe and lateralized influences. Journal of Clinical Psychology, 42, 845-860.

Joseph, R. (1986b). Reversal of language and emotion in a corpus callosotomy patient. Journal of Neurology, Neurosurgery, & Psychiatry, 49, 628-634.

Joseph, R. (1988a) The Right Cerebral Hemisphere: Emotion, Music, Visual-Spatial Skills, Body Image, Dreams, and Awareness. Journal of Clinical Psychology, 44, 630-673.

Joseph, R. (1988b). Dual mental functioning in a split-brain patient. Journal of Clinical Psychology, 44, 770-779.

Joseph, R. (1992) The Limbic System: Emotion, Laterality, and Unconscious Mind. The Psychoanalytic Review, 79, 405-455.

Joseph, R. (1996). Neuropsychiatry, Neuropsychology, Clinical Neuroscience, 2nd Edition. Williams & Wilkins, Baltimore.

Joseph, R. (1999). Frontal lobe psychopathology: Mania, depression, aphasia, confabulation, catatonia, perseveration, obsessive compulsions, schizophrenia. Psychiatry, 62, 138-172.

Joseph, R. (2000a). Limbic language/language axis theory of speech. Behavioral and Brain Sciences. 23, 439-441.

Joseph, R. (2000b). The evolution of sex differences in language, sexuality, and visual spatial skills. Archives of Sexual Behavior, 29, 35-66.

Joseph, R. (2000c). Fetal brain behavioral cognitive development. Developmental Review, 20, 81-98.

Joseph, R. (2001). The Limbic System and the Soul: Evolution and the Neuroanatomy of

Religious Experience. Zygon, the Journal of Religion & Science, 36, 105-136.

Joseph, R. (2003). Emotional Trauma and Childhood Amnesia. journal of Consciousness & Emotion, 4, 151-178.

Joseph, R. (2009). Quantum Physics and the Multiplicity of Mind: Split-Brains, Fragmented Minds, Dissociation, Quantum Consciousness, Journal of Cosmology, 3, 600-640.

Joseph, R. (2010). The neuroanatomy of free will: Loss of will, against the will "alien hand", Journal of Cosmology, 14, In press

Joseph, R., Gallagher, R., E., Holloway, J., & Kahn, J. (1984). Two brains, one child: Interhemispheric transfer and confabulation in children aged 4, 7, 10. Cortex, 20, 317-331.

Joseph, R., & Gallagher, R. E. (1985). Interhemispheric transfer and the completion of reversible operations in non-conserving children. Journal of Clinical Psychology, 41, 796-800.

Just, M. A., Carpenter, P. A., Keller, T. A., Eddy, W. F., & Thulborn, K. R. (1996). Brain activation modulated by sentence comprehension. Science, 274, 114-116.

Kaan, E., & Swaab, T. Y. (2002). The brain circuitry of syntactic comprehension. Trends in Cognitive Sciences, 6, 350-356.

Keller, T. A., Carpenter, P. A., & Just, M. A. (2001). The neural bases of sentence comprehension: A fMRI examination of syntactic and lexical processing. Cerebral Cortex, 11, 223-237.

Kobayashi, C. et al. (2007). Language in adult English speakers but not in English-speaking children

Kobayashi, C. et al. (2007) Cultural and linguistic effects on neural bases of 'Theory of Mind' in American and Japanese children. Brain Res. 1164, 95–107.

Kossyln, S. M., e t al. (1995) Topographical representations of mental images in primary visual cortex. Nature, 378: 496-498.

Lecours, A. R. (1975). Myelogenetic correlates of the development of speech and language. In E. Lenneberg & E. Lenneberg (Eds.), Foundations of language development. New York: Academic Press.

Lehmann, D., et al., (2009). Core networks for visual-concrete and abstract thought content: A brain electric microstate analysis, NeuroImage, 49, Issue 1, 1 January 2010, 1073-1079.

Mandler, G. (1975). Consciousness: Respectable, useful, and probably necessary. In R. Solso (Ed). Information processing and cognition. Erlbaum, NJ.

Mason OJ, Brady F. (2009). The psychotomimetic effects of short-term sensory deprivation. J Nerv Ment Dis. 197, 78378-78385.

Michael, E. B., Keller, T. A., Carpenter, P. A., & Just, M. A. (2001). fMRI investigation of sentence comprehension by eye and by ear: Modality fingerprints on cognitive processes. Human Brain Mapping, 13, 239-252.

Nakai, T., K. et al. (2005). An fMRI study of temporal sequencing of motor regulation guided by an auditory cue – a comparison with visual guidance. Cognitive Process 6.128–35.

Neisser, U. & Fivush, R. (2008) The Remembering Self: Construction and Accuracy in the Self-Narrative, Cambridge University

Neisser, U. (2006) The Perceived Self: Ecological and Interpersonal Sources of Self Knowledge, Cambridge U. Press.

Nelson, K. (1981). Individual differences in language development: Implications for development and language. Developmental Psychology, 17, 170-187.

Newman, S. D., Carpenter, P. A., Varma, S., & Just, M. A. (2003). Frontal and parietal participation in problem solving in the Tower of London: fMRI and computational modeling of planning and highlevel perception. Neuropsychologia, 41, 1668-1682.

Newman, S. D., Just, M. A., & Carpenter, P. A. (2002). Synchronization of the human cortical working memory network. NeuroImage, 15, 810-822.

Nisbett, R. E. & Wilson, T. D (1977). Telling More Than We Can Know: Verbal Reports on Mental processes, Psychological Review, 84, 231-259.

Ojemann, G. A., I. Fried, E. Lettich. (1989). Electrocorticographic (ECoG) correlates of language. I. Desynchronization in temporal language cortex during object naming. Electroencephalography and Clinical Neurophysiology 73.453–63.

Oller, D. K., & Lynch, M. P. (1992). Infant vocalizations and innovation in infraphonology: Toward a broader theory of development and disorders. In C. A. Ferguson, L. Menn, & C. Stoel-Gammon (Eds.), Phonological development: models, research, implications. Academic Press, New York.

Paulesu, E., Frith, C. D., & Frackowiak, R. S. J. (1993). The neural correlates of the verbal component of working memory. Nature, 362, 342-344.

Paulesu, P., et al., (2010). Neural correlates of worry in generalized anxiety disorder and in normal controls: a functional MRI study. Psychological Medicine (2010), 40: 117-124.

Penfield, W. (1952) Memory Mechanisms. Archives of Neurology and Psychiatry, 67, 178-191.

Penfield, W., & Perot, P. (1963) The brains record of auditory and visual experience. Brain, 86, 595-695.

Pernera, J., & Aichhorna, M. (2008) Theory of mind, language and the temporoparietal junction mystery Trends in Cognitive Sciences, 12, 123-126.

Petersen, S. E., Fox, P. R., Posner, M. I., et al. (1988). Positron emission tomographic studies of the cortical anatomy of single-word processing. Nature, 331, 585-589.

Petrides, M., & Pandya, D. N. (1999). Dorsolateral prefrontal cortex: comparative cytoarchitectonic analysis in the human and the macaque brain and corticocortical connection patterns. European Journal of Neuroscience 11.1011–1036.

Petrides, M., & Pandya, D. N. (2001). Comparative cytoarchitectonic analysis of the human and the macaque ventrolateral prefrontal cortex and corticocortical connection patterns in the monkey. European Journal of Neuroscience 16.291–310.

Piaget, J. (1952). The origins of intelligence in children. New York: Norton.

Piaget, J. (1962). Play, dreams and imitation in childhood. New York: Norton.

Piaget, J. (1974). The child and reality. New York: Viking Press.

Riesen, A. H. (1975). Developmental Neuropsychology of Sensory Deprivation, Academic Press.

Saxe, R., & Kanwisher, N. (2003). People thinking about thinking people – fMRI studies of Theory of Mind. Neuroimage, 19, 1835–1842.

Scheibel, A. B. (1991). Some structural and developmental correlates of human speech. In (K. R. Gibson & A. C. Petersen, Eds). Brain Maturation and Cognitive Development. New York,. De Gruyter.

Scheibel, A. B. (1993). Dendritic structure and language development. In B. de Boysson-Bardies, S., de Schonen, P. Jusczyk,

P, MacNeilage, and J. Morton (Eds.), Developmental neurocognitive speech and voice processing in the first year of life. Academic Press.

Schlosser, M. J., Aoyagi, N., Fulbright, R. K., Gore, J. C., & McCarthy, G. (1998). Functional MRI studies of auditory comprehension. Human Brain Mapping, 6, 1-13.

Scholz. J., et al., (2009). Distinct Regions of Right Temporo-Parietal Junction Are Selective for Theory of Mind and Exogenous Attention, PLoS ONE 4(3): e4869. doi:10.1371/journal.pone.0004869

Steinmetz H., and Seitz, R. J., (1991). Functional anatomy of language processing: neuroimaging and the problem of individual variability. Neuropsychologia 29 (12).1149–1161.

Szaflarski, J. P. et al., (2006) A longitudinal functional magnetic resonance imaging study of language development in children 5 to 11 years old, Annals of Neurology Volume 59, Issue 5, pages 796–807.

Tyler, L. K., & Marslen-Wilson, W. (2008). Fronto-temporal brain systems supporting spoken language comprehension, Phil. Trans. R. Soc. B., 363, 1037-1054.

Wilson, T. D. (2004). Strangers to Ourselves: Discovering the Adaptive Unconscious, Belknap Press, MA.

Vigneau, M., V. Beaucousin, P. Y. Hervé, H. Duffau, F. Crivello, O. Houdé, B. Mazoyer, and N. Tzourio-Mazoyer. 2006. Meta-analyzing left hemisphere language areas: Phonology, semantics, and sentence processing. Neuroimage 30(4).1414–1232.

Vygotsky, L. S. (1962). Thought and language. Cambridge: MIT Press.

Yakovlev, P. I., & Lecours, A. (1967). The myelogenetic cycles of regional maturation of the brain,. In A. Minkowski (Ed.), Regional development of the brain in early life, (pp.404-491). London: Blackwell.

Zubek, J. P. (1969). Sensory Deprivation: Fifteen years of research. Appleton-Century-Crofts.

29. Consciousness: A Direct Link to Life's Origins?
A. N. Mitra[1] and G. Mitra-Delmotte[2], Ph.D.

[1]Emeritus, Department of Physics, Delhi University, Delhi-110007; 244 Tagore Park, Delhi-110009, India.
[2]39 Cite de l'Ocean, Montgaillard, 97400 St.Denis, Reunion Island, France.

Abstract

Inspired by the Penrose-Hameroff thesis, we are intuitively led to examine an intriguing correspondence of 'induction' (by fields), with the complex phenomenon of (metabolism-sustained) consciousness: Did sequences of associated induction patterns in field-susceptible biomatter have simpler beginnings?

KEY WORDS: mind-matter, field-driven assembly, induction, environment, solitons, coherence, magnetism.

1. Introduction

Consciousness is a many-splendoured thing, whose anatomy has been under scrutiny almost since the birth of civilization. We have come a long way from the times when this term was associated with religion and spiritualism, to the present era when serious efforts are directed towards understanding it in the language of Science (Penrose 1995; Hameroff 2003). During this saga, physical science has progressed all the way from Newtonian mechanics (when Cartesian Partition ruled against such efforts) to the birth of relativity and quantum mechanics when sheer compulsions of logical self-consistency demanded that Cartesian Partition was no longer tenable and that mind and matter could no longer be divorced from each other. This was despite Bohr's Copenhagen Interpretation which had effectively decreed against difficult logical questions being asked about quantum mechanics. But Einstein could not accept this dictum and produced his EPR paper (Einstein et al 1935) ostensibly designed to demolish the tenets of quantum mechanics, but serendipitously treaded on a most fertile land as a logical consequence of the new paradigm, namely quantum entanglement and non-locality. This was directly at variance with the concept of local realism, the bedrock of the Copenhagen Interpretation. Since both could not be right at the same time, it took another half a century to decide on the issue: Alan Aspect, through his famous experiment (Aspect et al 1982), gave the verdict in favour of EPR's entanglement and non-locality and in so doing, ruled out Bohr's local causality. In the meantime Copenhagen had got rather outdated due to the emergence of decoherence, thanks to a two-decade-old development bearing on the very foundations of quantum mechanics (following the Aspect discovery) which gave an increasingly active role to the environment (see below for its fuller ramifications).

This episode offers a possible setting for bringing mind and matter on a common platform, since a direct touch with reality of these bizarre quantum concepts have willy-nilly got these two entities 'mutually entangled'! One may wonder how this 'Frankenstein' (read science with all its tools) which is the product of the human mind in the first place, has come to challenge its own Creator, and probe its `anatomy'. This essay is an attempt to string together some scientific advances designed to reduce the complex phenomena of consciousness (Sect.2-5), and to map (Sect.6-7) the resulting scenario to simpler ingredients that may have been available in the Hadean.

2. Bohm's Thesis: Integral Duality of Mind and Matter

For historical reasons, we start with a semi-intuitive model due to David Bohm (1990) who was led, by the conflict of quantum theory (discreteness) with GTR (continuous matter), to propose the existence of an undivided wholeness present in an implicate order which applies to both matter and mind, so that it can in principle access the relationship between these two different things. In this picture, matter and mind are seen as relative projections into an explicate order from the reality of the implicate order, with apparently no connection between them. Only at the deeper, fundamental level of the universe, does there exist an unbroken wholeness in which mind (consciousness?) merges with matter-- something akin to a holographic image of the brain (Pribram 1975).

Bohm also illustrates the idea of 'meaning' through the example of listening to music as a sequence of overlapping moments each with a short but finite interval of time. To produce the notes, one moment 'induces' the next, such that the content that was implicate in the immediate past, becomes explicate in the next interval (the immediate present). The sense of movement in music is thus the result of a sequence of overlapping transitions, thus producing consciousness from an implicate order. Consciousness is thus seen to be intimately related to the concept of 'time' -not merely a 'duration', in the sense of classical mechanics, but an active ingredient bearing on consciousness that reveals a world of continuous and unfolding events, a la Bergson (2001[1889]). Bohm (1990) further suggests that the tiny electron is an inseparable union of particle and (not or) field wherein the latter (like consciousness) organizes the movement of the former (like body). This is in line with Bohm's (1952) earlier thesis on quantum mechanics with 'hidden variables' wherein guiding waves determine the motion of the associated particles.

3. Quantum Coherence in Biology

Next, quantum coherence, which is naturally associated with quantum mechanics, is considered as a key ingredient in a more recent approach to consciousness studies, questioning the validity of purely algorithmic prescriptions for addressing phenomena like human insight. To deal with this issue of 'non-computability', Penrose (1995) suggests a new role for the environment, viz., as an 'external guide' influencing decisions in an algorithmic system, and for this a necessary condition is its quantum coherence. For a more concrete representation of such a picture in a biosystem, let us look at Frohlich's (1968) coherent excitations envisaged for cell membranes. Now, a stationary state is reached if the energy fed into an assembled material with polar vibrational modes, is sufficiently larger than that lost to bath degrees of freedom. Then, in the words of Frohlich (1968), 'The long-range Coulomb interaction then causes this energy to be shared with other dipoles.' The dipoles will tend to oscillate coherently provided the energy supply is sufficiently large compared with the energy loss. Nonlinear effects are likely to reduce this loss with increasing excitation and effectively transfer the system into a metastable state in which the energy supplied locally to dipolar constituents is channelled into a single longitudinal mode which exhibits long-range phase correlations'. Of course, the energy of the metabolic drive must be large enough, and the dielectric properties of the materials concerned need to have a matching capacity for maintaining (and withstanding) extremely high electric fields (Frohlich 1975). This is similar to Bose-Einstein condensation, in which a large number of particles participate in a single quantum state', i.e., behaving as one with a wave function applicable for a single particle, albeit scaled up appropriately. Despite their much higher than 'absolute zero' temperatures, it seems that coherence conditions are not only met in bio-systems, but that there is some direct experimental evidence too (Grundler and Keilmann 1983) of 10^{11} Hz oscillations predicted by Frohlich (1968). If this seems surprising vis-a-vis physical systems, note that it is because of (not despite!) the warm and soft sol-gel state that efficient nano-machines actually harness thermal fluctuations (Bustamante et al 2005, see below) in biology, where energy transformations occur under isothermal conditions.

4. Penrose-Hameroff Model: New Role for the Environment

In looking for appropriate biomaterials that on supplying with energy led to Frohlich-like excitations in the sub-nanosecond range, the Penrose-Hameroff model zeroes in on microtubules having the required ingredients of voltage effects and a geometry favoring "coupling among subunits" (Hameroff 2003) for acting as reservoirs of highly ordered energy. The early-evolved cytoskeleton forms the basic 'building block' in their new fractal perspective of the nervous system, which argues for a change in paradigm so that the sophisticated actions of animals down to single celled ones (all affected by general anaesthetics at about the same concentration) can be explained using only one basic control system (Penrose 1995). Now functional protein conformations appear to be correlated to such collective 'metastable states' mediated via the surroundings: "action of electric fields, binding of ligands or neurotransmitters, or effects of neighbor proteins' (Hameroff 2003). Thus consciousness can be partially inhibited using anaesthetics. (By reversibly binding to hydrophobic pockets within key neural proteins via weak forces, these can alter the environmental medium and thereby the electron mobility, in turn non-linearly coupled to mechanical movements).

A complementary approach to such long range cooperativity is due to Davydov (see Lomdahl et al 1984) who considers wave-like propagations --solitons-- for the spatial transfer of vibrational energy in ordered form, which also can be derived from the same type of non-linear effects leading to the coherent Frohlich ordered state. Tuszynski et al (1984) observe that while the latter lays its emphasis on time-independent dynamical ordering aspects, the former offers a plausible mechanism for not only localization but also transporting order through the system (time dependent aspects). (The unusual resilience of a soliton-- a quantum of energy that propagates as a traveling wave in nonlinear systems-- stems from two opposing tendencies as a result of which a dynamically stable entity emerges). Indeed, as the substrate for energy transfer in the cytoskeleton, just like electrons in computers, Hameroff (2003) considers Davydov's solitons. 'Objective reduction' (a self-collapse of the observer's wave function), then occurs in this algorithmic coherent system (see above), one in which the external (gravitational) field plays a key role. Here, the Penrose-Hameroff thesis makes a major departure from the conventional view, in that the (non-computable) field shows a new and active role for the environment, viz., as a concerned 'teacher' with a deep involvement in the system's decision-making- not merely a neutral examiner, assessing system variants (Penrose 1995).

Further, another study (Davia 2006) suggests the relation between the organism and the environment as one of mutual 'friendliness', in contrast to the reigning Darwinian perspective where, apart from being a source/sink, the environment presents itself to living systems as a sort of (potentially destructive) obstacle course to be negotiated; and the organism appears as a machine within an environment, with no causal relationship between the two. Briefly, Davia seeks to demonstrate that the question of how life maintains its organization through time is central to an understanding of the brain. To that end, he postulates life to be a scale- free (fractal) process of catalysis (which involves the fusion of energy and structure in the form of solitons). Then, rather than a hostile 'obstacle course', the environment becomes a willing partner in a set of transitions mediated by the living process via 'catalysis'.

5. Non-computable Bio-issues: External Control?

According to Goedel's incompleteness theorem, with any set of axioms, it is possible to produce a statement that is obviously true, but cannot be proved by means of the axioms themselves. Penrose (1995) took advantage of the Goedel theorem to claim that the functioning of the human brain also includes non-algorithmic processes, i.e. a system can be deterministic without being algorithmic. For this an excellent candidate is again quantum mechanics in its full glory of quantum coherence, in common with Bohm's (1990) semi-intuitive thesis. Now, mathematical induction is a well known concept which is akin to Goedel's theorem. Inspired by Penrose, we wish to extend this terminology to a physical level via the well known phenomena of (electric, magnetic) field-induced effects, which although conceived classically has a good promise of quantum extension. And with due respect to gravitation, the effects of other fields at different levels, from classical to quantum, should not be neglected in view of the properties of biomatter (Cope 1975).

We again return to the theme of non-computability for thought processes (Penrose 1995), looking to the environment as exerting external control. Now in fact, across biology we encounter instances where bio-solutions can include choices outside the space of existing possibilities. For instance, consider Bio-evolution (in particular the algorithmic complexity of sequences pointed out by Abel (2009)), and note that a similar Darwinian selection at the time scale of days--affinity maturation in B-cells-- can be found in higher vertebrates. So it is natural to ask how life itself must have emerged, (very likely from a set of not-as-yet-living systems), thus taking the problem to the door of life's origin!

6. The Environment as Guide; a Scaffold for Life (?)

Now in addition to the vital role for solitons in today's biology, they have strong implications for life's emergence owing to their fundamental association with both energy and information (Taranenko et al 2005). That is the boundary conditions offered by repeating structures could have been the answer to how energy and structure in biology got synonymous (Davia 2006), so that the patterns sustaining these quasiparticles could have been retained while gradually replacing the materials embodying them, en route to present day versions of metabolism and replication (c.f. Cairns Smith 1982; see below). In this context, Davydov's (1991) proposal for 'electrosolitons' (a plausible mechanism for electron transfers across distances with minimum energy losses, traditionally approached using tunneling effects) seems to be highly relevant for the hydrogenation of CO_2--seen as the basis for life's emergence (Nitschke and Russell 2009). Indeed, the quantum metabolism model of Demetrius (2008) approaches the issue of energy harnessing in the ATP-membrane proton pump-the most primitive of energy transduction mechanisms-using Frohlich's coherent excitations. Hence, revisiting Cairns- Smith's (1982) idea of a mineral scaffold for life 'taken over' by organic matter, in the light of these insights, prompts the question of whether the above non-linear interactions leading to coherent dynamics could have been achieved using simpler/less sophisticated substances that in turn got gradually replaced by the advanced ones of today with greater time-stability. Importantly, the new ingredient would be the environment having a penetrating influence in the coherent scaffold, and taking decisions a la Penrose.

A few years ago, we have drawn attention to another ubiquitous ingredient in terrestrial phenomena, viz., magnetism, which appears to exert its influence across kingdoms of life, and has a natural association with quantum coherence (see Merali 2007). A soft colloidal scaffold a la Russell and coworkers (1989) in terms of a field-induced assembly of magnetic dipoles (Mitra-Delmotte and Mitra 2010b), seems equipped with the potential to address symmetry-broken dynamics for primordial chemical reactions hosted within its 'layers'. A magnetic field can 'order' magnetic nanoparticles; the resulting structural order in natural assemblies could provide the boundary conditions needed for generating soliton-like structures. The synonymy between structure and energy across biology (Davia 2006) makes it compelling to speculate if magnetic solitons could not have been a primitive mechanism (c.f. Cairns- Smith 2008) for energy transport in a natural assembly (Russell et al 1990; Sawlowicz 1983), whose dynamical order was controlled by a field. To that end, it is encouraging to find studies using particles interacting via dipolar interactions (Ishizaka and Nakamura 2000) and indeed worth noting the recent field-modulated dynamics of magnetic nanoparticle ensembles by Casic et al. (2010). That solitons could be linked to transfer of order within field-induced colloidal structures, shows to what extent the analogies of energy landscapes for protein-folding and of disordered (solid) spin systems can be extended, and which thereby reduces the immense gap demarcating living and (considered as) non-living matter. Then it becomes tempting to cite a few other apparently disjointed features which fit into a bigger mosaic. For instance, there are intriguing analogies of conformational fluctuations of 'sophisticated' motor proteins carrying a load, with infinitesimal spin alignment changes of magnetic dipoles, ligand bound to organics, making their way through templates of head-to-tail aligned electric and magnetic dipoles, respectively (Mitra-Delmotte and Mitra 2010a). These can throw light on how thermal fluctuations can be harnessed in a simpler system with such life-like features, and which seems plausible to imagine in a Hadean setting.

Like in ATP-driven molecular motors, a gentle flux gradient (in a non-homogeneous rock magnetic field) can offer both detailed-balance breaking non-equilibrium as well as asymmetry to a magnetic dipole. Again, the correspondence of the local lowering of temperature (towards aiding coherence) theorized by Matsuno and Paton (2000) via the slow release mechanism of ATP hydrolysis, to the magnetic scenario comes in the form of an accompanying magnetocaloric effect, which allows interchange between system-entropy and bath temperature. And, not only does the matter-structuring role of a magnetic field gel with the boundary requirements of soliton-like structures, it provides a friendly background for a more dynamic role mediated by the soliton, besides being the same ingredient already found to be crucial to the Frohlich mechanism (1968).

7. Any Direct Link to the Origins?

Indeed, the above inductive form of reasoning by analogies, which is complementary to algorithmic deduction (c.f. Penrose 1995), matches the traditional 'pattern-recognition' approach to biology. Guided by the above, and the repetitively appearing phenomenon of sequential induction (in association with a self-referential character) across a hierarchy of life-processes, we propose that associated induction patterns could offer a richer 'simulation' of an 'active' experience as compared to a mere collection of data on a screen by a computer programmed to 'passively' mimic the same.

Quantum Physics, Evolution, Brain, Mind

Figure 1: The subjective experience

The picture above (Figure 1) (taken from the Web) depicts the said subjective experience. Indeed, the communication of the image via propagating patterns induced within a biological device could be what makes room for optical illusion effects and subjectivity. Now contrast this typical example of observation/measurement in biology with a detector (without outside field effects), say a camera, where the corresponding information gets quenched on the image plate. So this begs the question whether sequential induction as a measuring mode could throw light on the origins of the complex phenomena of consciousness, in view of the field-susceptible nature of biological matter. Note further that an environment--as a field-- does have the potential for induction, given access to any active d.o.f.s in matter. Such a scenario seems to gel with our proposal of a field-induced primitive scaffold for life.

8. Conclusion: Extra Scientific Dimensions?

We have chosen only a few samples of consciousness models, at the same time trying to extrapolate the environment-related issues to the emergence of life itself, yet they seem to go hardly beyond scratching only the outer surfaces of the problem so far. The huge gap is perhaps symbolized by a perspective taken from Whitehead if one substitutes 'consciousness' for his definition of 'religion' (Whitehead 1970):

'Religion is the vision of something which stands beyond, behind, and within the passing flux of immediate things; something which is real, and yet waiting to be realized; something which is a remote possibility, and yet the greatest of present facts; something which gives meaning to all that passes, and yet eludes apprehension; something whose possession is the final good, and yet is beyond all reach; something which is the ultimate ideal, and the hopeless quest.'

Acknowledgements: This essay is dedicated to the memory of ANM's mother, Rama Rani Mitra, on the occasion of her birth centenary (2011). We thank Prof. M.J. Russell for inspiration and constant help; and Dr.J.J. Delmotte for financial/infrastructural support.

References

Abel, D. (2009). The Capabilities of Chaos and Complexity. Intl J. Mod Sci, 10, 247-291.

Aspect, A., Grangier, P., Roger, G. (1982). Experimental realization of Einstein-Podolsky-Rosen-Bohm Gedankenexperiment: A new violation of Bell's inequalities. Phys. Rev. Lett., 48, 91-4.

Bergson, H. (2001[1889]). Time and free will. Sage Publication, New York.

Bohm, D. (1952). A Suggested Interpretation of the Quantum Theory in Terms of "Hidden" Variables. II , Phys. Rev. 85, 180Ð193.

Bohm, D. (1990). A new theory of the relationship of mind and matter. Philosophical Psychology, 3, 271-286.

Bustamante, C., Liphardt, J., Ritort, F. (2005) The nonequilibrium thermodynamics of small systems. Physics Today, 58, 43-48.

Cairns-Smith, A. G. (1982). Genetic Takeover and the Mineral Origins of Life. Cambridge University Press, Cambridge.

Cairns-Smith, A.G. (2008). Chemistry and the Missing Era of Evolution. Chem. Eur. J., 14, 3830-3839.

Casic, N., Schreiber, S., Tierno, P., Zimmermann, W., Fischer, T.M. (2010) Frictioncontrolled bending solitons as folding pathway toward colloidal clusters, EPL, 90, 58001

Cope, F.W. (1975). The solid-state physics of electron and ion transport In biology. A review of the applications of solid state physics concepts to biological systems. J. Biol. Phys., 3, 1-41.

Davia, C.J. (2006). Life, Catalysis and Excitable Media: A dynamic systems approach to metabolism and cognition. In: Tuszynski, J. (Ed.), The Emerging Physics of Consciousness, Heidelberg, Germany: Springer-Verlag.

Davydov, A.S. (1991). Solitons in Molecular Systems. Kluwer, Dordrecht, p115.

Demetrius, L. (2008). Quantum Metabolism and Allometric Scaling Relations in Biology, In: Abbott, D., Davies, P.C.W., Pati, A.K. (Eds.), Quantum aspects of life, Imperial College Press, London, pp. 127-146.

Einstein, A., Podolsky, P., Rosen, N. (1935). Can quantum-mechanical description of physical reality be considered complete? Phys. Rev. 47, 777-780. Frohlich, H. (1968). Long-range coherence and energy storage in biological systems. Int. J. Quant.Chem. 11, 641-649.

Frohlich, H. (1975). The extraordinary dialectric properties of biological molecules and the action of enzymes. Proc. Natl. Acad. Sci, 72(11), 4211-4215.

Grundler, W., Keilmann, F. (1983) Sharp resonances in yeast growth prove non thermal sensitivity to microwaves. Phys. Rev. Lett. 51(13), 1214-1216.

Hameroff, S. (2003). Ultimate Computing: Biomolecular consciousness and nanotechnology. Elsevier-North, Holland (1987), electronic ed.

Ishizaka, S., Nakamura, K. (2000) Propagation of solitons of the magnetization in magnetic nano-particle arrays. J. Magn. Magn. Mater., 210, 15.

Larter, R.C.L., Boyce, A. J., Russell, M.J. (1981). Hydrothermal pyrite chimneys from the Ballynoe baryte deposit, Silvermines, County Tipperary, Ireland. Mineralium Deposita, 16, 309-317.

Lomdahl, P.S., Layne, S.P., Bijio, I.J. (1984). Solitons in Biology. Los Alamos National Laboratory, Los Alamos, NM (2nd edn.), p. 3.

Matsuno, K., Paton, R.C. (2000). Is there a biology of quantum information? Biosystems 55, 39-46.

Merali, Z. (2007) Was life forged in a quantum crucible? New Sci., 196 (8 Dec.), 6-7.

Mitra-Delmotte, G., Mitra, A. N., (2010a). Magnetism, entropy, and the first nanomachines. Cent. Eur. J. Phys., 8(3), 259-272.

Mitra-Delmotte, G., Mitra, A. N. (2010b). Magnetism, FeS colloids, and the origins of life. In: Alladi, K., Klauder, J.R., Rao, C.R. (Eds.), The Legacy of Alladi Ramakrishnan in the Mathematical Sciences, Springer New York.

Nitschke, W., Russell, M.J. (2009). Hydrothermal Focusing of Chemical and Chemiosmotic Energy, Supported by Delivery of Catalytic Fe, Ni, Mo/W, Co, S and Se, Forced Life to Emerge. J. Mol. Evol., 69 (5), 481-96.

Penrose, R. (1995). Shadows of the Mind. Vintage, Random House, London.

Pribram, K.H. (1975). Towards a holonomic theory of perception. In: Ertel, S. (Ed.), Gestalttheorie in der modern psychologie, Erich Wengenroth, Koln, pp. 161-184.

Russell MJ, Hall AJ, Gize AP (1990) Pyrite and the origin of life. Nature 344, 387.

Russell, M.J., Hall, A.J., Turner, D. (1989). In vitro growth of iron sulphide chimneys: possible culture chambers for origin-of-life experiments. Terra Nova 1(3), 238-241.

Sawlowicz, Z. (1993). Pyrite framboids and their development: a new conceptual mechanism. Intl. J. Earth Sci 82, 148-156.

Tuszynski, J. A., Paul, R., Chatterjee, R., Sreenivasa, S.R. (1984). Relationship between Fršhlich and Davydov models of biological order. Phys. Rev. A, 30(5), 2666-2675.

Taranenko, V.B., Slekys, G., Weiss, C.O. (2005). Spatial Resonator Solitons, In: Akhmediev, N., Ankiewicz, A. (Eds.), Dissipative Solitons, Lect. Notes Phys., 661, Springer-Verlag, Heidelberg, 131-160.

Whitehead, A.N. (1970) Religion and Science, In: Gardner, M. (Ed.), Great Essays in Science, Washington Square Press, New York.

30. The Evolution of Human Consciousness: Reflections on the Discovery of Mind and the Implications for the Materialist Darwinian Paradigm

Martin Lockley

University of Colorado at Denver, Denver CO 80217-3364 USA

Abstract

Modern anthropology and paleontology generally look at hominid evolution through the 'physical' lens of anatomy and technological and cultural development. This approach may allow correlation between pre Homo sapiens species and their different anatomies with very rudimentary technology and art, presumed to indicate minimal self awareness. However, physical/anatomical-mind/culture correlations have limitations when applied to the ~150,000 year evolution of Homo sapiens, because without demonstrable changes in anatomy (brain architecture) there have been vast changes in cultural development (art, science and technology). These changes imply significant psychological evolution tied to an increase in individual self awareness. Just as distinct stages of psychological (emotional/intellectual) growth manifest in child development as s/he discovers the physical, sensory world and the emergent properties of mind/intellect/thought, so too stages of psychological development are manifest in the collective evolution of consciousness of our species as a whole. Given the robust biological evidence that evolution of development (evo-devo) involves convergent, recapitulating pathways, wherein ontogeny recapitulates phylogeny to varying degrees, there is equally compelling evidence that individual psychological development (ontogeny) is a complex and variable recapitulation of the evolution of consciousness at the species level (phylogeny). Among the best examples one may cite the development and co-evolution of language, religious sentiment, 'mind' and self-consciousness as traced in western literature since ~12th century BCE. Understanding these developments goes a long way towards explaining human fascination with our physical, psychological and spiritual origins.

KEY WORDS: human evolution, cultural evolution, self consciousness, ontogeny, phylogeny

1. Introduction

Modern evolutionary theory, like physical and cultural anthropology is barely two centuries old. However, 19th century views of human ancestors, and extant stone age (Paleolithic and Neolithic) cultures, as 'primitive' are today considered politically-incorrect, western, Eurocentric elitism, born of the evolutionary notion of 'progress' especially in science, technology and economics, but also in other areas such as law, ethics and morality. However, it may be equally misguided to view ancient cultures as 'the same' as ours or to dismiss the concept of progress entirely. Can we say the ancients employed rational analysis of the physical world, when the evidence overwhelmingly indicates that they produced cultures so manifestly different from ours?

At the present juncture, paleontology, anthropology and evolutionary psychology mostly focus on materialistic Darwinian evolutionary models that seek to understand developments in physical brain size, brain function and behavior in terms of adaptations and selective pressures that took place in the ice age: i.e. before 10,000 B.C. However, such approaches have yet to adequately explain the emergence of our most human 'psychological' traits - self-consciousness and language, even if it is now possible to understand which areas of the brain are 'activated' by language and other faculties in extant humans (Joseph, 2001). A more fruitful approach may be to consider how human development (ontogeny) is as much a psychological process as it is a phenomenon of physical growth, or as stated by Anderson (1995, p. 134-135) "psychological advances were not accomplished in the classic evolutionary sense. Rather they reflect movement....every bit as inherent as the physical one that creates the brain, the heart, and the muscles..." Thus, the evolution of our young (~150,000 year old) species may be better understood through a thoughtful understanding of our psychological maturation (sensu Anderson) or 'evolution of consciousness' expressed as much through art, language/literature and religion/spirituality, as through technological and economic manipulation of the physical environment. Whether physical evolution drives psychological evolution, or vice versa, is perhaps of less importance than to recognize that the two are intimately interrelated: see Trut (1999) for an exposition on the interplay of behavior and physical development in evolution.

Given that newborn infants lack language and self consciousness, and that humans consider themselves the only species manifesting well-developed self-consciousness the following conclusion is inescapable: humans became self conscious at some point in their history or prehistory (phylogeny) just as the individual becomes self conscious during ontogeny (Zahavi et al. 2004; Lockley 2010). This conclusion points to the importance of some psychological variant of the biogenetic law, that 'ontogeny recapitulates phylogeny' (Haeckel, 1866). Although Haeckel has been criticized for his errors and data manipuilation, in principle the biogenetic law can claim to be a pivotal precursor to the now- thriving field of evo-devo (Hobfeld and Olsson, 2003). In the 19th century the anthropological notion that our ancestors were child-like often carried the derogatory implication that they were primitive, emotional and lacking in reason and intellectual maturity (and children were often treated accordingly as 'incomplete,' immature adults).

However, as a species, humans manifest a number of remarkable juvenile traits, such as hairless bodies and extended periods of play (Verhulst, 2003). Ironically in the biological fields of 'heterochrony' and the 'evolution of development' (evo-devo), where such biological traits are well understood, such juvenilization traits, which 'retard' development, thereby preventing overspecialization (early aging or gerontomorphism) can be considered 'progressive' or novel evolutionary developments: see Trut (1999) for further observations. What lessons might we learn from changes in our own scientific paradigms? Obviously our scientific theories change as our consciousness evolves. Just as juvenile behavior is only juvenile, when viewed from the adult perspective, so ancient cultures cannot be labeled as advanced or primitive, without understanding the context in which they are viewed, and admitting our own prejudice for regarding ourselves as advanced and progressive. While present scientific hubris is often self-congratulatory on the subject of our technological achievements, our science fiction mythology readily speculates that future races (species) will regard us as primitive. In the short exposition which follows, we take these considerations into account as we look at the consciousness and culture of our ancestors from this psychological/consciousness perspective.

2. Before the 'World' Began

Humans are obsessed with origins, whether pertaining to the universe, the first unicellular life, vertebrates, mammals or our own human species. While our present self-consciousness allows us to experience existence, this same condition makes us wonder what it is like not to be self conscious. [Although this is rather silly, because one cannot experience 'non-experience,' one can nevertheless be curious about 'other' experience, especially spiritual epiphany, which humans have a long history of exploring and reporting, not least because spiritual gurus have made intriguing references to immortality]. The philosophical psychologist or theologian might observe that such obsessions with origins are a natural consequence of the mysteries of life and death, or the 'fact' that most of our evolutionary history pre-dates self consciousness. Given that our species had an origin (birth) and may die (go extinct), our interest in survival (also embedded in Darwinian evolutionary theory) is an understandable consequence (projection) of our perceived mortality. Perhaps, the only way round the perceived finality of mortal existence is to postulate immortality, a subject which has again preoccupied humanity for millennia. [Here one may distinguish between the psycho-spiritual and physical meanings of such intimations]. In biological terms immortality is not an altogether unrealistic, concept. Some biologists speak of immortality to characterize the asexual reproduction of countless identical clone generations. Even at the more complex level of sexual reproduction evolution, closely related generations and species transform one into another, so that while one may be lost another incarnates in its place. [Transformation and transmutation, are historical synonyms for evolution, and so could arguably be considered synonyms of reincarnation]. Are these preoccupations delusional, because in reality we are unequivocally and physically mortal? Or do they indicate an intuitive grasp of deeper non-material or spiritual realities that can be better appreciated by understanding the evolution of consciousness- or Consciousness in the Universe?

3. When Humans Became Self-Conscious and the 'World' Began

One might argue that it is almost self-evident that the emergence of human self consciousness, coincided with the origins of language and the first perception of what we call the 'world.' Abundant evidence from child development studies shows that this happens in most normal development (ontogeny). Rare cases of feral children and abnormal development severely compromise the quintessential human traits of language, self consciousness and their experience of the world (Newton, 2003). Normally, we speak of the child beginning to 'explore the world.' Such explorations involve self-referential language and emotional experience leading to cognitive understanding and meaning. Without language and self consciousness, the child's psychological experience would not be fully or healthily human. Indeed, as we have seen, the child's experience is not always easily conveyed to adults due to linguistic limitations. Likewise, the non-human experience of animals cannot be conveyed to humans linguistically, even if other non-self-conscious means of communication may be effective to varying degrees. Guldberg (2010), for example, points out that the ability of apes to use sign language, and recognize themselves in mirrors is very limited and too often seriously misinterpreted.

Commentators on the evolution of culture and language, have made the often compelling case that various ancestral cultures, through symbols, art and texts, have clearly recorded their experience of coming into the world of sensory perception and physical matter. By burying their dead such ancestral cultures also appear aware to have been aware of their mortality (Joseph, 2001). At such crucial stages in the evolution of consciousness and culture, there is a marked emphasis on fundamental spiritual questions pertaining to origins and the immortality of the soul. It is interesting that children very often also ask these same profoundly metaphysical life and death questions at a very early age (~3-4 years).

4. The Discovery of Mind

In a classic study of the evolution of Greek literature and culture Bruno Snell makes the compelling case that what we might call the modern European mind, evolved in a series of steps from the time of Homer (~12th century BCE) through the time of Socrates, Plato and Aristotle (6th-4th centuries BCE). During this span of time, roughly equivalent to the duration between the contributions of Thomas Aquinas and Albert Einstein, we witness the Discovery of the Mind (sensu Snell 1953). In less than a millennium, the Greeks discovered their inner souls and what it meant to be "individual" men and women. They carved proud statues of themselves, not just of the Gods. Put another way they developed a species of self-consciousness, discussed herein, that we would recognize as progressively-more characteristic of the modern human condition.

What evidence did Snell glean from his intimate knowledge of Greek literature? From the outset Snell (1953, p. v) stresses that the emergence of mind involved the "rise of thinking" (or intellect) which heralded "nothing less than a revolution" because "the existence of the intellect and the soul are dependent on man's awareness of his self" (op. cit. p. ix). Thinking of course involves the ability to make abstract internal representations of the world, and it is well known through Child Development studies that the thinking faculty 'arises' at a relatively late stage: i.e., at around 5- 7 years, with the advent of what Piaget (1976) calls the Concrete Operations stage.

Snell argues that we fail to grasp the ancient Greek mind if we 'think' people thought the way we do today. In a series of carefully argued steps Snell unravels what "the Greeks at any given time know about themselves." (Snell 1953, p. xi). Snell notes that Aristarchus was the first to notice that in Homer the word (soma), meaning body, is never used with reference to a living being (p. 5). Rather, Homer locates the vital "secret of life" in the mobile limbs, and the word thymos sometimes translated as 'soul' has, according to Snell, a more physical meaning of (e)motion: i.e. that which provides motion for the limbs. When used to describe "the escatological soul which flies off at the moment of death" (p. 11) it refers to the death of an animal. In contrast psyche often meaning "breath of life," and also translatable as 'soul,' is regarded as an animating force characteristic of humans that, at death, is seen to leave the body through the mouth. Noos, again having the connotation of 'soul' has yet another meaning closer to the faculty "in charge of intellectual matters," realization or seeing/perception (Snell, 1953, p. 12-13). Thus "Homeric man" has three 'organs' - thymos (motion/emotion), psyche (life) and noos (perception) - regarded as functions, not as physical entities.

By the time soma was used in reference to living bodies, Heraclitus (540-480 BCE) had used psyche to refer to the soul of a living person. Thus, arose the first distinction between the physical body and a radically different soul quality, which ever since has given us debates about mind-body duality. Before this, in the time of Homer, however, the soul had no depth or intensity and the "proper dimension of the spiritual receives no attention" (Snell, 1953, p. 19) nor has "Homeric man …yet awakened to the fact that he possesses in his own soul the source of his powers…" Thus, he does not regard himself as the source of his own decisions. Until he could do this he had not transplanted the external deity within himself as his own soul, and so the world could have little deep meaning. To the extent that individuals did this, they and their communities gained spiritual potential (potency) and, correspondingly, the Gods lost their power. This internalization of the divine, spiritual faculty is commented on by Jung in is interpretation of how the Old Testament Job "recognized God's inner antinomy…[it] allowed Job (man) to attain a divine numinosity" (Lockley and Morimoto, 2010 p., 260). When humans began to discern the moral and spiritual dimensions of events in their lives, they no longer attributed their fates to the Gods, and they began the philosophical process of evaluating morality (truth, beauty, goodness, virtue honor etc.,). They could as Snell puts it "toil for the sake of responsibility and justice" (op. cit. 102). This shift was often accompanied by a loss of awe and wonder, which various commentators have since alternately lamented or justified as necessary, perhaps even inevitable, as the capacity to distinguish good from evil arose. Clearly this distancing of the individual from the Gods (the invisible divine dimension of the universe) led to a greater burden of responsibility, loneliness, angst and the religious tradition of the "fall." So at first, according to Snell, we find a literary tendency to dialogs dealing with nostalgia for lost loved ones, but later, we encounter the monologue and the full expression of individual angst. Determining one's fate with no guidance from omniscient gods led almost inevitably to the wrenching drama of Greek tragedy arising from the dilemmas involved in making difficult decisions.

As noted by Rochat (2003) the development of "self awareness is a dynamic process" unfolding and "oscillating" not just in early life but throughout our ontogeny. In reading Greek literature and Snell's interpretations we see how our own decisions "oscillate" constantly as we apply moral, judgments to life situations. However, we must remember that the intellectual repertoire of the Greeks differed from our own, and that we would hardly find the justifications for their decisions satisfying to our morality. According to Barfield (1926, 1965, 1967) whose literary knowledge rivals, and whose interpretations, often parallel Snell's, it took time for thought to separate from perception, and so free the soul. Thus, Barfield holds there is no shred of evidence that the early Greeks interpreted life's events by rational or intellectual analysis, which requires the abstract and objective faculty we call "imagination." Likewise, Snell (1953, p.90) states that what we call myths were "accepted as reality' and that "what we would ascribe to the imagination, to an intellectual effort…. Homer… traces to actual experience." (op cit. p. 137).

Long (1984, p. 6) espouses much the same position when he states that " The Greek myths and Hindu traditions tell of gods and demi-gods walking the earth ... yet no one has ever seriously suggested that this was precisely how it was .." Jaynes (1976, p. 75) clearly echoes these interpretations when he states that Greek soldiers "did not have subjectivity as we do" [they] "had no awareness of [their] awareness of the world, no internal mind-space to introspect on." As a result they were "noble automatons who knew not what they did" fighting wars effectively "directed by hallucinations."

5. Self-Consciousness, Mind-Body and Heaven - Earth Dualism

> 'It is almost an absurd prejudice to suppose that existence can be only physical ... We might well say, on the contrary, that physical existence is a mere inference, since we know of matter [only] in so far as we perceive psychic images mediated by the senses'. -Carl Jung (1958)

Long (1984) should perhaps have cited Snell, Barfield and Jaynes on the authenticity of Greek experience. However he is on the same track when arguing that "neither the Greek myths, the bible or the early Sanskrit texts ... mentions that the accounts do not stand for exactly what they say. .. Central ... to all religious traditions are the unqualified references to the gods and demi-gods with miraculous powers participating in human affairs ..." (op. cit , p. 7).

In this vein of "heaven and earth dualism" we can cite from the much celebrated Nag Hammadi manuscripts: "the soul turned, at one time, toward matter: she fell in love with it, and, burning with desire to experience bodily pleasures, wishes no more to be separated from it. Thus the world was born.

From that moment the soul forgot herself; she forgot her original dwelling, her true center, her everlasting life..." (Chwolsohn, 1856; cited in Doressse, 2005)

Vitaliano (2000) is cogent in stating that: 'dualism is the act of severance, cutting (con-scire) the world into seer and seen, knower and known ... with the occurrence of the primary dualism, man's awareness shifts from the non-dual universal consciousness to his physical body."

The momentous implication of this conclusion is that, like infants, before they become self conscious, our early human ancestors not only had little concept of an inner self (soul) but they did not even recognize or identify with their physical, sensory bodies. Obvious as it may seem on reflection, we have to agree with Jung that self-awareness is an essential a pre-requisite for recognizing the physical existence of the body. Thus, as humans became aware of the physical world they still retained strong psychic ties to the non physical, non-sensory "spiritual" or psychic world from which they were just emerging. This in turn implies an abrupt emergence of sensory experience. Again biology and psychology confirm such experiences, not only at birth (when sight and other senses first function, but also with the advent of self consciousness at the toddler stage and in the many accounts of spiritual experience, including Near Death and Out of Body experience which, in the latter case, cross back from the world of physical to non-physical sensory experience: i.e. into the realm Vitaliano calls "non-dual universal consciousness."

Difficult as it may be for materialistic paradigms to accept that human evolution is not adequately or fully explained as a process of physical/anatomical evolution, caused by individual subjects interacting with objects in a dualistic world full of selective pressures, much evidence points to the fact our most characteristic attributes of our species (language, awareness of our existence in a physical, sensory world) came about as the result of a "fall" into self-consciousness quite late in our species history. These emergent psychological or consciousness faculties appear to be an inherent part of psychological maturation processes (sensu Anderson, 1995) with no simple, obvious or easily measured physical correlates such as changes in brain anatomy (brain waves and energy fields not withstanding). This is not to say that there are not intriguing expositions on the evolution of language in and of itself (Deutsher 2005), and in relation to neuroanatomy (Joseph, 2001). But again it seems the faculties of consciousness and language 'activate' or 'emerge' independent of any easily defined physical/anatomical changes. Thus, given the increasingly ambiguous meaning that modern physics gives to the concept of 'physical matter' it seems unlikely that explanations for the evolution of human consciousness will center on any strictly physical/material phenomena.

References

Anderson, C. (1995). The Stages of Life. Atlantic Monthly Press, New York.

Barfield, O. (1965) Saving the Appearances: Studies in Idolatry, (second ed. 1988) Wesleyan University Press, Hanover, NH.

Barfield, O. (1967). Speaker's Meaning, Wesleyan University Press, Hanover, NH.

Barfield, O. (1926/1988). History in English Words, Lindisfarne Press. Hudson NY.

Chwolsohn, D. (1856). Die Sabier und de Sabismus. St. Petersburg.

Deutscher, G. (2005) The unfolding of language Henry Holt and Co , New York.

Doresse, J. (2005). The discovery of the Nag Hammadi Texts, Inner Traditions, Rochester,

Guldburg, H. (2010). Just another Ape? Academic Imprint, Exeter, UK.

Haeckel, E. (1866) Generelle Morphologie der Organismen: allgemeine Grundzüge der organischen Formen-Wissenschaft, mechanisch begründet durch die von Charles Darwin reformirte Descendenz-Theorie, 2 vols., George Reimer, Berlin.

Hobfeld, U and Olsson, L. (2003). The Road from Haeckel: the jena tradition in Evolutionary Morhology and the Origins of "Evo-Devo." Biology and Philosophy 18: 285-307.

Jaynes, J. (1976) The Origin of Consciousness in the Breakdown of the Bicameral Mind, Houghton Mifflin, New York.

Joseph, R. (2001). The Limbic System and the Soul: Evolution and the Neuroanatomy of Religious Experience. Zygon, the Journal of Religion & Science, 36, 105-136.

Jung. C. (1958). Psychology and Religion:West and East, Pantheon New York.

Lockley, M. G. 2010. The evolutionary dynamics of consciousness. Journal of Consciousness Studies, 17, 266-116.

Lockley, M. G., and Morimoto, R. (2010). How Humanity Came into Being. Floris Books, Edinburgh.

Long, B. (1984). The Origins of Man and the Universe: The Myth that Came to Life, Routledge and Kegan Paul, London.

Newton, M. (2003). Savage Girls and wild boys: a history of Feral Children. MacMillan, London.

Piaget, J. (1976). The Grasp of Consciousness: Action and Concept In the Young Child, Harvard University Pres, Cambridge MA.

Rochat, P. (2003). Five levels of self-awareness as they unfold in early life. Consciousness and Cognition, 12: 717-731.

Snell, B. 1953. The Discovery of the Mind. Harvard University Press, Cambridge MA.

Trut L. N. (1999). Early Canid Domestication: the arm-Fox experiment. American Scientist, 87, 160-169.

Verhulst, J. (2003). Developmental Dynamics in Humans and other Primates: Discovering Evolutionary Principles Through Comparative Morphology, Adonis Press, New York.

Vitaliano, G. (2000) "A new integrative model for states of consciousness." NLP World 7: pp. 41-82.

Zahavi, D., Grunbaum, T. and Parnas, J. (Eds.) (2004). The Structure and Development of Self-Consciousness: Interdisciplinary Perspectives. J. Benjamins, Amsterdam.

31. Evolution of Paleolithic Cosmology and Spiritual Consciousness, and the Temporal and Frontal Lobes
Rhawn Joseph, Ph.D.

Emeritus, Brain Research Laboratory, California

Abstract

Complex mortuary rituals and belief in the transmigration of the soul, of a world beyond the grave, has been a human characteristic for at least 100,000 years. The emergence of spiritual consciousness and its symbolism, is directly linked to the evolution of the temporal and frontal lobes and to the Neanderthal and Cro-Magnon peoples, and then the first cosmologies, 20,000 to 30,000 years ago. These ancient peoples of the Upper and Middle Paleolithic were capable of experiencing love, fear, and mystical awe, and they carefully buried those they loved and lost. They believed in spirits and ghosts which dwelled in a heavenly land of dreams, and interned their dead in sleeping positions and with tools, ornaments and flowers. By 30,000 years ago, and with the expansion of the frontal lobes, they created symbolic rituals to help them understand and gain control over the spiritual realms, and created signs and symbols which could generate feelings of awe regardless of time or culture. Because they believed souls ascended to the heavens, the people of the Paleolithic searched the heavens for signs, and between 30,000 to 20,000 years ago, they observed and symbolically depicted the association between woman's menstrual cycle and the moon, patterns formed by stars, and the relationship between Earth, the sun, and the four seasons. These include depictions of 1) the "cross" which is an ancient symbol of the fours seasons and the Winter/Summer solstice and Spring/Fall equinox; 2) the constellations of Virgo, Taurus, Orion/Osiris, the Pleiades, and the star Sirius; 3) and the 13 new moons in a solar year. Although it is impossible to date these discoveries with precision, it can be concluded that spiritual consciousness first began to evolve over 100,000 years ago, and this gave birth to the first heavenly cosmologies over 20,000 years ago.

KEY WORDS: Consciousness, dreams, spirits, souls, evolution, amygdala, hippocampus, limbic system, temporal lobe, frontal lobes, Cro-Magnon, Neanderthals, constellations, Virgo, Taurus, Orion, Osiris.

1. THE TRANSMIGRATION OF THE SOUL

Belief in the transmigration of the soul, of a life after death, of a world beyond the grave, has been a human characteristic for at least 100,000 years, as ancient graves and mortuary rites attest (e.g., Belfer-Cohen & Hovers, 1992; Butzer, 1982; McCown, 1937; Rightmire, 1984; Schwarcz et al., 1988; Smirnov, 1989; Trinkaus 1986). Even ancient "archaic" humans, despite their small brains and primitive intellectual, linguistic, cognitive, and mental capabilities, and who wondered the planet over 120,000 years ago, carefully buried their dead (Butzer, 1982; Joseph 2000a; Rightmire, 1984). And like modern Homo sapiens, they prepared the recently departed for the journey to the Great Beyond: across the sea of dreams, to the land of the dead, the heavens, the realm of the ancestors and the gods.

Paleolithic burial in sleeping position.

Throughout the Middle and Upper Paleolithic it was not uncommon for tools and hunting implements to be placed beside the body, even 100,000 years ago, for the dead would need them in the next world (Belfer-Cohen & Hovers, 1992; McCown, 1937; Trinkaus, 1986). A hunter in life he was to be a hunter in death, for the ethereal world of the Paleolithic was populated by spirits and souls of bear, wolf, deer, bison, and mammoth (e.g., Campbell, 1988; Joseph 2001, 2002; Kuhn, 1955). Moreover, food and water might be set near the head in case the spirit hungered or experienced thirst on its long sojourn to the heavenly Hereafter. And finally, fragrant blossoming flowers and red ocher might be sprinkled upon the bodies (Solecki, 1971) along with the tears of those who loved them.

Given the relative paucity of cognitive, cultural, and intellectual development among Middle Paleolithic Neanderthal and "archaic" humans, and the likelihood that they had not yet acquired modern human speech (Joseph, 1996, 2000b), evidence of spiritual concerns among these peoples demonstrates the great antiquity of belief in an after-life and the soul. Humans began evolving a spiritual consciousness over 100,000 years ago. Seventy thousand years later, this consciousness would give birth to the first cosmologies.

2. MIDDLE PALEOLITHIC SPIRITUALITY

When humans first became aware of a "god" or "gods" cannot be determined. Nevertheless, the antiquity of religious and spiritual belief extends backwards in time to over 100,000 years. It is well established that Neanderthals and other Homo Sapiens of the Middle Paleolithic (e.g. 150,000 to 35,000 B.P.) and Upper Paleolithic (35,000 B.P. to 10,000 B.P.) engaged in complex ritualistic behavior. These rituals are evident from the manner in which they decorated their caves and the symbolism associated with death (Akazawa & Muhesen 2002; Conrad & Richter 2011; Harvati & Harrison 2010; Kurten 1976; Mellars 1996).

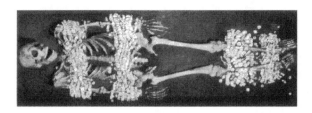

Cro-Magnon burial.

Neanderthals (a people who lived in Europe, Russia, Iraq, Africa, and China from around 150,000 to 30,000 B.P.), have been buried in sleeping positions with the body flexed and lying on its side. Some were laid to rest with limestone blocks placed beneath the head like a pillow—as if they were not truly dead but merely asleep (Akazawa & Muhesen 2002; Harvati & Harrison 2010).

Sleep and dreams have long been associated with the spirit world, and it is through dreams that gods including the Lord God worshipped by Jews, Christians, and Muslims, are believed to have communicated their thoughts, warnings, intentions, and commands. Throughout the ages (Campbell 1988; Freud, 1900; Jung 1945, 1964), and as repeatedly stated in the Old Testament and the Koran, dreams have been commonly thought to be the primary medium in which gods and human interact (Joseph 2002). Insofar as the ancients (and many moderns folks) were concerned, dreams served as a doorway, a portal of entry to the spirit world through which "God," His angels, or myriad demons could make their intentions known.

Paleolithic burial in sleeping position.

It is through dreams that one is able to come into direct contact with the spirit world and a reality so magical and profoundly different yet as real as anything experienced during waking. It is through dreams that ancient humans came to believe the spiritual world sits at the boundaries of the physical, where day turns to dusk, the hinterland of the imagination where dreams flourish and grow. And it is while dreaming that one's own soul may transcend the body, to soar like an eagle, or to commune with the spirits of loved ones who reside in heaven along side the gods.

Neanderthals prepared their dead for this great and final journey, by laying their loved ones to rest so that they would sleep with the spirits and dream of heavenly eternity.

Neanderthals have also been buried surrounded by goat horns placed in a circle, with reindeer vertebrae, animal skins, stone tools, red ocher, and in one grave, seven different types of flowers (Solecki 1971). In one cave (unearthed after 60,000 years had passed), a deep chamber was discovered which housed a single skull which was surrounded by a ring of stones (Harvati & Harrison 2010; Mellars 1996). Moreover, Neanderthals buried bears at a number of sites including Regourdou. At Drachenloch they buried stone "cysts" containing bear skulls (Kurten 1976); hence, "the clan of the cave bear."

Neanderthal burial.
This Neanderthal was buried with 7 different types of flowers.

Yet others were buried with large bovine bones above the head, with limestone blocks placed on top of the head and shoulders, and with heads severed coupled with evidence of ritual decapitation, facial bone removal, and cannibalism (Belfer-Cohen and Hovers 1992; Binford 1968; Harold 1980; Smirnov 1989; Solecki 1971). In one site, dated to over 100,000 years B.P., Neanderthals decapitated eleven of their fellow Neanderthals, and smashed their faces beyond recognition.

It therefore seems apparent that Neanderthals not only engaged in complex rituals, but they believed in spirits, ghosts, and a life after death. Hence the sleeping position, stone pillows, stone tools and food. They were preparing the departing spirit for the journey to the Hereafter and the land of dreams. However, they were also incapacitating their enemies, even after death, to prevent these souls from terrorizing the living or their dreams

3. SPIRITS, SOULS, GHOSTS & THE LAND OF DREAMS

The fact that so many of the Neanderthal dead were buried in a sleeping position implies an association between sleep and dreams. Since all vertebrates so far studied demontrate REM (dream) sleep, it can be assumed that Neanderthals dreamed. Among ancient (and even modern peoples) it was believed that souls and spirits could wonder about while people sleep and dream (Brandon 1967; Frazier 1950; Harris 1993; Jung 1945, 1964; Malinowkski 1990). Some believed the soul could escape the body via the mouth or nostrils while dreaming and that the spirit could leave the body and engage in various purposeful acts or interact with other souls including the soul or spirit of those who had died. The spirit and soul were believed to hover about in human-like, ghostly vestiges, at the fringes of reality, the hinterland where day turns into night (Campbell 1988; Frazier 1950; Jung 1964; Malinowski 1954; Wilson 1951); and it is at night when people dream.

Neanderthal burial.

And as is the case with modern day humans, it can be assumed the ancients, including Neanderthals, had dreams by which they were transported or exposed to a world of magic and untold wonders which obeyed its own laws of time, space, and motion. It is through dreams that humans came to believe the spiritual world sits at the boundaries of the physical, where day turns to dusk, the hinterland of the mind where imagination and dreams flourish and grow (Frazier, 1950; Jung, 1945, 1964; Malinowkski, 1954); hence the tendency to bury the dead in a sleeping position even 100,000 years ago.

It is also via dreams that humans came to know that spirits and lost souls populated the night. The dream was real and so too were the ghosts, gods and demons who thundered and condemned and the phantoms that hovered at the edge of night. Although but a dream, like modern humans, our ancient ancestors experienced this through the senses, much as the physical world is experienced. Dreams were real and they were taken seriously. Moreover, during dreams, both the living and the dead may be encountered. Thus, we can surmise that Neanderthals had similar dreams and may have dreamed about ghosts and wondering spirits,

It is also appears that they feared the dead, and were terrified by the prospect that certain souls might haunt the living. They were afraid of ghosts, and

frightened by the possibility that just as one might awake from sleep after visiting the land of the dead, the dead might also awake from this deathlike slumber. The dead, or at least their personal souls, had to be prevented from causing mischief among the living; especially dreaded enemies who had been killed. Hence, the ritual decapitation, facial bone removal, the smashing of faces, the removal of arms, hands, and legs, and placement of heavy stones upon the body.

It can be concluded, therefore, that almost 100,000 years ago, primitive humans had already come to believe in ghosts, souls, spirits, and a continuation of "life" after death. And, they also took precautions, in some cases, to prevent certain spirits and souls from being released from a dead body and returning to cause mischief among the living, which is why, in the case or powerful enemies, the Neanderthals would cut off heads and hands. They went to great lengths to obliterate all aspects of that dreaded individual's personal identity; e.g., smashing the face beyond recognition.

Of course, the fact that these Neanderthals were buried does not necessarily imply that they held a belief in "God." Rather, what the evidence demonstrates is that Neanderthals were capable of very intense emotions and feelings ranging from murderous rage to love to spiritual and superstitious awe. Although no god is implied, Neanderthals held spiritual and mystical beliefs involving the transmigration of the soul and all the horrors, fears, and hopes that accompany such feelings and beliefs. Although the Neanderthals had not discovered god, they stood upon the threshold.

4. THE NEANDERTHALS: A CHARACTER STUDY

Neanderthals were short, brutish, and an exceedingly violent, murderous people, as the remnants of their skeletons preserved for so many eons attests. Many of their fossils still betray the cruel ravages of deliberately and violently inflicted wounds (Conrad & Richter 2011; Harvati & Harrison 2010; Mellars 1996).

They also appear to have systematically engaged in female infanticide, and displayed a willingness to eat almost anything on four or two legs—including other Neanderthals. In one site, dated to over 100,000 years B.P., Neanderthals decapitated eleven of their fellow Neanderthals, and then enlarged the base of each skull (the foramen magnum) so the brains could be scooped out and presumably eaten. Even the skulls of children were treated in this fashion.

In fact, they would throw the bones and carcasses of other Neanderthals into the refuse pile. In one cave, a collection of over 20 Neanderthals were found mixed up with the remains of other animals and garbage. Presumably, these were enemies or just hapless strangers, innocent cave dwellers who were attacked and sometimes eaten after being brutally killed.

Hence, with the obvious exception of "friends," mates, and family, Neanderthals often saw one another as a potential meal, and had almost no regard for a stranger's innate humanness. These were a violent, murderous, ritualistic people, and strangers were often brutally killed and eaten.

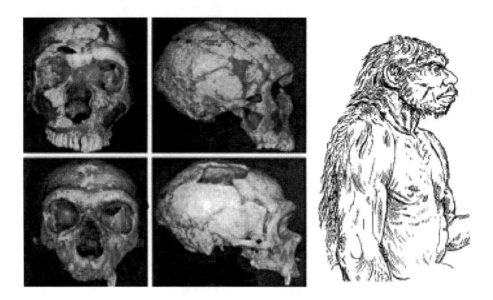

(Left) Two well preserved crania from northern European male Neanderthals. Reproduced from M. H. Wolpoff (1980), Paleo-Anthropology. New York, Knopf. (Right) Neanderthal Male. Reproduced from Howells, 1997. Getting Here. Compass Press, Washington D.C.

These characteristics are also associated with religious fervor. Among ancient and present day peoples, violence, murder, ritual cannibalism, and the sacrifice of children are common religious practices. The Five Books of Moses, are replete with stories of the mass murder and the genocide of non-Jews who were seen as subhuman, including pregnant women and children.

"...when you approach a town, you shall lay seizure to it, and when the Lord your god delivers it into your hand, you shall put all its males to the sword.... In the towns of the people which the Lord your god is giving you as a heritage, you shall not let a soul remain alive." Exodus 20:15-18; Deuterotomy 20:12-16.

"When Israel had killed all the inhabitants of Ai....and all of them, to the last man had fallen by the sword, all the Isrealites turned back to Ai and put it to the sword...until all the inhabitants of Ai had been exterminated... and the king of Ai was impaled on a stake and it was left lying at the entrance to the city gate." Deuteronomy 8:24-29.

I polluted them with their own offerings, making them sacrifice all their firstborn, which was to punish them, so that they would learn that I am Yahweh (Ezekiel 20:25-36. See also Ezekiel 22:28-29). "This very day you defile yourselves in the presentation of your gifts by making your children pass through the fire of all your fetishes." (Ezekiel 20:31). "A blessing on him who seizes your babies and dashes them against rocks." (Psalm 137:9).

Aztec and Indian natives were burnt alive in groups of 13 to honor Jesus Christ and his 12 disciples.

As is well known, the Spanish and Catholic missionaries, acting at the behest of the Catholic Popes (and their Spanish/Catholic Sovereigns), continued these genocidal practices once they invaded the Americas during the 1500's and up through the 19th century. As the Catholic Dominican Bishop Bartolom de Las Casas reported to the Pope: the Aztec and Indian natives were hung and burnt alive "in groups of 13... thus honoring our Savior and the 12 apostles."

5. The Limbic System: Love, Violence, & Spirituality.

Violence and murder are also under the control of the limbic system, the amygdala in particular (Joseph 1992, 1994, 1996, 1998). And it is the limbic system which mediates religious and spiritual experience and which provides much of the emotion and imagery which appears in dreams (Joseph 1988, 2000a, 2001, 2002).

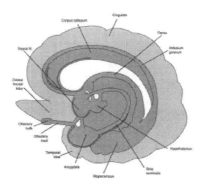

However, it is also the limbic system which subserves feelings of love and affection, and the ability to form long-term attachments. Thus we see that Neanderthals provided loving care for friends and family who had been injured or maimed, enabling them to live many more years despite their grievous injuries. For example, the skeleton of one Neanderthal male, who was about age 45 when he died, had been nursed for a number of years following profoundly crippling injuries. His right arm had atrophied, and his lower arm and hand had apparently been ripped or bitten off, and his left eye socket, right shoulder, collarbone, and both legs were badly injured. Obviously someone loved and tenderly cared for this man. He was no doubt a father, a husband, a brother, and son, and someone in his family not only provided long term loving care to make him comfortable in this life, but prepared him for the next life as well (Mellars 1996).

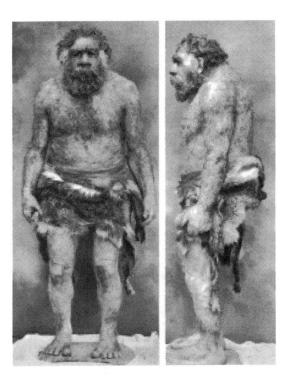

The ability to feel love is a function of the limbic system, the amygdala in particular which is buried in the depths of the temporal lobe.

6. THE NEANDERTHAL BRAIN & TEMPORAL LOBES

An examination of Neanderthal skulls and endocasts of the inner skull provides a gross indication of the size and configuration of their brains. Based on physical indices, the temporal lobe was well developed and little different from that of modern humans.

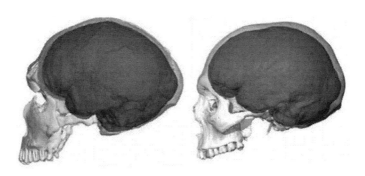

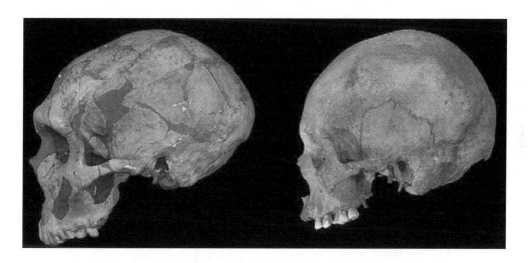

(Right)Neanderthal. (Left) Modern human.

Likewise, given the great antiquity of the limbic system, it can be surmised that the Paleolithic human limbic system was well developed, and similar to the limbic system of modern humans (Joseph 2000a, 2001; 2002).

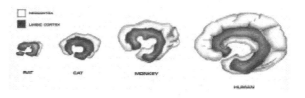

Consciousness and the Universe

The Neanderthals were not a very intelligent or tidy people and were unable to fashion complex tools which, along with other indices, suggests they were unable think complex thoughts. Yet they were people of passion who experienced profound emotions and love; made possible by the limbic system and temporal lobe. In fact, it is because they had the limbic capacity to experience love, spiritual awe, and religious concerns, that these expressions of love continued following death of those they loved, as it has been conclusively demonstrated that these brain structures mediate these functions (Joseph 1992, 1994, 1998a, 2001, 2002). Thus the Neanderthals carefully buried their dead, providing them with food and even sprinkling the bodies with seven different types of blooming, blossoming, fragrant flowers (Belfer-Cohen & Hovers, 1992; McCown, 1937; Solecki 1971; Trinkaus, 1986).

In overall size, the posterior portion of the Neanderthal brain, i.e. the occipital and superior parietal lobes, were slightly larger in length and breadth than the modern human brain on average (Joseph 1996, 2000b; Wolpoff, 1980); a reflection of the environment in which they dwelled and the neural capacities their life style required-- the body moving in visual space as Neanderthals spent most of their non-sleeping hours searching for food. The occipital and superior parietal areas are directly concerned with visual analysis and positioning the body in space (Joseph 1986, 1996). As male and female Neanderthals spent a considerable amount of their time engaged in hunting activities, scanning the environment for prey and running and throwing in visual space were more or less ongoing concerns. A large occipital and superior parietal lobe would reflect these activities.

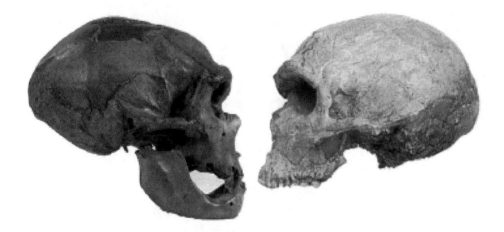

By contrast, concerns about the dead, and attendant mortuary rituals are activities linked to the temporal lobes. The temporal lobes and underlying limbic structures (amygdala, hippocampus), could be likened the seat of the soul and the senior executive of the personality. It is the temporal lobes and the amygdala and hippocampus which have been directly implicated in the generation of religious feelings and supernatural experiences including visions of floating above the body, seeing angels and devils, and what has been described as the after-death and near-death experiences (Joseph 1996, 1998b, 1999a,b, 2000a, 2001, 2002).

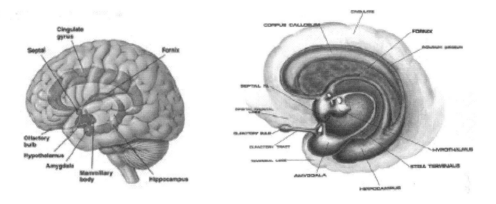

The amygdala (which is buried in the depths of the anterior temporal lobe) enables us to hear "sweet sounds," recall "bitter memories," or determine if something is spiritually significant, sexually enticing, or good to eat and makes it possible to experience the spiritually sublime. It is concerned with the most basic animal emotions, and allows us to store affective experiences in memory or even to reexperience them when awake or during the course of a dream in the form of visual, auditory, or religious or spiritual imagery. The amygdala also enables an individual to experience emotions such as love and religious rapture, as well as the ecstasy associated with orgasm and the dread and terror associated with the unknown.

In fact, the amygdala (in conjunction with the hippocampus and overlying temporal lobe) contributes in large part to the production of very bizarre, unusual and fearful mental phenomenon including dissociative states, feelings of depersonalization, and hallucinogenic and dream-like recollections involving threatening men, naked women, the experience of god, as well as demons and ghosts and pigs walking upright dressed as people (Daly 1958; Gloor 1997; Halgren 1992; Horowitz et al. 1968; MacLean 1990; Penfield and Perot 1963; Schenk, and Bear 1981; Slater and Beard 1963; Subirana and Oller-Daurelia 1953; Trimble 1991; Weingarten, et al. 1977; Williams 1956). Moreover, some individuals report communing with spirits or receiving profound knowledge from the Hereafter, following amygdala stimulation or abnormal activation (Penfield and Perot 1963; Subirana and Oller-Daurelia, 1953; Williams 1956).

Intense activation of the temporal lobe, hippocampus, and amygdala has been reported to give rise to a host of sexual, religious and spiritual experiences; and chronic hyperstimulation can induce an individual to become hyper-religious or visualize and experience ghosts, demons, angels, and even "God," as well as claim demonic and angelic possession or the sensation of having left their body (Bear 1979; Gloor 1992; Horowitz et al. 1968; MacLean 1990; Penfield and Perot 1963; Schenk, and Bear 1981; Weingarten, et al. 1977; Williams 1956).

Much of the visual, emotional, and hallucinatory aspects of dream sleep, have their source in the temporal lobe and underlying limbic system structures (Joseph 1992, 1994, 1996, 1998, 2001, 2002). It is the evolution of the temporal lobes, this "transmitter to god" which also explains why even primitive humanity likely believed in spirits, souls, and ghosts, and practiced complex mortuary rites for those they feared or loved.

7. THE BIG BANG IN SYMBOLIC THINKING: THE CRO-MAGNON FRONTAL LOBES

There is considerable evidence that over the course of human history, the temporal lobe evolved at a faster and earlier rate than the frontal lobe (Joseph 1996, 2000b, Gloor, 1997). Likewise, the temporal lobes mature more rapidly than the frontal lobes over the course of human ontological development (Joseph, 1982, 1996, 1998b, 1999a, 2000b,c).

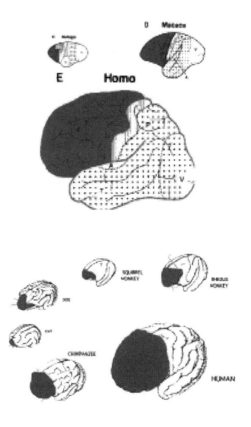

Comparison of the frontal lobes (red) in different species

The Neanderthals were blessed with a well developed temporal lobe, whereas more anterior regions of the brain, the frontal lobes, remained little different from more ancient ancestral primate species. However, with the evolution of the Cro-Magnon people, the brain mushroomed in size, with much of that development in the frontal lobes.

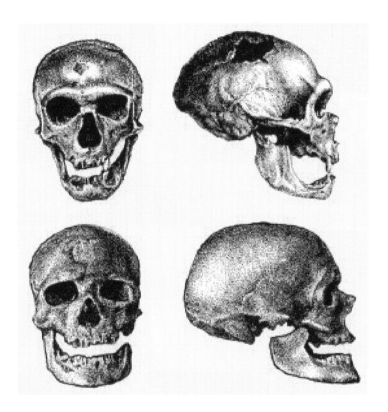

(Top) Neanderthal skull. (Bottom) Cro-Magnon Skull

As based on cranial comparisons and endocasts of the inside of the skull, and using the temporal and frontal poles as reference points, it has been demonstrated that the brain has tripled in size over the course of human evolution, and that the frontal lobes significantly expanded in length and height during the Middle to Upper Paleolithic transition (Blinkov and Glezer 1968; Joseph 1993; MacLean 1990; Tilney 1928; Weil 1929; Wolpoff 1980).

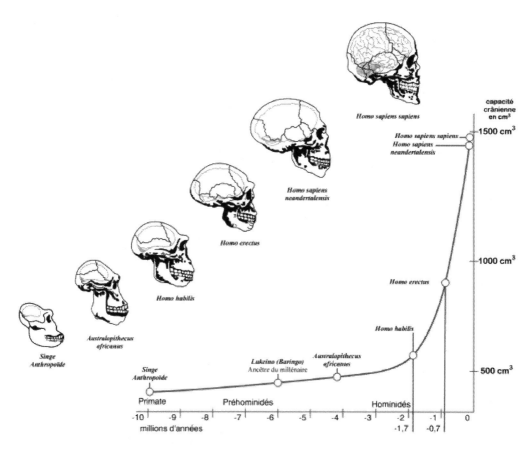

It is obvious that the height of the frontal portion of the skull is greater in the six foot tall, anatomically modern Upper Paleolithic H. sapiens (Cro-Magnon) versus Neanderthal and archaic H. sapiens (Joseph 1996, 2000b; Tilney, 1928; Wolpoff 1980). The evolution and expansion of the frontal lobe is also evident when comparing the skills and creative and technological ingenuity of the Cro-Magnons, vs the Neanderthals (Joseph 1993, 1996, 2000b).

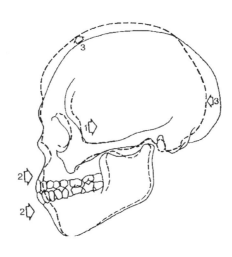

Figure 126. A modern (dotted line) mesolithic cranium compared with a more ancient cranium (solid line). Arrows indicate the main average changes in skull structure including a reduction in the length of the occiput and an increase and upward expansion in the frontal cranial vault. Reproduced from M. H. Wolpoff (1980), Paleo- Anthropology. New York, Knopf.

Therefore, whereas the temporal, occipital and parietal lobes were well developed in archaic and Neanderthals, the frontal lobes would increase in size by almost a third in the transition from archaic humans to Cro-Magnon (Joseph 1996, 2000a,b, 2001).

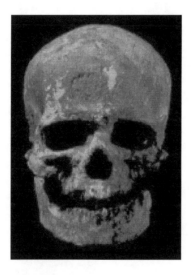

It is the evolution of the frontal lobes which ushered in a cognitive and creative big bang which gave birth to a technological revolution and complex spiritual rituals and beliefs in shamans and goddesses and their relationship to the heavens, and thus the moon and the stars.

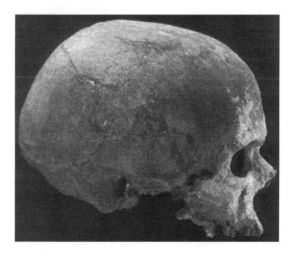

Neanderthals died out as a species around 30,000 years ago; but for at least 10,000 years they shared the planet with the Cro-Magnon people. The Cro-Magnon men stood 6ft tall on average and the males and females were very handsome and beautiful, with thin hips, aquiline noses, prominent chins, small even perfect teeth, and high rounded foreheads. There was nothing ape-like or Neanderthal about these people.

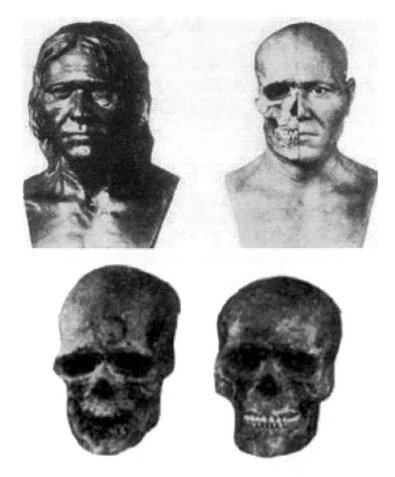

The Cro-Magnon cerebrum was also significantly larger than the Neanderthal brain, with volumes ranging from around 1600 to 1880 cc on average compared with 1,033 to 1,681 cc for Neanderthals (Blinikov & Glezer 1968; Clark 1967; Day 1996; Holloway 1985; Roginskii & Lewin 1955; Wolpoff 1980). In fact, the Cro-Magnon brain is one third larger than the modern human brain, i.e. 1800 cc vs 1350 ccs. However, a distinguishing characteristic of the Cro-Magnon brain, was the massively developed frontal lobes.

The frontal lobes are the senior executive of the brain and are responsible for initiative, goal formation, long term planning, the generation of multiple alternatives, and the consideration of multiple consequences (Joseph, 1986a, 1999a). The frontal lobes are the source of creativity, imagination, and what has been described as free will (Joseph 1996, 2011). Through interactional pathways maintained with brainstem, limbic system, thalamus and the primary receiving and association areas in the neocortex (Petrides & Pandya 1999, 2001)), the frontal lobes have access to every stage of information analysis, and are able to coordinate and regulate attention, memory, personality, and information processing throughout the brain so as to direct intellectual, creative, artistic, symbolic, and cognitive processes (Joseph, 1986a, 1999a).

It is well established that the frontal lobes enable humans to think symbolically, creatively, imaginatively, to plan for the future, to consider the consequences of certain acts, to formulate secondary goals, and to keep one goal in mind even while engaging in other tasks, so that one may remember and act on those goals at a later time (Joseph 1986, 1990b, 1996, 1999c). Selective attention, planning skills, and the ability to marshal one's intellectual resources so as to to anticipate the future rather than living in the past, are capacities clearly associated with the frontal lobes.

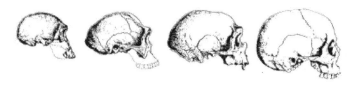

Comparison of the frontal cranium over the course of "evolution:" from Australopithecus, to H. Habilis, to H. erectus. to modern humans. Note obvious expansion of the anterior portion of the skull frontal lobe.

The frontal lobes are associated with the evolution of "free will" (Joseph 1986, 1996, 1999c, 2011b) and the Cro-Magnon were the first species on this planet to exercise that free will, shattering the bonds of environmental/genetic determinism by doing what had never been done before: After they emerged upon the scene over 35,000 years ago, they created and fashioned tools, weapons, clothing, jewelry, pottery, and musical instruments that had never before been seen. They created underground Cathedrals of artistry and light, adorned with magnificent multi-colored paintings ranging from abstract impressionism to the surreal and equal to that of any modern master (Breuil, 1952; Leroi-Gourhan 1964, 1982). And they used their skills to carve the likeness of their female gods.

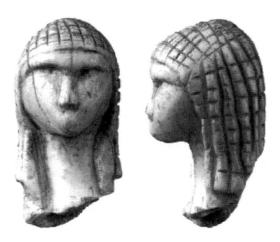

Paleolithic Goddess: Venus de Brassempouy.

Thirty five thousand years ago, Cro-Magnon were painting animals not only on walls but on ceilings, utilizing rich yellows, reds, and browns in their paintings and employing the actual shape of the cave walls so as to conform with and give life-like dimensions, including the illusion of movement to the creature they were depicting (Breuil, 1952; Leroi-Gourhan 1964, 1982). Many of their engraving on bones and stones also show a complete mastery of geometric awareness and they often used the natural contours of the cave walls, including protuberances, to create a 3-dimensional effect (Breuil, 1952; Leroi-Gourhan 1964, 1982).

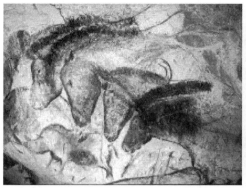
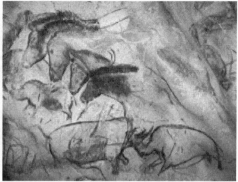

The drawing or carving often became a harmonious or rather, an organic part of the object, wall, ceiling, or tool upon which it was depicted. The Cro-Magnon drew and painted scenes in which animals mated, defecated, fought, charged, and/or were fleeing and dying from wounds inflicted by hunters. The Cro-Magnon cave painters were exceedingly adept at recreating the scenes of everyday life. Moreover, most of the animals were drawn to scale, that is, they were depicted in their actual size; and all this, 30,000 years ago (e.g. Chauvet, et al., 1997).

They created art that was meant to be looked at, owned and admired, and for trade, as jewelry and household decorations, and as highly prized possessions as well as for religious reverence. Picasso was awestruck by these Paleolithic masterpieces, and complained that although 30,000 years had elapsed, "we have learned nothing new. We have invented nothing."

With the evolution of the Cro-Magnon people, the frontal lobes mushroomed in size and there followed an explosion in creative thought and technological innovation. The Cro-Magnon were intellectual giants. They were accomplished artists, musicians, craftsmen, sorcerers, and extremely talented hunters, fishermen, and highly efficient gatherers and herbalists. And they were the first to contemplate the heavens and the cosmos which they symbolized in art.

From the time of Homo Erectus (1.9 million to 500,000 B.P), humans had utilized fire to keep warm, to provide light, to cook their food, and to ward off animals. However, the Cro-Magnon learned over 30,000 years ago how to make fire using the firestone; iron pyrite which when repeatedly struck with a flint makes sparks which can easily ignite brush. They also created the first rudimentary blast furnaces which were capable of emitting enormous amounts of heat, so as to fire clay. This was accomplished by digging a tiny tunnel into the bottom of the hearth which allowed air to be drawn in. Indeed, 30,000 years ago these people were making fire hardened ceramics and clay figures of animals and females with bulging buttocks and breasts—which are presumed to be the first goddesses and fertility fetishes(Breuil, 1952; Leroi-Gourhan 1964).

Many of these female figurines were shaped so that they tapered into points so they could be stuck into the ground or into some other substance either for ornamental or supernatural purposes, e.g., household goddesses, fertility figures, and earth mothers (Breuil, 1952). In fact, much of the art produced, be it finely crafted "laurel leafs" or other artistic masterpieces, served ritual, spiritual, and esthetic functions.

It is the evolution of the Cro-Magnon and their massive frontal lobes which ushered in a cognitive and creative big bang which gave birth to complex spiritual rituals and beliefs in shamans, goddesses, and the cosmos (Joseph 2001, 2001, 2002).

By contrast, Neanderthals, archaics, and other peoples of the Middle Paleolithic were not very smart, and lived in the "here and now." They had little capacity for creative or abstract thought, and constructed and made and used the same simple stone tools over and over again for perhaps 200,000 years, until around 35,000 B.P., with little variation or consideration of alternatives (Binford, 1982; Gowlett, 1984; Mellars, 1989, 1996).

Neanderthal tools.

As neatly summed up by Hayden (1993, p. 139), "as a rule, there is no evidence of private ownership or food storage, no evidence for the use of economic resources for status or political competition.... no ornaments or other status display items, no skin garments requiring intensive labor to produce, no tools requiring high energy investments, no intensive regional exchange for rare items like sea shells or amber, no competition for labor to produce economic surpluses and no corporate art or labor intensive rituals in deep cave recesses to impress onlookers and help attract labor."

Neanderthals greatly lacked in creativity, initiative, imagination, and tended to create simple stone tools that served a single purpose. They tended to live in the immediacy of the present, with little ability to think about or consider the distant future or engage in creative or abstract thought (Binford, 1973, 1982; Dennell, 1985; Mellars, 1989, 1996). These are capacities associated with the frontal lobes.

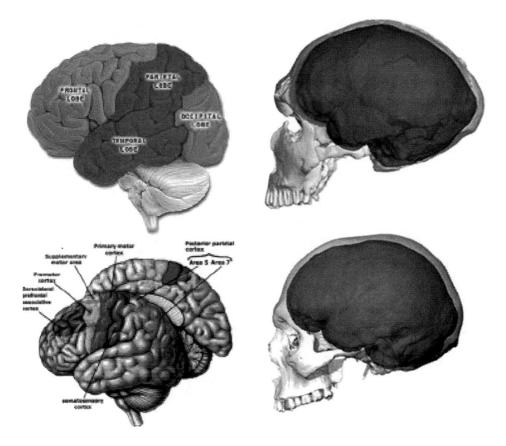

(Left & Top Left) Modern human brain. (Lower Right) Modern human (Upper Right) Neanderthal cranium with endocasts of their brains superimposed

About one third of the frontal lobe, i.e. the motor areas, are concerned with initiating, planning, and controlling the movement of the body and fine motor functioning (Joseph 1990b, 1999c). It is this part of the "archaic" and Neanderthal frontal lobe that appears to be most extensively developed. However, the more anterior frontal lobes are concerned with wholly different functions ranging from creative thought to analytical and planning skills (Joseph 1986, 1996, 1999c), and it is this region of the brain which exploded in size with the evolution of the Cro-Magnon people.

Thus, whereas mortuary rites and primitive spirituality can be associated with the temporal lobes and the Neanderthals, it was not until the evolutionary expansion of the non-motor regions of the frontal lobes that spirituality and concepts of the soul could be expressed through abstract symbolism. Therefore, the evolution of spirituality preceded abstract concepts which could be associated with religiosity; all of which in turn are directly linked to the differential evolution of the frontal and temporal lobes.

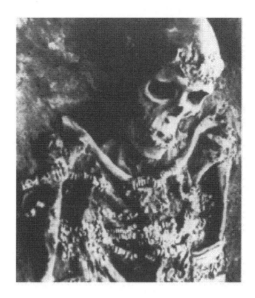

Upper Paleolithic Cro-Magnons buried with tools, ornaments, hunting implements, and other essentials.

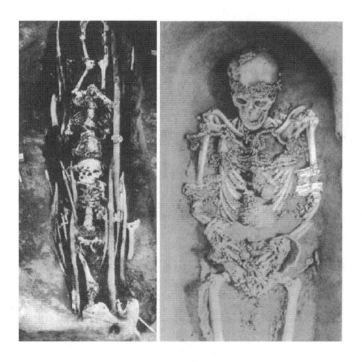

(Left) Two boys buried together. (Right) Cro-Magnon adult. Note tools, ornaments, hunting implements, and other essentials.

8. CRO-MAGNON UPPER PALEOLITHIC SPIRITUAL

CONSCIOUSNESS: THE BIRTH OF THE GODS

The Cro-Magnon practiced complex religious rituals and apparently were the first peoples to have arrived at the conception of "god." However, it was not a male god who they worshipped but female goddesses who were attended by animals and shaman.

Beginning over 30,000 years ago the Cro-Magnon were painting, drawing, and etching bear and mammoth, dear and horse, and pregnant females and goddesses in the recesses of dark and dusky caverns (Bandi 1961; Breuil, 1952; Chauvet et al., 1996; Leroi-Gourhan 1964, 1982; Prideaux 1973). The pregnant females include Venus statuettes, many of which were fertility goddesses.

Paleolithic fertility rites. Dancing Paleolithic Goddess surrounded by female dancers.

A pregnant woman is a symbol of fertility. However, so to is a slim, big busted female. The Cro-Magnon were able to draw both. In fact, these were the first people to paint and etch what today might be considered Paleolithic porn: slim, shapely, naked and nubile young maidens in various positions of repose.

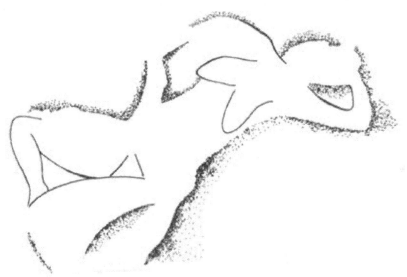

These naked females were not drawn for the sake of prurient interests. These were fertility gods associated with the heavens and the stars.

The Cro-Magon paid homage to a number of goddesses who was associated with the fertility of the earth, as well as the moon and the stars. One great goddess, linked to the moon was carved in limestone over the entrance to an underground cathedral in Laussel, France, perhaps 20,000 years ago. She was painted in the colors of life and fertility: blood red. Her left hand still rests upon her pregnant belly whereas in her right hand she holds the horned crescent of the moon which is engraved with thirteen lines, the number of new moon cycles in a solar year. She was a goddess of life, linked to the mysteries of the heavens and the magical powers of the moon whose 29 day cycle likely corresponded with the Cro-Magnon menstrual cycle which issues from a woman's life-giving womb. The Cro-Magnon believed in gods. God was a woman linked to the Moon, and the earth was her womb from which life would spring anew.

A mother goddess, holding in her hand the symbol of the moon (or a bison's horn) with 13 lines, which is the number of menstrual/lunar cycles in a solar year. Her other hand rests upon her pregnant belly. This goddess was carved outside the entrance to an underground Paleolitic Cathedral, in Laussel.

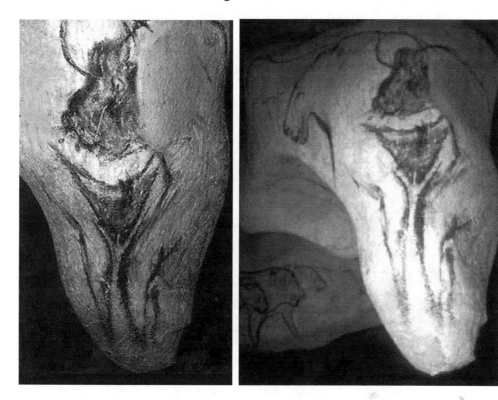

A bull head shaman, the legs and vaginal area of a woman, the head/body of a lion, facing a cross, and painted on a breast-like protrusions. The shaman appears to be mating with the female.

These great underground cathedrals may have also served as the Earth-womb of the goddess, where souls were reborn as men, women, and animals. Specific locations within the Earth-womb were of were of ritualistic, mystical, and spiritual significance in that many paintings were in out of the way places where one had to crawl long distances through tiny spaces and along rather tortuous routes to get to them (Leroi-Gourhan 1964). Moreover, for almost 20,000 years, subsequent generations of Cro-Magnon artists crawled to these same difficult to reach locations to repaint or paint over existing drawings which were hidden away in deep recesses of these dark underground caverns that were extremely difficult to find (Leroi-Gourhan 1964). This indicates that the location within the cave was of particular mystical and ritualistic importance. And not just the location but the journey to these hidden recesses may have been of mystical significance perhaps relating to birth, or rebirth following death.

9. SPIRITS, SOULS, AND SORCERERS

As is evident from their cave art and symbolic accomplishments, the nether world of the Cro- Magnon and other peoples of the Upper Paleolithic, was haunted by the spirits and souls of the living, the dead, and those yet to be born (Brandon 1967; Campbell 1988; Prideaux 1973).

Upper Paleolithic peoples apparently believed these souls and spirits could be charmed and controlled by hunting magic, and through the spells of sorcerers and shamans. Hence, in conjunction with the worship of the goddess, the Cro-Magnons also relied on shamans dressed as animals.

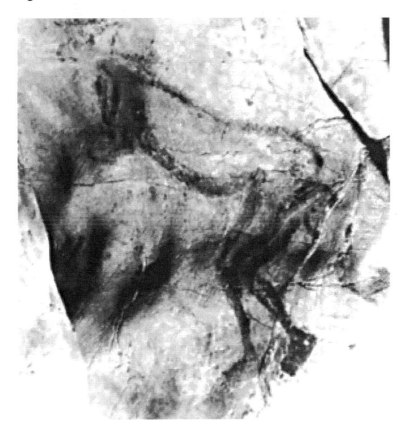

Sorcerer/Shaman: Half bull / half human

Hundreds of feet beneath the earth, the likeness of one ancient shaman attired in animal skins and stag antlers, graces the upper wall directly above the entrance to the 20,000-25,000 year-old grand gallery at Les Trois-Freres in southern France (Breuil, 1952; Leroi-Gourhan 1964). Galloping, running, and swirling about this ancient sorcerer are bison, stag, horse, deer, and presumably their spirits and souls. Images of an almost identical "sorcerer" appear again in ancient Sumerian and Babylonian inscriptions fashioned four to six thousand years ago (Joseph 2000a). The "sorcerer" has a name: "Enki"-the god of the double helix.

(Left) "The Shaman." Dressed in animal skins and stag antlers. (Right) A Sumerian/Babylonian god/shaman dressed as a bull.

10. UNDERGROUND CATHEDRALS: EMBRACED BY THE LIGHT

In order to view these Cro-Magnon paintings, statues and shrines, one has to enter the hidden entrance of an underground cave, and crawl a considerable distance, sometimes hundreds of yards, through a twisting, narrowing, pitch black tunnel before reaching these Upper Paleolithic underground cathedrals which were shrouded in darkness (Breuil, 1952). Here the Cro-Magnon would light candles and lamps, performing magical and spiritual rituals as the painted animals and spirits wavered in the cave light.

Lascaux cave, France.

The nature and location of the Cro-Magnon cathedrals, which have been found throughout Europe, and the nature of the tortuous routes to get to them, and the effect of cave light bringing these paintings to life, is significant as it embraces features associated with after death experiences as retold by present day (as well as ancient) peoples.

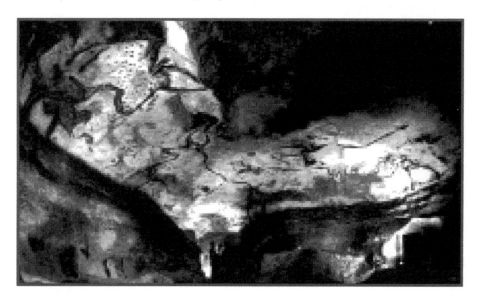

Lascaux cave, France.

In the ancient Egyptian and Tibetan Books of the Dead, and has been reported among many of those who have undergone a "near death" or "life after death" experience, being enveloped in a dark tunnel is commonly experienced soon after death. It is only as one ascends the tunnel that one will see in the decreasing distance, a light, the "light" of "Heaven" and of paradise. Once embraced by the light "the recently dead" may be greeted by the souls of dead relatives, friends, and/or radiant human or animal-like entities (Eadie, 1992; Rawling 1978; Ring 1980).

However, emerging from the tunnel and mouth of the cave, would also be a symbolic rebirth through the birth canal and womb of the earth....

11. DREAMS, ANIMAL SPIRITS AND LOST SOULS

Across time and culture, people have believed that not just humans but animals, plants and trees were sensitive, sentient, intelligent, and the abode of spirits including the souls of dead ancestors (Campbell 1988; Frazier 1950; Harris, 1993; Jung 1964; Malinowski 1948). Before hunting and killing an animal, its spirit sometimes had to be conjured forth so as to not harm it, or to ask forgiveness (Campbell 1988; Frazier 1950). The great hunters respected and paid homage to the souls and spirits of the animals they killed, and the Cro-Magnon were great hunters.

A dead hunter and a birds head. A disembowled bison stands above him. Presumably this scene depicts the death of a hunter and the flight of his soul as symbolized by the bird. Bird heads were commonly employed by ancient peoples including the Egyptians to depict the ascension to heaven. Eventually, bird heads were replaced by creatures with wings, e.g. angels. However, the symbolism of the birds also refers to flight, and the spirits of the dead were believed to ascend to the heavens which were filled with stars.

Be it human, animal or plant, souls were also believed capable of migrating to new abodes, and that souls could migrate from humans, to animals or plants and then back again (Campbell, 1988; Frazier, 1950; Harris, 1993; Jung, 1964; Malinowkski, 1948). The spirit left the body at death, and the body was buried in the womb of the earth, from which new life would emerge. And the liberated soul might ascend to starry vault of heaven, sometimes taking the shape and form of a bird.

Goddesses with bird heads. From Libya, approximately 10,000 B.P. The bird head symbolizes the capacity for flight and thus the ability to ascend to heaven.

Be it following death, or during a dream, sometimes the soul was believed to take on another form, such as a bird, or deer, fox, rabbit, wolf, and so on. The spirit and the soul could also hover about in human- or animal-like, ghostly vestiges, at the fringes of reality, the hinterland where day turns into night (Campbell 1988; Frazier 1950; Jung 1964; Malinowski 1954; Wilson 1951). The souls of animal's such as a wolf or eagle, could also leave the body and take on various forms including that of a woman or Man. Not just men but animals too had souls that had to be respected.

Even after death souls continued to interact with the living, and every living being possessed a soul. Hence, the ancients believed that these souls could be influenced, their behavior controlled, and, in consequence, a good hunt insured or with the assistance of a soul. And thus, deep within the womb of the Earth, the Cro-Magnon painted and paid homage to the spirits of the animal world.

Lascaux cave, France.

Souls were also believed by ancient humans to wonder about while people sleep and dream (Brandon 1967; Frazier 1950; Harris 1993; Jung 1945, 1964; Malinowkski 1990). That is, among many different cultures and religions the soul is believed to sometimes wonder away from the body, especially while dreaming, and may engage in certain acts or interact with other souls including those of the dear but long dead and departed. These peoples believed in an afterlife and a spirit world which could be entered through a doorway of dreams. Thus, at death, the soul or spirit would be completely liberated from the body.

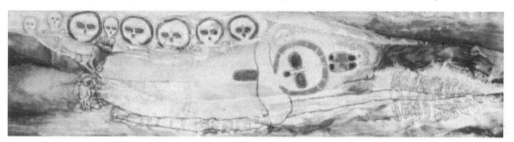

Spirits and the souls (wondjinas) of the dead

Paleolithic spirits ascending to the heavens.

According to the ancients, the soul could exit the body following death and thus we see that the peoples of the Paleolithic peoples often buried their dead in sleeping positions. And, because the Cro-Magnons obviously believed in an after-life, they buried their dead with food, weapons, flowers, jewelry, clothing, pendants, rings, necklaces, multifaceted tools, head bands, beads, bracelets and so on. The Cro-Magnon were a profoundly spiritual people and they fully prepared the dead for the journey to the spirit world, equipping them so that they could live for all eternity in the land of the ancestors and the gods.

12. THE COSMOLOGY OF ANCIENT SPIRITUALITY

When humans first turned their eyes to the sun, moon, and stars to ponder the nature of existence and the cosmos, is unknown. The Cro-Magnon people were keen observers of the world around them, which they depicted with artistic majesty. The heavens were part of their world and they searched the skies for signs and observed the moon, the patterns formed by clusters of stars, and perhaps the relationship between the Earth, the sun, and the changing seasons. Although it is impossible to date cave paintings with precision, the first evidence of this awareness of the cosmic connection between Sun, Moon, Woman, Earth and the changing seasons are from the Paleolithic; symbolized in the creations of the Cro-Magnon.

12.1. GODDESS OF THE MOON Among the ancients, the Sun and the Moon were of particular importance and the Cro-Magnon observed the relationship between woman and the lunar cycle. Consider, the pregnant goddess, the Venus of Laussel, who holds the crescent moon in her hand (though others say it is a bison's horn). Although the length of a Cro-Magnon woman's menstrual cycle is unknown, it can be assumed that like modern woman she menstruated once every 28 to 29 days, which corresponds to a lunar month 29 days long, and which averages out to 13 menstrual cycles in a solar year. And not just menstruation, but pregnancy is linked to the phases of the moon.

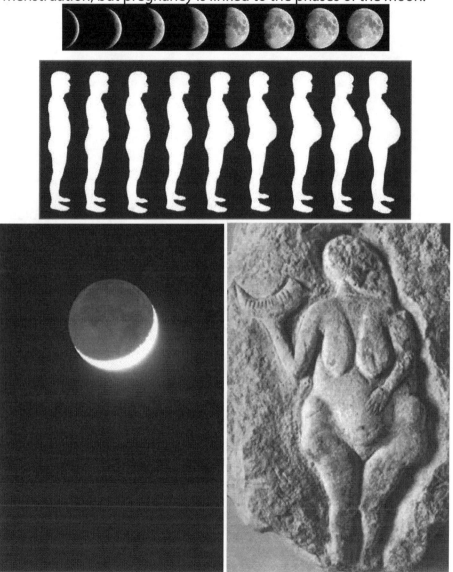

12.2. THE FOUR CORNERS OF THE SOLAR CLOCK. When the Cro-Magnon turned their eyes to the heavens, seeking to peer beyond the mystery that separated this world from the next, they observed the sun. With a brain one third larger than modern humans, and given their tremendous power of observation, it can be predicted these ancient people would have associated the movement of the sun with the changing seasons which effected the behavior of animals, the growth of plants, and the climate and weather; all of which are directly associated with cyclic alterations in the position of the sun and the length of a single day over the course of a solar year which is equal to 13 moons.

The four seasons, marked by two solstices and the two equinoxes have been symbolized by most ancient cultures with the sign of the cross, e.g. the "four corners" of the world and the heavens. The "sign of the cross" generally signifies religious or cosmic significance. The Cro-Magnon also venerated the sign of the cross, the first evidence of which, an engraved cross, is at least 60,000 years old (Mellars, 1989). Yet another cross, was painted in bold red ochre upon the entryway to the Chauvet Cave, dated to over 30,000 years ago (Chauvet et al., 1996).

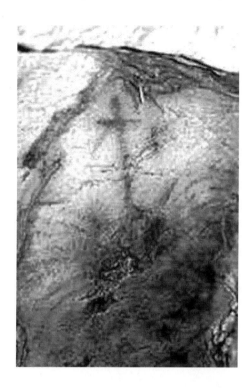

The entrance to the underground Upper Paleolithic cathedral. The Chauvet cave. Note the sign of the cross. Reprinted from Chauvet et al., (1996). Dawn of Art: The Chauvet Cave. Henry H. Adams. New York.

The illusion of movement of the Sun, from north to south, and then back again, in synchrony with the waxing and waning of the four seasons, is due to the changing tilt and inclination of the Earth's axis, as it spins and orbits the sun. Thus over a span of 13 moons, it appears to an observer that the days become shorter and then longer and then shorter again as the sun moves from north to south, crosses the equator, and then stops, and heads back north again, only to stop, and then to again head south, crossing the equator only to again stop and head north again. The two crossings each year, over the equator (in March and September) are referred to as equinoxes and refers to the days and nights being of equal length. The two time periods in which the sun appears to stop its movement, before reversing course (June and December), are referred to as solstices—the "sun standing still."

The sun was recognized by ancient astronomer priests, as a source of light and life-giving heat, and as a keeper of time, like the hands ticking across the face of a cosmic clock. Because of the scientific, religious, and cosmological significance of the sun, ancient peoples, in consequence, often erected and oriented their religious temples to face and point either to the rising sun on the day of the solstice (that is, in a southwest—northeast axis), or to face the rising sun on the day of the equinox (an east-west axis). For example, the ancient temples and pyramids in Egypt were oriented to the solstices, whereas the Temple of Solomon faced the rising sun on the day of the equinox.

Thus the sign of the cross is linked to the heavens and to the sun. Understanding the heavens and the sun, has been been a common astronomical method of divining the the will of the gods, and for navigation, localization, and calculation: these celestial symbols have heavenly significance.

Regardless of time and culture, from the Aztecs, Mayans, American Indians, Romans, Greeks, Africans, Christians, Cro-Magnons, Egyptians (the key of life), and so on, the cross consistently appears in a mystical context, and/or is attributed tremendous cosmic significance (Budge,1994; Campbell, 1988; Joseph, 2000a; Jung, 1964). The sign of the cross was the ideogram of the goddess "An", the Sumerian giver of all life from which rained down the seeds of life on all worlds including the worlds of the gods. An of the cross gave life to the gods, and to woman and man.

The God Seb supporting the Goddess Nut who represents heaven. Note the repeated depictions of the key of life; i.e. a ring with a cross at the end.

The symbol of the cross is in fact associated with innumerable gods and goddesses, including Anu of the ancient Egyptians, the Egyptian God Seb, the Goddess Nut, the God Horus (the hawk), as well as Christ and the Mayan and Aztec God, Quetzocoatl. For example, like the Catholics, the Mayas and Aztecs adorned their temples with the sign of the cross. Quetzocoatl, like Jesus, was a god of the cross.

Quetzocoatl the Mayan and Aztec god of the cross. The round shield encircling the cross represents the sun.

In China the equilateral cross is represented as within a square which represents the Earth, the meaning of which is: "God made the Earth in the form of a cross." It is noteworthy that the Chinese cross-in-a-box can also be likened to the swastika—also referred to as the "gammadion" which is one of the names of the Lord God: "Tetragammadion." The cross, in fact forms a series of boxes when aligned from top to bottom or side by side, and cross-hatchings such as these were carved on stone over 60,000 years ago.

Ochre etched with crosses, forming a series of cross-hatchings, dating to 77,000 years ago.

Among the ancient, the sign of the cross, represented the journey of the sun across the four corners of the heavens. The Cro-Magon adorned the entrance and the walls of their underground cathedrals with the sign of the cross, which indicates this symbol was of profound cosmic significance. However, that some of the Cro-Magnon depictions of animal-headed men have also been found facing the cross, may also pertain to the heavens: the patterns formed by stars, which today are refereed to as "constellations."

Consciousness and the Universe

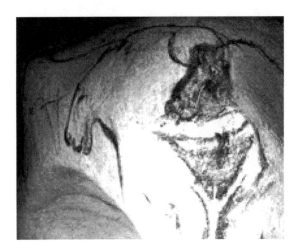

12.3. THE CONSTELLATION OF VIRGO There is nothing "virginal" about the constellation of Virgo. The pattern can be likened to a woman in lying on her back with an arm behind her head, and this may have been the visage which stirred the imagination of the Cro-Magnon.

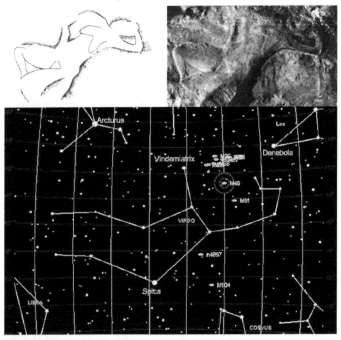

Cro-Magnon goddess, depicting the constellation of Virgo. La Magdelain cave.

12.4. THE PLEIADES AND THE CONSTELLATIONS OF TAURUS AND OSIRIS

It would be unreasonable to assume that the Cro-Magnon would not have observed the heavens or the illusory patterns formed by the alignment of various stars. Depictions of the various constellations, such as Taurus and Orion, and "mythologies" surrounding them, are of great antiquity, and it appears that similar patterns were observed by the Cro-Magnon people.

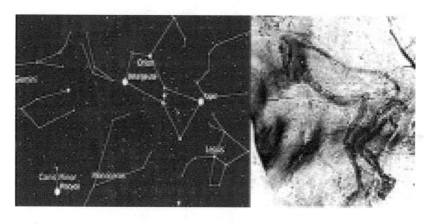

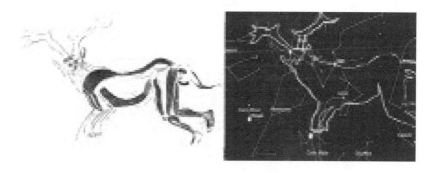

(Upper Right / Lower Left) The "Sorcerer" Trois-Frères cave. (Upper Left / Lower Right) Constellation of Orion/Osiris.

Consciousness and the Universe

Consider, for example the "Sorcerers" or "Shamans" wearing the horns of a bull, and possibly representing the constellation of Taurus; a symbol which appears repeatedly in Lascaux, the "Hall of the Bulls" and in the deep recesses of other underground cathedrals dated from 18,000 to 30,000 B.P. And above the back of one of these charging bulls, appears a grouping of dots, or stars, which many authors believe may represent the Pleiades which is associated with Taurus. These Paleolithic paintings of the bull appear to be the earliest representation of the Taurus constellation.

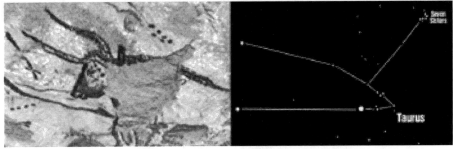

(Top) The main freeze of the bulls in the Lascaux Cave in Dordogne. There is a group of dots on the back of the great bull (Taurus) which may represent six of the seven stars of the Pleiades (the seven sisters). As stars are also in motion, not all would be aligned or as bright or dim today, as was the case 20,000 to 30,000 years ago.

In the "modern" sky, the constellation of Orisis/Orion the hunter, faces Taurus, the bull; and these starry patterns would not have been profoundly different 20,000 to 30,000 years ago. In ancient Egypt, dating back to the earliest dynasties (Griffiths 1980), Osiris was the god of death and of fertility and rebirth, who wore a a distinctive crown with two horns (later symbolized as ostrich feathers at either side). He was the brother and husband of Isis. According to myth, Orisis was killed by Set (the destroyer) and dismembered. Isis recovered all of his body, except his penis. After his death she becomes pregnant by Orisis. The Kings of Egypt were believed to ascend to heaven to join with Osiris in death and thereby inherit eternal life and rebirth, symbolized by the star Sirius (Redford 2003). The Egyptian "King list" (The Turin King List) goes backward in time, 30,000 years ago to an age referred to as the "dynasty of gods" which was followed by a "dynasty of demi-gods" and then dynasties of humans (Smith 1872/2005).

Over 20,000 years ago, the 6ft tall Cro-Magnon, with their massive brain one third larger than modern humans, painted a hunter with two horns who had been killed. And just as the constellation of Orion the hunter faces Taurus, so too does the dead Cro-Magnon hunter who has dismembered/disemboweled the raging bull. And below and beneath the dead Cro-Magnon hunter, another bird, symbol of rebirth, and perhaps symbolizing the star Sirius.

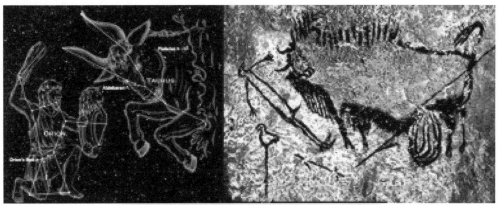

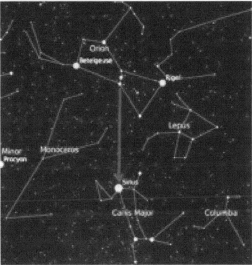

The constellation of Osiris (Orion the hunter) in Egyptian mythology is the god of the dead who was dismembered; but also represents resurrection and eternal life as signified by the star Sirius. (Upper Right) Constellation of Osiris/Orion and Taurus. (Upper Left) Cave painting. Lascaux. The dead (bird-headed or two horned) hunter killed by a bull whom he disemboweled. (Bottom) Constellation of

Orion/Osiris in relation to Sirius.

13. CONCLUSIONS

Complex mortuary rituals and belief in the transmigration of the soul, of a world beyond the grave, has been a human characteristic for at least 100,000 years. The emergence of spiritual consciousness and its symbolism, is directly linked to the evolution of the temporal and frontal lobes and to the Neanderthal and Cro-Magnon.

These ancient peoples were capable of experiencing love, fear, and mystical awe, and they carefully buried those they loved and lost. They believed in spirits and ghosts which dwelled in a heavenly land of dreams, and interned their dead in sleeping positions and with tools, ornaments and flowers. By 30,000 years ago, and with the expansion of the frontal lobes, they created symbolic rituals to help them understand and gain control over the spiritual realms, and created signs and symbols which could generate feelings of awe regardless of time or culture.

Because they believed souls ascended to the heavens, the people of the Paleolithic searched the heavens for signs. They observed and symbolically depicted the association between woman and the moon, patterns formed by stars, and the relationship between Earth, the sun, and the four seasons.

The ancestry and origins of the Cro-Magnon peoples, are completely unknown. There are no transitional forms that link them with Neanderthals or the still primitive "early modern" peoples of the Middle Paleolithic who were decidedly more archaic in appearance as compared to Cro-Magnons. Neanderthals did not evolve into Cro-Magnons, and they coexisted for almost 15,000 years, until finally the Neanderthals disappeared from the face of the Earth, around 30,000 years B.P. (Mellars, 1996). Indeed, the Neanderthals were of a completely different race; and not just physically, but genetically, for when they died out, so too did their genetic heritage and almost all traces of their DNA (Conrad & Richter 2011; Harvati & Harrison 2010).

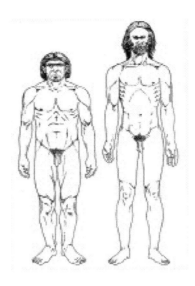

(Left) Neanderthal. (Right) Cro-Magnon.

Since Cro-Magnons shared the planet with Neanderthals during overlapping time periods it certainly seems reasonable to assume that the technologically and intellectually superior Cro-Magnons probably engaged in wide spread ethnic cleansing and exterminated the rather short (5ft 4in.), sloped-headed, heavily muscled Neanderthals, eradicating all but hybrids from the face of the Earth, some 35,000 to 28,000 years ago.

Presumably, the evolutionary lineage of the Cro-Magnon is linked to the evolution of "early modern" archaic humans who first appeared in what is now the Middle East around 100,000 years ago. In contrast, to Neanderthals, the frontal portions of the craniums of "early modern" archaics, were rounded, indicating an expansion of the frontal lobes (Joseph 1996). In addition, these archaics began engaging in mortuary rites before the Neanderthals. For example, archaic H. sapiens and "early moderns" were carefully buried in Qafzeh, near Nazareth and in the Mt. Carmel, Mugharetes-Skhul caves on the Western coast of the Middle East over 90,000 to 98,000 years ago (McCown 1937; Smirnov 1989; Trinkaus 1986). This includes a Qafzeh mother and child who were buried together, and an infant who was buried holding the antlers of a fallow deer across his chest. In a nearby site equally as old (i.e. Skhul), yet another was buried with the mandible of a boar held in his hands, whereas an adult had stone tools placed by his side (Belfer-Cohen and Hovers 1992; McCown 1937). "Early modern," and other "archaic" Homo sapiens commonly buried infants, children, and adults with tools, grave offerings, and animal bones.

However, it was not until the evolution of the Cro-Magnon and the expansion of the frontal lobes that symbolic representations of religious and spiritual feelings literally became an art. The spiritual belief systems of the Cro-Magnon and other peoples of the Upper Paleolithic, completely outstripped those of their predecessors in complexity, originality, and artistic and symbolic expression. Hence, the Cro-Magnon conception of, and ability to symbolically express the spirit world, became much more complex as well, undergoing what has been described as a "symbolic explosion" (Bandi 1961; Kuhn 1955; Leroi-Gourhan 1964, 1982; Prideaux 1973). As the brain and man and woman evolved, so too did their spiritual beliefs.

The Cro-Magnon were taller and had a larger brain than Neanderthals and modern humans.

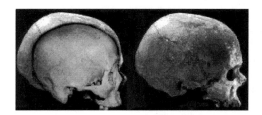

Comparison of modern human skull superimposed on Cro-Magnon skull (Left). Cro-Magnon skull (Right). The Cro-Magnon brain was 1/3 larger on average, than the modern human brain.

With a massive frontal lobe and a brain one third larger than modern humans, the 6ft tall Cro-Magnons were intellectual giants, as the remnants of their creations attest. What they might have been capable of mentally, what they may have achieved is unknown to us, except indirectly, through what today is classified as "myth."

What is known for fact is the people of the Paleolithic were among the greatest hunters, craftsmen, and artists to have walked this Earth. It is these people who were the first to develop complex beliefs involving spirits, souls, sorcerers, shamans, goddesses, and the moon, and sun, and the stars which shine in the darkness of night.

Spiritual consciousness first began to evolve 100,000 years ago. It is this consciousness of the spirit, and belief in the transcendence of the soul, which gave birth to the first heavenly cosmologies over 20,000 years ago.

References

Akazawa, T & Muhesen, S. (2002). Neanderthal Burials. KW Publications Ltd.

Amaral, D. G., Price, J. L., Pitkanen, A., & Thomas, S. (1992). Anatomical organization of the primate amygdaloid complex. In J. P. Aggleton (Ed.). The Amygdala. (Wiley. New York.

Bandi, H. G. (1961). Art of the Stone Age. New York, Crown PUblishers, New York.

Bear, D. M. (1979). Temporal lobe epilepsy: A sydnrome of sensory-limbic hyperconnexion. Cortex, 15, 357-384.

Belfer-Cohen, A., & E.Hovers, (1992). In the eye of the beholder: Mousterian and Natufian burials in the levant. Current Anthropology 33: 463-471.

Breuil. H. (1952). Four hundred centuries of cave art. Montignac.

Budge, W. (1994). The Book of the Dead. New Jersey, Carol.

Butzer, K. (1982). Geomorphology and sediment stratiagraphy, in The Middle Stone Age at Klasies River Mouth in South Africa. Edited by R. Singer and J. Wymer. Chicago: University of Chicago Press.

Binford, L. (1981). Bones: Ancient Men & Modern Myths. Academic Press, NY

Binford, S. R. (1973). Interassemblage variability--the Mousterian and the 'functional' argument. In The explanation of culture change. Models in prehistory. edited by C. Renfrew. Pittsburgh: Pittsburgh U. Press.

Binford S. R. (1982). Rethinking the Middle/Upper Paleolithic transition. Current Anthropology 23: 177-181.

Blinkov, S. M., & Glezer, I. I. (1968). The human brain in figures and tables. New York: Plenum.

Campbell, J. (1988) Historical Atlas of World Mythology. New York, Harper & Row.

Cartwright, R. (2010) The Twenty-four Hour Mind: The Role of Sleep and Dreaming in Our Emotional Lives. Oxford University Press.

Chauvet, J-M., Deschamps, E. B. & Hillaire, C. (1996) Dawn of Art: The Chauvet Cave. H.N. Abrams.

Clark, G. (1967) The stone age hunters. Thames & Hudson.

Clark, J. D., & Harris, J. W. K. (1985). Fire and its role in early hominid lifeways. African Archaeology Review, 3, 3-27.

Conrad, N. J., & Richter, J. (2011). Neanderthal Lifeways, Subsistence and Technology. Springer.

Daly, D. (1958) Ictal affect. American Journal of Psychiatry, 115, 97-108.

Day, M. H. (1996). Guide to Fossil Man. University of Chicago Press, Chicago.

Dennell, R. (1985). European prehistory. London, Academic Press.

Eadie, B. J. (1992). Embraced by the light. California, Gold Leaf Press.

Frazier, J. G. (1950). The golden bough. Macmillan, New York.

Freud, S. (1900). The interpretation of dreams. Standard Edition (Vol 5). London: Hogarth Press.

Gloor, P. (1997). The Temporal Lobes and Limbic System. Oxford University Press. New York.

Gowlett, J. (1984). Ascent to civlization. New York: Knopf.

Gowlett, J.A. (1981). Early archaeological sites, hominid remains and traces of fire from Chesowanja, Kenya. Nature, 294, 125-129.

Griffiths, J. G. (1980). The Origins of Osiris and His Cult. Brill.

Halgren, E. (1992). Emotional neurophysiology of the amygdala within the context of human cognition. In J. P. Aggleton (Ed.). The Amygdala. New York, Wiley-Liss.

Halgren, E., Walter, R. D., Cherlow, D. G., & Crandal, P. H. (1978). Mental phenomenoa evoked by electrical stimualtion of the human hippocampal formation and amygdala, Brain, 101, 83-117.

Harold, F. B. (1980). A comparative analysis of Eurasian Palaeolithic burials. World Archaeology 12: 195-211.

Harold, F. B. (1989). Mousterian, Chatelperronian, and Early Aurignacian in Western Europe: Continuity or disconuity?" In P. Mellars & C. B. Stringer (eds). The human revolution: Behavioral and biological perspectives on the origins of modern humans, vol 1.. Edinburgh: Edinburgh University Press.

Harris, M. (1993) Why we became religious and the evolution of the spirit world. In Lehmann, A. C. & Myers, J. E. (Eds) Magic, Witchcraft, and Religion. Mountain View: Mayfield.

Hasselmo, M. E., Rolls, E. T., & Baylis, G. C. (1989). The role of expression and identity in the face-selective responses of neurons in the temporal visual cortex of the monkey. Behavioral Brain Research, 32, 203-218.

Harvati, K., & Harrison, T. (2010). Neanderthals Revisited. Springer.

Hayden, B. (1993). The cultural capacities of Neandertals: A review and re-evaluation. Journal of Human Evolution 24: 113-146.

Holloway, R. L. (1988) Brain. In: Tattersall, I., Delson, E., Van Couvering, J. (Eds.) Encyclopedia of human evolution and prehistory. New York: Garland.

Joseph, R. (1982). The Neuropsychology of Development. Hemispheric Laterality, Limbic Language, the Origin of Thought. Journal of Clinical Psychology, 44 4-33.

Joseph, R. (1986). Confabulation and delusional denial: Frontal lobe and lateralized influences. Journal of Clinical Psychology, 42, 845-860.

Joseph, R. (1988) The Right Cerebral Hemisphere: Emotion, Music, Visual-Spatial Skills, Body Image, Dreams, and Awareness. Journal of Clinical Psychology, 44, 630-673.

Joseph, R. (1990a) The temporal lobes. In A. E. Puente and C. R. Reynolds (series editors). Critical Issues in Neuropsychology. Neuropsychology, Neuropsychiatry, Behavioral Neurology. Plenum, New York.

Joseph, R. (1990b) The frontal lobes. In A. E. Puente and C. R. Reynolds (series editors). Critical Issues in Neuropsychology. Neuropsychology, Neuropsychiatry, Behavioral Neurology. Plenum, New York.

Joseph, R. (1992) The Limbic System: Emotion, Laterality, and Unconscious Mind. The Psychoanalytic Review, 79, 405-456.

Joseph, R. (1993) The Naked Neuron. Evolution and the languages of the body and the brain. Plenum. New York.

Joseph, R. (1994) The limbic system and the foundations of emotional experience. In V. S. Ramachandran (Ed). Encyclopedia of Human Behavior. San Diego, Academic Press.

Joseph, R. (1996). Neuropsychiatry, Neuropsychology, Clinical Neuroscience, 2nd Edition. 21 chapters, 864 pages. Williams & Wilkins, Baltimore.

Joseph, R. (1998a). The limbic system. In H.S. Friedman (ed.), Encyclopedia of Human health, Academic Press. San Diego.

Joseph, R. (1998b). Traumatic amnesia, repression, and hippocampal injury due to corticosteroid and enkephalin secretion. Child Psychiatry and Human Development. 29, 169-186.

Joseph, R. (1999a). Environmental influences on neural plasticity, the limbic system, and emotional development and attachment, Child Psychiatry and Human Development. 29, 187-203.

Joseph, R. (1999b). The neurology of traumatic "dissociative" amnesia. Commentary and literature review. Child Abuse & Neglect. 23, 715-727.

Joseph, R. (1999c). Frontal lobe psychopathology: Mania, depression, aphasia, confabulation, catatonia, perseveration, obsessive compulsions, schizophrenia. journal of Psychiatry, 62, 138-172.

Joseph, R. (2000a). The Transmitter to God. University Press California.

Joseph, R. (2000b). The evolution of sex differences in language, sexuality, and visual spatial skills. Archives of Sexual Behavior, 29, 35-66.

Joseph, R. (2000c). Fetal brain behavioral cognitive development. Developmental Review, 20, 81-98.

Joseph, R. (2001). The Limbic System and the Soul: Evolution and the Neuroanatomy of Religious Experience. Zygon, the Journal of Religion & Science, 36, 105-136.

Joseph, R. (2002). NeuroTheology: Brain, Science, Spirituality, Religious Experience. University Press.

Joseph, R. (2011a). Dreams and Hallucinations: Lifting the Veil to Multiple Perceptual Realities, Cosmology, 14, In press.

Joseph, R. (2011). The neuroanatomy of free will: Loss of will, against the will "alien hand", Journal of Cosmology, 14, In press.

Jung, C. G. (1945). On the nature of dreams. (Translated by R.F.C. Hull.), The collected works of C. G. Jung, (pp.473-507). Princeton: Princeton University Press.

Jung, C. G. (1964). Man and his symbols. New York: Dell.

Kawashima, R., Sugiura, M., Kato, T., et al., (1999). The human amygdala plays an important role in gaze monitoring. Brain, 122, 779-783.

Kling. A. S. & Brothers, L. A. (1992). The amygdala and social behavior. In J. P. Aggleton (Ed.). The Amygdala. New York, Wiley-Liss.

Kurten, B. (1976). The cave bear story. New York: Columbia University Press.

Leroi-Gourhan, A. (1964.) Treasure of prehistoric art. New York: H. N. Abrams.

Leroi-Gourhan, A. (1982). The archaeology of Lascauz Cave. Scientific American 24: 104-112.

MacLean, P. (1990). The Evolution of the Triune Brain. New York, Plenum.

Malinowski, B. (1954) Magic, Science and Religion. New York. Doubleday.

McCown, T. (1937). Mugharet es-Skhul: Description and excavation, in The stone age of Mount Carmel. Edited by D. A. E. Garrod and D. Bate. Oxford: Clarendon Press.

Mellars, P. (1989). Major issues in the emergence of modern humans. Current Anthropology 30: 349-385.

Mellars, P. (1996) The Neanderthal legacy. Princeton University Press.

Mellars, P. (1998). The fate of the Neanderthals. Nature 395, 539-540.

Mesulam, M. M. (1981) Dissociative states with abnormal temporal lobe EEG: Multiple personality and the illusion of possession. Archives of General Psychiatry, 38, 176-181.

Moody, R. (1977). Life after life. Georgia, Mockingbird Books.

Morris, J. S., Frith, C. D., Perett, D. I., Rowland, D., Young, A. W., Calder, A. J., & Colan, R. J. (1996). A differential neural response in the human amygdala to fearful and happy facial expression. Nature, 383, 812-815.

Mullan, S., & Penfield, W. (1959). Epilepsy and visual halluciantions. Archives of Neurology and Psychiatry, 81, 269-281.

Neihardt, J. G. & Black Elk, (1979). Black Elk speaks. Lincoln. U. Nebraska Press.

emchin, A. A., Whitehouse, M.J., Menneken, M., Geisler, T., Pidgeon, R.T., Wilde, S. A. (2008). A light carbon reservoir recorded in zircon-hosted diamond from the Jack Hills. Nature 454, 92-95.

Noyes, R., & Kletti, R. (1977) Depersonalization in response to life threatening danger. Comprehensive Psychiatry, 18, 375-384.

Parson, E. R. (1988). Post-traumatic self disorders (PTsfD): Theoretical and practical considerations in psychotherapy of Vietnam War Veterans. In J. P. Wilson, Z. Harel, & B. Kahana (Eds). Human Adaptation to Extreme Stress. New York, Plenum.

Pena-Casanova, J., & Roig-Rovira, T. (1985). Optic aphasia, optic apraxia, and loss of dreaming. Brain and Language, 26, 63-71.

Penfield, W., & Perot, P. (1963). The brains record of auditory and visual experience. Brain, 86, 595-695.

Perryman, K. M., Kling, A. s., & Lloyd, R. L. (1987). Differential effects of inferior temporal cortex lesions upon visual and auditory-evoked potentials in the amygdala of the squirrel monkey. Behavioral and Neural Biology, 47, 73-79.

Petrides, M., & Pandya, D. N. (1999). Dorsolateral prefrontal cortex: comparative cytoarchitectonic analysis in the human and the macaque brain and corticocortical connection patterns. European Journal of Neuroscience 11.1011–1036.

Petrides, M., & Pandya, D. N. (2001). Comparative cytoarchitectonic analysis of the human and the macaque ventrolateral prefrontal cortex and corticocortical connection patterns in the monkey. European Journal of Neuroscience 16.291–310.

Prideaux, T. (1973). Cro-Magnon. New York: Time-Life.

Rawlings, M. (1978). Beyond deaths door. London, Sheldon Press.

Redford, D. B. (2003). The Oxford Guide: Essential Guide to Egyptian Mythology, Berkley.

Rightmire, G. P. (1984). Homo sapiens in Sub-Saharan Africa, In F. H. Smith and F. Spencer (eds). The origins of modern humans: A world survey of the fossil evidence. New York: Alan R. Liss.

Ring, K. (1980). Life at death. New York, Coward, McCann & Geoghegan.

Roginskii Y. Y., & Lewin S. S. (1955). Fundamentals of Anthropology. Moscow: Moscow University Press.

Rolls, E. T. (1984). Neurons in the cortex of the temporal lobe and in the amygdala of the monkey with responses selective for faces. Human Neurobiology, 3, 209-222.

Rolls, E. T. (1992). Neurophysiology and functions of the primate amygdala. In J. P. Aggleton (Ed.). The Amygdala. New York, Wiley-Liss.

Sabom, M. B. (1982). Recollections on death. New York, Harper & Row.

Sawa, M., & Delgado, J. M. R. (1963). Amygdala unitary activity in the unrestrained cat. Electroencephalography and Clinical Neurophysiology, 15, 637-650.

Schenk, L., & Bear, D. (1981) Multiple personality and related dissociative phenomenon in patients with temporal lobe epilepsy. American Journal of Psychiatry, 138, 1311-1316.

Schutze, I., Knuepfer, M. M., Eismann, A., Stumpf, H., & Stock, G. (1987). Sensory input to single neurons in the amygdala of the cat. Experimental Neurology, 97, 499-515.

Slater, E. & Beard, A.W. (1963). The schizophrenia-like psychoses of epilepsy. British Journal of Psychiatry, 109, 95-112.

Schwarcz, A. et al. (1988). ESR dates for the hominid burial site of Qafzeh. Journal of Human Evolution 17: 733-737.

Smirnov, Y. A. (1989). On the evidence for Neandertal burial. Current Anthropology 30: 324.

Smith, G. A. (1872/2005). Chaldean Account of Genesis (Whittingham & Wilkins, London, 1872). Adamant Media Corporation (2005).

Solecki, R. (1971). Shanidar: The first flower people. New York: Knopf.

Subirana, A., & Oller-Daurelia, L. (1953). The seizures with a feeling of paradisiacal happiness as the onset of certain temporal symptomatic epilepsies. Congres Neurologique International. Lisbonne, 4, 246-250.

Tarachow, S. (1941). The clinical value of hallucinations in localizing brain tumors. American Journal of Psychiatry, 99, 1434-1442.

Taylor, D. C. (1972). Mental state and temporal lobe epilepsy. Epilepsia, 13, 727-765.

Tilney, F. (1928). The brain from ape to man. New York: P. B. Hoeber.

Tobias, P. V. (1971). The Brain in Hominid Evolution. Columbia University Press, New York.

Trimble, M. R. (1991). The psychoses of epilepsy. New York, Raven Press.

Trinkaus, E. (1986). The Neanderthals and modern human origins. Annual Review of Anthropology 15: 193-211.

Turner, B. H. Mishkin, M. & Knapp, M. (1980). Organization of the amygdalopetal projections from modality-specific cortical association areas in the monkey. Journal of Comparative Neurology, 191, 515-543.

Ursin H., & Kaada, B. R. (1960). Functional localization within the amygdaloid complex in the cat. Electroencephalography and Clinical Neurophysiology, 12, 1-20.

Weil, A. (1929). Measurements of cerebral and cerebellar surfaces. American Journal of Physical Anthropology 13: 69-90.

Weingarten, S. M., Cherlow, D. G. & Holmgren. E. (1977). The relationship of hallucinations to depth structures of the temporal lobe. Acta Neurochirugica 24: 199-216.

Williams, D. (1956). The structure of emotions reflected in epileptic experiences. Brain, 79, 29-67.

Wilson, I. (1987). The after death experience. New York, Morrow.

Wilson, J. A. (1951) The culture of ancient Egypt. Chicago, U. Chicago Press.

Wolpoff, M. H. (1980), Paleo-Anthropology. New York, Knopf.

32. Evolution of Modern Human Consciousness
Ian Tattersall

Division of Anthropology, American Museum of Natural History,
New York NY 10024, USA

Abstract

Humans are symbolic, mentally processing information in a radically different way even from our closest living relatives, which are unable to reason about the unobservable. Trawling through the human fossil and archaeological records reveals that the unique human cognitive style was not gradually acquired; for even though the diverse hominids of the last 2-3 million years showed strong tendencies toward increasing brain size and technological and behavioral complexity, symbolic reasoning itself was a very recent acquisition of the modern human lineage alone. Indeed, our cognitive faculty appears to have been acquired well within the tenure of anatomically-recognizable Homo sapiens, as an emergent property. The most straightforward scenario is that, after many millions of years of accretionary history, the human brain had by some 200 kyr ago evolved to a point at which a structurally minor addition or modification was able to produce a structure with an entirely new potential. This structural modification was plausibly a passive byproduct of the major developmental event that gave rise to our species as the highly distinctive anatomical entity it is. The new cognitive potential was subsequently "released" by a cultural stimulus, most plausibly the invention of language.

KEY WORDS: Human consciousness; human evolution; emergence; exaptation; Homo sapiens

1. The Symbolic Species

The list of unique human features is almost endless, mostly involving physical alterations that stem directly or indirectly from our unusual form of upright locomotion. But the attribute that makes us feel so different from all the other living organisms around us is an intangible one, involving the way we process information in our brains. Alone among all creatures on the Earth today, we human beings are symbolic, meaning that we mentally disassemble the world around us into a vocabulary of abstract symbols - which we can then recombine in our minds to create alternative versions of reality. We cannot entirely escape reality, as any surviving victim of a natural disaster will tell you; but unlike other organisms which respond, with greater or lesser sophistication, to stimuli impinging upon them, we human beings are able to live for much of the time in the world as we re-create it in our heads. The fact that each one of us will re-create it slightly - or even radically - differently is, of course, at the root of the unusual complexities of human society, and of the need every society exhibits to have elaborate rules and procedures governing the behaviors of its members.

Today, then, we are most strikingly symbolic (see, for example, Lock and Peters, 1996). But the way in which we otherwise fit so seamlessly into the great Tree of Life on Earth indicates unequivocally that we are descended from a non-symbolic ancestor. Unique features are famously hard to explain; and indeed, the only reason we have for thinking that any earthbound form could ever have become symbolic, is that we so obviously did. But if we are ever to form a proper understanding of ourselves, and of just how we fit into Nature, we will have to understand exactly how it was that we arrived at our entirely unprecedented cognitive state.

This is complicated because not only cognitive states, but also the behaviors they generate, do not leave direct tangible traces. All we have by which to trace the history of our precursors' transformation from a more or less run-of-the-mill primate to the totally unmatched physical and cognitive entity that Homo sapiens is today, is our fossil and archaeological records. The first of these consists of the mineralized bones and other direct traces of themselves left by earlier members of our zoological family Hominidae (or subfamily Homininae; it makes no practical difference); the second consists principally of the artifacts these hominids made, and the ways in which they were strewn both around the landscape and the specific spots at which they were found. And in each case, the indicators to hand must stand as indirect proxies for behaviors and cognitive states that must themselves necessarily be inferred. Such inferences are often arguable, and in many cases they are not directly testable. Still, the overall pattern of change over most of hominid evolution is pretty clear, and is briefly outlined below, following an attempt to clarify an initial cognitive state.

2. The Starting Point

One of the problems that paleoanthropologists face that paleontologists of other stripes do not, is that Homo sapiens is the lone representative of its family in the world today. We no longer have anything close to compare ourselves with. But while our closest living relatives, the chimpanzees and bonobos, are actually fairly remotely related to us, they nonetheless bear sufficient cranial and brain-size resemblances to our earliest hominid ancestors that we can reasonably view them as approximate cognitive models for those ancestors.

In this context, we can concede right away that cognitively the apes are very impressive. Indeed, barely a week seems to pass without the discovery that one or another of them displays yet another behavior that we had thought unique to ourselves. An impressive recent example is the use by savanna chimpanzees of wooden "spears" to impale sleeping bushbabies (Pruetz and Bertolani, 2007). What's more, "ape language" experiments have also shown that these primates are capable of recognizing and responding not only to individual visual and auditory symbols, but to limited combinations of them (Cohen, 2010). Still, there can be no doubt that apes are nonsymbolic in the human sense, and the cognitive scientist Daniel Povinelli suggests one limiting factor is that "chimpanzees rely strictly upon observable features of others to forge their social concepts … they do not realize that there is more to others than their movements, facial expressions, and habits of behavior" (Povinelli, 2004: 33). From these and other observations, Povinelli concludes that very early hominids did not "reason about unobservable things" and lacked any "notion of causation" (Povinelli, 2004: 34). And while this negative deduction involves a long chain of inference, it nonetheless seems to be as robust a statement as we can currently achieve about the cognitive condition of our earliest ancestors.

3. Patterns of Change in Human Evolution

The very earliest hominid fossils known are in the seven to four million year (myr) age range. Mostly fragmentary, they make a quite oddly assorted group, and perhaps their most important function is to demonstrate that from the very beginning the hominid family tree has been bushy, much like that of any other successful mammalian family. It is highly unusual for Homo sapiens to be the only hominid on Earth (Fig. 1). Among other things, this means that we need to resist the temptation to reconstruct the history of our species by extrapolating it back into the past to reveal a sort of singleminded slog from primitiveness to perfection, a process of fine-tuning over the eons. Earlier hominids were emphatically not simply junior-league versions of us.

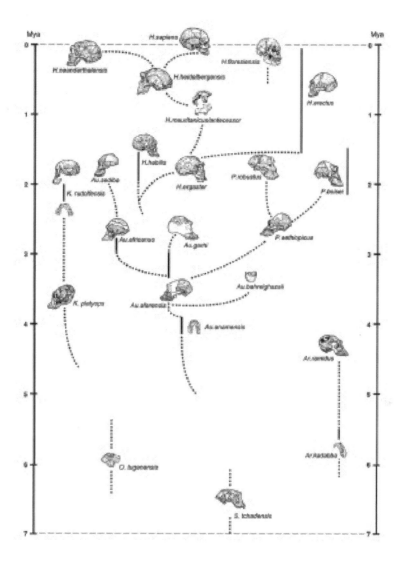

Figure 1. Highly provisional evolutionary tree of the family Hominidae, sketching in some possible relationships among species and showing how multiple hominid species have typically coexisted - until the appearance of Homo sapiens. ©Ian Tattersall.

The earliest well-documented hominids are classified as Australopithecus afarensis. Fossils of this species are known from several eastern African sites in the 3.6 to 3.0 myr time range, during which the African climate was drying and dense forests were giving way to woodland. When on the ground these small-bodied hominids walked bipedally, presumably because they were already most comfortable holding their trunks upright while climbing arboreally. Still, they were broad-hipped and short-legged, and many features of their upper bodies indicate retained agility in the trees. Similarly, their brain sizes and cranial proportions remained comparable to those of chimpanzees, hence their frequent description as "bipedal apes." Still, ecologically there was a significant difference. For while chimpanzees living in open environments today still eat basically the same fruit-based diet as their forest-dwelling fellows, it is clear that the early hominids had expanded their dietary repertoire to include the new resources offered by woodland and savanna environments. These probably included not only tough tubers and rhizomes, but probably also small mammals such as hyraxes, as indicated by analysis of stable carbon isotopes preserved in their teeth (Sponheimer and Lee-Thorp, 2007).

The pursuit of such resources along the forest-edges and in the expanding woodlands exposed the "bipedal apes" to significant hazards of predation; and the potency of this new factor suggests that in reconstructing their social organization we should turn not to today's forest-dwelling apes, but to other higher primates that live in similarly dangerous habitats, namely, the baboons and macaques. These highly social primates live not in small groups as apes do, but in extremely large ones that have complex social hierarchies, and are spatially organized to protect the crucially important but vulnerable core of females and infants (Hart and Sussman, 2009). These groups are, however, sufficiently flexible to break up into subunits for daily foraging out in the open as necessary, while coming together in trees or on rock-faces for nocturnal protection. Physically the bipedal apes were supremely well-equipped for a lifestyle of this kind that spanned grassland to forest - and we know that theirs was not a "transitional" adaptation, but a stable and successful one that remained essentially unaltered for millions of years, even as new species came and went.

We have no reason to suspect that the earliest A. afarensis differed significantly in their underlying cognitive processes from what we see among the social great apes today - at least until, at about 2.5 myr ago, we find evidence of the first stone tools - simple sharp flakes knocked off rock "cores." Here is evidence of a huge cognitive leap that not only involved appreciation of the qualities of materials, but that also demanded planning and foresight: appropriate rocks were carried long distances before being made into tools (Klein, 2009). Although some claimed "early Homo" fossils are known at 2.5 myr, the earliest stone tools are almost certainly the work of Australopithecus; and there are even intimations - in the form of butchery marks on animal bones - that such tools were being used substantially earlier (McPherron et al., 2010). Thus, right from the beginning, a consistent theme is established: the fossil and archaeological records are out of phase. Still, the new ability to cut and dismember carcasses must have hugely affected the lives of the first toolmakers, and it may signal a significant acceleration of the tendency to include animal fats and proteins in the diet. Sadly, the evidence doesn't allow us to say anything about how this technological advance might have reflected change relative to the apes in the toolmakers' subjective experience of the world.

The first unequivocal members of our genus Homo, the earliest hominids with basically modern body size and structure, are known in East Africa at around 1-9 to 1.6 myr ago. They had reduced faces in concert with brains that were larger than those of bipedal apes, although they were not much more than half the size of our brains today. And while their species, Homo ergaster, is initially found in association with stone tools of archaic type, these hominids must have undergone a huge lifestyle change.

This is because they were physically committed to the open savanna, where the range of dietary resources available, and the ambient conditions, were radically different from anything experienced by their forebears. Indeed, it has been argued not only that an increase in animal protein intake was mandated at this point by the energy-hungry larger brain, but that meat - and plant foods, as well - must have been cooked to increase the availability of the nutrients contained (Wrangham, 2010). It is even possible that the reduction in metabolically-expensive gut tissue typical of the genus Homo was an inevitable price of increasing brain size. However this may be, domesticated fire would certainly have been valuable in discouraging a daunting carnivore fauna, especially at night; and it would also have provided a physical focus for the group, something that would in turn have had huge social ramifications. All this makes an attractive argument, but for the time being it remains a circumstantial one, since there is no tangible evidence for the domestication of fire before about 800 thousand years (kyr) ago (Goren-Inbar et al., 2004). All we can say with any confidence at present is that with the new body form we are seeing hominids transitioning from prey species to a (still vulnerable) role as secondary predators. Like us the new hominids were slow; but they had enormous endurance under the hot sun, which may have enabled them to obtain fleeter prey by wearing them down. And, as predators, they would necessarily have been thin on the ground, suggesting that earlier large group sizes had become drastically reduced and perhaps intensifying social bonds among group members.

A new kind of stone tool appears at about 1.5 myr ago. "Handaxes" were large implements deliberately formed to a standard symmetrical teardrop shape, implying for the first time the existence of a "mental template" in the minds of the toolmakers (Schick and Toth, 1993). Clearly, another cognitive advance is implied; but alas this innovation tells us little about Homo ergaster's wider experience of the world. Still, and probably very significantly, we have nothing at all at this stage to suggest any kind of symbolic behavior. Indeed, there is nothing in any aspect of Old Stone Age technology, right up until the very last Ice Age, into which we can confidently read symbolic input. Craftsmanship, yes; craftiness, yes; even an intuitive aesthetic appreciation. But not symbolism.

This continued to be true with the advent of Homo heidelbergensis, the first cosmopolitan hominid species. Appearing in Europe and Africa at about 600 kyr ago, this hominid spread as far east as China, and its tenure witnessed the first building of artificial shelters, the regular domestication of fire, the first hafting of stone tools into compound implements, and the first fabrication of javelin-like throwing spears (see overview in Tattersall, 2009a), all documented at about 400 kyr ago. Somewhat later, in the 300-200 kyr span, came the next radical invention in stone tool-making. This was the "prepared-core" tool, produced by fashioning a stone core on both sides until a single blow could knock off a flake of predetermined form. Brains of the robustly-built Homo heidelbergensis were within the modern size-range, though a little under the modern average; but again, this species bequeathed us nothing that we can unequivocally identify as the product of a symbolic mind. Homo heidelbergensis was clearly an accomplished hunter and exploiter of its environment; but the most we can currently say of it is that it was doing what its predecessors had done, but a bit better. No doubt these hominids had a very sophisticated intuitive intelligence, and complex ways of communicating involving vocalization, gesture, and body language; but we have no grounds for believing they communicated or mentally processed information in the ways that we do - even in rudimentary form. It is thus particularly frustrating that we have such extreme difficulty in seeing through the eyes of any form that does or did not share our particular human mindset.

Homo heidelbergensis is often reckoned to be the progenitor of the lineages that lead to Homo neanderthalensis on one hand (in Europe), and to Homo sapiens on the other (in Africa). Whether or not this is the case, in the later part of the Pleistocene epoch we can identify several independent hominid lineages in which brain size was independently increasing. It was happening in Europe, in the Neanderthal lineage; in Africa, in the lineage that eventuated in Homo sapiens; and in eastern Asia, in the Homo erectus lineage. Clearly, there was something about the genus Homo that predisposed its members to metabolically-expensive brain expansion, regardless of geography or environment; and understanding just what this something was, will be crucial to comprehending how we human beings became the extraordinary creatures we are. But it was clearly not linked to the specific and unusual way in which we process information in our brains, because for neither the Neanderthal nor the Homo erectus lineage is there any firm evidence that symbolic cognition was ever achieved.

The Neanderthals are particularly fascinating because they are the best-known of all extinct hominids, and thus provide the most reliable yardstick we have by which to measure our own uniqueness. Homo neanderthalensis occupied Europe and a wide swath of western Asia between about 200 kyr and 30 kyr ago, and had a brain as big as ours. Physically distinctive (from us; from the neck down it was probably a fairly standard later Homo species: Tattersall, 2009b), this hominid was clearly a skilled hunter of large animals and exploiter of its (often severe) environments (see discussion and references in Tattersall, 2009a). Neanderthals were accomplished practitioners of the prepared-core tool-making method, and they buried their dead (at least occasionally, and without fanfare) while at the same time practicing a chillingly prosaic form of cannibalism (Carbonell et al., 2010). Clearly, the Neanderthals communicated subtly and effectively with each other, and had a keen understanding of the difficult world around them. But, again, over the vast expanse of time and space they inhabited these hominids left no unequivocal proxy evidence of symbolic thought processes; and it is probably telling that they were entirely eliminated, in a few short thousand years, by the fully symbolic Homo sapiens who began trickling into their territory at about 40 kyr ago. These were the Cro-Magnons, the creators of such stupendous cave art sites as Chauvet, Lascaux and Altamira: beings whose lives were totally drenched in symbol.

4. Becoming Human

Modern Homo sapiens fossils begin appearing in Africa (without clear fossil antecedents) at about 200-160 kyr ago. Archaeological associations are scanty and modest (Klein, 2009). Better archaeological contexts (Klein, 2009) exist for the first anatomically modern humans outside Africa. These are found at around 100 kyr ago in Israel which also, at least intermittently, hosted Neanderthals between about >130 kyr and 45 kyr. And throughout this period, the stone tool assemblages associated with the two kinds of hominid are virtually identical. There is no reason whatever to suspect any cognitive difference between the two in this period. Neither was symbolic. This actually agrees with the record in Africa, because the first intimations of symbolic behavior in that continent are found only subsequent to about 100 kyr ago, in the form of geometrically-engraved ochre plaques and complex multi-stage technologies reported from sites on the South African coast in the period around 75 kyr ago (Henshilwood et al., 2004; Brown et al., 2009). Even earlier intimations exist in the form of pierced shell "beads" from sites both in the north and the south of the continent; but such expressions are more arguably symbolic, and are still in the <100 kyr range. Thus, modern human anatomy appears to have been achieved significantly before modern symbolic behavior patterns.

So what happened? Most plausibly, after something like 400 myr of vertebrate brain evolution, the neural substrate permitting symbolic thought had been produced, in an already near-enabled brain, as a byproduct of the apparently radical developmental reorganization that gave rise to Homo sapiens as an anatomically distinctive entity (Tattersall, 2009b). Whatever its exact nature, this innovation lay fallow until its new use was "discovered" by its possessors, through the effect of some necessarily cultural stimulus. However radical its result, the evolutionary mechanism of exaptation that must have been involved here is not in the least unusual; after all, the ancestors of birds had feathers for million of years before using them to fly. So what was that cultural stimulus? The most plausible candidate is the invention of language, which is almost synonymous with what we experience as thought today. Like thought, language involves creating symbols in the mind, and rearranging them according to rules to generate an infinite number of possible statements from a finite vocabulary. What's more, it is known that structured language can be readily and spontaneously invented (Kegl et al., 1999); and, as an externalized attribute, language is highly likely to spread rapidly within populations already biologically predisposed for it. Once symbolic thought had been "kicked" into existence in this way, newly symbolic Homo sapiens was able to envision alternative worlds, and thus to plan in an unprecedentedly complex manner. It acquired that sense of "self" that Povinelli found so conspicuously lacking in chimpanzees. This opened up a huge cognitive gulf between our species and even its closest contemporaneous relatives (the Neanderthals, Homo erectus, its own ancestor). Complex as they undoubtedly were, these "old-style" hominids were cognitively limited to responding, in however sophisticated a manner, to the outside world as it was presented to them by Nature. They were unable to remake the world in their minds, as Homo sapiens now could. At some point after acquiring symbolic thought in its parent continent Homo sapiens left Africa, and its rapid spread through the Old World, and eventually to the New has been exhaustively recorded by molecular biologists (see overview by DeSalle and Tattersall, 2008). Its nearest (and apparently not so dearest) residing in those territories never stood a chance.

References

Brown, K. S., Marean, C. W., Herries, A. I. R., Jacobs, Z., Tribolo, C., Braun, D., Roberts, D. L., Meyer, M. C., Bernatchez, J. (2009). Fire as an engineering tool of early modern humans. Science, 325, 859-862.

Carbonell, E., I. Caceres, M. Lizano, P. Saladie, J. Rosell, C. Lorenzo, J. Vallverdu, R. Huguet, A. Canals, J. M. Bermudez de Castro. (2010). Cultural cannibalism as a paleoeconomic system in the European lower Pleistocene.

Curr. Anth., 51, 539-549.

Cohen, J. (2010). 2010. Almost chimpanzee. Times Books, New YorkCity.

DeSalle, R., Tattersall, I. (2008). Human origins: What bones and genomes tell us about ourselves. Texas A&M Press, College Station, TX.

Goren-Inbar, N., Alperson, N., Kislev, M.

E., Simchoni O., Melamed, Y., Ben-Nun, A., Werker, E. (2004). Evidence of hominid control of fire at Gesher Benot Ya'aqov, Israel. Science, 304, 725-727.

Hart, D., Sussman, R. W. (2009). Man the hunted: Primates, predators, and human evolution. Expanded edition. Westview Press, Boulder, CO.

Kegl, J., Senghas, A., Coppola, M. (1999). Creation through contact: Sign language emergence and sign language change in Nicaragua. In: M. deGraaf (Ed.), Comparative grammatical change: The intersection of language acquisition, Creole genesis and diachronic syntax. MIT Press, Cambridge MA, pp. 179-237.

Klein, R. (2009). The human career: Human biological and cultural origins, 3rd ed. University of Chicago Press, Chicago.

Lock, A., Peters, C. E. (1996). Handbook of human symbolic evolution. Oxford University Press, Oxford, UK.

Henshilwood, C. S., d'Errico, F., Yates, R., Jacobs, Z., Tribolo, C., Duller, G. A. T., Mercier, N., and 4 others. (2002). Emergence of modern human behavior: Middle Stone Age engravings from South Africa. Science, 295, 1278-1280.

McPherron, S., Alemseged, Z. Marean, C. W., Wynne, J. G. Reed, D. Geraads, D. Bobe R. Béarat, H. A. (2010). Evidence for stone-tool-assisted consumption of animal tissues before 3.39 million years ago at Dikika, Ethiopia. Nature, 466, 857-860.

Povinelli, D. J. (2004). Behind the ape's appearance: Escaping anthropocentrism in the study of other minds. Daedalus, 133 (1), 29-41.

Pruetz, J. D., Bertolani, P. (2007). Savanna chimpanzees, Pan troglodytes verus, hunt with tools. Curr. Biol., 17, 412-417.

Schick, K. D., Toth, N. (1993). Making silent stones speak. Simon and Schuster, New York City.

Sponheimer, M. and J. Lee-Thorp. (2007). Hominin paleodiets: The contribution of stable isotopes. In: Henke W., I. Tattersall, I. (Eds), Handbook of paleoanthropology. Springer, Heidelberg, pp. 555-585.

Tattersall, I. (2009a). The fossil trail: How we know what we think we know about human evolution, 2nd ed. Oxford University Press, New York.

Tattersall, I. (2009b). Human origins: Out of Africa. Proc. Natl. Acad. Sci. USA, 106, 16018-16021.

Wrangham, R. 2009. Catching Fire: How Cooking Made Us Human. Basic Books, New York.

33. Evolution's Gift: Subjectivity and the Phenomenal World
Arnold Trehub

Department of Psychology, University of Massachusetts at Amherst, Amherst, Massachusetts 01003

Abstract

A particular system of brain mechanisms, called the retinoid system, is proposed as the evolutionary adaptation responsible for the existence of subjectivity and our sense of being here in a surrounding 3D world. The structural and dynamic properties of the retinoid system successfully predict a novel conscious experience in which the brain constructs a vivid visual representation of an object moving in space without a corresponding image projected to the retinas. Implications of the retinoid system for human creativity and our scientific understanding of the universe are suggested.

KEY WORDS: consciousness, 3-D world, brain, retina, retinoid

1. INTRODUCTION

We are each born with a system of brain mechanisms that constitute the full scope of our entire phenomenal universe. The structure and dynamics of this crucial brain system actually construct the world of our experience. What we call consciousness is the presence of this world for us. Consciousness did not exist before a creature was able to represent something somewhere in a perspectival relationship to a fixed internal "point" of origin – a transformative evolutionary event that ushered in the dawn of the first-person perspective – subjectivity and consciousness. To this extraordinary biological development we owe the possibility of all of our past and present conceptions of everything that is in our cosmos. Here is the key question: What biological mechanism emerged that enabled subjectivity and set us on a path by which we can now engage in our present discourse about consciousness? My proposal is that a unique brain system with a particular kind of neuronal structure and dynamics – the retinoid system (Trehub 1977, 1991, 2007) -- is the essential generator of our conscious experience. The structural and dynamic properties of the retinoid system enable it to register and appropriately integrate disparate stimuli into an egocentric representation of one's volumetric surround – the world around us. This system of neuronal mechanisms composes an internal egocentric space that receives Input from all sensory modalities and, in recurrent fashion, projects its excitatory neuronal patterns back to each sensory modality. It organizes its multi-sensory features into a coherent global representation of 3D space. All the sensory features of objects and events in retinoid space are organized in proper spatiotemporal register, called feature binding. For example, if a red car were to travel from left to right in your field of view, the retinoid representation of its color would be contained within the contour of its shape, and both color and shape would move in concert left to right within your egocentrically organized retinoid space. Each sensory modality is served by a particular kind of neuronal mechanism called a synaptic matrix (Trehub 1991). The synaptic matrix is a self-organizing brain mechanism that has the capacity to learn and classify complex stimulus patterns, store them in long-term memory, and recall images of them in the absence of external stimulation. Synaptic matrices are specialized processors; each serves only one kind of sensory feature; e.g., for visual shape, for color, for motion, for taste, etc. It is the role of the retinoid system to integrate signals from these disparate sensory modalities, widely separated in the brain, into a single coherent representation of the surrounding world.

The rich content of our sense of the world around us is provided by reciprocal evocation among sensory images and their neuronal tokens embodied in the recurrent loops of the synaptic matrices in parallel coupling with the retinoid system. If we think in metaphorical terms of a theater of consciousness within the brain, then retinoid space would correspond to the bright stage of the "theater" on which we are a participant. This stage is our conscious content. The synaptic matrices in the various sensory modalities and the other cognitive mechanisms which categorize and evaluate the patterns of neuronal activity (the objects and events) presented on the retinoid "stage" would correspond to something like a critical observing audience in the dark (unconscious) part of the theater (Trehub 2007).

Most organisms on this earth are able to adapt reasonably well with nothing but mechanisms that we would relegate to the unconscious. They do not respond to an internal global representation of the world they live in. They have no such internal representation and respond instead only to those isolated stimuli that reach their sensory transducers. Other creatures, mankind in particular, respond to a phenomenal world full of events and contingencies that are far beyond what impacts the lower organisms.

While the retinoid system is assumed to have first appeared in the brains of lower animals, its evolutionary development, together with the development of related cognitive brain mechanisms, peaked with the emergence of mankind.

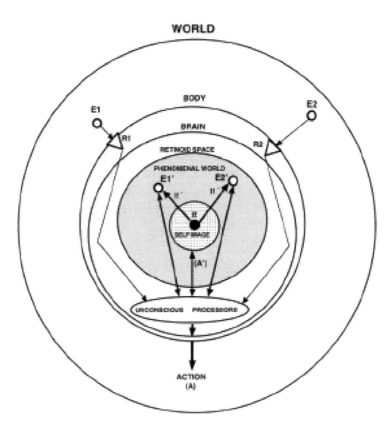

Figure 1. A. Non-conscious creatures. E1 and E2 are discrete events in the physical world. R1 and R2 are sensory transducers in the body that selectively respond to E1 and E2. R1 and R2 signal their response to unconscious processing mechanisms within the brain. These mechanisms then trigger adaptive actions. B. Conscious creatures. In addition to the mechanisms described in A, the brain of a conscious creature has a retinoid system that provides a holistic volumetric representation of a spatial surround from the privileged egocentric perspective of the self-locus -- the core self (I!). For example, in this case, there is a perspectival representation of E1 and E2 (shown as E1' and E2') within the creature's phenomenal world. Neuronal activity in retinoid space is the entire current content of conscious experience.

Notice in Fig. 1 that that there are neuronal pathways from the body's sensory receptors (R1 and R2 in the diagram) to the brain's unconscious sensation-toaction processors, then from these neuronal mechanisms to the neuromuscular structures for overt action. It is assumed that all sentient and motile preconscious organisms have this kind of sensory-motor system. However, with the emergence of the retinoid system in the course of creature evolution, an entirely new kind of brain mechanism with new possibilities for adaptive action appeared in the world.

Some investigators have claimed that consciousness depends on particular kinds of quantum events in the brain's neuronal circuits. On the basis of our present understanding of quantum electrodynamics we should expect quantum events to be relevant to all biophysical processes at a fundamental level. However, I would argue that if particular kinds of quantum events are selectively determinate for conscious content, they must conform in some way to the structural and dynamic properties of the retinoid system. With a nod to the anthropic principle, I suggest that the entire conceptual edifice of modern science is a product of biology and is necessarily constrained by the conscious-cognitive structure and dynamics of the human brain. The cosmos as it is subjectively represented in the retinoid system of the human brain is the only cosmos we can think about.

2. Neuronal Activity in the Retinoid System Constitutes Our Phenomenal World

A key feature of the representational space within the retinoid system is that it is organized around a fixed cluster of cells which constitute the neuronal origin – the 0,0,0 (X, Y, Z) coordinate -- of its 3D spatiotopic neuronal structure. All phenomenal representations are constituted by patterns of neuronal excitation on the Z-planes of retinoid space. I have proposed that the fixed spatial coordinate of origin in the Z-plane structure can be thought of as one's self-locus in one's phenomenal world, and I designate this central cluster of neurons as the core self (I!) (Trehub 1991, 2007).

Our consciousness is no more nor less than the current content of our phenomenal world, and on this I base my working definition of consciousness as follows:

> Consciousness is a transparent brain representation of the
> world from a privileged egocentric perspective.

Since, in this theoretical model, retinoid space is the space of all of our conscious experience, vision should be understood as only one of the sensory modalities that project content into our egocentrically organized phenomenal world. A blind person can have as keen a phenomenal sense of a surrounding volumetric space as a fully sighted person. All of our sensory modalities that serve both external and internal sensations can contribute to our phenomenal experience, as shown in Figure 1. The minimal neuronal structure and dynamics of the retinoid system and the non-conscious sensory and cognitive brain mechanisms, primarily with respect to the visual system, have been detailed in Trehub 1991. On the question of the retinoid system as the biophysical foundation of consciousness, we might ask: Is there a critical experiment that provides strong support to the theoretically formulated properties of the mechanisms in the retinoid system? The Seeing-more- than-is-there experiment described in the following section shows that there is such evidence.

3. Complementary Neuronal and Phenomenal Properties

In the development of the physical theory of light, the double-slit experiment was critical in demonstrating that light can be properly understood as both particle and wave. Similarly, I believe that a particular experiment – a variation of the seeing-more- than-is-there (SMTT) paradigm – is a critical experiment in demonstrating that consciousness can be properly understood as a complementary relationship between the activity of a specialized neuronal brain mechanism, having the neuronal structure and dynamics of the retinoid system, and our concurrent phenomenal experience.

Seeing-More-Than-is-There (SMTT) If a narrow vertically oriented aperture in an otherwise occluding screen is fixated while a visual pattern is moved back and forth behind it, the entire pattern may be seen even though at any instant only a small fragment of the pattern is exposed within the aperture. This phenomenon of anorthoscopic perception was reported as long ago as 1862 by Zöllner. More recently, Parks (1965), McCloskey and Watkins (1978), and Shimojo and Richards (1986) have published work on this striking visual effect. McCloskey and Watkins introduced the term seeing-more-than- is-there to describe the phenomenon and I have adopted it in abbreviated form as SMTT. The following experiment was based on the SMTT paradigm (Trehub 1991).

Procedure:

1. Subjects sit in front of an opaque screen having a long vertical slit with a very narrow width, as an aperture in the middle of the screen. Directly behind the slit is a computer screen, on which any kind of figure can be displayed and set in motion. A triangular-shaped figure in a contour with a width much longer than its height is displayed on the computer. Subjects fixate the center of the aperture and report that they see two

tiny line segements, one above the other on the vertical meridian. This perception corresponds to the actual stimulus falling on the retinas (the veridical optical projection of the state of the world as it appears to the observer).

2. The subject is given a control device which can set the triangle on the computer screen behind the aperture in horizontal reciprocating motion (horizontal oscillation) so that the triangle passes beyond the slit in a sequence of alternating directions. A clockwise turn of the controller increases the frequency of the horizontal oscillation. A counter-clockwise turn of the controller decreases the frequency of the oscillation. The subject starts the hidden triangle in motion and gradually increases its frequency of horizontal oscillation.

Results:

As soon as the figure is in motion, subjects report that they see, near the bottom of the slit, a tiny line segment which remains stable, and another line segment in vertical oscillation above it. As subjects continue to increase the frequency of horizontal oscillation of the almost completely occluded figure there is a profound change in their experience of the visual stimulus.

At an oscillation of ~ 2 cycles/sec (~ 250 ms/sweep), subjects report that they suddenly see a complete triangle moving horizontally back and forth instead of the vertically oscillating line segment they had previously seen. This perception of a complete triangle in horizontal motion is strikingly different from the tiny line segment oscillating up and down above a fixed line segment which is the real visual stimulus on the retinas.

As subjects increase the frequency of oscillation of the hidden figure, they observe that the length of the base of the perceived triangle decreases while its height remains constant. Using the rate controller, the subject reports that he can enlarge or reduce the base of the triangle he sees, by turning the knob counterclockwise (slower) or clockwise (faster).

3. The experimenter asks the subject to adjust the base of the perceived triangle so that the length of its base appears equal to its height.

Results:

As the experimenter varies the actual height of the hidden triangle, subjects successfully vary its oscillation rate to maintain approximate base-height equality, i.e. lowering its rate as its height increases, and increasing its rate as its height decreases.

This experiment demonstrates that the human brain has internal mechanisms that can construct accurate analog representations of the external world. Notice that when the hidden figure oscillated at less than 2 cycles/sec, the observer experienced an event (the vertically oscillating line segment) that corresponded to the visible event on the plane of the opaque screen. But when the hidden figure oscillated at a rate greater than 2 cycles/sec., the observer experienced an internally constructed event (the horizontally oscillating triangle) that corresponded to the almost totally occluded event behind the screen. The experiment also demonstrates that the human brain has internal mechanisms that can accurately track relational properties of the external world in an analog fashion. Notice that the observer was able to maintain an approximately fixed one-to-one ratio of height to width of the perceived triangle as the height of the hidden triangle was independently varied by the experimenter.

These and other empirical findings obtained by this experimental paradigm were predicted by the neuronal structure and dynamics of a putative brain system (the retinoid system) that was originally proposed to explain our basic phenomenal experience and adaptive behavior in 3D egocentric space (Trehub, 1991). It seems to me that these experimental findings provide conclusive evidence that the human brain does indeed construct phenomenal representations of the external world and that the detailed neuronal properties of the retinoid system can account for our conscious content.

4. The Heuristic Self-Locus (I!*)

The ability to move excitation from the source point of the self-locus (I!) to selected regions within the depth (Z-planes) of retinoid space also provides an important means of selective attention. The projection of neuronal excitation from the excitatory source of the core self to a target of interest in 3D retinoid space constitutes a selective shift of attention which is realized by an excursion of the heuristic self-locus (I!*; see Fig.2). Neurons in regions of a retinoid that are stimulated by the added local excitation of a heuristic self-locus excursion are preferentially primed and marked relative to other cells in retinoid space. Cells in a primed region respond more quickly and vigorously than those in unprimed regions.

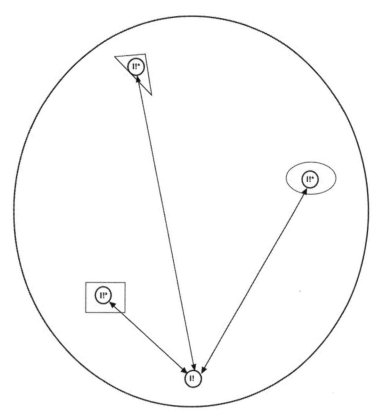

Figure 2. Excursions of the heuristic self-locus (I!*) are the biophysical equivalent of selective attention. In this example, attention (I!*) is directed at a square, a triangle, and an oval in the phenomenal world of retinoid space (egocentrically arranged with respect to the core self (I!).

An important feature of the heuristic self-locus is its property of inducing excitatory traces or patterns of cellular activity in accordance with its movement through retinoid space. We can think of these excitatory images as similar to the patterns drawn by a stylus on a display board. These self-locus traces might represent unimpeded travel routes between regions of interest, they might represent barriers to be avoided, or they might be retinoid images of our own imaginative construction serving us as cognitive maps or internal sketches to be used for many different purposes.

5. Creativity

The capability for invention, trivial and great, is arguably the most consequential characteristic that distinguishes humans from all other creatures. Our cognitive brain is especially endowed with neuronal mechanisms that can model within their biological structures all conceivable worlds as well as the world that we directly perceive or know to exist. External expressions of an unbounded diversity of brain-created models constitute the arts and sciences and all the artifacts and enterprises of human society (Trehub, 1991).

Our retinoid system together with a specialized preconscious brain mechanism for learning, long-term memory, and imaging, which I call a synaptic matrix, make creative modeling possible. A detailed description of the basic neuronal design of the synaptic matrix, as well as the structure and dynamics of the retinoid system, is given in Trehub 1991 (The Cognitive Brain). Synaptic matrices in all of our sensory modalities serve learning, memory, recognition, recall, and imaging of sensory features and events, while the retinoid system receives feedback signals from the imaging matrix of each modality to organize a coherent global layout of objects and events in our egocentrically organized retinoid space.

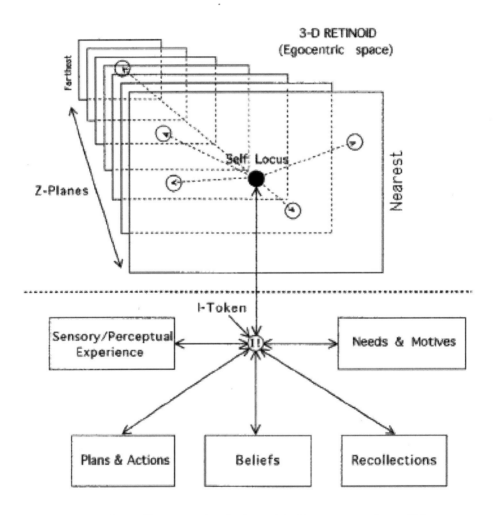

Figure 3. The self system (Trehub 2007). Neurons at the self-locus anchor the I-token (I!) to the retinoid origin of egocentric space. I! has reciprocal synaptic links to sensory and cognitive processes.

Damage to the neuronal mechanisms below the dotted line results in cognitive impairment. Interruption of the synaptic link between the neurons at the origin of retinoid space (the self locus) and I! results in loss of consciousness.

Previously learned images and layouts can be recalled from the preconscious cognitive mechanisms, then decomposed and recomposed by the neuronal mechanisms of the retinoid system to form novel images or models. The content of retinoid space may consist of such imaginative constructions as well as the veridical makeup of a current environment. External expressions of novel phenomenal objects and events in retinoid space comprise our creative productions.

The conscious brain constructs the world of our experience by inserting and arranging on the Z-planes of retinoid space selected patterns of exteroceptive and interoceptive sensory patterns together with images recalled from the memory stores of synaptic matrices. The retinoid mechanisms are able to create phenomenal representations of novel objects and events by parsing objects or their parts in existing representations and rearranging these excitatory neuronal patterns in new combinations projected into retinoid space. The analytic mechanisms of the cognitive brain (shown below the dotted line in Fig.3) can then evaluate the novel retinoid images in terms of their practical or theoretical utility. These putative brain mechanisms have been described in detail (Trehub 1991).

On the basis of Fig. 3, we can say that damage to any of the mechanisms below the dotted line would result in cognitive impairment with a sparing of consciousness. But if the synaptic link between the self-locus cells in the retinoid's Z-plane and its neuronal token (I!) below the dotted line were broken or damaged we would expect a loss of consciousness. Taking this into account, we might conjecture that the reason a sharp blow to the head can cause a loss of consciousness is precisely because the jolt can effectively interrupt synaptic communication between the self-locus neurons in the retinoid structure and I!, which is in an excitatory feedback loop with the retinoid's self locus (the core self), and which provides a gateway to the rest of the cognitive system.

6. Discussion

Can the retinoid model answer the daunting questions that have long perplexed the search for a scientific understanding of consciousness? An important consideration in making an assessment is whether the theoretical model enables the scientific community to perform reasonable empirical tests of its implications. In this respect, I have suggested that a science of consciousness requires the adoption of a bridging principle between the first-person subjective description of conscious content and a third-person objective description of conscious content (Pereira Jr, A. et al 2010). To this end, I have proposed the following principle:

> For any instance of conscious content there is a corresponding analog in the biophysical state of the brain.

The objective, then, is to formulate brain mechanisms that can generate proper analogs of conscious content. Application of this bridging principle led to successful predictions about subjects' conscious experiences in the SMTT experiment on the basis of the detailed structure and dynamics of the retinoid system. It should be added that there are many more previously puzzling subjective phenomena that are straightforwardly explained by the causal properties of the retinoid system, among them, the moon illusion, size constancy, and motion after-effects (see Trehub 1991, pp. 89-93 and pp. 239-255). A positive aspect of the retinoid theory, beside its explicit account of subjectivity and phenomenal consciousness, is its ability to explain human creativity on the basis of the normal operation of plausible brain mechanisms. This, together with the well founded supposition that the retinoid structure of the brain is an advanced evolutionary adaptation, adds credence to the theoretical model. Moreover, the implications of the retinoid model for understanding the source of our scientific concepts lends substance to the weak version of the anthropic principle (Barrow and Tipler 1986). In addition to extending the retinoid model and further testing its implications, one would want to see others pursue the scientific challenge of formulating an alternative testable model that can do a better job of explaining subjectivity and the brain mechanisms that present us with our phenomenal world.

7. Conclusion

Our phenomenal world, the world of everyday experience, the world in which we try to thrive and probe for understanding, is an amazing product of biological evolution. The big questions that science now tries to answer could not be posed before the evolutionary emergence of the brain's retinoid system. It is this biological system which gives us subjectivity -- a sense of a self centered in a surrounding universe.

References

Barrow, J. D. and Tipler, F. J. (1986). The Anthropic Cosmological Principle. Oxford University Press.

McCloskey, M. and Watkins, J.W. (1978). The seeing-more-than-is-there phenomenon: Implications for the locus of iconic storage. Journal of Experimental Psychology: Human Perception and Performance 4: 553-564.

Parks, T. (1965). Post-retinal visual storage. American Journal of Psychology 78: 145-147.

Pereira Jr., A., Edwards, J. C. W., Lehmann, D., Nunn, C., Trehub, A., and Velmans, M. (2010). Understanding Consciousness:

A Collaborative Attempt to Elucidate Contemporary Theories. Journal of Consciousness Studies 17: 213-219.

Shimojo, S. and Richards, W. (1986). Seeing shapes that are almost totally occluded: A new look at Park's camel. Perception and Psychophysics 39: 418-426.

Trehub, A. (1977). Neuronal models for cognitive processes: Networks for learning, perception, and imagination. Journal of Theoretical Biology 65: 141-169.

Trehub, A. (1991). The Cognitive Brain. MIT Press.

Trehub. A. (2007). Space, self, and the theater of consciousness. Consciousness and Cognition 16: 310-330.

Zollner, F. (1862). Uber einer neuer Art anorthoscopischer Zerrbilder. Annalen der Physik und Chemie 27: 477-484.

34. Intention and Attention in Consciousness: Dynamics and Evolution

Hans Liljenström

Biometry and Systems Analysis Group, Energy & Technology, SLU, Box 7013, S-750 07 Uppsala, Sweden, Agora for Biosystems, Box 57, S-193 22 Sigtuna, Sweden

Abstract

All through the history of the universe there is an apparent tendency for increasing complexity, with the organization of matter in evermore elaborate and interactive systems. The living world in general, and the human brain in particular, provides the highest complexity known. Presumably, the neural system with its complex dynamics has evolved to cope with the complex dynamics of the environment, where it is embedded. The evolution of a nervous system constitutes a major transition in biological evolution and allows for an increasing capacity for information storage and processing. Neural knowledge processing, cognition, shows the same principal features as non-neural adaptive processes. Similarly, consciousness might appear, to different degrees, at different stages in evolution. Both cognition and consciousness seem to depend critically on the organization and complexity of the organism. Different states of consciousness can apparently be associated with different levels of neural activity, in particular with different oscillatory modes at the mesoscopic level of cortical networks. Transitions between such modes could also be related to transitions between different states of consciousness. For example, a transition from an awake to an anaesthetized state, or sleep, is accompanied by a transition from high frequency oscillations to low frequency oscillations in the cortical neurodynamics. In this article, I will briefly discuss some general aspects on the evolution of the nervous system and its complex neurodynamics, which provides organisms with ever increasing capacity for complex behaviour, cognition and consciousness. Consciousness and cognition apparently evolve through interaction with the environment, where the organism is embedded. Such exploration of the environment requires both attention and intention. I will discuss these dual and complementary aspects of consciousness, and their effects as perception and action. Finally, I will speculate on consciousness related to life, and how it may be regarded as a driving force in the exploration of our world.

KEY WORDS: Attention, intention, free will, neurodynamics, cortical networks, EEG, evolution

1. Introduction

When we open our eyes after a good night's sleep, we again become conscious of the world around us. A few moments before, we had closed our senses to the external world, resting in an inner world, which only in the dreaming state had some degree of consciousness. We were unable to attend to our physical environment, or move voluntarily in it. When we wake up, we make a transition into a state, where we suddenly can perceive and act intentionally. If we wake up in a familiar environment, we can act almost automatically in it, but otherwise, we may start to explore our environment, by attending to it and interact with it intentionally. However, in our wake conscious state, we may shift between several sub-states, corresponding to different levels of alertness or wakefulness.

During sleep, our cognitive activity is mostly unconscious. We may wake up from this state spontaneously (by some internal "clock" or other internal event), or as a result of a sounding alarm clock, or the whispering of our name, or a weak smell of smoke, or any other significant signal, that is below threshold for any insignificant sensory input. Obviously, there is a certain level of "awareness" also in the so-called unconscious state.

Even in our wake state, most of our cognitive activity may be unconscious. While it is important to distinguish between cognition and consciousness, (which is often mixed up in the literature), both seem to linked to system and process complexity. In fact, the cognitive and conscious capacity seems to increase with an increasing level of complexity, or interconnectedness (Århem &

Liljenström, 2007).

The brain, like other biological systems, has presumably evolved to efficiently increase the probability of survival in a constantly changing environment. This would imply, among other things, a rapid and appropriate response to external (and internal) events. We may assume that the neural system with its complex dynamics has evolved to cope with the complex dynamics of the environment, by constantly interacting with it.

In the following, I attempt to relate the neurodynamics of the brain with our cognitive and conscious activity, and also set it in an evolutionary context. Specifically, I will consider the dual aspects of consciousness, intention and attention, and its effective counterparts of the neural system as perception and action. Finally, I will speculate on these aspects as universal to all life, and on consciousness as a driving force in nature.

2. Neural Dynamics

Brain structures are characterized by their complexity in terms of organization and dynamics. This complexity appears at many different spatial and temporal scales, which in relative terms can be considered as micro, meso, and macro scales. The corresponding dynamics may range from ion channel kinetics, to spike trains of single neurons, to the neurodynamics of cortical networks and areas. The high complexity of this neurodynamics is partly a result of the web of non-linear interrelations between levels and parts, including positive and negative feedback loops. (Freeman, 2000; Århem & Liljenström, 2001; Århem et al, 2005).

Very little is still known about the functional significance of the neural dynamics at the various organizational levels, and even less about the relation between activities at these levels. However, it is reasonable to assume that different dynamical states correlate with different functional or conscious states. Transitions within the brain dynamics at some level would correlate with transitions in the "consciousness dynamics", involving various cognitive levels and conscious states, e.g. when going from sleep to awake states, or from drowsiness to alertness, etc. In general, there is a spectrum of states, each one with its own characteristic neurodynamical (oscillatory) mode.

Presumably, phase transitions at the mesoscopic level of neural systems, i.e. at levels between neurons and the entire brain, are of special relevance to mental state transitions. The mesoscopic neurodynamics of cortical neural networks typically occurs at the spatial order of a few millimetres to centimetres, and temporally on the order of milliseconds to seconds. This type of dynamics can be measured by methods, such as electrocorticography (ECoG), electroencephalography (EEG), or magnetoencephalography (MEG).

Mesoscopic brain dynamics with its transitions is partly depending on thresholds and the summed activity of a large number of microscopic elements (molecules and cells) interconnected with positive and negative feedback. Some of this activity is spontaneous and can be considered as noise (which may have a functional significance, see e.g. Århem et al., 2005).

In addition, the mesoscopic dynamics is influenced and regulated by various hormones and neuromodulators, such as acetylcholine and serotonin, as well as by the state of arousal or motivation, relating to more macroscopic processes of the brain. Hence, mesoscopic neurodynamics can be seen as resulting from the dynamic balance between opposing processes at several scales, from the influx and efflux of ions, inhibition and excitation etc. Such interplay between opposing processes often results in (transient or continuous) oscillatory and chaoticlike behaviour (Skarda & Freeman, 1987; Freeman, 2000; Liljenström, 2010). Clearly, the brain activity is constantly changing, due to neuronal information processing, internal fluctuations, neuromodulation, and sensory input.

An essential feature of the mesoscopic brain dynamics is spatio-temporal patterns of activity, appearing at the collective level of a very large number of neurons. Waves of activity move across the surface of sensory cortices, with oscillations at various frequency bands (Freeman, 1975, 2000). In general, low frequencies correspond to low mental activity, drowsiness or sleep, whereas higher frequencies are associated with alertness and higher conscious activity. Fig. 1 shows the simulated effect of certain anaesthetics, which are presumed to cause a transition from high frequency (gamma) to low frequency (theta) oscillations, mimicking a transition from an awake to an anaesthetized state (Halnes et al., 2007).

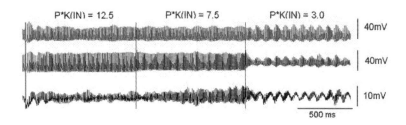

Fig. 1. Simulated effect of anaesthetics, which may shift the mesoscopic neurodynamics from high frequency (gamma) oscillations to low frequency (theta) oscillations for three different concentration levels of potassium ion channel blocking, increasing from left to right. These shifts in neurodynamics would presumably correspond to shifts in conscious states, going from alert (left) to anaestetized/ sleep (right). The two upper time series show the activity of a single excitatory and inhibitory neuron, respectively, while the lower time series is the network mean. (Adopted from Halnes et al, 2007).

In particular, oscillations in the gamma frequency band, around 40 Hz, has been associated with (visual) attention, based on experiments on cats about two decades ago (Eckhorn et al., 1988; Gray & Singer, 1989). It was this phenomenon that triggered a lot of studies on neural correlates of consciousness, as e.g. suggested by Crick and Koch (1990; Koch, 2004) and contributed to opening the field of neuroscience to consciousness studies. This experimental and theoretical research on neural oscillations has been complemented by computational models, which attempt to relate structure to dynamics and function. (I have dealt with these issues in more detail elsewhere, e.g. in Liljenström 1991, 1995, 1997, 2010).

3. Consciousness Dynamics

The complex neurodynamics at a mesoscopic level of the brain seem significant for cognitive functions and conscious activity. It has been related to perception, attention and associative memory, but also to volition and activity in the sensory motor areas of the brain. Even though many details are still unknown, there is an interplay between the neurodynamics of the sensory and motor pathways, which seem essential for the interaction with our environment.

We explore our world in a perception-action cycle (Freeman, 2000). The development of our cognitive and conscious abilities depend on an appropriate interaction with the complex and changing environment, in which we are embedded. Our perceptions and actions develop and are refined to effectively deal with our external (and internal) world.

However, prior to perception is attention, and prior to action is intention. Some believe that attention and intention are at the core of our experience of existence (Popper et al., 1993). While attention has been shown to correlate with certain neural oscillations, intention has not been demonstrated to the same extent. Attention is primarily related to the sensory/perceptual pathways and brain areas, whereas intention would be more related to the motor areas and pathways. In particular, the supplementary motor area (SMA), but also the parietal cortex, show early signs of intentional motor activity (Eccles, 1982; Libet 1985; Desmurget et al., 2009).

Attention could be extended, but also be focused to some specific part of our internal or external worlds. We can turn our attention to some specific memory, thought, or sensory experience, while neglecting or inhibiting the rest. Naturally, and perhaps more commonly, our attention is drawn to some object or event in our environment, something that is of special relevance or interest to us, for our survival or for our curiosity.

We may also intentionally turn our attention to some area of interest. Intention can be viewed as a precursor to volition and will, as an "urge" or "desire" to act in a certain direction, to attain a certain goal. Voluntary movement, or more generally, behavior, is based on perception and past experience (memory), which are required for prediction of internal and external interactions. Attention may provide information about the internal and external worlds, but intention guides our actions. We can attend to our intentions, but we can also intentionally guide our attention.

While attention and intention would be most effective in the wake conscious state, they could also be present, to lesser degree, in "unconscious" states (e.g. visual attention in the original 40 Hz experiments were made on anesthetized cats (Eckhorn et al., 1988; Gray & Singer, 1989)). They could be present at different levels of intensity; where we become conscious awake above a certain threshold.

Attention and intention could be seen as complementary aspects of consciousness, with their effective neural counterparts of perception and action, respectively, when implemented through the interactions within sensory and motor hierarchies of the nervous system. They correspond to an "inward" and an "outward" going activity of consciousness (see Fig. 2). Both of these aspects seem essential for exploring our external (and internal) world. In other words, we play the dual role of actor and observer in the drama of existence (Joseph, 1982, 2011).

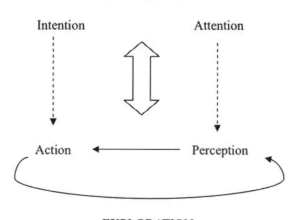

Fig. 2. The relation between the "implicate" order of consciousness, as in the dual aspect of attention and intention, and the "explicate order" of exploration, as in the perception-action cycle.

4. Consciousness and Life

Attention and intention, as complementary aspects of consciousness are presumably not only part of the human experience, but could be much more universal. For example, (visual) attention has been studied in insects, such as fruit-flies (van Swinderen, 2008) and honey bees (Srinavasan 2009), and can be expected to be found also in other invertebrates. Likewise, intentional behaviour can be attributed to many types of (in particular social) animals, including crustaceans (Hauser & Nelson, 1991; Mather 2011; Nani et al., 2011).

Indeed, these aspects of consciousness might be intimately linked to life at all levels, albeit not in the same sense as we use the terms for humans (or other mammals). Naturally, in organisms without a nervous system, attention and intention could never be implemented through any sensory or motor hierarchy of such a system, but would rather have to be implemented through other molecular or cellular structures.

This view is akin to that of Delbrück (1986), who considers consciousness to include all types of "awareness" of external and internal events, from a crude form of perception in the simplest organisms to self-consciousness in Man. The more complex the cellular organization of an organism is the more this organism can be aware, or conscious, of the external and internal world. In this sense, consciousness has evolved together with other properties of life. In his book, "Mind from Matter?" (1986), Delbrück regards perception as the basic form of consciousness and he states (p. 43) that,

> "The unity and continuity of life is equally manifest in its psychic aspects. Perception in plants and animals is a familiar phenomenon, but the beginnings of perception are also clearly present in microorganisms, in which adaptive behavior demonstrates that they can detect and evaluate signals from the environment and respond appropriately".

This is not a dominating view today, but it is appealing in the sense that it relates processes in simple organisms with those in higher. It avoids the difficult task of finding a possible origin of consciousness in evolution, and points to a continuity in the living world. A similar view is shared by Margulis and Sagan (1995), who suggest that also intentionality may be attributed to simple organisms, including bacteria.

The electro-chemical (or other) processes that take place in the intricate neural networks of the brain can be assumed to be associated to various internal states and "mental" processes of an animal. In lower animals the nervous system may allow only for "primitive mental" processes, such as perception, learning, recalling, drives, and emotions (Mather 2011; Nani et al., 2011). To a large extent the animal is presumably unconscious of these internal processes, and it is only aware (conscious, in the Delbrück sense) of the external environment. As the nervous system evolves and attains a higher degree of complexity and organization, its processes become more advanced and sophisticated, including cognition, as an act of knowing, thinking, reasoning, believing, and willing. Eventually, (symbolic) language becomes an important and integrated part of the cognitive functions (Joseph 2000).

In birds and mammals, and most articulate in man, there is an advanced ability to "internalize" the world through the formation of spatial and temporal patterns. In particular, temporal binding allows us to understand the relation between cause and effect and to experience "the arrow of time". For humans, the internal model of the world also includes other individuals and their minds. Models of other individuals are important for determining the significance of their behavior, and for predicting the "next step" in that behavior.

Self-consciousness, a sense of an "I", may develop when experienced events are given a sequential context in the internal model of the world. It is obvious that many cognitive functions (but not consciousness in general) depend upon memory, but the "I" needs memories well structured in time, in order to be perceived. Obviously, also planning for the future, involving attention as well as intention, is an essential component in the making of an "I" (Ingvar 1994). Our subjective experience of a decision to act, as well as of a sensory stimulation, seems to require a time span of a few to several hundred milliseconds (Libet 1985, Crick 1994, Soon et al., 2008).

Thus, the capabilities to predict future events have been extended to the anticipation and expectation of a continuous individual life. Also, if one has a (good) model of oneself one can infer ones own thinking, feelings and behaviors also to others, thus allowing for a more efficient interaction with other individuals, and better prediction of their behavior. The ultimate pattern in our understanding of the world results in an experience of meaning, where our interrelationships can be set in a context (attention), and be seen as purposeful (intention). Fig. 3 shows an attempt to link the evolution of cognitive and conscious interaction with the external environment to an increasing complexity of living organisms.

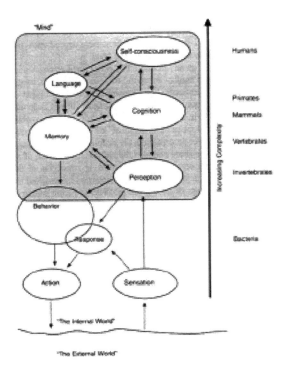

Fig. 3. A crude attempt to map various cognitive functions and relate them to each other on an "evolutionary scale" that is at the same time a measure of the degree of complexity. Intention would correspond to the downward direction, and attention to the upward direction in this figure. (Adopted from Liljenström, 1997).

Consciousness and cognition apparently evolve through interaction with the environment, where the organism is embedded. Such exploration of the environment requires both attention and intention. The evolution of consciousness presumably occurs through larger leaps when new species appear, and through smaller steps within a species and even smaller steps within an organism. A great leap of consciousness must have happened when we became aware of ourselves and our own cognitive activity. That leap should also include an awareness of our intentions, a sense of a free will. This more advanced consciousness is presumably experienced, or at least reflected on, by humans alone.

5. Discussion

In previous sections, I have given an outline of a view on consciousness that is based on its complementary aspects of intention and attention, and more universal than commonly considered. It implies that both of these aspects of consciousness are essential for exploring our internal and external world. As suggested by William James (1890), consciousness could be viewed as a constantly changing process, interacting with a constantly changing environment in which it is embedded. Another view is that consciousness can be seen as a "force field", similar to an electromagnetic field (Popper et al., 1993; Lindahl & Århem, 1994). These views imply that we cannot reduce consciousness to neural or electrochemical processes. Indeed, that would have little meaning to us, even if we claimed we had succeeded. Concepts at one level are in many cases not transferable in a meaningful way to another level. New qualities and properties emerge at each new level in the hierarchical organization of matter, qualities which are irrelevant at lower levels. Such a "holistic" view was also given by Delbrück, who compared mind with quantum reality:

> "The mind is not a part of the man-machine but an aspect of its entirety extending through space and time, just as, from the point of view of quantum mechanics, the motion of the electron is an aspect of its entirety that cannot be unambiguously dissected into the complementary properties of position and momentum."

Penrose (1989, 1994) brings this analogy further, believing that mind indeed may need some quantum mechanical description. He argues that there are some aspects of mind, or mental phenomena, like e.g. understanding and insight, that are non-computable in nature, and thus, for example, can never be simulated on a computer. Such phenomena may even require a new physics, new laws and principles that are not mechanistically derivable from lower levels. With this perspective, consciousness seems to fundamentally transcend physics, chemistry and any mechanistic principle of biology.

If intentionality is a fundamental aspect of consciousness, and if intentionality corresponds to our sense of free will, it is difficult to fit it within the framework of current science, which is based on deterministic laws and chance. In my view, consciousness provides a freedom beyond the indeterminism of quantum physics or the determinism of classical physics. Indeed, consciousness should perhaps not at all be viewed as a phenomenon, that is observable within the framework of current scientific approaches (even though science entails observation and experiment, as an advanced form of attention and intention). Instead, consciousness could be considered a meta-phenomenon (or noumenon, in a Kantian sense), which can only be "observed" indirectly (by a conscious observer) through its effects on the material world.

The effects of consciousness is most clearly linked to intention, as its more active aspect, which has to be considered if we are to understand its role in evolution and in the universe. Exploration, through attention-perception and intention-action, is fundamental to all levels of existence. In this sense, consciousness is inseparable from life, expanding and evolving together. It is expressed and evolving through inter-relations (attention) and inter-actions (intention). Hence, consciousness can be regarded as a driving force in nature.

References

Århem, P., Liljenström, H. (2001) Fluctuations in neural systems: From subcellular to network levels. In: Moss, F. & Gielen, S. (Eds.) Handbook of Biological Physics, Vol. 4: Neuroinformatics, Neural Modelling, Elsevier, Amsterdam, pp. 83-125.

Århem, P., Liljenström, H. (2007). Beyond cognition - on consciousness transitions. In: Liljenström, H. and Århem, P., (Eds.), Consciousness Transitions - Phylogenetic, Ontogenetic and Physiological Aspects. Elsevier, Amsterdam, pp. 1-25.

Århem, P., Braun, H., Huber, M. and Liljenström, H. (2005). Non-Linear state transitions in neural systems: From ion channels to networks. In: Liljenström, H. & Svedin, U. (Eds.) Micro - Meso - Macro: Addressing Complex Systems Couplings, World Scientific Publ. Co., Singapore, pp. 37-72.

Crick, F. (1994). The Astonishing Hypothesis - the Scientific Search for the Soul. Simon & Schuster, London.

Crick, F., Koch, C. (1990). Towards a neurobiological theory of consciousness, Seminars Neurosci. 2, 263-275.

Delbrück, M. (1986). Mind from Matter? Blackwell Scientific Publ., Palo Alto, CA.

Desmurget, M., Reilly, K. T., Richard, N., Szathmari, A., Mottolese, C., Sirigu, A. (2009). Movement intention after parietal stimulation in humans. Science 324, 811-813.

Eccles, J. C. (1982). The initiation of voluntary movements by the supplementary motor area. Arch. Psychiatr. Nervenkr. 231, 423-441.

Eckhorn, R., Bauer, R., Jordon, W., Brosch, M., Kruse, W., Monk, M., Reitboeck, H.J. (1988). Coherent oscillations: A mechanism of feature linking in the in the visual cortex? Biol. Cybern. 60, 121-130.

Freeman, W. J. (1975) Mass Action in the Nervous System. Academic Press, New York.

Freeman, W. J. (2000). Neurodynamics: An Exploration in Mesoscopic Brain Dynamics. Springer, London.

Gray, C. M., Singer, W.(1989). Stimulus-specific neuronal oscillations in orientation columns of cat visual cortex. Proc. Natl. Acad. Sci. USA 86, 1698-1702.

Halnes, G., Liljenström, H., Århem, P. (2007). Density dependent neurodynamics. Biosystems 89, 126-134.

Hauser, M. D., Nelson, D. A. (1991). 'Intentional' signaling in animal communication. TREE 6, 186-189.

Ingvar, D. H. (1994). The will of the brain: Cerebral correlates of willful acts, J. theor. Biol. 171, 7-12.

James, W. (1890). The Principles of Psychology. Harvard Univ. Press, Cambridge, MA.

Joseph, R. (1982). The neuropsychology of development. Hemispheric laterality, limbic language, the origin of thought. Journal of Clinical Psychology, 44 4-33.

Joseph, R. (2000). The evolution of sex differences in language, sexuality, and visual spatial skills. Archives of Sexual Behavior, 29, 35-66.

Joseph, R. (2011). Consciousness, language, and the development and origins of thought, Journal of Cosmology, 14, In press.

Koch, C. (2004). The Quest for Consciousness - A Neurobiological Approach. Roberts & Co. Publ., Greenwood Village, CO.

Libet, B. (1985). Unconscious cerebral initiative and the role of conscious will in voluntary action. Behav. Brain Sc. 8, 529-566.

Liljenström, H. (1991). Modeling the dynamics of olfactory cortex using simplified network units and realistic architecture, Int. J. Neural Syst. 2:1-15

Liljenström, H. (1995). Autonomous learning with complex dynamics. Int. J. Intell. Syst. 10, 119-153

Liljenström, H. (1997). Cognition and the efficiency of neural processes. In: Århem, P., Liljenström, H., Svedin, U. (Eds.), Matter Matters? On the Material Basis of the Cognitive Activity of Mind, Springer, Heidelberg, pp. 177-213.

Liljenström, H. (2010). Inducing transitions in mesoscopic brain dynamics. In: Steyn-Ross, D. A., Steyn-Ross, M. (Eds.) Modeling Phase Transitions in the Brain, 149, Springer Series in Computational Neuroscience 4, pp. 149-179.

Lindahl, B. I. B., Århem, P. (1994). Mind as a force field: Comments on a new interactionistic hypothesis. J. theor. Biol. 171, 111-122.

Margulis, L., Sagan, D. (1995). What is Life? Widenfeld & Nicolson, London.

Mather, J. (2011). Consciousness in Cephalopods? Journal of Cosmology, 14, In press.

Nani, A., Clare M. Eddy, C. M., Cavanna, A. E. (2011). The quest for animal consciousness. Journal of Cosmology, 14, In press.

Penrose, R. (1989). The Emperor's New Mind. Oxford University Press, GB.

Penrose, R. (1994). Shadows of the Mind. Oxford University Press, GB.

Popper, K., Lindahl, B. I. B., Århem, P. (1993). A discussion of the mind-brain problem. Theor. Medicine 14, 167-180.

Skarda, C.A., Freeman, W.J. (1987). How brains make chaos in order to make sense of the world. Behav. Brain Sci. 10, 161-195.

Soon, C. S., Brass, M., Heinze, H-J., Haynes, J-D. (2008). Unconscious determinants of free decisions in the human brain. Nature Neuroscience 11, 543-545.

Srinavasan, M. V. (2009). Honey bees ans a model for vision, perception, and cognition. Annu. Rev. Entomol. 55, 267-284.

van Swinderen, B. (2008). The remote roots of consciousness in fruit-fly selective attention? In: Liljenström, H. & Århem, P. (Eds.), Consciousness Transitions. Phylogenetic, Ontogenetic and Physiological Aspects. Elsevier, Amsterdam, pp. 27-44.

35. The Ecological Cosmology of Consciousness
Tom Lombardo, Ph.D.
Center for Future Consciousness

Abstract

Four fundamental mysteries regarding consciousness and its relationship with the physical world are identified. Various classical philosophical and scientific solutions to these mysteries are described. Subsuming the first four mysteries, a fifth deeper mystery is proposed, providing a new theoretical scheme of inquiry—"ecological reciprocalism"—for understanding the relationship of consciousness and the physical world. Ecological reciprocalism posits that that the physical universe and consciousness are interdependent realities, and it is this reciprocity that is the fundamental ontological truth and mystery of consciousness. Consciousness is always embodied, supported, and locally embedded within the physical universe, and the very existence and nature of the physical universe is always revealed and conceptualized through consciousness. Some of the essential features of ecological reciprocalism are described, demonstrating how the self, the physical body and environment, technology, and awareness of other conscious minds fit into this theoretical framework. The temporal and evolutionary dimensions of consciousness are then described, which leads into a discussion of the cosmological foundations and directionality of the future evolution of consciousness.

KEY WORDS: Consciousness, reciprocity, ecological, embodied mind, cosmology, evolution, perception

> "For things are things because of mind,
> as mind is mind because of things."
> Hsin Hsin Ming

1. Introduction

In this paper, four fundamental mysteries regarding consciousness and its relationship with the physical world are identified. Various classical philosophical and scientific solutions to these mysteries are described. Subsuming the first four mysteries, a fifth deeper mystery is proposed, providing a new theoretical scheme of inquiry--"ecological reciprocalism"--for understanding the relationship of consciousness and the physical world. Some of the essential features of ecological reciprocalism are described, demonstrating how the self, the physical body and environment, technology, social networks, and awareness of other conscious minds fit into this theoretical framework. The temporal and evolutionary dimensions of consciousness are then described, leading into a discussion of the cosmological directionality of the future evolution of consciousness.

2. The Mysteries of Consciousness

There are (at least) four fundamental mysteries connected to consciousness:

What is consciousness? Is it energy, spirit, an activity or process, a form of illumination, the interiority of all being, an ethereal or refined kind of substance, or something else? This is a perplexing question, since our very essence is consciousness, and yet we can't seem to grasp what this very essence is (Velmans and Schneider, 2006).

How is consciousness connected with biological and physical beings (like ourselves) who possess it? The existence of consciousness seems to clearly depend upon a supportive physical world, including an active brain and functioning body (Edelman, 2006; Edelman and Tonini, 2002; Koch, 2007), but the puzzle is that the qualities of consciousness seem very different than the qualities of the physical world. For example--to pose a question often referred to as the "hard problem"--how can a physical brain of electro-chemical impulses produce conscious sensations, emotions, and thoughts?

Third, how can there be consciousness of a surrounding and yet physically distal environment that extends beyond the body and brain of the perceiver? If consciousness is somehow located in the brain, how can consciousness reach out and make epistemic contact with a physical world? Perhaps perceptual consciousness is awareness of brain states rather than the external physical world (Frith, 2007). This is the classical philosophical problem of the perception of an external world (Lombardo, 1987).

The fourth mystery--a natural implication of the first three--has to do with what the physical world is. This question might seem strange since not only do we appear intimately conscious of it through perception (the third mystery) but we also have a deep and intricate knowledge of the physical world through the sciences. Yet, it is not clear whether our present understanding of the physical world suggests any understandable way that it could be connected with consciousness. In fact, the two above puzzles regarding how consciousness is connected with the physical world largely derive from perceived incompatibilities between consciousness and the physical world. Perhaps some significant limitation in our understanding of the physical world contributes to these philosophical quagmires, as much as our lack of understanding of the nature of consciousness (Feigl, 1967).

3. The solutions to consciousness and the physical world

Answers to these mysteries invariably involve general ontologies of the nature of reality. It is a deep and profound point about the mysteries of consciousness that answers seem to require the need to address the nature of reality as a whole; the nature of consciousness has cosmological significance.

Idealism posits that the universe is, in its entirety, a mental reality or manifestation of consciousness. Since our conceptual understanding (including all of the theories of science and the very idea of an independently existing physical world) and the phenomenological manifestation of the physical world within perception arise within consciousness, one can argue that the physical world as experienced and understood depends upon (or is a manifestation of) consciousness (Berkeley, 1713; Kant, 1781).

Materialism posits that the universe is entirely physical. Since consciousness depends upon (at the very least) active states of the physical brain, perhaps consciousness is nothing but states of the brain, or of the body as a totality. There is no separate ontological realm of consciousness (Churchland, 1986; Dennett, 1991; Ryle, 1949).

Monistic theories, such as materialism and idealism, explain the two mysteries of how consciousness is related to the physical world by attempting to derive one ontological realm from the other, in fact, to reject as an independent reality either consciousness or physical matter. Either consciousness doesn't "really" exist as something separate from the physical world (the materialist thesis), or physical matter doesn't "really" exist as something separate from consciousness (the idealist thesis).

Neutral monism posits that the mental (consciousness included) and the physical are two manifestations of the same underlying reality (Spinoza, 1677). Perhaps, as in the identity theory of consciousness and brain, what we have are two different perspectives, interior and exterior respectively, on the same reality (Feigl, 1967).

Dualism posits that there are two types of being, physical matter and consciousness. Dualism, which doesn't reduce one realm to the other, may or may not posit interactivity between the realms; conscious states and brain states may parallel each other, or brain states may "cause" conscious states, and conscious states, such as intentions and desires, may "cause" physical actions of the body. But dualism does not provide a convincing explanation for how the two realms could interact since it is assumed that the two realms are qualitatively different. For example, how can electro-chemical activity in a brain generate a conscious thought or feeling (Chalmers, 1996), or how can a conscious intention move a muscle?

Evolutionism posits that physical reality is primordial and both mind and consciousness progressively emerge across time out of this physical substrate (Morowitz, 2002; Kauffman, 2008); as the physical world evolved in complexity, consciousness emerged. Perhaps physical reality was even "primed" at the start (as in the "strong anthropic principle") for this evolutionary emergence (Tipler, 1994). But evolutionism can sound dualistic and magical, since consciousness appears to pop into existence within a physical world at some point in time. Alternatively, one could argue that consciousness has always been there within the physical world, and as the physical world has evolved in complexity, consciousness and mind concurrently have evolved in complexity as well.

4. Ecological Reciprocalism

Two fundamental--and what would seem contradictory--theses are contained in the above solutions: First, consciousness depends on a physical support system. Second, the phenomenal manifestation and meaning of the physical world depends upon consciousness. Both these theses seem true. But how is this possible? How can each depend on the other? And doesn't this reciprocity of consciousness and the physical world contradict our deep intuitive sense that the physical world exists independent of conscious minds that know it? Yet, I will take this conundrum as my starting point. Ecological reciprocalism posits that the physical universe and consciousness are interdependent realities, and it is this reciprocity that is the fundamental ontological truth and mystery.

Ecological reciprocalism goes beyond simple interactivity between the physical and conscious realms (this is the position of interactive dualism). It is not simply that consciousness and the physical world causally affect each other. Rather, each realm literally requires the other for its existence. Further, instead of supposing that the puzzles described above can be solved by explaining how one of the two realms can be derived from the other, as in monistic solutions, the deepest puzzle is explaining how the two realms are inextricably interdependent. Monistic solutions deny (or explain away) the existence of one of the two realms; ecological reciprocalism does not. Hence, reality is neither a reductionistic monism nor an incompatible dualism of two independent separate realms, but a reciprocity of consciousness and the physical world. In essence, ecological reciprocalism reformulates the mind-body and consciousness-physical matter problems (Lombardo, 1987, 2009).

The two puzzles regarding the nature of consciousness and the nature of the physical world can be subsumed within this reciprocal framework. Though distinct, these two realms must possess some underlying incompleteness and relational dimension that requires the other. One can't answer the questions of what is consciousness and what is the physical world independent of each other. In adopting this ontological position, I am rejecting the view that "substance" (or whatever primordial concepts we use to identify the ground or essence of being) is an independent reality; that is, the primary ontological realities of the cosmos are not individually self-sufficient.

I propose a symmetry or balance of the initial four puzzles: What is consciousness? What is the physical world? How does consciousness depend upon the physical world? And how does the physical world depend on consciousness? The fundamental puzzle, subsuming these four would be: What is it about consciousness and the physical world such that they are interdependent with each other? One can conceptualize this ontological interdependency in terms of the Taoist Yin-Yang. There are two realities--Yin and Yang--that are distinct yet interdependent with each other; neither can exist without the other. The Tao, which is symbolized as the sine wave defining the interface of Yin and Yang, represents the underlying reality of the interdependency of Yin and Yang. Consciousness and the physical world form such a reciprocal Yin-Yang.

5. The reciprocities of consciousness and the physical universe

The thesis of ecological reciprocalism can be made more precise: Consciousness is always embodied, relative to a point of view, surrounded and locally situated within the physical universe, and the meaningful manifestation of the physical universe is always an integrated and selective differentiation relative to an embodied consciousness.

Beginning with consciousness, at the most basic level there is perceptual awareness of (what appears to be) an ambient, structured, and dynamic physical world surrounding and spreading outward from a proprioceived localized body. The body is experienced at the center of this ambient surround and consciousness is experienced as situated and manifested within this body. Further, consciousness is felt (though tactual, kinesthetic, and articular awareness) throughout the body. Consciousness also accompanies, to varying degrees, purposeful actions of the body within the surrounding environment; Synthesizing these points, the basic configuration of consciousness is of an embodied and active conscious being within an ambient environment.

Through movement and changes in bodily position the appearance of the perceived physical world changes. Phenomenologically what is revealed are physically situated perspectives of the surrounding physical environment. Hence, perceptual consciousness of the world is relative to a point of view (the position of the body) within the world. Reciprocally, the position and configuration of the body is proprioceived relative to the perceived layout of the environment (Lombardo, 1987). Conscious purposeful actions are guided and informed by this reciprocal awareness of the configuration of the body relative to the surround. Hence, bodily behavior and perceptual-proprioceptual awareness form a fundamental reciprocal loop of interdependency.

The self may in fact be grounded in this proprioceptual awareness of the body and sense of distinctive point of view. The self is likely based on the conscious sense of identity, configured, distinguished, and localized within a physical and social environment. Therefore the self is experienced as both the agency possessing consciousness and an object of consciousness (Baars, 1997).

Aside from perceptual-proprioceptual consciousness and conscious action, at least some conscious beings experience emotional states, conscious desires, sequences of conscious thoughts, and sensory-like imagery and memories. All these additional conscious states are experienced as localized within and frequently volitionally created by the embodied conscious being. Thoughts and emotions are experienced as situated within an embodied consciousness, itself within the perceived ambient physical world, rather than in some second separate ontological realm or space. Also, when one conscious being encounters another conscious being within the perceived environment, consciousness clearly seems to be manifested and expressed through the body of the other conscious being; the consciousness of the other, through attention, purposeful behavior, emotional expressions, and efforts to connect and communicate, shows itself. The consciousness of the other is not entirely private and hidden away.

Through the physical sciences, invariably guided by abstract thought and facilitated through technologies that heighten sensory and behavioral capacities, it has been discovered that there are many scales and dimensions of structure within the physical world beyond what is revealed directly through perception. Psychophysiological research has repeatedly demonstrated that perceptual systems are highly selective, only reacting to certain forms, scales, and patterns of energy. The experienced perceptual world is highly selective. Science has also explored the intricate bodily systems (including the nervous system) that seem necessary for the realization of consciousness and its complex structure and dynamics; these biological systems, through elaborate integrative and differentiating processes, constrain and configure the make-up of consciousness and the selective apprehension of the world (Koch, 2007). Further, as a general principle, perceptual awareness is selectively attuned to those environmental features that are meaningful, apprehending the relational properties or environmental "affordances" relevant to the ways of life and purposeful behaviors of a conscious being (Gibson, 1979).

In the broadest sense, even our scientific explorations into the physical world, which expand and deepen our understanding of reality, are realized through embodied consciousness, selective abstract thinking, and selective technologies that augment our capacities. Following Smolin (1997), physical reality can only be consciously apprehended from within an ambient universe from a point of view (or a series or collection of points of view). There may not be such a thing as a detached, absolute, or non-relative realization of consciousness--consciousness may be manifested from a point of view within an ambient universe. And complementarily, there may not be such a thing as a non-relative, non-selective revelation of the physical world.

Science has also discovered that the physical universe, at all levels of magnitude or scale, structures the forms of physical energy within it, and these patterns of structured energy permeate through it. Hence, physical realities produce specific structured energetic effects within the universe providing information about their existence and make-up, affording the possibility of being known. Further, it is possible that at each and every location within the universe there is a structured energetic array deterministically specific to the location and the relational configuration of its surround (Gibson, 1979). (Gibson, 1966; Gibson, 1979; Lombardo, 1987). The fundamental perceptual-proprioceptual experience of center-surround is informed and grounded in stimulus information specific to the physical configuration of an embodied and localized conscious being within a world.

But because energy converging to any local area is structured at multiple levels of magnitude and complexity, across all forms and variable ranges of energy, containing an indeterminately rich and immense set of differences and patterned relationships, often in mixtures and interference patterns with each other, the function of a highly selective embodied conscious being is to differentiate and integrate out of this indeterminately rich plenum a distinctive conscious apprehension of itself in relationship with a physical environment.

Conversely, the physical world can only manifest itself relative to a selective, localized, and integrated perspective; this active and relational process of selecting, differentiating, and integrating may be the nature and fundamental function of consciousness. Still, the actual existential content of perceptual consciousness is a particular relational manifestation of the physical world, rather than some constructed representation hidden away in either the brain or mind.

And though locally grounded in a biological system, the embodiment and relational reach of consciousness is not fixed. The central nervous system forms the physical nexus of coordinated activity within a conscious being, but instrumentalities (such as tools and technologies) can be attached or functionally coupled with the biological body, extending its functional integration and reach, both perceptually and behaviorally (Joseph 2000). Consciousness (and embedded psychological processes such as purposeful behaviors, perceptions, and even thinking) is realized and experienced throughout both the conscious body and functionally integrated instrumentalities (Clark, 2003; Clark, 2008: Noe, 2009; Shapiro, 2011).

6. The Cosmological Evolution of Consciousness

There is a temporal dimension and directionality to consciousness. Conscious beings are aware of duration, relative stability, and patterns of change; of becoming and passing away; and of an experiential direction to time. The conscious now--which is inherently transformative and not instantaneous--may be anchored at the level of perception, and contextualized within consciousness of the past (memories) and conscious anticipation of the future, all three phenomenologically blurring together at the "edges" (Johnson and Sherman, 1990; Lombardo, 2006a, 2007).

Consciousness includes what could be described as a sequential directional flow (Fraser, 1978; Fraser, 1987; Carroll, 2010). Though conscious states pulsate through an ongoing series of relatively distinct, highly selective and integrated apprehensions (Baars, 1997), conscious time is always opening into the future or looking backwards into the past. Our conscious anticipatory thoughts and emotions, our desires or motives, and purposeful behaviors are directed toward the future (Lombardo, 2006a) and are often based on the past. Perhaps the primary function of our cognitive-cerebral processes is anticipation and guidance of the future (Frith, 2007; Hawkin, 2004).

It has been frequently argued that the function of the conscious mind is to create order within a chaotic physical world (Lombardo, 1987). But the physical universe, as noted above, also possesses an unimaginably immense amount of order that must be differentiated out by a conscious being. In fact, the direction of evolution appear to be toward both increasing order and complexity

(Prigogine and Stengers, 1984; Morowitz, 2002). Building upon the evolved complexities of physical systems, the evolution of future-focused conscious minds selectively and purposefully creates even more order and complexity within the physical universe (Gell-Mann, 1994; Kurzweil, 1999; Kurzweil, 2005).

Since populations of conscious minds organize into coordinative agencies, the physical embodiment and reach of consciousness spreads outward into greater expanses and more complex networks. With the ongoing evolution of technologies that amplify capacities to perceive and manipulate the surrounding environment, the conscious reach of minds evolves and expands. Many argue that technologically facilitated social networking is leading to the emergence of a "global consciousness" that potentially could expand into a "cosmic consciousness" (Lombardo, 2006b; Stock, 1993; Tipler, 1994). Consciousness will more pervasively and coherently drive the evolution of its own cosmic embodiment. In this evolutionary process, consciousness, through advancing theoretical abstractions and enhanced technologies, will progressively differentiate and integrate more of the cosmos.

7. Summary and Conclusion

The universe affords the possibility of being known but it is indeterminately rich in information. Grounded in localized and embodied center-surround relationships, consciousness arises as perspectival and ecological. Conscious beings extract and integrate meaningful perspectives or manifestations of the cosmos in relationship to themselves. Consciousness is relational with respect to the cosmos because it is selective and integrated apprehensions and purposeful manipulations of the universe; and the universe is relational with respect to consciousness because it reveals itself as differentiated and unique perspectives and opportunities of action to consciousness. Consciousness is temporal and future-directed and evolves through increasing mental complexity and the ongoing functional integration of social networks and embodied instrumentalities. In essence, the universe is evolving an embodied conscious mentality that progressively differentiates and integrates and brings under its volitional control more of its primordial substratum.

References

Baars, B. J. (1997). In the Theatre of Consciousness: The Workplace of the Mind. Oxford University Press, New York.

Berkeley, G. (1713). Three Dialogues between Hylas and Philonous. In: Smith, T.V. and Grene, M. (Eds.), (1957) Berkeley, Hume, and Kant. The University of Chicago Press, Chicago.

Carroll, S. (2010). From Eternity to Here: The Quest for the Ultimate Theory of Time. Dutton, New York.

Chalmers, D. (1996). The Conscious Mind: In Search of a Fundamental Theory. Oxford University Press, Oxford.

Churchland, P. (1986). Neurophilosophy: Toward a Unified Theory of the Mind-Brain. Cambridge University Press, Cambridge, MA.

Clark, A. (2003). Natural-Born Cyborgs: Minds, Technologies, and the Future of Human Intelligence. Oxford University Press, Oxford.

Clark, A. (2008). Supersizing the Mind: Embodiment, Action, and Cognitive Extension. Oxford University Press, Oxford.

Dennett, D. C. (1991). Consciousness Explained. Little, Brown, and Co., Boston.

Edelman, G. (2006). Second Nature: Brain Science and Human Knowledge. Yale University Press, New Haven, CT.

Edelman, G. and Tononi, G. (2000). A Universe of Consciousness: How Matter Becomes Imagination. Basic Books, New York.

Feigl, H. (1967). The "Mental" and the "Physical". University of Minnesota Press, Minneapolis.

Fraser, J.T. (1978). Time as Conflict. Birkhauser Verlag, Basel and Stuttgart.

Fraser, J. T. (1987). Time, the Familiar Stranger. Tempus, Redmond, WA.

Frith, C. (2007). Making Up the Mind: How the Brain Creates Our Mental World. Blackwell, Maiden, MA.

Joseph, R. (2000). The evolution of sex differences in language, sexuality, and visual spatial skills. Archives of Sexual Behavior, 29, 35-66.

Gell-Mann, M. (1994). The Quark and the Jaguar: Adventures in the Simple and the Complex. W.H. Freeman and Company, New York.

Gibson, J. J. (1966). The Senses Considered as Perceptual Systems. Houghton Mifflin, Boston.

Gibson, J. J. (1979). The Ecological Approach to Visual Perception. Houghton Mifflin, Boston.

Hawkins, Jeff (2004). On Intelligence. Times Books, New York.

Kant, I. (1781). Critique of Pure Reason. (1990) Prometheus Books, Amherst, NY.

Kauffman, S. (2008). Reinventing the Sacred: A New View of Science, Reason, and Religion. Basic Books, New York.

Koch, C. (2007) The Quest for Consciousness: A Neurobiological Approach. Roberts and Company, Englewood, CO.

Kurzweil, R. (1999) The Age of Spiritual Machines: When Computers Exceed Human Intelligence. Penguin Books, New York.

Kurzweil, R. (2005). The Singularity is Near: When Humans Transcend Biology. Viking Press, New York.

Lombardo, T. (1987). The Reciprocity of Perceiver and Environment: The Evolution of James J. Gibson's Ecological Psychology. Lawrence Erlbaum Associates, Hillsdale, NJ.

Lombardo, T. (2006a). The Evolution of Future Consciousness: The Nature and Historical Development of the Human Capacity to Think about the Future. AuthorHouse, Bloomington, IN.

Lombardo, T. (2006b). Contemporary Futurist Thought: Science Fiction, Future Studies, and Theories and Visions of the Future in the Last Century. AuthorHouse, Bloomington, IN.

Lombardo, T. (2007). The evolution and psychology of future consciousness. Journal of Future Studies, Vol. 12, No. 1, 1-23.

Lombardo, T. (2009). The future evolution of the ecology of mind. World Future Review, Vol. One, No. 1, 39-56.

Morowitz, H. (2002). The Emergence of Everything: How the World Became Complex. Oxford University Press, Oxford.

Noe, A. (2009). Out of Our Heads: Why You are not Your Brain, and Other Lessons from the Biology of Consciousness. Hill and Wang, New York.

Prigogine, I. and Stengers, I. (1984). Order out of Chaos: Man's New Dialogue with Nature. Bantam, New York.

Ryle, G. (1949). The Concept of Mind. Barnes and Noble, New York.

Searle, J. (1997). The Mystery of Consciousness. New York Review Book, New York.

Shapiro, L. (2011). Embodied Cognition. Routledge, New York.

Smolin, L. (1997). The Life of the Cosmos. Oxford University Press, Oxford.

Spinoza, B. (1677). On the Improvement of Understanding, Ethics, Correspondence. (1955) Dover Publications, New York.

Stock, G. (1993) Metaman: The Merging of Humans and Machines into a Global Superorganism. Simon and Schuster, New York.

Tipler, F. (1994). The Physics of Immortality: Modern Cosmology, God, and the Resurrection of the Dead. Doubleday, New York.

Velmans, M. and Schneider, S. (Eds.) (2006) The Blackwell Companion to Consciousness. Blackwell, New York.

36. Consciousness, Dissociation and Self-Consciousness
Ellert R.S. Nijenhuis, Ph.D.

Top Referent Trauma Center, Mental Health Care
Drenthe, Assen, The Netherlands

Abstract

The philosophy and psychology of dissociative disorders and the fragmenting of consciousness and personality are discussed in reference to self-consciousness.

KEY WORDS: Consciousness, dissociation, self-consciousness

1. Introduction

Our conscious personality constitutes a well integrated, dynamic biopsychosocial system that is associated with characteristic mental and behavioral actions but which remains open to change through experience and maturation which may be integrated into the core personality (Edelman & Tononi, 2000; Fuster, 2003; Janet, 1889, 1935; Van der Hart, Nijenhuis, & Steele, 2006). However, the same is not true of the personality and consciousness of patients with dissociative disorders, a condition associated with emotional trauma so profound the conscious personality shatters (American Psychiatric Association, 1994). In dissociative disorders, the conscious personality is fragmented, and is insufficiently integrated and consists of overly rigid biopsychosocial subsystems or dissociative parts (Nijenhuis, 2011; Nijenhuis & Van der Hart, 2011); it is as if several different conscious personalities exist in the same head, but with each separate from the other. A crucial feature is that each dissociated fragment of the conscious personality has their own sense of self, world, and self-in-the-world (Nijenhuis, 2011). Each dissociated consciousness also believes their dissociative ways of experiencing and knowing themselves and the world reflect 'objective' reality. The different, often conflicting feelings, perceptions, and ideas of dissociative parts cause struggles among them, as well as fears of each other (Van der Hart et al., 2006).

More specifically, dissociative fragmentation of the conscious mind can be manifested in the expression of multiple distinct identities or personalities, each with its own unique pattern of thinking, feeling, and behaving, and with their own memories. This is not some epiphenomenon. It has been demonstrated that EEGs (Hughes et al., 1990; Lapointe et al., 2006), and cerebral blood flow (Matthew et al., 1985; Reinders et al., 2003, 2006) significantly differ when different aspects of the fragmented conscious mind come to the fore, and that the parietal lobe differs from normals (Garcia-Campay et al., 2009); a region of the brain associated with the body image (Joseph 2009).

Dissociative patients characteristically display shifts in dominance of consciousness and behavior among their parts, as well as intrusions of one part into the conscious domain of one or more other parts (Nijenhuis, 2004; Dell, 2006; Kluft, 1987; Ross et al., 1990; Van der Hart et al., 2006).

Dissociation of the personality manifests in dissociative symptoms, that can be categorized as psychoform and somatoform, and as negative and positive (Nijenhuis, 2004; Van der Hart et al., 2006; see Table 1). For example, dissociative parts of the conscious mind can have negative symptoms such as analgesia (i.e., insensitivity for pain), anesthesia (e.g., visual, kinesthetic, auditory), lack proprioceptis, as well as sense of ownership and agency with respect to particular body parts, inability to move, or to remember recent or remote episodes. They may also have positive dissociative symptoms such as pain, hearing voices of other dissociative parts; that is, one part of the dissociated mind may receive information from another region of the mind, in a manner similar to split-brain patients (Joseph 1988a,b; 2009). Negative and positive dissociative symptoms are generally linked. Thus, one part of the dissociated mind may be analgesic, but an other part may suffer pain in particular parts of their body. Dissociative parts may hear each the voices of the other fragments of consciousness, or intrude upon each other in different ways. They may reenact traumatic memories, or be intruded by a different part's traumatic memories (Van der Kolk & Van der Hart, 1989); that is, one dissociated conscious personality will remember events it did not experience, but which were experienced and committed to the memory of a different dissociated consciousness which had split off from the rest of the mind. These memories are not narratives like normal autobiographical memories, but sensorimotor and highly emotionally charged experiences (Van der Hart et al., 2006). Although they may not realize it, they are reenacting the past, and what is being being remembered, is not a memory, but is happening here and now.

Table 1. Dissociative symptoms: A 2 x 2 classification

	Psychoform	Somatoform
Negative	Amnesia Depersonalization Emotional anesthesia	Bodily anesthesia (visual, auditory, et cetera) Analgesia Motor inhibitions
Positive	Intrusions (not personified experiences and perceptions), such as hearing voices, undergoing emotions, thoughts, images, and (traumatic) memories of other dissociative parts	Intrusions (not personified bodily feelings and body movements), such as feeling physically touched by someone else, having pain, bodily movements, bodily components of traumatic memories, that all stem from other dissociative parts

These and related differences among different dissociative parts suggest that their sense of self, world, and self-in-the-world are mental contents generated by particular epistemic actions, that is, constructions developed in interaction with a material and social environment (Metzinger, 2003; Nijenhuis, 2011). This action perspective on (self-)consciousness is also grounded in the observation that 'fusion' of formerly dissociated parts is associated with instant loss of the former and rapid generation of a more or less different sense of self--including the body--world, and self-in-the-world.

Like normal individuals, dissociative parts experience and believe their dissociated self to an independent entity or substance that could live by itself, and that it consists of an invariant set of intrinsic properties, as well as a unique and indivisible unity (cf. Metzinger, 2003). The phenomena of dissociative disorders clearly challenge the reality of this view even when applied to the normal conscious personality. The way in which dissociative parts experience and know themselves and the world involve the epistemic contents of their ongoing constructive actions, the same can be said of anyone's self and world--as the Buddha contended, and as some Western philosophers contend (e.g., Kant, 1998; Schopenhauer, 1958; Metzinger, 2003).

In this light, dissociative disorders may provide a window of opportunity to examine what mental actions and biopsychosocial underpinnings are involved in what has been described as "self-consciousness"; what individuals must do to be consciously aware of themselves and their environment.

2. Parts and Wholes

Does it make sense to study parts of a whole? This question obviously applies to dissociative disorders. However, it is not any less relevant to the study of (self-)consciousness in health. This topic is therefore addressed first at some length.

As human beings, we, as a whole organism, are conscious of a world and of ourselves, and we have goals and engage in actions. Although it is true that the brain is related to consciousness and different regions of the brain subserve different functions (Joseph 1988a,b; 2009), ascribing psychological predicates to functional parts of a whole system confuses relationships between wholes and parts and could be considered a mereological fallacy (Bennett & Hacker, 2003). It is "[o]nly of a human being and what resembles (behaves like) a living human being can one say: it has sensations; it sees; hears; is deaf; is conscious or unconscious" (Wittgenstein, 1953, entry 281).

Body and Mind The 17th Century witnessed the start of a major mereological discussion on the mind-body relationship. As Spinoza (1996) emphasized, Descartes (1960) erred to distinguish between a body and a mind. Ascribing thoughts, feelings, beliefs, fantasies, and other psychological predicates to the mind rather than to a living human being or his personality as an embodied functional entity, is a fallacy. Lacking such faculties, mind is rather a term that denotes a range of human powers, the exercise of these powers, and a range of characteristic human character traits. Whereas most contemporary neuroscientists repudiate Descartes' body-mind dualism, many also assign psychological attributes to parts of a whole—i.e., the brain or parts of the brain--that are characteristic features of functions of the individual as a whole biopsychosocial system (Joseph 1988a,b; 2009). This could be construed as a mereologic error (Bennett & Hacker, 2003).

Brain/body and Environment Another mereological error is to regard the brain in separation of the environment in which it exists. The brain/body as the necessary apparatus for mental contents is dependent on the environment with respect to these contents, otherwise, the brain might remain 'empty' (Northoff, 2003). And an environment can only exist for a living brain/body because, according to some philosophers, there is no observer-independent environment (Kant, 1998; Schopenhauer, 1958). Environment and brain are bilaterally dependent, they are necessary conditions for each other (Northoff, 2003; Thompson, 2007). According to various theories of quantum physics (Bohr 1958; Heisenberg, 1958; Neuman & Tamir, 2009; Stapp 2009), they also determine each other, they interact, which also applies to the relationship between the body and the environment.

Individuals as whole organisms must adapt to and can creatively change their environment by selecting particular actions to achieve their goals. In this context, individuals at the brain level select particular dynamic patterns, and at the brain/body level engage in particular patterns of motor actions, that match and mold the environment. Conversely, the environment in which an individual lives influences the dynamic brain states and motor actions that will be selected. There is embedment, that is, intrinsic relationships between brain, body, and environment in ontological regard (Northoff, 2003; see Figure 1). Brain, body, and environment are intrinsically coupled (van Gelder, 1995; Gallagher, 2005), and this coupling is dynamic, selective, adaptive, and creative.

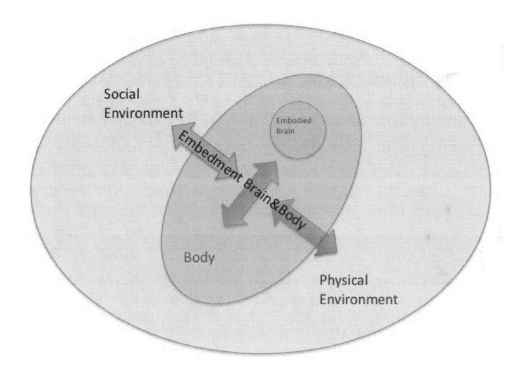

Figure 1. The ontological trinity: Intrinsic relationships between brain, body, environment

For example, the smell of food activates in us the motivational or action system (Lang, Bradley, & Cuthbert, 1998; Panksepp, 1998) of energy management when our organism has a shortage of calories; a function of the hypothalamus and brainstem and other neurological structures which continually sample the internal and external environment. Threat to our physical integrity awakens the action system of defense (Fanselow & Lester, 1988) mediated by, for example, the limbic system which is responsible for emotional and motivational functioning (Joseph 1992). Depending on an assessment of the actual environmental conditions, we will startle, freeze, flee, fight, or play dead.

What odors will foster eating, and what conditions present 'threat' is also dependent on phylogenetic and ontogenetic features of the subject as a whole organism. Thus, the intrinsic relationship between the embrained and embodied individual as a whole biopsychosocial system and the environment is also characterized by this system's biological, psychological, and social history. For example, this relationship will be influenced by prior and anticipated experiences, past social relations, and the individual's developmental phase. Brain, body, and environment, thus, mutually constitute and determinate each other, and they are mutually dependent (Northoff, 2003).

This ontological determination also applies to dissociative subsystems of personality. A dissociative part involves a co-constitution and co-determination of a subsystem that, as clinical observations suggest, strives to realize particular goals (see Figure 2). This subsystem consists in a particular intrinsic relationship, that is, in ongoing dynamic configurations of an individual's brain, body, and environment, as well as an intrinsic relationship of one dissociative part with other existing dissociative parts. These parts do not exist in a void, but constitute and determine each other in ontological regard. For example, as one part, a patient may not eat. Whether this part realizes it or not, he or she can only exist because there are other parts that do eat.

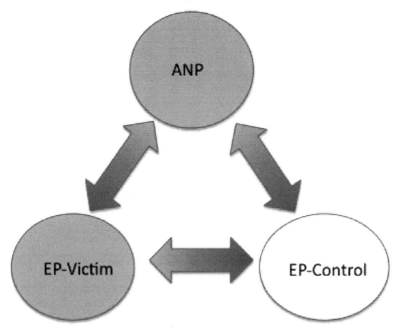

Figure 2. The trinity of dissociative parts of the personality: Three essential types:

* ANP = Apparently Normal Part of the Personality. As ANP, the patient focuses primarily on functioning in daily life. ANP is foremost mediated by action systems for daily life functioning such as energy management, exploration, and care taking. ANP typically fears and avoids EP-Victim, as well as EP-controlling. In some cases, ANP may verbally fight with EP-Control.

* EP-Victim = Emotional Part of the Personality that is fixated in the 'there and then,' that they often experience and regard as the actual 'here and now.' As EP-Victim, the patient's phenomenal self-model and world-model is that of an endangered child. As EP, the patient focuses primarily on surviving actual or perceived threat. For space, it is not discussed here that complex dissociation is a condition that relate to chronic traumatization, including attachment disruptions in early childhood-related condition (see, e.g., Chu, Frey, Ganzel, & Matthews, 1999; Diseth, 2006; Dutra, Bureau, Holmes, Lyubchik, & Lyons-Ruth, 2009; Nijenhuis & Den Boer, 2009; Ogawa, Sroufe, Weinfield, Carlson, & Egeland, 1997; Van der Hart et al., 2006. EP is foremost mediated by action systems for defense (flight, freeze, fight, playing dead) and attachment cry. EP-Victim's fear of ANP and EP-Controlling is proportional to ANP's and EP-controlling's rejection, as well as verbal and physical attacks.

* EP-Control = Emotional Part of the Personality that tends to imitate perpetrators in an effort to control the uncontrollable (i.e., traumatization, including attachment disruptions). EP-Control regards EP-Victim as weak, bad, and undeserving, and regards ANP in a similar fashion. EP-Control is afraid to realize that he or she was also hurt and abandoned, thus fears to realize that he or she is intimately related, that is, belongs to EP-Victim and ANP. EP-Control may be mediated by the action system of social dominance

Where is the mind? If there are intrinsic relationships of brain, body, and environment, where does this leave the mind? The tempting (Cartesian) belief that the mind could exist in separation of the body involves an epistemic illusion caused by our inability to know how we generate our mental contents, including our sense of self. This inability is known as transparency (McGinn, 1989, 1998; Metzinger, 2003) or autoepistemic limitation (Northoff, 2003). Whereas there is no mind without a brain, the mind does not equal the brain. We tend to experience that our mind is in our head, but no one has found the mind in the brain. According to Alva Noë (2009, p. xiii):

Human experience is a dance that unfolds in the world and with others. You are not your brain. We are not locked up in a prison of our own ideas and sensations. The phenomenon of consciousness, like that of life itself, is a world-involving dynamic process. We are already at home in the environment. We are out of our heads.

"Consciousness isn't something that happens inside us: it is something that we do, actively, in our dynamic interaction with the world around us" (Noë, 2009, p. 24).

A feature that distinguishes dissociative parts from other subsystems of personality (e.g., the sympathetic nervous system) is that they are conscious and self-conscious (Nijenhuis, 2011). If the mind in mental health cannot be found in the brain, the mind of disociative parts cannot be found in the brain or brain/body either (though not all would agree with this statement, e.g. Joseph 2009). The mind unfolds in dynamic configurations of brain, body, and environment, particularly in embedded actions that are ongoing when these parts are awake or dreaming.

Cause and Effect The fact that cause and effect are inseparable in the case of co-occurrence and co-constitution does not imply that there is no causation at all. Aristotle distinguished apart from efficient causation three other causes. Final cause involves the goal toward which something aims, hence pertains to the integration of the brain and body within the environment. For example, mediated by the limbic system of attachment, individuals seek protection by a caretaker in situations that they perceive as calling for this form of protection (Joseph 1992). These and other action systems bias individuals as a whole system toward particular perceptions, affects, and conceptions, and toward the effectuation of particular goals through action.

As Northoff (2003) puts it, goals are about observable and to-be effectuated events in the environment. Observable events pertain to sensory perception, and to-be effectuated events include more motor action. Goal-orientation involves sensorimotor integration, which cannot be captured in terms of efficient causation. Sensations/perceptions are not the linear and physical cause of motor actions, and motor action is not the linear and physical cause of sensations/perceptions. Sensations/perceptions and motor actions rather have an ongoing effect on each other (Hurley, 1998). Moreover, physical activity such as patterns of brain activity does not reveal what the owner and agent of that activity was sensing and doing, and why. Sensorimotor integration or synthesis, more complex forms of integration, and goal-orientation (e.g., the aim to attach or defend) cannot be explained in physical terms, but involve final causes.

Formal cause describes what makes a particular 'thing' one sort of 'thing' and no other, and describes the kind of organization of the 'thing.' We thus do not simply respond to physical 'stimuli' and do not, like computers, 'process' 'information' understood as discrete, encapsulated, symbolic, and predefined, thus context-independent and context-isolated instructional codes. Rather, we are enactive complex systems that find or generate meaning in a dynamic and self-organizing fashion (Edelman & Tononi, 2000; Kelso, 1995; Thompson, 2007; Varela, Thompson, & Rosch, 1993).

As whole embodied and embedded biopsychosocial systems, we are associated with all four causes. As dynamic, self-organizing systems (formal cause), we are oriented toward observable events regarding which we aim to realize particular effects as guided by our goals (final cause). The material cause involves the physical material we are made of (e.g., cells). With regard to the brain, the efficient cause includes 'the force of the neural mechanisms by means of which different parts of the brain are specifically related to each other with the ultimate realization of self-organisation and dynamic pattern formation" (Northoff, 2003, p. 291).

Dissociative parts are also associated with all four causes. They are goal-oriented (final cause) and self-organizing (formal cause), they have an organic basis (material cause) and involve physical and neural mechanisms (efficient cause). Mediated by their goals, each dissociative part synthesizes its own kinds of perceptions, sensations, emotional feelings, thoughts, memories, and motor actions. Alternations among different parts thus go along with profound shifts in the quality and kind of syntheses, hence, shifts in consciousness.

Like mentally healthy individuals, dissociative parts also engage in the mental action of personification regarding particular perceptions, sensations, feelings, thoughts, memories, and motor actions (Van der Hart et al., 2006). Personification is generating a phenomenal self-model (the raw feeling of being someone) and a more reflective self-model (Metzinger, 2003; Nijenhuis, 2011). While each dissociative part generates a self-model, they do not personify each other. They do not feel or grasp that they are parts of a whole, that they, so to speak, are the other parts as well (in an ontological sense), and should integrate these parts' mental and behavioral contents (e.g., that the memories 'of other parts' should also be(come) their memories) in an experiential and epistemological sense.

Each dissociative part involves its own kinds of dynamic configuration of brain/body/environment. For example, as one part but not as a different part, the patient may have personified the right arm of the body that the first part has mutilated. Shifts among these parts are thus associated with profound shifts in personification, hence, in the patient's sense of self, world, and self-in-the-world. Personification includes the actions of experiencing and acception ownership ('this is my hand') and agency ('I move my hand').

Another mental action is presentification (Van der Hart et al., 2006), the demanding action of integrating one's past, present and future, such that the present is experienced as the most real. More than knowing, presentification includes taking the content of one's experience and knowledge into account with respect to one's current and future actions. The extent to which dissociative parts can and will engage in presentification is quite different. They may confuse past and present (as in re-enactment of traumatic experiences), fail to presentify past episodes (as in dissociative amnesia), or overlook empending dangers.

Does it make sense to study dissociative parts? In sum, ontologicallly, dissociative parts constitute interrelated but insufficiently integrated embrained, embodied, and environmentally embedded biopsychosocial structures, that is, subsystems of personality. They include their own phenomenal and epistemic sense of self, world, and self-in-the-world (Nijenhuis, 2011; Nijenhuis & Van der Hart, 2011). As conscious and self-conscious goal-directed subsystems, they are subject to all four Aristotelian causes.

Considering these features and their degree of relative autonomy, it makes logical sense to maintain that dissociative parts of a patient decide, think, feel, want, or do something, provided we realize that (1) they are subsystems of a higher-order system that constitutes the patient's personality, (2) they are not totally separated from each other, (3) a full understanding of them involves an analysis of their mutual relations, and (4) a full understanding of the patient requires a structural and functional analysis of his or her personality as a whole biopsychosocial system.

3. Four Interrelated Epistemic Perspectives

Different dynamical configurations of brain, body, and environment account for our different epistemic abilities and inabilities. The first-person, quasi second-person, second-person, and third-person perspective are such configurations.

The first-person perspective (FPP) is about our phenomenal experience, and involves experiencing 'raw', prereflective sensations and other feelings that are subjective, private, and internal. This perspective subserves living and experiencing of present events (Northoff, 2003). FPP involves our body as our spatial center, and our 'I' as the center of our mental states. In other words, FPP pertains to the action of generating the subjective feeling of being someone with a point of view, that is, of being an acting and experiencing self with a subjectively experienced outward perspective on his or her perceived environment, and an inward perspective regarding himself or herself (Metzinger, 2003). This someone involves our 'I' or 'self', or phenomenal self-model. How this model is generated by some component of us as a whole system remains hidden, because the brain does not tell us how it generates its own states, including our 'I'. Due to this autoepistemic inability, we experience our phenomenal self and world as given.

Dissociative parts of the personality include their own FPP. They feel they are/have a self, and experience that their world is given. Subject to the described autoepistemic limitation, they are unable to experience that their syntheses, personifications and presentifications are personal constructions. However, for the observer, the constructive nature of these contents can be clear. For example, whereas dissociative parts experience they have their own body, observers will agree this is not the case.

Mentally healthy individuals can detect, recognize, and compare, that is, judge their our different phenomenal experiences and states (Chalmers, 1996). In Northoff's (2003) terms, this phenomenal judgment involves our Second-Person Perspective (SPP), or, as I say, our Quasi SPP (QSPP). QSPP is a form of intra-subjective communication, whereas FPP is about intra-subjective experience. This intra-subjective communication serves, among others, for the development of a 'relation of mineness' (Metzinger, 1993). For example, in QSPP, we can say "I know this is my body." QSPP therefore involves a more objective self-model than the phenomenal self-model in FPP in which we might experience 'being cold' which may but need not be expressed as "I am cold." In FPP, personification--i.e., the mental actions of owning experience and agency--is prereflective. In QSPP, personification involves at least a phenomenal judgment of ownership (e.g., "My body hurts") and agency (e.g., "I run").

SPP pertains to the relationship between the 'I' of the own person and the 'I' of another person, the 'Thou' (Buber, 1983). Access to someone else's phenomenal experience exists, but is indirect, and involves actions such as empathy, joint perspective taking, and imitation. These actions include the activity of particular brain structures and functions, e.g., mirror and canonical neurons (Pineda, 2008), that mediate our phenomenal sense of being in mental/experiential touch with another individual.

QSPP and SPP are epistemic intermediates between phenomenal experience in FPP and physical judgment in TPP. In FPP, we are experientially and spatially centered in our body, provided we, for example, do not experience out of body phenomena, or somatoform anesthesia, as can happen in dissociative disorders. In QSPP, we are also centered in our body, but since we phenomenally judge our phenomenal experience, we are linked with, but also at some distance of our mental and bodily state. This bridge between phenomenal experience and physical judgment in QSPP allows us to modulate our emotional experiences.

Many dissociative parts can reach a level of functioning high enough to engage in phenomenal judgment regarding themselves and others. Thus, they have an explicit sense of self to the extent that they engage in personification, and they are oriented in phenomenal space and time to the extent that they engage in presentification. However, what healthy individuals phenomenally judged in terms of QSPP (e.g., this is my arm, memory, motor action), dissociative parts may phenomenally judge in terms of SPP (e.g., this is not my arm, memory, motor action, but it belongs to [an other dissociative part; I am not the others [i.e., the other dissociative parts]). Confusion between QSPP and SPP is a core feature of dissociative disorders.

A dissociative part may also regard an other part in Third-Person Perspective (TPP). TPP does not include experiential and bodily centeredness, because in this epistemic perspective, we are not experientially--i.e., mentally and bodily--linked with ourselves, other individuals, other organisms, and immaterial objects. We engage in physical judgment in TPP, and cannot experience perceptions or perform motor actions (Northoff, 2003). For example, some neuroscientists tend to study the brain and its workings as physical objects in TPP. Clinicians assess patients' mental disorders in TPP, but engage in SPP when they are empathically attuned to patients in therapeutic encounters. As private persons, we may notice in TPP that our leg is broken and needs treatment, or that we are depressed or happy. In TPP, we thus observe and judge ourselves, other individuals, and objects as physical objects.

At least some dissociative parts can function at a mental level that is high enough to engage in TPP. For example, as such a part, a patient may be able to hold a job that requires the ability to engage in physical judgments. While this TPP is adaptive, dissociative parts also engage in maladaptive TPP. For example, when a dissociative part does not personify a different part that 'did not resist abuse,' he or she may physically judge that the other part is bad, and deserves punishment.

The different perspectives and the involved epistemic abilities and inabilities are interrelated. TPP is only possible on the basis of FPP, and in many cases QSPP and SPP as well. For example, scientists' TPP regarding their object of study can only exist in the framework of their FPP, QSPP, and SPP. Science would not exist without human prereflective (FPP) and reflective (QSPP/SPP) consciousness and self-consciousness. As discussed above, objects only exist in the awareness of subjects. Furthermore, QSPP/SPP depend on FPP because phenomenal judgment involves a judgment of phenomenal experience. Whereas FPP can exist without QSPP, QSPP can influence FPP because individuals' phenomenal experience tends to be affected by their phenomenal judgment.

Summary There are no selves, objects and events without experiencing and knowing subjects. Subjects include individuals whose personality is integrated, as well as conscious and self-conscious dissociative parts of an individual. Any 'self,' 'world,' and 'event' depends on an embrained, embodied, and embedded subject. Brain, body, and environment cannot exist by themselves, but exist in intrinsic relationships and dynamic configurations of brain, body, and environment. How an individual and dissociative parts of an individual experience, perceive, and conceive these embedded selves, objects and events depends on their particular epistemic perspective: FPP, QSPP, SPP, and TPP. In epistemic regard, dissociative parts engage in underinclusive and/or overinclusive acts of synthesis, personification, and presentification.

4. Research of Consciousness and Self-Consciousness

Under the formulated restrictions, the study of dynamic configurations of brain, body, and environment in dissociation can contribute to an understanding of consciousness and self-consciousness. For example, this research includes comparisons between different dissociative parts with their own goal-orientation and epistemological features such as their different syntheses, and degrees and kinds of personification, and presentification. It also encompasses comparisons between dissociative parts and mentally healthy individuals.

Comparisons of different (types of) dissociative parts indicate what neural activity, brain structures, and behavioral patterns are associated with experiencing, perceiving, and knowing oneself, other selves, objects, and events in FPP, QSPP, SPP, or TPP. For example, these studies can find differences between dynamic configurations of brain, body, and environment for dissociative parts that do not synthesize ('There is no arm') or personify particular body parts ('This is not my arm'), and those that do. Comparisons of dissociative parts and mentally healthy individuals can also document such correlates, which research is similar to comparisons of patients with structural brain defects and healthy individuals (e.g., Gallagher, 2005). For example, by comparing the neural activity of dissociative parts with a deficient phenomenal self-model and mentally healthy individuals with an adequate phenomenal self-model, it may be found what neural activity is associated with the generation of an adequate FPP and QSPP. Both subtypes of research can contribute to an understanding that the normal phenomenal sense of self, world, and self-in-the-world involve ongoing constructions, developed in interaction with a material and social environment.

Studies of dissociative parts are possible and fruitful. For example, different resting state patterns of neural activation patterns have been found for DID patients, functioning as ANP, and healthy controls (Sar, Unal, & Ozturk, 2007). This kind of research can document what brain structures and functions are associated with normal and abnormal resting states of (self)consciousness. The study of task-related states of (self)consciousness in dissociation and normality is also within reach. We had two different types of dissociative parts in women with DID listen to neutral and trauma memory scripts (Reinders et al., 2003, 2006, 2008). These types involved a part oriented to achieving goals of daily life, and a part oriented to physical defense in the form of foremost sympathetically mediated tendencies to flight and freeze. Theoretically, these types are described as Apparently Normal Parts (ANP) and Emotional Parts (EP) (Van der Hart et al., 2006). For ANP only the neutral memory script, and for EP both scripts pertained to personal experiences. ANP and EP had indistinguishable subjective, psychophysiological and neural reaction patterns for listening to the neutral memory script, but their respective reactions to the trauma memory script were, as was theoretically predicted, very different. Broadly speaking, EPs but not ANPs had strong phenomenal and psychophysiological reactions, as well as more subcortical (caudate, amygdala) and less cortical activation than ANPs. These differences were not due to role-playing, suggestion, and fantasy proneness, as some scientists have suggested (Giesbrecht, Lynn, Lilienfeld, & Merckelbach, 2008; Lilienfeld et al., 1999; Merckelbach, Rassin, & Muris, 2000; Piper & Merskey, 2004): neither high nor low fantasy prone mentally healthy women instructed to simulate ANP and EP had psychophysiological and neural reaction patterns that characterized authentic ANPs and EPs (Reinders et al., 2008).

Studies of phenomenal self-models and lack of personification in dissociative disorders and health are feasible as well. Thus, we currently prepare a study of biopsychosocial reactions of ANP and EP to audiotaped ANP's, EP's, a friend's, and a stranger's self-spoken self-descriptions. Healthy controls will listen to audiotaped self-spoken self-descriptions, and self-descriptions of a friend and a stranger. This work will allow us to explore what neural structures and patterns are involved in correct and incorrect discriminations of QSPP (I-me), SPP (I-other dissociative part/I-friend), and TPP (I-stranger).

5. Conclusion

Theoretical considerations and pioneering empirical work suggest that studies of dissociation can contribute to an understanding of consciousness and self-consciousness as phenomena implied in particular ongoing actions. Ontologically, this study examines dynamic configurations of the intrinsic relationships of brain, body, and environment, that is, links between mental, behavioral, (psycho)physical, and neuronal states, and the actions that generate these states (Joseph 2009; Northoff, 2003). Epistemically, the focus of this work is on the differences and interdependency of FPP, QSPP, SPP, and TPP. The theory of dissociation of the personality—for space only minimally introduced in this article-- can serve as a solid heuristic with regard to these explorations.

With some exceptions (e.g., Edelman & Tononi, 2000; Metzinger, 2003), the possibility of this study is hardly realized to date. For example, there are only a handful neuroimaging studies of DID, whereas there are about a thousand neuroimaging studies of schizophrenia, a disorder with a similar prevalence. Realization of the relevance of complex dissociative disorders for the study of consciousness and self-consciousness would be most rewarding for philosophy, science, and the wellfare of patients with these severe conditions alike.

References

American Psychiatric Association (1994). DSM-IV. Washington, DC: American Psychiatric Press.

Bennett, M.R. & Hacker, P.M.S. (2003). Philosophical foundations of neuroscience. Oxford: Blackwell.

Bohr, N. (1958/1987), Essays 1932-1957 on Atomic Physics and Human Knowledge, reprinted as The Philosophical Writings of Niels Bohr, Vol. II, Woodbridge: Ox Bow Press.

Buber, M. (1983). Ich und Du. 11. durchgesehene Auflage. Heidelberg:

Lambert Schneider.

Chalmers, D.J. (1996). The conscious mind: In search of a fundamental theory. New York: Oxford University Press.

Dell, P.F. (2006). A new model of dissociative identity disorder. Psychiatric Clinics of North America, 29, 1–26.

Descartes, R. (1960). Discourse on Method and Meditations. New York: The Liberal Arts Press.

Edelman, G. M., & Tononi, G. (2000). A universe of consciousness: How matter becomes imagination. New York: Basic Books.

Fanselow, M.S. & Lester, L.S. (1988). A functional behavioristic approach to aversively motivated behavior: Predatory imminence as a determinant of the topography of defensive behavior. In R.C. Bolles & M.D. Beecher (Eds.), Evolution and learning (pp.185-212). Hillsdale, NJ: Erlbaum.

Fuster, J.M. (2003). Cortex and mind: Unifying cognition. New York: Oxford University Press.

Gallagher, S. (2005). How the body shapes the mind. New York: Oxford University Press.

Garcia-Campayo, J., Fayed, N., Serrano-Blanco, A., Roca, M., (2009). Brain dysfunction behind functional symptoms: neuroimaging and somatoform, conversive, and dissociative disorders. Current Opinion in Psychiatry 22 (2): 224.

Giesbrecht, T., Lynn, S.J., Lilienfeld, S.O., & Merckelbach, H. (2008). Cognitive processes in dissociation: An analysis of core theoretical assumptions. Psychological Bulletin,134, 617-647.

Hermans, E.J., Nijenhuis, E.R.S., Van Honk, J., Huntjens, R., & Van der Hart, O. (2006). State dependent attentional bias for facial threat in dissociative identity disorder. Psychiatry Research, 141, 233-236.

Heisenberg, W. (1958), Physics and Philosophy: The Revolution in Modern Science, London: Goerge Allen & Unwin.

Hughes, J.R., et al., (1990). Brain mapping in a case of multiple personality". Clinical EEG, 21 (4): 200–209.

Hurley, S.L. (1998). Consciousness in action. Cambridge MA: Harvard University Press. Janet, P. (1889). L'automatisme psychologique. Paris: Félix Alcan.

Janet, P. (1935). Réalisation et interpretation [Realization and interpretation]. Annales Médico-Psychologiques, 93, 329-366.

Joseph, R. (1988a) The Right Cerebral Hemisphere: Emotion, Music, Visual-Spatial Skills, Body Image, Dreams, and Awareness. Journal of Clinical Psychology, 44, 630-673.

Joseph, R. (1988b). Dual mental functioning in a split-brain patient. Journal of Clinical Psychology, 44, 770-779.

Joseph, R. (1992) The Limbic System: Emotion, Laterality, and Unconscious Mind. The Psychoanalytic Review, 79, 405-455.

Joseph, R. (2009). Quantum Physics and the Multiplicity of Mind: Split-Brains, Fragmented Minds, Dissociation, Quantum Consciousness, Journal of Cosmology, 3, 600-640.

Kant, I. (1998). Critique of pure reason. Cambridge, MA: Cambridge University Press.
Kelso, J.A.S. (1995). Dynamic patterns: The self-organization of brain and behavior. Cambridge, MA: MIT Press.

Kluft, R.P. (1987). First-rank symptoms as a diagnostic clue to multiple personality disorder. American Journal of Psychiatry,144(3), 293–298.

Lang, P.J., Bradley, M.M., & Cuthbert, B.N. (1998). Emotion, motivation, and anxiety: Brain mechanisms and psychophysiology. Biological Psychiatry, 44, 1248-1263.

Lapointe, A.R., et al., (2006). Similar or disparate brain patterns? The intra-personal EEG variability of three women with multiple personality disorder". Clinical EEG and Neuroscience, 37 (3): 235–242.

Lilienfeld, S.O, Lynn, S.J., Kirsch I, Chaves JF, Sarbin TR, Ganaway GK, & Powell RA (1999). Dissociative identity disorder and the sociocognitive model: recalling the lessons of the past. Psychological Bulletin, 125, 507-523.

Mathew, R.J., Jack, R.A., West, W.S. (1985). Regional cerebral blood flow in a patient with split personality". The American Journal of Psychiatry 142 (4): 504–505.

McGinn, C. (1989). Can we solve the mind-body problem? Mind, 98, 349-366.

McGinn, C. (1999). The mysterious flame: Conscious minds in a material world (1st ed.). Oxford, UK; Cambridge, MA, USA: B. Blackwell.

Merckelbach, H., Rassin, E., & Muris, P. (2000). Dissociation, schizotypy, and fantasy proneness in undergraduate students. Journal of Nervous and Mental Disease, 188, 428-431.

Metzinger, T. (2003). Being no one: The self-model theory of subjectivity. Cambridge, MA: MIT Press.

Neuman, Y., and Tamir, B. (2009). On Meaning, Consciousness and Quantum Physics. Journal of Cosmology, 2009, 3, 540-547.

Northoff, G. (2003). Philosophy of the brain: The brain problem. Amsterdam/Philadelphia: John Benjamins.

Nijenhuis, E.R.S. (2004). Somatoform dissociation: Phenomena, measurement, and theoretical issues. New York: Norton.

Nijenhuis, E.R.S. & Den Boer, J.A. (2009). Psychobiology of traumatization and trauma- related structural dissociation of the personality. In P. Dell & J. O'Neil (Eds.) Dissociation and dissociative disorders: DSM-IV and beyond. Oxford: Routledge.

Nijenhuis, E.R.S. & Van der Hart (2011). Dissociation in trauma: A new definition and comparison with previous formulations. Journal of Trauma and Dissociation. In press.

Nijenhuis, E.R.S., Van der Hart, & Steele, K. (2002). The emerging psychobiology of trauma- related dissociation and dissociative disorders. In H. D'Haenen, J.A. Den Boer, & P. Willner (Eds.), Biological Psychiatry (pp. 1079-1098). London: Wiley.

Nijenhuis, E.R.S. (2011). Consciousness and self-consciousness in dissociative parts of the personality. In V. Sinason (Ed.), Attachment, trauma, and multiplicity. London: Routledge.

Noë, A. (2009). Out of our heads. New York: Hill and Wang.

Panksepp, J. (1998). Affective neuroscience: The foundations of human and animal emotions. New York: Oxford University Press.

Pineda, J.A. (2008). Sensorimotor cortex as a critical component of an 'extended' mirror neuron system: Does it solve the development, correspondence, and control problems in mirroring? Behav Brain Funct, 4, 47.

Piper, A., & Merskey, H. (2004). The persistence of folly: Critical examination of dissociative identity disorder. Part II. The defence and decline of multiple personality or dissociative identity disorder. Canadian Journal of Psychiatry, 49, 678-683.

Reinders, A.A.T.S., Nijenhuis, E.R.S., Paans, A.M.J., Korf, J., Willemsen, A.T.M., & Den Boer, J.A. (2003). One brain, two selves. NeuroImage, 20, 2119-2125.

Reinders, A.A.T.S., Nijenhuis, E.R.S., Quak, J., Korf, J., Paans, A.M.J., Haaksma, J., Willemsen, A.T.M., & Den Boer, J. (2006). Psychobiological characteristics of dissociative identity disorder: A symptom provocation study. Biological Psychiatry, 60, 730-740.

Reinders, A.A.T.S., van Eekeren, M., Vos, H., Haaksma, J., Willemsen, A., den Boer, J., Nijenhuis, E. (2008). The dissociative brain: Feature or ruled by fantasy? Proceedings of the First International Conference of the European Society of Trauma and Dissociation. Amsterdam, April 17-19, p. 30.

Reinders, A.A.T.S., Willemsen, A.T.M., Vos, H.P.J., den Boer, J.A., & Nijenhuis, E.R.S. (submitted). Is dissociative identity disorder fake or fact? A psychobiological study of simulated and pathological dissociative identity states.

Ross, C.A., Miller, S.D., Reagor, P., et al. (1990). Schneiderian symptoms in multiple personality disorder and schizophrenia. Comprehensive Psychiatry, 31 (2),111–118.

Sacks, O. (1984). A leg to stand on. London: Duckworth.

Sar, V., Unal, S.N., & Ozturk, E. (2007). Frontal and occipital perfusion changes in dissociative identity disorder. Psychiatry Research, 156(3), 217-223.

Schopenhauer, A. (1958). The world as will and representation. Vol. I and II. Clinton, MA: The Falcon's Wing Press.

Schopenhauer, A. (2007). On the fourfold root of the principle of sufficient reason. New York: Cosima.

Spinoza, B. de (1996). Ethics. London: Penguin Books.

Stapp, H. P. (2009). Quantum Reality and Mind. Journal of Cosmology, 3, 570-579.

Thompson, E. (2007). Mind in life: Biology, phenomenology, and the sciences of mind. Cambridge, MA: Belknap Harvard.

Van der Hart, O., Nijenhuis, E.R.S., & Steele, K. (2006). The haunted self: Structural dissociation and the treatment of chronic traumatization. New York: Norton.

Van der Kolk, B.A., & Van der Hart, O. (1989). Pierre Janet and the breakdown of adaptation in psychological trauma. American Journal of Psychiatry, 146, 1530–1540.

Van Gelder, T. (1995). What might cognition be, if not computation? Journal of Philosophy, 91, 345-381.

Varela, J., Thompson, E., & Rosch, E. (1993). The embodied mind: Cognitive science and human experience. Cambridge, MIT Press.

Wittgenstein, L. (1953). Philosophical investigations. New York: The MacMillan Company.

37. Science and the Self-Referentiality of Consciousness

Michel Bitbol, Ph.D.[1], and Pier-Luigi Luisi, Ph.D.[2],

[1]CREA, CNRS/Ecole Polytechnique, ENSTA, 32, Boulevard Victor, 75015 Paris, France. [2]Department of Biology, University of Roma 3, Viale G. Marconi 446, 00146 Rome, Italy

Abstract

In this paper, we study the extent and limits of scientific inquiry about consciousness. We first insist on the exceptional status of conscious experience, which is no proper object of investigation, but rather an actual presence and a precondition of any investigation. To better characterize this status, we develop the concept of "radical self-reference". Questioning about consciousness is radically self-referential in so far as it is itself an act of consciousness. This suggests that consciousness is existentially primary; a kind of primacy which clearly departs from the ontological primacy advocated by property dualists or panpsychists. We then notice that, accordingly, science has some basic features which hinder in principle its approach of consciousness: it distantiates from its objet, whereas consciousness is at no distance; it tends to formulate truths that do not depend on one's situation, whereas consciousness is what it is like to be situated; it claims that physical explanations are self-sufficient, thus threatening to reduce consciousness to an epiphenomenon. These remarks tend to increase the "hardness" of the "hard problem" of the origin and existence of consciousness. Yet, we also point out that scientific inquiries are able to clarify a host of interesting issues about the forms and development of consciousness. We thus present a discussion about the significance of evolutionary arguments about consciousness. We wonder whether higher order features of consciousness such as reflectivity or self-knowledge can be considered as "spandrels" of evolution (mere side consequence of other features), or rather as adaptative features that were selected in the same way as love behavior. We also discuss the adaptative value of an important byproduct of the higher-order features of consciousness, namely belief in supernatural powers. We finally examine the evolutionary relevance of a conception according to which the brain is a modulator of consciousness rather than its source.

KEY WORDS: consciousness, phenomenology, first-person experience, objectivity, hard problem, darwinism, evolution,

1. INTRODUCTION

This paper is the coordinate effort of a philosopher and a scientist to tackle the question of consciousness. Its aim is to investigate whether and to what extent a harmony or even better a synergy of view is possible on the subject. From the beginning of this endeavor, we were well aware of the difficulty of the task, our advantage being that we also had a common background. Both of us tend to criticize classical reductionist science, and to advocate the systemic conceptions of life and knowledge developed by Maturana and Varela under the names of autopoiesis and enaction. We also accept the fact of elementary consciousness, or pure lived experience, as primary, although with small (but profound) nuances. And we both share interest in Husserl's (1970) and Heidegger's (2010) phenomenology, that we consider as a pioneering attempt at exploring the field of consciousness in the first person. Thus, we were close to each other and nevertheless not ready to completely yield to the other. In particular, the scientist was not ready to accept that science has nothing really significant to say about the origin of consciousness; and the philosopher was somehow reluctant to complement his views about the lived immediacy of conscious experience with empirical inquiry, that he saw as a distraction from the truly central issue. So, let's see where this debate leads us in the field of consciousness studies.

Before we come to the heart of the subject, some precisions about words are necessary, since when they use the term "consciousness" people mean the most disparate things. We are not concerned here with acts of volition, possibly connected with the moral problem of discriminating the good from the evil – we leave that to St Augustine (2009); and we acknowledge the restriction of this field of meaning to the parent term "conscience". We are rather interested in the actuality of lived experience attending to any act of feeling, thinking, doing, and more generally being. Within it, we make a difference between three layers of this experience: (a) pure non-reflective experience (the mere "feel" of sensing and being, without awareness of this feel); (b) reflective consciousness, namely the second-order awareness of having the experience of something; (c) self-consciousness, in the sense of being aware of one's own identity and local embodiment. Of course, we admit a certain amount of blending between the three layers. Firstly, in human life, non-reflective experience is usually associated with reflective consciousness (Joseph 1982). It is precisely reflection that gives us access to experience, thus allowing to speak of it. Secondly, self-consciousness is a special case of reflection: it is reflection about the experience of one's body and biographical memory. However, despite this large amount of interdependence between the levels of consciousness, making the distinction proves useful to understand many familiar situations. For instance, the connection of the two first layers appears to break in cases of displaced attention: when a person is skilfully driving a car while being distracted by talking to somebody, one can easily understand this as a case of nonreflective experience of the road associated with reflective consciousness of the discussion.

We then start, in section 2, with a strong statement of the givenness of pure experience. In section 3 we draw some consequences from this universal condition of scientific inquiry. With this epistemological proviso in mind, we turn to science in section 4 and discuss the evolutionary background of the elaborate forms of consciousness, namely reflective consciousness, full self-consciousness, and the spiritual quest that arises from them. In section 5, we address the relevance of quantum mechanics to consciousness. Our conclusive remarks are devoted to embedding the philosophical and scientific aspects of our investigation in a broader picture.

2. ON THE RADICAL SELF-REFERENTIALITY OF CONSCIOUSNESS

According to Maurice Merleau-Ponty (1964), philosophy is "(…) the set of questions in which the one who questions is himself implicated in the question". Any question about consciousness (here understood as pure experience revealed by reflection) is then utterly philosophical. For when we raise a question about consciousness, we are not only implicated in it in abstracto, timelessly, as generic human beings ; we are fully implicated in it by what we are at this precise moment. We are fully and presently implicated in it because formulating a question about consciousness is an act of consciousness; understanding a question about consciousness is an act of consciousness ; figuring out how we could answer a question about consciousness is yet another act of consciousness. In short, questions about consciousness are radically self-referential.

Let us ponder about this notion of radical self-referentiality, because it may bring us closer to the heart of the issue of consciousness. A sentence such as "this sentence uses five words" is self-referential.

It is easy to see that it indeed refers to itself, provided one shifts attention from the meaning of the sentence to its lexicon, from what it says to the graphemes it is made of. Here, the required attention shift goes from one object of consciousness (what is meant) to another object of consciousness (the written words). By contrast, the self-referential character of a question about consciousness is seen only if attention shifts from the meaning of the question to present conscious experience as the background of this very act of attention. In this case, the second focus of attention is no "object" at all ; rather, it is the condition for anything to be taken as an object. This is why questions about consciousness are more than self-referential : they are radically self-referential. In fact, the radical self-referentiality of questions about consciousness is no abstract circumstance: it can be recognized as a fundamental feature of our lived experience in certain educated practices such as the phenomenological "suspension of judgment", or Zen meditation (Bertossa & Ferrari, 2004).

The question "where does consciousness come from ?" provides us with a good illustration of how misguided one can be if this kind of self-referentiality is ignored. When we ask the question "where ?", we usually prepare ourselves to focus our attention on some restricted region of our conscious experience : right or left, up or down, nearby or far away, inside or outside the skull, in this or that part of the brain cortex or nuclei, in this or that spatio-temporal phase of neural firings, in cognitive functions or in their material substrate. And when we think we have got the answer, after a deep speculative reflection or after a long experimental inquiry, this answer often consists in pointing towards an object or a process that we can describe, think about, or even sometimes imagine. Only advanced spiritual thinkers are ready to endow their "where" with a cosmic dimension, and ascribe consciousness a locus which is much broader than any object of possible perception or standard conceptualization (e.g. the well-known Indian metaphor which represents our personal conscious life as a small ripple on the global ocean of being). In our culture, answering a question about the origin of consciousness is usually tantamount to singling out a given content of our consciousness, and encouraging others to modulate their own consciousness accordingly. Everything looks as if we were trying to ascribe consciousness as a whole to some part of it ; as if conscious experience, this all-pervasive fact that constitutes our lives, were tentatively encapsulated in a fraction of it. This sounds awkward indeed! There is an easy way to alleviate this feeling of awkwardness, though.

Turning our attention to the background condition of any act of attention (in line with radical self-referentiality), we are bound to reply that "consciousness comes from nowhere else than ... here". True, "here" does not look like a serious answer, because it does not refer to a special place, a special object, or a special process. But should we dismiss it so quickly ? Let us think a little further. "Here" is an indexical (or deictic, or demonstrative) term, like "I", "now", and "this". As any indexical term, it fully commits the person who utters it. It thereby invites other persons to figure out how things appear from the standpoint of the utterer, or more generally how things appear from the standpoint of any utterer whatsoever. "Here" is a verbal operator that brings each one of us back to one's own situation. Saying that "consciousness comes from here" then means that consciousness has no other obvious origin than the present situation. Consciousness is the name we give to the astounding realization of immediate existence, even before its more intricate connotations such as reflective consciousness, self-consciousness, or moral conscience. Consciousness, in this very elementary sense, is existentially primary.

These obvious (yet destabilizing) remarks are not derived from any argument. They rather arise when we suspend any judgment, and just state the elementary features of what we are living. They express what E. Husserl (1913/1931) called a phenomenological description; a plain statement of what is immediately experienced, irrespective of any interpretation of the contents of experience in naturalistic terms. So, asserting that consciousness is "existentially primary" was no metaphysical doctrine. Asserting the existential primacy of consciousness was no idealist, property dualist (Chalmers, 1996), or panpsychist (Strawson, 2007) doctrine of the ontological primacy of consciousness to be contrasted with a doctrine of the ontological primacy of matter. Unlike Chalmers, who claims that consciousness is a fundamental (yet non-physical) property of information processors, and unlike Strawson who spreads experience (or proto-experience) on matter in general, we refrain from any such doctrine. We just invite our readers to be faithful to their own lived experience in its most pristine form.

3. THE SCIENTIFIC APPROACH TO CONSCIOUSNESS: GENERAL REMARKS

What about science? What about the physics that physicalist doctrines of consciousness summon? What about the neurobiology that reductionism or eliminativism put forward? To begin with, we must ponder about the fact that the physical sciences cannot avoid using language and symbols. This use is not innocent; it has momentous consequences. Two features of language have an especially strong impact on what science can say about conscious experience (Joseph 1982). Two features that are so elementary that one usually does not even pay serious attention to them: language means and discriminates (Joseph 1982, 1988a,b).

Meaning is tantamount to displacing attention. It displaces attention from the sound of a word to what it signifies, from the pointed finger to what it aims at showing. Meaning thereby pushes us outwards, towards the future, towards something that is not close at hand. When we use a word for "consciousness", we are then automatically led astray, because conscious experience is not something over there to be meant in any way. Once again consciousness is plainly here; this "here" that submerges us ; this "here" that is presupposed by any location in space. Trying to mean consciousness is self-defeating, since what is allegedly meant is not beyond the very act of meaning it. It is radically self-referring.

The same holds for the discriminative power of language (Joseph 1982). How can we discriminate present conscious experience from anything else? This can be done in everyday use for making a difference between somebody else's apparent states of wakefulness and sleep ; but not in the proper existential (and radically self-referential) sense, since at this precise moment that contains in it all the memories of the past and all the projects for the future, there is nothing that can be contrasted with it.

Classical science can be biased about consciousness in the same way as language itself: it attempts to put a distance to what is at no distance from us, and to discriminate what can be contrasted with nothing. Drawing from language, many scientists tend to treat consciousness as a property of human organisms. However, when ascribing a "property" to something it must be based on reliable criteria bearing on this thing, whereas any external bodily consciousness-criterion, be it presence or absence of verbal report, or presence or absence of certain waves on an electroencephalogram, is weak and ambiguous. One proof of consciousness is that we are living at this precise moment (an embodied living as opposed to a contemplated body); this testifies to no property, but rather to the most basic condition for ascribing properties at all.

More specifically, classical science was born from the decision to objectify, namely to select the elements of experience that are invariant across persons and situations. Its aim is to formulateuniversal truths, namely truths that can be accepted by anyone irrespective of one's situation. Therefore, the kind of truths science can reach is quite peculiar: they take the form of universal and necessary connections between phenomena (the so-called scientific laws). This epistemological remark has devastating consequences. It means that in virtue of the very methodological presupposition on which it is based, classical science has little to say about the mere fact that there are phenomena (ultimately construed as appearances experienced by someone), let alone on the qualitative content of these phenomena/appearances (Wright, 2008).

Let us give a few examples. Physics establishes laws about phenomena that are characterized as electromagnetic. It classifies the waves that give rise to the perception of colours according to their wavelengths. But classical physics has little to say about the very existence of an experience of colour and even less about its lived quality. Psychophysics and neurology of occipital cortex areas add more and more precise knowledge about the structure of colour perception in humans, about the mutual relations of various perceived colours, about the physiological states in which colour perception is reported to be altered. But often those who study this phenomenon remain mute about how and why there should be any lived experience of colour at all when neuronal activity occurs in these brain areas, and about what it is like to experience blue or red. More generally, we have witnessed amazing advances of neurophysiology about how the brain stores information, binds its maps and programs for action, and even elaborates self-mapping (Joseph 2009, 2011). These discoveries have also been actively correlated with human subjects' descriptions of their own conscious experience, thus allowing some scientists to speak of memory or unified consciousness instead of information binding, and of self-awareness instead of self-mapping. But only a few neuroscientists have sought to explain why and how these neuronal processes should generate anything like conscious experience (Joseph 1982, 1988a,b, 2009). In other terms, borrowed from David Chalmers, physical and neurological sciences have shown their ability to solve an unlimited number of "easy problems" about the structure and neural correlates of conscious events, but most remain silent about the "hard problem" of the existence, origin, and "feel" of conscious experience itself. This is just a consequence of the methodological decision by some scientists to objectify that has been taken at the very foundation of science. Objectification automatically pushes situated lived experience in the "blind spot" of research.

This fundamental limitation of the classical scientific inquiry about consciousness has serious consequences. One of them arises from the so-called "causal closure" of physical and physiological explanations (the conviction that the system of causes of physics is complete, that it is sufficient to explain any event without resorting to any non-physical cause). Nothing prevents one from offering a purely physical or physiological account of the chain of causes operating from a sensory input received by an organism to the behaviour of this organism. At no point does one need to invoke the circumstance that this organism is perceiving and acting consciously; that it has a feel. The fact of consciousness here appears as irrelevant or incidental. This is what Max Velmans (2009, p. 300) calls the "causal paradox" of consciousness. Therefore, the more a scientific/objective account of mental processes is complete (namely the more it complies with the norm of causal closure), the more some believe conscious experience to be bound or to be construed as a mere epiphenomenon. A confirmation of this impression can be found in a recent about turn concerning the meaning of the word "mind".

Formerly, this word implied consciousness in a tacit way, be it pure non-reflective experience or reflective consciousness. But later on, the word "mind" became, for some, a synonym of "system of cognitive functions". One insisted that many mental operations are unconscious, and that only a few of them give rise to conscious experience. So much so that the question as to whether or not consciousness is a property of our mind arose, thus suggesting that consciousness might be somehow subsidiary with respect to mind.

The same is true in principle of evolutionist arguments, which are documented in the next section. Evolution can select certain useful functions ascribed to consciousness (such as unification of information, or behavioral emotivity of the organism, or reflectivity), but not the mere fact that there is something it is like to implement these functions. In other terms, borrowed from N. Block et al. (1997), evolution can select features of access consciousness but not phenomenal consciousness itself. Some scientists (e.g. Hardcastle, 1996) have indirectly aknowledged this fundamental lacuna by arguing that scientists should be allowed to completely ignore the "hard problem" of the existence and origin of elementary consciousness, and just proceed with the many interesting "easy" problems about the structure and correlates of consciousness it is able to clarify.

4. AN EVOLUTIONARY VIEW OF CONSCIOUSNESS

We have just outlined a series of cautious remarks on why a strictly third person approach to consciousness, in particular to the hard problem of the physical origin of experience, is bound to fail. These remarks actually point towards the necessity of introducing and taking seriously first person reports in science (Joseph 1982, 1988a,b; Petitmengin & Bitbol, 2009).

This does not mean that we claim that no scientific inquiry should be carried out on the subject, but only that either one should be permanently aware of the scope and limits of such inquiry, or that one should include first person approaches in a broadened definition of science, as Varela (1996) did.

Even though strictly classical science is often mute about the stubborn and isolated "hard problem", it is able to highlight and elucidate a growing number of the so-called "easy" aspects of consciousness, which are remarkably rich and important. An objective scientific approach is therefore complementary to the view of the radical self-referentiality discussed above. Actually, it may be interesting to see what this complementary aspect is – and this is one of the main aims of this section.

We can start from a very general, apparently harmless statement: consciousness is connected to our biology of human beings. This rather innocent statement becomes less trivial, however, as soon as we recall, and attach to it, the famous 1973 predicament of Theodosius Dobzhansky, "Nothing in Biology Makes Sense Except in the Light of Evolution". Once we accept this corollary, namely a link between consciousness and evolution, we are almost automatically bound to ask questions such as: "is consciousness, or at least some crucial aspects of it, then part of biological evolution?" "Does it mean that it is somehow based on evolving genes?" "What is consciousness good for, in terms of reproductive advantage?" And also "where was consciousness before mankind arose?"

The main point here is the acceptance of the notion that some crucial features of consciousness, such as, say, reflectivity or self-awareness are part of the biological evolution and therefore important for mankind development and reproduction (see e.g. Joseph 1982, about how "extended consciousness", or reflective consciousness, emerges out of primary consciousness).

But is there any scientific way to establish a relation between evolution and consciousness without falling into the machinery of the Darwinian natural selection theory? Something that comes to mind is the notion of "spandrels" as advocated by Jay Stephen Gould and Lewontin (1979) after having visited San Marco' cathedral in Venice. There they noticed beautiful curved area of masonry which were not originally planned by the architects, but arose as necessary architectural byproducts. Gould and Lewontin thus defined spandrels in evolutionary biology to mean any biological feature that arises as a necessary side consequence of other features, a type of feature which is not directly detemined by natural selection.

Should we then consider the subjective experience of being and knowing it, namely reflective consciousness, as a spandrel of evolution? This idea would have the advantage that we do not have to care about evolving genes for consciousness. Of course, all this is difficult to prove, and it is difficult to argue against or in favour of it. So, let us go back for a moment to the main stream, and ask: what would really mean to consider consciousness as part of the Darwinian evolutionary pathway?

To understand this question somehow, let us make the analogy with another important genetic determinant of the animal kingdom, love (or at least love behavior). We can say that love behavior is an important mechanism of evolution, in the sense that without the love of the mother for her puppies, or without sexual attraction between man and woman, there would be no reproduction and therefore no survival of the species. So, with all the great beauty and poetic respect for this most important and noblest element of mankind, we should also say that love is an animal instinct, somehow linked to genetic determinants. Are we ready to say that the same is true for consciousness?

Clearly, things are not so simple as in the case of love behavior. In particular, the reproductive advantage of the highest forms of consciousness (especially reflectivity, self-awareness, and even social awareness) is far from being obvious. But one can make the point that there is indeed something like that, once we consider an important "spin off" of consciousness, which is spirituality/religiosity.

There is by now a vast literature in anthropology about men being "born to believe" as a Darwinian evolutionary trait (Boyer 1994, Girotto et al., 2008, Feierman 2009). To see this relation in a kind of pictorial/poetic way, let us go back 6 million year ago, when the common ancestors homo and chimpanzee separated from each other, and homo, moving in the savanna, eventually started to walk on two legs and developing a bi-pedal consciousness.

Imagine then the awe, and fear, of our early ancestors looking at the immensity of the sky, feeling on the one hand the inner experience of being, and at the same time strongly believing that the motion of the sun and the terrifying lightening in the storm is likely to be due to the presence of some superior powers that also have inner experience and will – the Gods (Joseph 2001). The gods to whom human beings first devoted rituals which enabled the tribe to have more internal cohesion and strength – thus enabling reproductive advantage, and later erected altars and temples.

The equation consciousness (as subject's reflective feeling of being) = inclination for religiosity (desire of contact with the mysteries of the universe), may be a valid and important one. Can we then also assume their mutual co-evolution?

Difficult to say, difficult to prove. We have no way of testing the evolution of consciousness. Actually, at this point, when we talk about evolution of consciousness, we may depart from the notion of consciousness as subjective experience – to embrace a broader notion of consciousness which also includes the dimensions we have eliminated from the start: decision making and moral code.

The hard-core classical scientist at this point would certainly intervene, to say that what has been evolving and growing is the brain – from the 500 cm or so of the first homo to the final value of 1500 cm (with an apparent maximum at the time of the Cro-Magnon people 30,000 years ago) – and with that, certainly, the optimization of certain human skills and intelligence (Joseph, 2000). Then, would it not be the easiest thing to say – he would add – that by increasing the brain size a critical value was reached, by which consciousness arose? Isn't this precisely what Crick was advocating with his "Astonishing Hypothesis" over 15 years ago?

According to us, this does not have to follow. These relevant evolutionary theories must not be mistaken for a true reductionist scientific account of the radical origin of conscious experience. Rather, they can easily be understood in terms of a non-reductionist conception of the relation between the lived and the living, between conscious experience and biological processes. One such conception was remarkably expressed by Beauregard and O'leary (2007) : "More than a century ago, William James proposed that the brain may serve as a permissive / transmissive / expressive function rather than a productive one, in terms of the mental events and experiences it allows (just as a prism, which is not the source of the light, changes the incoming light to form the coloured spectrum). Following James, Bergson and Huxley speculated that the brain acts as a filter or reducing valve by blocking out much of, and allowing registration and expression of only a narrow band of, perceivable reality. They believed that over the course of evolution, the brain has been trained to eliminate most of those perceptions that do not directly aid our everyday survival. This outlook implies that the brain normally limits the human capacity to have spiritual experiences". Here, the brain is no longer seen as the "cause and place" of experience in general, but as the cause and place of its narrowed down variety which includes reflection and self-knowledge. Such vision of the brain and body as modulators makes sense of many scientific theories and experiments about the many-layered functionalities and amplifications of consciousness, without weakening our knowledge that raw experience is our most immediate given, a feature that is so basic and pervasive that nothing more basic is to be found to account for it. This vision even helps us to gain a better understanding of a true wonder: that there is not only "something it is like to be", but that there is also at this very moment reflective realization of being.

In terms of evolution of consciousness this hypothesis contains something potentially mischievous: it could mean that mankind becomes less and less capable of having peak spiritual experiences, that it is less and less capable to connect with what Rilke (2009), in his eighth Duino elegy, called "openness" – while the brain evolves in order to develop the skill and intelligence for everyday life-hunting, housing and making better weapons. A strange conclusion indeed: a negative evolution of spiritual capability and a positive evolution for technicalities.

All this relates to the old question, whether there is, or whether there has been, instead, a perceivable evolution of human consciousness. Can we say that human consciousness has made some progress in the course of evolution, or not? Is there any hope, expectancy, that mankind will develop towards more tolerance, less aggressivity, no war, etc?

Biological evolution moves at a slow pace and the distance that separates us from Lucy or our Neanderthal relatives is probably too small – a few million years – to clearly detect a difference. And for the same reason, we cannot prove the hypothesis of Bergson and Huxley. What can we conclude from this brief analysis using a scientific approach based on evolution? Certainly we did not learn anything new about the ultimate nature of consciousness - as we had foreseen. We have perceived the limit of this approach by aknowledging that we are not able to provide "proofs" for or against the idea of the evolution of consciousness in general. Yet, an evolutionary approach of certain important features of our consciousness (such as self-awareness and spiritual experience) has opened up and enriched our knowledge.

5. SOME LESSONS OF QUANTUM PHYSICS

Can't at least quantum physics, which clearly departs from classical science, provide us with some unconventional understanding of the status of consciousness in nature? In past literature, there have been two major, but diametrically opposite, ways of seeing the putative connection between consciousness and quantum mechanics: consciousness as the cause of state vector reduction (Von Neumann, 1955, London and Bauer, 1939, Wigner, 1967, French, 2002), and state vector reduction as the physical basis of consciousness (Penrose, 1994, Hameroff & Penrose, 1996, Stapp, 2007). To understand why such connections were assumed, let us first remember that in Heisenberg's (1958) and Popper's (1982) realist reading, the state vector represents a complex of potentialities (or propensities) of physical systems. Now, an event has to be actualized out of the many potentialities, in order to enter into a univocal description of what is the case. According to a straightforward view, actualization should occur as soon as an interaction between the system and a measuring apparatus has taken place. Unfortunately, when this interaction is accounted for quantum mechanically, a new (entangled) state vector is to be ascribed to the compound system that includes the measuring apparatus, thus extending the domain of potentiality and apparently leaving no room to actuality. This is the well-known measurement problem of quantum mechanics. Decoherence only half-solves the problem by showing how the structure of quantum potentialities quickly tends towards a classical probabilistic structure as the interaction between the measurement chain and its environment develops (Joos et al., 2003). So, it is quite natural that many physicists have summoned consciousness as their last resource. On the one hand, conscious experience is precisely what is realized here and now as actual: it is the very paradigm of actuality and it then automatically stops the apparently infinite regress of potentialities as soon as it enters the scene. On the other hand, conversely, actualization by state vector reduction can be construed as what is needed in order to give birth to consciousness as a paradigm of actuality. Alternative challenging conceptions have insisted that, in order to account for what is witnessed in laboratories and in nature, it is not necessary to suppose that a single actuality is realized. It may be the case that several actualities co-exist and evolve in parallel, yet self-interpreting themselves as unique. This is the case in the many-worlds (Everett, 1957) or many-minds (Lockwood, 1989) interpretations, as well as in a recent interpretation that cogently points out the multiplicity of functions and mental states processed by various areas of the brain (Joseph, 2009).

One must pause at this point and ponder about the hidden assumptions of the whole debate. What has been assumed until now is a realist interpretation of quantum mechanical state vectors as intrinsic propensities of physical systems. But state vectors can also be interpreted anti-realistically as mere bearers of presently available experimental information, used for predicting outcomes of future experiments (Bruckner & Zeilinger, 2005; Bitbol, 1996, 1998, 2010). State vector reduction then only represents a convenient way of updating the available information. Along with this anti-realist interpretation, the relations between quantum mechanics and consciousness are seen in a very different light. One no longer has to believe that consciousness bursts in the outer world to reduce the state vector, nor to figure out that the outer world generates consciousness by a spontaneous process of actualization. Here, consciousness is no longer an object of the theory, but rather its most basic presupposition : it is that for which information makes sense.

This does not mean, however, that quantum mechanics interpreted in this non-realist way has nothing to teach us about consciousness. In fact, this theory has important clarifications in store for the philosophy of mind and consciousness (Bohr, 1987). The first clarification arises from the contextuality of quantum phenomena. Quantum phenomena adhere to the experimental situation in which they arise ; they are not independent of the measurement context in which they manifest ; accordingly, they cannot be said to "reveal" an intrinsic property. Despite this, quantum physicists were able to build a theory which holds independently of particular experimental situations. They obtained intersituational consent without detachment of an object.

Now, the epistemological status of a science of mind and consciousness is remarkably isomorphic to the epistemological status of quantum physics (Bitbol, 2000, 2002). Just as microphysical phenomena cannot be detached from measuring apparatuses, conscious experiences cannot be detached from the very state of being aware. Just as there are no true "properties" in quantum physics but only "observables", there are no properties when consciousness is at stake, but only "livables" (first-person experiences). Just as in quantum physics one formulated an intersituational theory about "observables", one can look for ways of obtaining intersubjective consensus about "livables" (Bitbol, 2008a,b). The latter strategy is similar to the one advocated by Varela (1996) in his neurophenomenology.

The second clarification to be obtained from quantum physics has to do with our concept of matter. Indeed, the conception of matter favored by quantum physics, in my opinion, hardly supports the materialist view (Bitbol, 2007). To sum up, this conception is that matter in the classical sense (namely bodies extended in space) is nothing else than a phenomenon shaped by our sense-organs or experimental apparatuses. What is left is only an abstract pattern of field-like dispositions out of which matter-like appearances manifest at our scale. As a consequence, ordinary matter can no longer be taken as the fundamental stuff out of which anything else emerges, including consciousness. At most, one could figure out that matter (qua macroscopic appearance) and consciousness (for which appearances make sense) are both immediately implied by the same background vaguely indicated through the concept of a dispositional quantum field. The latter view is speculative, but it is at least consistent with the former evolutionary picture of the brain as a mere modulator, amplifier, and self-monitoring device of ubiquitous elementary consciousness.

6. CONCLUDING REMARKS

The exercise of using the evolutionary approach was interesting in many ways. It did not elucidate the absolute origin of consciousness (we left untouched most of the questions listed at the beginning of the former section) but it was instrumental to bring out the shortcomings and the advantages of the scientific method in this area.

Let us take an exemple of shortcoming, which echoes our initial insistence on the radical self-referentiality of questions about consciousness. As any scientific approach should be, our evolutionary considerations were based on a typical combination of experiments and reasoning.

But, strangely enough for those who consider reasoning and rationality in general as their highest value, the use of reasoning turns out to be a major loophole when consciousness is at stake. Indeed, as E. Schrödinger (1964, p. 19) noticed, when the issues of mind and consciousness are dealt with, the reasoning is part of the overall phenomenon to be explained, not a tool for any genuine explanation. As any reasoning, a reasoning about consciousness involves conscious experience ; aknowledging the validity of a personal reasoning or of a mechanical inference performed by a Turing machine, is still a conscious experience. To sum up, any reasoning bearing on consciousness is included in what is reasoned about. So, when consciousness is presented as an object (rather than a background condition) of reasoning, this can only be in a fake sense.

This being granted, the most promising approach to fundamental questions about consciousness seems to be phenomenology, a method that (as we noticed in section 1) does not rely on reasoning, but on pure description of what is lived in the first person. Far from being a defect of the phenomenological approach, this bracketing of reasoning might well be its major quality.

The implication of what has been said until now is on the one hand that traditional science cannot help to shed light on the ultimate nature of consciousness; and on the other hand that pursuing the scientific inquiry, as indicated even by the coarse exercise carried out in section 4, is capable to enrich us immensely with complementary notions about the successive forms, restrictions, or ability to self-representation, of consciousness. Thinking of consciousness in terms of genes, in the same way as we do when love behavior is dealt with, is a hard pill to swallow for many philosophers and cognitive scientists, and because of that, possibly, a very useful exercise. The bold hypothesis of the development of the brain in two opposite directions (improvement of technical and logical abilities, yet loss of contemplative/spiritual potential) is something challenging from all kinds of perspectives. And asking questions about the origin of human forms of consciousness in the broad scenario of a development of our species is another useful intellectual enterprise.

To sum up, we emphasize the radical self-referentiality of consciousness which may be beyond the reach of objective science, and also beyond any ontological categorization unlike Chalmers' "fundamental properties". Yet, we also add that the scientific approach is the appropriate avenue to investigate and clarify the paraphernalia of conscious life.

References

Beauregard M. and O'Leary D. (2007), The spiritual brain, London: HarperOne.

Bertossa F., Ferrari R., & Besa M. (2004), Matrici senza uscita. Circolarità della conoscenza oggettiva e prospettiva buddhista, in: M. Cappuccio (ed.), Dentro la Matrice, scienza e filosofia di The Matrix, Milan: Alboversorio.

Bitbol M. (1996), Mécanique quantique : une introduction philosophique, Paris : Flammarion.

Bitbol M. (1998), Some steps towards a transcendental deduction of quantum mechanics, Philosophia naturalis, 35, 253-280.

Bitbol M. (2000), Physique et philosophie de l'esprit, Paris: Flammarion.

Bitbol M. (2002), Science as if situation mattered. Phenomenology and the Cognitive Science 1: 181-224.

Bitbol M. (2007), Materialism, Stances, and Open-Mindedness, in: Monton B. (ed.), Images of Empiricism: Essays on Science and Stances, with a Reply from Bas C. van Fraassen, Oxford : Oxford University Press.

Bitbol M. (2008a), Is Consciousness Primary?, NeuroQuantology, 6, 53-71.

Bitbol M. (2008b), Consciousness, Situations, and the Measurement Problem of Quantum Mechanics, NeuroQuantology, 6, 203-213.

Bitbol M. (2010), Downward causation without foundations, Synthese, DOI 10.1007/s11229-010-9723-5.

Block N., Flanagan O., Güzeldere G. (eds.), (1997), The Nature of Consciousness, Cambridge: M.I.T Press.

Bohr N. (1987), Atomic theory and the description of nature, Ox Bow Press.

Boyer P. (1994), The Naturalness of Religious Ideas, Berkeley: University of California Press.

Brukner C. & Zeilinger A. (2005), Quantum Physics as a Science of Information, in: Elitzur A., Dolev S., Kolenda N., Quo Vadis Quantum Mechanics? Berlin : Springer.

Chalmers, D. (1996), The Conscious Mind, Oxford: Oxford University Press.

Feierman JR. (Ed.), (2009), The Biology of Religious Behavior: the evolutionary origins of faith and religion. Santa Barbara: Praeger.

French S. (2002), A phenomenological solution to the measurement problem? Husserl and the foundations of quantum mechanics, Studies In History and Philosophy of Science Part B: Studies In History and Philosophy of Modern Physics, 33, 467-491.

Girotto,V., Pievani T., and Vallortigara, G. (2008), Nati per credere, Torino: Codice edizioni. Gould, S. J. and Lewontin, R. (1979), "The spandrels of San Marco and the Panglossian paradigm: a critique of the adaptationist programme", Proc. R. Soc. Lond., B, Biol. Sci. 205, 581-598.

Hameroff S.R., Penrose R. (1996), Conscious events as orchestrated spacetime selections, Journal of Consciousness Studies 3, 36-53.

Hardcastle V. (1996), The why of consciousness: a non-issue for materialists, Journal of Consciousness Studies, 3, 7-13.

Heidegger M. (1935/1998), Einfuehrung in die Metaphysik, Berlin: Niemeyer.

Heidegger M. (2010), Being and Time, New York: SUNY Press.

Husserl E. (1913/1931), Ideas: General Introduction to Pure Phenomenology, London: George Allen & Unwin.

Husserl E. (1970), The Crisis of European Sciences and Transcendental Phenomenology, Evanston (IL): Northwestern University Press.

Joos E., Zeh H.D., Kiefer C., Giulini D., Kupsch J. and Stamatescu I.O. (2003), Decoherence and the Appearance of Classical World in Quantum Theory, Berlin : Springer.

Joseph, R. (1982). The Neuropsychology of Development. Hemispheric Laterality, Limbic Language, the Origin of Thought. Journal of Clinical Psychology, 44 4-33.

Joseph, R. (1988a) The Right Cerebral Hemisphere: Emotion, Music, Visual-Spatial Skills, Body Image, Dreams, and Awareness. Journal of Clinical Psychology, 44, 630-673.

Joseph, R. (1988b). Dual mental functioning in a split-brain patient. Journal of Clinical Psychology, 44, 770-779.

Joseph, R. (2000). The evolution of sex differences in language, sexuality, and visual spatial skills. Archives of Sexual Behavior, 29, 35-66.

Joseph, R. (2001). The Limbic System and the Soul: Evolution and the Neuroanatomy of Religious Experience. Zygon, the Journal of Religion & Science, 36, 105-136.

Joseph, R. (2009). Quantum Physics and the Multiplicity of Mind: Split-Brains, Fragmented Minds, Dissociation, Quantum Consciousness, Journal of Cosmology, 3, 600-640.

Joseph, R. (2011). The neuroanatomy of free will. Loss of will, against the will, "alien hand". Journal of Cosmology, 14, 6000-6045.

London F., Bauer E. (1939), La Théorie de l'Observation en Mécanique Quantique, Paris : Hermann.

Merleau-Ponty M. (1964), Le visible et l'invisible, Paris: Gallimard.

Penrose R. (1994), Shadows of the Mind, Oxford : Oxford University Press.

Petitmengin C. & Bitbol M. (2009), "The validity of first-person descriptions as authenticity and coherence", Journal of Consciousness Studies, 16, 363-404.

Popper, K. (1982), The Postscript to the Logic of Scientific Discovery, II. The Open Universe, London : Hutchinson.

Rilke R.M. (2009), Duino Elegies, London: Vintage.

Schrödinger E. (1964), My view of the world, Cambridge: Cambridge University Press.
Stapp H. (2007), Mindful Universe : Quantum Mechanics and the Participating Observer, Berlin : Springer.

St Augustine (2009), Confessions, Oxford: Oxford University Press.

Strawson G. (2007), "Why physicalism entails panpsychism", in: A. Freeman (ed.), Consciousness and its Place in Nature, Exeter: Imprint Academic.

Varela F. (1996), "Neurophenomenology, a methodological remedy for the hard problem", Journal of Consciousness Studies, 3, 330-350.

Velmans M. (2009), Understanding Consciousness, London: Routledge.

Von Neumann J. (1955), Mathematical Foundations of Quantum Mechanics, Princeton : Princeton University Press.

Wigner E.P. (1967), Remarks on the mind-body question, in: Symmetries and Reflections, Bloomington : Indiana University Press.

Wittgenstein L. (1982), Notes on private experience and sense-data, Paris: T.E.R.

Wright E. (2008), The Case for Qualia, Cambridge: MIT Press.

38. Cosmological Implications of Near-Death Experiences
Bruce Greyson

University of Virginia, University of Virginia Health System, 210 10th Street, NE, Suite 100, Charlottesville, VA 22902-4754

Abstract

"Near-death experiences" include phenomena that challenge materialist reductionism, such as enhanced mentation and memory during cerebral impairment, accurate perceptions from a perspective outside the body, and reported visions of deceased persons, including those not previously known to be deceased. Complex consciousness, including cognition, perception, and memory, under conditions such as cardiac arrest and general anesthesia, when it cannot be associated with normal brain function, requires a revised cosmology anchored not in 19th-century classical physics but rather in 21st-century quantum physics that includes consciousness in its conceptual formulation. Classical physics, anchored in materialist reductionism, offered adequate descriptions of everyday mechanics but ultimately proved insufficient for describing the mechanics of extremely high speeds or small sizes, and was supplemented a century ago by quantum physics. Materialist psychology, modeled on the reductionism of classical physics, likewise offered adequate descriptions of everyday mental functioning but ultimately proved insufficient for describing mentation under extreme conditions, such as the continuation of mental function when the brain is inactive or impaired, such as occurs near death.

KEY WORDS: materialism, reductionism, near-death experience, mind-body problem, consciousness

1. Introduction to Near-death Experiences

Many but not all neuroscientists, physicists, and psychologists believe the mind and consciousness are produced by, or are subjective concomitants of, brain states (Crick, 1994; Damasio & Meyer, 2009; Searle, 2000). This theory receives considerable support from the correlation between brain changes and mental changes: inhibiting brain activity generally inhibits mental activity (Churchland, 1986; Jeeves & Brown, 2009; Tononi & Laurys, 2009). However, correlation is not the same as causation. For example, it is well established that the frontal lobes control or mediate perceptual and other cognitive activities through inhibition (Joseph 1988, 1999a). It is in this manner that concentration and attention may be maintained, so that the mind remains focused. On the other hand, it is also well established that injuries to specific regions of the brain disrupt various aspects of consciousness and mental activity, including thinking, speech, and awareness of the body image (Joseph 1986, 1996, 1999b).

Thus the production model posits that the brain generates the mind (Churchland, 1986; Crick, 1994; Searle, 2000), whereas the filter or transmission model posits that the brain may permit or mediate the mind (Broad, 1953; Burt, 1968; Huxley, 1954; James, 1898; Kelly, 2007; Schiller, 1891). By contrast, the evidence compiled by Joseph (1996, 1999b, 2001) could be said to support both views. Likewise, it is the opinion of this author that the observed correlation between brain states and mind states is compatible with the "production" theory that mind is produced by the brain, but it is also compatible with the "filter" or "transmission" theory that the mind is filtered, focused, limited, constrained, or received by the brain; i.e. that the brain may be a vehicle which receives, transports, and transmits, but is not synonymous with the mind. These and other models of the brain-mind relationship have been debated for centuries, with discussions appearing even in the Platonic dialogues (Kelly, 2007).

One major clue to the nature of the mind-brain relationship may be found in descriptions of near-death, or after death experiences. If the mind, if consciousness is retained during clinical death, this would indicate the mind may only be dependent on the brain much as a radio transmission is dependent upon a receiver and broadcast unit.

Dozens of case reports in the medical literature spanning centuries have documented the phenomenon of "terminal lucidity," the unexplained return of mental clarity and memory shortly before death in patients who had suffered years of chronic schizophrenia or dementia (Brierre de Boismont, 1862; Burdach, 1826; Marshall, 1815; Nahm & Greyson, 2009; Turetskaia & Romanenko, 1975). Beyond this paradoxical enhanced mental clarity while brain function deteriorates, considerable research in the past several decades has delineated parameters of what have come to be called "near-death experiences", where those who appear to have died report dissociative experiences where they are separate from their bodies and can observe and are conscious of their surroundings. These profound subjective experiences that many people report when they are near death pose challenges to the materialist mind-brain production model (Greyson, 2003; Parnia et al., 2001; Schwartz et al., 2005; van Lommel et al., 2001).

Experiences of heightened or mystical consciousness on the threshold of death have been described sporadically in the Western medical literature since the 19th century and have been studied systematically for the past 30 years (Holden et al., 2009). Recent research suggests that near-death experiences (NDEs) are reported by 12% to 18% of cardiac arrest survivors (Parnia et al., 2001; Greyson, 2003; van Lommel et al., 2001). These near-death experiences include feelings of peace and joy; a sense of being out of one's physical body and watching events from an out-of-body perspective; a cessation of pain; seeing an unusually bright light, sometimes experienced as a "Being of Light" that radiates love and may speak or otherwise communicate with the person; encountering other beings, often deceased people; experiencing a revival of memories or even a full life review, seeing some "other realm," often of great beauty; sensing a barrier or border beyond which the person cannot go; and returning to the physical body, often reluctantly.

A number of hypotheses have been proposed to explain NDEs within the mind-brain production theory, attributing them to psychopathology (Noyes & Kletti, 1976; Pfister, 1930), unique personality traits (Gow, Lane, & Chant, 2003; Lynn & Rhue, 1988), altered blood gases (Whinnery, 1997), neurotoxic metabolic reactions (Carr, 1982; Jansen, 1997), or alterations in brain activity (Morse, Venecia, & Milstein, 1989; Saavedra-Aguilar & Gómez-Jeria, 1989).

Joseph (1996, 1999b, 2001), for example, has provided considerable data of near-death like dissociative experiences following abnormal activation or electrode stimulation of temporal lobe structures which include the amygdala and the hippocampus. According to Joseph, these latter structures play important roles in emotional and cognitive memory, and may explain the "life review"; the phenomenon where victims see their "life flash before their eyes." He also argues that since the hippocampus and overlying temporal lobe structures are involved in vision, the cognitive mapping of the visual environment, and the processing of the upper visual field, they may contribute to the dissociative experience of floating above the body during near death and when these structures are activated by electrode. Although he raises the possibility these are hallucinations, he also reports on those who died and who were able to correctly described their surroundings including doctors and nurses who attended to them while they were dead.

Although the theories and evidence provided by Joseph (1996, 1999b, 2001a) and others is intriguing, they only provide empirical support and address only selected aspects of the phenomena (Greyson et al., 2009). The most important objection to the adequacy of all such reductionistic hypotheses is that mental clarity, vivid sensory imagery, a clear memory of the experience, and a conviction that the experience seemed more real than ordinary consciousness are the norm for NDEs. They occur even in conditions of drastically altered cerebral physiology under which the production theory would deem consciousness impossible.

2. Physiology of General Anesthesia and Cardiac Arrest

NDEs are typically triggered when patients are clinically near death, such as following catastrophic physical traumas, or during cardiac arrest or some other, usually sudden, loss of vital functions. In one study of 1595 consecutive admissions to a cardiac care unit, NDEs were reported 10 times more often by patients who had survived definite cardiac arrest than by patients with other serious cardiac incidents (Greyson, 2003). The incompatibility of NDEs with the mind-brain production theory is particularly evident in connection with experiences that occur under two conditions, namely, general anesthesia and cardiac arrest.

In our collection at the University of Virginia, 22% of our NDE cases occurred under anesthesia, and they include the same features as other NDEs, such as out-of-body experiences that involved watching medical personnel working on the body, an unusually bright or vivid light, meeting deceased persons, and thoughts, memories, and sensations that were clearer than usual.

John et al. (2001) identified reliable electroencephalographic (EEG) correlates of loss and recovery of consciousness during general anesthesia. Their results confirmed the standard thinking about anesthesia and EEG, namely, that unconsciousness is associated with a profound reduction in brain activity under anesthesia. Additional results supportive of this conclusion derive from other recent functional imaging studies that have looked at blood flow, glucose metabolism, or other indicators of cerebral activity under general anesthesia (Alkire, 1998; Alkire et al., 2000; Shulman et al., 2003; White & Alkire, 2003). In these studies, brain areas essential to the global workspace are consistently greatly reduced in activity individually and may be decoupled functionally, thereby providing considerable evidence against the possibility that the anesthetized brain could produce clear thinking, perception, or memory.

The situation is even more dramatic with regard to NDEs occurring during cardiac arrest, many of which in fact occur also in conjunction with major surgical procedures involving general anesthesia. In four published studies alone, more than 100 cases of NDEs occurring during cardiac arrest were reported (Greyson, 2003; Parnia et al., 2001; Sabom, 1982; van Lommel et al., 2001). Like

NDEs that occur with general anesthesia, those that occur in connection with cardiac arrest include the typical features associated with NDEs, including enhanced sensation and mentation, out-of-body experiences, and visions of deceased acquaintances.

However, in cardiac arrest, cerebral functioning shuts down within a few seconds. With circulatory arrest, blood flow and oxygen uptake in the brain plunge to near-zero levels. EEG signs of cerebral ischemia are detectable within 6-10 seconds, and progress to isoelectricity (flat-line EEGs) within 10-20 seconds (DeVries et al., 1998; Vriens et al., 1996). In sum, full arrest leads rapidly to the three major clinical signs of death: absence of cardiac output, absence of respiration, and absence of brainstem reflexes (Parnia & Fenwick, 2002; van Lommel et al., 2001).

Defenders of the mind-brain production theory might object that even in the presence of a flat-lined EEG there still could be undetected brain activity going on; current scalp-EEG technology detects only activity common to large populations of neurons, mainly in the cerebral cortex. However, the issue is not whether there is brain activity of any kind whatsoever, but whether there is brain activity of the specific form agreed upon by contemporary neuroscientists as the necessary condition of conscious experience. Activity of this form is eminently detectable by current EEG technology, and it is abolished either by general anesthesia or by cardiac arrest.

In cardiac arrest, even neuronal action-potentials, the ultimate physical basis for coordination of neural activity between widely separated brain regions, are rapidly abolished (Kelly et al., 2007). Moreover, cells in the hippocampus, the region thought to be essential for memory formation, are especially vulnerable to the effects of anoxia (Vriens et al., 1996). In short, it is not credible to suppose that NDEs occurring under conditions of general anesthesia, let alone cardiac arrest, can be accounted for in terms of some hypothetical residual capacity of the brain to process and store complex information under those conditions.

A second defense of the mind-brain production theory for NDEs is to suggest that these experiences do not occur during the actual episodes of brain insult, but before or just after the insult, when the brain is more or less functional (Augustine, 2007; Rodabaugh, 1985).

However, unconsciousness produced by cardiac arrest characteristically leaves patients amnesic and confused for events immediately preceding and following these episodes (Aminoff et al., 1988; Parnia & Fenwick, 2002; van Lommel et al., 2001). Furthermore, a substantial number of NDEs contain apparent time "anchors" in the form of verifiable reports of events occurring during the period of insult itself. For example, a cardiac-arrest victim described by van Lommel et al. (2001) had been discovered lying in a meadow 30 minutes or more prior to his arrival at the emergency room, comatose and cyanotic, and yet days later, having recovered, he was able to describe accurately various circumstances occurring in conjunction with the ensuing resuscitation procedures in the hospital.

3. Relevance of Near-Death Experiences to Cosmology

Until the early 20th century, it was plausible to base a scientific cosmology on materialist reductionism, the idea that any complex phenomenon could be understood by reducing it to its individual components, and eventually down to elementary material particles. This worldview implied that all complex psychological phenomena could ultimately be understood in material terms. This materialist cosmology influenced psychology as it did other sciences, even though this reductionism required mainstream academic psychologists to ignore consciousness as the subject was viewed as "not scientific." Materialist psychology was epitomized by Watson, who asserted: "Psychology, as the behaviorist views it, is a purely objective, experimental branch of natural science which needs consciousness as little as do the sciences of chemistry and physics" (1914, p. 27). However, while Watson was aligning behaviorist psychology with classical mechanics, physicists were already moving beyond that model with a quantum physics that could not be formulated without reference to consciousness.

Classical dynamics adequately described the motion of macroscopic objects moving at everyday speeds; it was only the investigation of extraordinary circumstances, involving objects moving with velocities approaching the speed of light or the behavior of microscopic wave-particles, that revealed the limits of the classical model and the need for additional explanatory paradigms. So too with the question of the mind-brain relationship: it is the exploration of extraordinary circumstances of mental function that reveal the limitations of the production theory and the need for a more comprehensive theory of mind-brain interaction.

4. Near-Death Experiences and the Mind-Brain Production Theory

Although many 20th-century physicists, psychologists and neuroscientists accepted the reductionistic model that brain produces mind, or indeed is the mind (Churchland, 1986; Crick, 1994; Damasio, 1999; Pinker, 1997), several features of NDEs call into question whether materialist reductionism will ever provide a full explanation of mind, including and most notably enhanced mental processes, accurate out-of-body perception, and visions of deceased relatives and close friends (Greyson, 2010a).

Perhaps the most important of these features, because it is so commonly reported in NDEs, is the occurrence of enhanced mental activity at times when, according to the mind-brain production model, such activity should be diminishing, if not impossible. Individuals reporting NDEs often describe their mental processes during the NDE as remarkably clear and lucid and their sensory experiences as unusually vivid, surpassing those of their normal waking state.

A recent analysis of several hundred NDE cases showed that 80% of experiencers described their thinking during the NDE as "clearer than usual" or "as clear as usual" (Kelly et al., 2007, p. 386). An analysis of the medical records of people reporting NDEs showed that, in fact, people reported enhanced mental functioning significantly more often when they were actually physiologically close to death than when they were not (Owens et al., 1990).

An example of enhanced mental functioning during an NDE is a rapid revival of memories that sometimes extends over the person's entire life. An analysis of several hundred NDEs showed that in 24% of them there was a revival of memories during the NDE (Kelly et al., 2007, p. 386). Moreover, in contrast to the isolated and often just single brief memories evoked during cortical stimulation, memories revived during an NDE are frequently described as an almost instantaneous "panoramic" review of the person's entire life (Noyes & Kletti, 1977; Stevenson & Cook, 1995).

Another important feature of NDEs that the mind-brain production theory cannot adequately account for is the experience of being out of the body and perceiving events that one could not ordinarily have perceived. A recent analysis of several hundred cases showed that 48% of near-death experiencers reported seeing their physical bodies from a different visual perspective. Many of them also reported witnessing events going on in the vicinity of their body, such as the attempts of medical personnel to resuscitate them (Kelly et al., 2007). The mind-brain production theory could attribute the belief that one has witnessed events going on around one's body to a retrospective imaginative reconstruction based on a persisting ability to hear, even when unconscious, or to the memory of objects or events that one might have glimpsed just before losing consciousness or while regaining consciousness, or to expectations about what was likely to have occurred (Saavedra-Aguilar & Gómez-Jeria, 1989; Woerlee, 2004).

Such explanations are inadequate, however, for several reasons. First, memory of events occurring just before or after loss of consciousness is usually confused or completely absent (Aminoff et al., 1988; Parnia & Fenwick, 2002; van Lommel et al., 2001). Second, anecdotal reports that adequately anesthetized patients retain a significant capacity to be aware of or respond to their environment in more than rudimentary ways—let alone to hear and understand—have not been substantiated by controlled studies (Ghoneim & Block, 1992,

1997).

The phenomenology of awakenings under anesthesia is altogether different from that of NDEs, and often extremely unpleasant, frightening, and even painful, typically brief and fragmentary, and primarily auditory or tactile, but not visual (Osterman et al., 2001; Spitellie et al., 2002). There is no convincing evidence that memories of complex sensory experiences occurring during general anesthesia could have been acquired by the impaired brain itself during the period of unconsciousness.

Furthermore, any such explanatory claims are even less credible when, as commonly happens, the specific sensory channels involved in the reported experience have been blocked as part of the surgical routine—for example, when visual experiences are reported by patients whose eyes were taped shut during the relevant period of time.

Sabom (1982) carried out a study specifically to examine whether claims of out-of-body perceptions could be attributed to retrospective reconstruction. He interviewed patients who reported NDEs in which they seemed to be watching what was going on around their body, most of them cardiac patients who were undergoing cardiopulmonary resuscitation (CPR) at the time of their NDE. He also interviewed "seasoned cardiac patients" who had not had an NDE during their cardiac-related crises, and asked them to describe a cardiac resuscitation procedure as if they were watching from a third-person perspective. He found that 80% of the comparison patients made at least one major error in their descriptions, whereas none of the NDE patients made any (pp. 87–115). Sartori (2008) recently replicated Sabom's findings in a five-year study of hospitalized intensive care patients, in which patients who reported leaving their bodies during cardiac arrests described their resuscitations accurately, whereas every cardiac arrest survivor who had not reported leaving the body described incorrect equipment and procedures when asked to describe their resuscitation.

An even more difficult challenge to the mind-brain production theory comes from NDEs in which experiencers report that, while out of the body, they became aware of events occurring at a distance or that in some other way would have been beyond the reach of their ordinary senses even if they had been fully and normally conscious. Clark (1984) and Owens (1995) each published a case of this type, and we have reported on 15 cases, including seven cases previously published by others and eight from our own collection (Cook et al., 1998; Kelly et al., 2000), including accurate perceptions of highly unexpected or unlikely details. Additionally, Ring and Cooper (1997, 1999) reported 31 cases of blind individuals, nearly half of them blind from birth, who experienced during their NDEs quasi-visual and sometimes veridical perceptions of objects and events. One criticism of these reports of perception of events at a distance from the body is that they often depend on the experiencer's testimony alone. However, many cases have in fact been corroborated by independent witnesses (Clark, 1984; Hart, 1954; Ring & Lawrence, 1993; van Lommel et al., 2001; Cook et al., 1998).

In a recent review of 93 published reports of potentially verifiable out-of-body perceptions during NDEs, Holden (2009) found that 43% had been corroborated to the investigator by an independent informant, an additional 43% had been reported by the experiencer to have been corroborated by an independent informant who was no longer available to be interviewed by the investigator, and only 14% relied solely on the experiencer's report. Of these out-of-body perceptions, 92% were completely accurate, 6% contained some error, and only 1% was completely erroneous. Even among those cases corroborated to the investigator by an independent informant, 88% were completely accurate, 10% contained some error, and 3% were completely erroneous. The cumulative weight of these cases is inconsistent with the conception that purported out-of-body perceptions are nothing more than hallucinations.

Many people who approach death and recover report that, during the time they seemed to be dying, they met deceased relatives and friends (Cook et al., 1998; Kelly et al, 2000; Osis & Haraldsson, 1977). In a recent analysis of several hundred NDEs, 42% of experiencers reported meeting one or more recognizable deceased acquaintances during the NDE (Kelly, 2001). In the mind-brain production theory, such experiences are widely viewed as being hallucinations, caused by drugs or other physiological conditions or by the person's expectations or wishes to be reunited with deceased loved ones at the time of death. However, a closer examination of these experiences indicates that such explanations are not adequate.

People close to death are more likely to perceive deceased persons than are people who are not close to death: the latter, when they have waking hallucinations, are more likely to report seeing living persons (Osis & Haraldsson, 1977). For example, Whinnery (1997) reported that healthy fighter pilots exposed to acceleration-induced anoxia to the point of loss of consciousness (G-LOC) typically report hallucinations of living friends and family. One 20-year-old pilot reported his G-LOC experience: "I was home . . . saw my mom and my brother. . . . I got to go home [by dreaming] without taking [military] leave!" (Whinnery, 1997, p. 245; brackets in original). Near-death experiencers whose medical records show that they really were close to death also were more likely to perceive deceased persons than experiencers who were ill but not close to death, even though many of the latter thought they were dying (Kelly, 2001).

People more often perceive deceased individuals with whom they were emotionally close, but in one-third of the cases the deceased person was either someone with whom the experiencer had a distant or even poor relationship or someone whom the experiencer had never met, such as a relative who died long before the experiencer's birth (Kelly, 2001). Van Lommel (2004,

p. 122) reported the case of a man who had an NDE during cardiac arrest in which he saw his deceased grandmother and an unknown man. Later shown a picture of his biological father, whom he had never known and who had died years ago, he immediately recognized him as the man he had seen in his NDE.

There is one particular kind of vision of the deceased that calls into question even more directly their dismissal as subjective hallucinations: cases in which the dying person apparently sees, and often expresses surprise at seeing, a person whom he or she thought was living, who had in fact recently died. Reports of such cases were published in the 19th century (Cobbe, 1882; Gurney & Myers, 1889; Johnson, 1899; Sidgwick, 1885) and have continued to be reported in recent years (Greyson, 2010b; Osis & Haraldsson, 1977; Sartori, 2008; van Lommel, 2004). In one recent case, a 9-year-old boy with meningitis, upon awakening from a 36-hour coma, told his parents he had been with his deceased grandfather, aunt, and uncle, and also with his 19-year-old sister who was, as far as his family knew, alive and well at college 500 miles away. Later that day, his parents received news from the college that their daughter had died in an automobile accident early that morning (Greyson, 2010b). Because in these cases the experiencers had no knowledge of the death of recently deceased person, the vision cannot plausibly be attributed to the experiencer's expectations.

5. Conclusion

In sum, the challenge of NDEs to the mind-brain production theory lies in asking how complex consciousness, including mentation, sensory perception, and memory, can occur under conditions in which current neurophysiological models deem it impossible. This conflict between a materialist model of brain producing mind and the occurrence of NDEs under conditions of general anesthesia and/or cardiac arrest is profound and inescapable. Only when we expand models of mind to accommodate extraordinary experiences such as NDEs will we progress in our understanding of consciousness and its relation to brain. The predominant contemporary models of consciousness are based on principles of classical physics that were shown to be incomplete in the early decades of the 20th century. However, the development of post-classical physics over the past century offers empirical support for a new scientific conceptualization of the interface between mind and brain compatible with a cosmology in which consciousness is a fundamental element (Schwartz et al., 2005; Stapp, 2007).

References

Alkire, M. T. (1998). Quantitative EEG correlations with brain glucose metabolic rate during anesthesia in volunteers. Anesthesiol, 89, 323-333.

Alkire, M. T., Haier, R. J., & Fallon, J. H. (2000). Toward a unified theory of narcosis. Consciousness & Cognition, 9, 370-386.

Aminoff, M. J., Scheinman, M. M., Griffin, J. C., & Herre, J. M. (1988). Electrocerebral accompaniments of syncope associated with malignant ventricular arrhythmias. Ann Int Med, 108, 791-796.

Augustine, K. (2007). Does paranormal perception occur in near-death experiences? J Near-Death Stud, 25, 203-236.

Brierre de Boismont, A. (1862). Des hallucinations. Paris: Germer Baillière.

Broad, C. D. (1953). Religion, philosophy, and psychical research. New York: Harcourt Brace.

Burdach, K. F. (1826). Vom Baue und Leben des Gehirns, Vol. 3. Leipzig: Dyk'sche Buchhandlung.

Burt, C. (1968). Psychology and psychical research. London: Society for Psychical Research.

Carr, D. (1982). Pathophysiology of stress-induced limbic lobe dysfunction: A hypothesis for NDEs. J Near-Death Stud, 2, 75-89.

Churchland, P. S. (1986). Neurophilosophy: Toward a unified science of the mind/brain. Cambridge, MA: MIT Press.

Clark, K. (1984). Clinical interventions with near-death experiencers. In B. Greyson & C. P. Flynn (Eds.), The near-death experience (pp. 242-255). Springfield, IL: Charles C Thomas.

Cobbe, F. P. (1882). The peak in Darien. Boston: George H. Ellis.

Cook, E. W., Greyson, B., & Stevenson, I. (1998). Do any near-death experiences provide evidence for the survival of human personality after death? J Sci Exploration, 12, 377-406.

Crick, F. (1994). The astonishing hypothesis: The scientific search for the soul. London: Simon & Schuster.

Damasio, A. (1999). How the brain creates the mind. Sci Amer, 281, 112 – 117.

Damasio, A., & Meyer, K. (2009). Consciousness: An overview of the phenomenon and of its possible neural basis. In S. Laurys & G. Tononi (Eds.), The neurology of consciousness: Cognitive neuroscience and neuropathology (pp. 3-14). Amsterdam: Elsevier.

DeVries, J. W., Bakker, P. F. A., Visser, G. H., Diephuis, J. C., & van Huffelin, A. C. (1998). Changes in cerebral oxygen uptake and cerebral electrical activity during defibrillation threshold testing. Anesthesiol Analgesia, 87, 16-20.

Ghoneim, M. M., & Block, R. I. (1992). Learning and consciousness during general anesthesia. Anesthesiol, 76, 279-305.

Ghoneim, M. M., & Block, R. I. (1997). Learning and memory during general anesthesia. 18 Anesthesiol, 87, 378-410.

Gow, K., Lane, A., & Chant, D. (2003). Personality characteristics, beliefs, and the near-death experience. Austral J Clin Exper Hypn, 31, 128-152.

Greyson, B. (2003). Incidence and correlates of near-death experiences in a cardiac care unit. Gen Hosp Psychiatry, 25, 269-276.

Greyson, B. (2010a). Implications of near-death experiences for a postmaterialist psychology. Psychol Relig Spirituality, 2, 37-45.

Greyson, B. (2010b). Seeing deceased persons not known to have died: "Peak in Darien" experiences. Anthropology & Humanism, 35, 159-171.

Greyson, B., Kelly, E. W., & Kelly, E. F. (2009). Explanatory models for near-death experiences. In J. M. Holden, B Greyson, & D. James (Eds.), The handbook of near-death experiences (pp. 213-234). Santa Barbara, CA: Praeger/ABC-CLIO.

Gurney, E., & Myers, F. W. H. (1889). On apparitions occurring soon after death. Proc Soc Psychical Res, 5, 403-485.

Hart, H. (1954). ESP projection. J Amer Soc Psychical Res, 48, 121-146.

Holden, J. M. (2009). Veridical perception in near-death experiences. In J. M. Holden, B Greyson, & D. James (Eds.), The handbook of near-death experiences (pp. 185-211). Santa Barbara, CA: Praeger/ABC-CLIO.

Holden, J. M., Greyson, B., & James, D. (Eds.). (2009). The handbook of near-death experiences. Santa Barbara, CA: Praeger/ABC-CLIO.

Huxley, A. (1954). The doors of perception. London: Chatto & Windus.

James, W. (1898). Human immortality: Two supposed objections to the doctrine. Boston: Houghton Mifflin.

Jansen, K. L. R. (1997). The ketamine model of the near-death experience: A central role for the N-methyl-D-aspartate receptor. J Near-Death Stud, 16, 5-26.

Jeeves, M., & Brown, W. S. (2009). Neuroscience, psychology, and religion: Illusions, delusions, and realities about human nature. West Conshohocken, PA: Templeton Foundation Press.

John, E. R., Prichep, L. S., Kox, W., Valdés-Sosa, P., Bosch-Bayard, J., Aubert, E., Tom, M., diMichele, F., & Gugino, L. D. (2001). Invariant reversible QEEG effects of anesthetics. Consciousness & Cognition, 10, 165-183.

Johnson, A. (1899). Coincidences. Proc Soc Psychical Res, 14, 158-330.

Joseph, R. (1986). Confabulation and delusional denial: Frontal lobe and lateralized influences. Journal of Clinical Psychology, 42, 845-860.

Joseph, R. (1996). Neuropsychiatry, Neuropsychology, Clinical Neuroscience, 2nd Edition. Williams & Wilkins, Baltimore.

Joseph, R. (1999a). Frontal lobe psychopathology: Mania, depression, aphasia, confabulation, catatonia, perseveration, obsessive compulsions, schizophrenia. Psychiatry, 62, 138-172.

Joseph, R. (1999b). The neurology of traumatic "dissociative" amnesia. Commentary and literature review. Child Abuse & Neglect. 23, 715-727.

Joseph, R. (2001). The Limbic System and the Soul: Evolution and the Neuroanatomy of Religious Experience. Zygon, the Journal of Religion & Science, 36, 105-136.

Kelly, E. F. (2007). A view from the mainstream: Contemporary cognitive neuroscience and the consciousness debates. In E. F. Kelly, E. W. Kelly, A. Crabtree, A. Gauld, M. Grosso, & B. Greyson, Irreducible mind (pp. 1-46). Lanham, MD: Rowman & Littlefield.

Kelly, E. W. (2001). Near-death experiences with reports of meeting deceased people. Death Stud, 25, 229-249.

Kelly, E. W., Greyson, B., & Kelly, E. F. (2007). Unusual experiences near death and related phenomena. In E. F. Kelly, E. W. Kelly, A. Crabtree, A. Gauld, M. Grosso, & B. Greyson, Irreducible mind (pp. 367-421). Lanham, MD: Rowman & Littlefield.

Kelly, E. W., Greyson, B., & Stevenson, I. (2000). Can experiences near death furnish evidence of life after death? Omega, 40, 513-519.

Lynn, S., & Rhue, J. (1988). Fantasy proneness: Hypnosis, developmental antecedents, and psychopathology. Amer Psychologist, 43, 35-44.

Marshall, A. (1815). The morbid anatomy of the brain in mania and hydrophobia. London: Longman.

Morse, M. L. Venecia, D., & Milstein, J. (1989). Near-death experiences: A neurophysiological explanatory model. J Near-Death Stud, 8, 45-53.

Nahm, N., & Greyson, B. (2009). Terminal lucidity in patients with chronic schizophrenia and dementia: A survey of the literature. J Nerv Ment Dis, 197, 942-944.

Noyes, R., & Kletti, R. (1976). Depersonalization in the face of life-threatening danger: An interpretation. Omega, 7, 103-114.

Noyes, R., & Kletti, R. (1977). Panoramic memory: A response to the threat of death. Omega, 8, 181-194.

Osis, K., & Haraldsson, E. (1977). At the hour of death (3rd ed.). New York: Avon.

Osterman, J. E., Hopper, J., Heran, W. J., Keane, T. M., & van der Kolk, B. A. (2001). Awareness under anesthesia and the development of posttraumatic stress disorder. Gen Hosp Psychiatry, 23, 198-204.

Owens, J. E. (1995). Paranormal reports from a study of near-death experience and a case of an unusual near-death vision. In L. Coly & J. D. S. McMahon (Eds.), Parapsychology and thanatology (pp. 149-167). New York: Parapsychology Foundation.

Owens, J. E., Cook, E. W., & Stevenson, I. (1990). Features of "near-death experience" in relation to whether or not patients were near death. Lancet, 336, 1175-1177.

Parnia, S., & Fenwick, P. (2002). Near death experiences in cardiac arrest. Resuscitation, 52, 5-11.

Parnia, S., Waller, D. G., Yeates, R., & Fenwick, P. (2001). A qualitative and quantitative study of the incidence, features and aetiology of near death experiences in cardiac arrest survivors. Resuscitation, 48, 149-156.

Pfister, O. (1930). Shockdenken und Shockphantasien bei höchster Todesgefahr. Internat Zeitschr Psychoanalyse, 16, 430-455.

Pinker, S. (1997). How the mind works. New York: Norton.

Ring, K., & Cooper, S. (1997). Near-death and out-of-body experiences in the blind: A study of apparent eyeless vision. J Near-Death Stud, 16, 101-147.

Ring, K., & Cooper, S. (1999). Mindsight: Near-death and out-of-body experiences in the blind. Palo Alto, CA: William James Center/Institute of Transpersonal Psychology.

Ring, K., & Lawrence, M. (1993). Further evidence for veridical perception during near-death experiences. J Near-Death Stud, 11, 223-229.

Rodabaugh, T. (1985). Near-death experiences: An examination of the supporting data and alternative explanations. Death Stud, 9, 95-113.

Saavedra-Aguilar, J. C., & Gómez-Jeria, J. S. (1989). A neurobiological model for near-death experiences. J Near-Death Stud, 7, 205-222.

Sabom, M. B. (1982). Recollections of death. New York: Harper & Row.

Sartori, P. (2008). The near-death experiences of hospitalized intensive care patients: A five year clinical study. Lewiston, UK: Edward Mellen Press.

Schiller, F. C. S. (1894). Riddles of the sphinx: A

study in the philosophy of evolution. London: Swan Sonnenschein.

Schwartz, J. M., Stapp, H. P., & Beauregard, M. (2005). Quantum physics in neuroscience and psychology. Phil Trans Roy Soc B, 360, 1309-1327.

Searle, J. R. (2000). Consciousness. Ann Rev Neurosci, 23, 557-578.

Shulman, R. G., Hyder, F., & Rothman, D. L. (2003). Cerebral metabolism and consciousness. Comptes Rendus Biologies, 326, 2532-273.

Sidgwick, E. M. (1885). Notes on the evidence, collected by the Society, for phantasms of the dead. Proc Soc Psychical Res, 3, 69-150.

Spitellie, P. H., Holmes, M. A., & Domino, K. B. (2002). Awareness under anesthesia. Anesthesiol Clin N Amer, 20, 555-570.

Stapp, H. P. (2007). Mindful universe: Quantum mechanics and the participating observer. Berlin: Springer-Verlag.

Stevenson, I., & Cook, E. W. (1995). Involuntary memories during severe physical illness or injury. J Nerv Ment Dis, 183, 452-458.

Tononi, G., & Laurys, S. (2009). The neurology of consciousness: An overview. In S. Laurys & G. Tononi (Eds.), The neurology of consciousness: Cognitive neuroscience and neuropathology (pp. 375-412). Amsterdam: Elsevier.

Turetskaia. B. E., & Romanenko, A. A. (1975). Agonal remission on the terminal stages of schizophrenia. Korsakov J Neuropathol Psychiat, 75, 559-562.

Van Lommel, P. (2004). About the continuity of our consciousness. Adv Exper Med Biol, 550, 115-132.

Van Lommel, P., van Wees, R., Meyers, V., & Elfferich, I. (2001). Near-death experience in survivors of cardiac arrest. Lancet, 358, 2039-2045.

Vriens, E. M., Bakker, P. F. A., DeVries, J. W., Wieneke, G. H., & van Huffelin, A. C. (1996). The impact of repeated short episodes of circulatory arrest on cerebral function. EEG Clin Neurophysiol, 98, 236-242.

Watson, J. B. (1914). Behavior. New York: Holt.

Whinnery, J. E. (1997). Psychophysiologic correlates of unconsciousness and near-death experiences. J Near-Death Stud, 15, 473-479.

White, N.S., & Alkire, M. T. (2003). Impaired thalamocortical connectivity in humans during general-anesthetic-induced unconsciousness. NeuroImage, 19, 402-411.

Woerlee, G. M. (2004). Cardiac arrest and near-death experiences. J Near-Death Stud, 22, 235-249

39. Near Death Experiences and the 5th Dimensional Spatio-Temporal Perspective
Jean-Pierre Jourdan, M.D.

IANDS-France President - Director of medical research, International Association for Near-Death Studies, 28 Av. Flourens Aillaud, 04700 Oraison, France,

Abstract

The cognitive and perceptive characteristics of 70 cases of Near Death Experiences have been studied. The detailed analysis of the unusual modes and characteristics of spatial and temporal perception during these experiences reveals a "hidden" logic for which I propose a model where the point of perception would be in an extra dimension. The appropriateness of such a model is analyzed and shown to be consistent with the NDE accounts in the study. In contrast, those interpretations of such perceptions as being purely hallucinatory are undermined. Whatever its meaning, the underlying logic shown in this study suggests that NDEs seem to follow precise rules. Since these experiences can be viewed as an unusual but consistent behavior of consciousness, they deserve further pluridisciplinary study.

KEY WORDS: NDEs; consciousness; perception; information; modelling; large extra dimensions; 5D; space-time.

1. Introduction

The perceptions we are going to study here come from persons who, having survived a cardiac arrest or some other life-threatening circumstances, report a Near-Death-Experience (NDE). Although they were totally unconscious for any witness around them, these patients frequently report clear memories and can describe accurately their vicinity, including details that can be checked afterwards and scenes that allow clarifying the "moment" of the experience, all that seemingly perceived from a point external to their body.

The number of NDE cases with objective perceptions is far from anecdotal: of the 70 cases included in this study, 48 (68.5%) involve what one calls an "Out of Body" (OBE) stage (at the very least, all declare having "seen" their bodies and the surrounding activity from a perspective outside the body). Of these 48 cases, 23 (47.9%) report precise perceptions corresponding to verified details in the environment and/or to scenes which took place just as described, thus giving a precise idea of when the actual experience occurred.

Many cases of this type (Sabom 1993-1998, Ring,1980, Ring & Valarino1997, Van Lommel 2001, Jourdan 2006) have been reported, and two recent cases have been the subject of thorough investigations (Sartori and Al., 2008). These apparently nonphysical veridical NDE perceptions (AVPs) are extensively discussed in (Holden and Al. 2009, Holden 2009).

An important point needs to be specified here: despite a superficial similarity, these experiences have to be properly differentiated from well known and explained phenomena like sleep paralysis or lucid dreams (Jourdan 2006).

A classical NDE comprises several stages (OBE stage, tunnel, perception of a brilliant light, life review, encountering a " being of light" and/or deceased relatives, feeling of a limit not to be trespassed…), the majority of which, although very similar within different accounts, are for the time being beyond the reach of objective study. Only, the OBE stage shows some apparently objective elements, and is therefore a good candidate for a scientific investigation.

But sometimes, as we are about to see, research in a specific field may have surprising results.

2. A Strange Account

In the late 90s, while he was telling me his experience, he stressed how his consciousness had "seen" his surroundings while it was happening:

> "I was able to see the sofa and my body simultaneously from all directions. I saw my body through the sofa, I could see the top of my head and in the same time I saw my left and right sides and the sofa from below and from above, and all the room like that, I was everywhere at the same time!". Dashing off several sketches, front and side-on views as well as views from

above, below, etc., he repeated "I saw all that".

Trying to schematize a perception "from everywhere", I drew a circle around a person lying on a sofa and mused aloud: "in fact, it's as if you had "seen" simultaneously from all points of a sphere surrounding the scene". At that moment, I realized that the drawing allowed me to see the couch as well as the body from all angles simultaneously. Moreover, the back of the sofa did not need to be transparent to allow seeing the left side of the body lying on it … Essentially, I was looking at a two-dimensional universe (the drawing) from the third dimension, which allowed me to perceive simultaneously everything in it.

Nothing magic about all that, but was it that simple? Even if we do not go further than Euclidian geometry, considering only spatial dimensions for the time being, the scene described by the experiencer was three-dimensional. Everything happened as though during his NDE he had perceived from a point external to our universe, thus logically… from a fourth dimension! Of course, this interpretation seemed somewhat speculative (!) but nevertheless seemed compatible with some old accounts I had not paid attention to, thinking they were just curiosities. Going back to these latter, I found some similar evidences which persuaded me to investigate in detail the perceptual particularities in NDEs.

Without making any assumptions as to the reality of these paradoxical perceptions, be it as it may it was interesting as a first step to define their modalities by asking the patients to specify how they perceived things during the experience.

One could have expected to find various eccentric features, they could have described only more or less vague outlines, black and white, fanciful, soft or garish colours, distorted shapes, imaginary details or scenes, etc. But none of this was to be found in their answers. All that they described was perfectly banal considering the circumstances. All, except for the way they perceived the environment! Apart from those who were unable to give precisions, the answers were quite strange, while remaining confined in several precise categories.

3. Overview: Out of Body Multi-Dimensional Consciousness

An overview of the testimonies and extracts cited in this paper (recorded at the IANDS-France association and are taken from Jourdan 2000, 2001 and 2006).

> I was surprised that I could see at an angle of 360°, I could see in front and behind, I could see underneath, from far, I could see up close and also transparently. I remember seeing a stick of lipstick in one of the nurses' pockets. If I wanted to see inside the lamp which illuminated the room, I would manage to, and all of this instantly, as soon as I wanted to. I could

say how people were dressed, I could see the sandstone wall, and also the stone slabs of the floor. I was able to verify their presence in a photograph later on since I thought it strange and anachronistic to have such slabs in an operating room. It was surprising and I could see, all at once, a green plaque with white letters saying "Manufacture de Saint Etienne". The plaque was under the edge of the operating table, covered up by the sheet I was lying on. I could see with multiple axes of vision, from many places at once. This is the reason why I saw this plaque under the operating table, from a completely different angle, since I was up there by the ceiling and I still managed to see this plaque located under the table, itself covered by a sheet. When I wanted to check this, we realized the plaque really was there and read "Manufacture d'armes de Saint Etienne". (J.M.).

Table 1. The most frequently reported perceptual features are summarized in Table.

Perceptual features	Out of 48	
Global perception	17	35,4%
360°/spherical perception	15	31,3%
Perception "from everywhere"	12	25%
Perception by transparency	18	37,5%
Zoom/ instantaneous displacement	15	31,3%

4. Establishing a Model

Rather than a theory, what I will now propose is a model that allows to understand every seemingly irrational feature we have just outlined. We will now try to understand how the perception of a n-dimensional universe from a vantage point situated in an n+1 universe differs from its usual perception from "inside".

If we are able to see everything in the plan below, looking at it as a whole or focusing our attention on each of its details, it is because our point of perception is situated away from it. This "global perception" obviously disappears when the vantage point belongs to the observed universe, as is the case with an imaginary being who lives in the diagram: Tweedee (he only possesses Two Dimensions and is visible in his bedroom). In his universe, he is subject to the same limitations as us. He can only see what is within his visual field, he cannot cross the walls neither see inside or through them. In short, all that is quite normal.

Yet you and I see his house in its entirety at a glance, we do not need to move around to see the damaged car in the garage or his step-mother in front of the fridge. We see all the stools simultaneously, as we do the paintings -even the hidden one-, without the need to turn around or see through the walls. Which is also normal... Depending on the spot from where we perceive, these two totally different viewpoints are both equally logical. Comparing them will allow us to understand the apparently irrational perceptions described in the NDEs.

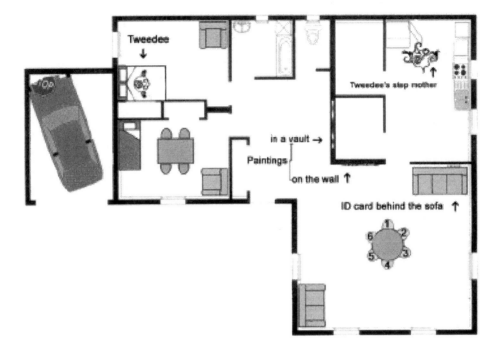

Figure 1. Tweedee's house.

For that, let us suppose that Tweedee suffers a cardiac arrest and "lives" an NDE. We will now assume that for the duration of his experience his perception will be similar to ours. But what is perfectly normal and common for us is not for him. When he returns, how will he manage to describe the particular features of this perception, which is all but ordinary for him?

Here is an example that describes perceiving a scene or the environment in a way that is unusual, but nevertheless frequent during a NDE:

> I was at ceiling level in the emergency room, above my head. I could see myself lying down on the bed. A (male) doctor was working the resuscitation machines on my left. There was ringing everywhere. It was quite surreal. A nurse was close to me, adjusting perfusions and other tubes. Another nurse was running back and forth between the doctor and my bed, leaving the room and then coming back, running, all the time The nurse next to me was talking to me, "Stay with us, this isn't the time to leave". I saw her slapping me. I was fine. I no longer felt pain. I said, "Why do you want me to come back, when it's not hurting for once". Then I added, "Oh all right,"

but I really wasn't happy about it. What could be verified was the number of people in the room and what they were doing and saying.

Jourdan: "Did the notions of up and down, left and right make any sense?"

Yes. For the resuscitation scene I was at the top of the room at ceiling level. It is a notion of the orientation of things, in relation to each other. Not a real geographical, physical, position. It's hard to explain.

Jourdan: "Is there a difference (or a contradiction) between what you perceive (things stay in their usual positions with respect to one another) and the fact that the scene is perceived in a global way, which means that the position of objects in relation to where you are is impossible to define?"

No, for me there is no contradiction. People and things are properly oriented in respect to one another in three dimensions. However, we see the scene globally, by this I mean in its entirety. It's difficult to explain. For example, if you have a person on the ground placed in front of an object, you do not see this object. To see it you need to move. In this instance, it's different. The person is really in front of the object. The way things are oriented stays the same. Despite this, you can still see the object. You don't need to move. You see the entire scene. On the ground, you need to move, to change your angle of vision to see everything. There, you see everything without having to move. But people and things are placed normally in respect to one another.

Jourdan: "Did you have the impression of having a larger angle of vision, to see in front and behind you simultaneously?"

Yes. Total, instant vision.

Jourdan: "Did you have the impression you were seeing an object or scene from multiple locations at once?"

Yes, absolutely. This is what I meant before when speaking of an orientation but not of a real physical location. Once again, a total, instantaneous vision. (C.P.)

Tweedee is accustomed to live within a 2D plane but is now watching it from our vantage point. Although this could sound completely surrealistic, the patient describes exactly what this latter could say: not a real geographical location. Indeed, his point of perception cannot be located anywhere inside what he perceives. He "sees the scene globally", which is not easy to explain. Tweedee's step-mother is really in front of the fridge but she cannot hide it from him. Indeed, he has no need to move or change his angle of vision in order to "see" anything in his house.

This perception of his universe is as natural for us as it is strange for him! Let us now detail some peculiarities.

4. Instantaneous Moving and Zoom

Some NDErs report having explored their surroundings. A curious feature appears when one asks them to be more specific about the way they moved around: in fact, most of them hesitate between a displacement and a zooming sensation:

> What must also be understood is that it works like a zoom and a displacement all at once. When we take an interest in something, it's as if we zoomed in. It is the displacement and perception occurring simultaneously which allows this to happen. It is hard to separate them, in the sense that there is no notion of time, thus no time spent moving. However, there is a certain notion of space, but not of space with limits and boundaries like in usual space. In the same way there is no compartmentalization or delineated directions, the notion of time and space aren't compartmentalized. It's hard to explain. (A.S.)

> Jourdan: "How did this displacement happen? Instantly or not, sensation of moving, sense of speed, more of a zooming sensation without real displacement?"

> This displacement was instantaneous - but the question "More of a zooming sensation without real displacement" bothers me - and it seems more or less that this is how it happened...What (maybe) makes me understand why, is a PC game that I've enjoyed recently where the hero "Predator" often used this "Zoom" function to move about. Every time I pressed the "Zoom" button I was both disturbed and happy to use this function...which in fact reminded me of this state. Maybe this is why I felt so light... (maybe I wasn't moving about after all, maybe it was my highly developed vision which gave me the feeling of moving about...). (F.E.)

In everyday life, we unconsciously associate being in a spot with perceiving from it. If Tweedee's attention is attracted to a detail, he will focus on it, just as our gaze can instantly switch from the step-mother to the garage where it will be rapidly attracted to the incongruous "STOP" sign.

The only thing which has changed is the direction of our gaze, but for Tweedee, who is accustomed to see his universe from within, the feeling is that of being able to move instantly and/or to "zoom" on to a detail:

> My displacements were subject to my will with instantaneous effect. In-

stant zooming of my vision, without any displacement on my part. When I was on the outside in the park at tree level, I remember experiencing this zoom effect very clearly since I could see inside a tree without having moved. (J.M.)

I see everywhere at once, except when I target an object towards which I am "hurled" at great speed, as if I was zooming onto it. (...) It is like a rapid zoom to be there where I am looking. Like a very fast zoom, I cannot recall if it really is a displacement, or simply a zoom, but I was where I was aiming at, so there is a displacement...actually I don't know but it is very pleasant and fun. (Be.N.)

My consciousness, like a beam of light, can move around very fast, nearly instantly. This gaze, just like thought, can in addition move about very rapidly from points that are distant to one another. (P.C.)

Moving around is done as if time does not exist anymore (or nearly). We "think" about where we wish to be and we make a volitional effort and we get there instantaneously (or nearly, since there is a sensation of movement, but very fast). (D.U.) A sort of sliding, moving by zooming. (J-Y.C.) Feeling of displacement, but ultrafast. (M.L.) Displacement: in one go (F.U.)

5. Crossing the Walls

Many NDErs claim having been capable to pass through walls or ceilings:

Everything was like in reality, exactly in the same place, with the same appearance except that I could see the elements of matter and pass through it. (D.U.)

Usually, Tweedee is of course unable to cross any wall and is used to pass through the doors. But if we move our gaze from the garage to the kitchen quite slowly, paying attention to everything we can see between them, our look will pass through several 2D walls without any difficulty. These latter, which are for us only black lines, are usually perfectly real for him. How will Tweedee interpret that?

At some point, I told myself "I know this surgery but as I didn't know the rest of it, I would like to look a little more", then I went through the wall, I wonder how I moved and then I looked, but the next room was dark and black. (C-A.D.)

That's to say that during my transfer to A. Hospital in the fire-fighters' vehicle I felt my mind off from my body and hear the conversations of these people. Then, realizing this faculty of elevation I got out of the cabin of the

> vehicle through the ceiling … (R.H.)
>
> A sudden upward attraction seized me up, making me go through all the physical structures of the building, without the slightest difficulty, neither any jolt. (F.I.)
>
> I went through the ceiling of my room, through the roof, without pain (I was very surprised ...) (Cl.N.)
>
> (...) Since I passed through the wall. (H.C.)

7. "360°" Perception Without Limiting the Field of Vision

> A "360° spherical" perception is very common during NDEs:
>
> I was surprised that I could see at an angle of 360°, I could see in front, I could see behind, I could see underneath… (J.M.)
>
> I had a 360° spherical-like vision. (X.S.) I had a 360° angle of vision. … (N.D.)

Tweedee is usually as limited as we are : he can only see what is in front of him. If he looks at the painting on the living room's wall, the table and the stools surrounding it are behind him, therefore out of his visual field. If he now moves to our own external visual perspective, he will "see" everything around him –or more precisely around his point of projection without any movement. This "360°" perception can be interpreted as a "centrifugal" global perception where the subject is interested in what surrounds him:

> It is not like the view from here, which is limited in its field of vision and acuity, which is dazzled or impressed by different things such as light or darkness. The vision is only focused when we move, as if we were aiming at target. Otherwise, we see everywhere.
>
> Jourdan: "Did you have the impression you had a bigger angle of vision, that you were seeing in front and behind you simultaneously? "
>
> Yes, without physically turning around, I could see everywhere. (Be.N.)
>
> I had a global perception of the room, like a sphere. My field of vision was larger than usual. Global vision without the need to turn my head left/right or even to turn around. I did not need all of this. (F.E.)

In addition, we can notice the perspective that "changes a little" while the vision remains complete. This is exactly what happens when we watch the layout from a close and moving point of view.

> I could see the whole room under me despite this position, and in particular a physical body lying on the bed, detached from me, far away, which

I knew was my body, but which I wasn't inhabiting at that moment (...) It was then that I noticed I could move around in space voluntarily. This was interesting, new, and so I made a few trial movements. It seems to me that although my view of the room was always complete, panoramic, that the perspective changed a little according to my movements and my position in that space. (F.I.)

8. Perception "From Everywhere Simultaneously"

Our visual perception usually originates in a single vantage point. Obviously, this restriction disappears during many NDEs:

> I visited various places I managed to identify afterwards. I remember a window in a village, a building with very white plaster, sand-carved windows. My curiosity was attracted to details. This is quite important, since we cannot do this normally, like seeing inside and outside at the same time, an impression of a quasi holographic vision... Not a panoramic view, but seeing in front, behind, all details simultaneously which is completely different from ordinary sight. It is very rich. (A.S.)

Tweedee's attention is now attracted by his own body, his point of perception being at a few centimeters from the plan. He will obviously be surprised to simultaneously "see" both his hands, the back of his head as well as the tip of his nose. It is a "centripetal" global perception now. For him, the sole way of understanding that will be the impression of "seeing" from everywhere at once:

> I could see from above, 360° and from all sides, all at once. I see/am/feel this matter with my "sight from above" I see/am/feel this matter from underneath, I see/am/feel this matter which fills the room more and more, I see/am sideways, in profile, underneath, in front, behind, from everywhere, I am the spectator/actor/scene. I had a 360° spherical-like vision. I saw everything and had different points of observation: above, sideways, from the front, underneath, it was really extraordinary to see and be all of it at the same time. When I saw the sofa, the furniture and the room in which I was, I was simultaneously above, sideways, from profile, facing forwards... it was very clear. (X.S.)

> I noticed : "we see everything from all sides simultaneously!", I could see everywhere. (N.D.)

As the distance between him and the observed universe is increasing, he can also, just as we do, "see" simultaneously different places:

> And then I realized that I was both in that space and out of my body. I saw myself lifeless on the bed, I felt my body very heavy, I (my mind? my soul?) got to float around the room. On the one hand, I saw my friends who had gone to play cards, on the other I saw the basement window that attracted me. (K.E.)

Remembering that our vantage point is usually confused with the place we find ourselves in allows us to understand the NDErs who claim having "been" everywhere, J.M.P. who reports having seen an entire clinic and having "been" everywhere at once, and finally A.L. who describes an "enlarged vision" and "was" simultaneously at home and at her grand mother's:

> Here you see this, then elsewhere you see something else, you know everything, from one place to another from the spot where you are. For example, if I want to go to the window, I have to move. But there you don't move, you're everywhere. Unbelievable, but it's great! (J-M.W.)

> ...because I found myself above, on the bottom, everywhere all at once in the clinic… I told you earlier on that I was in the bedroom or more precisely in Mrs.E's bathroom. So, to tell you whether I used the stairs or not, no, I think not; hmm… This displacement cannot be explained since I was downstairs and upstairs, and everywhere all at once actually. (J.M.P.)

> The funny thing is that we have a greatly enlarged vision of things. It was as if I was in many places at once. After their shower my children went to the village, to my grandmother's who lived in a house opposite ours. It was on the other side of a big valley about 800 meters away. She often watched what was happening at our place with her binoculars. So, at the same time, I also found myself at my grandmother's, who was saying: "Oh, something must have happened over at the parents' place because the ambulance is there…" She was watching with her binoculars, the children watched with her through the window and I, was behind them! It is a very strange impression, everything I saw was very luminous, very clear and my senses were sharpened, a much sharper perception of things, I saw and heard everything, all the while being pretty much in a coma. (A.L.)

9. Perception by Transparency

The cases of perception by transparency are recurrent in NDE accounts:

> I could see everything at once and if I focused on one thing, I could see this thing through any obstacle and in every detail, from its surface to how its

atoms were organized, truly a global and detailed vision. (M.L.)

I could see in front and behind oneself simultaneously, through objects, a holographic view. (A.R.)

This means that I did see the entire accident, I left the car and could see myself from above, thus the car's roof was transparent. (P.F.)

When I left my body, I could see through all objects. (C.C.) A very wide vision, through the walls if I wanted to. (K.E.)

This seemingly strange perceptual feature becomes easily understandable within our extra-dimensional model. Indeed, could Tweedee hide anything from our eyes?

Let us return to his house, in which there are at least two things he cannot usually see: a masterpiece painting in a vault placed inside the wall separating the hall and kitchen, and the identity card he has been looking for several days lying behind the couch.

Just like X.S who had the impression of seeing through the couch, so Tweedee will also have the impression of seeing the front and back of the sofa, his ID card, the two sides of the wall and the painting hidden within it, all at once. In our world as much as in his, when we can see both the front and the back of an object as well as that which stands behind it, it simply means that this object is transparent:

I also saw everything that was happening around me. It was in a tent and thus very dark. I went out of the tent very quickly, but the funny thing was that everything was transparent to me. It was very fast because I was rising rapidly, and I saw through the tent. (P.T.)

I was in a global vision, I could see as if I were using my eyes, with clarity, I saw everything at once, I could see everything at once, the impression of seeing backwards and transparently. Sometimes as if I was inside my own eyes. (J-M.M.)

I saw myself exit, since I saw my body on the operating table. I was above and could see everything, everywhere, even through the surgeon. We see everything. -Could you see behind yourself or through objects ? Yes, through objects. (H.C.)

If the NDEr's attention is attracted to a detail, this latter will be perceived without anything being able to hide it. In fact, anything which could stand as an obstacle for normal vision such as a pocket, a curtain, a vault or a wall, is not so anymore:

From the memory of where my bed was placed -in the old people's ward- and of how it was situated in the corner of the room : the door and the coming and going of the personnel were to the right of my head... Also, I think my normal sight should have been cut off and limited due to the wall stopping at my bed. This wall should have limited my visual capacity - and I feel troubled remembering that... (F.E.)

I felt like a soap bubble with eyes strolling about above at ceiling level, in a space which seemed a little "closer" than real space. Behind a wall was a woman dying in the resuscitation room. I saw the instruments, the doctors' gestures and their conversation, I could see through the curtains which hung in front of the glass partition. (J-P-L.)

I could see up close and also transparently. I remember seeing a stick of lipstick in one of the nurses' pockets. If I wanted to see inside the lamp which illuminated the room, I'd manage, and all of this instantly, as soon as I wanted to. (J.M.)

If we look closely at Tweedee and his step-mother, we will see a spiral pattern representing their internal organs, which, for them, are usually as hidden as ours. But now that Tweedee perceives his own body from our vantage point, he is able to "see" inside or through his body... exactly like NDErs do:

I saw all around me, I saw the inside of my body. (M.H.)

After a while I see my "hand made out of crystal" and I tell myself ,"There I can see all the small blood-vessels, but how can it be, there I see all the small blood-vessels in my hand?". It amused me, nothing more. (D.J.)

When I was at ceiling level I could see through myself. (Be.N.)

During the excision of a vesicle in September 1972 and during the anesthesia I found myself...well, floating on the left of the ceiling and looking down on the people operating. I was surrounded by medical personnel, at least 6 people who seemed to be working on my body. I had the time to see, to see...well...I had a very sharp vision and saw through a section of the table. I saw through the operating cloth which is around the operation... the shoes of...of one of the resuscitation team probably...One of them had untied shoe laces. Well, I went through the fabric and so...I came to the conclusion that they were actually resuscitating me. I therefore had the time to realize that it was my body and off I went into the tunnel. (F.U.)

10. Dynamic Perspective Effects

Until now, we have detailed only "static" effects. But it seems that in some cases the point of perception is likely to move, that translates as some new "dynamic perspective" effects.

Let us remember that Tweedee has only notions of moving along length and width, but not at all in height. How can he interpret his perceptions when the distance between his point of perception and his body gradually increases in a third dimension? Going up "perpendicularly", he moves away from his body and everything in its vicinity, while remaining above it. M.M. uses a nice expression to describe this feeling of moving away from her body in a direction perpendicular to our universe:

> After the crash, I left my body in a "geographically total" manner. I was everywhere at once, with a panoramic view. (M.M)

Here is another way to describe the same perception:

> After what I call the familiarization with the "walking around" state, I "went away" while I'm sure I stayed at the same place, I would say that my "sight" got "wider". (K.E.)

As Tweedee can see (and then "be") everywhere while being nowhere in his own universe, he can also describe this perceptual conflict as being "everywhere and nowhere at once":

> I spoke again to the people involved and they were very surprised that I knew exactly what they told me during these times. Like the nurse to which I gave the names of those who came to see me, who thought that someone had told me ... It's as if all my senses were magnified. I could "feel" the people, "guess" them (something that still often happens to me). My vision seemed wider, I could have heard, understood and followed several conversations without difficulty if I wanted ... It was very strange because I seemed to be "everywhere and nowhere at once" and with such clarity ... (M.Q.)

If he goes up even higher, he will see his body getting smaller and smaller. In his own universe, a thing will look smaller only if he moves away from it, but at a certain point the walls will prevent him to move back farther. Here is another perceptual conflict : considering the long distance from which he perceives now, he obviously ought to have crossed a wall, which then "seems to move back with him":

> Very slow moving, diagonally to the ceiling, upwards from the table that was there, as if I was there. But the ceiling seemed extremely high, it seemed that as far as I went upwards to this spot, the ceiling went up the same time as me. (C.F.)

Another way of describing the conflict is to "see through the concrete, and at the same time it is not the same thing":

> First, I watched all this bustle. Then I realized there was a body. I did not say, but when I say "up" is not a sight from two to three meters. I was much higher, much higher. I had a perception, an overview. I was not at three meters up. It was a holistic, panoramic view into the room. But from very

high as if I could have seen through the concrete and at the same time it is not the same thing. I would rather say that I was in another dimension of space where I had another vision capability as if I were both very close and very far because I could see very fine details, every detail. I saw this body and at first I did not realize it was my body. (P.B.)

11. Peculiarities of Time Perception

Until now, we have considered the perception of a 2D universe from a third dimension that allowed us to understand the particularities reported by the NDErs. These latter seem extremely similar to those which would appear when a 3D space is perceived from an external vantage point, thus situated in an additional spatial dimension.

This notion of a fourth dimension is far from new. As a purely spatial extension, it has been evoked in the 19th (Bork 1964) and early 20th century as mathematical and geometric speculations (Bragdon 1913, Manning 1914, Durrel 1938, Rucker 1977, 1984), within metaphysics (Willink 1893, Zollner 1901, Gardner 1981), philosophy (Kant 1783) as well (remember some paintings from Picasso showing a model seen from several different places simultaneously!) in fine arts (Dalrymple Henderson 1983).

As it concerns for the moment only the perception of space, the modelling I have set out until now can be considered as a purely geometrical analogy in an Euclidian affine extended space. A similar hypothesis has been proposed independently in 2003 by another NDE researcher (Brumblay 2003). But is it sufficient? Since Minkowski, Einstein and Poincaré's works, it is well known that we live in a four-dimensional world, in which space and time are closely linked within a spatio-temporal continuum. Thus there already exists a fourth time-like dimension and if we want to be rigorous we have now to talk at least of a fifth dimension.

11.1. No time The NDErs' answers to questions about their perception of time during their experience led us to enlarge the 4D spatial model to a 5D spatio-temporal one. For a start, during a NDE the notions of time or of duration may disappear:

Feeling that time no longer existed. (S.D.)

In fact, there was no time, it was like a moment of eternity. (K.E)

Time did not exist. Now it's a real knowledge for me, time does not exist! (M.M.)

On the other side, time does not exist. One truly realizes it. Time is a completely mental concept. A thousand years may be instantaneous. (M-P.S.)

There I had the distinct impression of finding myself in a familiar place, a place I had known. As if I was gone for some time and then back home Some time ... But what does it mean: "some time"? The concept of duration to which it usually refers was absent from this story. All I can say, even if I am unable to explain it, is that I existed in what might be called a kind of absolute timelessness. For this entire trip out of my body also unfolded outside of time. No body: no time! So I wonder whether our perception of temporal flow could not be an enormous performance, a joke. (M.N.)

Notion of time? No, I think we totally lose track of time. Maybe faster. But in reality I do not know, because it happens like flashes by, you see things, you hear, you see, it seems that everything happens at once. (A.L.)

11.2. An eternal present In other cases, a less drastic disappearance of time may be translated by various expressions. Many speak about an "eternal present", an "omnitime", or of a "time that no longer unfolds":

I could tell some facts that were going on at places where I was not supposed to be, since I was strapped to a bed in ICU. They checked, they found it surprising that this was true, as this seems surprising to some that I talk to them about what happened. About their lives. They call it the past because they reason in terms of time, but there is no time outside the body. There is no past. There is no present, no future. There is an eternal present. (P.M.)

I had a horrible feeling of eternity. I had an experience where time no longer unfolded. Furthermore, no past, no future, just an eternal present. I had the feeling that all that was real, the feeling of "living" in eternity. (I.H.)

Their experience led C.N. to study Einstein, and H.R. to speak spontaneously of space-time as a whole:

I had no notion of time during the experiment. It's just another time, in fact we are no longer in time, it's omnitime, that is to say, the eternal present ... One is truly in the eternal present. There is no more time. But having said that, after my return, I had a very very big problem with time, I was very obsessed by time, space-time, that is what made me study Einstein, the fourth dimension, etc. Because at the same time I was very anxious at the idea of having no time, no time enough to do what I had to do, it's funny; I had a very very big problem with time having been out of time gave me a problem with the chronological time that we live on earth. At least I know I was in that omnitime and omnispace. (C.N.) No notion of time and no limit. To my knowledge there is no possibility of comparing the earthly time and that of this dimension. The whole makes up this space-time, a form of totality, comprehensiveness. (H.R.)

There is an apparent contradiction in living "out of time": to experience something, one way or another last. In fact, some expressions used by NDErs suggest that, during their experience, there remains at least a present allowing them to continue to exist, but it is also clear that this present of their own is no longer subject to the time arrow. Like an astronaut who is no longer submitted to gravity but who can watch the Earth with a telescope and see the fall of an apple, in their experience they "observe" a time which is no longer their own.

11.3. A second form of time We could consider a purely psycho-physiological explanation, as this "other form of time" might be a reconstruction by our brain, which is used to run sequentially, in particular with regard to memory, not to mention that the narrative of the experience can only be done sequentially too. But NDErs are adamant that, just like any of the other perceptual particularities we have reviewed so far, this second time was experienced during their NDE and is registered in their memory as well as the rest, as shown by some pertinent remarks:

> It is a profound conviction that I do not explain, displacements are infinitely fast but there is still a before and an after, a chronology and a souvenir of the action that just took place, so there is a form time but I cannot explain it. (Be.N.)

> All this took place "outside time" - or in a time that has no earthly reference. I had the impression of being outside of time, and yet there was some sort of time (it was another time). (A.T.)

> As I said, in the absence of time there is still time. I know it sounds absurd, but I can not explain more. (J.X.)

> The notion of time has nothing to do with ordinary life, that's for sure. Physical, material time does not exist. Time does not flow. To say that there is another " time system", I do not know. If there was a complete "timelessness," all emotions would be simultaneous. For me, anyway, I had various feelings. Knowledge is complete and simultaneous. Emotions not. We react emotionally to what we see. In my opinion, there must exist another form of time, anyway. (C.P.)

> Jourdan: "Did you have a notion of time?"

> Yes and no. Yes because events followed events. I feel they did not occur at the same time. No, because the concept of time is not the same. There is no yesterday, today and tomorrow. I would say that events are instantaneous but emotions come and go. And then maybe I say that the events follow one another because the emotions are, themselves, quite distinct

from each other. Maybe it has nothing to do with the concept of temporality within ordinary life. No, because my own chronology doesn't match that of the ordinary world. Events that I placed before actually occurred afterwards when I asked for confirmation. And vice versa. (P.C.)

12. A Spatialized Time

Now, let us follow our 5D hypothesis through, supposing that during an NDE our 4D universe could really be perceived as a whole from an extra-dimension. If this hypothesis more or less reflects some reality, we ought to find some accounts reporting several kinds of temporal perspective.

12.1. Past, present, future, all confused If we walk on a lane, one part of the way is behind us, another is ahead. But if we see this lane from above, no more walking on it, not only are we nowhere on it but also, as they were relative to us, the notions of behind and ahead logically disappear. In the same way, if we are "out of time" we are no more subject to its arrow. Thus the notions of past, present and future can disappear or merge together. In fact, almost every feature we are about to review seems to translate a spatialization of time:

> Past, present future are merged in a single concept, that's what I experienced (X.S.) Time no longer existed, past, present, future, all confused. (M.O.)

> No sensation of duration, neither of waiting. No sense of past, present, future, as if all that was away from me. (F.E.)

During NDEs, frequently following the OBE stage, most patients describe a "life review". They report having been able to "see" or sometimes "live again" some moments or only significant scenes of their life, in chronological order or in reverse order. During this stage some additional anomalies and perspective effects concerning time will appear, strengthening the hypothesis that our universe could be perceived from an external vantage point.

12.2. Flying over time The non-locality of the observer in relation to the observed universe, which has helped us understanding the spatial 4D perspective effects appears to concern also time: to be "outside" the space-time would give the same impression of "being" everywhere at once compared to the latter. The expressions are various, but translate the same strange feelings of perceiving time from outside or flying above it:

> I had access to both past, present, future and any place in space. (M.Z.)

> I had no access to the future, I don't think so, but to the past yes, exactly so, as well as to present since I was seeing myself. It seems to me that I could move around. (P.B.)

I felt I could fly over time. (J-M.M.)

It seems to me that the time is no longer valid. That is, I don't take place within time. There is no longer any past neither future, everything is within the same plane. I got out of the timeline and I can contemplate it AS A WHOLE. But thirty years later, I am still unable to define accurately, using common words, this perceived lack of time ... and both its presence. When you move from one place to another in a flash, when one sees multiple views of the same situation, physically and temporally, that's not "every day life."

Jourdan: "Have you had the impression of "flying over" time, as one can fly over a landscape, or see it from above?"

Yes, in some ways, move forward or backward at the same time. "Time" no longer appears as fragmented, but as a one and single moment: a "continuum" related to will and free will. (D.S.)

There, the time does not seem to unfold as here. I would say it's "above", a place from where you can "govern" the events and the destinies of the earthly world. Neither was there any space. (A.T.)

I wonder about the word "time". I had the impression of "flying over" a certain portion of time, to fly so quickly but the time seemed at once long and short. That's funny. I felt able to move in time. (F.N.)

When I saw my life, it was like an accelerating videotape, somewhat as if I could fly over it, it goes fast enough to review one's life and yet it lasts forever, I can't explain. (Be.N.)

13. Spatio-temporal Perspective

What could be the predictable consequences of a hypothetical perception from outside our spatio-temporal continuum? Within the framework of this 5D model, everything happens as though NDErs were able to take enough distance to see in perspective not only the immediate vicinity of their body, but also their whole life. Then we could now expect some precise temporal perspective effects.

In our everyday experience, the concepts of time and space are fundamentally different. It is surprising to find several accounts of a uniqueness that has nothing natural nor intuitive for us. Even if he finds it difficult to explain (what we will readily admit), J-Y.C. briefly summarizes relativity with some expressions that would have pleased Einstein. Even better, the way he watches his own life as what we could call a "4D spatiotemporal object" is amazing : a 3D form under his eyes, with an "integrated time" which doesn't unfold, a life he can see from every angle, get more or less closer or change his angle of view, focusing on one part or another... exactly as we do in our everyday life when examining a banal 3D object.

> Indeed, at the time I receive this new form of intelligence, I find before me ... my life. I look at this 3D thing that is my life and which does not unfold. The time is integrated in it, it is no more linear. All of one's life is visible and this "global" intelligence can read it, understand it. (…)
>
> I saw my entire life, in relief, with all the details, people, situations. But in a time that does not unfold, life being seen from every angle with this universal or global understanding. My life was a form under my eyes, which contained everything and that I consulted.
>
> (…) My whole senses were concentrated or condensed in a single understanding concept. The ability to understand and develop ALL, in its wholeness and in its detail. Should I have watched a car, I had known in one thought its mileage, fuel quantity, the wear of spark plugs, how many times it had turned left or right, the condition of all its parts, etc... It is very difficult to share the encompassing of the three dimensions with the fourth, which merge in a concept that can be easily read when one gets this form of over-intelligence.
>
> (…) Time is no longer linear. Your own life is in 3D and the fourth dimension is fully integrated. At that time, if I had watched a man, I could have known everything about him. His age, height, blood type, his siblings, the amount of all his taxes, his diseases, etc.. etc.. ALL in a single concept.
>
> Jourdan: "Did you feel yourself moving?"
>
> Yes.
>
> Jourdan: "When?"
>
> To get closer to my life.
>
> Jourdan: "How did it happen?"
>
> A sort of sliding, zoom displacement.
>
> The only "thing" that I was able to contemplate was my own life. An oblong shape, three-dimensional pink-orange hue (always "metallic" as having its own light). I could see inside, seeing-through my entire life course, including time without unfolding time. To see another part of this life I just had to °change my angle of view. (J-Y.C.)

This second example is less spectacular, but we find again a "frozen time", a "whole life spread before (the NDEr's) eyes, its slices being seen instantly":

> Totally calm and in a state of unimaginable bliss, I continued to float in a world of breathtaking clarity where the notion of time, that seemed frozen, defies understanding. In tune with this inexplicable timelessness the slices of my life were seen instantly, without any sense of duration. That's quite difficult to explain with "earthly words." My past life did not just appear to me like images following one another in a reverse chronology, as might be suggested by my previous comments. The events unfolded in accordance with the original script, but their succession went backward over the course of my life. Sometimes, and here it's even more difficult to explain, I felt like my whole life was spread before my eyes, undifferentiated in its stages, and still without the sequence of events being linked to time. I know it's crazy, totally incomprehensible, but that's the way it happened. (M.N.)

14. Discussion

NDE are frequently viewed as hallucinatory experiences. Indeed, in spite of numerous confirmed accounts reporting precise details and scenes in the immediate vicinity of unconscious patients, for the time being we have no irrefutable proof about AVPs.

On another hand, the hyperdimensional model I propose allows to understand very simply every seemingly strange perceptions, implying that these experiences could follow definite rules.

So we are faced with several possibilities. The first one is an "inner" hypothesis : NDE are purely subjective experiences, the AVPs being the result of brain activity, this latter having "rebuilt" scenes and details very close to the reality from various elements gathered after the experience. Joseph (1996, 2001) provides evidence which he believes demonstrates it is the hippocampus which is responsible for the hallucinations of floating above the body. As detailed by Joseph (2001):

> "The hippocampus contains "place" neurons which are able to encode one's position and movement in space. The hippocampus, therefore, can create a cognitive map of an individuals environment and their movements within it. Presumably it is via the hippocampus that an individual can visualize themselves as if looking at their body from afar, and can remember and thus see themselves engaged in certain actions, as if one were an outside witness (Joseph, 1996). However, under conditions of hyperactivation (such as in response to extreme fear) it appears that the hippocampus may create a visual hallucination of that "cognitive map" such that the individual may "ex-

perience" themselves as outside their body, observing all that is occurring. In fact, it has been repeatedly demonstrated that hyperactivation or electrical stimulation of the amygdala-hippocampus-temporal lobe, can cause some individuals to report they have left their bodies and are hovering upon the ceiling staring down. That is, their ego and sense of personal identity appears to split off from their body, such that they may feel as if they are two different people, one watching, the other being observed."

If this hypothesis proves to be the correct one, that would at least lead us to explore the hypothesis of some 5D-like brain organisation, which could present some interest for cognitive neurosciences, neurology, psychology, and all those disciplines which generally seek to explore the nature of consciousness and the functioning of our brain. Like radioactivity at its very beginning, what appears to be only an oddity can conceal major avenues of research.

Another possibility is that of an "outer" hypothesis. Nobody, at present, can clearly define consciousness. We can at the very most safely say that it is part of our world, follows the laws of nature and has been for a long time our only tool to try to puzzle over it. Could it, in some unusual circumstances, show us the first evidences of additional dimensions?

14.1. Four or five dimensions? It is important to clarify some potential confusions if we envisage an extra dimension. We have seen that the first proposals about this subject date back several centuries. At that time, scientists reasoned within an Euclidean space -which comprises only spatial dimensions-, envisaging a fourth spacelike dimension that was a virtual mathematical or geometrical concept. Nowadays, the mathematical setting of relativity is a Minkowski space comprising three spacelike dimensions and a timelike fourth dimension. Then, as our visible universe is a 4D space-time continuum, and considering the multiple accounts reporting spatio-temporal perspective effects, the extra dimension giving an accurate background to the model I have set out would be a fifth one.

14.2. Recent extra-dimensions theories The first proposal of a fifth dimension, in order to unify electromagnetism and gravity, comes from the German mathematician Theodor Kaluza (1921) and the Swedish physicist Oskar Klein (1926). This theory was abandoned, but after a few decades appeared superstrings and strings theories, largely initiated by Peter Freund (1982,1985), who introduced extra dimensions of space in physics and found the mechanisms by which these extra dimensions curl up. These theories involve 10, 11 or up to 26 extra dimensions, which are compactified, curled up at each point of our universe with a finite minuscule size (about the Planck length, i.e 10^{-33} cm in the K.K. theory). Obviously, this tiny size does not offer a sufficient distance to allow the perspective effects that we have reviewed.

Derived from string theories, which concern essentially particle physics, brane cosmology is based upon brane theories, which attempts to understand the weakness of gravity within our visible universe. In brane theory, a string is a 1-brane, a "membrane" is a 2-brane. In general a p-brane (p is the number of spatial dimensions, therefore a p-brane is in practice a (p+1) space-time) is viewed as a slice inside a (p+1) brane. Thus, according to this theory, our four-dimensional universe is confined in a 3-brane within a 4-brane, a "super"universe endowed with (4+1) dimensions.

Following a first proposal (Antoniadis & al 1998, Arkani-Hamed & al 1998, 2000), Lisa Randall and Raman Sundrum established in 1999 two models of brane cosmology. In the first one (Randall and Sundrum 1999-a), the size of the extra dimension is finite, about 1mm, which is far better than the Planck length but still insufficient.

On the other hand, in the second one (Randall and Sundrum 1999-b) the extra-dimension might be infinite, which is perfectly suitable for the extradimensional modelling that I propose.

Another model, elaborated by Laurent Nottale (Nottale 1993, 2010, Nottale and Timar 2008), is scale relativity. Within it appear two interesting characteristics: a fifth topological dimension and a spatialization of time, which could explain the particularities that we have reviewed about the perception of this latter.

14.3. The Time issue During NDEs, our universe seems to be perceived not only as spatial but indeed as a whole space-time. Several testimonies seem to report some sort of time spatialization, and the main issue is to understand how that could be possible. Saying "our (3+1)D universe is a subset of a (4+1)D universe" implies that we have merely added a spacelike dimension, the time dimension remaining the same. Concerning this particular point, the status of time within extra-dimensional theories is not clear and above all I am not qualified to go further.

At the very most, I could perhaps say that we might understand the particularities described by NDERs such as "no time", "eternal present", "being out of time" by remembering that, according to relativity, an object or particle is subject to time – and therefore has a duration of its own - only if it has a mass, and therefore suppose that "that which perceives" during an NDE is massless.

Be that as it may, I hope one more time that qualified scientists will accept to think about that according to the accounts we have reviewed.

14.4 A predictive modelling? The modelling I propose, like every self-respecting model, must be predictive and lend itself to experimentation. We have seen that, to an observer whose vantage point is situated in a (n+1)D universe, nothing can be hidden within a (n)D universe. Thus a very simple test could be proposed, consisting of a hidden target (for example a colored drawing enclosed in a sealed envelope) put in the vicinity of places where NDEs are likely to occur (ICUs, surgery, etc.). Provided it is unusual and interesting enough, this target could attract the attention of an NDEr, who would be able to describe it after resuscitation.

15. Conclusion

In this short paper, I hope to have given the reader enough information so that he or she can make up his or her own mind about the hyperdimensional interpretation of perceptual particularities in NDEs. The fact that the perspective effects concern time as well as space, and that some patients without any training or education in physics were able to describe with their own words a spatio-temporal continuum seems to me particularly interesting.

To summarize, the particularities that we have reviewed could lead one to suppose that consciousness could be the result of some interactions between 4D and 5D phenomena and/or universes, an hypothesis we cannot simply dismiss and that is considered very seriously by some neuroscientists (Smythies 1994, 2003) and cosmologists (Carr 2008) as well as philosophers (Droulez 2010).

The look we have on a screen, a sketch, a painting or any 2D-like universe allows an instantaneous global information. Waiting for further research and results, the analysis of the perceptual particularities in NDEs in terms of global perception/acquiring of information, a concept that is coherent with our model, should allow us to conduct research calmly and in a purely scientific way. In addition, it should be free from all metaphysical a priori and use concepts which are already within our reach.

Whether the logic revealed by this analysis reflects a particular cerebral function, a new phenomenon or a combination of both, it casts doubt on purely hallucinatory interpretations of these experiences and constitutes an argument in favor of scientific research into NDEs, justifying a multidisciplinary approach gathering physicians, neuroscientists, cognitive scientists, philosophers, psychologists, anthropologists, and now maybe, mathematicians, physicists and cosmologists.

ACKNOWLEDGMENTS: It is a pleasure to thank the editors of the Journal of Cosmology for inviting this paper, hoping that the evidences set out in it will represent some food for thought for its readers. I am grateful to the confidence NDErs graced me with, and for the time they spent answering my numerous and sometimes strange questions. Without them this study would never have seen the light. I am also eager to thank John Smythies and Thomas Droulez for their encouragements and for a careful proofreading of this paper.

References

Antoniadis I. & al (1998) ., New dimensions at a millimeter to a fermi and superstring at a Tev, Phys Lett B., 436, 257,.

Arkani-Hamed N., Dimopoulos S., Dvali G. (1998). "The Hierarchy problem and new dimensions at a millimeter". Phys. Lett. B 436: 263–272.

Arkani-Hamed N., Dimopoulos S., Dvali G. (August 2000). "The Universe's Unseen Dimensions". Scientific American 283 (2): 62–69.

Bork A. (1964), The Fourth Dimension in Nineteenth-Century Physics, Isis 181.

Bragdon C.F.(1913) A Primer of Higher Space (The Fourth Dimension), Cosimo Classics, 2005.

Brumblay, R.J. (2003) Hyperdimensional perspectives in Out-of-Body and Near-Death Experiences. Journal of Near-Death Studies, 21, 201-221.

Carr, B. (2008). Worlds apart? Proceedings of the Society for Psychical Research, 59, 1–96.

Dalrymple Henderson L. (1983) The Fourth Dimension and Non-Euclidean Geometry in Modern Art, Princeton University Press.

Droulez T. (2010), Conscience, espace, réalité. Implications d'une critique du réalisme perceptuel direct, Cahiers Philosophiques de Strasbourg, n°28 .

Durrel F. (1938), The fourth Dimension : an Efficiency Picture, in Mathematical Adventures, Boston, Bruce Humphries.

Freund, P.G.O. (1982). Kaluza-Klein Cosmologies. Nucl.Phys.B 209 (1): 146.

Freund, P.G.O. (1985). Superstrings from Twenty-six-Dimensions?. Phys.Lett.B 151 (5-6): 387–390

Gardner M. (1981), Parapsychology and quantum Mechanics, in Abell G. & Singer B, Science and the Paranormal: Probing the Existence of the Supernatural, New York, Charles Scribner's Sons.

Greene, B. (2003) The Elegant Universe: Superstrings, Hidden Dimensions, and the Quest for the Ultimate Theory. W.W. Norton & Co.

Hawking, S. (2001), The Universe in a Nutshell (London: Bantam Press).

Holden J. M., (2009) Veridical perception in Near-Death Experiences: A Comprehensive, Critical Review of the Professional Literature. . In J. M. Holden, B. Greyson, & D. James (Eds.), The handbook of near-death experiences: Thirty years of investigation (pp. 185-212). Santa Barbara, CA: Praeger/ABC-CLIO.

Holden J. M., Greyson B., & James D. (Eds.), The handbook of near-death experiences: Thirty years of investigation, Santa Barbara, CA (2009): Praeger/ABC-CLIO.

Holden, J. M., & Joesten, L. (1990). Near-death veridicality research in the hospital setting. J.Near-Death Studies 9 (1): 45–54.

Joseph, R. (1996). Neuropsychiatry, neuropsychology, clinical neuroscience. Lippincott Williams & Wilkins.

Joseph, R. (2001). The limbic system and the soul. Zygon, 36, 105-136.

Jourdan J.P., Juste une dimension de plus. Cahiers scientifiques de IANDSFrance. Hors Série N°1 Février 2000. English version: « Just an extra dimension » http://iands-france.org.pagesperso-orange. fr/SRC/PDF/justextra.

pdf

Jourdan J.P., Les dimensions de la conscience. Les Cahiers de IANDSFrance. N°7 Janvier 2001.

Jourdan J.P., (2006) Deadline - Dernière limite. Les 3 Orangers, Paris 2006; Pocket Paris 2010.

Kaluza T. (1921), On the problem of unity in physics, Sitzungsber. Preuss. Akad. Wiss. Berlin. (Math. Phys.) 966-972.

Kant E. 1783, Prolégomènes à toute métaphysique future, Paris, Librairie Philosophique J. Vrin, 1968.

Keeton, C. R., and Petters, A. O. (2005). Formalism for testing theories of gravity using lensing by compact objects. III. Braneworld gravity. Physical Review D 73:104032.

Klein O. (1926), « Quantum theory and five dimensional theory of relativity », Z. Phys. 37 895-906.

Manning H.P. (1914), Geometry Of Four Dimensions, New York, Dover 1956.

Nottale, L., (1993) "Fractal Space-Time and Microphysics: Towards a Theory of Scale Relativity" (World Scientific).

Nottale L., (2010). "Scale Relativity and Fractal Space-Time. A New Approach to Unifying Relativity and Quantum Mechanics" (Imperial College Press, in press).

Nottale L. and Timar P., (2008), in « Simultaneity : Temporal Structures and Observer Perspectives », Susie Vrobel, Otto E. Rossler, Terry Marks- Tarlow, Eds., (World Scientific, Singapore), Chap. 14, p. 229-242.

Parnia S., Waller D.G., Yeates R., Fenwick P., A qualitative and quantitative study of the incidence, features and aetiology of near death experiences in cardiac arrest survivors. Resuscitation 48 (2001); 149-156.

Randall, L., and Sundrum, R. (1999a). Large Mass Hierarchy from a Small Extra Dimension. Phys. Rev. Lett. 83, 3370–3373.

Randall L., Sundrum R (1999b), An alternative to compactification, Phys. Rev. Lett.83, 4690-4693,.

Ring K. (1980) Life at Death: A scientific investigation of Near-Death Experience. New York,NY: Coward,McCann and Geoghegan.

Ring K., Cooper S., (1997). Near-Death and Out-of-Body Experiences in the Blind: a study of apparent eyeless vision. Journal of Near Death Studies, (2), 101-147.

Rucker R. (1977), Geometry, Relativity and the Fourth Dimension , Dover.

Rucker R. (1984), The Fourth Dimension, Houghton Mifflin.

Sabom M.B., (1981) Recollections of Death: A Medical Investigation, Bantam.

Sabom M.B., (1998) Light and Death: One doctor's fascinating account of neardeath experiences. Grand Rapids, MI; Zondervan.

Schwaninger J. & al., (2002). A prospective analysis of Near-Death Experiences in cardiac arrest patients. Journal of Near Death Studies, 20 (4), 215-232.

Sartori P. (2004) A prospective study of NDEs in an intensive therapy unit. Christian parapsychologist, 16, 34-40.

Sartori P. (2006) A long-term prospective study to investigate the incidence and phenomenology of near-death experiences in a Welsh intensive therapy unit. Network review, 90, 23-25.

Sartori P., et al., (2006) A Prospectively Studied Near-Death Experience with Corroborated Out-of-Body Perceptions and Unexplained Healing. Journal of Near Death Studies Vol. 25 (2), Winter.

Smit H.R., (2008). Corroboration of the

dentures anecdote involving veridical perception in a Near-Death Experience, Journal of Near-Death Studies, 27 (1), Fall.

Smythies J. (1994). The Walls of Plato's Cave: The Science and Philosophy of Brain. Consciousness and Perception. Avebury Press.

Smythies J. (2003). Space, time and consciousness. Journal of Consciousness Studies, 10, 47–56.

Smythies J. (2009) Brain and Consciousness: The Ghost in the Machines. Journal of Scientific Exploration, Vol. 23, No. 1, pp. 37-50.

Van Lommel Pim & al., (2001). Near-Death Experience in survivors of cardiac arrest : a prospective study in the Netherlands. The Lancet, vol 358, décembre.

Willink A. (1893). Th World of the unseen : An Essay on the Relation of Higher Space to Things Eternal, New York, Macmillan.

Zollner J.C.F. (1901). Transcendental Physics, Boston, Beacon of Light Publishing.

40. In the Borderlands of Consciousness and Dreams: Spirituality Rising from Consciousness in Crisis

Kevin R. Nelson, MD

Professor of Neurology, University of Kentucky,
Lexington, KY 40536

Abstract

Through understanding consciousness in crisis, we see that spirituality is inextricably bound to our primal brain. During crisis, the border between conscious states and the border between these states and unconsciousness can blur, creating borderlands of consciousness. Emerging from these borderlands comes some of our most powerful experiences.

KEY WORDS: Near-death, consciousness, REM consciousness, "fight-or-flight", syncope, locus coeruleus, arousal, consciousness borderlands

"And his pure brain, which some suppose the soul's frail dwelling-house" -William Shakespeare (King John, Act VI Scene VII)

"All mental processes, even the most complex psychological processes, derive from operations of the brain" - Eric R. Kandel (1998) Nobel Laureate

1. Introduction

The brain is our most glorious organ. To survey the majesty of human achievement is to survey the brain's majesty. Plato's philosophy, Shakespeare's plays, Beethoven's symphonies, and Einstein's theories are but a few of the brain's triumphs that come readily to mind. And in our most sublime moments, spirituality must be touching the brain if we seriously consider the words of Shakespeare and Kandel. The brain's grandeur and power is dazzling, yet at the same time this splendor blinds us to the brain's prime purpose-to keep us alive.

Key to this primal role, the brain tightly regulates it's own blood supply each second of life. The brain depends on aerobic metabolism, and so must control its blood flow at rest, in exercise, as well as during physiological and emotional stress. Cerebral blood flow is maintained through the arterial baroreflex (Benarroch, 2008) that in turn relies on the yoked opposition of cholinergic and adrenergic neurons in the peripheral and central nervous systems, as well as mechanisms intrinsic to the cerebrovasculature (Deegan et al., 2010).

Fading cerebral blood flow with looming unconsciousness signals a crisis to the brain. When the brain finds itself in crisis, we will see how it calls upon crucial impulses that have guided our forbearers' survival for tens upon tens of millions of years.

2. Consciousness in Crisis

One such crisis is near-death experience, whose frequent cause includes transient cardiac dysrhythmia. For reasons beyond biology alone, near-death has become a social stereotype of going through a tunnel, being enveloped by "the light", and floating above one's body, often combined with transcendental or mystical elements. Each near-death experience is colored by the person's life experiences and psychology. Philosopher Sir Alfred Ayer recounts, while near-death, that he had crossed the River Styx, writing afterwards "I have not wholly put my classical education behind me" (Ayer, 1988). Carl Jung visualized burning lamps surrounding a door leading to the inside of a stone temple during his near-death, and the psychiatrist later remarked that "I had once actually seen this when I visited the Temple of the Holy Tooth at Kandy in Ceylon" (Jung and Jaffé, 1989).

Whereas some Americans have sighted Elvis in their near-death experiences (Moody, 1987), Elvis is not featured in experiences of children nor in adults of other cultures (Morse et al., 1986, Pasricha and Stevenson, 1986).

Although there is little mention of near-death experience by William James, near-death does meet his conception of a spiritual experience whereby "feelings, acts and experiences" touch "whatever they may consider the divine" (James and Marty, 1982).

All the clamor over near-death has blinded us, until recently, to the fact that first and foremost near-death experience is consciousness in crisis (Nelson et al., 2006).

3. The Brain's Three Conscious States

There are three mental states possible for the brain: waking, REM sleep and non-REM sleep. We expect in crisis to be awake and attentive so we can meet the danger head on. This is so intuitively obvious that it's rarely given thought. But the brain must not take for granted that it will be in the right conscious state during crisis. Waking consciousness must immediately orient attention to whatever is required for survival, and so consciousness must be in lock step with "flight-or-fight" action. It is the brain (specifically the limbic system) that orchestrates the "fight-or-flight" to survive (figure 1).

Figure 1. On Friday January 20, 1911 Walter B. Cannon wrote three lines in his laboratory journal: "Tried experiment on rabbit-but no success. Got idea that the adrenals in excitement serve to affect muscular power and mobilize sugar for muscular use-thus in wild state readies for fight or run!" (Cannon, 1923-2003) with permission. Cannon's critics soon turned "run" into "flight", making the phrase "fight-or-flight" similar to one used earlier by Cannon's most renown teacher, William James.

4. The First Borderland of Consciousness '

The border between consciousness and unconsciousness is not always abrupt and absolute. Between the hazy edges of consciousness and unconsciousness lies the first borderland of consciousness entered by someone whose brain is ischemic, starving for blood. The brain is in crisis if blood flow drops below the threshold of 23cc/100 grams of brain/minute, whereupon the cerebral cortex fails (Jones et al., 1981), and consciousness is lost after ten seconds or so (Brenner, 1997).

Consciousness can come and go if cerebral blood flow waxes and wanes across this threshold, which often happens in clinical settings.

Neuronal death begins within minutes after cerebral blood flow completely ceases. But even with sustained flow there is a second threshold below which neurons die. This threshold increases over hours to eventually reach a plateau of 17 to 18 ml/100 grams of brain/minute (Jones et al., 1981).

In the initial seconds of failing blood flow and fading consciousness, there is no reason to expect that the brain reacts differently between simple syncope and cardiac dysrhythmia. This explains why in a series of near-death subjects, otherwise harmless syncope was the most common event leading to near-death (Nelson et al., 2006). Even in the controlled safety of the laboratory, the syncope experience can be nearly indistinguishable from near-death (Lempert et al., 1994a), which reinforces the finding that only half of those experiencing near-death are actually medically threatened (Owens et al., 1990). Together these observations clash with the popular misconception that the near-death experience happens only in the face of truly eminent death. Upwards of one third of people faint within their lifetime, often while feeling endangered, and this makes syncope potentially fertile ground for spiritual experience.

5. A Metronome of Waking Consciousness

Epinephrine (adrenaline) in the body, and the corresponding neurotransmitter nor-epinephrine (nor-adrenaline) in the brain serve vital functions when peril confronts us. Elemental to "fight-or-flight" behavior is the brain's source of norepinephrine, the locus coeruleus (LC). From the brainstem pons, LC neurons project widely throughout the brain to help regulate consciousness and actively promote behaviors critical to survival (figure 2).

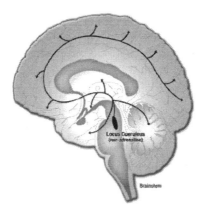

Figure 2. The locus coeruleus is a miniscule cluster of approximately 16 thousand neurons in the right and left pons. Through it's diffuse projections, the locus coeruleus is the nearly exclusive source of the neurotransmitter nor-epinephrine (nor-adrenalin) for almost every region of the brain (Nelson, 2011).

The pontine arousal system is the fulcrum of a reciprocal swing between colossal neurochemical systems that sweep through the brain (Hobson et al., 1975). The REM promoting cholinergic pedunculopontine (PPT) and laterodorsal tegmental (LDT) nuclei (figure 3), are in counterbalance to the waking actions of the serotonergic dorsal raphe (DR) and noradrenergic LC nuclei.

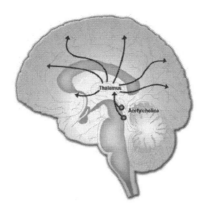

Figure 3. The pontine arousal system for REM consciousness uses the neurotransmitter acetylcholine (Nelson, 2011).

The locus coeruleus constantly discharges during waking consciousness. In primates, LC discharges are tightly linked to specific behaviors. There is some evidence to suggest that the firing patterns of this tiny nucleus anticipate certain primate behaviors. Low discharge rates correspond to low arousal when the animal is inattentive to the world around it. Moderate rates (with synchronized bursts) are seen with focused attention. High LC discharge rates correlate to the animal visually scanning the environment and rapidly shifting its attentiveness (Aston-Jones et al., 2000). Swiftly directing attention to meet the demands of the outside world is a fundamental role for the LC (Benarroch, 2009) (figure 4).

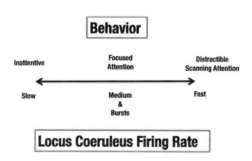

Figure 4. Locus coeruleus firing patterns anticipate primate behavior (derived from Aston- Jones et al., 2000).

During waking consciousness the activity of the LC is extremely important to maintaining vigilance in response to stress (Rajkowski et al., 1994, Abercrombie and Jacobs, 1987). Feelings of fear, hypoxia, hypotension and hypercarbia, often present during near-death, all vigorously stimulate the LC, increasing its tonic discharge rates (Kaehler et al., 1999, Valentino et al., 1991, Bodineau and Larnicol, 2001).

6. What Becomes of the Activated Locus Coeruleus?

Physiologic mechanisms do not function unchecked in isolation. A classic example is the reciprocal action between the cholinergic and adrenergic portions of the peripheral autonomic nervous system. The LC could be central to an arousal system predisposed to blending REM and waking consciousness. Although factors activating the LC have been extensively investigated, much less is known about how LC activity is tempered. Conditions promoting REM consciousness powerfully inhibit the LC. Only during REM consciousness is the LC relatively silent (Reiner, 1986). Since LC suppression anticipates both behavior and REM consciousness (Foote et al., 1980), conceivably if scanning attention becomes maladaptive, or focused attention necessary in crisis, then counterbalances like the cholinergic REM system could act on the adrenergic LC.

7. An Inopportune Consciousness

The functioning of the LC provides a clue to how the brain might, counter-intuitively, shift from waking to REM consciousness at a most inopportune moment for survival. Further exploring how the brain could tilt to REM consciousness, when all consciousness is about to cease, requires us to know more about REM consciousness itself.

8. The Consciousness of Light

The visual system comes to the fore during REM consciousness. Light and visual impressions are among the many sensations experienced during REM (McCarley and Hoffman, 1981), and give rise to what can be called REM consciousness. Rapid eye movements (REM) during the dreaming stage of sleep is accompanied by visual system activation. Ponto-geniculo-occipital (PGO) waves, as the name indicates, travel widely from the pons in the brainstem to the (lateral) geniculate nucleus in the thalamus, and then to the visual cortex, thus triggering powerful visual sensations when dreaming. REM consciousness is also characterized by cerebral cortical activation and atonia of nonrespiratory muscles so that individuals do not act on their dreams (Lu et al., 2006, Anaclet et al., 2010). Dreaming, another hallmark of REM consciousness, takes place in cortical areas physically far removed from the brainstem (Bischof and Bassetti, 2004).

It is believed all mammals enter REM consciousness, which testifies to this conscious state's importance. Yet the biologic purpose of dreams remains elusive. The limbic system contributes emotions to our dreams, and sometimes that emotion is fear (Merritt et al., 1994). Limbic structures appeared early in vertebrate evolutionary development, before the primate neocortex formed (Crosby and Schnitzlein, 1982). Dreams have long been considered important to instinctual behavior, and it is argued that one role of our dreams is to simulate and then rehearse solutions to threats (Jouvet, 1973, Valli and Revonsuo, 2009).

9. Into a Second Borderland of Consciousness

Different REM elements can be independently expressed. When consciousness transitions between REM and waking, REM can fragment and leave its traces in waking consciousness. REM intrusions into waking consciousness can take the form of complex visual and auditory hallucinations, as well as the atonia of sleep paralysis or cataplexy. Although a person can get stuck in the borderland between REM and waking, this borderland is unstable--within seconds or minutes consciousness reverts to a more stable REM or waking state. How does someone get caught in this second borderland, the one between REM and waking consciousness? The REM consciousness switch, located near the LC, shifts us between these two states (Lu et al., 2006). The switch is made up of several components. Some of these components tilt consciousness to REM, and others tilt us awake. The switch operates in an all or none flip-flop fashion, and most but not all of the time, it crisply moves the brain between REM and waking.

10. The REM Consciousness Switch in Moments of Crisis

A critical component of the REM switch is the ventrolateral portion of the periaqueductal grey (vlPAG) (Lu et al., 2006). When it activates, consciousness tilts towards waking and away from REM. The vlPAG is fundamental to suppressing REM consciousness (Sapin et al., 2009).

It is fascinating what the vlPAG region does or doesn't do in response to hypotension (with low cerebral blood flow). The vlPAG activates during pain, hypoxia, or moderate blood loss (Keay et al., 2002). This is also when the LC is very active. But something profoundly changes when blood pressure becomes profoundly low. Here vlPAG neurons cause the peripheral adrenergic response to recede, bringing the cholinergic system to dominance (Vagg et al., 2008). The heart, instead of beating rapidly to maintain blood pressure, slows and blood pressure falls even further.

Why would the vlPAG do that?

As the vlPAG retracts the adrenergic nervous system, the once agitated animal with shifting attention becomes quiet and inattentive (Persson and

Svensson, 1981), disengaging from its surroundings. Remaining quiet and still when an injury is severe or inescapable may be an effective survival strategy (Bandler et al., 2000). Whatever its advantage, this response to crisis is effective enough to have become imbedded within our evolutionarily ancient and conserved brainstem. Presumably the LC shifts from high discharge rates to the slow pulse of low arousal.

What brings the LC to this sluggish beat?

Normally when the vlPAG subsides, REM consciousness immediately follows. Is the REM system, the powerful brake on the LC, engaged in crisis?

It is uncertain if the vlPAG neurons responsible for diminishing peripheral adrenergic nervous system activity in the face of hypotension are the same, or even if they are functionally related to the neurons within the REM consciousness switch. That question lies before us. Yet it strains probability to dismiss as simply coincidence the adrenergic withdrawal in two physiological domains coordinated by the same brainstem region.

The known connections between the REM switch and LC emphasizes that successful behaviors like "fight-or-flight", or lying quietly must be coupled to the right conscious state. Although these connections indicate how our visceral responses might bring us into REM consciousness during crisis, what evidence supports this notion?

11. Blending REM and Waking Consciousness is not a Fluke

Blending REM into waking consciousness frequently occurs but is infrequently recognized. The REM atonia of sleep paralysis happens in the lifetime of roughly six percent of people (Ohayon et al., 2002, Aldrich, 1996), often combined with visual or auditory hallucinations (Ohayon et al., 1999, Fukuda et al., 1987, Cheyne et al., 1999, Takeuchi et al., 1992). Cataplexy is less common, occurring in 1.2% to 3.2% (Ohayon et al., 2002, Hublin et al., 1994). REM visual activation during waking consciousness is found in 19 to 28 % of people (Aldrich, 1996, Ohayon et al., 2002, Cheyne et al., 1999).

The borderland between REM and waking consciousness underlies other clinical conditions, and nowhere is this truer than for narcolepsy. Here the cardinal abnormality is an inability to control the boundaries between REM and waking consciousness (Broughton et al., 1986). Patients with narcolepsy have a hypocretin deficiency causing their REM switch, including the vlPAG, to tilt rapidly and frequently between waking and REM consciousness (Lu et al., 2006, Kaur et al., 2009).

It is reasonable to expect that the brains are different in the 6.3% to 12% who survive cardiac dysrhythmia with a near-death experience (Parnia et al., 2001, Greyson, 2003, van Lommel et al., 2001). In fact, persons with a near-death experience have an arousal system predisposed to entering the borderland of REM and waking consciousness (Nelson et al., 2006). Sixty percent had in their life some form of REM blending into waking consciousness. This is greater than the twenty-four percent of age and gender matched controls. Remarkably, the forty-six percent of those near-death who also had sleep paralysis compares similarly to the fifty percent of narcoleptics with sleep paralysis (Aldrich, 1996).

12. An Origin of Near-Death Experience

The experience of near-death requires a confluence of events. Oftentimes one key factor is the psychological reaction to danger. Survivors of danger can feel detached or psychologically dissociated from the world or their body (Noyes et al., 1977). They may also experience heightened arousal with thoughts speeded up, sharp or lucid. A sense of greater control is common, and all of this likely reflects neural pathways that improve survival by diminishing panic. These pathways may also be utilized during syncope in the absence of danger (Lempert et al., 1994b).

Near-death experiences can rise from danger alone and be nearly indistinguishable from those emerging during physiological crisis like cardiac dysrhythmia (Owens et al., 1990). One of the few discerning features is the appearance of "enhanced light" during genuine physiological threat.

13. The Tunnel and Light

At the beginning of syncope, a tunnel-like peripheral to central visual loss develops over 5 to 8 seconds from retinal ischemia, while brain cortical functions remain (Lambert and Wood, 1946). Ambient light is at the end of the tunnel. The eyes are kept open during syncope (Lempert and von Brevern, 1996), allowing light to strike an ischemic and failing retina. Smudges of outside light at the end of the tunnel may be all the brain is capable of seeing while on the threshold of unconsciousness. Light is the prime sensation in REM consciousness, and the often cited light of near-death could also arise from visual system activation brought on by REM mechanisms. In the absence of retinal input, pontine REM mechanisms are the dominant influence over the visual relay to the cerebral cortex (McCarley et al., 1983).

Cortical ischemia alone would not prevent REM light and visions. The cortically blind are capable of visual dream imagery (Solms, 1997), although during REM the primary visual (striate) cortex is deactivated (Braun et al., 1997), the extrastriate visual cortex is activated (Braun et al., 1998). Moreover, simple and complex visual hallucinations are reported by a majority during the brain ischemia of syncope (Lempert et al., 1994b).

14. Nerves of the Heart Draw Us into the Borderland of Consciousness

Undoubtedly the near-death conditions of danger, imperiled cerebral blood flow and cardio-respiratory crisis heighten cardio-respiratory afferent nerve activity. Autonomic nervous system afferent fibers transmit information from stretch, pressure, mechanical, and chemical receptors located within the heart, vascular and pulmonary systems. These impulses are conveyed to the brainstem principally by the vagus, but also by the glossopharygeal and trigeminal nerves. The cervical portion of the vagus is made up of approximately 80% visceral afferents (Agostoni et al., 1957). Vagal afferents alone robustly bring the brain into REM consciousness.

Electrical stimulation of the vagus in animals triggers the visual, cortical and atonic physiological facets of REM consciousness (Puizillout and Foutz, 1976, Valdes- Cruz et al., 2002, Fernandez-Guardiola et al., 1999, Puizillout and Foutz, 1977, Foutz et al., 1974). The transition from waking to REM can be so brisk that it spawned the terms "reflex REM narcolepsy" (Puizillout and Foutz, 1977) and "narcoleptic reflex" (Valdes-Cruz et al., 2002). This rapid transition between conscious states is the very condition that in humans causes REM and waking consciousness to merge (Takeuchi et al., 1992).

When the vagus nerve is stimulated in humans to treat epilepsy, REM intrudes into non-REM consciousness (Malow et al., 2001). Furthermore, the cardio-respiratory instability arising from an autoimmune attack on cardiac, vascular and respiratory autonomic nerve fibers in patients with Guillain-Barré syndrome leads to florid intrusion of REM consciousness (Cochen et al., 2005).

How can the vagus nerve shift consciousness?

Vagal afferents project upwards to synapse within the medullary nucleus tractus solitarius. From here, neural fibers rise to the pontine parabrachial nuclear complex (PBN) that is the principal relay for ascending cardio-respiratory afferents to the forebrain. In addition, the nucleus tractus solitarius and the PBN reciprocally connect with cholinergic REM structures (Semba and Fibiger, 1992, Quattrochi et al., 1998). The PBN region forms an intersection where neurons promoting and functioning specifically during REM consciousness (Datta et al., 1992, Datta and Hobson, 1994) intermingle with neurons participating in cardio-respiratory function (Chamberlin and Saper, 1992, Chamberlin and Saper, 1994).

Although these relationships lead to a connection between fight-or-flight, consciousness, and cardio-respiratory crisis, the full story of vagal afferents and REM consciousness remains to be seen. What is clear is that through its nerves the heart can cause REM consciousness, thereby transporting consciousness to unexpected places.

15. Out-of Body Experience and REM Consciousness

Out-of body experience (autoscopy) is an astonishingly common and normal experience. In a survey of over 13,000 people, 5.8% reported at least one autoscopic experience (Ohayon, 2000). Autoscopy is a common feature of near-death, that occurs with danger alone (Noyes and Kletti, 1976), and surprisingly does not distinguish between those who are or are not medically near death (Owens et al., 1990). This is consistent with the observation that syncope in the safe laboratory provokes autoscopy about 10% of the time (Lempert et al., 1994b).

Autoscopy has a long and established relationship with REM consciousness. Narcoleptics are prone to autoscopy (Mahowald and Schenck, 1992, Overeem et al., 2001), and its frequency wanes as the narcolepsy is treated. Autoscopy may appear in lucid dreams, a special expression of dreaming when the dreamer possesses the insight they are dreaming (LaBerge et al., 1988). In young healthy adults autoscopy accompanies the sleep paralysis of REM consciousness (Cheyne and Girard, 2009).

Autoscopy is directly produced by stimulating the temporoparietal region (Blanke et al., 2002), probably by disturbing sensory integration into the coherent self. Other temporoparietal disruptions cause autoscopy as well (Blanke et al., 2004). The temporoparietal region is also selectively inactive during REM consciousness, directly suggesting how REM consciousness and autoscopy are related (Maquet et al., 2005) (figure 5).

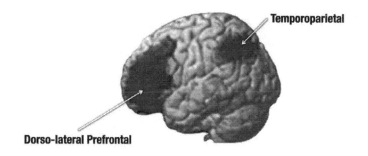

Figure 5. A combined 207 PET scans during REM consciousness show selective metabolic inactivity of the dorso-lateral prefrontal and temporoparietal regions. Adopted from Maquet et al., (2005) with permission

Autoscopy is particularly forceful during REM consciousness. Persons with a near-death experience are as likely to have autoscopy transitioning between REM and waking consciousness as they are to have autoscopy during near-death itself (Nelson et al., 2007). Often their autoscopy occurs during sleep paralysis.

16. Heaven-like Rewards

The brain's reward system could underlie the feelings of rapture, peace or euphoria often present during near-death. The REM consciousness promoting PPT and LDT nuclei are also instrumental to promoting reward behavior (Yeomans et al., 1993). Pathways from these REM structures project to an integral part of the reward system, the brainstem ventral tegmental region (Oakman et al., 1995). During REM consciousness ventral tegmental neurons vigorously discharge (Dahan et al., 2007). In animals, PPT injury reduces the reward seeking behavior for many strong stimuli including food (Alderson et al., 2002) and self administered heroin (Olmstead et al., 1998). In humans, the limbic and paralimbic regions active in REM sleep are also important in the reward system (Nofzinger et al., 1997). Pleasant or positive feelings are common during syncope (Lempert et al., 1994b), suggesting these reward pathways have been activated.

17. Divine-like Dreams

REM consciousness during peril provides a mechanism for activating limbic and paralimbic structures believed to underpin the narrative, ineffable, transcendental and paranormal qualities of near-death. In REM sleep, amygdala and anterior cingulate gyrus activity is detected on PET scan (Nofzinger et al., 1997, Braun et al., 1997, Maquet et al., 1996), and PGO waves propagate to the basolateral amygdala, cingulate gyrus, and hippocampus (Calvo and Fernandez-Guardiola, 1984). REM consciousness could also underlie the "dreaming" during syncope that pilots report (Lambert and Wood, 1946, Forster and Whinnery, 1988).

Many ancient and modern cultures regard dreams an augur of the future, and connection to the divine and deceased. There are many shared narrative qualities of dreaming and near-death, and a fuller comparison is found elsewhere (Nelson, 2011). One such example is sensing "someone's" presence. This happens in 9% of people as REM intrudes into waking consciousness (Ohayon, 2000), and bears similarity to the presences sensed during near-death. These presences, like autoscopy, can arise directly by stimulating the temporoparietal cortex (Arzy et al., 2006).

Obviously dreams and near-death differ. Near-death is recalled with an intense sense of realness that sharply contrasts to dreams. Near-death narratives lack the bizarreness of dreams. Yet near-death can be almost identical to lucid dreams (LaBerge and Rheingold, 1990), when the dorso-lateral prefrontal region, instrumental to logical executive cognition, may remain active during REM consciousness(Hobson, 2009).

Why some experiences seem real and others do not is a compelling ambiguity. But one thing is certain; our sense of reality profoundly shifts in unexpected ways in REM consciousness. Our dreams seem so real at the time regardless of how strange we find them on awaking. Inactivating the dorsolateral prefrontal brain during REM consciousness might contribute to suspending waking reality. How does this brain region function as REM blends with waking consciousness in crisis?

REM consciousness blurs the borders of reality. We must expect an unfamiliar and baffling reality when REM and waking consciousness merge to form a borderland of consciousness.

18. Epilogue

The notion that REM consciousness physiologically contributes to the experience of near-death falls within the realm of reasonable neurological probability. To understand the brain during near-death experience we needn't resort to extraordinary supernatural explanations when ordinary natural explanations suffice. Nonetheless, even if we know how the brain works during spiritual experience, the mystery of why lives on.

References

Abercrombie, E. D. & Jacobs, B. L. (1987). Single-unit response of noradrenergic neurons in the locus coeruleus of freely moving cats. II. Adaptation to chronically presented stressful stimuli. J.Neurosci., 7, 2844-2848.

Agostoni, E., Chinnock, J. E., Daly, M. D. B. & Murray, J. G. (1957). Functional and histological studies of the vagus nerve and its branches to the heart, lungs and abdominal viscera in the cat. J.Physiol, 135, 182-205.

Alderson, H. L., Brown, V. J., Latimer, M. P., Brasted, P. J., Robertson, A. H. & Winn, P. (2002). The effect of excitotoxic lesions of the pedunculopontine tegmental nucleus on performance of a progressive ratio schedule of reinforcement. Neuroscience, 112, 417-425.

Aldrich, M. S. (1996). The clinical spectrum of narcolepsy and idiopathic hypersomnia. Neurology, 46, 393-401.

Anaclet, C., Pedersen, N. P., Fuller, P. M. & Lu, J. (2010). Brainstem circuitry regulating phasic activation of trigeminal motoneurons during REM sleep. PLoS One, 5, e8788.

Arzy, S., Seeck, M., Ortigue, S., Spinelli, L. & Blanke, O. (2006). Induction of an illusory shadow person. Nature, 443, 287.

Aston-Jones, G., Rajkowski, J. & Cohen, J. (2000). Locus coeruleus and regulation of behavioral flexibility and attention. Prog Brain Res, 126, 165-82.

Author. (1988). What I saw when I was dead... The Sunday Telegraph, August 28, p.5.

Bandler, R., Keay, K. A., Floyd, N. & Price, J. (2000). Central circuits mediating patterned autonomic activity during active vs. passive emotional coping. Brain Res Bull, 53, 95- 104.

Benarroch, E. E. (2008). The arterial baroreflex: functional organization and involvement in neurologic disease. Neurology, 71, 1733-8.

Benarroch, E. E. (2009). The locus ceruleus norepinephrine system: functional organization and potential clinical significance. Neurology, 73, 1699-704.

Bischof, M. & Bassetti, C. L. (2004). Total dream loss: A distinct neuropsychological dysfunction after bilateral PCA stroke. Ann Neurol, 56, 583-586.

Blanke, O., Landis, T., Spinelli, L. & Seeck, M. (2004). Out-of-body experience and autoscopy of neurological origin. Brain, 127, 243-58.

Blanke, O., Ortigue, S., Landis, T. & Seeck, M. (2002). Neuropsychology: Stimulating illusory own-body perceptions. Nature, 419, 269-270.

Bodineau, L. & Larnicol, N. (2001). Brainstem and hypothalamic areas activated by tissue hypoxia: Fos-like immunoreactivity induced by carbon monoxide inhalation in the rat. Neuroscience, 108, 643-653.

Braun, A. R., Balkin, T. J., Wesensten, N. J., Gwadry, F., Carson, R. E., Varga, M., Baldwin, P., Belenky, G. & Herscovitch, P. (1998). Dissociated pattern of activity in visual cortices and their projections during human rapid eye movement sleep. Science, 279, 91-5.

Braun, A. R., Balkin, T. J., Wesenten, N. J., Carson, R. E., Varga, M., Baldwin, P., Selbie, S., Belenky, G. & Herscovitch, P. (1997). Regional cerebral blood flow throughout the sleep-wake cycle. An H2(15)O PET study. Brain, 120, 1173-1197.

Brenner, R. P. (1997). Electroencephalography in syncope. J.Clin Neurophysiol., 14, 197-209.

Broughton, R., Valley, V., Aguirre, M., Roberts, J., Suwalski, W. & Dunham, W. (1986). Excessive daytime sleepiness and the pathophysiology of narcolepsy-cataplexy: a laboratory perspective. Sleep, 9, 205-215.

Calvo, J. M. & Fernandez-Guardiola, A. (1984). Phasic activity of the basolateral amygdala, cingulate gyrus, and hippocampus during REM sleep in the cat. Sleep, 7, 202-210.

Cannon, W. B. 1923-2003. Bradford Cannon Papers, 1911. Boston, MA: Harvard Medical Library, Francis A. Countway Library of Medicine.

Chamberlin, N. L. & Saper, C. B. (1992). Topographic organization of cardiovascular responses to electrical and glutamate microstimulation of the parabrachial nucleus in the rat. J.Comp Neurol., 326, 245-262.

Chamberlin, N. L. & Saper, C. B. (1994). Topographic organization of respiratory responses to glutamate microstimulation of the parabrachial nucleus in the rat. J.Neurosci., 14, 6500- 6510.

Cheyne, J. A. & Girard, T. A. (2009). The body unbound: vestibular-motor hallucinations and out-of-body experiences. Cortex, 45, 201-15.

Cheyne, J. A., Rueffer, S. D. & Newby-Clark, I. R. (1999). Hypnagogic and hypnopompic hallucinations during sleep paralysis: neurological and cultural construction of the nightmare. Conscious.Cogn, 8, 319-337.

Cochen, V., Arnulf, I., Demeret, S., Neulat, M. L., Gourlet, V., Drouot, X., Moutereau,

S., Derenne, J. P., Similowski, T., Willer, J. C., Pierrot-Deseiligny, C. & Bolgert, F. (2005). Vivid dreams, hallucinations, psychosis and REM sleep in Guillain-Barre syndrome. Brain, 128, 2535-45.

Crosby, E. C. & Schnitzlein, H. N. (1982). Comparative correlative neuroanatomy of the vertebrate telencephalon, New York, Macmillan.

Dahan, L., Astier, B., Vautrelle, N., Urbain, N., Kocsis, B. & Chouvet, G. (2007). Prominent burst firing of dopaminergic neurons in the ventral tegmental area during paradoxical sleep. Neuropsychopharmacology, 32, 1232-41.

Datta, S., Calvo, J. M., Quattrochi, J. & Hobson, J. A. (1992). Cholinergic microstimulation of the peribrachial nucleus in the cat. I. Immediate and prolonged increases in pontogeniculo- occipital waves. Arch.Ital. Biol., 130, 263-284.

Datta, S. & Hobson, J. A. (1994). Neuronal activity in the caudolateral peribrachial pons: relationship to PGO waves and rapid eye movements. J.Neurophysiol., 71, 95-109.

Deegan, B. M., Devine, E. R., Geraghty, M. C., Jones, E., Olaighin, G. & Serrador, J. M. (2010). The relationship between cardiac output and dynamic cerebral autoregulation in humans. J Appl Physiol, 109, 1424-31.

Fernandez-Guardiola, A., Martinez, A., Valdes-Cruz, A., Magdaleno-Madrigal, V. M., Martinez, D. & Fernandez-Mas, R. (1999). Vagus nerve prolonged stimulation in cats: effects on epileptogenesis (amygdala electrical kindling): behavioral and electrographic changes. Epilepsia, 40, 822-829.

Foote, S. L., Aston-Jones, G. & Bloom, F. E. (1980). Impulse activity of locus coeruleus neurons in awake rats and monkeys is a function of sensory stimulation and arousal. Proc Natl Acad Sci U S A, 77, 3033-7.

Forster, E. M. & Whinnery, J. E. (1988). Recovery from Gz-induced loss of consciousness: psychophysiologic considerations. Aviat. Space Environ.Med., 59, 517-522.

Foutz, A. S., Ternaux, J. P. & Puizillout, J. J. (1974). Les stades de sommeil de la preparation "encephale isole": II. Phases paradoxales. leur declenchement par la stimulation des afferences baroceptives. Electroencephalogr. Clin.Neurophysiol., 37, 577-588.

Fukuda, K., Miyasita, A., Inugami, M. & Ishihara, K. (1987). High prevalence of isolated sleep paralysis: kanashibari phenomenon in Japan. Sleep, 10, 279-286.

Greyson, B. (2003). Incidence and correlates of near-death experiences in a cardiac care unit. Gen Hosp Psychiatry, 25, 269-76.

Hobson, J. A. (2009). REM sleep and dreaming: towards a theory of protoconsciousness. Nat Rev Neurosci, 10, 803-13.

Hobson, J. A., Mccarley, R. W. & Wyzinski, P. W. (1975). Sleep cycle oscillation: reciprocal discharge by two brainstem neuronal groups. Science, 189, 55-58.

Hublin, C., Kaprio, J., Partinen, M., Koskenvuo, M., Heikkila, K., Koskimies, S. & Guilleminault, C. (1994). The prevalence of narcolepsy: an epidemiological study of the Finnish Twin Cohort. Ann.Neurol, 35, 709-716.

James, W. & Marty, M. E. 1982. The varieties of religious experience : a study in human nature, Harmondsworth, Middlesex, England ; New York, N.Y., Penguin Books.

Jones, T. H., Morawetz, R. B., Crowell, R. M., Marcoux, F. W., Fitzgibbon, S. J., Degirolami, U. & Ojemann, R. G. (1981). Thresholds of focal cerebral ischemia in awake monkeys. J Neurosurg, 54, 773-82.

Jouvet, M. (1973). [Sleep study]. Arch Ital Biol, 111, 564-76.

Jung, C. G. & Jaffé, A. 1989. Memories, dreams, reflections, New York, Vintage

Books. Kaehler, S. T., Singewald, N. & Philippu, A. (1999). The release of catecholamines in hypothalamus and locus coeruleus is modulated by peripheral chemoreceptors. Naunyn Schmiedebergs Arch.Pharmacol., 360, 428-434.

Kandel, E. R. (1998). A new intellectual framework for psychiatry. Am.J.Psychiatry, 155, 457- 469.

Kaur, S., Thankachan, S., Begum, S., Liu, M., Blanco-Centurion, C. & Shiromani, P. J. (2009). Hypocretin-2 saporin lesions of the ventrolateral periaquaductal gray (vlPAG) increase REM sleep in hypocretin knockout mice. PLoS One, 4, e6346.

Keay, K. A., Clement, C. I., Matar, W. M., Heslop, D. J., Henderson, L. A. & Bandler, R. (2002). Noxious activation of spinal or vagal afferents evokes distinct patterns of fos-like immunoreactivity in the ventrolateral periaqueductal gray of unanaesthetised rats. Brain Res, 948, 122-30.

Laberge, S., Levitan, L., Brylowski, A. & Dement, W. (1988). "Out-of-body" experiences occurring in REM sleep. Sleep Res, 17, 115.

Laberge, S. & Rheingold, H. 1990. Exploring the World of Lucid Dreaming, New York, NY, Ballantine.

Lambert, E. H. & Wood, E. H. (1946). The problem of blackout and unconsciousness in aviators. Med.Clinic.North Am., 30, 833-844.

Lempert, T., Bauer, M. & Schmidt, D. (1994a). Syncope and near-death experience. Lancet, 344, 829-830.

Lempert, T., Bauer, M. & Schmidt, D. (1994b). Syncope: a videometric analysis of 56 episodes of transient cerebral hypoxia. Ann. Neurol., 36, 233-237.

Lempert, T. & Von Brevern, M. (1996). The eye movements of syncope. Neurology, 46, 1086-1088.

Lu, J., Sherman, D., Devor, M. & Saper, C. B. (2006). A putative flip-flop switch for control of REM sleep. Nature, 441, 589-94.

Mahowald, M. W. & Schenck, C. H. (1992). Dissociated states of wakefulness and sleep. Neurology, 42, 44-51.

Malow, B. A., Edwards, J., Marzec, M., Sagher, O., Ross, D. & Fromes, G. (2001). Vagus nerve stimulation reduces daytime sleepiness in epilepsy patients. Neurology, 57, 879-884.

Maquet, P., Peters, J., Aerts, J., Delfiore, G., Degueldre, C., Luxen, A. & Franck, G. (1996). Functional neuroanatomy of human rapid-eye-movement sleep and dreaming. Nature, 383, 163-166.

Maquet, P., Ruby, P., Maudoux, A., Albouy, G., Sterpenich, V., Dang-Vu, T., Desseilles, M., Boly, M., Perrin, F., Peigneux, P. & Laureys, S. (2005). Human cognition during REM sleep and the activity profile within frontal and parietal cortices: a reappraisal of functional neuroimaging data. Prog Brain Res, 150, 219-27.

Mccarley, R. W., Benoit, O. & Barrionuevo, G. (1983). Lateral geniculate nucleus unitary discharge in sleep and waking: state- and rate-specific aspects. J.Neurophysiol., 50, 798- 818.

Mccarley, R. W. & Hoffman, E. (1981). REM sleep dreams and the activation-synthesis hypothesis. Am J Psychiatry, 138, 904-12.

Merritt, J. M., Stickgold, R., Pace-Schott, E., Williams, E. F. & Hobson, J. A. (1994). Emotion Profiles in the Dreams of Young Adult Men and Women. Conscious Cogn, 3, 46-60.

Moody, R. 1987. Elvis After Life, Atlanta, GA, Peachtree Publishers, Ltd. Morse, M., Castillo, P., Venecia, D., Milstein, J. & Tyler, D. C. (1986). Childhood near-death experiences. Am.J.Dis. Child, 140, 1110-1114.

Nelson, K. R. 2011. The Spiritual Doorway in the Brain: A Neurologist's Search for the God

Experience, New York, NY, Penguin Group.

Nelson, K. R., Mattingly, M., Lee, S. A. & Schmitt, F. A. (2006). Does the arousal system contribute to near death experience? Neurology, 66, 1003-9.

Nelson, K. R., Mattingly, M. & Schmitt, F. A. (2007). Out-of-body experience and arousal. Neurology, 68, 794-5.

Nofzinger, E. A., Mintun, M. A., Wiseman, M., Kupfer, D. J. & Moore, R. Y. (1997). Forebrain activation in REM sleep: an FDG PET study. Brain Res., 770, 192-201.

Noyes, R., Jr., Hoenk, P. R., Kuperman, S. & Slymen, D. J. (1977). Depersonalization in accident victims and psychiatric patients. J Nerv Ment Dis, 164, 401-7.

Noyes, R., Jr. & Kletti, R. (1976). Depersonalization in the face of life-threatening danger: a description. Psychiatry, 39, 19-27.

Oakman, S. A., Faris, P. L., Kerr, P. E., Cozzari, C. & Hartman, B. K. (1995). Distribution of pontomesencephalic cholinergic neurons projecting to substantia nigra differs significantly from those projecting to ventral tegmental area. J.Neurosci., 15, 5859-5869.

Ohayon, M. M. (2000). Prevalence of hallucinations and their pathological associations in the general population. Psychiatry Res., 97, 153-164.

Ohayon, M. M., Priest, R. G., Zulley, J., Smirne, S. & Paiva, T. (2002). Prevalence of narcolepsy symptomatology and diagnosis in the European general population. Neurology, 58, 1826-1833.

Ohayon, M. M., Zulley, J., Guilleminault, C. & Smirne, S. (1999). Prevalence and pathologic associations of sleep paralysis in the general population. Neurology, 52, 1194-1200.

Olmstead, M. C., Munn, E. M., Franklin, K. B. & Wise, R. A. (1998). Effects of pedunculopontine tegmental nucleus lesions on responding for intravenous heroin under different schedules of reinforcement. J.Neurosci., 18, 5035-5044.

Overeem, S., Mignot, E., Van Dijk, J. G. & Lammers, G. J. (2001). Narcolepsy: clinical features, new pathophysiologic insights, and future perspectives. J Clin Neurophysiol, 18, 78-105.

Owens, J. E., Cook, E. W. & Stevenson, I. (1990). Features of "near-death experience" in relation to whether or not patients were near death. Lancet, 336, 1175-1177.

Parnia, S., Waller, D. G., Yeates, R. & Fenwick, P. (2001). A qualitative and quantitative study of the incidence, features and aetiology of near death experiences in cardiac arrest survivors. Resuscitation, 48, 149-156.

Pasricha, S. & Stevenson, I. (1986). Near-death experiences in India. A preliminary report. J.Nerv.Ment.Dis., 174, 165-170.

Persson, B. & Svensson, T. H. (1981). Control of behaviour and brain noradrenaline neurons by peripheral blood volume receptors. J Neural Transm, 52, 73-82.

Puizillout, J. J. & Foutz, A. S. (1976). Vago-aortic nerves stimulation and REM sleep: evidence for a REM-triggering and a REM-maintenance factor. Brain Res., 111, 181-184.

Puizillout, J. J. & Foutz, A. S. (1977). Characteristics of the experimental reflex sleep induced by vago-aortic nerve stimulation. Electroencephalogr.Clin.Neurophysiol., 42, 552-563.

Quattrochi, J., Datta, S. & Hobson, J. A. (1998). Cholinergic and non-cholinergic afferents of the caudolateral parabrachial nucleus: a role in the long-term enhancement of rapid eye movement sleep. Neuroscience, 83, 1123-1136.

Rajkowski, J., Kubiak, P. & Aston-Jones, G. (1994). Locus coeruleus activity in monkey: phasic and tonic changes are associated with altered vigilance. Brain Res.Bull., 35, 607-616.

Reiner, P. B. (1986). Correlational analysis of central noradrenergic neuronal activity and sympathetic tone in behaving cats. Brain Res., 378, 86-96.

Sapin, E., Lapray, D., Berod, A., Goutagny, R., Leger, L., Ravassard, P., Clement, O., Hanriot, L., Fort, P. & Luppi, P. H. (2009). Localization of the brainstem GABAergic neurons controlling paradoxical (REM) sleep. PLoS One, 4, e4272.

Semba, K. & Fibiger, H. C. (1992). Afferent connections of the laterodorsal and the pedunculopontine tegmental nuclei in the rat: a retro- and antero-grade transport and immunohistochemical study. J.Comp Neurol., 323, 387-410.

Solms, M. 1997. The neuropsychology of dreams: A clinico-anatomical study., Mahwah, NJ, Erlbaum.

Takeuchi, T., Miyasita, A., Sasaki, Y., Inugami, M. & Fukuda, K. (1992). Isolated sleep paralysis elicited by sleep interruption. Sleep, 15, 217-225.

Vagg, D. J., Bandler, R. & Keay, K. A. (2008). Hypovolemic shock: critical involvement of a projection from the ventrolateral periaqueductal gray to the caudal midline medulla. Neuroscience, 152, 1099-109.

Valdes-Cruz, A., Magdaleno-Madrigal, V. M., Martinez-Vargas, D., Fernandez-Mas, R., Almazan-Alvarado, S., Martinez, A. & Fernandez-Guardiola, A. (2002). Chronic stimulation of the cat vagus nerve: effect on sleep and behavior. Prog. Neuropsychopharmacol.Biol.Psychiatry, 26, 113-118.

Valentino, R. J., Page, M. E. & Curtis, A. L. (1991). Activation of noradrenergic locus coeruleus neurons by hemodynamic stress is due to local release of corticotropin-releasing factor. Brain Res., 555, 25-34.

Valli, K. & Revonsuo, A. (2009). The threat simulation theory in light of recent empirical evidence: a review. Am J Psychol, 122, 17-38.

Van Lommel, P., Van Wees, R., Meyers, V. & Elfferich, I. (2001). Near-death experience in survivors of cardiac arrest: a prospective study in the Netherlands. Lancet, 358, 2039-2045.

Yeomans, J. S., Mathur, A. & Tampakeras, M. (1993). Rewarding brain stimulation: role of tegmental cholinergic neurons that activate dopamine neurons. Behav.Neurosci., 107, 1077-1087.

41. Dreams and Hallucinations: Lifting the Veil to Multiple Perceptual Realities

Rhawn Joseph, Ph.D.

Emeritus, Brain Resarch Laboratory, California

Abstract

Regardless of culture or antiquity, humans have similar dreams and experience hallucinations. There are numerous reports of individuals who claim to have dreamed about future events which then took place, including deaths, tragedies, mass murders, horrific accidents, and environmental catastrophes. Yet others have experienced hallucinations where they have left their bodies, or have seen entities that appear to be from other realities. We should ask: is it really a hallucination if someone experiences a dissociation of consciousness and floats above their body and can later accurately described what was taking place and the appearance of those around them? Is it really a hallucination if an individual can see inside his hand and watch the blood cells swishing through his blood vessel? Abraham Lincoln, one of the great presidents in the history of the United States dreamed of his death 13 days before he was assassinated. Was it just a dream? Based on the evidence marshaled here and elsewhere, it can be inferred that not all dreams and hallucinations are dreams and hallucinations. There are specific neuroanatomical structures within the brain, including the amygdala, hippocampus, and temporal lobe, which make these experiences possible. There are also neurotransmitters, such as serotonin, which inhibit these tissues of the mind, so as to suppress and prevent the reception of much of the information they are able to perceive. Why would the brain evolve capabilities which are suppressed? Why did neural structures evolve which can process multiple sensations simultaneously, but then come to be inhibited by serotonin?

Why would activation of specific brain areas result in the sensation of having left the body, or being privy to cosmic wisdom, or witnessing events and entities which appear to be from other dimensions or realities? It is concluded that these are mental capabilities which are de-evolving or still evolving, and which are or were destined to serve a specific purpose: Lifting the veil so we can gaze deeply into the past, the future, and the unknown.

KEY WORDS: Consciousness, dreams, hallucinations, alternate realities, near death, out of body experiences, dissociation, evolution, amygdala, hippocampus, limbic system, temporal lobe, serotonin, LSD.

"In sleep and in dreams we pass through the whole thought of earlier humanity.....as a man now reasons in dreams, so humanity reasoned for many thousands of years when awake....This atavistic element in man's nature continues to manifest itself in our dreams, for it is the foundation upon which the higher reason has developed and still develops in every individual. Dreams carry us back to remote conditions of human culture and afford us a means of understanding it better." --Friedrich Nietzsche (1879).

1. Dreams of Life's Past

When I was a boy of 3, and for many years until around age 7, I had dreams of a previous life... and in that dream of a previous life I had also been a little boy... playing by the sea shore... by the ocean... and there were crowds of people...and it was a warm summer day... and then...in the dream... the ocean began to recede... the ocean waters drew back back back... and I could see shells and fish flopping on the wet sand where moments before there had been ocean... and I ran to where the ocean had been, on the wet sand, picking up shells... and many other people also ran onto the wet sand, laughing and talking in amazement that the ocean had pulled back for miles and miles.... and then... and then... and then...

I walked further and further out to where the ocean had been, picking up giant shells some with wiggling living creatures still inside, and gazing in wonder at what the ocean had hidden but which was now revealed... and then, I heard screams... women and men and children were screaming... and in my dream, they were all running from the wet sand toward the dry shore...and people on the shore were also running... everyone was running away and screaming... and when I looked to see why they were running, I could see the ocean... it was still miles away--but it was a WALL OF OCEAN.. a WALL OF WATER looming up maybe 100 yards into the sky... and in my dream the wall of ocean was rushing forward, to where the ocean had been minutes before, toward where I was standing with sea shells in my hands... and I started running... like everyone else, running running running... and I could see, over my shoulder, behind me, the wall of ocean water coming closer, and closer... and faster faster faster... and I kept running... everyone was running and screaming...trying to get away... and then the towering WALL OF WATER was just behind me... then looming over me... and then it crashed down upon me... and the little boy that I was, in this previous life, drowned.... and then I awoke in my bed... the same boy who drowned, but a different boy...me...

I had this dream over and over... for years. The same dream...

Twenty years later, I learned, for the first time, about Tsunamis--- what I first dreamed about when I was 3 years old... was in fact exactly what happens if there is a giant Tsunami... the ocean pulls back and recedes... and people foolishly run out to where the ocean was... and then... the ocean comes rushing back as a wall of water drowning everyone who did not immediately run away...

How could I have dreamed so vividly about something 3-year old me, knew nothing about?

I have been here before... you have been here before... we will be here again...

According to Carl Jung (Jung, 1945, 1964), not all dreams are related to wish fulfillments or impressions from one's personal life. Some dreams contain very archaic elements which seem to have absolutely no bearing on the dreamer's personal experience. Instead these dreams consists of ancestral memories and archetypes, the residue of ancient impressions and profound experiences that somehow became litterally engraved into the mind and brain of humanity; ancestral memories which are recalled even thousands of years later in a dream.

Yet others have argued, and have presented considerable evidence to back up their claims, that some children in fact dream of a previous life and a previous death (Dossey, Greyson, Sturrock, Tucker 2011). The dream is thus, not a dream at all, but a personal memory, an experiential reincarnation, that is passed on through mechanisms as yet unknown.

2. Dreams of Genetic Destiny

Science of the future, would be perceived as magic, today. The science of today, would have been magic to those who lived just a few hundred years ago. Therefore, what seems to stretch the boundaries of science, and which then become confined to the realms of "the supernatural", may be explained by a science as yet unknown.

If a cutting of a plant is placed in water, takes root and blooms, is it a re-incarnation? Or a continuation? Our genetic ancestry stretches backwards in time to the first creatures to take root on Earth. Memories, too, are presumably stored in DNA. And these genetic memories need not be passed down strictly from father/mother to daughter/son. Genes may be horizontally transferred between species, along with the information, the genetic memories they store.

3. But What of Those Who Dream of the Future?

Dreams serve a number of purposes, and at times are highly improbable and bizarre. However, they often reflect something significant about the mental and emotional life of the dreamer, as well as other issues of concern. For example, when subjects are awakened repeatedly over the course of several days when physiological indices indicated they were dreaming, often an evolving thematic pattern, an unfolding story, can be discerned (Cartwright et al., 1980). These patterns frequently reflect mental-emotional activity concerned with the solution of particular problems (Cartwright 2010; Freud, 1900; Joseph 1988a, 1992a; Jouvet 2001; Jung, 1945, 1964).

For example, one subject, a student, noted that "after being woken many times and seeing three or four dreams a night, I could realize there was a certain problem being worked out, like coping with responsibilities that were thrust upon me, but that weren't necessarily my own but I took on anyway. It was working out the feelings of resentment of taking somebody else's responsibility, but I met them well in my dreams. A good thing about spending time in the sleep lab was you could relate a common bond to some of the dreams" (Cartwright et al., 1980, p. 277). Similar patterns were, of course, recognized by Freud (1900) and Jung (1945) many years ago.

Given all the multiple forms of information one is exposed to on a daily basis, coupled with the personal concerns of the dreamer, not surprisingly this information may be analyzed during the dream, in dream-language, and the resulting dream may reflect not just the past, but one's future intentions. Dreams often instruct the dreamer, much in the way thinking serves the conscious mind (Joseph 2011a), and dreams may be predict the future based on what it has been perceived thereby creating probabilistic scenarios which can serve as rehearsals for future behaviors.

4. Abraham Lincoln Dreams of His Assassination

In some cases, dreams do no just fore tell the future, but may predict the death of the dreamer:

In April of 1865, the commanding general of the Confederate Army, Robert E. Lee, had surrendered to General Ulysses S. Grant, and the days of the South were numbered. John Wilkes Booth, an actor and southern sympathizer hoped to rally the remaining Confederate troops to continue fighting and plotted with several other men in a conspiracy to kill President Abraham Lincoln.

On April 2, 1865, President Abraham Lincoln dreamed of his own death by assassination. The dream troubled him deeply, and on April 11, 1865, three days prior to his assassination, Abraham Lincoln shared this dream with his wife and a few friends which included Ward Hill Lamon (1865/1994):

> About ten days ago, I retired very late. I had been up waiting for important dispatches from the front. I could not have been long in bed when I fell into a slumber, for I was weary. I soon began to dream. There seemed to be a death-like stillness about me. Then I heard subdued sobs, as if a number of people were weeping. I thought I left my bed and wandered downstairs. There the silence was broken by the same pitiful sobbing, but the mourners were invisible. I went from room to room; no living person was in sight, but the same mournful sounds of distress met me as I passed along. I saw light in all the rooms; every object was familiar to me; but where were all the people who were grieving as if their hearts would break? I was puzzled and alarmed. What could be the meaning of all this? Determined to find the cause of a state of things so mysterious and so shocking, I kept on until I arrived at the East Room, which I entered. There I met with a sickening surprise. Before me was a catafalque, on which rested a corpse wrapped in funeral vestments. Around it were stationed soldiers who were acting as guards; and there was a throng of people, gazing mournfully upon the corpse, whose face was covered, others weeping pitifully. 'Who is dead in the White House?' I demanded of one of the soldiers, 'The President,' was his answer; 'he was killed by an assassin.' Then came a loud burst of grief from the crowd, which woke me from my dream. I slept no more that night; and although it was only a dream, I have been strangely annoyed by it ever since." -Abraham Lincoln

On April 14, 1865, President Lincoln was shot in the back of the head while watching the play "Our American Cousin" at Ford's Theatre in Washington, D.C. with his wife, Mary Todd Lincoln. He died the next morning.

5. Dreams of the Future

Among ancient societies dreams were seen as extremely important sources of information, not just regarding the past, but the future (Joseph, 1992a,b,

1996, 2001, 2002; Jung 1945). As possible harbingers of the future they had to be observed carefully interpreted.

In the ancient world, be it Greek, Rome, Egypt, or Babylon, it was believed that some dreams contain important information regarding not only the individual, but his friends, family, and even the entire clan, village, city, or nation (Freud, 1900; Joseph 1992a, 2001, 2002; Jung 1945, 1964): The "big dream" of a child, woman, or man, were taken seriously by highly sophisticated and cultured ancient societies and were even announced two thousand years ago in the Roman Senate. On one occasion a senator's daughter had a dream in which Minerva the Goddess, appeared and complained that her temple was being neglected by the Roman people. The dream was announced to the Senate which in turn voted funds for restoration of the temple.

The dreams of generals, kings, queens, emperors, and Pharaohs, were commonly scrutinized and their symbolism interpreted as they were believed to foretell the future:

>...Now Israel loved Joseph more than all his children, and he made him a coat of many colors... ...and Joseph dreamed a dream and he told it to his brethren... for behold we were binding sheaves in the field and lo, my sheaf arose, and stood upright: and, behold, your sheaves stood round about, and made obeisance to my sheaf. And his brethren said to him. Shalt thou indeed reign over us? And they hated him yet the more for his dreams...

>-Genesis 37

But the predictions of these dreams came to pass.

>...And... the Pharaoh dreamed: and behold, he stood by the river and seven well favored kine and flatfish came up out of the river and they fed in a meadow. And behold, seven other kine came up after them out of the river, ill favoured and leanfleshed. And the ill favoured and leanfleshed kine did eat up the seven well favoured and fat kine. So Pharaoh awoke. And he slept and dreamed the second time....

>...And it came to pass in the morning that his spirit was troubled and he sent and called for all the magicians and wise men of Egypt...but there was none that could interpret the dream...

>....And Joseph said unto Pharaoh, God hath showed Pharaoh what he is about to do. The seven good kine are seven years.... and the seven thin and ill favoured kine that came up after them are seven years... Behold, there come seven years of great plenty throughout all the land of Egypt: And there shall arise after them seven years of famine: and all the plenty shall be forgotten and the famine shall consume the land....

-Genesis 37

6. The Quantum Future is Now?

Joseph was not the first to dream of the future, nor would he be the last. How can this be? Why did Lincoln dream of his assassination?

It may also be that the past, present, and future are a simultaneity but which are located in different regions of space-time, within the 4th dimension. And it may be that it is the mind which journeys along the dimension, journeying across space-time and encountering what is experienced as the ever present now which slips away as quickly as it is grasped, only to be replaced by a future which becomes the now--just as a DVD or CD contains the beginning and ending of a film or song simultaneously, but which is encoded in different locations within the medium.

And just as the ripples of a pond may strike distant shores, the quantum states of the future may also effect the distant shores occupied by what the mind experiences as the now.

If there is a quantum continuum, then why should it be confined to what our minds define as the present? If the 4th dimension is space-time, and if differences in time are related to movement through space and thus distance between locations, then within the quantum continuum, everything is connected: stars, planets, dogs, cats, and the future and the past.

If correct, this would imply that there may be something unique about the process of dreaming, which enables some dreamers to enter this quantum state and dream of an ancestral past, or to see what may lie ahead in the future.

Or, it may be that that those regions of the mind which subserve dreams, analyze the myriad details commonly experienced to make predictions about the future. Could this explain Lincoln's dream of his own death? Did his dream simply explain his own realistic fears? Or did he gain access to information about the future?

7. Genetic Destiny

The future can be predicted by the past. The future is in fact engraved into our genes. Have we not inherited our genes which determined what we have become, and does not this genetic ancestry leads interminable into the past? And so, too, life in the future, may also be encoded into ancestral genes, for it is these ancestral genes, passed on from mother/father, which code for what will be: genetic destiny.

Certainly genes interact with the environment. However, much of the information contained in our genome is hardwired. Dogs behave like dogs, cats like cats, and human like humans, not because of free will, but genetic destiny. The hardware which supports the software is hardwired into our genes.

Our lives and the future are shaped, at least in part, by genetic destiny and our genetic ancestry is certainly much older than this Earth (Joseph 2011b). Indeed, there is evidence of life in this planet's oldest rocks, dated to 4.2 bya (Nemchin et al. 2008; O'Neil et al. 2008), indicating life was present on this planet from the very beginning. Let us engage in a thought experiment and imagine life on Earth came from other planets, and these seeds of life contained the genes for the tree of life which took root on Earth. However, if the same genetic seeds landed on other Earth-like planets, the same trees of life may have evolved, such that, humans just like the humans of Earth may populate innumerable worlds; and just as dogs behave like dogs, and cats like cats, those humans of other worlds may act just like us; because their genes and our genes have a common source.

Or let us say there are multiple dimensions, a multiverse with multiple worlds, many of which are just like our own and where our own cosmic quantum twins may have evolved, and where they behave just like us--my multiverse twin writing this article, and your multiverse twin reading it. We behave the same on these multiple worlds, because we have the same genes. Consider the often reported instances of twins separated at birth, but who go on to lead nearly identical lives, even marrying women who look alike and have the same first names.

If the future is engraved within our genomes, then this would imply there is something unique about dreams, which unlock these genetic codes, revealing to the conscious mind what had had been concealed.

In fact, gene expression is enhanced during dream sleep which in turn has been correlated with increased activity and plasticity within neurons in the neocortex and limbic system (Ribeiro et al. 1999, 2002, 2008). Therefore, dream sleep is not only associated with the activation of otherwise silent genes which change the shapes of neural interconnections, but may unlock the secrets of changes yet to come.

Be it genetics, the quantum continuum, a multiverse, the unconscious dream analysis of myriad details, or the opening of windows to sensory capabilities which had been inhibited or suppressed, it appears that during dreams the veils are lifted, thereby enabling the mind to see and reveal what had been concealed.

8. The Neuroanatomy of Dreams: Overview

Sleep consists of five distinct stages, one of which is closely associated with the appearance of dreams (Hobson 2004, Monti et al., 2008; Steriade & McCarley 2005). It is during the course of the dream that the eyes begin to move quite rapidly as if the dreamer were observing some action. This is referred to as REM (rapid eye movement). The appearance of REM during sleep has been found to occur in a rhythmical fashion in all terrestrial mammals so far studied.

REM occurs during a sleep stage referred to as "paradoxical sleep." It is called paradoxical (or active sleep), for electrophysiologically the brain is aroused and quite active, similar to its condition during waking. However, the body musculature is paralyzed and motor functioning is all but abolished except in certain regions which control respiration, and eye movements (Hobson 2004, Hobson et al. 1986; Jouvet 2001; Monti et al., 2008; Steriade & McCarley 2005). This prevents the dreamer from acting out their dreams. The ability to perceive outside sensory events, normally received through the five senses, is also greatly attenuated (Hobson 2004, Monti et al., 2008; Steriade & McCarley 2005; Vertes 1990). In addition during the course of a dream, temperture control is lost, pain sensation is rare, and males tend to have an erection.

These REM dream cycles occur every 90 to 100 minutes. By contrast, non-REM (N-REM) periods occur during a stage referred to as "slow-wave" or synchronized sleep.

Thus, during REM dream sleep, the brain is in a state of heightened activity and arousal, indicating that considerable processing of information is taking place. Yet, simultaneously, the normal routes of sensory reception have been restricted. Yet, the brain of the sleeper does in fact receive and process sensory information.

Most individuals awakened during REM report dream activity approximately 80% of the time. REM dreams involve a considerable degree of visual imagery, emotion, and tend to be distorted and implausible to various degrees (Foulkes, 1962; Hobson 2004; Steriade & McCarley 2005).

When awakened during the N-REM period, dreams are reported approximately 20% of the time (Foulkes, 1962; Jouvet 2001; Monroe et al. 1965). However, the type of dreaming that occurs during N-REM is quite different from REM. For example, N-REM dreams (when they occur) are often quite similar to thinking and speech (i.e. lingusitic thought), such that a kind-of rambling verbal monologue is experienced in the absence of imagery (Foulkes 1962; Hobson 2004; Jouvet 2001; Monroe et al. 1965). It is also during N-REM in which an individual is most likely to talk in his or her sleep (Kamiya, 1961).

9. Right Hemisphere Dreams

REM is characterized by high levels of activity within the pons of the brainstem, the lateral geniculate nucleus of the thalamus, and occipital lobes; referred to as PGO waves (Hobson 2004; Monti et al., 2008; Steriade & McCarley 2005). It also has been reported that electrophysiologically the right hemisphere becomes highly active during REM, whereas, conversely, the left half of the brain becomes more active during N-REM (Goldstein et al. 1972; Hodoba, 1986). This may account for the striking differences in the content of dreams, with left hemisphere dreams being more "thought-like" and verbal, and right hemisphere dreams more emotional, vivid, and visual-spatial.

Measurements of cerebral blood flow have shown an increase in the right temporal and parietal regions during REM sleep and in subjects who upon wakening report visual, hypnagogic, hallucinatory and auditory dreaming (Meyer et al., 1987). Moreover, abnormal and enhanced activity in the right temporal and temporal-occipital area acts to increase dreaming and REM sleep for an atypically long time period (Hodoba, 1986). Similarly, REM sleep is associated with increased activity in this same region much more than in the left hemisphere (Hodoba, 1986). These findings indicate that there is a specific complementary relationship between REM sleep and right temporal-occipital electrophysiological activity.

Conversely, there have been reports of patients with right cerebral damage who have ceased dreaming altogether or to dream only in words (Humphrey & Zangwill, 1951; Kerr & Foulkes, 1978, 1981). For example, defective dreaming, deficits that involve visual imagery, and loss of hypnagogic imagery have been found in patients with focal lesions or hypoplasia of the posterior right hemisphere and abnormalities in the corpus callosum which would prevent transfer from the right to left hemisphere (Botez et al. 1985; Kerr & Foulkes, 1981; Murri et al. 1984).

An absence or diminished amount of dreaming during sleep also has been reported after split-brain surgery; i.e., as reported by the disconnected left hemisphere (Bogen & Bogen, 1969; Hoppe & Bogen, 1977). Similarly, a paucity of REM episodes have been noted in other callosotomy patients, although these particular individuals continued to report some dream activity (Greenwood, Wilson, & Gazzaniga, 1977).

On the other hand it has been reported that when the left hemisphere has been damaged, particularly the posterior portions (i.e. aphasic patients), the ability to verbally report and recall dreams also is greatly attenuated (Murri et al., 1984; Pena-Casanova & Roig-Rovira, 1985; Schanfald et al. 1985). Of course, aphasics have difficulty describing much of anything, let alone their dreams.

The differential activation of the right and left hemisphere during REM vs N-REM, is a major factor in the visual-emotional hallucinatory-mosaic experienced during the dream (Joseph 1988a, 1996). As has been well established, the right hemisphere is dominant for most aspects of non-verbal and visual-spatial perceptual activity as well as the expression and comprehension of social-emotional nuances. It is for this reason that the right hemisphere is sometimes thought to be the more intuitive half of the cerebrum.

As demonstrated in individuals who have had the two hemispheres surgically separated, the right half of the brain is able to draw conclusions, make predictions, selectively store certain images and experiences in memory, and can call on and act on these information at will (Joseph 1988a,b; 1996). Moreover, given its sensitivity to a host of non-social environmental variables (Joseph 1988a), it is able to assimilate and draw conclusions if not make predictions about this material which, conversely, the left hemisphere has difficulty processiing. In this regard, it is not at all surprising that during the course of a dream, when the right half of the brain is at a peak level of activity, that it may draw upon these capacities to arrive at certain conclusions or to make predictions regarding events, people, or the future, and that those aspects of consciousness associated with the language-dependent aspects of the mind (Joseph 1988a,b, 2011a) would view these cognitions as bizarre and inexplicable.

However, other factors may also be involved, including the perception of sensory and other information which is normally filtered out and suppressed. That is, the right hemisphere may be perceiving stimuli during the course of the dream, which during waking is not normally perceived.

10. REM-On, REM-Off & Serotonin

The visual-emotional hallucinatory aspects of dreaming occur during REM, and the activation of a variety of brain regions are involved, i.e. the amygdala, hippocampus, right temporal lobe, right occipital lobe, the lateral geniculate nucleus of the thalamus, and brainstem nuclei located in the lateral and medial pons. In addition, the production of REM sleep is mediated by cyclic fluctuations in the levels of various neurotransmitters, including, and especially serotonin which serves an inhibitory function and when at high levels suppresses REM sleep and the activity of neurons which contribute to the generation of dreams.

Specifically, cholinergic (ACh) neurons located in the lateral pons, and neurons located in the medial pontine reticular formation appear to be the locus for REM-on neurons which initiate and/or maintain the production of REM sleep and which produce muscle atonia so that dreamers do not act on their dreams (Lydic, et al., 1991; Monti et al., 2008; Steriade & McCarley, 2005; Vertes 1990). That is, during the production of REM and paradoxical sleep, there is increased cholinergic activity and the production of pontine, lateral geniculate, occipital activity; i.e. PGO waves. Whereas with the termination of REM, and with the onset of slow wave N-REM sleep these same neurons greatly reduce their activity. As ACh is also implicated in memory, this may well explain why recent memories tend to become incorporated in dreams.

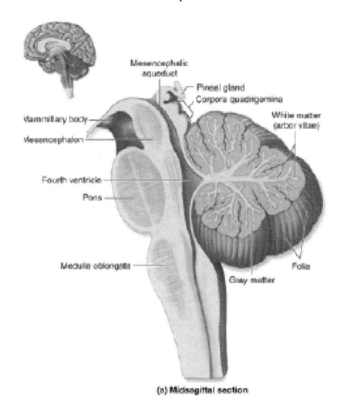

(a) Midsagittal section

In contrast, REM-off neurons, which tend to be located in the medial raphe nucleus (which contain 5HT neurons) and in the locus coeruleus (located at the midbrain-pons junction and which contain NE neurons) are highly active during waking but then significantly decrease their activity with the initition and onset of REM and the production of pontine, lateral geniculate, occipital activity; i.e. PGO waves (Hobson 2004; Monti et al., 2008; Steriade & McCarley 2005).

Hence, these REM-Off neurons appear to suppress REM and PGO activity and interfere with the onset of dreaming and paradoxical sleep, whereas REM-on neurons initiate the opposite sleep phase in which case the brain appears to be highly active and the individual begins to dream.

REM-on and REM-off neurons, therefore, appear to oscillate in a rhythmic fashion thus inducing sleep, dreaming, and waking. When the REM-off neurons cease to fire, the REM-on neurons (which are predominately cholinergic) become highly active until a REM episode is produced. However, as REM-on neuron activity decreases, REM-off activity increases, thus setting into motion a continuous 90 minute cycle of REM - Non-REM sleep.

11. Serotonin & Sensory Filtering

Low levels of 5HT are associated with REM sleep and dreaming (Hobson 2004; Monti et al., 2008) and thus with increased activity in the amygdala, hippocampus, and right hemisphere. 5HT in fact exerts inhibitory influences on a variety of brain structures, thereby suppressing incoming sensory input and the processing of sensory information from a variety of modalities simultaneously (Applegate, 1980; Jacobs & Azmita 1992; Soubrie, 1986; Spoont, 1992). That is 5HT restricts perceptual and information processing and in fact increases the threshold for neural responses to occur at both the neocortical and limbic level. In this way, attention can be focused and everything considered irrelevant may be filtered out. In fact, 5HT appears to be involved in learning not to respond to stimuli (Benninger, 1989). These signals are filtered out and suppressed.

It has also been demonstrated that 5HT acts to suppress activity in the lateral (visual) geniculate nucleus of the thalamus and synaptic functioning in the visual cortex as well as the amygdala (Jacobs & Azmita 1992; Soubrie, 1986; Spoont, 1992). By contrast, substances which block 5HT transmission -such as LSD- results in increased activity in the amygdala (Chapman & Walter, 1965; Chapman, Walter, Ross et al., 1963) and in the sensory pathways to the neocortex (Purpura 1956), which induces complex hallucinatory experiences.

12. The Amygdala: Gateway to Multiple Perceptual Realities

The amygdala is exceedingly responsive to social and emotional stimuli as conveyed vocally, visually, through touch, or body language including the face (Gloor, 1997; Halgren, 1992; Kling & Brothers 1992; Morris et al., 1996; 1992) and contains neurons which respond selectively to smiles and to the eyes, and which can differentiate between male and female faces and the emotions they convey (Hasselmo, Rolls, & Baylis, 1989, Kawashima, et al., 1999; Rolls, 1984; Morris et al., 1996). Single neurons in the amygdala, in fact, can respond to multiple sensory modalities, simultaneously (Amaral et al. 1992; O'Keefe & Bouma, 1969; Perryman, Kling, & Lloyd, 1987; Rolls 1992; Sawa & Delgado, 1963; Schutze et al. 1987; Turner et al. 1980; Ursin & Kaasa, 1960). Overall, because emotional, motivational, and multimodal assimilation of various sensory impressions occurs in this region, it is also involved in attention, learning, and memory (Gloor, 1997; Halgren, 1992; Halgren et al., 1978).

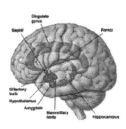 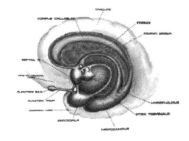

The right amygdala (as well as the right hippocampus, and the right hemisphere in general) is also involved in the production of dream imagery as well as REM sleep (Broughton, 1982; Goldstein et al., 1972; Hodoba, 1986; Humphrey & Zangwill, 1961; Kerr & Foulkes, 1978; Meyer et al. 1987). Simulation of the amygdala triggers and increases ponto-geniculo-occipital paradoxical activity during sleep (Calvo, et al. 1987), which in turn is associated with REM and dreaming.

Presumably, during REM, the amygdala (and hippocampus) serve as a reservoir from which various images, emotions, faces, words, and ideas are drawn and incorporated into the matrix of dream-like activity being woven by the right hemisphere. Essentially, this increased activity, resulting in REM sleep and the productions of dreams is due in in large part to cyclic reductions in serotonin.

Serotonin, inhibits these multi-modal amygdala neurons, thereby suppressing their activity and preventing the reception and processing of a vast array of sensory impressions. However, if serotonin release or uptake is prevented, the amygdala and other structures are released from their inhibitory sensory prison, and in consequence, single amygdala neurons will process multi-modal properties simultaneously, such that individuals will dream.

13. Dreaming Backwards

During the dream state, the dominant sensory streams are suppressed prior to transmission to the neocortical receiving areas. However, subcortical and limbic structures continue to receive and process sensory information, and not uncommonly these sensations, experienced during sleep, will be incorporated into the dream and may even trigger the dream, in which case, the dreamer may dream backwards

"Julie" dreams she is walking in San Francisco lugging large bags of gifts. Feeling tired she sets them down on the sidewalk. She looks for a bus and see a cable car coming. As it pulls up the conductor begins to ring its bell. The sound of the bell grows louder and then jolts her awake. Fully awake she realizes someone is ringing her doorbell. In this regard, the hearing of the bell seemed to be a natural part of the dream, and it is. What seems paradoxical, however, is that the dream seemed to lead up to the bell so that its ringing made sense in the context of the dream.

The dream did not lead up to the bell; the bell initiated the dream. The dream was produced, via the unique language of the right hemisphere during sleep (as well as amygdala activation), so as to explain the sound of the bell. The bell was heard and the dream was instantly produced in explanation and association. The bell stimulated the dream which may have only last a second.

One individual (described by Freud, 1900) dreamt he was in 18th Century France in the midst of the French Revolution. After a trial in which he was been found guilty, he was being led down a street lined with yelling and cursing Frenchmen and women. At the end of the street he could see the gallows where the heads of various political criminals were being chopped off at the neck. Mad with fright he felt and saw himself led up the stairs and his head being placed in the yoke of the chopping block. The executioner gave the signal, the crowd screamed its approval, he could hear and sense the blade falling, and with a loud crack it struck him across the neck. Indeed, it struck him with such a jolt that he awoke to find that his poster bed had broke and that a railing had fallen and struck him across the side and back of his neck.

Thus, we see that although the perception of external stimuli by the five senses is greatly attenuated, external stimuli can still trigger a complex dream. The dream explains this external stimuli. However, what if this stimuli is not transmitted via the dominant sensory channels, and is conveyed through electro-magentic radiation, pheromones, DNA, quantum field fluctuations, or the dreams of other dreamers?

14. Damon Wells: Son of the Devil

The following is based on court, psychiatric, police, and investigative records, and interviews with many of the principles directly involved.

District Attorney (D.A.): "Did you join the United States Army in 1980?

Witness (Stephen M): "Yes, sir."

D.A.: "You were stationed in Germany?"

S.M.: "Yes."

D.A.: "And while in Germany, did you come to know a person by the name of Damon Wells?"

S.M.: "Yes, sir."

D.A.: "Do you recall when you first met Damon Wells?"

S.M.: "Yes. It was 1981."

D.A.: "How did he introduce himself?"

S.M.: "He said, 'My name is Damon Wells, son of the devil.'"

D.A.: "Did you become friends with Damon Wells?"

S.M.: "We were both soldiers, and it was a pretty isolated site. Pretty secluded on top of a mountain, so I had a lot of contact with him."

D.A.: "What type of duties did you perform?"

S.M. "I was a satellite technician. Wells was a mechanic, a motor mechanic."

D.A.: "What kind of worker was he?"

S.M. "He was a fantastic worker. One of the best mechanics we ever had up there. He'd fix vehicles and the electric generators which powered our site. We didn't have any electricity up there."

D.A. "Did you ever notice anything odd about Damon?"

S.M. "Sure. Many times."

D.A. "Such as?"

S.M. "Sometimes he would stop what he was doing and would say, The voices are here. Can you hear them?"

D.A. "Anything else?"

S.M. "Damon used to sit for hours on top of this mountain, looking down into the valley. It was totally remote and isolated and heavily forested, but sometimes you could see him from the communications tower. Sometimes he would meditate by these ruins on the side of this mountain.

"He'd sit for hours, like in a trance. His legs would be crossed. He'd be rocking, and chanting, mumbling about demons and angels from Hell. Even when it began to rain, or even snow, Damon would just sit there, trance-like. He'd be soaking wet or freezing, sometimes in just shirt sleeves. I had to go get him a few times."

D.A. "Did Damon ever show you a cave?"

S.M. "Yes, he did. He had found this cave, hidden deep in the woods, on the back side of a mountain across the valley. I don't know how he ever found it. He had a rope tied to a tree on the mountain top, and you had to climb, to rappel down the rope in order to get to this cave. It was really kind of creepy."

D.A. "Did you ever go inside this cave?"

S.M. "Yes, I did. There was a large stone slab, and some candles and gun powder, and a knife and some books inside. It was obvious he had spent a lot of time there."

D.A. "Did Damon ever tell you why he visited this cave?"

S.M. "Yeah. He told me he went there to talk to the voices. The voices, from his point of view, came from the devil. He said he could communicate with the devil, and that the devil would appear in this cave and speak with him."

D.A. "Did he every talk to you about sacrifices?"

S.M. "Yes, he did. Especially after I pointed out what looked like animal bones, and blood stains on this big slab of rock. In fact, he showed me this book of black magic and it had a picture of a naked woman lying spread eagle on a rock. Standing over her was a hooded man with a big knife. Damon said that was how the devil would sacrifice his victims. Damon said, that since he was a slave of the devil, he was supposed to sacrifice victims the same way, on this rock."

D.A. "And what did you tell him after he said and showed these things to you?"

S.M. "I told him he had a serious problem, that he needed help. In fact, I suggested a few times that maybe he should see a psychiatrist or a priest."

D.A. "And what did he say when you told him he should see a priest?"

S.M. "He laughed and began talking about voodoo and demons. About how he could project his thoughts, and with the help of the demons, he could harm his enemies. He said that priests were enemies of the devil, they couldn't help him. Besides, he said, I have everything under control. The Demons do as I tell them.'"

D.A. "Did Damon ever tell you he had committed a serious crime? That he had sacrificed anyone to the devil?"

S.M. "Yes sir, he did. A few months later. He said he picked up a hitchhiker, drove her to the woods, and then killed her because the devil had told him to. He said the devil took control over him, and he killed her and had sex with her."

D.A. "Did you believe him?"

S.M. "No. He was really mixed up, confused like he was two people. He seemed possessed. I thought he was flipping out. When I asked if he had really killed someone, he just said, They're taking me over, Steve. They are taking me over."

On the afternoon of 8/27/84, Damon Wells had taken LSD and was again communing with the devil. Damon was to be sent on a mission. The devil wanted a woman. Damon was to sacrifice this woman for the devil. That woman would be Tanya Z.

That same afternoon, Tanya Z, a very attractive, dark haired, 21-year old woman, had just left work at a Santa Cruz bank and was accompanied to her car by a female friend and coworker. Tanya was chattering away and showing her friend a birthday card for her fiance and explaining that she was going shopping for a birthday present. However, first she would have to drive over the Santa Cruz mountains, on Highway 17, to stop at her house in Santa Clara County.

At the same time, Damon was also walking to his car, talking with the Devil who was urging him on, explaining what to do. A neighbor in the next yard overhead the mumble speech and thought Damon might have been talking to him? "What did you say?" he asked?

Damon looked at him: "Don't you hear them?"

"What?"

"The voices are here. Can you hear them?"

"Hear who?"

"The Devil," Damon replied. He got into his car and drove into the Santa Cruz mountains.

Tanya had just made it over the summit, the highest point on Highway 17, when her car began to stall, and then it died. She pulled over to the side of the busy highway, and then repeatedly attempted to get it to start, but to no avail. She had run out of gas.

It was then that Damon drove by. According to Damon, the voice of the Devil told him that this woman was to be the sacrifice that He required. Damon pulled up, introduced himself, and offered her a ride to the nearest phone. She agreed, and got in. Instead, he took her deep into the mountains, attacked her, beat her until she was nearly unconscious, and then dragged her down a heavily wooded mountainside, and then pushed away some branches which hid a trail leading to a huge rock. Damon dragged her to the rock, laid her out sacrificial style, killed her, sexually assaulted her, and then left the dead body lying naked on the rock.

That evening Tanya's father and fiancee began searching for her. They found only the abandoned car near the summit of highway 17. They contacted the police, who, however, could find no evidence of a crime. There was no body. No blood. No signs of a struggle. Just an abandoned car.

That evening, Damon, still experiencing the after-affects of the LSD, could not stop visualizing the murder. It played over and over in his mind, even after he fell asleep. But once he began to dream, the victim, Tanya, did not stay dead, but appeared to him, accusing him, showing him what he had done. His victim had become an angry avenging spirit.

That same night a woman named "Sunshine" a nudist who lived in a nudist colony, Lupin Lodge, situated in the Santa Cruz Mountains, had a nightmare: A woman was being brutally murdered. The next day, Sunshine read the story of Tanya's disappearance in the local newspaper, and that night, and the night following, she had the dream again, but this time the victim appeared to her. It was Tanya. But she was no avenging spirit. According to Sunshine, Tanya showed her the road off highway 17 where Damon had taken her, and then the spot where Damon had parked his car and attacked her. Next she led Sunshine down a rather steep incline, and then along a trail, and pointed out her naked body, lying spread eagle on this huge slab of rock. Sunshine had this same dream repeatedly.

Sunshine was not the only dreamer dreaming of the murder. Damon dreamed of the horrible crime night after night. But his victim would not stay dead. She haunted his dreams, accusing him. It was an unending nightmare for Damon Wells, who called himself: Son of the Devil.

On the morning of 9/15/84, having dreamed of the murder, and with the victim's help, Sunshine was convinced she knew where Tanya's body lay hidden. She contacted Tanya's family, told them of her dreams, and that same day led them and the police to the the side road Tanya had showed her, and then to the very spot where she had dreamed of the murder. The police climbed down the steep incline, and just as Sunshine had dreamed, they found the trail. But, there was no body.

That night Sunshine had another dream and this time Tanya took her to the same spot, down the same trail, then pointed at and emphasized a little trail that forked off to the right and which led directly to the body. The next day, Sunshine and the family met again, and then climbed down the incline, took the trail to the right, and there was Tanya's body laid out exactly as revealed to Sunshine when dreaming.

The murder remained unsolved, however, until 2/7/88 when Damon Wells, beset by horrible nightmares, sought psychiatric treatment. He confessed and hoped the psychiatrist could help him escape the dreams and visions which tormented him.

14.1 Dreams and Wondering Spirits. Souls and spirits were believed by ancient humans to wonder about while people sleep and dream (Brandon 1967; Frazier 1950; Harris 1993; Jung 1945, 1964; Malinowkski 1990). Some believed the soul could escape the body via the mouth or nostrils while dreaming and that the soul could wonder away from the body and engage in various purposeful acts or interact with other souls including the soul or spirit of those who had died. The spirit and soul were believed to hover about in human-like, ghostly vestiges, at the fringes of reality, the hinterland where day turns into night (Campbell 1988; Frazier 1950; Jung 1964; Malinowski 1954; Wilson 1951). It was also believed that even after death souls continued to interact with the living, and could haunt their dreams, and that the spirits of the dead might visit the dreams of the living, to convey knowledge and important information. Of course these are all silly superstitions. Certainly it is not possible Tanya Z's spirit was in fact haunting the dreams of Damon Wells and Sunshine.

In 1970, Ullman and Krippner reported statistically significant findings from a dream lab where dreamers were targeted with specific images from art prints, such as "School of the Dance" by Degas, depicting a dance class of several young women. The subject's dream reports included phrases as "I was in a class made up of maybe half a dozen people; it felt like a school." "There was one little girl that was trying to dance with me" (Krippner 1993). Could it be that Damon's dreams effected the dreams of Sunshine? Or perhaps it was all a coincidence.

15. LSD, 5HT, Dreams & Hallucinations

Damon Wells had consumed LSD on the afternoon of the murder. LSD blocks the release and uptake of 5HT (Bennett & Snyder, 1976) including in the amygdala (Gresch et al., 2002). LSD also acts directly on the amygdala and hippocampus (Bennett & Snyder, 1976; Gresch et al,, 2002), the right amygdala (and hippocampus) in particular. In consequence, once serotonin release or uptake is reduced or blocked, the amygdala will process multi-modal stimuli simultaneously, and the person will see sound, taste colors, and all aspects of perceptual processing will be greatly enhanced:

It was 1967, the "summer of love" and I am my friends made the trek to Haight-Ashbury where I acquired pure LSD created by Stanley Owsley. I was about to take my first "trip."

About half hour after I ingested the drug I forgot I had taken it, and instead I noticed how all the colors of the trees and plants were much brighter, more colorful and luminous. I was walking toward a park and stopped to touch a green leaf which was sparkling with emerald light, and I could feel the life inside the leaf, I could taste it's "greenness" through my fingers, and then my eyes became like a microscope and I could see the fine cellular structure of the leaf and then inside the leaf... and my attention turned to my hand... and I could see the fine cellular structure of my skin, then beneath my skin, and then I could see a blood vessel and then inside the vessel and I could see the red and white blood cells swishing past inside the vessel--and I was totally amazed and kept wondering: How come I never noticed this before! It was as if my eyes had become a tunneling microscope--and this was 15 years before its invention!

This was the first hour of my experience on LSD.... as the experience wore on I could see sound, I could see sound waves. I could taste colors. And I was able to see through the Santa Cruz mountains to what was on the otherside: the ocean and a jet plane and then the jet flew over the top of the mountain. And no, I do not think I was hallucinating per se. LSD blocks 5HT, which is an inhibitor. Structures such as the amygdala are inhibited by 5HT, and many amygdala neurons are multi-sensory, a single amygdala neuron can process sound, touch, taste, and vision, simultaneously---but this information is inhibited and filtered out as we would be overwhelmed if we were constantly tasting colors or seeing inside our skin...So LSD blocks 5HT which turns off the filtering... I was 17 years old when I took LSD. At the end of the experience I felt as if my intelligence had increased by 20 IQ points and my consciousness, understanding, and awareness of the world was certainly much greater. My mind had expanded and from the day forward I saw the world with open eyes.

As is well known, LSD can elicit profound hallucinations involving all spheres of experience. Following the administration of LSD, high-amplitude slow waves (theta) and bursts of paroxysmal spike discharges occur in the hippocampus and amygdala (Chapman & Walter, 1965; Chapman et al., 1963), but with little cortical abnormal activity. In both humans and chimpanzees in whom the temporal lobes, amygdala, and hippocampus have been removed, LSD ceases to produce hallucinatory phenomena (Baldwin et al., 1955, 1959; Serafetinides, 1965). Moreover, LSD-induced hallucinations are significantly reduced when the right versus left temporal lobe has been surgically ablated (Serafetinides, 1965). Dreaming is sometimes abolished with right but not left temporal lobe removals (Kerr & Foulkes, 1981). Likewise, Penfield and Perot (1963) report that the most vivid hallucination tend to be triggered from the right not the left temporal lobe.

LSD is structurally similar to serotonin, but acts as a serotonin antagonist and acts both pre-synaptically (Montigny and Aghajanian 1977) and post-synaptically (Bennett & Snyder, 1976) by blocking 5-HT secretion and 5-HT receptors (Bennett & Snyder, 1975), thereby preventing serotonin from exerting its normal inhibitory effects on sensory reception and multi-modal sensory processing. Further, LSD acts on the brainstem raphe nucleus (Strahlendorf, et al., 1982) which produces serotonin, thereby preventing this structure from exerting inhibitory influences not just on the amydgala, but in the pons, lateral geniculate, and visual cortex--structures which become highly active during dreaming.

Moreover, LSD acts on the frontal lobes (Gresch et al., 2002), which exert controls over the rest of the brain and sensory processing in the neocortex through inhibition (Joseph 1996, 1999a). Thus, following adminstration of LSD and suppression of 5-HT influences, sensory inhibition is signficantly attenuated throughout the brain, such that neurons which are normally supressed begin processing information normally filtered out, all of which is then experienced by the conscious mind.

16. Day Dreams, Hallucinations, Out-of-Body Consciousness, and Alternate Realities

Hallucinations are typically defined as the "Perception of visual, auditory, tactile, olfactory, or gustatory experiences in the absence of an external stimulus coupled with a compelling sense of their reality." The Diagnostic and Statistical Manual of the American Psychiatric Association also defines hallucinations as occurring "without external stimulation of the relevant sensory organ."

According to definition, hallucinatory experiences under LSD, insofar as they are based on external stimulation and perceptions freed of inhibitory restraint, are not necessarily hallucinations. Of course, this supposition could be dismissed by attributing the experience to the LSD. However, this dismissal fails as it is not the LSD which induces the experience, but the reductions in 5-HT, as also occurs during REM sleep. Therefore, it could be said that strictly speaking they are not hallucinating. Just as external stimuli may trigger a dream during REM sleep, that during waking and under reduced 5-HT external stimulation also produced a dream, what is experienced could be likened to a day-dream.

In fact, There is some evidence to suggest that during the course of the day and night the two cerebral hemispheres oscillate in activity every 90 to 100 minutes and are 180 degrees out of phase --a cycle that corresponds to changes in cognitive efficiency, the appearance of day dreams, REM (dream sleep), and, conversely, N-REM sleep (Bertini et al. 1983; Broughton, 1982; Hodoba, 1986; Klein & Armitage, 1979; Kripke & Sonnenschein, 1973). That is, like two pistons sliding up and down, it appears that when the right cerebrum is functionally at its peak of activity, the left hemisphere is correspondingly at its nadir. Day dreams also correspond to this cycle, such that when dreaming at night, or during the day, the dream is association with increased right hemisphere and reduced left hemisphere activity (Joseph 1988a, 1992a, 1996).

Day dreams, like night dreams, and LSD, are all associated with differential and increased right hemisphere activation. Likewise, in studies of hallucinations secondary to cerebral tumors or seizure activity, although simple hallucinations are likely following damage to either hemisphere, complex hallucinations are usually associated with right rather than left cerebral lesions (Teuber et al., 1960; Mullan & Penfield, 1959; Hecaen & Albert, 1978; Joseph 1996).

Moreover, direct stimulation of the neocortex and the amygdala, also produce complex hallucinations . For example, electrical stimulation of visual association areas 18 and 19 can produce complex and vivid images of men, animals, various objects and geometric figures, liliputian-type individuals (Hecaen & Albert, 1978; Joseph 1996; Tarachow, 1941). Those who experience these hallucinations may also have the illusion that their vision has telescoped such that they can see objects and people which are exceedingly far away, or existing in another reality or dimension. However, just as often, what they see may appear right before then and may not be unusual and no different from any other perception. For example, one patient saw a butterfly then attempted to catch it when area 19 of the visual cortex was elecrically stimulated. Another hallucinated a dog and then called to it, denying the possibility that it was not real (Joseph 1996).

However, under conditions of exceedingly heightened activity or other disturbances in the temporal lobes and underlying amygdala and hippocampus, the hallucination may become exceedingly vivid and unusual and include images of threatening men, naked women, sexual intercourse, demons and ghosts and pigs walking upright dressed as people (Bear 1979; Daly 1958; Gloor 992; Halgren 1992; Horowitz et al. 1968; Joseph 1999, 2002; Penfield & Perot 1963; Slater & Beard 1963; Taylor 1972; Trimble 1991; Weingarten, et al. 1977; Williams 1956). Often the experience could best be described as "other worldly" as if seeing into another supernatural dimension or entering into another reality.

Some individuals report communing with spirits, angels or gods, or receiving profound knowledge from the Hereafter, following temporal lobe activation (Daly 1958; MacLean 1990; Penfield & Perot 1963; Williams 1956). Some have visualized and have seen ghosts, demons, angels, and even God, or claim to have left their body (Bear 1979; Daly 1958; Gloor 1992; Horowitz et al. 1968; MacLean 1990; Mesulam 1981; Penfield & Perot 1963; Schenk & Bear 1981; Slater & Beard 1963; Subirana & Oller-Daurelia, 1953; Trimble 1991; Weingarten, et al. 1977; Williams 1956).

Some individuals have described feelings such as elation, security, eternal harmony, immense joy, paradisiacal happiness, euphoria, completeness. Between .5 and 20% of such patients claim such feelings (Williams, 1956; Daly, 1958). One patient of Williams (1956) claimed he was overwhelmed by "sudden feeling of extreme well being involving all my senses. I see a curtain of beautiful colors before my eyes and experience a pleasant but indescribable taste in my mouth. Objects feeling pleasurably warm. the room assumes vast proportions, and I feel as if in anothe world."

A patient described by Daly (1958) claimed his seizure felt like "a sunny day when your friends are all around you." He then felt disociated from his body, as if he were looking down upon himself and wathcing his actions.

Penfield and Perot (1963) describe several patients who claimed they could see themselves outside their body engaging in various activity. One woman stated that "it was though I were two persons, one watching, and the other having this happen to me," and that it was she who was doing the watching as if she was completely separated from her body. According to Penfield, "It was as though the patient were attending a play and was both actor and audience.

Williams (1956) describes a patient who claimed that during an aura she experienced a feeling that she was being lifted up out of her body, coupled with a very pleasant sensation of eleation and the sensation that she was "just about to find out knowlede no one else shares, something to do with the line between life and death."

Subirana and Oller-Daurelia (1953) described two patients who experienced ecastic feelings of either "extrraordinary beatitude" or of paradise as if they had gone to heaven and noted that his fantastic feelings lasted for hours.

Other patients have noted that feelings and things suddenly become "cyrstal clear" or that they have a feeling of clairvoyance, or of having the truth revealed to them, or opf having achieved a sense of greater awareness and of a new awarness such that sounds, smells and visual objects seemd to have a greater meaning and sensibility (Joseph 2001, 2002).

It has frequently been reported that as compared to other cortical areas, the most complex and most forms of hallucination occur secondary to temporal lobe involvement (Malh et al., 1964; Horowitz et al., 1968; Penfield & Perot, 1963; Tarachow, 1941) and that the hippocampus and amygdala (in conjunction with the temporal lobe) appear to be the responsible agents (Gloor 1992, 1997; Horowitz et al., 1968; Halgren et al., 1978). Thus, the same brain regions implicated in the generation of the dream are also linked to hallucinations following the administration of LSD or abnormal activation or direct electrode stimulation.

17. Out-of-Body and Near Death Experience

Some children and adults who have been declared "clinically" dead but who subsequently return to life, have reported that after "dying" they left their body and floated above the scene (Eadie 1992; Joseph, 1996; Rawling 1978; Ring 1980). Typically they become increasingly euphoric as they float above their body, after which they may float away, become enveloped in a dark tunnel and then enter a soothing radiant light. And later, when they come back to life, they may even claim conscious knowledge of what occurred around their body while they were dead and floating nearby. Similar experiences are detailed in the Egyptian funery texts and "book of the dead," written almost 6000 years ago (Budge 1994) as well as by otherwise completely "modern" and sophisticated humans.

"Lisa" for example, was a 22 year old college coed with no religious background or spiritual beliefs, who was badly injured in an auto accident when the windshield collapsed and all but completely severed her arm. According to Lisa, when she got out of the car her was blood spraying everywhere and she only walked a few feet before collapsing. Apparently an ambulance arrived in just a few minutes. However, the next thing she noticed was that part of the time she was looking up from the ground, and part of the time she was up in the air looking down and could see the ambulance crew working, picking up her body, placing it on a gurney and into the ambulance.

According to Lisa, during the entire ride to the hospital it was like she was half in and half out of the ambulance, as if she were running along outside or just extending out of the vehicle watching the cars and tress go by. When they got to the hospital she was no longer attached to her body but was floating up and down the halls, watching the doctors and nurses and attendants. One doctor in particular drew her attention because he had a big belt buckle with his name written on it. She could even read it and it said "Mike."

According to Lisa, she was "tripping out, bobbing up and down the halls, just checking everything out" when she noticed a girl lying on a gurney with several doctors and nurses working frantically. When she floated over and peered over the shoulder of one of the doctors to take a look she suddenly realized the girl was her and that her hair and face were very bloody and needed to be washed. At that point she realized she was floating well above her body and the doctors and that she looked to be "dead." However, according to Lisa she did not feel afraid or upset, although the fact that her hair was dirty bothered her.

As detailed by Lisa, she soon floated up and outside the Emergency room and was enveloped in a total blackness, "like I was passing through a tunnel at the end of which was a vague light which became brighter and more biliant, radiating outward." The light soon enveloped her body which made her feel exceedingly happy and very warm. A few moments later she heard the voice of her grandmother who had died when Lisa was a young girl. Although Lisa had no memory of this grandmother, she nevertheless recognized her and felt exceedingly happy. However, as Lisa approached, her grandmother very sorrowfully told her it was "too soon", she would have to "go back." Lisa didn't want to go back, but had no choice. She was drawn away from the light and felt herself falling only to land with a painful thump in her own body. At this point she moved her hand which alerted one of the emergency room staff that Lisa was no longer dead.

It is noteworthy that Lisa had never heard of "near death experiences" (she was injured in 1982) and that after returning to life she only reluctantly explained what had happened when she was questioned by one of her doctors. Lisa also claimed that while she was dead and floating about the emergency room that she saw, heard and is able to recall everything that occurred up to the point when she was enveloped in darkness. She was able to accurately describe "Mike" as well as some of the staff who first attended her, the conversations that occurred around her as well as some of the other patients. Indeed, similar "after death" claims of leaving and floating above the body, and seeing everything occurring below, are common (Eadie 1992; Moody 1977; Rawling 1978; Ring 1980; Sabom 1982; Wilson 1987), and, as noted, are even reported in the 6,000 year old Egyptian Book of the Dead (Budge 1994), as well as the Tibetan Book of the Dead (the Bardo Thodol) which was composed over 1,300 years ago. Approximately 37% of patients who are resuscitated report similar "out of body" experiences (Ring 1980).

Consider for example, the case of Army Specialist J. C. Bayne of the 196th Light Infantry Brigade. Bayne was "killed" in Chu Lai, Vietnam, in 1966. He was simultaneously machine gunned and struck by a mortar. According to Bayne, when he opened his eyes he was floating in the air, looking down on his crumpled, burnt, and bloody body, and he could see a number of Vietcong who were searching and stripping his him: "I could see me... it was like looking at a manikin laying there... I was burnt up and there was blood all over the place... I could see the Vietcong. I could see the guy pull my boots off. I could see the rest of them picking up various things... I was like a spectator... It was about four or five in the afternoon when our own troops came. I could hear and see them approaching... I could see me... It was obvious I was burnt up. I looked dead... they put me in a bag... transferred me to a truck and then to the morgue. And from that point, it was the embalming process. I was on that table and a guy was telling a couple of jokes about those USO girls... all I had on was bloody undershorts... he placed my leg out and made a slight incision and stopped... he checked my pulse and heartbeat again and I could see that too...It was about that point I just lost track of what was taking place.... [until much later] when the chaplain was in there saying everything was going to be all right.... I was no longer outside. I was part of it at this point" (reported in Wilson, 1987, pp 113-114; and Sabom, 1982, pp 81-82).

Moreover, some surgery patients also claim to "leave their bodies" and recall seeing not just the events occurring below, but in one case, dirt on top of a light fixture (Ring 1980). "It was filthy. And I remember thinking, 'Got to tell the nurses about that."

Did the above surgical patient or Lisa or Army Specialist Bayne really float above and observe their bodies and the events taking place below? Or did they merely transpose what they heard (e.g. conversations, noises, etc.) and then visualize, imagine, or hallucinate an accompanying and plausible scenario? This seems likely, even in regard to the "filthy" light fixture. On the other hand, not all those who have an "out of body" hear conversations, voices, or even sounds. Rather, they may be enveloped in silence.

"I was struck from behind...That's the last thing I remember until I was above the whole scene viewing the accident. I was very detached. This was the amazing thing about it to me... I could see my shoe which was crushed under the car and I thought: Oh no. My new dress is ruined... I don't remember hearing anything. I don't remember anybody saying anything. I was just viewing things...

like I floated up there..." (Sabom, 1982; p. 90). Moreover, even individuals born blind experience these "near death" hallucinations.

18. Fear and Out-of-Body Experiences

"I was shooting down the freeway doing about 100 or more in my Mustang when a Firebird suddenly cut me off. As I switched lanes to avoid him, he also switched lanes at which point I hit the breaks and began to lose control. The Mustang began to slide and spin... I felt real terror.... I was probably going to be killed... I was trying to control the Mustang and avoid turning over, or hitting any of the surrounding cars or the guard rail.... time seemed to slow down and then I suddenly realized that part of my mind was a few feet outside the car looking all around; zooming above it and then beside it and behind it and in front of it, looking at and analyzing the respective positions of my spinning Mustang and the cars surrounding me. Simultaneously I inside trying to steer and control it in accordance with the multiple perspectives I was suddenly given by that part of my mind that was outside. It was like my mind split and one consciousness was inside the car, while the other was zooming all around outside and giving me visual feedback that enabled me to avoid hitting anyone or destroying my Mustang."

The prospect of being terribly injured or killed in an auto accident or fire fight between opposing troops, or even dying during the course of surgery, are often accompanied by feelings of extreme fear. It is also not uncommon for individuals who experience terror to report perceptual and hallucinogenic experiences, including dissociation, depersonalization and the splitting off of ego functions such that they feel as if they have separated from their bodies and floated away, or were on the ceiling looking down (Campbell 1988; Courtois 1995; Grinker and Spiegel 1945; James 1958; Neihardt and Black Elk 1932/1989; Noyes and Kletti 1977; Parson 1988; Southard 1919; Terr 1990). Consider the following:

"The next thing I knew I wasn't in the truck anymore; I was looking down from 50 to 100 feel in the air.""I had a clear image of myself... as though watching it on a television screen.""I had a sensation of floating. It was almost like stepping out of reality. I seemed to step out of this world" (Noyes and Kletti 1977).

19. Hippocampal Hyperactivation and Astral Projection

Feelings of fear and terror are mediated by the amygdala, whereas the capacity to cognitively map, or visualize one's position and the position of other objects and individuals in visual-space is dependent on the hippocampus (Nadel, 1991; Joseph, 1996; O'Keefe, 1976; Wilson and McNaughton, 1993). The hippocampus contains "place" neurons which are able to encode one's position and movement in space.

The hippocampus, therefore, can create a cognitive map of an individuals environment and their movements within it. Presumably it is via the hippocampus that an individual can visualize themselves as if looking at their body from afar, and can remember and thus see themselves engaged in certain actions, as if one were an outside witness (Joseph, 1996). However, under conditions of hyperactivation (such as in response to extreme fear) it appears that the hippocampus may create a visual hallucination of that "cognitive map" such that the individual may "experience" themselves as outside their body, observing all that is occurring.

Again, it has been repeatedly demonstrated that hyperactivation or electrical stimulation of the amygdala-hippocampus-temporal lobe, can cause some individuals to report they have left their bodies and are hovering upon the ceiling staring down (Daly 1958; Jackson and Stewart 1899; Joseph, 1996; Penfield 1952; Penfield and Perot 1963; Williams 1956). That is, their ego and sense of personal identity appears to split off from their body, such that they may feel as if they are two different people, one watching, the other being observed.

20. The Evolution of Dream Consciousness

That so many people, regardless of culture or antiquity, have similar dreams and hallucinations, is presumably due to all possessing a limbic system and temporal lobe that is organized similarly. Of course, many of these experiences are also colored by one's cultural background and differences in thinking patterns.

However, we should ask: is it really a hallucination if someone experiences a dissociation of consciousness and floats above their body and can later accurately described what was taking place and the appearance of those around them? Is it really a hallucination if an individual can see inside his hand and watch the blood cells swishing through his blood vessel? If one of the great presidents in the history of the United States dreams of his death 13 days before he is assassinated, was it just a dream?

There are numerous reports of individuals who claim to have dreamed about future events which then took place, including deaths, tragedies, mass murders, horrific accidents, and environmental catastrophes (Barker 1967; Jung 1945, 1964; Wiseman 2011). Yet, given that six billion people dream multiple dreams every night, it could be argued that we should not be surprised that a few of these dreams just happen to accurately coincide with what takes place. These are just chance coincidences, which, when considering the billions of dreams dreamed nightly, should be dismissed as meaningless and of no significance (Wiseman 2011). Since so few people have these dreams, it should not be concluded that these dreams represent an important cognitive capacity. However, if we apply the same criteria to Einstein's theory of relativity, or the music of Mozart or Beethoven, then the absurdity of this position becomes clear. Certain cognitive capacities are well developed in just a few people.

Another question should also be asked: Why would the brain evolve capabilities which are suppressed? Why did neural structures evolve which can process multiple sensations simultaneously, but then come to be inhibited by serotonin? Why would activation of specific brain areas result in the sensation of having left their body, or being privy to cosmic wisdom, or witnessing events and entities which appear to be from other dimensions or realities?

These experiences are made possible by the brain. They must serve an adaptive function. These capabilities must have evolved following natural selection. Why would we evolve the ability to hallucinate or dream about that which supposedly does not exist? Why should we have evolved the ability to hallucinate that we have left our bodies, or can see inside our hands, or dream that a friend will tragically die days before his accidental death?

One possibility is the human brain is de-evolving, and has lost or is losing capabilities which served a more adaptive purpose in ancient humans 30,000 to 10,000 years ago--ancient humans (the Cro-Magnon) whose brain was 1/3 larger in size than the modern brain! The Cro-Magnon men stool 6 ft tall on average, and the Cro-Magnon people created the magnificent underground cathedrals of art which first began to appear 30,000 years ago.

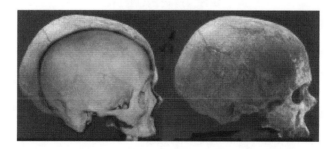

Comparison of modern human skull superimposed on Cro-Magnon skull (Left). Cro-Magnon skull (Right). The Cro-Magnon brain was 1/3 larger on average, than the modern human brain.

Conversely, it may be that these are capacities which are still evolving, but which at the present stage of human evolution, the human brain is unable to utilize adaptively. Consider, the evolution of language. Certainly, Australopithecus, Homo habilis, Homo erectus, and Neaderthals did not engage in complex conversations (Joseph 1996, 2000, 2001). We can surmise based on indirect evidence, that the language capabilities of these ancient hominids and humans may have been limited to grunts, groans, screams, and a variety of emotional sounds (Joseph 2000). Whatever language abilities ancestral species evolved prior to 100,000 years ago were primitive manifestations of what was yet to evolve, i.e. grammatical speech, reading, writing, and associated modes of abstract thinking, such as math. Although the rudimentary foundations had been laid long ago, these were only primitive steps toward what was later to more fully evolve.

It can be surmised that if humans continue to evolve, that maybe 100,000 years from now they will possess a brain which is able to fully utilize capacities which the modern brain is, as yet, unable to master. Or, just as the brain has shrunk by 1/3 in size since the ending of the Paleolithic, that it may continue to lose capacities due to advances in technology which will increasingly render reading, writing, speaking, or creative endeavors obsolete.

Based on the evidence marshaled here and elsewhere, it can be inferred that not all dreams and hallucinations are dreams and hallucinations. They are mental capabilities which are de-evolving or still evolving, and which are or were destined to serve a specific purpose: Lifting the veil so we can gaze deeply into the past, the future, and the unknown.

References

Amaral, D. G., Price, J. L., Pitkanen, A., & Thomas, S. (1992). Anatomical organization of the primate amygdaloid complex. In J. P. Aggleton (Ed.). The Amygdala. (Wiley. New York.

Applegate, C. D. (1980). 5,7,-dihydroxytryptamine-induced mouse killing and behavioral reversal with ventricular administation of serotonin in rats. Behavioral and Neural Biology, 30, 178-190.

Baldwin, M., Lewis, S. A., & Bach, S. A. (1959). The effects of lysergic acid after cerebral ablation. Neurology, 469-474

Baldwin, M., Lewis, S.A., & Bach, S.A. (1959). The effects of lysergic acid after cerebral ablation.Neurology, 9, 469-474.

Barker, J. (1967). Premonitions of the Aberfan Disaster, JSPR, 44, 169-181

Bear, D. M. (1979). Temporal lobe epilepsy: A sydnrome of sensory-limbic hyperconnexion. Cortex, 15, 357-384.

Benninger, R. J. (1989). The role of serotonin and dopamine in learning to avoid aversive stimuli. In T. Archer & L-G Nilsson (Eds) Aversion, Avoidance and Anxiety. New Jersey, Erlbaum.

Bennett, Jr. J. P., and Snyder, S. H. (1975). Stereospecific binding of d-lysergic acid diethylamide (LSD) to brain membranes: Relationship to serotonin receptors. Brain Research, 94, 523-544.

Bennett, Jr. J. P., and Snyder, S. H. (1976). Serotonin and Lysergic Acid Diethylamide Binding in Rat Brain Membranes: Relationship to Postsynaptic Serotonin Receptors. Molecular Pharmacology, 12, 373-389.

Bertini, M., Violani, C., Zoccolotti, P., Antonelli, A., & DiStephano, L. (1983). Performance on a unilateral tactile test during waking and upon awakenings from REM and NREM. In W. P. Koella (Ed.), Sleep, (pp. 122-155). Basel: Karger.

Bogen J.,& Bogen,C. (1969).The other side of the brain: III. The corpus callosum and creativity. Bulletin of the Los Angeles Neurological Society, 34, 191-220.

Botez, M. I., Olivier, M., Vezina, J.-L., Botez, T., & Kaufman, B. (1985). Defective revisualization: Dissociation between cognitive and imagistic thought case report and short review of the literature. Cortex, 21, 375-389.

Brandon, S.G. F. (1967) The Judgment of the Dead. New York, Scribners.

Broughton, R. (1982). Human consciousness and sleep/waking rhythms: A review and some neuropsychological considerations. Journal of Clinical Neuropsychology, 4, 193-218.

Budge, W. (1994). The Book of the Dead. New Jersey, Carol.

Calvo, J. M., Badillo, S., Morales-Ramirez, M., Palacios-Salas, P. (1987) The role of the temporal lobe amygdala in ponto-geniculo-occipital activity and sleep organization in cats. Brain Research, 403, 22-30.

Campbell, J. (1988) Historical Atlas of World Mythology. New York, Harper & Row.

Cartwright , R. (2010) The Twenty-four Hour Mind: The Role of Sleep and Dreaming in Our Emotional Lives. Oxford University Press.

Cartwright, R. D., Tipton, L. W., & Wicklund, J. (1980). Focusing on dreams. Archives of General Psychiatry, 37, 275-288.

Chapman, L. F., & Walter, R. D. (1965). Actions of lysergic acid dienthalamid on averaged human cortical evoked rsposnes to light flash. Recent Advances in Biological Psychiatry, 7, 23-36.

Chapman, L. F., Walter, R. D., Ross, W., et al. (1963). Altered electrical activity of human hippocampus and amygdala induced by LSD-25. Physiologist, 5, 118.

Courtois, C.A. (1995). Healing the Incest Wound. New York, Norton.

Daly, D. (1958) Ictal affect. American Journal of Psychiatry, 115, 97-108.

Dossey, L., Greyson, B., Sturrock, P. A., Tucker, J. B. (2011). Consciousness -- What Is It? Shared Consciousness, Twin Consciousness, Near Death, Journal of Cosmology, Vol 14. In press.

Eadie, B. J. (1992). Embraced by the light. California, Gold Leaf Press.

Foulkes, W. D. (1962). Dream reports from different stages of sleep. Journal of Abnormal and Social Psychology, 65, 14-25.

Frazier, J. G. (1950). The golden bough. Macmillan, New York.

Freud, S. (1900). The interpretation of dreams. Standard Edition (Vol 5). London: Hogarth Press.

Gloor, P. (1997). The Temporal Lobes and Limbic System. Oxford University Press. New York.

Goldstein, L., Stoltzfus, N. W., & Gardocki, J. F. (1972). Changes in interhemispheric amplitude relationships in the EEG during sleep.

Physiology and Behavior, 8, 811-815.

Greenwood, P., Wilson, D. H., & Gazzaniga, M. S. (1977). Dream report following commissurotomy. Cortex, 13, 311-316.

Gresch, P. J., Strickland, L. V. and Sanders-Bush, E. (2002). Lysergic acid diethylamide-induced Fos expression in rat brain: role of serotonin-2A receptors. Neuroscience, 114, 707-713.

Grinker, R. R., Spiegel, J. P. (1945) Men Under Stress. McGraw-Hill, New York.

Halgren, E. (1992). Emotional neurophysiology of the amygdala within the context of human cognition. In J. P. Aggleton (Ed.). The Amygdala. New York, Wiley-Liss.

Halgren, E., Walter, R. D., Cherlow, D. G., & Crandal, P. H. (1978). Mental phenomenoa evoked by electrical stimualtion of the human hippocampal formation and amygdala, Brain, 101, 83-117.

Harris, M. (1993) Why we became religious and the evolution of the spirit world. In Lehmann, A. C. & Myers, J. E. (Eds) Magic, Witchcraft, and Religion. Mountain View: Mayfield.

Hasselmo, M. E., Rolls, E. T., & Baylis, G. C. (1989). The role ofexpression and identity in the face-selective responses of neurons in thetemporal visual cortex of the monkey. Behavioral Brain Research, 32,203-218.

Hecaen, H., & Albert, M. L.(1978). Human Neuropsychology, New York: John Wiley.

Hodoba, D. (1986). Paradoxic sleep facilitation by interictal epileptic activity of right temporal origin. Biological Psychiatry, 21, 1267-1278.

Hobson, J. A. (2004) Dreaming: An Introduction to the Science of Sleep. Oxford University Press.

Hoppe, K. D., & Bogen, J. E. (1977). Alexithymia in twelve commissurotomized patients. Psychotherapy and Psychosomatics, 28, 148-155.

Horowitz, M. J., Adams, J. E., & Rutkin, B. B. (1968). Visual imagery on brain stimulation. Archives of General Psychiatry, 19, 469-486.

Humphrey M. E., & Zangwill, O. L. (1911). Cessation of dreaming after brain injury. Journal of Neurology, Neurosurgery, and Psychiatry, 14, 322-325.

Jacobs, B. L., & Azmita, E. C. (1992). Structure and function of the brain Serotonin System. Physiological Reviews, 72, 165-245.

Joseph, R. (1988a) The Right Cerebral Hemisphere: Emotion, Music, Visual-Spatial Skills, Body Image, Dreams, and Awareness. Journal of Clinical Psychology, 44, 630-673.

Joseph, R. (1988b). Dual mental functioning in a split-brain patient. Journal of Clinical Psychology, 44, 770-779.

Joseph, R. (1992a) The Limbic System: Emotion, Laterality, and Unconscious Mind. The Psychoanalytic Review, 79, 405-456.

Joseph, R. (1992b). The Right Brain and the Unconscious. New York, Plenum.

Joseph, R. (1996). Neuropsychiatry, Neuropsychology, Clinical Neuroscience, 2nd Edition. 21 chapters, 864 pages. Williams & Wilkins, Baltimore.

Joseph, R. (1999a). Frontal lobe psychopathology: Mania, depression, aphasia, confabulation, catatonia, perseveration, obsessive compulsions, schizophrenia. journal of Psychiatry, 62, 138-172.

Joseph, R. (1999b). The neurology of traumatic "dissociative" amnesia. Commentary and literature review. Child Abuse & Neglect. 23, 715-727.

Joseph, R. (2000). The evolution of sex differences in language, sexuality, and visual spatial skills. Archives of Sexual Behavior, 29,

35-66.

Joseph, R. (2001). The Limbic System and the Soul: Evolution and the Neuroanatomy of Religious Experience. Zygon, the Journal of Religion & Science, 36, 105-136.

Joseph, R. (2002). NeuroTheology: Brain, Science, Spirituality, Religious Experience. University Press.

Joseph, R. (2011a). Origins of Thought: Consciousness, Language, Egocentric Speech and the Multiplicity of Mind Journal of Cosmology, 14.

Joseph, R. (2011b). Life on Earth Came from Other Planets. Cosmology Science Publishers, Cambridge.

Jouvet, M. (2001). The Paradox of Sleep: The Story of Dreaming. MIT press.

Jung, C. G. (1945). On the nature of dreams. (Translated by R.F.C. Hull.), The collected works of C. G. Jung, (pp.473-507). Princeton: Princeton University Press.

Jung, C. G. (1964). Man and his symbols. New York: Dell.

Kamiya, J. (1961). Behavioral, subjective and physiological aspects of drowsiness and sleep. In D. W. Fiske, & S. R. Maddi (Eds.). Function of varied experience, (pp. 145-174). Homewood, IL: Dorsey Press.

Kawashima, R., Sugiura, M., Kato, T., et al., (1999). The human amygdala plays an important role in gaze monitoring. Brain, 122, 779-783.

Kerr N. H., & Foulkes, D (1978). Reported absence of visual dream imagery in a normally sighted subject with Turner's syndrome. Journal of Mental Imagery, 2, 247-264.

Kerr, N. H., & Foulkes, D. (1981). Right hemisphere mediation of dream visualization: A case study. Cortex, 17, 603-611.

Klein, R., & Armitage, R. (1979). Rhythms in human performance: 11/2 hour oscillations in cognitive style. Science, 204, 1326-1328.

Kling. A. S. & Brothers, L. A. (1992). The amygdala and social behavior. In J. P. Aggleton (Ed.). The Amygdala. New York, Wiley-Liss.

Kripke, D. F., & Sonnenschein, D. (1973). A 90 minute daydream cycle. Sleep Research, 2, 187-188.

Krippner, S (1993). The Maimonides ESP-dream studies - Maimonides Medical Center, Journal of Parapsychology, 57, 279-319.

Lamon, W. H. (1865/1994). Recollections of Abraham Lincoln 1847-1865, by Ward Hill Lamon, University of Nebraska Press.

Lydic, R., Baghdoyan, H. A., & Lorinc, Z. (1991). Microdialysis of cat pons reveals enhanced ACh release during state-dependent respiratory depression. American Journal of Physiology, 261, 766-770.

MacLean, P. (1990). The Evolution of the Triune Brain. New York, Plenum.

Malh, G. F., Rothenberg, A., Delgado, J. M. R., & Hamlin, H. (1964). Psychological response in the human to intracerebral electrical stimulation. Psychosomatic Medicine, 26, 337-368.

Malinowski, B. (1954) Magic, Science and Religion. New York. Doubleday.

Meyer, J. S., Ishikawa, Y., Hata, T., & Karacan, I. (1987). Cerebral blood flow in normal and abnormal sleep and dreaming. Brain and Cognition, 6, 266-294.

Mesulam, M. M. (1981) Dissociative states with abnormal temporal lobe EEG: Multiple personality and the illusion of possession. Archives of General Psychiatry, 38, 176-181.

Monti, J. Pandi-Permal, S. R., Sinton, C. M. (2008). Neurochemistry of Sleep and Wakefulness, Cambridge U. Press.

Montigny, C. de, and Aghajanian G.K. (1977). Preferential action of 5-methoxytryptamine and 5-methoxydimethyltryptamine on presynaptic serotonin receptors: A comparative iontophoretic study with LSD and serotonin. Neuropharmacology, 16, 811-818.

Moody, R. (1977). Life after life. Georgia, Mockingbird Books.

Morris, J. S., Frith, C. D., Perett, D. I., Rowland, D., Young, A. W., Calder, A. J., & Colan, R. J. (1996). A differential neural response in the human amygdala to fearful and happy facial expression. Nature, 383, 812-815.

Mullan, S., & Penfield, W. (1959). Epilepsy and visual halluciantions. Archives of Neurology and Psychiatry, 81, 269-281.

Murri, L., Arena, R., Siciliano, G., Mazzotta, R., & Muratorio, A. (1984). Dream recall in patients with focal cerebral lesions. Archives of Neurology, 41, 183-185.

Nadel, L. (1991). The hippocampus and space revisited. Hippocampus, 1, 221-229.

Neihardt, J. G. & Black Elk, (1979). Black Elk speaks. Lincoln. U. Nebraska Press.

emchin, A. A., Whitehouse, M.J., Menneken, M., Geisler, T., Pidgeon, R.T., Wilde, S. A. (2008). A light carbon reservoir recorded in zircon-hosted diamond from the Jack Hills. Nature 454, 92-95.

Noyes, R., & Kletti, R. (1977) Depersonalization in response to life threatening danger. Comprehensive Psychiatry, 18, 375-384.

O'Keefe, J. (1976). Place units in the hippocampus of the freely moving rat. Experimental Neurology, 51, 78-109.

O'Keefe, J., & Bouma, H. (1969). Complex sensory properties of certain amygdala units in the freely moving cat. Experimental Neurology, 23, 384-398.

O'Neil, J., Carlson, R. W., Francis, E., Stevenson, R. K. (2008). Neodymium-142 Evidence for Hadean Mafic Crust Science 321, 1828 - 1831.

Parson, E. R. (1988). Post-traumatic self disorders (PTsfD): Theoretical and practical considerations in psychotherapy of Vietnam War Veterans. In J. P. Wilson, Z. Harel, & B. Kahana (Eds). Human Adaptation to Extreme Stress. New York, Plenum.

Pena-Casanova, J., & Roig-Rovira, T. (1985). Optic aphasia, optic apraxia, and loss of dreaming. Brain and Language, 26, 63-71.

Penfield, W., & Perot, P. (1963). The brains record of auditory and visual experience. Brain, 86, 595-695.

Perryman, K. M., Kling, A. s., & Lloyd, R. L. (1987). Differential effects of inferior temporal cortex lesions upon visual and auditory-evoked potentials in the amygdala of the squirrel monkey. Behavioral and Neural Biology, 47, 73-79.

Purpura, D. P. (1956). Electrophysiological analysis of psychotogenic drug action. I & II. Archives of Neurology & Psychiatry, 40, 122-143.

Rawlings, M. (1978). Beyond deaths door. London, Sheldon Press.

Ribeiro, S., Goyal, V., Mello, C. & Pavlides, C. (1999). Brain gene expression during REM sleep depends on prior waking experience. Learning & Memory, 6: 500-508.

Ribeiro, S., Mello, C., Velho, T., Gardner, T., Jarvis, E., & Pavlides, C. (2002). Induction of hippocampal long-term potentiation during waking leads to increased extra hippocampal zif-268 expression during ensuing rapid-eye-movement sleep. Journalof Neuroscience, 22(24), 10914-10923.

Riberio, S., Simões, C. & Nicolelis, M. (2008). Genes, Sleep and Dreams. In Lloyd & Rossi (Eds.) Ultradian rhythms from molecule to mind. Springer. N.Y., 413-430.

Ring, K. (1980). Life at death. New York, Coward, McCann & Geoghegan.

Rolls, E. T. (1984). Neurons in the cortex of the temporal lobe and in the amygdala of the monkey with responses selective for faces. Human Neurobiology, 3, 209-222.

Rolls, E. T. (1992). Neurophysiology and functions of the primate amygdala. In J. P. Aggleton (Ed.). The Amygdala. New York, Wiley-Liss.

Sabom, M. B. (1982). Recollections on death. New York, Harper & Row.

Sawa, M., & Delgado, J. M. R. (1963). Amygdala unitary activity in the unrestrained cat. Electroencephalography and Clinical Neurophysiology, 15, 637-650.

Schenk, L., & Bear, D. (1981) Multiple personality and related dissociative phenomenon in patients with temporal lobe epilepsy. American Journal of Psychiatry, 138, 1311-1316.

Serafetinides, E. A. (1965). The significance of the temporal lobes and of hemisphere dominance in teh production of the LSD-25 symptomology in man. Neuropsychologia, 3, 69-79.

Schutze, I., Knuepfer, M. M., Eismann, A., Stumpf, H., & Stock, G. (1987). Sensory input to single neurons in the amygdala of the cat. Experimental Neurology, 97, 499-515.

Slater, E. & Beard, A.W. (1963). The schizophrenia-like psychoses of epilepsy. British Journal of Psychiatry, 109, 95-112.

Soubrie, P. (1986). Reconciling the role of central serotonin neurons human and animal behavior. Behavioral and Brain Sciences, 9, 319-364.

Spoont, M. R. (1992). Modulatory role of serotonin in neural information processing: Implications for human psychopathology. Psychological Bulletin, 112, 330-350.

Steriade, M. M. & McCarley, R. W. (2005) Brain Control of Wakefulness and Sleep, Springer.

Strahlendorf, J. C. R., et al., (1982). Differential effects of LSD serotonin and l-tryptophan on visually evoked responses. Pharmacology Biochemistry and Behavior, 16, 51-55.

Subirana, A., & Oller-Daurelia, L. (1953). The seizures with a feeling of paradisiacal happiness as the onset of certain temporal symptomatic epilepsies. Congres Neurologique International. Lisbonne, 4, 246-250.

Tarachow, S. (1941). Tjhe clinical value of hallucinations in localizing brain tumors. American Journal of Psychiatry, 99, 1434-1442.

Taylor, D. C. (1972). Mental state and temporal lobe epilepsy. Epilepsia, 13, 727-765.

Teuber, H. L., Battersfy, W. S., & Bender, M. B. (1960). Visual field defects after penetrating missile wounds of the brain. Cambridge: Harvard University Press.

Trimble, M. R. (1991). The psychoses of epilepsy. New York, Raven Press.

Turner, B. H. Mishkin, M. & Knapp, M. (1980). Organization of the amygdalopetal projections from modality-specific cortical association areas in the monkey. Journal of Comparative Neurology, 191, 515-543.

Ullman, M., & Krippner, S. (1970). Dream studies and telepathy; An experimental approach. New York: Parapsychology Foundation.

Ursin H., & Kaada, B. R. (1960). Functional localization within the amygdaloid complex in the cat. Electroencephalography and Clinical Neurophysiology, 12, 1-20.

Vertes, R. P. (1990). Brainstem mechanisms of slow-wave sleep and REM sleep. In W. R. Klemm, & R. P. Vertes (Eds.). Brainstem mechanisms of behavior. Wiley. New York.

Weingarten, S. M., Cherlow, D. G. & Holmgren. E. (1977). The relationship of hallucinations to depth structures of the temporal lobe. Acta Neurochirugica 24: 199-216.

Williams, D. (1956). The structure of emotions reflected in epileptic experiences. Brain, 79, 29-67.

Wilson, I. (1987). The after death experience. New York, Morrow.

Wilson, J. A. (1951) The culture of ancient Egypt. Chicago, U. Chicago Press.

Wilson, M. A., & McNaughton, B. L. (1993). Dynamics of the hippocampal ensemble for space. Science, 261, 1055-1058.

Wiseman, R. (2011) Paranormality, Macmillan.

42. Altered Consciousness Is A Many Splendored Thing
Etzel Cardeña, Ph. D.

Thorsen Professor, Department of Psychology, Center for Research on Consciousness and Anomalous Psychology (CERCAP), Lund University, P.O. Box 213 SE-221 00, Lund, Sweden

Abstract

Contrary to the notion that altering our ordinary state of consciousness necessarily produces delusional beliefs and is generally deleterious, various findings in psychology and other disciplines suggest that 1) the ordinary state of consciousness does not provide an accurate mapping of reality and 2) alterations of consciousness can have various positive functions. Among the most important ones are to: a) mildly alter our consciousness to be more effective at our tasks, b) obtain greater pleasure than we can achieve ordinarily, 3) explore nonfactual possibilities that may be actualized through creative ways, 4) enhance the sense of meaningfulness in life and enhance psychological and medical healing, and 5) obtain valid knowledge to comprehend better ourselves and perhaps the universe at large.

KEY WORDS: Altered states of consciousness, epistemology, mysticism.

A discussion of altered states of consciousness (ASC), like one on politics or religion, invites strong emotional reactions. In this paper I question some assumptions about the accuracy and benevolence of the typical ordinary state of consciousness (OSC) while discussing some of the main functions of ASC. Three of the main arguments against ASC are that they: 1) go against what is normal and rational, 2) wreak havoc at personal and social levels, and 3) produce a delusional account of reality, as compared with the OSC. Because we live in a monophasic (rather than polyphasic, see Laughlin, McManus, & d'Aquili, 1992; Whitehead, in press) society that primarily values our ordinary state of consciousness to the detriment of other states, these assumptions are rarely questioned.

1. Ordinary and altered states of consciousness

Regarding the first argument, we are immediately confronted by the relativity of what is "normal" and "rational." As various anthropologists have pointed out, what we consider "normal" in post-industrial, Western societies differs markedly from the experiences of other groups. Turnbull (1993, p. 74) gives a lucid example of how he could not even begin to understand the Mbuti of Congo until he transformed his consciousness to fully participate in their world:

> "But the more it happened the more other things happened. Not only did seemingly incontrovertible oppositions disappear, such as joy and grief, noise and quietness… somehow the differentiation between my senses seemed to disappear and I began touching moonlight, smelling the sound of the songs, hearing the scent of the various kinds of woods blazing away… and seeing the truth, even if I could not understand what I saw."

As to rationality, Richard Shweder (1986) has cogently discussed how holding such ideas as reincarnation, which may at first blush strike the reader as irrational, may be based on a rational consideration of empirical evidence, although parting from different metaphysical axioms than those held by many in the secular West who hold different ones (and by definition axioms are not the result of rational consideration but a-priori assumptions).

As to the second issue, undoubtedly the search for and consequences of ASC can be destructive, as in the personal and social costs of drug addiction, which is why various traditional societies provide training on and ritualize the use of psychoactive drugs, which then cause no harm, in contrast with what occurs in our midst (Dobkin de Rios, 1984). Let me be clear that despite my cheery title it is not my contention that ASC are necessarily beneficent. Although evidence has accumulated that just having unusual experiences and ASC is not per se a sign of dysfunction (Cardeña, Lynn, & Krippner, 2000; Moreira-Almeida & Cardeña, in press), this does not deny that hellish ASC are also encountered in the ravines of a schizophrenic or otherwise seriously disordered mind (Cardeña, 2011), or that ritually-induced ASC to form in-group cohesion may not be used for horrible purposes as in the Nazi Nuremberg rallies. The other side of the coin, however, is that the vast majority of atrocities have been planned while in an OSC, from decisions to wage unnecessary wars and genocides to the socially accepted mistreatment of non-human sentient beings to save some money.

The final point in this section concerns the presumption that ASC "create phenomenal contents of consciousness that misrepresent or create delusional beliefs of the surrounding world and oneself" and the OSC does not (Kallio & Revonsuo, 2003, pp. 141-142; a re-statement of a position previously advanced by Natsoulas, 1983). This position goes against not only centuries old critiques of naïve realism in both the West and the East, but a vast amount of research on perception, cognition, and personality. For instance, perceptual illusions show that our senses routinely distort stimuli in various ways, physiological habituation evidences that repeated exposure to a stimulus may make it imperceptible, and studies of other species show how limited our perceptual channels are as compared to theirs (Balcombe, 2010). To these we can add research on how the OSC is geared to detect certain aspects of the environment and disregard others (Ornstein, 1986), and on how emotions and cognitions routinely have a self-serving bias and distort how we evaluate and remember events (Greenwald, 1980). Thus, Western and Eastern spiritual ascriptions that state that we perceive the world through a glass darkly or the veil of maya (or illusion) are consistent with psychological and neuroscientific research. As to the argument that evolution should produce beings that do not create delusional beliefs or misrepresentations, research shows that a mere approximation to what may actually be "there" is selected for rather than a resources-taxing very accurate representation (Hoffman, 2009).

Instead of considering ASC as necessarily leading to delusions and misrepresentations, I concur with Mishara & Schwartz (in press) that ASC barely suspend or disrupt our consensual common sense ways of constructing reality, and that we cannot state a-prior what is the ontological status of these alternatives. A discussion of the main purposes to alter our consciousness constitutes the gist of this paper, inspired by multidisciplinary contributions to the study of ASC (Cardeña & Winkelman, in press a, b).

That most cultures throughout history (around 90%, see Bourguignon, 1973) until very recently have established forms to attain ASC (often during well-developed rituals) suggests that they have considered them positive rather than as the purveyors of misinformation. For example, the phenomenon known as shamanism, found in hunting-gathering societies across the globe, is characterized by the controlled practice of altering the shaman's (and often his or her audience's) state of consciousness or, in less secular terms, experiencing other realities (Cardeña & Krippner, 2010; Winkelman, in press). Eastern spiritual traditions have described and codified meditation techniques that aim to alter one's consciousness in the short- and long-term (Shear, in press), as have performance disciplines (Zarrilli, in press), martial arts, and sexual practices

(Maliszewski, Vaughan, Krippner, Holler, & Fracasso, in press).

The West has also had a plethora of techniques to induce ASC, from the use of caves in prehistoric and Greek times, to drugs, dances, mystery rituals, hypnotic and other suggestions, meditative practices, and so on (Cardeña & Alvarado, in press; Sluhovsky, in press; Ustinova, in press). In our times, various groups have continued or modified previous techniques and enlisted technology with the goal of altering consciousness (St John, in press), but we have barely gone beyond the cogitations of Plato on altered consciousness (Cardeña, 2009).

2. Why Do We Seek to Alter Our Consciousness?

A visitor from another planet would surely find it puzzling that humans spend about one third of their lives in sleep states yet pay little attention to them. Besides this inbuilt alteration of consciousness, humans (and other species, see Siegel, 1989) seek to affect their consciousness in other ways. Some psychoactive drugs have been synthesized in the laboratory but non-urbane inhabitants have had at their disposal for millennia compounds with remarkable abilities to produce ASC. For instance, N,Ndimethyltryptamine (DMT) is found in many plants and animals, and its molecular structure is remarkably similar to that of the neurotransmitter serotonin (Mishor, McKenna, & Callaway, in press). Indeed, the most ubiquitous drug supplier we have is our own nervous system, whose function is based on various neurotransmitters with very similar structures to those of psychoactive drugs (the latter can affect us precisely because they mimic our endogenously produced substances; Presti, in press).

Contrary to what many of us were told about the artificiality of affecting our states of consciousness, it turns out that we have been using all along our biology given resources. But why do we try to alter our state of consciousness? Ludwig (1966) posited that ASC can have healing, social, or epistemological functions. In the remaining of the paper, I will expand that list and discuss how ASC can help us: A) make small adjustments so we can be more effective at our tasks, B) enhance the experience of life, C) explore nonfactual possibilities that may be actualized through creative ways, D) enhance the sense of meaningfulness in life and heal in different ways, and E) obtain alternate epistemological routes to comprehend better ourselves and the universe at large.

A) Making small adjustments in everyday life Because they are taken routinely and produce minor modifications, we generally disregard how we routinely affect our consciousness by using caffeine to make us more alert, chamomile tea to produce a gentle sedative affect, or nicotine as either a stimulant or a relaxant. Feeling low we may choose to initiate an energetic walk or, being tired of thinking about and writing this paper, I may decide to give my mind a break and let a good film director invite me into another reality or try to achieve a flow state (i.e., full immersion in an activity, Csikszentmihalyi, 1988) while playing tennis. I do not claim that these various activities generally bring about ASC (although flow states have a number of commonalities, such as time changes, with more radical alterations of consciousness, Cardeña, 2006), but at the very least they exemplify our propensity to affect our states of consciousness in subtle ways throughout the day.

B) Enhance the experience of life We do not need to embrace hedonism and declare pleasure the only intrinsic good to realize that one of the major joys of life is the experience with which our senses regale us. Throughout our development, however, we substitute more and more the expanded, open consciousness of the baby and infant for an ego-centered, focused, conceptual mediation of the world (Gopnik, 2009). Although pragmatic in many ways, this form of apprehension superposes conceptual layers on direct, sensuous experience. That many people find something amiss in a low intensity existence is borne by the myriad activities through which they seek to enhance sensuous experiences, from culinary to bungee-jumping adventures. These pale, however, when compared with the re-enchantment of the world which some find in the use of psychedelics or "esoteric" erotic techniques. While the term "hallucinogen" suggests that the main effect of a drug will be some kind of hallucinations, users may specifically look for an enhancement of their sensual experiences (e.g., Tart, 1971), perhaps akin to what some artists experience and seek to convey (Durr, 1970). We may thus induce ASC to become more childlike and immerse us in sounds, tastes, or the skin of the beloved (cf. Granqvist, Reijman, & Cardeña, in press).

Some excerpts from an account by psychologist Stanley Krippner (1970) after ingesting psilocybin will give a taste of what this experience is like:

> "I opened my eyes to find the living room vibrating with brilliant colors… I seemed to be in the middle of a three-dimensional Vermeer painting… I was astounded by the extraordinarily delicious taste and perfection of [an apple]… My exploration of the softness of the sweater and the warmth of her flesh was an ecstatic sensual experience… I was hearing the music as I had never heard it before… Only the sheer beauty of each individual tone mattered. I was listening to music vertically rather than horizontally."

C) Discover novel, creative ways to look at or express reality A recent finding in developmental psychology is that the creation and experience of fantasy during infancy and childhood is not only not opposed to analytical, rational development, but may be necessary to it (cf. Gopnik, 2009). The initial flowing spurts of game and created worlds tend to be overwhelmed later on by pragmatic endeavors, yet during ASC not only can we experience more intensely our sense data but become fully immersed in experiences in which ordinary physical or psychological constraints do not apply, have unusual associations, or try different solutions that could not be implemented during our OSC. It is not surprising that the propensity to easily transit among states of consciousness is positively related to creativity (Hartmann, 1991). Artists (Levy, in press), and writers (Cousins, in press), have explored dreaming, shamanism, meditation, hypnosis, and other ASC to develop their art in various ways or as a source of inspiration. For example, Van Gogh wrote that he would often look at a landscape for hours without a break until he would arrive to a special state of "lucidity," which may partly explain the intensity of his paintings (in Levy, in press).

Some eminent scientists have also linked ASC, brought deliberately or not (Koestler, 1964), with technical invention and scientific breakthrough, as in the case of dreams providing a guide in the work of Loewi, Elias Howe, and others (Krippner & Hughes, 1970). This does not mean, of course, that being in an ASC will ipso facto allow someone to produce a great work of art or have a scientific breakthrough, but that entering an ASC may generate novel ideas, associations, and experiences that may then be developed in that or an OSC.

D) Enhancing the meaningfulness of life/improving medical and psychological health Anomalous experiences overall are experienced as enhancing creativity and the meaning and purpose of life, in contrast with obtaining wealth (Kennedy, Kanthamani, & Palmer, 1994). While it is arguable that near-death-experiences (NDEs) provide evidence for the survival of consciousness after death, they are a primary illustration of how ASC can improve the experients' life. The majority of those having undergone an NDE seem to become more harmonious in the long-term (in the short term they may have to adapt themselves and others to their experience), caring more for others and the environment and being less materialistic and selfcentered (Greyson, 2000). Similarly, an experience with psilocybin in a supervised setting was not only described as extremely meaningful by experients, but made them better people as judged by those close to them (Griffiths, Richards, McCann, & Jesse, 2006).

That a psychoactive drug is not required can be seen in research with very high hypnotizables individuals who reported positive changes after experimental sessions without specific suggestions. Previously they had reported such ongoing experiences as "the overwhelming serenity of it," "all the feelings that are good… There's a bright light, I'm in it… It feels like I've been here before" (Cardeña, 2005, unpublished data). The ultimate ontology of these and other ASC can be contested, but not that they increase the perceived meaning of experients' lives.

Healing is a multivocal concept that includes experiences of health improvement that may or may not be related to actual improvement in health indexes, besides being used to denote finding greater meaning in life irrespective of changes in health. Ever since shamanism have ASC been associated with the ability to induce or experience healing (Winkelman, in press), and often that purported ability has followed real or symbolic injury and illness, as in this fragment of a North American shaman:

> "I was torn and torn... I accepted it all... And then I began to be put back together again... Something was there wasn't there before... it made the whole thing more than it had been... that song gave me new strength and power" (in Halifax, 1979, p. 155).

Accounts of some form of healing after ASC do not occur only in traditional societies as illustrated by an account of an NDE in which after the person experienced "the kind of love that cures, heals, regenerates" his brain tumor disappeared (in Dossey, 2011, p. 59). We require far more systematic research on the connection of ASC and such "spontaneous remissions" (Krippner & Achterberg, 2000). Techniques that can produce alterations in consciousness such as hypnosis and meditation have been found to have a therapeutic effect in a variety of psychological and medical conditions (Horowitz, 2010; Mendoza & Capafons, 2009), but we are still a long way from being able to determine what factors, including possible alterations in consciousness, effect positive outcomes. Mishara and Schwartz (in press) propose that the alternative self-world representations created by ASC may disrupt maladaptive psychobiological patterns and bring about improvement in symptoms or, at least, in the way they are experienced.

E) Provide valid knowledge about reality The final issue goes to the core of what is the most contested issue about ASC, namely whether they can provide alternative and valid knowledge to that offered by our OSC. The import of ASC for ontology, epistemology, and metaphysics has been discussed by thinkers of the stature of Plato and Descartes, and continue to intrigue contemporary philosophers (Windt, in press). Let me start with less controversial claims and then move to more challenging ones.

A mainstream theory of the functions of sleep maintains that by keeping new learning to a minimum during that state we can consolidate and reorganize recently acquired information (Hobson, 1995); this seems to be the case especially with respect to emotional memories (Wagner, Gais, & Born, 2001). A less investigated purpose for ASC is that they may provide insight into aspects of cognition that generally remain opaque. Besides the insights into one's psyche that various drugs and other forms to alter consciousness may provide, seeking to affect ordinary consciousness through meditation may reduce the effect of perceptual illusions (e.g., Brown, Forte, & Dysart, 1984) and reveal the architecture of cognition (Hunt, 1995). That ASC may provide access to valid information at least with regard to the experient is borne out by a study in which traumatic dreams provided a significant indicator of the extent of the, until then unascertained, heart damage of the dreamer (Smith, 1987).

Yet the most daring claim is that the alternative perspectives about reality engendered by some (by no means all!) ASC is a more real or at least a complementary view than that provided ordinarily by our senses and reason. The clearest instance is the view enunciated by mystics throughout the centuries that we are interconnected within an all-pervading unity (Wulff, 2000). Mystical reports include sensual experiences such as "I was immersed in sweetness words cannot express. I could hear the singing of the planets, and wave after wave of light washed over me... I was the light as well" (in Wulff, 2000, p. 398) or instances of pure awareness: "[In my meditations]I just remain in the Absolute for the entire sitting and nothing else seems to happen, other than the feeling of bliss permeating me completely (in Shear, in press). Regardless of these differences these and other reports imply a general sense of unity with all there is. Serious consideration for this view of reality has come not only from mystics and artists of various stripes but from eminent psychologists and neuroscientists (cf. Cardeña & Winkelman, in press a, b), and quantum physicists such as David Bohm. Just compare Bohm describing his theory of implicate order "We must learn to view everything as part of "Undivided Wholeness in Flowing Movement"... On this stream, one may see an ever-changing pattern of vortices, ripples, waves, splashes, etc., which evidently have no independent existence as such. Rather, they are abstracted from the flowing movement, arising and vanishing in the total process of the flow" (Bohm, 1980, pp. 11, 48), with a fragment from Lord Shelley's poem Adonais:

> The One remains, the many change and pass;
> Heaven's light forever shines, Earth's shadows fly;
> Life, like a dome of many-coloured glass,
> Stains the white radiance of Eternity

The growing research database on parapsychological phenomena also suggests that not only do we seem to be affected independently of time and space constraints, but that this happens more often (or at least we become aware that it happens more often) during ASC (Luke, in press).

Thus, not only consciousness as the capacity to have experiences and observe events, but also ASC must be taken into consideration as we ponder the nature of the universe. This suggests that the world is found not only in a grain of sand but in the state of consciousness that intuits that connection.

References

Balcombe, J. (2010). Second Nature. The Inner Lives of Animals. New York: Macmillan.

Bohm, D. & Hiley, B. J. (1993). Wholeness and the Implicate Order. London: Routledge & Kegan Paul.

Bourguignon, E. (1973). Religion, Altered States of Consciousness, and Social Change. Columbus: Ohio State University Press.

Brown, D., Forte, M., & Dysart, M. (1984a). Differences in visual sensitivity among mindfulness meditators and non-meditators. Perceptual and Motor Skills, 58, 727– 733.

Cardeña, E. (2005). The phenomenology of deep hypnosis: Quiescent and physically active. International Journal of Clinical & Experimental Hypnosis, 53, 37-59.

Cardeña, E. (2006). Flow and anomalous experiences. Stanford University Nissan Salon, Invited address, March, 2006.

Cardeña, E. (2009). Beyond Plato? Toward a science of alterations of consciousness. In C. A. Roe, W. Kramer, & L. Coly (Eds.). Utrecht II: Charting the Future of Parapsychology (pp. 305-322). New York, NY: Parapsychology Foundation, Inc.

Cardeña, E. (in press). Altered consciousness in emotion and psychopathology. In E. Cardeña, & M. Winkelman (Eds.), Altering Consciousness. A Multidisciplinary Perspective. Volume II. Biology and Clinical Sciences. Praeger Publishers.

Cardeña, E., & Alvarado, C. (in press). Altered consciousness from the Age of Enlightenment through mid-20th century. In E Cardeña, M Winkelman (Eds.), Altering Consciousness. A Multidisciplinary Perspective. Volume I. History, Culture, and the Humanities. Praeger Publishers.

Cardeña, E., Lynn, S. J., & Krippner, S. (Eds.). (2000). Varieties of Anomalous Experience: Examining the Scientific Evidence. Washington, DC: American Psychological Association.

Cardeña, E., & Winkelman, M. (in press, a). Altering Consciousness. A Multidisciplinary Perspective. Volume I. History, Culture, and the Humanities. Praeger Publishers.

Cardeña, E., & Winkelman, M. (in press, b). Altering Consciousness. A Multidisciplinary Perspective. Volume II. Biology and Clinical Sciences. Praeger Publishers.

Cousins, W. (in press). Colored inklings: Altered states of consciousness and literature. In E Cardeña, M Winkelman (Eds.), Altering Consciousness. A Multidisciplinary Perspective. Volume I. History, Culture, and the Humanities. Praeger Publishers.

Csikszentmihalyi, Mihaly (1988), Optimal experience: Psychological studies of flow in consciousness, Cambridge, NY: Cambridge University Press.

Dobkin de Rios, M. (1984). Hallucinogens: Cross-cultural perspectives. Albuquerque: University of New Mexico Press.

Dossey, L. (2011). Dying to heal: A neglected aspect of NDEs. Explore: The Journal of Science and Healing, 7(2), 59-62.

Durr, R.A. (1970). Poetic vision and the psychedelic experience. New York: Syracuse University Press.

Gopnik, A. (2009). The Philosophical Baby. New York: Farrar, Straus and Giroux. Granqvist, P., Reijman, S., & Cardeña, E. (in press). Altered consciousness and human development. In E Cardeña, M Winkelman (Eds.), Altering Consciousness. A Multidisciplinary

Perspective. Volume II. Biology and Clinical Sciences. Praeger Publishers.

Greyson, B. (2000). Near-death experiences. In E. Cardeña, S. J. Lynn, & S. Krippner (Eds.), Varieties of Anomalous Experience: Examining the Scientific Evidence (pp. 315–351). Washington, DC: American Psychological Association.

Griffiths, R. R., Richards, W. A., McCann, U., & Jesse, R. (2006). Psilocybin can occasion mystical-type experiences having substantial and sustained personal meaning and spiritual significance. Psychopharmacology, 187, 268–283.

Halifax, J. (1979). Shamanic Voices: The Shaman as Seer, Poet and Healer. Middlesex, UK: Penguin.

Hartmann, E. (1991). Boundaries of the Mind: A New Psychology of Personality. New York: Basic Books.

Hobson, J. A. (1995). Sleep. New York: Scientific American Library.

Hoffman, D. D. (2009). Nature and consciousness. Mindfield. The Bulletin of the Parapsychological Association, 1(1), 6–7.

Horowitz, S. (2010). Health benefits of meditation: What the newest research shows. Alternative and Complementary Therapies 16(4), 223-228.

Hunt, H. (1995). On the nature of consciousness. New Haven, CT: Yale University Press.

Kallio, S., & Revonsuo, A. (2003). Hypnotic phenomena and altered states of consciousness: A multilevel framework of description and explanation. Contemporary Hypnosis, 20, 111–164.

Kennedy, J. E., Kanthamani, H., & Palmer, J. (1994). Psychic and spiritual experiences, health, well-being, and meaning in life. Journal of Parapsychology, 58, 353-383.

Koestler, A. (1964). The act of creation. Middlesex, UK: Penguin.

Krippner, S. (1970). An adventure in psilocybin. In B. Aaronson & H. Osmond (Eds.), Psychedelics. New York: Anchor Books.

Krippner, S., & Achterberg, J. (2000). Anomalous healing experiences. In E. Cardeña, S. J. Lynn, & S. Krippner (Eds.), Varieties of anomalous experience: Examining the scientific evidence (pp. 353–395). Washington, DC: American Psychological Association.

Krippner, S., & Hughes, W. (1970). Dreams and human potential. Journal of Humanistic Psychology, 10, 1-20.

Laughlin, C. D., McManus, J., & d'Aquili, E. G. (1992). Brain, Symbol and Experience: Toward a Neurophenomenology of Human Consciousness. New York: Columbia University Press.

Levy, M. (in press). Altered consciousness and modern art. In E Cardeña, M Winkelman (Eds.), Altering Consciousness. A Multidisciplinary Perspective. Volume I. History, Culture, and the Humanities. Praeger Publishers.

Ludwig, A. M. (1966). Altered states of consciousness. Archives of General Psychiatry, 15, 225–234.

Luke, D. (in press). Anomalous phenomena, psi, and altered consciousness. In E Cardeña, M Winkelman (Eds.), Altering Consciousness. A Multidisciplinary Perspective. Volume II. Biology and Clinical Sciences. Praeger Publishers.

Maliszewski, M., Vaughan, B., Krippner, S., Holler, G., & Fracasso, C. (in press). Altering consciousness through sexual activity. In E Cardeña, M Winkelman (Eds.), Altering Consciousness. A Multidisciplinary Perspective. Volume II. Biology and Clinical Sciences. Praeger Publishers.

Mendoza, M. E., & Capafons, A. (2009).

Efficacy of clinical hypnosis: A summary of its empirical evidence. Papeles del Psicólogo, 38, 98-116.

Mishara, A. L., & Schwartz, M. A. (in press). Altered states of consciousness as paradoxically healing: An embodied social neuroscience perspective. In E Cardeña, M Winkelman (Eds.), Altering Consciousness. A Multidisciplinary Perspective. Volume II. Biology and Clinical Sciences. Praeger Publishers.

Mishor, Z., McKenna, D. J., & Callaway, J. C. (in press). DMT and human consciousness. In E Cardeña, M Winkelman (Eds.), Altering Consciousness. A Multidisciplinary Perspective. Volume II. Biology and Clinical Sciences. Praeger Publishers.

Moreira-Almeida, A., & Cardeña, E. (in press). Differential diagnosis between nonpathological psychotic and spiritual experiences and mental disorders: A contribution from Latin American studies to the ICD-11. Revista Brasileira de Psiquiatria Natsoulas, T. (1983). Concepts of consciousness. Journal of Mind and Behavior, 4, 13–59.

Ornstein, R. (1986). The psychology of consciousness (rev. ed.). New York: Penguin.

Presti, D. E. (in press). Neurochemistry and altered consciousness. In E Cardeña, M Winkelman (Eds.), Altering Consciousness. A Multidisciplinary Perspective. Volume II. Biology and Clinical Sciences. Praeger Publishers.

Shear. J. (in press). Eastern approaches to altered states of consciousness. In E Cardeña, M Winkelman (Eds.), Altering Consciousness. A Multidisciplinary Perspective. Volume I. History, Culture, and the Humanities. Praeger Publishers.

Shweder, R.A. (1986). Divergent rationalities. In D.W. Fiske & R.A. Shweder (Eds.), Metatheory in social science: Pluralisms and subjectivites (pp. 163-196). Chicago: University of Chicago Press.

Siegel, R. K. (1989). Intoxication. New York: E. P. Dutton.

Sluhovsky, M. (in press). Spirit possession and other alterations of consciousness in the Christian Western tradition. In E Cardeña, M Winkelman (Eds.), Altering Consciousness. A Multidisciplinary Perspective. Volume I. History, Culture, and the Humanities. Praeger Publishers.

Smith, R. C. (1987). Do dreams reflect a biological state? Journal of Nervous and Mental Disease, 147, 587-604.

St John, G. (in press). Spiritual technologies and altering consciousness and altering consciousness in contemporary counterculture. In E Cardeña, M Winkelman (Eds.), Altering Consciousness. A Multidisciplinary Perspective. Volume I. History, Culture, and the Humanities. Praeger Publishers.

Ustinova, Y. (in press). Consciousness alteration practices in the West from Prehistory to Late Antiquity. In E Cardeña, M Winkelman (Eds.), Altering Consciousness. A Multidisciplinary Perspective. Volume I. History, Culture, and the Humanities. Praeger Publishers.

Tart, C. T. (1971). On Being Stoned: A Psychological Study of Marijuana Intoxication. Palo Alto, CA: Science and Behavior Books.

Turnbull, C. (1993). Liminality: A synthesis of subjective and objective experience. In: R. Schechner & W. Appel (Eds). By means of performance (pp. 50-81). Cambridge: Cambridge University Press.

Wagner, U., Gais, S., & Born, J. (2001). Emotional memory formation is enhanced across sleep intervals with high amounts of rapid eye movement sleep. Learning and Memory, 8, 112-119.

Whitehead, C. (in press). Altered

consciousness in society. In E Cardeña, M Winkelman (Eds.), Altering Consciousness. A Multidisciplinary Perspective. Volume II. Biology and Clinical Sciences. Praeger Publishers; in press.

Windt, J. M. (in press). Altered consciousness in philosophy. In E Cardeña, M Winkelman (Eds.), Altering Consciousness. A Multidisciplinary Perspective. Volume I. History, Culture, and the Humanities. Praeger Publishers.

Winkelman, M. (in press). Shamanism and the alteration of consciousness. In E Cardeña, M Winkelman (Eds.), Altering Consciousness. A Multidisciplinary Perspective. Volume II. Biology and Clinical Sciences. Praeger Publishers.

Wulff, D. M. (2000). Mystical experiences. In E. Cardeña, S. J. Lynn, & S. Krippner (Eds.), Varieties of Anomalous Experience: Examining the Scientific Evidence (pp. 397–440). Washington, DC: American Psychological Association.

Zarrilli, P. B. (in press). Altered consciousness in performance. West and East. In E Cardeña, M Winkelman (Eds.), Altering Consciousness. A Multidisciplinary Perspective. Volume II. Biology and Clinical Sciences. Praeger Publishers.

43. Quantum Physics and the Multiplicity of Mind: Split-Brains, Fragmented Minds, Dissociation, Quantum Consciousness

R. Joseph, Ph.D.

Emeritus, Brain Research Laboratory, Northern California

Abstract

Quantum physics and Einstein's theory of relativity make assumptions about the nature of the mind which is assumed to be a singularity. In the Copenhagen model of physics, the process of observing is believed to effect reality by the act of perception and knowing which creates abstractions and a collapse function thereby inducing discontinuity into the continuum of the quantum state. This gives rise to the uncertainty principle. Yet neither the mind or the brain is a singularity, but a multiplicity which include two dominant streams of consciousness and awareness associated with the left and right hemisphere, as demonstrated by patients whose brains have been split, and which are superimposed on yet other mental realms maintained by the brainstem, thalamus, limbic system, and the occipital, temporal, parietal, and frontal lobes. Like the quantum state, each of these minds may also become discontinuous from each other and each mental realm may perceive their own reality. Illustrative examples are detailed, including denial of blindness, blind sight, fragmentation of the body image, phantom limbs, the splitting of the mind following split-brain surgery, and dissociative states where the mind leaves the body and achieves a state of quantum consciousness and singularity such that the universe and mind become one.

1. Introduction

In 1905 Albert Einstein published his theories of relativity, which promoted the thesis that reality and its properties, such as time and motion had no objective "true values", but were "relative" to the observer's point of view (Einstein, 1905a,b,c). However, what if the observer is not a singularity and has more than one point of view and more than one stream of observing consciousness? And what if these streams of consciousness were also relative?

Quantum physics, as exemplified by the Copenhagen school (Bohr, 1934, 1958, 1963; Heisenberg, 1930, 1955, 1958), also makes assumptions about the nature of reality as related to an observer, the "knower" who is conceptualized as a singularity. Because the physical world is relative to being known by a "knower" (the observing consciousness), then the "knower" can influence the nature of the reality which is being observed. In consequence, what is known vs what is not known becomes relatively imprecise (Heisenberg, 1958).

For example, as expressed by the Heisenberg uncertainty principle (Heisenberg, 1955, 1958), the more precisely one physical property is known the more unknowable become other properties, whose measurements become correspondingly imprecise. The more precisely one property is known, the less precisely the other can be known and this is true at the molecular and atomic levels of reality. Therefore it is impossible to precisely determine, simultaneously, for example, both the position and velocity of an electron.

However, we must ask: if knowing A, makes B unknowable, and if knowing B makes A unknowable, wouldn't this imply that both A and B, are in fact unknowable? If both A and B are manifestations of the processing of "knowing," and if observing and measuring can change the properties of A or B, then perhaps both A and B are in fact properties of knowing, properties of the observing consciousness, and not properties of A or B.

In quantum physics, nature and reality are represented by the quantum state. The electromagnetic field of the quantum state is the fundamental entity, the continuum that constitutes the basic oneness and unity of all things.

The physical nature of this state can be "known" by assigning it mathematical properties (Bohr, 1958, 1963). Therefore, abstractions, i.e., numbers, become representational of a hypothetical physical state. Because these are abstractions, the physical state is also an abstraction and does not possess the material consistency, continuity, and hard, tangible, physical substance as is assumed by Classical (Newtonian) physics. Instead, reality, the physical world, is created by the process of observing, measuring, and knowing (Heisenberg, 1955).

Consider an elementary particle, once this positional value is assigned, knowledge of momentum, trajectory, speed, and so on, is lost and becomes "uncertain." The particle's momentum is left uncertain by an amount inversely proportional to the accuracy of the position measurement which is determined by values assigned by the observing consciousness. Therefore, the nature of reality, and the uncertainty principle is directly affected by the observer and the process of observing and knowing (Heisenberg, 1955, 1958).

The act of knowing creates a knot in the quantum state; described as a "collapse of the wave function;" a knot of energy that is a kind of blemish in the continuum of the quantum field. This quantum knot bunches up at the point of observation, at the assigned value of measurement.

The process of knowing, makes reality, and the quantum state, discontinuous. "The discontinuous change in the probability function takes place with the act of registration…in the mind of the observer" (Heisenberg, 1958).

Reality, therefore, is a manifestation of alterations in the patterns of activity within the electromagnetic field which are perceived as discontinuous. The perception of a structural unit of information is not just perceived, but is inserted into the quantum state which causes the reduction of the wave-packet and the collapse of the wave function.

Knowing and not knowing, are the result of interactions between the mind and concentrations of energy that emerge and disappear back into the electromagnetic quantum field.

However, if reality is created by the observing consciousness, and can be made discontinuous, does this leave open the possibility of a reality behind the reality? Might there be multiple realities? And if consciousness and the observer and the quantum state is not a singularity, could each of these multiple realities also be manifestations of a multiplicity of minds?

Heinserberg (1958) recognized this possibility of hidden realities, and therefore proposed that the reality that exists beyond or outside the quantum state could be better understood when considered in terms of "potential" reality and "actual" realities. Therefore, although the quantum state does not have the ontological character of an "actual" thing, it has a "potential" reality; an objective tendency to become actual at some point in the future, or to have become actual at some point in the past.

Therefore, it could be said that the subatomic particles which make up reality, or the quantum state, do not really exist, except as probabilities. These "subatomic" particles have probable existences and display tendencies to assume certain patterns of activity that we perceive as shape and form. Yet, they may also begin to display a different pattern of activity such that being can become nonbeing and thus something else altogether.

The conception of a deterministic reality is therefore subjugated to mathematical probabilities and potentiality which is relative to the mind of a knower which effects that reality as it unfolds, evolves, and is observed (Bohr 1958, 1963; Heisenberg 1955, 1958). That is, the mental act of perceiving a non-localized unit of structural information, injects that mental event into the quantum state of the universe, causing "the collapse of the wave function" and creating a bunching up, a tangle and discontinuous knot in the continuity of the quantum state.

Einstein ridiculed these ideas (Pais, 1979): "Do you really think the moon isn't there if you aren't looking at it?"

Heisenberg (1958), cautioned, however, that the observer is not the creator of reality: "The introduction of the observer must not be misunderstood to imply that some kind of subjective features are to be brought into the description of nature. The observer has, rather, only the function of registering decisions, i.e., processes in space and time, and it does not matter whether the observer is an apparatus or a human being; but the registration, i.e., the transition from the "possible" to the "actual," is absolutely necessary here and cannot be omitted from the interpretation of quantum theory."

Shape and form are a function of our perception of dynamic interactions within the continuum which is the quantum state. What we perceive as mass (shape, form, length, weight) are dynamic patterns of energy which we selectively attend to and then perceive as stable and static, creating discontinuity within the continuity of the quantum state. Therefore, what we are perceiving and knowing, are only fragments of the continuum.

However, we can only perceive what our senses can detect, and what we detect as form and shape is really a mass of frenzied subatomic electromagnetic activity that is amenable to detection by our senses and which may be known by a knowing mind. It is the perception of certain aspects of these oscillating patterns of continuous evolving activity, which give rise to the impressions of shape and form, and thus discontinuity, as experienced within the mind.

This energy that makes up the object of our perceptions, is therefore but an aspect of the electromagnetic continuum which has assumed a specific pattern during the process of being sensed and processed by those regions of the brain and mind best equipped to process this information.

Perceived reality, therefore, becomes a manifestation of mind.

However, if the mind is not a singularity, and if we possessed additional senses or an increased sensory channel capacity, we would perceive yet other patterns and other realities which would be known by those features of the mind best attuned to them. If the mind is not a singularity but a multiplicity, this means that both A and B, may be known simultaneously.

2. Duality vs Multiplicity

In the Copenhagen model, the observer is external to the quantum state the observer is observing, and they are not part of the collapse function but a witness of it (Bohr, 1958, 1963; Heisenberg 1958). However, if the Copenhagen model is correct, and as the cosmos contains observers, then the standard collapse formulation can not be used to describe the entire universe as the universe contains observers (von Neumann, 1932, 1937).

Further, reality becomes, at a minimum, a duality (observer and observed) with the potential to become a multiplicity.

As described by DeWitt and Graham (1973; Dewitt, 1971), "This reality, which is described jointly by the dynamical variables and the state vector, is not the reality we customarily think of, but is a reality composed of many worlds. By virtue of the temporal development of the dynamical variables the state vector decomposes naturally into orthogonal vectors, reflecting a continual splitting of the universe into a multitude of mutually unobservable but equally real worlds, in each of which every good measurement has yielded a definite result and in most of which the familiar statistical quantum laws hold."

The minimal duality is that aspect of reality which is observed, measured, and known, and that which is unknown.

However, this minimal duality is an illusion as indicated not only by the potential to become multiplicity, but by the nature of mind which is not a singularity (Joseph, 1982, 1986a; 1988a,b).

Even if we disregard the concept of "mind" and substitute the word "brain", the fact remains that the brain is not a singularity. The human brain is functionally specialized with specific functions and different mental states localized to specific areas, each of which is capable of maintaining independent and semi-independent aspects of conscious-awareness (Joseph 1986a,b, 1988a,b, 1992, 1999a). Different aspects of the same experience and identical aspects of that experience may be perceived and processed by different brain areas in different ways (Gallagher and Joseph, 1982; Joseph 1982; Joseph and Gallagher 1985; Joseph et al., 1984).

Therefore, although it has been said that orthodox quantum mechanics is completely concordant with the defining characteristics of Cartesian dualism, this is an illusion. Cartesian duality assumes singularity of mind, when in fact, the overarching organization of the mind- and the brain- is both dualistic and multiplistic.

If quantum physics is "mind-like" (actual/operational at the quantum level, but mentalistic on the ontological level) then quantum physics, or rather, the quantum state (reality, the universe) is not a duality, but a multiciplicity. Indeed, the entire concept of duality is imposed on reality by the dominant dualistic nature of the brain and mind which subordinates not just reality, but the multiplicity of minds maintained within the human brain (Joseph, 1982).

Like the Copenhagen school, Von Neumann's formulation of quantum mechanics (1932, 1937), fails to recognize or understand the multiple nature of mind and reality. Von Neumann postulated that the physical aspects of nature are represented by a density matrix. The matrix, therefore, could be conceptualized as a subset of potential realities, and that by averaging the values of these evolving matrices, the state of the universe and thus of reality, can be ascertained as a unified whole. However, in contrast to the Copenhagen interpretation, Von Neumann shifted the observer (his brain) into the quantum universe and thus made it subject to the rules of quantum physics.

Ostensibly and explicitly, Von Neumann's conceptions are based on a conception of mind as a singularity acting on the quantum state which contains the brain. Von Neumann's mental singularity, therefore, imposes itself on reality, such that each "event" that occurs within reality, is associated with one specific experience of the singularity-mind. Thus, Von Neumann assumes the brain and mind has only "one experience" which corresponds with "one event;" and this grossly erroneous misconception of the nature of the brain and mind, unfortunately, is erroneously accepted as fact by most cosmologists and physicists. Further, he argues that in the process of knowing, the quantum state of this singularity brain/mind also collapses, or rather, is reduced in a mathematically quantifiable manner, just as the quantum universe is collapsed and reduced by being known (Von Neumann, 1932, 1937).

However, the brain and mind are not a singularity, but a multiplicity (Joseph, 1982, 1988a,b, 1999a). Nevertheless, Von Neumann's conceptions can be applied to the multiplicity of mind/brain when each mental realm is considered individually as an interactional subset of the multiplicity.

3. The Multiplicity of Mind and Perception

According to Von Neumann (1932), the "experiential increments in a person's knowledge" and "reductions of the quantum mechanical state of that person's brain" corresponds to the elimination of all those perceptual functions that are not necessary or irrelevant to the knowing of the event and the increase in the knowledge associated with the experience.

If considered from the perspective of an isolated aspect of the mind and the dominating stream of consciousness, Von Neuman's conceptions are essentially correct. However, neither the brain nor the mind function in isolation but in interaction with other neural tissues and mental/perceptual/sensory realms (Joseph 1982, 1992, 1999a). Perceptual functions are not "eliminated" and removed from the brain. Instead, they are prevented from interfering with the attentional processes of one aspect of the multiplicity of mind which dominates during the knowing event (Joseph, 1986b, 1999a).

Consider, by way of example, you are sitting in your office reading this text. The pressure of the chair, the physical sensations of your shoes and clothes, the musculature of your body as it holds one then another position, the temperature of the room, various odors and fragrances, a multitude of sounds, visual sensations from outside your area of concentration and focus, and so on, are all being transmitted to the brainstem, midbrain, and olfactory limbic system. These signals are then relayed to various subnuclei within the thalamus.

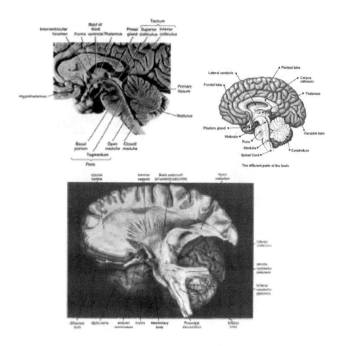

The neural tissues of the brainstem, midbrain, limbic system and thalamus are associated with the "old brain." However, those aspects of consciousness we most closely associated with humans are associated with the "new brain" the neocortex (Joseph, 1982, 1992). Therefore, although you may be "aware" of these sensations while they are maintained within the old brain, you are not "conscious" of them, unless a decision is made to become conscious or they increase sufficiently in intensity that they are transferred to the neocortex and forced into the focus of consciousness (Joseph, 1982, 1986b, 1992, 1999a).

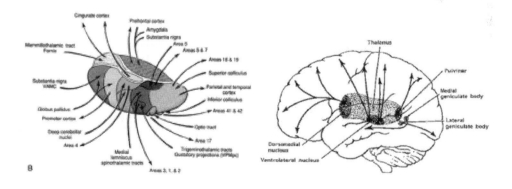

The old brain is covered by a gray mantle of new cortex, neocortex. The sensations alluded to are transferred from the old brain to the thalamus which relays these signals to the neocortex. Human consciousness and the "higher" level of the multiplicity of mind, are associated with the "new brain."

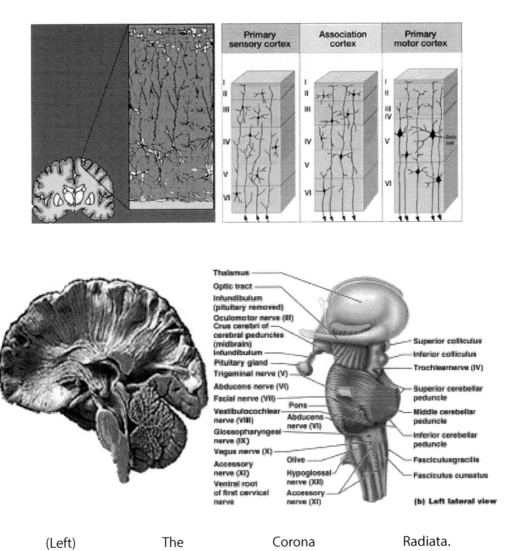

(Left) The Corona Radiata.
(Right) The human Brainstem & Thalamus

For example, visual input is transmitted from the eyes to the midbrain and thalamus and is transferred to the primary visual receiving area maintained in the neocortex of the occipital lobe (Casagrande & Joseph 1978, 1980; Joseph and Casagrande, 1978). Auditory input is transmitted from the inner ears to the brainstem, midbrain, and thalamus, and is transferred to the primary auditory receiving area within the neocortex of the temporal lobe. Tactual-physical stimuli are also transmitted from the thalamus to the primary somatosensory areas maintained in the neocortex of the parietal lobe. From the primary areas these signals are transferred to the adjoining "association" areas, and simple percepts become more complex by association (Joseph, 1996).

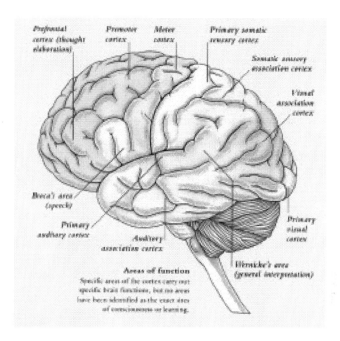

Monitoring all this perceptual and sensory activity within the thalamus and neocortex is the frontal lobes of the brain, also known as the senior executive of the brain and personality (Joseph 1986b, 1999a; Joseph et al., 1981). It is the frontal lobes which maintain the focus of attention and which can selectively inhibit any additional processing of signals received in the primary areas.

There are two frontal lobes, a right and left frontal lobe which communicate via a bridge of nerve fibers. Each frontal lobe, and subdivisions within each are concerned with different types of mental activity (Joseph, 1999a).

The left frontal lobe, among its many functions, makes possible the ability to speak. It is associated with the verbally expressive, speaking aspects of consciousness. However, there are different aspects of consciousness associated not only with the frontal lobe, but with each lobe of the brain and its subdivisions (Joseph, 1986b; 1996, 1999a).

4. Knowing Yet Not Knowing: Disconnected Consciousness

Consider the well known phenomenon of "word finding difficulty" also known as "tip of the tongue." You know the word you want (the "thingamajig") but at the same time, you can't gain access to it. That is, one aspect of consciousness knows the missing word, but another aspect of consciousness associated with talking and speech can't gain access to the word. The mind is disconnected from itself. One aspect of mind knows, the other aspect of mind does not.

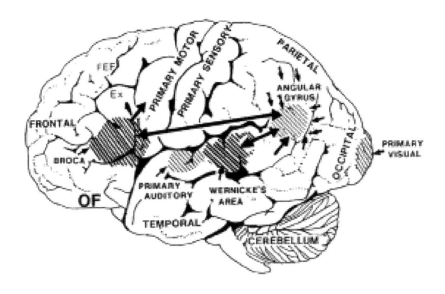

This same phenomenon, but much more severe and disabling, can occur if the nerve fiber pathway linking the language areas of the left hemisphere are damaged. For example, Broca's area in the frontal lobe expresses humans speech. Wernicke's area in the temporal lobe comprehends speech. The inferior parietal lobe associates and assimilates associations so that, for example, we can say the word "dog" and come up with the names of dozens of different breeds and then describe them (Joseph, 1982; Joseph and Gallagher 1985; Joseph et al., 1984). Therefore, if Broca's area is disconnected from the posterior language areas, one aspect of consciousness may know what it wants to say, but the speaking aspect of consciousness will be unable to gain access to it and will have nothing to say. This condition is called "conduction aphasia."

Or consider damage which disconnects the parietal lobe from Broca's area. If you place an object, e.g., a comb, out-of-sight, in the person's right hand, and ask them to name the object, the speaking aspect of consciousness may know something is in the hand, but will be unable to name it. However, although they can't name it, and can't guess if shown pictures, if the patient is asked to point to the correct object, they will correctly pick out the comb (Joseph, 1996).

Therefore, part of the brain and mind may act purposefully (e.g. picking out the comb), whereas another aspect of the brain and mind is denied access to the information that the disconnected part of the mind is acting on.

Thus, the part of the brain and mind which is perceiving and knowing, is not the same as the part of the brain and mind which is speaking. This phenomenon occurs even in undamaged brains, when the multiplicity of minds which make up one of the dominant streams of consciousness, become disconnected and/or are unable to communicate.

5. The Visual Mind: Denial of Blindness

All visual sensations first travel from the eyes to the thalamus and midbrain. At this level, these visual impressions are outside of consciousness, though we may be aware of them. These visual sensations are then transferred to the primary visual receiving areas and to the adjacent association areas in the neocortex of the occipital lobe. Once these visual impressions reach the neocortex, consciousness of the visual word is achieved. Visual consciousness is made possible by the occipital lobe.

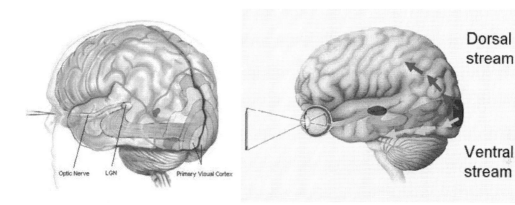

Destruction of the occipital lobe and its neocortical visual areas results in cortical blindness (Joseph, 1996). The consciousness mind is blinded and can not see or sense anything except vague sensations of lightness and darkness. However, because visual consciousness is normally maintained within the occipital lobe, with destruction of this tissue, the other mental systems will not know that they can't see. The remaining mental system do not know they are blind.

Wernicke's area in the left temporal lobe in association with the inferior parietal lobe comprehends and can generate complex language. Normally, visual input is transferred from the occipital to the inferior parietal lobe (IPL) which is adjacent to Wernicke's area and the visual areas of the occipital lobe. Once these signals arrive in the IPL a person can name what they see; the visual input is matched with auditory-verbal signals and the conscious mind can label and talk about what is viewed (Joseph, 1982, 1986b; Joseph et al., 1984). Talking and verbally describing what is seen is made possible when this stream of information is transferred to Broca's area in the left frontal lobe (Joseph, 1982, 1999a). It is Broca's area which speaks and talks.

Therefore, with complete destruction of the occipital lobe, visual consciousness is abolished whereas the other mental system remain intact but are unable to receive information about the visual world. In consequence, the verbal aspects of consciousness and the verbal-language mind does not know it can't see because the brain area responsible for informing these mental system about seeing, no longer exists. . In fact the language-dependent conscious mind will deny that it is blind; and this is called: Denial of blindness.

Normally, if it gets dark, or you close your eyes, the visual mind becomes conscious of this change in light perception and will alert the other mental realms. These other mental realms do not process visual signals and therefore they must be informed about what the visual mind is seeing. If the occipital lobe is destroyed, visual consciousness is destroyed, and the rest of the brain cannot be told that visual consciousness can't see. Therefore, the rest of the brain does not know it is blind, and when asked, will deny blindness and will make up reasons for why they bump into furniture or can't recognize objects held before their eyes (Joseph, 1986b, 1988a).

For example, when unable to name objects, they might confabulate an explanation: "I see better at home." Or, "I tripped because someone moved the furniture."

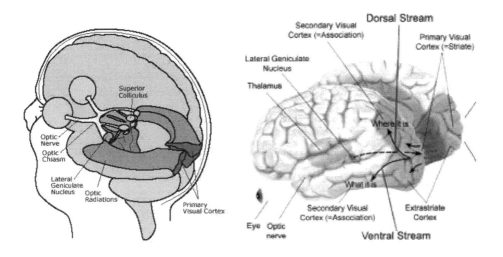

Even if you tell them they are blind, they will deny blindness; that is, the verbal aspects of consciousness will claim it can see, when it can't. The Language-dependent aspects of consciousness does not know that it is blind because information concerning blindness is not being received from the mental realms which support visual consciousness.

The same phenomenon occurs with small strokes destroying just part of the occipital lobe. Although a patient may lose a quarter or even half of their visual field, they may be unaware of it. This is because that aspect of visual consciousness no longer exists and can't inform the other mental realms of its condition.

6. "Blind Sight"

The brains of reptiles, amphibians, and fish do not have neocortex. Visual input is processed in the midbrain and thalamus and other old-brain areas as these creatures do not possess neocortex or lobes of the brain. In humans, this information is also received in the brainstem and thalamus and is then transferred to the newly evolved neocortex.

As is evident in non-mammalian species, these creatures can see, and they are aware of their environment. They possess an older-cortical (brainstem-thalamus) visual awareness which in humans is dominated by neocortical visual consciousness.

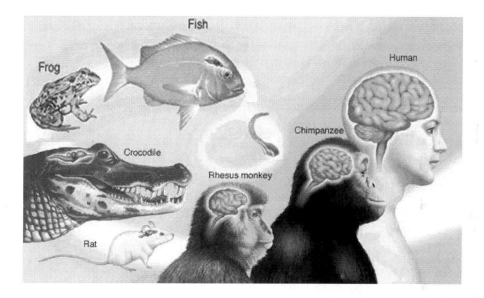

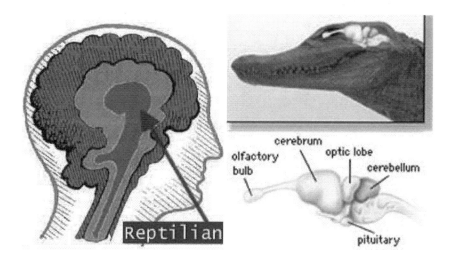

(Left) Human Brain. Reptile Brain (Right)

Therefore, even with complete destruction of the visual neocortex, and after the patient has had time to recover, some patients will demonstrate a nonconscious awareness of their visual environment. Although they are cortically blind and can't name objects and stumble over furniture and bump into walls, they may correctly indicate if an object is moving in front of their face, and they may turn their head or even reach out their arms to touch it--just as a frog can see a fly buzzing by and lap it up with its tongue. Although the patient can't name or see what has moved in front of his face, he may report that he has a "feeling" that something has moved.

Frogs do not have neocortex and they do not have language, and can't describe what they see. However, humans and frogs have old cortex that process visual impressions and which can control and coordinate body movements. Therefore, although the neocortical realms of human consciousness are blind, the mental realms of the old brain can continue to see and can act on what it sees; and this is called: Blind sight (Joseph, 1996).

7. Body Consciousness: Denial of the Body, and Phantom Limbs

All tactile and physical-sensory impressions are relayed from the body to the brainstem and the thalamus, and are then transferred to the primary receiving and then the association area for somatosensory information located in the neocortex of the parietal lobe (Joseph, 1986b, 1996). The entire image of the body is represented in the parietal lobes (the right and left half of the body in the left and right parietal lobe respectively), albeit in correspondence with the sensory importance of each body part. Therefore, more neocortical space is devoted to the hands and fingers than to the elbow.

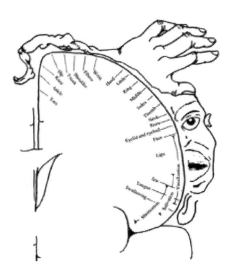

It is because the body image and body consciousness is maintained in the parietal area of the brain, that victims of traumatic amputation and who lose an arm or a leg, continue to feel as if their arm or their leg is still attached to the body. This is called: phantom limbs. They can see the leg is missing, but they feel as if it is still there; body-consciousness remains intact even though part of the body is missing (Joseph, 1986b, 1996). They may also continue to periodically experience the pain of the physical trauma which led to the amputation, and this is called "phantom limb pain."

Thus, via the mental system of the parietal lobe, consciousness of what is not there, may appear to consciousness as if it is still there. This is not a hallucination. The image of the body is preserved in the brain and so to is consciousness of the body; and this is yet another example of experienced reality being a manifestation of the brain and mind. In this regard, reality is literally mapped into the brain and is represented within the brain, such that even when aspects of this "reality" are destroyed and no longer exists external to the brain, it nevertheless continues to be perceived and experienced by the brain and the associated realms of body-consciousness.

Conversely, if the parietal lobe is destroyed, particularly the right parietal lobe (which maintains an image of the left half of the body), half of the body image may be erased from consciousness (Joseph, 1986b, 1988a). The remaining realms of mind will lose all consciousness of the left half of the body, which, in their minds, never existed.

Doctor: "Give me your right hand!" (Patient offers right hand). "Now give me your left!" (The patient presents the right hand again. The right hand is held.) "Give me your left!" (The patient looks puzzled and does not move.) "Is there anything wrong with your left hand?"

Patient: "No, doctor."

Doctor: "Why don't you move it, then?"

Patient: "I don't know."

Doctor: "Is this your hand?" (The left hand is held before her eyes.)

Patient: "Not mine, doctor."

Doctor: "Whose hand is it, then?"

Patient: "I suppose it's yours, doctor."

Doctor: "No, it's not; I've already got two hands. look at it carefully." (The left hand is again held before her eyes.)

Patient: "It is not mine, doctor."

Doctor: "Yes it is, look at that ring; whose is it?" (Patient's finger with marriage ring is held before her eyes)

Patient: "That's my ring; you've got my ring, doctor. You're wearing my ring!"

Doctor: "Look at it—it is your hand."

Patient: "Oh, no doctor."

Doctor: "Where is your left hand then?"

Patient: "Somewhere here, I think." (Making groping movements near her left shoulder).

Because the body image has been destroyed, consciousness of that half of the body is also destroyed. The remaining mental systems and the language-dependent conscious mind will completely ignore and fail to recognize their left arm or leg because the mental system responsible for consciousness of the body image no longer exists. If the left arm or leg is shown to them, they will claim it belongs to someone else, such as the nurse or the doctor. They may dress or groom only the right half of their body, eat only off the right half of their plates, and even ignore painful stimuli applied to the left half of their bodies (Joseph, 1986b, 1988a).

However, if you show them their arm and leg (whose ownership they deny), they will admit these extremities exists, but will insist the leg or arm does not belong to them, even though the arm or the leg is wearing the same clothes covering the rest of their body. Instead, the language dependent aspects of consciousness will confabulate and make up explanations and thus create their own reality. One patient said the arm belonged to a little girl, whose arm had slipped into the patient's sleeve. Another declared (speaking of his left arm and leg), "That's an old man. He stays in bed all the time."

One such patient engaged in peculiar erotic behavior with his left arm and leg which he believed belonged to a woman. Some patients may develop a dislike for their left arms, try to throw them away, become agitated when they are referred to, entertain persecutory delusions regarding them, and even complain of strange people sleeping in their beds due to their experience of bumping into their left limbs during the night (Joseph, 1986b, 1988a). One patient complained that the person sharing her bed, tried to push her out of the bed and then insisted that if it happened again she would sue the hospital. Another complained about "a hospital that makes people sleep together." A female patient expressed not only anger but concern least her husband should find out; she was convinced it was a man in her bed.

The right and left parietal lobes maintain a map and image of the left and right half of the body, respectively. Therefore, when the right parietal lobe is destroyed, the language-dependent mental systems of the left half of the brain, having access only to the body image for the right half of the body, is unable to become conscious of the left half of their body, except as body parts that they then deduce must belong to someone else.

However, when the language dominant mental system of the left hemisphere denies ownerhip of the left extremity these mental system are in fact telling the truth. That is, the left arm and leg belongs to the right not the left hemisphere; the mental system that is capable of becoming conscious of the left half of their body no longer exist.

When the language axis (Joseph, 1982, 2000), i.e. the inferior parietal lobe, Broca's and Wernicke's areas, are functionally isolated from a particular source of information, the language dependent aspect of mind begins to make up a response based on the information available. To be informed about the left leg or left arm, it must be able to communicate with the cortical area (i.e. the parietal lobe) which is responsible for perceiving and analyzing information regarding the extremities. When no message is received and when the language axis is not informed that no messages are being transmitted, the language zones instead rely on some other source even when that source provides erroneous input (Joseph, 1982, 1986b; Joseph et al., 1984); substitute material is assimilated and expressed and corrections cannot be made (due to loss of input from the relevant knowledge source). The patient begins to confabulate. This is because the patient who speaks to you is not the 'patient' who is perceiving- they are in fact, separate; multiple minds exist in the same head.

8. Split-Brains and Split-Minds.

The multiplicity of mind is not limited to visual consciousness, body consciousness, or the language-dependent consciousness. Rather the multiplicity of mind include social consciousness, emotional consciousness, and numerous other mental realms linked with specific areas of the brain (Joseph, 1982, 1986a,b, 1988a,b, 1992, 1999a), such as the limbic system (emotion), frontal lobes (rational thought), the inferior temporal lobes (memory) and the two halves of the brain where multiple streams of mental activity become subordinated and dominated by two distinct realms of mind; consciousness and awareness (Joseph, 1982, 1986a,b, 1988a,b).

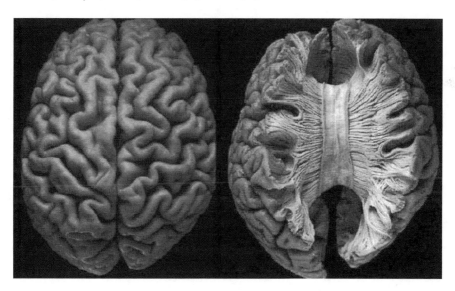

The brain is not a singularity. This is most apparent when viewing the right and left half of the brain which are divided by the interhemispheric fissure and almost completely split into two cerebral hemispheres. These two brain halves are connected by a rope of nerve fibers, the corpus callosum, which enables them to share and exchange some information, but not all information as these two mental realms maintain a conscious awareness of different realities.

For example, it is well established that the right cerebral hemisphere is dominant over the left in regard to the perception, expression and mediation of almost all aspects of social and emotional functioning and related aspects of social/emotional language and memory. Further, the right hemisphere is dominant for most aspects of visual-spatial perceptual functioning, the comprehension of body language, the recognition of faces including friend's loved ones, and one's own face in the mirror (Joseph, 1988a, 1996).

Recognition of one's own body and the maintenance of the personal body image is also the dominant realm of the right half of the brain (Joseph, 1986b, 1988a). The body image, for many, is tied to personal identity; and the same is true of the recognition of faces including one's own face.

The right is also dominant for perceiving and analyzing visual-spatial relationships, including the movement of the body in space (Joseph, 1982, 1988a). Therefore, one can throw or catch a ball with accuracy, dance across a stage, or leap across a babbling brook without breaking a leg.

The perception of environmental sounds (water, wind, a meowing cat) and the social, emotional, musical, and melodic aspects of language, including the ability to sing, curse, or pray, are also the domain of the right hemisphere mental system (Joseph, 1982, 1988a). Hence, it is the right hemisphere which imparts the sounds of sarcasm, pride, humor, love, and so on, into the stream of speech, and which conversely can determine if others are speaking with sincerity, irony, or evil intentions.

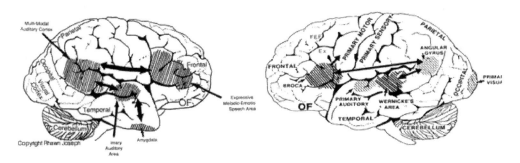

By contrast, expressive and receptive speech, linguistic knowledge and thought, mathematical and analytical reasoning, reading, writing, and arithmetic, as well as the temporal-sequential and rhythmical aspects of consciousness, are associated with the functional integrity of the left half of the brain in the majority of the population (Joseph, 1982, 1996). The language-dependent mind is linked to the left hemisphere.

Certainly, there is considerable overlap in functional representation. Moreover, these two mental system interact and assist the other, just as the left and right hands cooperate and assist the other in performing various tasks. For example, if you were standing at the bar in a nightclub, and someone were to tap you on the shoulder and say, "Do you want to step outside," it is the mental system of the left hemisphere which understands that a question about "outside" has been asked, but it is the mental system of the right which determines the underlying meaning, and if you are being threatened with a punch in the nose, or if a private conversation is being sought.

However, not all information can be transferred from the right to the left, and vice versa (Gallagher and Joseph, 1982; Joseph, 1982, 1988a; Joseph and Gallagher, 1985; Joseph et al., 1984). Because each mental system is unique, each "speaks a different language" and they cannot always communicate. Not all mental events can be accurately translated, understood, or even recognized by the other half of the brain. These two major mental systems, which could be likened to "consciousness" vs "awareness" exist in parallel, simultaneously, and both can act independently of the other, have different goals and desires, and come to completely different conclusions. Each mental system has its own reality.

The existence of these two independent mental realms is best exemplified and demonstrated following "split-brain" surgery; i.e. the cutting of the corpus callosum fiber pathway which normally allows the two hemisphere's to communicate.

As described by Nobel Lauriate Roger Sperry (1966, p. 299), "Everything we have seen indicates that the surgery has left these people with two separate minds, that is, two separate spheres of consciousness. What is experienced in the right hemisphere seems to lie entirely outside the realm of awareness of the left hemisphere. This mental division has been demonstrated in regard to perception, cognition, volition, learning and memory."

Consciousness and the Universe

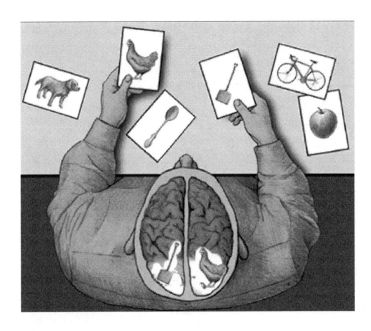

The right half of the brain controls and perceives the left half of the body and visual space, whereas the right half of the body and visual space is the domain of the left hemisphere. Therefore, following split-brain surgery, if a comb, spoon, or some other hidden object is placed in the left hand (out of sight), the left hemisphere, and the language-dependent conscious mind, will not even know the left hand is holding something and will be unable to name it, describe it, or if given multiple choices point to the correct item with the right hand (Joseph 1988a,b; Sperry, 1966). However, the right hemisphere can raise the left hand and not only point to the correct object, but can pantomime its use.

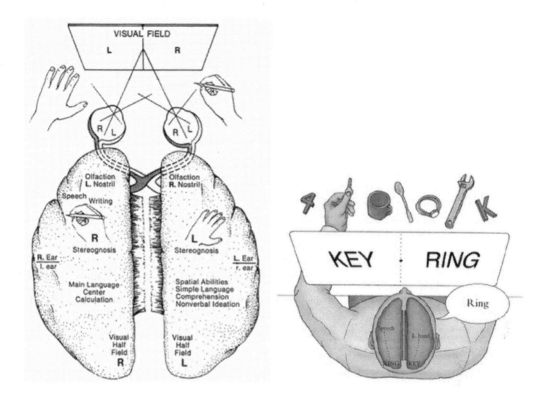

If the split-brain patient is asked to stare at the center of a white screen and words like "Key Ring" are quickly presented, such that the word "Key" falls in the left visual field (and thus, is transmitted to the right cerebrum) and the word "Rings" falls in the right field (and goes to the left hemisphere), the language dependent conscious mind will not see the word "Key." If asked, the language-dependent conscious mind will say "Ring" and will deny seeing the word "Key." However, if asked to point with the left hand, the mental system of the right hemisphere will correctly point to the word "Key."

Therefore, given events "A" and "B" one half of the brain may know A, but know nothing about B which is known only by the other half of the brain. In consequence, what is known vs what is not known becomes relatively imprecise depending on what aspects of reality are perceived and "known" by which mental system (Joseph 1986a; Joseph et al., 1984). There is no such thing as singularity of mind. Since the brain and mind is a multiplicity, "A" and "B" can be known simultaneously, even when one mind is knows nothing about the existence of A or B.

In that the brain of the normal as well as the "split-brain" patient maintains the neuroanatomy to support a multiplicity of mind, and the presence of two dominant psychic realms, it is therefore not surprising that "normal" humans often have difficulty "making up their minds," suffer internal conflicts over love/hate relationships, and are plagued with indecision even when staring into an open refrigerator and trying to decide what to eat. "Making up one's mind" can be an ordeal involving a multiplicity of minds.

However, this conflict becomes even more apparent following split-brain surgery and the cutting of the corpus callosum fiber pathway which links these two parallel streams of conscious-awareness.

Akelaitis (1945, p. 597) describes two patients with complete corpus callosotomies who experienced extreme difficulties making the two halves of their bodies cooperate. "In tasks requiring bimanual activity the left hand would frequently perform oppositely to what she desired to do with the right hand. For example, she would be putting on clothes with her right and pulling them off with her left, opening a door or drawer with her right hand and simultaneously pushing it shut with the left. These uncontrollable acts made her increasingly irritated and depressed."

Another patient experienced difficulty while shopping, the right hand would place something in the cart and the left hand would put it right back again and grab a different item.

A recently divorced male patient complained that on several occasions while walking about town he found himself forced to go some distance in another direction by his left leg. Later (although his left hemisphere was not conscious of it at the time) it was discovered that this diverted course, if continued, would have led him to his former wife's new home.

Geschwind (1981) reports a callosal patient who complained that his left hand on several occasions suddenly struck his wife--much to the embarrassment of his left (speaking) hemisphere. In another case, a patient's left hand attempted to choke the patient himself and had to be wrestled away.

Bogen (1979, p. 333) indicates that almost all of his "complete commissurotomy patients manifested some degree of intermanual conflict." One patient, Rocky, experienced situations in which his hands were uncooperative; the right would button up a shirt and the left would follow right behind and undo the buttons. For years, he complained of difficulty getting his left leg to go in the direction he (or rather his left hemisphere) desired. Another patient often referred to the left half of her body as "my little sister" when she was complaining of its peculiar and independent actions.

Another split-brain patient reported that once when she had overslept her left hand began slapping her face until she (i.e. her left hemisphere) woke up. This same patient, in fact, complained of several instances where her left hand had acted violently toward herself and other people (Joseph, 1988a).

Split brain patient, 2-C, complained of instances in which his left hand would perform socially inappropriate actions, such as striking his mother across the face (Joseph, 1988b). Apparently his left and right hemisphere also liked different TV programs. He complained of numerous instances where he (his left hemisphere) was enjoying a program, when, to his astonishment, the left half of his body pulled him to the TV, and changed the channel.

The right and left hemisphere also liked different foods and had different attitudes about exercise. Once, after 2-C had retrieved something from the refrigerator with his right hand, his left took the food, put it back on the shelf and retrieved a completely different item "Even though that's not what I wanted to eat!" On at least one occasion, his left leg refused to continue "going for a walk" and would only allow him to return home.

In the laboratory, 2-C's left hemisphere often became quite angry with his left hand, and he struck it and expressed hate for it. Several times, his left and right hands were observed to engage in actual physical struggles, beating upon each other. For example, on one task both hands were stimulated simultaneously (while out of view) with either the same or two different textured materials (e.g., sandpaper to the right, velvet to the left), and he was required to point (with the left and right hands simultaneously) to an array of fabrics that were hanging in view on the left and right of the testing apparatus. However, at no time was he informed that two different fabrics were being applied.

After stimulation he would pull his hands out from inside the apparatus and point with the left to the fabric felt by the left and with the right to the fabric felt by the right.

Surprisingly, although his left hand (right hemisphere) responded correctly, his left hemisphere vocalized: "Thats wrong!" Repeatedly he reached over with his right hand and tried to force his left extremity to point to the fabric experienced by the right (although the left hand responded correctly! His left hemisphere didn't know this, however). His left hand refused to be moved and physically resisted being forced to point at anything different. In one instance a physical struggle ensued, the right grappling with the left with the two halves of the body hitting and scratching at each other!

Moreover, while 2-C was performing this (and other tasks), his left hemisphere made statements such as: "I hate this hand" or "This is so frustrating" and would strike his left hand with his right or punch his left arm. In these instances there could be little doubt that his right hemisphere mental system was behaving with purposeful intent and understanding, whereas his left hemisphere mental system had absolutely no comprehension of why his left hand (right hemisphere) was behaving in this manner (Joseph, 1988b).

These conflicts are not limited to behavior, TV programs, choice of clothing, or food, but to actual feelings, including love and romance. For example, the right and left hemisphere of a male split-brain patient had completely different feelings about an ex-girlfriend. When he was asked if he wanted to see her again, he said "Yes." But at the same time, his left hand turned thumbs down!

Another split-brain patient suffered conflicts about his desire to smoke. Although is left hemisphere mental system enjoyed cigarettes, his left hand would not allow him to smoke, and would pluck lit cigarettes from his mouth or right hand and put them out. He had been trying to quit for years.

Because each head contains multiple minds, similar conflicts also plague those who have not undergone split-brain surgery. Each half of the brain and thus each mental system may have different attitudes, goals and interests. As noted above, 2-C experienced conflicts when attempting to eat, watch TV, or go for walks, his right and left hemisphere mental systems apparently enjoying different TV programs or types of food (Joseph 1988b). Conflicts of a similar nature plague us all. Split-brain patients are not the first to choke on self-hate or to harm or hate those they profess to love.

Each half of the brain is concerned with different types of information, and may react, interpret and process the same external experience differently and even reach different conclusions (Joseph 1988a,b; Sperry, 1966). Moreover, even when the goals are the same, the two halves of the brain may produce and attempt to act on different strategies.

Each mental system has its own reality. Singularity of mind, is an illusion.

9. Dissociation and Self-Consciousness

The multiplicity of mind is not limited to the neocortex but includes old cortical structures, such as the limbic system (Joseph 1992). Moreover, limbic nuclei such as the amygdala and hippocampus interact with neocortical tissues creating yet additional mental systems, such as those which rely on memory and which contribute to self-reflection, personal identity, and even self-consciousness (Joseph, 1992, 1998, 1999b, 2001).

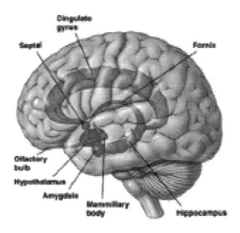

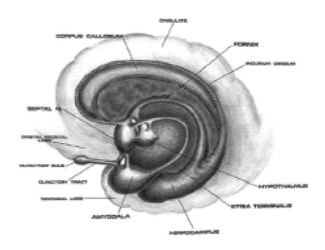

For example, both the amygdala and the hippocampus are implicated in the storage of long term memories, and both nuclei enable individuals to visualize and remember themselves engaged in various acts, as if viewing their behavior and actions from afar. Thus, you might see yourself and remember yourself engage in some activity, from a perspective outside yourself, as if you are an external witness; and this is a common feature of self-reflection and self-memory and is made possible by the hippocampus and overlying temporal lobe (Joseph, 1996, 2001).

The hippocampus in fact contains "place neurons" which cognitive map one's position and the location of various objects within the environment (Nadel, 1991; O'Keefe, 1976; Wilson & McNaughton, 1993). Further, if the subject moves about in that environment, entire populations of these place cells will fire. Moreover, some cells are responsive to the movements of other people in that environment and will fire as that person is observed to move around to different locations or corners of the room (Nadel, 1991; O'Keefe, 1976; Wilson and McNaughton, 1993).

Electrode stimulation, or other forms of heightened activity within the hippocampus and overlying temporal lobe can also cause a person to see themselves, in real time, as if their conscious mind is floating on the ceiling staring down at their body (Joseph, 1998, 1999b, 2001). During the course of electrode stimulation and seizure activity originating in the temporal lobe or hippocampus, patients may report that they have left their bodies and are hovering upon the ceiling staring down at themselves (Daly, 1958; Penfield, 1952; Penfield & Perot 1963; Williams, 1956). That is, their consciousness and sense of personal identity appears to split off from their body, such that they experience themselves as as a consciousness that is conscious of itself as a conscious that is detached from the body which is being observed.

One female patient claimed that she not only would float above her body, but would sometimes drift outside and even enter into the homes of her neighbors. Penfield and Perot (1963) describe several patients who during a temporal lobe seizure, or neurosurgical temporal lobe stimulation, claimed they split-off from their body and could see themselves down below. One woman stated: "it was though I were two persons, one watching, and the other having this happen to me." According to Penfield (1952), "it was as though the patient were attending a familiar play and was both the actor and audience."

Under conditions of extreme trauma, stress and fear, the amygdala, hippocampus and temporal lobe become exceedingly active (Joseph, 1998, 1999b). Under these conditions many will experience a "splitting of consciousness" and have the sensation they have left their body and are hovering beside or above themselves, or even that they floated away (Courtois, 2009; Grinker & Spiegel, 1945; Noyes & Kletti, 1977; van der Kolk 1987). That is, out-of-body dissociative experiences appear to be due to fear induced hippocampus (and amygdala) hyperactivation.

Likewise, during episodes of severe traumatic stress personal consciousness may be fragmented and patients may dissociate and experience themselves as splitting off and floating away from their body, passively observing all that is occurring (Courtois, 1995; Grinker & Spiegel, 1945; Joseph, 1999d; Noyes & Kletti, 1977; Southard, 1919; Summit, 1983; van der Kolk 1987).

Noyes and Kletti (1977) described several individuals who experienced terror, believed they were about to die, and then suffered an out-of body dissociative experience: "I had a clear image of myself... as though watching it on a television screen." "The next thing I knew I wasn't in the truck anymore; I was looking down from 50 to 100 feet in the air." "I had a sensation of floating. It was almost like stepping out of reality. I seemed to step out of this world."

One individual, after losing control of his Mustang convertible while during over 100 miles per hour on a rain soaked freeway, reported that "time seemed to slow down and then... part of my mind was a few feet outside the car zooming above it and then beside it and behind it and in front of it, looking at and analyzing the respective positions of my spinning Mustang and the cars surrounding me. Simultaneously I was inside trying to steer and control it in accordance with the multiple perspectives I was given by that part of my mind that was outside. It was like my mind split and one consciousness was inside the car, while the other was zooming all around outside and giving me visual feedback that enabled me to avoid hitting anyone or destroying my Mustang."

Numerous individuals from adults to children, from those born blind and deaf, have also reported experiencing a dissociative consciousness after profound injury causing near death (Eadie 1992; Rawling 1978; Ring 1980). Consider for example, the case of Army Specialist J. C. Bayne of the 196th Light Infantry Brigade. Bayne was "killed" in Chu Lai, Vietnam, in 1966, after being simultaneously machine gunned and struck by a mortar. According to Bayne, when he opened his eyes he was floating in the air, looking down on his burnt and bloody body: "I could see me... it was like looking at a manikin laying there... I was burnt up and there was blood all over the place... I could see the Vietcong. I could see the guy pull my boots off. I could see the rest of them picking up various things... I was like a spectator... It was about four or five in the afternoon when our own troops came. I could hear and see them approaching... I looked dead... they put me in a bag... transferred me to a truck and then to the morgue. And from that point, it was the embalming process. I was on that table and a guy was telling jokes about those USO girls... all I had on was bloody undershorts... he placed my leg out and made a slight incision and stopped... he checked my pulse and heartbeat again and I could see that too... It was about that point I just lost track of what was taking place.... [until much later] when the chaplain was in there saying everything was going to be all right.... I was no longer outside. I was part of it at this point" (reported in Wilson, 1987, pp 113-114; and Sabom, 1982, pp 81-82).

Therefore, be it secondary to the fear of dying, or depth electrode stimulation, these experiences all appear to be due to a mental system which enables a the conscious mind to detach completely from the body in order to make the body an object of consciousness (Joseph, 1998, 1999b, 2001).

10. Quantum Consciousness

It could be said that consciousness is consciousness of something other than consciousness. Consciousness and knowledge of an object, such as a chair, are also distinct. Consciousness is not the chair. The chair is not consciousness. The chair is an object of consciousness, and thus become discontinuous from the quantum state.

Consciousness is consciousness of something and is conscious of not being that object that is is conscious of. By knowing what it isn't, consciousness may know what it is not, which helps define what it is. This consciousness of not being the object can be considered the "collapse function" which results in discontinuity within the continuuum.

Further, it could be said that consciousness of consciousness, that is, self-consciousness, also imparts a duality, a separation, into the fabric of the quantum continuum. Therefore this consciousness that is the object of consciousness, becomes an abstraction, and may create a collapse function in the continuun.

However, in instances of dissociation, this consciousness is conscious of itself as a consciousness that is floating above its body; a body which contains the brain. The dissociative consciousness is not dissociated from itself as a consciousness, but only from its body. That is, there is an awareness of itself as a consciousness that is floating above the body, and this awareness is simultaneously one with that consciousness, as there is no separation, no abstractions, and no objectification. It is a singularity that is without form, without dimension, without shape.

Moreover, because this dissociated consciousness appears to be continuous with itself, there is no "collapse function" except in regard to the body which is viewed as an object of perception.

Therefore, in these instances we can not say that consciousness has split into a duality of observer and observed or knower and known, except in regard to the body. Dissociated consciousness is conscious of itself as consciousness; it is self-aware without separation and without reflecting. It is knowing and known, simultaneously.

In fact, many patients report that in the dissociated state they achieve or nearly achieved a state of pure knowing (Joseph, 1996, 2001).

A patient described by Williams (1956) claimed she was lifted up out of her body, and experienced a very pleasant sensation of elation and the feeling that she was "just about to find out knowledge no one else shares, something to do with the link between life and death." Another patient reported that upon leaving her body she not only saw herself down below, but was taken to a special place "of vast proportions, and I felt as if I was in another world" (Williams, 1956).

Other patients suffering from temporal lobe seizures or upon direct electrical activation have noted that things suddenly became "crystal clear" or that they had a feeling of clairvoyance, of having the ultimate truth revealed to them, of having achieved a sense of greater awareness and cosmic clarity such that sounds, smells and visual objects seemed to have a greater meaning and sensibility, and that they were achieving a cosmic awareness of the hidden knowledge of all things (Joseph, 1996, 2001).

Although consciousness and the object of consciousness, that which is known, are not traditionally thought of as being one and the same, in dissociative states, consciousness and knowing, may become one and the same. The suggestion is of some type of cosmic unity, particularly as these patients also often report a progressive loss of the sense of individuality, as if they are merging into something greater than themselves, including a becoming one with all the knowledge of the universe; a singularity with god.

Commonly those who experience traumatic dissociative conscious states, not only float above the body, but report that they gradually felt they were losing all sense of individuality as they became embraced by a brilliant magnificent whiteness that extended out in all directions into eternity.

Therefore, rather than the dissociated consciousness acting outside the quantum state, it appears this mental state may represent an increasing submersion back into the continuity that is the quantum state, disappearing back into the continuum of singularity and oneness that is the quantum universe.

Dissociated consciousness may be but the last preamble before achieving the unity that is quantum consciousness, the unity of all things.

11. Conclusion: Quantum Consciousness and the Multiplicity of Mind

What is "Objective Reality" when the mind is a multiplicity which is capable of splitting up, observing itself, becoming blind to itself, and becoming blind to the features of the world which then cease to exist for the remaining mental realms?

Each mental system has its own "reality." Each observer is a multiplicity that engages in numerous simultaneous acts of observation. Therefore, non-local properties which do not have an objective existence independent of the "act of observation" by one mental system, may achieve existence when observed by another mental system. The "known" and the "unknown" can exist simultaneously and interchangeably and this may explain why we don't experience any macroscopic non-local quantum weirdness in our daily lives.

This means that quantum laws may apply to everything, from atoms to monkeys and woman and man and the multiplicity of mind. However, because of this multiplicity, this could lead to seemingly contradictory predictions and uncertainty when measuring macroscopically objective systems which are superimposed on microscopic quantum systems. Indeed, this same principle applies to the multiplicity of mind, where dominant parallel streams of conscious awareness may be superimposed on other mental systems, and which may be beset by uncertainty.

Because of the multiplicity of mind, as exemplified by dissociative states, the observer can also be observed, and thus, the observer is not really external to the quantum state as is required by the standard collapse formulation of quantum mechanics. This raises the possibility that the collapse formulation can be used to describe the universe as a whole which includes observers observing themselves being observed.

The multiplicity of mind also explains why an object being measured by one mental system therefore becomes bundled up into a state where it either determinately has or determinately does not have the property being measured. Measurements performed by one mental system are not being performed by others, such that the same object can have an initial state and a post-measurement state and a final state simultaneously as represented in multiple minds in parallel, or separate states as represented by each mental realm individually.

The collapse dynamics of observation supposedly guarantees that a system will either determinately have or determinately not have a particular property. However, because the observer is observing with multiple mental systems the object can both have and not have specific properties when it is being measure and not measured, simply because it is being measured and not measured, or rather, observed and not observed or its different features observed simultaneously by multiple mental systems. Therefore, it can be continuous and discontinuous in parallel, and different properties can be known and not known in parallel simultaneously.

And the same rules apply to the mental systems which exist in multiplicity within the head of a single observer. Mental systems can become continuous or discontinuous, and can be known and not known simultaneously, in parallel. Thus, the standard collapse formulation can be used to describe systems that contain observers, as the mind/observer can be simultaneously internal and external to the described system.

The mind is a multiplicity and there is no such thing as a "single observer state." Therefore, each element may be observed by multiple observer states which perceive multiple object systems thereby giving the illusion that the object has been transformed during the collapse function. What this also implies is that contrary to the standard or Copenhagen interpretations, states may have both definite position and definite momentum at the same time.

Each mental system perceives a different physical world giving rise to multiple worlds and multiple realities which may be subordinated by one or another more dominant stream of conscious awareness.

Moreover, as the multiplicity of mind can also detach and become discontinuous with the body, whereas dissociative consciousness is continuous with itself, this indicates that the mind is also capable of becoming one with the continuum, and can achieve singularity so that universe and mind become one.

References

Akelaitis, A. J. (1945). Studies on the corpus callosum. American Journal of Psychiatry, 101, 594-599.

Bogen, J. (1979). The other side of the brain. Bulletin of the Los Angeles Neurological Society. 34, 135-162.

Bohr, N. (1934/1987), Atomic Theory and the Description of Nature, reprinted as The Philosophical Writings of Niels Bohr, Vol. I, Woodbridge: Ox Bow Press.

Bohr, N. (1958/1987), Essays 1932-1957 on Atomic Physics and Human Knowledge, reprinted as The Philosophical Writings of Niels Bohr, Vol. II, Woodbridge: Ox Bow Press.

Bohr, N. (1963/1987), Essays 1958-1962 on Atomic Physics and Human Knowledge, reprinted as The Philosophical Writings of Niels Bohr, Vol. III, Woodbridge: Ox Bow Press.

Casagrande, V. A. & Joseph, R. (1978). Effects of monocular deprivation on geniculostriate connections in prosimian primates. Anatomical Record, 190, 359.

Casagrande, V. A. & Joseph, R. (1980). Morphological effects of monocular deprivation and recovery on the dorsal lateral geniculate nucleus in prosimian primates. Journal of Comparative Neurology, 194, 413-426.

Courtois, C. A. (2009). Treating Complex Traumatic Stress Disorders: An Evidence-Based Guide Daly, D. (1958). Ictal affect. American Journal of Psychiatry, 115, 97-108.

DeWitt, B. S., (1971). The Many-Universes Interpretation of Quantum Mechanics, in B.

D.'Espagnat (ed.), Foundations of Quantum Mechanics, New York: Academic Press. pp. 167–218.

DeWitt, B. S. and Graham, N., editors (1973). The Many-Worlds Interpretation of Quantum Mechanics. Princeton University Press, Princeton, New-Jersey.

Eadie, B. J. (1993) Embraced by the light. New York, Bantam

Einstein, A. (1905a). Does the Inertia of a Body Depend upon its Energy Content? Annalen der Physik 18, 639-641.

Einstein, A. (1905b). Concerning an Heuristic Point of View Toward the Emission and Transformation of Light. Annalen der Physik 17, 132-148.

Einstein, A. (1905c). On the Electrodynamics of Moving Bodies. Annalen der Physik 17, 891-921.

Einstein, A. (1926). Letter to Max Born. The Born-Einstein Letters (translated by Irene Born) Walker and Company, New York.

Gallagher, R. E., & Joseph, R. (1982). Non-linguistic knowledge, hemispheric laterality, and the conservation of inequivalance. Journal of General Psychology, 107, 31-40.

Geschwind. N. (1981). The perverseness of the right hemisphere. Behavioral Brain Research, 4, 106-107.

Grinker, R. R., & Spiegel, J. P. (1945). Men Under Stress. McGraw-Hill.

Heisenberg. W. (1930), Physikalische Prinzipien der Quantentheorie (Leipzig: Hirzel). English translation The Physical Principles of Quantum Theory, University of Chicago Press.

Heisenberg, W. (1955). The Development of the Interpretation of the Quantum Theory, in W. Pauli (ed), Niels Bohr and the Development of Physics, 35, London: Pergamon pp. 12-29.

Heisenberg, W. (1958), Physics and Philosophy: The Revolution in Modern Science, London: Goerge Allen & Unwin.

Joseph, R. (1982). The Neuropsychology of Development. Hemispheric Laterality, Limbic Language, the Origin of Thought. Journal of Clinical Psychology, 44 4-33.

Joseph, R. (1986). Reversal of language and emotion in a corpus callosotomy patient. Journal of Neurology, Neurosurgery, & Psychiatry, 49, 628-634.

Joseph, R. (1986). Confabulation and delusional denial: Frontal lobe and lateralized influences. Journal of Clinical Psychology, 42, 845-860.

Joseph, R. (1988) The Right Cerebral Hemisphere: Emotion, Music, Visual-Spatial Skills, Body Image, Dreams, and Awareness. Journal of Clinical Psychology, 44, 630-673.

Joseph, R. (1988). Dual mental functioning in a split-brain patient. Journal of Clinical Psychology, 44, 770-779.

Joseph, R. (1992) The Limbic System: Emotion, Laterality, and Unconscious Mind. The Psychoanalytic Review, 79, 405-455.

Joseph, R. (1996). Neuropsychiatry, Neuropsychology, Clinical Neuroscience, 2nd Edition. Williams & Wilkins, Baltimore.

Joseph, R. (1998). Traumatic amnesia, repression, and hippocampal injury due to corticosteroid and enkephalin secretion. Child Psychiatry and Human Development. 29, 169-186.

Joseph, R. (1999a). Frontal lobe psychopathology: Mania, depression, aphasia, confabulation, catatonia, perseveration, obsessive compulsions, schizophrenia. Psychiatry, 62, 138-172.

Joseph, R. (1999b). The neurology of traumatic "dissociative" amnesia. Commentary and literature review. Child Abuse & Neglect. 23, 71-80.

Joseph, R. (2000). Limbic language/language axis theory of speech. Behavioral and Brain Sciences. 23, 439-441.

Joseph, R. (2001). The Limbic System and the Soul: Evolution and the Neuroanatomy of Religious Experience. Zygon, the Journal of Religion & Science, 36, 105-136.

Joseph, R., & Casagrande, V. A. (1978). Visual field defects and morphological changes resulting from monocular deprivation in primates. Proceedings of the Society for Neuroscience, 4, 1978, 2021.

Joseph, R. & Casagrande, V. A. (1980). Visual field defects and recovery following lid closure in a prosimian primate. Behavioral Brain Research, 1, 150-178.

Joseph, R., Forrest, N., Fiducia, N., Como, P., & Siegel, J. (1981). Electrophysiological and behavioral correlates of arousal. Physiological Psychology, 1981, 9, 90-95.

Joseph, R., Gallagher, R., E., Holloway, J., & Kahn, J. (1984). Two brains, one child: Interhemispheric transfer and confabulation in children aged 4, 7, 10. Cortex, 20, 317-331.

Joseph, R., & Gallagher, R. E. (1985). Interhemispheric transfer and the completion of reversible operations in non-conserving children. Journal of Clinical Psychology, 41, 796-800.

Nadel, L. (1991). The hippocampus and space revisited. Hippocampus, 1, 221-229.

Neumann, J. von, (1937/2001), "Quantum Mechanics of Infinite Systems. Institute for Advanced Study; John von Neumann Archive, Library of Congress, Washington, D.C.

Neumann, J. von, (1938), On Infinite Direct Products, Compositio Mathematica 6: 1-77.

Neumann, J. von, (1955), Mathematical Foundations of Quantum Mechanics, Princeton, NJ: Princeton University Press.

Noyes, R., & Kletti, R. (1977). Depersonalization in response to life threatening danger. Comprehensive Psychiatry, 18, 375-384.

O'Keefe, J. (1976). Place units in the hippocampus. Experimental Neurology, 51-78-100. Pais, A. (1979). Einstein and the quantum theory, Reviews of Modern Physics, 51, 863-914.

Penfield, W. (1952) Memory Mechanisms. Archives of Neurology and Psychiatry, 67, 178-191.

Penfield, W., & Perot, P. (1963) The brains record of auditory and visual experience. Brain, 86, 595-695.

Rawlins, M. (1978). Beyond Death's Door. Sheldon Press.

Ring, K. (1980). Life at Death: A Scientific Investigation of the Near-Death Experience. New York: Quill.

Sabom, M. (1982). Recollections of Death. New York: Harper & Row.

Southard, E. E. (1919). Shell-shock and other Neuropsychiatric Problems. Boston.

Sperry, R. (1966). Brain bisection and the neurology of consciousness. In F. O. Schmitt and F. G. Worden (eds). The Neurosciences. MIT press.

van der Kolk, B. A. (1985). Psychological Trauma. American Psychiatric Press.

Williams, D. (1956). The structure of emotions reflected in epileptic experiences. Brain, 79, 29-67.

Wilson, I. (1987). The After Death Experience. Morrow.

Wilson, M. A., & McNaughton, B. L. (1993). Dynamics of the hippocampus ensemble for space. Science, 261, 1055-1058.

44. Consciousness and Quantum Physics: A Deconstruction of the Topic
Gordon Globus, M.D.

Professor Emeritus of Psychiatry and Philosophy,
University of California Irvine, Irvine, CA

Abstract

The topic of "consciousness and quantum physics" is deconstructed. Consciousness is not only a vague concept with very many definitions but is entangled with unresolved controversies, notably the notorious measurement problem in quantum physics and the qualia problem in philosophy, so the problematic calls for redefinition. The key issue is that of primary "closure"--the nonphenomenality of quantum physical reality--and the action that brings "dis-closure." Dis-closure of the phenomenal world can be understood within the framework of dissipative quantum thermofield brain dynamics without any reference to consciousness. Some unsettling monadological consequences of this view are brought out in the discussion.

KEY WORDS: consciousness, quantum physics, dissipative quantum thermofield brain dynamics, measurement problem, between-two, monadology

1. Introduction

At present there is no agreed upon definition of consciousness--Vimal (2009) identified over forty!--so how could we discuss it in the same breath as the scientific field of quantum physics? A recent issue of The Journal of Consciousness Studies devoted ten articles to the topic of defining consciousness, the editor concluding that we should all try harder to both specify what we mean when referring to "consciousness" and pay more attention to the contexts within which that meaning applies. And we should embrace the resultant diversity … . (Nunn 2009, p.7) Might a science-based discussion of consciousness and quantum physics be barking up the wrong tree? And if so, how might discussion be redeployed away from consciousness without falling into the dullness of crass materialism?

Since "consciousness" is so ill-defined, we are at the mercy of tacit subtextual meanings. Yet clues might be found in the term's etymology, which gains us a certain detachment. The very term "consciousness" already carries profound biases. It derives from the Latin con-scieri, which is to know-together, and accordingly cognitive-social. Conscientia--conscience--is an internalization of social knowing, viz. a self-knowing and judging. But today consciousness has a much broader meaning than this historical cognitive emphasis. "Consciousness" in present usage is not just a socialized knowing but is also perceptual. We say that we are conscious of the phenomenal world. The supplement to knowing found in the contemporary meaning of "consciousness" encompasses "qualia" too--the conscious experience of colors, sounds, odors, etc. --but this is precisely where there is hot debate. Despite enormous discussion (e.g. Kazniak 2001; Wright 2008) there is no philosophical consensus regarding the qualia problem. So the extension of the original cognitive meaning of "consciousness" to include qualitative experiences brings complications to any engagement with an unsuspecting physics, which ought to make us suspicious.

It should not be thought that substituting "awareness" or "experience" for "consciousness" improves the situation. "Aware" comes from an old English term meaning cautious (cf. "wary"), which is a cognitive activity. "Experience" (cf. "experiment") comes from the Latin experio, to try out, which is cognitive-behavioral. It is noteworthy that the term 'consciousness' does not even appear until the 17th century. If this is something so fundamental as to be related to quantum physics, why should it take so long to be distinguished in philosophical discourse? Were the philosophers of ancient Greece so unwise?

As if the ambiguity and subtext of "consciousness" were not enough, consciousness is already party to a long-standing and still highly contentious problem within quantum physics itself: the "measurement problem," which is typified by Schrödinger's notorious cat. Consistent with the principles of quantum physics Schrödinger's cat seems to be in a superposition state of being dead and being alive, until a conscious observation is made.

Even if it is supposed that consciousness has nothing to do with the outcome--supposed that the wave function readily collapses on its own (Gihradi 2007; Hameroff and Penrose 1996)--the result is not a phenomenal cat, dead or alive as the case may be. Wave function collapse is to near-certainty of location, not to a phenomenal cat. The wave function is a wave of probability and its collapse is still expressed in probabilistic terms ("exceedingly near 1.0" at some point). Any phenomenality depends on the observer's consciousness, and so the measurement problem drags in some meta-physics. Since the tradition of metaphysics runs back through Kant and Descartes to Plato--a tradition that uncommonsensical quantum physics otherwise challenges at every turn--the very topic of consciousness and quantum physics cries for deconstruction.

There is no place for the phenomenal cat in quantum physics itself--that cat right there in the box, dead or alive, after you open it. Quantum field theory is well capable of describing macroscopic objects with sharp boundary structures (Umezawa 1993, Chapter 6). Scale is not an issue for quantum field theory. (There is no need to get to the macroscopic by the fiat of letting Planck's constant go to zero.) But a macroscopic quantum object is not of the same kind as a cat-in-a-box: that would be a colossal category mistake to equate them. Might quantum brain theory ride to the rescue? Many theories have been proposed for how quantum brain mechanisms might generate consciousness and thereby rescue from metaphysics the principle of the causal closure of the physical domain. (Quantum physicists in general have been slow to recognize the importance of quantum brain theory to their endeavor.) But these attempts run into the wasps nest already emphasized: the definability problem for consciousness, the qualia problem, and unyielding controversy over the measurement problem. A fresh start is tried here and the brutal consequences are faced.

2. Dis-Closure

The world presences, "is," exists, has Being, appears … . Instead of saying that we are "conscious of the world" it is less prejudicial to say that we always find ourselves already amidst one, waking and dreaming both. Being is no easy replacement for consciousness, to be sure … and the term "Being" may sound more problematic than "consciousness" to the physics ear. For reasons to be brought out, I shall use instead the term "dis-closure."

To speak of the quantum "world," as is sometimes loosely done, is self-contradictory. As already emphasized above, the quantum realm lacks phenomenality, has no appearance, is closed. What is needed is an account of the appearance of Being, presence, or since the fundamental property is closedness, what is called for is dis-closedness.

Dis-closure entails an action--an action on closure--a dis-closure in which Being--a cat in a box--appears. What is called for by the above deconstruction of consciousness in quantum physics is an account of dis-closure. Thermofield brain dynamics provides an explanation of dis-closure that may get free from the problematical consciousness.

3. Thermofield Brain Dynamics

The origins of thermofield brain dynamics go back to the quantum brain dynamics of Umezawa and coworkers in the late sixties (Umezawa 1995). It was recognized that symmetry-breaking in the ground state of the brain--the vacuum state of a water electric dipole field--offers a mechanism for memory. Sensory inputs fall into the ground after dissipating their energy and break the dipole symmetry. The broken symmetry is preserved by boson condensation (Nambu-Goldstone condensates). When the sensory input is repeated, the condensate-trace is excited from the vacuum state and becomes conscious. Jibu and Yasue (1995, 2004) worked this idea into a full-fledged quantum brain dynamics of consciousness.

Vitiello (1995, 2001, 2004) greatly extended quantum brain dynamics to a thermofield brain dynamics by bringing in dissipation. Now the brain is a dissipative system and its vacuum state has two modes: a system mode and an environment mode. The system mode contains the boson memory traces and the environment mode expresses ongoing input. ("Input" should be understood here as both sensory signals and signals the brain generates on its own, that is, other-generated signals and self-generated signals.) Neither mode exists without the other. The vacuum state is "between-two," between other-generated and self-generated signals on the one hand and memory traces on the other. Vitiello proposed that consciousness is not the quantum brain dynamics model of excitation of memory traces from the vacuum but is the state of match between dual modes. Consciousness for Vitiello is between-two.

Notable in the Vitiello model is a new version of ontological duality. The traditional dualities are the two substances of Descartes or the two aspects of a "neutral reality" proposed by Spinoza. Mitigated dualisms include Sperry's (1969) emergent level which is more than the sum of its interacting parts and Huxley's (1898) epiphenomenalism in which the mental is a nomological dangler without causal influence. The between-two is a new idea. The two are indissolubly coupled--no one without the other--and unlike ontological substances or aspects of a neutral substance, what is primary is their between.

Since Vitiello wedded thermofield brain dynamics to consciousness, the same difficulties already detailed arise. But we may alternatively reinterpret thermofield brain dynamics as a theory of dis-closure (Globus 2003, 2009). Dis-closure is between-two.

Phenomenal world appears when the dual modes belong-together, in the sense that a complex number "belongs-to" its complex conjugate. The match between such dual modes is real. Phenomenal world is a function of the brain's quantum vacuum state, dis-closed in the match between other-action, self-action and memory traces.

4. Paying the Dues for Giving-Up Consciousness

We should not expect to get off so easily after jettisoning consciousness for action-cum-disclosure. Something dear to our hearts is exploded: the world-in-common that faithful observers might by and large agree on. Not only do we lose our consciousness but we lose the world too. Now every quantum thermofield brain is dis-closing worlds in parallel. There is no world-in-common that we each represent (re-present) in our own way, nor do we each pick up a common world's sensory offerings according to our individual predilection (Gibson 1979; Neisser 1976), not even a common world that is selected by sensory input from the possibilities we variously bring to it (Edelman 1987). For the present proposal there is no physical reality of an external world, only unworldly macroscopic and microscopic quantum objects. All worlds are between-two.

So Being, which is to replace consciousness, can be precisely specified. Ontologically primary is distinctionless abground, and Being is secondarily dis-closed--appears--in virtue of an action on the abground, an action that unfolds. That is, the primary ontological condition is closure and Being requires its undoing: dis-closure. The deconstructed topic of "consciousness and quantum physics" is succeeded by the topic of "Being and quantum physics."

This view should not be confused with multiple worlds theory (Greene 2011), which just compounds the difficulties surrounding consciousness already discussed. In multiple worlds theory every possible result of wave function collapse is realized. In one world the Schrödinger cat is found alive by a consciousness and in another world found dead by a different consciousness. For the present account, in contrast, different observers perceive the same result. The worlds dis-closed are multiple yet in agreement. There is consensus to the extent that other-action, self-action and memory are comparable across brains.

The present view might be called "monadological" but in a distinct sense from that of Leibniz. Leibniz did not doubt that there is in fact a transcendent world bestowed through God's love. "God produces substances from nothing," Leibniz (1952, sect. 395) states in the Theodicy. The worlds in parallel of monads are in "pre-established harmony" with the transcendent world God thinks into being. But there is no transcendent world according to the view developed here. There is closure--a distinctionless "abground" (Heidegger 1999) or even "holomovement" (Bohm 1980; Bohm and Hiley 1993)--and multiple parallel phenomenal world disclosures. This is a rather scary thought, to lose the quotidian world-in-common, when you really think about it ...

Other macroscopic quantum objects besides brains also have a matching state of their between-two but the disclosures of, say, an old cabbage, barely change from moment to moment. To use the Kantian phrase, such a macroscopic quantum object "is in itself" but its Being doesn"t amount to much. The brain's genius is that the between-two comes under exquisite control by three influences: other-action, self-action and memory. Complex dis-closures flowingly evolve in waking life.

Three distinct types of brain state bring out some implications of the model. In the case of well-formed slow wave sleep, the between-two disclosure is closer to that of the cabbage. During active REM sleep, however, the between-two revives, though with the participation of other-action (sensory input) strongly inhibited. The between-two becomes a function of only residual self-actions (mainly from the preceding day, Freudian "day residues") and retraces (typically of emotional significance): the dream life is dis-closed. (See Globus (2010) for a detailed illustration of how this works.) Monadological disclosure accordingly varies dramatically in content across waking, dreaming and sleeping but in no case can be transcended. We are windowless monads in parallel and best get on with it! Such a counter-intuitive conclusion is founded in a relentless deconstruction of the role of consciousness in quantum physics. The successor concept to consciousness is "world-thrownness" in virtue of between-two dis-closure.

References

Bohm, D. (1980). Wholeness and the implicate order. Routledge & Kegan Paul, London.

Bohm, D. and Hiley, B. (1993). The Undivided Universe. Routledge & Kegan Paul, London.

Edelman, G. (1987). Neural Darwinism. Basic Books, New York.

Ghiradi, G. (2007). Collapse theories. In Stanford Encyclopedia of Philosophy. http://Plato.Stanford.edu/entries/qm-collapse/

Gibson, J. (1979). The ecological approach to visual perception. Houghton- Mifflin, Boston.

Globus, G. (2003). Quantum Closures and Disclosures: Thinking together Postphenomenology and Quantum Brain Dynamics. John Benjamins, Amsterdam.

Globus, G. (2009). The transparent becoming

of world: A crossing between process thought and quantum neurophilosophy. John Benjamins, Amsterdam.

Green, B. (2010). The Hidden Reality. Knopf, New York.

Hameroff, S. Penrose, R. (1996). Conscious events as orchestrated space-time selections. Journal of consciousness studies 3, 36-53.

Heidegger, M. (1999). Contributions to philosophy (from Enowning). P. Emad & K. Maly, trans. Indiana University Press, Bloomington, Indiana.

Huxley, D.H. (1898). Methods and results: Essays by Thomas H. Huxley. D. Appleton, New York.

Jibu, M., Yasue, K. (1995). Quantum brain dynamics and consciousness. John Benjamins, Amsterdam.

Jibu, M., Yasue, K. (2004). Quantum brain dynamics and quantum field theory. In: Brain and being. Globus, G., Pribram, K., Vitello, G. eds. John Benjamins, Amsterdam.

Kaszniak, A. (ed.) (2001). Emotions, qualia and consciousness. World Scientific: Singapore.

Leibniz, G.W. (1952). Theodicy: Essays on the Goodness of God, the Freedom of Man, and the Origin of Evil. A. Farrer, ed., E. M. Huggard, trans. Yale University Press, New Haven.

Neisser, U. (1976). Cognition and Reality. W.H. Freeman, San Francisco.

Nunn, C. (2009). Editors introduction: Defining consciousness. J. of consciousness studies 16, 5-8.

Sperry, R. (1969). A modified concept of consciousness. Psychological review 77, 585-590.

Umezawa, H. (1993). Advanced Field Theory: Micro, Macro, and Thermal Physics, American Institute of Physics, New York.

Umezawa, H. (1995). Development in concepts in quantum field theory in half century. Mathematica Japonica, 41, 109-124.

Vimal, R.L.P. (2009) Meanings attributed to the term "Consciousness". J. of consciousness studies, 16: 9-27.

Vitiello G. (1995). Dissipation and memory capacity in the quantum brain model. Int J of Modern Physics B 9, 973-989.

Vitiello, G. (2001). My Double Unveiled. John Benjamins, Amsterdam.

Vitiello, G. (2004). The dissipative brain. In: G. Globus, K. Pribram & G. Vitiello, eds. Brain and being. Amsterdam: John Benjamins.

Wright, E. (ed.) (2008). The case for qualia. MIT Press, Cambridge.

45. Logic of Quantum Mechanics and Phenomenon of Consciousness
Michael B. Mensky

P.N. Lebedev Physical Institute, Russian Academy of Sciences,
Leninsky prosp. 53, 119991 Moscow,

Abstract

The phenomenon of consciousness, including its mystical features, is explained on the basis of quantum mechanics in the Everett's form. Everett's interpretation (EI) of quantum mechanics is in fact the only one that correctly describes quantum reality: any state of the quantum world is objectively a superposition of its classical counterparts, or "classical projections". They are "classically incompatible", but are considered to be equally real, or coexisting. We shall call them alternative classical realities or simply classical alternatives. According to the Everett's interpretation, quantum reality is presented by the whole set of alternative classical realities (alternatives). However, these alternatives are perceived by humans separately, independently from each other, resulting in the subjective illusion that only a single classical alternative exists. The ability to separate the classical alternatives is the main feature of what is called consciousness. According to the author's Extended Everett's Concept (EEC), this feature is taken to be a definition of consciousness (more precisely, consciousness as such, not as the complex of processes in the conscious state of mind). It immediately follows from this definition that turning the consciousness off (in sleeping, trance or meditation) allows one to acquire access to all alternatives. The resulting information gives rise to unexpected insights including scientific insights.

KEY WORDS: Quantum mechanics; quantum reality; Everett's interpretation; consciousness; unconscious; insights; miracles; probability

1 INTRODUCTION

Strange as this may seem, we do not know what is the nature of consciousness, especially of the very strange features of consciousness which resemble mystical phenomena. The most familiar examples of these mysterious phenomena are scientific insights (of course, we mean only "great insights" experienced by great scientists). Some people supposed that the mystery of consciousness may be puzzled out on the basis of quantum mechanics, the science which is mysterious itself.

This viewpoint has been suggested, as early as in 20th years of 20th century, by the great physicist Wolfgang Pauli in collaboration with the great psychologist Carl Gustav Jung. They supposed particularly that quantum mechanics may help to explain strange psychic phenomena observed by Jung and called "synchronisms". Jung told of a synchronism if a series of the events happened such that these events were conceptually close but their simultaneous (synchronous) emergence could not be justified causally. For example he observed causally unjustified, seemingly accidental, appearance of the image of fish six times during a single day. The work of Pauli and Jung on this topic was not properly published and was later completely forgotten, but it became popular in last decades (see about this in Enz, 2009).

The idea of connecting consciousness with quantum mechanics was suggested by some other authors, mostly without referring Pauli and Jung. In the last three decades this idea was supported by Roger Penrose (Penrose, 1991), (Penrose, 1994). He particularly remarked that people manage to solve such problems which in principle cannot be solved with the help of computers because no algorithms exist for their solving. Penrose suggested that quantum phenomena should be essential for explaining the work of brain and consciousness.

Usually attempts to explain consciousness on the base of quantum mechanics follow the line of consideration that is natural for physicists. Everything must be explained by natural sciences, may be with accounting quantum laws. Therefore, in order to explain consciousness, one has to apply quantum mechanics for analysis of the work of brain. For example, the work of brain may be explained as the work of quantum computer instead of classical one. Thus, it is usually assumed, explicitly or implicitly, that consciousness must be derived from the analysis of the processes in brain.

The approach proposed by the author in 2000 does not include this assumption. This approach is based on the analysis of the logical structure of quantum mechanics, and the phenomenon of consciousness is derived from this purely logical analysis rater than from the processes in brain. The actual origin of the concept of consciousness is, according to this approach, specific features of the concept of reality accepted in quantum mechanics (contrary to classical physics) and often called quantum reality.

Quantum reality has its adequate presentation in the so-called Everett's interpretation (EI) of quantum mechanics known also under name of many-worlds interpretation (Everett III, 1957). The approach of the author is based on the Everett's form of quantum mechanics and called Extended Everett's Concept (EEC).

Some physicists, having in mind purely classical concept of reality, consider the Everett's interpretation of quantum mechanics too complicated and "exotic". However, it is now experimentally proved that reality in our world is quantum, and the conclusions based on classical concept of reality, are not reliable. The comprehension of the concept of quantum reality was achieved after long intellectual efforts of genius scientists. Unfortunately, the ideas of Pauli and Jung were not properly estimated and timely used. The first whose thoughts about quantum reality became widely know was Einstein who, in the work with his coauthors, suggested so-called Einstein-Podolski-Rosen paradox (Einstein et al., 1935). Much later John Bell formulated now widely known Bell's theorem (Bell, 1964), (Bell, 1987) which provided an adequate tool for direct quantum-mechanical verification of the concept of quantum reality, the Bell's inequality. Less than in 20 years the group of Aspect experimentally proved (Aspect et al., 1981) that the Bell's inequality is violated in some quantum processes, and therefore reality in our world is quantum.

Most simple and convenient formulation of quantum reality was given even earlier that the Bell's works, in 1957, by the Everett's interpretation of quantum mechanics. It was enthusiastically accepted by the great physicists John Archibald Wheeler and Brice Dewitt, but was not recognized by the wide physical community until last decades of 20th century, when the corresponding intellectual base was already prepared. From this time the number of adepts of the Everett's interpretation grows permanently. Results of this difficult but very important process of conceptual clarification of quantum mechanics justify the appreciation of the Everett's interpretation as the only correct form of quantum mechanics. It is exciting that, as an additional prize, the Everett's interpretation explains the mysterious phenomenon of consciousness.

Quantum mechanics in the Everett's form implies coexisting "parallel worlds", or parallel classical realities. This clearly expresses the difference of quantum reality from the common classical reality. According to the author's Extended Everett's Concept (Menskii, 2000), consciousness is the ability to perceive the Everett's parallel worlds separately, independently from each other. An immediate consequence of this assumption is that the state of being unconscious makes all parallel realities available without the separation. This leads to irrational insights and other "mystical" phenomena. This is the central point of the theory making it plausible. Indeed, it is well known (from all spiritual schools and from deep psychological researches) that the strange abilities of consciousness arise just in the states of mind that are close to being unconsciousness (sleeping, trance or meditation).

2 PARALLEL WORLDS

According to the Everett's "many-worlds" interpretation of quantum mechanics, quantum mechanics implies coexistence of "parallel classical worlds", or alternative classical realities. This follows from the arguments (see details below) including the following points:

• The very important specific feature of quantum systems is that their states are vectors. This means that a state of any quantum system may be a sum (called also superposition) of other states of the same system. All the states which are the counterparts of this sum, are equally real, i.e. they may be said to coexist. This is experimentally proved for the states of microscopic systems (such as elementary particles or atoms).

• However, this should be valid also for macroscopic systems (consisting of many atoms). It follows from the logical analysis of the measurements of microscopic systems. Indeed, let a microscopic system S be measured with the help of a macroscopic measuring device M. If the state of S is a sum of a series of states, then, after the measurement, the state of the combined system (S and M) is also a sum (each its term consisting of a state of S and the corresponding state of M).

• Different states of the measuring device, by the very definition of a measuring device, have to be macroscopically distinct. Therefore, a macroscopic system may be in the state which is a sum (superposition) consisting of the states which are incompatible (alternative to each other) from the point of view of classical physics. However, quantum mechanics requires them to "coexist" (in the form of a sum, or superposition).

• Therefore, classically incompatible states of our world (alternative classical realities) must coexist as a sort of "parallel worlds" which are called Everett's worlds.

Let us now give some details of these arguments. What does it mean that the states of a quantum system are vectors? If vpi are states of some quantum system, then $\psi = \psi_1 + \psi_2 + \cdots$ is also a state of the same system, called a superposition of the states ψ_i (counterparts of the superposition).

This feature is experimentally proved for microscopic systems (such as elementary particles or atoms), but it has to be valid also for macroscopic systems. This follows from the analysis of measurements with microscopic measured systems and macroscopic measuring devices.

The conclusion following from this analysis is that, even for a macroscopic system, its state ψ may be a superposition of other states of this system, which have evident classical interpretation (are close to some classical states of the system) while the state ψ has no such an interpretation:

$$\psi = \psi_1 + \psi_2 + ... + \psi_n ... \quad (1)$$

It is important that the states ψ here may be macroscopically distinct, therefore, from classical point of view incompatible, alternative to each other, presenting alternative classical realities. Nevertheless, it follows from quantum theory that even such macroscopically distinct states ψ_i may be in superposition, i.e. may coexist. Let us formulate the above situation a little bit more precisely. The quantum system is in the state denoted by the state vector and only this state objectively exists (however, taken as a whole, has no classical interpretation). The counterparts ^ of the superposition are in fact classical projections of the objectively existing quantum state ψ. These classical projections describe images of the quantum system which arise in consciousness of an observer (therefore, they concern the subjective aspect of quantum reality). This status of the classical projections will be made more transparent below.

In the following we shall use ψ for the state of our (quantum) world as a whole. The components ψi of the superposition will be alternative classical states of this world (more precisely, quasiclassical, i.e. the states as close to classical as is possible for the quantum world).

In the Everett's interpretation of quantum mechanics the states ψ are called Everett's worlds. We shall use also the terms "parallel worlds", "alternative classical realities", "classical alternatives" or simply "alternatives". In case if such alternatives are superposed (as in Eq. (1)), we shall say that they "coexist". Of course this word is nothing else than a convenient slang, meaning in fact "to form a superposition" or "to be in superposition". The status of the "coexistence" is connected with the consciousness and subjective perception of the world, which will be explained below.

3 EXTENDED EVERETT'S CONCEPT (EEC)

We see that the alternative classical realities in our quantum world may coexist (as components of a superposition presenting the state of the quantum world). Subjectively however each observer perceives only a single "classical alternative". These two assertions seem to contradict to each other. Are they in fact compatible? We shall show how this seeming contradiction is resolved in the Everett's interpretation (EI) and how it may be taken as a basis for the theory of consciousness and the unconscious if the EI is properly extended.

3.1 Everett's "Many-Worlds" Interpretation One may naively think that the picture of the world arising in his consciousness (the picture of a single classical alternative) is just what objectively exists. However, EI of quantum mechanics (unavoidably following from the logics of quantum mechanics applied to the phenomenon of quantum measurements) claims that it is only the superposition of all alternatives (as in Eq. (1)) that objectively exists.

The seemingly strange and counter-intuitive presentation of objective reality (in the Everett's form of quantum mechanics) as the set of many objectively coexisting classical realities adequately expresses quantum character of reality in our (quantum) world.

The single alternatives (components of the superposition) present various subjective perceptions of this quantum reality in an observer's consciousness. The natural question arises how the multiplicity of the "classical pictures" that may arise in our consciousness may be compatible with the subjective sensory evidence of only a single such picture.

This is the most difficult point of the EI and the reason why this interpretation has not been readily accepted by the physicists. This point may be made more transparent if it is presented in the terms of "Everett's worlds" as it has been suggested by Brice DeWitt.

Thus, all of the separate components of the superposition (classical alternatives, or Everett's worlds) are declared by Everett to be "equally real". No single alternative may be considered to be the only real, while the others being potentially possible but not actualized variants of reality (this might be accepted in classical physics, but not in quantum mechanics, because of the special character of quantum reality).

To make understanding of the EI easier, Brice DeWitt proposed (De-Witt & Graham, 1973) to think that each observer is present in each of the Everett's worlds. To make this even more transparent, one may think that a sort of twins ("clones") of each observer are present in all Everett's worlds. Subjective perception is the perception of a single twin. Objectively the twins of the given observer exist in all Everett's worlds, each of them perceiving corresponding alternative pictures.

Each of us subjectively perceives around him a single classical reality. However, objectively the twins, or clones, of each of us perceive all the rest realities. It is important that all twins of the given observer are equally real. It is incorrect to say that there is "I" and there are my twins, which are not "I". All twins embody me as an observer, each of them can be called "I".

Thus, the concept of "Everett's world" allows to make the Everett's presentation of quantum reality more transparent. We prefer to verbalize the same situation in another way (Menskii, 2000). We shall say that all alternative classical projections of the quantum world's state objectively exist, but these projections are separated in consciousness. Subjective perception of the quantum world by human's consciousness embraces all these classical pictures, but each picture is perceived independently of the rest.

Regardless of the way of wording, the Everett's assumption of objectively coexisting classical alternatives implies that all these alternatives may be accessible for an observer in some way or another. Yet it is not clear how the access to "other alternatives" (different from one subjectively perceived) may be achieved for the given observer. It is usually claimed that the EI does not allow observing "other alternatives". This makes the interpretation non-falsifiable and thus decreases its value.

It turns out however that the EI may be improved in such a way that that the question "How one can access to other alternatives?" is answered in a very simple and natural way. This improvement is realized in the author's Extended Everett's Concept. The (improved) EI becomes then falsifiable, although in a special sense of the word (see below Sect. 3.2).

3.2 Extension of the Everett's Interpretation Starting with the above mentioned formulation (that classical alternatives are separated in consciousness), the present author proposed (Menskii, 2000) to accept a stronger statement: consciousness is nothing else than separation of the alternatives.

This seemingly very small step resulted in important consequences. Finally the so-called Extended Everett's Concept (EEC) has been developed (Menskii, 2005), (Menskii, 2007), (Mensky, 2007), (Mensky, 2005), (Mensky, 2010).

The first advantage of EEC is that the logical structure of the quantum theory is simplified as compared with the EI.

The point is that the formulation "alternatives are separated in consciousness" (accepted in one of the possible formulations of EI) includes two primary (not definable) concepts, "consciousness" and "alternative separation". These concepts have no good definitions. One may object that many different definitions have been proposed for the notion "consciousness". This is right, but all these definitions concern in fact mental and sensual processes in brain rather than "consciousness as such", while the latter (more fundamental) notion has no good definition.

After the notions "consciousness" (more precisely, "consciousness as such") and "separation of alternatives" are identified (as it is suggested in EEC), only one of these concepts remains in the theory. Therefore, EEC includes only one primary concept instead of two such concepts in the EI. The logical structure of the theory is simplified after its extension.

Much more important is that EEC gives a transparent indication as to how an observer may obtain access to "other alternatives" (different from the alternative subjectively perceived by him). This very important question remains unanswered (or answered negatively) in the original form of the EI. This is seen from the following argument.

If consciousness = separation (of the alternatives from each other), then absence of consciousness = absence of separation.

Therefore, turning off consciousness (in sleeping, trance or meditation) opens access to all classical alternatives put together, without separation between them. Of course, the access is realized then not in the form of visual, acoustic or other conscious images or thoughts. Nothing at all can be said about the form of this access. However, if we accept EEC, then we may definitely conclude that the access is possible in the unconscious state of mind.

This of course has very important consequences. The access to the enormous "data base" consisting of all alternative classical realities enables one to acquire valuable information, or rather knowledge. This information is unique in the sense that it is unavailable in the conscious state when only a single alternative is subjectively accessible. One may suppose that a part of this unique information may be kept on returning to the usual conscious state of mind and recognized in the form of usual conscious images and thoughts.

Thus, when going over to the unconscious state, one obtains the information, or knowledge, which is in principle unavailable in the usual conscious state.

This information is unique first of all because it is taken from "other" classical alternatives (different from one subjectively observed). There is however something more that makes this information unique and highly valuable. All alternatives together form a representation of the quantum state of the world (vector $\wedge$ in Eq. (1)). Time evolution of this state vector, according to quantum laws, is reversible. This means that, given at some time moment, this vector is known also at all other times. Therefore, information about "all alternatives together" (i.e. about the state vector of the world) includes information from any time moments in future and past. This information may be thought of as being obtained with the help of a "virtual time machine".

Evidently, this makes "irrational" inspirations (including scientific insights) possible. Here is a simple example. Let a scientist be confronted with a scientific problem and consider a number of hypotheses for solving this problem. Going, by means of the above mentioned virtual time machine, into future and backward, the scientist may find out what of these hypotheses will be confirmed by future experiments or proved with the help of the future theories. Then, on returning to the conscious state, he will unexpectedly and without any rational grounds get certain about which of these hypotheses has to be chosen.

Remark that it is not necessary, for making use of such a virtual time machine, to turn off consciousness completely (although it is known that some important discoveries were made in sleeping, or rather after awakening). It is enough to disconnect it from the problem under consideration. This is why solutions of hard problems are sometimes found not during the work on these problems but rather during relaxation.

Preliminary "rational" work on the problem is however necessary. The deep investigation of all data concerning the problem enables the consciousness to form a sort of query (clear formulation of the problem and its connections with all relevant areas of knowledge). Then the query will be worked out in the unconscious state (during relaxation) and will result in an unexpected insight.

It is clear that not only scientific problems can be solved in this way, but also problems of general character. Quite probable that, besides ordinary intuitive guesses, we meet in our experience examples of super-intuitive insights of the type described above. Anybody knows that many efficient solutions come in the morning just after awakening. This fact may be an indirect confirmation of the ability of super-intuition.

4 PROBABILISTIC MIRACLES

Thus, Extended Everett's Concept (EEC) leads to the conclusion that unconscious state of mind allows one to take information "from other alternatives" that reveals itself as unexpected insights, or direct vision of truth.

Another consequence is feasibility of even more weird phenomenon looking as arbitrary choice of reality. Let us describe this ability in a special case of what can be called "probabilistic miracles".

Consider an observer who subjectively perceives one of the alternative classical realities at the present time moment t_0. Let in a certain future moment $t > t_0$ some event E may happen, but with a very small probability p. Call it the objective probability of the event E and suppose that this probability is small.

According to the Everett's interpretation of quantum mechanics, at time t two classes of alternative classical realities will exist so that the event E happens in each alternative of the first class and does not happen in the alternatives of the second class. The twins of our observer will be objectively present in each of the alternatives (this is the feature of Everett's worlds, see Sect. 3.1). However, subjectively our observer will feel to be in one of them. With some probability p it will turn out to be the alternative of the first class. The probability p may be called subjective probability of the event E for the given observer.

It is accepted in the Everett's interpretation that subjective and objective probabilities coincide, p = p. However, in the context of EEC we may assume that they may differ and, moreover, the observer may influence the value of subjective probability p'. Let us assume that the observer prefers the event E to happen. Then he can enlarge the subjective probability of this event, i.e. the probability to find himself subjectively at time t just in that classical reality in which this rare event actually happens.

Thus, besides the objective probability of any event, there is a subjective probability of this event for the given observer, and the observer may in principle influence the subjective probability. In the above mentioned situation, an event under consideration can happen according to usual laws of the natural sciences, but with small probability. This means that the objective probability of this event is small, it may seem even negligible. It is important that the objective probability is non-zero. One may say that among all alternatives at the moment t few of the alternatives correspond to the pictures of the world in which the event happens, and much more alternatives correspond to the pictures of the world where the event does not happen.

However, according to EEC, an observer can, simply by the force of his consciousness, make the subjective probability of this event close to unity. Then very likely he will find himself at the moment t in one of those classical realities where the event does happen.

The subjective experience of such an observer will evidence that the objectively improbable event may be realized by the effort of his will. This looks like a miracle. However, this is a miracle of a special type, which may be called "probabilistic miracle".

Probabilistic miracles essentially differ from "absolute" miracles that happen in fairy tales. The difference is that the event realized as a probabilistic miracle (i.e. "by the force of consciousness") may in principle happen in a quite natural way, although with a very small probability. This small but nonzero probability is very important. Particularly, because of fundamental character of probabilistic predictions in quantum mechanics, it is in principle impossible to prove or disprove the unnatural (miraculous) character of the happening.

Indeed, if the objectively rare event happens, the person who has strongly desired for it to happen is inclined to consider the happening as a result of his will. Yet any skeptic may insist in this situation that the event occurred in a quite natural way: what happened, was only a rare coincidence. The secret is in the nature of the concept of probability: if the probability of some stochastic event is equal p, then in a long series consisting of N tests the event will happen pN times (very rarely for small p). But it is in principle impossible to predict in which of these tests the event will happen. Particularly, it may happen even in the very first test from the long series of them.

The latter is a very interesting and general feature of the phenomena "in the area of consciousness" as they are treated in EEC. These phenomena in principle cannot be unambiguously assigned to the sphere of natural events (obeying the laws of natural sciences) or to the sphere of spiritual or psychic phenomena (which are treated by the humanities and spiritual doctrines). Impossibility to do this may be called relativity of objectiveness.

Synchronisms studied by Jung may be considered to be probabilistic miracles. One who observes a subject or event which somehow attracts his special attention, involuntarily thinks about it (often even not clearly fixing his thoughts). According to the above said, he may increase the subjective probability of immediately observing something similar or logically connected.

Some Biblical miracles can also be explained as probabilistic miracles. An example is the miracle at the Sea of Galilee where Jesus calmed the raging storm (Matthew 8:23-27). This was completely natural event. Wonderful was only the fact that the storm ceased precisely at that moment when this was necessary for Jesus and his disciples. The probability of "timely" occurring this natural event was of course very small. The miracle was probabilistic.

5 CONCLUDING REMARKS

We shortly followed in this paper the main ideas of Extended Everett's Concept (EEC) about nature of consciousness. Let us briefly comment on the further development of this theory.

All that has been discussed above, makes sense for humans (possessing consciousness) and may be partly for higher animals. However, the theory may be generalized to give the quantum concept of life in a more general aspect. Thus modified theory is meaningful not only for humans, but for all living beings (belonging to the type of life characteristic for Earth). The idea of the generalization is follows (see (Mensky, 2010) for details).

The main point of EEC is the identification of the "separation of classical alternatives" with the human's consciousness. Now we have to identify this quantum concept with the ability of the living beings to "perceive the quantum world classically". This is an evident generalization of consciousness but for all living beings. Instead of "consciousness" in EEC we have now "classical perceiving of quantum reality" which means that the alternative classical realities (forming the state of the quantum world) are perceived separately from each other.

Similarly to what we told about consciousness (in case of humans), this ability of living beings to "classically perceive the quantum world" is necessary for the very phenomenon of life (of local type). The reason is the same: elaborating efficient strategy of surviving is possible only in a classical world which is "locally predictable". Existing objectively in the quantum world, any creature is living in each of the classical realities separately from all the rest classical realities. Life is developing parallely in the Everett's parallel worlds.

Remark by the way that from this point of view "existing" and "living" are different concepts. Important difference is that existing (evolution in time) of the inanimate matter is determined by reasons while living of the living beings is partly determined also by goals (first of all the goal of survival). Let us make some other remarks concerning philosophical or rather meta-scientific aspects of EEC.

This theory shows that a conceptual bridge between the material (described by natural sciences) and the ideal (treated by the humanities and spiritual doctrines) does exist.

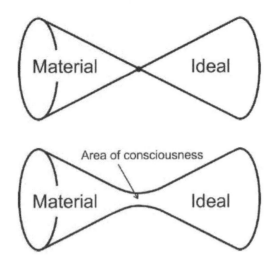

Fig 1: Carl Jung illustrated the relation between sphere of the material and sphere of the ideal as two cones with common vertex (above). In EEC the "area of consciousness" is common for both spheres (below). Some of the phenomena in the area of consciousness cannot be unambiguously assigned to exclusively one of the spheres: objectiveness is relative.

This bridge is determined in EEC in a concrete way, but the idea of such a bridge is not novel (see Fig. 1). The creators of quantum mechanics from the very beginning needed the notion of the "observer's consciousness" to analyze conceptual problems of this theory (the "problem of measurement"). In fact, the difficulties in solving these problems were caused by the insistent desire to construct quantum mechanics as a purely objective theory. Nowadays it becomes clear that there is no purely objective quantum theory. Objectiveness is relative (see Sect. 4).

There is a very interesting technical point in relations between the material and the ideal. We see from the preceding consideration that the description of the ideal, or psychic (consciousness and the unconscious), arises in the interior of quantum mechanics when we consider the whole world as a quantum system. This provides the absolute quantum coherence which is necessary for the conclusions that are derived from EEC. Usually only restricted systems are considered in quantum mechanics. The resulting theory is purely material. Ideal (psychic) elements arise as the specific aspects of the whole world. The unrestricted character of the world as a quantum system is essential for this (cf. the notion of microcosm).

All these issues demonstrate the specific features of the present stage of quantum theory. Including theory of consciousness (and the unconscious) in the realm of quantum mechanics (starting by Pauli and Jung and now close to being accomplished) marks a qualitatively new level of understanding quantum mechanics itself. The present stage of this theory can be estimated as the second quantum revolution. When being completed, it will accomplish the intellectual and philosophical revolution that accompanied creating quantum mechanics in the first third of 20th century.

References

Aspect, A., Grangier, P., Roger, G. (1981). Phys. Rev. Lett., 47, 460.

Bell, J. S. (1964). Physics, 1, 195. Reprinted in (Bell, 1987).

Bell, J. S. (1987). Speakable and Unspeakable in Quantum Mechanics. Cambridge Univ. Press, Cambridge.

DeWitt, B. S., Graham, N. (Eds.) (1973). The Many-Worlds Interpretation of Quantum Mechanics. Princeton Univ. Press, Princeton, NJ.

Einstein, A., Podolsky, B., Rosen, N. (1935). Phys. Rev., 47, 777.

Enz, C. P. (2009). Of Matter and Spirit. World Scientific Publishing Co., New Jersey etc. Everett III, H. (1957). Rev. Mod. Phys., 29, 454. Reprinted in (Wheeler & Zurek, 1983).

Menskii, M. B. (2000). Quantum mechanics: New experiments, new applications and new formulations of old questions. Physics-Uspekhi, 43, 585-600.

Menskii, M. B. (2005). Concept of consciousness in the context of quantum mechanics. Physics-Uspekhi, 48, 389-409.

Menskii, M. B. (2007). Quantum measurements, the phenomenon of life, and time arrow: Three great problems of physics (in Ginzburg's terminology). Physics-Uspekhi, 50, 397-407.

Mensky, M. (2007). Postcorrection and mathematical model of life in Extended Everett's Concept. NeuroQuantology, 5, 363-376. www.neuroquantology.com, arxiv:physics.gen-ph/0712.3609.

Mensky, M. (2010). Consciousness and Quantum Mechanics: Life in Parallel Worlds (Miracles of Consciousness from Quantum Mechanics). World Scientific Publishing Co., Singapore.

Mensky, M. B. (2005). Human and Quantum World (Weirdness of the Quantum World and the Miracle of Consciousness). Vek-2 publishers, Fryazino. In Russian.

Penrose, R. (1991). The Emperor's New Mind: Concepting Computers, Minds, and the Laws of Physics. Penguin Books, New York.

Penrose, R. (1994). Shadows of the Mind: a Search for the Missing Science of Consciousness. Oxford Univ. Press, Oxford.

46. A Quantum Physical Effect of Consciousness

Shan Gao

Unit for HPS & Centre for Time, SOPHI,
University of Sydney, Sydney, NSW 2006, Australia

Abstract

It is shown that a conscious being can distinguish definite perceptions and their quantum superpositions, while a physical measuring system without consciousness cannot distinguish such nonorthogonal quantum states. This result may have some important implications for quantum theory and the science of consciousness. In particular, it implies that consciousness is not emergent but a fundamental feature of the universe.

KEY WORDS: consciousness, quantum superposition, quantum-to-classical transition, panpsychism

1. Introduction

The relationship between quantum measurement and consciousness has been studied since the founding of quantum mechanics (von Neumann 1932/1955; London and Bauer 1939; Wigner 1967; Stapp 1993, 2007; Penrose 1989, 1994; Hameroff and Penrose 1996; Hameroff 1998, 2007; Gao 2004, 2006b, 2008b). Quantum measurement problem is generally acknowledged as one of the hardest problems in modern physics, and the transition from quantum to classical is still a deep mystery. On the other hand, consciousness remains another deep mystery for both philosophy and science, and it is still unknown whether consciousness is emergent or fundamental. It has been conjectured that these two mysteries may have some intimate connections, and finding them may help to solve both problems (Chalmers 1996).

There are two main viewpoints claiming that quantum measurement and consciousness are intimately connected. The first one holds that the consciousness of an observer causes the collapse of the wave function and helps to complete the quantum measurement or quantum-to-classical transition in general (von Neumann 1932/1955; London and Bauer 1939; Wigner 1967; Stapp 1993, 2007). This view seems understandable. Though what physics commonly studies are insensible objects, the consciousness of observer must take part in the last phase of measurement. The observer is introspectively aware of his perception of the measurement results, and consciousness is used to end the infinite chains of measurement here. The second view holds that consciousness arises from objective wavefunction collapse (Penrose 1989, 1994; Hameroff and Penrose 1996; Hameroff 1998, 2007). One argument is that consciousness is a process that cannot be described algorithmically, and the gravitation-induced wavefunction collapse seems non-computable as a fundamental physical process, and thus the elementary acts of consciousness must be realized as objective wavefunction collapse, e.g., collapse of coherent superposition states in brain microtubules. Though these two views are obviously contrary, they both insist that a conscious perception is always definite and classical, and there are no quantum superpositions of definite conscious perceptions.

Different from these seemingly extreme views, it is widely thought that the quantum-to-classical transition and consciousness are essentially independent with each other (see, e.g. Nauenberg (2007) for a recent review). At first sight, this common-sense view seems too plain to be intriguing. However, it has been argued that, by permitting the existence of quantum superpositions of different conscious perceptions, this view will lead to an unexpected new result, a quantum physical effect of consciousness (Gao 2004, 2006b, 2008b). In this article, we will introduce this interesting result and discuss its possible implications.

2. The Effect

Quantum mechanics is the most fundamental theory of the physical world. Yet as to the measurement process or quantum-to-classical transition process, the standard quantum mechanics provides by no means a complete description, and the collapse postulate is just a makeshift (Bell 1987). Dynamical collapse theories (Ghirardi 2008), many-worlds theory (Everett 1957) and de Broglie-Bohm theory (Bohm 1952) are the main alternatives to a complete quantum theory. The latter two replace the collapse postulate with some new structures, such as branching worlds and Bohmian trajectories, while the former integrate the collapse postulate with the normal Schrödinger evolution into a unified dynamics. It has been recently shown that the dynamical collapse theories are probably in the right direction by admitting wavefunction collapse (Gao 2011). Here we will mainly discuss the possible quantum effects of consciousness in the framework of dynamical collapse theories, though the conclusion also applies to the other alternatives. Our analysis only relies on one common character of the theories, i.e., that the collapse of the wave function (or the quantum-to-classical transition in general) is one kind of objective dynamical process, essentially independent of the consciousness of observer, and it takes a finite time to finish.

It is a well-known result that nonorthogonal quantum states cannot be distinguished (by physical measuring device) in both standard quantum mechanics and dynamical collapse theories (see, e.g. Wootters and Zurek 1982; Ghirardi et al 1993; Nielsen and Chuang 2000). However, it has been argued that a conscious being can distinguish his definite perception states and the quantum superpositions of these states, and thus when the physical measuring device is replaced by a conscious observer, the nonorthogonal states can be distinguished in principle in dynamical collapse theories (Gao 2004, 2006b, 2008b). The distinguishability of nonorthogonal states will reveal a distinct quantum physical effect of consciousness, which is lacking for physical measuring systems without consciousness. In the following, we will give a full exposition of this result.

Let $v1$ and $v2$ be two definite perception states of a conscious being, and $v1 + v2$ is the quantum superposition of these two definite perception states. For example, $v1$ and $v2$ are triggered respectively by a small number of photons with a certain frequency entering into the eyes of the conscious being from two directions, and $v1 + v2$ is triggered by the superposition of these two input states. Assume that the conscious being satisfies the following slow collapse condition, i.e., that the collapse time of the superposition state $v1 + v2$, denoted by tc, is longer than the normal conscious time tp of the conscious being for definite states, and the time difference is large enough for him to identify. This condition ensures that consciousness can take part in the process of wavefunction collapse; otherwise consciousness can only appear after the collapse and will surely have no influence upon the collapse process. Now we will explain why the conscious being can distinguish the definite perception states $v1$ or $v2$ and the superposition state $v1 + v2$.

First, we assume that a definite perception can appear only after the collapse of the superposition state $v1 + v2$. This assumption seems plausible. Then the conscious being can have a definite perception after the conscious time tp for the states $v1$ and $v2$, but only after the collapse time tc can the conscious being have a definite perception for the superposition state $v1 + v2$. Since the conscious being satisfies the slow collapse condition and can distinguish the times tp and tc, he can distinguish the definite perception state $v1$ or $v2$ and the superposition state $v1 + v2$. Note that a similar argument was first given by Squires (1992).

Next, we assume that the above assumption is not true, i.e., that the conscious being in a superposition state can have a definite perception before the collapse has completed. We will show that the conscious being can also distinguish the states v1 + v2 and v1 or v2 with non-zero probability.

(1). If the definite perception of the conscious being in the superposed state v1 + v2 is neither v1 nor v2 (e.g. the perception is some sort of mixture of the perceptions v1 and v2), then obviously the conscious being can directly distinguish the states v1 + v2 and v1 or v2.

(2). If the definite perception of the conscious being in the superposed state v1+v2 is always v1, then the conscious being can directly distinguish the states v1+v2 and v2. Besides, the conscious being can also distinguish the states v1 + v2 and v1 with probability 1/2. The superposition state v1 + v2 will become v2 with probability 1/2 after the collapse, and the definite perception of the conscious being will change from v1 to v2 accordingly. But for the state v1, the perception of the conscious being has no such change.

(3). If the definite perception of the conscious being in the superposed state v1 + v2 is always v2, the proof is similar to (2).

(4). If the definite perception of the conscious being in the superposed state v1 + v2 is random, i.e., that one time it is v1, and another time it is v2, then the conscious being can still distinguish the states v1 + v2 and v1 or v2 with non-zero probability. For the definite perception states v1 or v2, the perception of the conscious being does not change. For the superposition state v1 + v2, the perception of the conscious being will change from v1 to v2 or from v2 to v1 with non-zero probability during the collapse process.

In fact, we can also give a compact proof by reduction to absurdity. Assume that a conscious being cannot distinguish the definite perception states v1 or v2 and the superposition state v1 + v2. This requires that for the superposition state v1 + v2 the conscious being must have the perception v1 or v2 immediately after the conscious time tp, and moreover, the perception must be exactly the same as his perception after the collapse of the superposition state v1 + v2. Otherwise he will be able to distinguish the superposition state v1 + v2 from the definite state v1 or v2. Since the conscious time tp is shorter than the collapse time tc, the requirement means that the conscious being knows the collapse result beforehand. This is impossible due to the essential randomness of the collapse process. Note that even if this is possible, the conscious being also has a distinct quantum physical effect, i.e., that he can know the random collapse result beforehand.

To sum up, we have shown that if a conscious being satisfies the slow collapse condition, he can readily distinguish the nonorthogonal states v1 + v2 (or v1 - v2) and v1 or v2, which is an impossible task for a physical measuring system without consciousness.

3. The Condition

The above quantum physical effect of consciousness depends on the slow collapse condition, namely that for a conscious being the collapse time of a superposition of his conscious perceptions is longer than his normal conscious time. Whether this condition is available for human brains depends on concrete models of consciousness and wavefunction collapse. For example, if a definite conscious perception involves less neurons such as several thousand neurons, then the collapse time of the superposition of such perceptions will be readily in the same level as the normal conscious time (several hundred milliseconds) according to some dynamical collapse models (Gao 2006a, 2006b, 2008a, 2008b). This result is also supported by the Penrose-Hameroff orchestrated objective reduction model (Hameroff and Penrose 1996; Hagan, Hameroff and Tuszynski 2002). In the model, if a conscious perception involves about 109 participating tubulin, then the collapse time will be several hundred milliseconds and in the order of normal conscious time. When assuming that 10% of the tubulin contained becomes involved, the conscious perception also involves about one thousand neurons (there are roughly 107 tubulin per neuron). In addition, even though the slow collapse condition is unavailable for human brains, it cannot be in principle excluded that there exist some small brain creatures in the universe who satisfy the slow collapse condition (see also Squires 1992).

A more important point needs to be stressed here. The collapse time estimated above is only the average collapse time for an ensemble composed of identical superposition states. The collapse time of a single superposition state is an essentially stochastic variable, which value can range between zero and infinity. As a result, the slow collapse condition can always be satisfied for some collapse events with a certain probability. For these random collapse processes, the collapse time of the single superposition state is much longer than the average collapse time and the normal conscious time, and thus the conscious being can distinguish the nonorthogonal states and have the distinct quantum physical effect. As we will see, this ultimate possibility will have important implications for the nature of consciousness.

Lastly, we note that the slow collapse condition is also available in the many-worlds theory and de Broglie-Bohm theory (Gao 2004). For these two theories, the collapse time will be replaced by the decoherence time. First, since a conscious being is able to be conscious of its own state, he can always be taken as a closed self-measuring system in theory. In both many-worlds theory and de Broglie-Bohm theory, the state of a closed system satisfies the linear Schrödinger equation, and thus no apparent collapse happens or the decoherence time is infinite for the superposition state of a closed conscious system. Therefore, the slow collapse condition can be more readily satisfied in these theories when a conscious system has only a very weak interaction with environment. By comparison, in most dynamical collapse theories, the superposition state of a closed system also collapses by itself. Secondly, a conscious system (e.g. a human brain or neuron groups in the brain) often has a very strong interaction with environment in practical situations. As a result, the decoherence time is usually much shorter than the collapse time, and the slow collapse condition will be less readily satisfied in many-worlds theory and de Broglie-Bohm theory than in the dynamical collapse theories. This difference can be used to test these different quantum theories.

4. Implications

Consciousness is the most familiar phenomenon. Yet it is also the hardest to explain. The relationship between objective physical process and subjective conscious experience presents a well-known hard problem for science (Chalmers 1996). It retriggers the recent debate about the long-standing dilemma of panpsychism versus emergentism (Strawson et al 2006; Seager and Allen-Hermanson 2010). Though emergentism is currently the most popular solution to the hard problem of consciousness, many doubt that it can bridge the explanation gap ultimately. By comparison, panpsychism may provide an attractive and promising way to solve the hard problem, though it also encounters some serious problems (Seager and Allen-Hermanson 2010). It is widely believed that the physical world is causally closed, i.e., that there is a purely physical explanation for the occurrence of every physical event and the explanation does not refer to any consciousness property (see, e.g. McGinn 1999). But if panpsychism is true, the fundamental consciousness property should take part in the causal chains of the physical world and should present itself in our investigation of the physical world. Then does consciousness have any causal efficacy in the physical world?

As we have argued above, a conscious observer can distinguish two nonorthogonal states, while the physical measuring system without consciousness cannot. Accordingly, consciousness does have a causal efficacy in the physical world when considering the fundamental quantum processes. This will provide a strong support for panpsychism. In fact, we can argue that if consciousness has a distinct quantum physical effect, then it cannot be emergent but be a fundamental property of substance. Here is the argument.

If consciousness is emergent, then the conscious beings should also follow the fundamental physical principles such as the principle of energy conservation etc, though they may have some distinct high-level functions. According to the principles of quantum mechanics, two nonorthogonal states cannot be distinguished. However, a conscious being can distinguish the nonorthogonal states in principle. This clearly indicates that consciousness violates the quantum principles, which are the most fundamental physical principles. Therefore, the consciousness property cannot be reducible or emergent but be a fundamental property of substance. It should be not only possessed by the conscious beings, but also possessed by atoms as well as physical measuring devices. The difference only lies in the conscious content. The conscious content of a human being can be very complex, while the conscious content of a physical measuring device is probably very simple. In order to distinguish two nonorthogonal states, the conscious content of a measuring system must at least contain the perceptions of the nonorthogonal states. It might be also possible that the conscious content of a physical measuring device can be complex enough to distinguish two nonorthogonal states, but the effect is too weak to be detected by present experiments.

On the other hand, if consciousness is a fundamental property of substance, then it is quite natural that it violates the existing fundamental physical principles, which do not include it at all. It is expected that a complete theory of nature must describe all properties of substance, thus consciousness, the new fundamental property, must enter the theory from the start. Since the distinguishability of nonorthogonal states violates the linear superposition principle, consciousness will introduce a nonlinear element to the complete evolution equation of the wave function. The nonlinearity is not stochastic but definite. It has been argued that the nonlinear quantum evolution introduced by consciousness has no usual problems of nonlinear quantum mechanics (Gao 2006b).

Lastly, it should be noted that the above argument for panpsychism depends on the assumption that the wavefunction collapse or the quantum-to-classical transition in general is an objective physical process. However, the conclusion is actually independent of the origin of the wavefunction collapse. If the wavefunction collapse results from the consciousness of observer, then consciousness will also have the distinct quantum effect of collapsing the wave function, and thus consciousness should be a fundamental property of substance too. In addition, we stress that this conclusion is also independent of the interpretations of quantum mechanics. It only depends on two firm facts: one is the existence of indefinite quantum superpositions, and the other is the existence of definite conscious perceptions.

5. Conclusions

It is widely thought that the quantum-to-classical transition and consciousness are two essentially independent processes. But this does not mean that the result of their combination must be plain. In this article, we have shown that a conscious being can have a distinct quantum physical effect during the quantum-to-classical transition. A conscious system can measure whether he is in a definite perception state or in a quantum superposition of definite perception states, while a system without consciousness cannot distinguish such nonorthogonal states. This new result may have some important implications for quantum theory and the science of consciousness. In particular, it may provide a quantum basis for panpsychism.

References

Bell, J. S. (1987). Speakable and Unspeakable In Quantum Mechanics. Cambridge: Cambridge University Press.

Bohm, D. (1952). A suggested interpretation of quantum theory in terms of "hidden" variables, I and II. Phys. Rev. 85, 166-193.

Chalmers, D. (1996). The Conscious Mind. Oxford: University of Oxford Press.

Everett, H. (1957). "Relative state" formulation of quantum mechanics, Rev. Mod. Phys. 29, 454-462.

Gao, S. (2004). Quantum collapse, consciousness and superluminal communication, Found. Phys. Lett, 17(2), 167-182.

Gao, S. (2006a). A model of wavefunction collapse in discrete space-time, Int. J. Theo. Phys. 45 (10), 1943-1957.

Gao, S. (2006b). Quantum Motion: Unveiling the Mysterious Quantum World. Bury St Edmunds: Arima Publishing.

Gao, S. (2008a). God Does Play Dice with the Universe. Bury St Edmunds: Arima Publishing.

Gao, S. (2008b). A quantum theory of consciousness. Minds and Machines 18 (1), 39-52.

Gao, S. (2011). Meaning of the wave function, Int. J. Quant. Chem. Article first published online: http://onlinelibrary.wiley.com/doi/10.1002/qua.22972/abstract.

Ghirardi, G. C., Grassi, R., Butterfield, J., and Fleming, G. N. (1993). Parameter dependence and outcome dependence in dynamic models for state-vector reduction, Found. Phys., 23, 341.

Ghirardi, G. (2008). Collapse Theories, The Stanford Encyclopedia of Philosophy (Fall 2008 Edition), Edward N. Zalta (ed.), http://plato.stanford.edu/archives/fall2008/entries/qm-collapse/.

Hagan, S., Hameroff, S. R., and Tuszynski, J. A. (2002). Quantum computation in brain microtubules: decoherence and biological feasibility, Phys. Rev. E 65, 061901.

Hameroff, S. R. (1998). Funda-Mentality: Is the conscious mind subtly linked to a basic level of the universe? Trends in Cognitive Sciences 2(4):119-127.

Hameroff, S. R. (2007). Consciousness, neurobiology and quantum mechanics: The case for a connection, In: The Emerging Physics of Consciousness, edited by Jack Tuszynski, New York: Springer-Verlag.

Hameroff, S. R. and Penrose, R. (1996), Conscious events as orchestrated space-time selections, Journal of Consciousness Studies, 3 (1), 36-53.

London, F., and Bauer, E. (1939). La théorie de l'observation en mécanique quantique.

Hermann, Paris. English translation: The theory of observation in quantum mechanics. In Quantum Theory and Measurement, ed. by J.A. Wheeler and W.H. Zurek, Princeton University Press, Princeton, 1983, pp. 217-259.

McGinn, C. (1999). The Mysterious Flame: Conscious Minds in a Material World. New York: Basic Books.

Nauenberg, M. (2007). Critique of "Quantum enigma: Physics encounters consciousness". Foundations of Physics 37 (11), 1612–1627.

Nielsen, M. A. and Chuang, I. L. (2000). Quantum Computation and Quantum Information. Cambridge: Cambridge University Press. Section 1.6.

Penrose, R. (1989). The Emperor's New Mind. Oxford: Oxford University Press.

Penrose, R. (1994). Shadows of the Mind. Oxford: Oxford University Press.

Seager, W. and Allen-Hermanson, S. (2010). Panpsychism, The Stanford Encyclopedia of Philosophy (Fall 2010 Edition), Edward N. Zalta (ed.), http://plato.stanford.edu/archives/fall2010/ entries/panpsychism/.

Squires, E. (1992). Explicit collapse and superluminal signaling, Phys. Lett. A 163, 356-358.

Stapp, H. P. (1993). Mind, Matter, and Quantum Mechanics. New York: Springer-Verlag.

Stapp, H. P. (2007). Mindful Universe: Quantum Mechanics and the Participating Observer. New York: Springer-Verlag.

Strawson, G. et al. (2006). Consciousness and its Place in Nature: Does Physicalism entail Panpsychism? (ed. A. Freeman). Exeter, UK: Imprint Academic.

von Neumann, J. (1932/1955). Mathematical Foundations of Quantum Mechanics. Princeton: Princeton University Press. German original Die mathematischenGrundlagen der Quantenmechanik. Berlin: Springer-Verlag, 1932.

Wigner, E. P. (1967). Symmetries and Reflections. Bloomington and London: Indiana University Press, 171-184.

Wootters, W. K. and Zurek, W. H. (1982). A single quantum cannot be cloned. Nature 299, 802-803.

47. The Conscious Observer in the Quantum Experiment

Fred Kuttner and Bruce Rosenblum

Physics Department, University of California, Santa Cruz,
1156 High Street, Santa Cruz, CA 95064

Abstract

A quantum-theory-neutral version of the two-slit experiment displays the intrusion of the conscious observer into physics. In addition to the undisputed experimental results, only the inescapable assumption of the free choice of the experimenter is required. In discussing the experiment in terms of the quantum theory, other aspects of the quantum measurement problem also appear.

KEY WORDS: Consciousness, quantum, experiment, free-will, John Bell

1. INTRODUCTION

The intrusion of the observer into physics appeared at the inception of quantum theory eight decades ago. With this "quantum measurement problem," the physics discipline encountered something apparently beyond "physics" (Greenstein & Zajonc 1997).

Photons and electrons manifested "wave-particle duality": They exhibited wave properties or particle properties depending on the experimental technique used to observe them. Wave properties imply a spread-out entity, while particle properties imply a not spread-out entity. The contradiction was perhaps acceptable for these not-quite-real objects seen only as effects on macroscopic measuring apparatus (Heisenberg 1971).

Today, quantum weirdness is demonstrated in increasingly large systems (Haroche & Raimond 2006), and interpretations of "what it all means" proliferate (Elitzur et. al. 2006). Essentially every interpretation ultimately requires the intrusion of the conscious observer to account for the classical-like world of our experience (Squires 1994, Penrose, R. 2005).

The quantum measurement problem is often considered a problem of the quantum theory: How to explain the collapse of the multiple possibilities of the wavefunction to a single observed actuality. This is indeed unresolved (Squires, 1993). However, the measurement problem also arises directly from the quantum-theory-neutral experiment, and depends crucially on the assumption of free will of the experimenter. We present a version of the archetypal quantum experiment illustrating the intrusion of the conscious observer into the experiment.

2. THE ARCHETYPAL QUANTUM EXPERIMENT

According to Richard Feynman, "[The two-slit experiment] contains the only mystery. We cannot make the mystery go away by "explaining" how it works. . . In telling you how it works we will have told you about the basic peculiarities of all quantum mechanics" (Feynman, et. al. 2006).

In the two-slit experiment one can choose to demonstrate either of two contradictory things: that each object was a compact entity coming through a single slit or that each was a spread out entity coming through both. Similar experiments have been done with photons, electrons, atoms, large molecules, and are being attempted with yet larger objects such as live viruses (Clauser 2010). We will just refer to "objects."

We present an equivalent version of the two-slit experiment in which one can choose to show that an object was wholly in a single box (Rosenblum & Kuttner 2002). But one could have chosen to show that that same object was not wholly in a single box. By telling the story with objects captured in boxes, one can decide at leisure which of the two contradictory situations to demonstrate for each isolated object. This displays the quantum challenge to our intuition that an observer-independent physical reality exists "out there." We describe quantum-theory-neutral experiments, by telling only what could be directly observed.

The experimenter is presented with a set of box pairs, say twenty pairs of boxes. Each pair of boxes contains a single object. How the box pairs were prepared is irrelevant for these quantum-theory-neutral experiments. Since it's easiest to describe the preparation in quantum language, we do that in the next section.

The "Which Box" Experiment The experimenter is instructed to determine, for this set of box pairs, which box of each pair contains the object. He does this by placing each box pair in turn in the same position in front of a screen that an object would mark on impact. He then cuts a narrow slit first in one box of the pair, and then the other. For some box pairs, an impact occurs only on opening the first box, and then not on opening the second box. For others, the impact occurs only on the second opening. In this "which box" experiment, the experimenter thus determines which box of each pair contained the object, and which box was empty. The experimenter establishes that for this set of box pairs, each object had been wholly in a single box of its pair.

Repeating this with box pairs placed in the same position in front of the screen, the experimenter notes a more or less random spread of marks on the screen.

The "Interference" Experiment The experimenter is presented with second set of box pairs. This time he is instructed to cut slits in both boxes of each pair at about the same time. He does so, positioning each box pair in the same position in front of the screen. (It's an interference experiment, and we so name it. But we make no reference to waves.)

This time the objects do not impact randomly. There are regions where many objects land, and regions where none land. Each object followed a rule specifying the regions in which it was allowed to land.

To investigate the nature of this rule, the experimenter repeats the experiment with different spacings of the boxes of a pair from each other. He finds that the rule each object follows depends on the spacing of its box pair. Each object "knows" the spacing of its box pair. Something of each object therefore had to have been in each box of its pair. The experimenter establishes that–unlike the previous set, for which objects were each wholly in a single box–for this set of box pairs, objects were not wholly in a single box.

The Free Choice Of Experiment The experimenter is reminded that he established that the objects in the first set of box pairs were wholly in a single box, while the objects in the second set of box pairs were not wholly in a single box. Now presented with a third set of box pairs, he is asked to establish whether the objects in this set of box pairs are, or are not, wholly in a single box.

The experimenter arbitrarily chooses to do the "which box" experiment, and thus establishes that this box-pair set had contained objects wholly in a single box. Given another set of box pairs and similar instructions, he chooses the "interference" experiment, and establishes that the objects in this set were not wholly in a single box.

Offered further sets of box pairs, the experimenter finds that each time he chooses to do a "which box" experiment, objects were wholly in a single box. Each time he chooses an "interference" experiment, he establishes a contradictory physical situation, that objects were not wholly in a single box. His free choice of experiment seemed to create the prior history of what had been in the boxes. He's baffled.

If the experimenter's choice of experiment were predetermined to match what was actually in each box pair set, he would see no problem. He recognizes this, but he is certain his choices were freely made. His conscious certainty of his free will causes him to experience a measurement problem with the archetypal quantum experiment. In fact, the free choice of the observation set is generally recognized as an essential aspect of any inductive science.

No experiment in classical physics raised the issue of free will. In classical physics, questions of free will arose only out of an ignorable aspect of the deterministic theory.

We emphasize that in describing the intrusion of the observer into the archetypal quantum experiment we never referred to quantum theory, wavefunctions, or waves of any kind. Even were quantum theory never invented, one could do these experiments, and the results would present an inexplicable enigma (Greenstein & Zajonc 1997).

3. THE BOX-PAIR EXPERIMENT IN QUANTUM THEORY

The Preparation Of The Box Pairs Objects are sent one at a time, at a known speed, toward a "mirror" that equally transmits and reflects their wavefunction. In Figure 1, a wavefunction is shown at three successive times. The reflected part is subsequently reflected so that each part is directed into one of a pair of boxes. The doors of the boxes are closed at a time when the wavefunction is within the boxes.

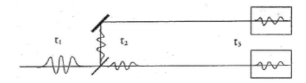

Figure 1. Schematic diagram of the preparation of the box pairs. The object's wavefunction is shown at three successive times.

Dividing a wavefunction into well-separated regions is part of every interference experiment. Holding an object in a box pair without disturbing its wavefunction would be tricky, but doable in principle. Capturing an object in physical boxes is not actually required for our demonstration. A sufficiently extended path length would be enough. The box pair is a conceptual device to emphasize that a conscious choice can be made while an object exists as an isolated entity.

Observation In the box-pairs version of the two-slit experiment, the wavefunction spreads widely from the small slit in a box. In the "which box" experiment, it emerges from each single box and impinges rather uniformly on the detection screen. In the "interference" experiment, parts of the wavefunction emerge simultaneously from both boxes and combine to form regions of maxima and minima on the detection screen.

The Born postulate has the absolute square of the wavefunction in a region giving the probability of an object being "observed" there. In the Copenhagen interpretation of quantum mechanics, observation takes place, for all practical purposes, as soon as the microscopic quantum object encounters the macroscopic screen (Griffiths 1995, Stapp 1972). Macroscopic objects are then assumed to be classical-physics objects viewed by the experimenter. Other interpretations of quantum mechanics, attempting to go beyond practical purposes, consider observation to be more involved with the actual conscious experience of the experimental result.

History Creation Finding an object in a single box means the whole object came to that box on a particular single path after its earlier encounter with the semi-transparent mirror. Choosing an interference experiment would establish a different history: that aspects of the object came on two paths to both boxes after its earlier encounter with the semi-transparent mirror. (As noted above, the question of history creation also arose in the quantum-theory-neutral experiment.) Quantum cosmologist John Wheeler (1980) suggested that quantum theory's history creation be tested. He would have the choice of which experiment to do delayed until after the object made its "decision" whether to come on a single path or whether to come on both at the semi-transparent mirror. The experiment was done with photons and a mirror arrangement much like our Figure 1. Getting the same results as in the usual quantum experiment would imply that the relevant history was indeed created by the later choice of experiment.

For a human to make a conscious choice of which experiment to do takes perhaps a second, in which a photon travels 186,000 miles. Therefore the actual "choice" of experiment was made by a fast electronic switch making random choices. The most rigorous version of the experiment was done in 2007 (Jacques et al., 2007), when reliable single-photon pulses could be generated, and fast enough electronics were available. The result (of course?) confirmed quantum theory's predictions. Observation created the relevant history.

Non-locality and Connectedness When an object is observed to be in a particular location, its probability of being elsewhere becomes zero. Its wavefunction elsewhere "collapses" to zero, and to unity (a certainty) in the location in which the object was found. If an object is found to be in one box of its pair, its wavefunction in the other box instantaneously becomes zero—no matter how far apart the boxes are.

In its usual interpretation, quantum theory does not include an object in addition to the wavefunction of the object. The wavefunction is, in this sense, the physical entity itself. Thus the wavefunction being affected by observation everyplace at once is problematic in the light of special relativity, which prohibits any matter, or any message, to travel faster than the speed of light. The non-local, instantaneous collapse of the wavefunction on observation poses the quantum measurement problem as viewed from quantum theory.

The instantaneous, non-local collapse of the wavefunction provoked Einstein to challenge the completeness of quantum theory with the famous EPR paper (Einstein, et al. 1935). To avoid what Einstein later derided as "spooky action at a distance," EPR held that there must be properties at the microscopic level that quantum theory did not include. However, since EPR provided no experimental challenge to quantum theory, it was largely ignored by physicists for three decades as merely arguing a philosophical issue.

John Bell (1964) proved a theorem allowing experimental tests establishing the existence of an instantaneous connectedness, Einstein's "spooky action." The experiments (Freedman & Clauser, 1972; Aspect et al. 1984) showed that if objects had interacted, what an observer chose to observe about one of them would instantaneously influence the result that an arbitrarily remote observer chose to observe for the other object.

The Bell results included the assumption of the free will of the observers, that their choices of what to observe were independent of each other and independent of all prior physical events. Denying that assumption would be "more mind boggling" than the connectedness the denial attempts to avoid. Such denial would imply, Bell wrote: "Apparently separate parts of the world would be conspiratorially entangled, and our apparent free will would be entangled with them" (Bell 1981).

4. THE ROBOT FALLACY

The most common argument that consciousness is not involved in the quantum experiment is that a not-conscious robot could do the experiment. However, for any experiment to be meaningful, a human must eventually evaluate it. A programmed robot sees no enigma. Consider the human evaluation of the robot's experiment:

The robot presents a printout to the human experimenter. It shows that with some sets of box pairs the robot chose a which-box experiment, establishing that the objects were wholly in a single box. With other sets of box pairs, choosing the interference experiment, it established that the objects were not wholly in a single box.

On the basis of this data, the human experimenter could conclude that certain box-pair sets actually contained objects wholly in a single box, while others contained objects not wholly in a single box. However, a question arises in the mind of the experimenter: How did the robot choose the appropriate experiment with each box-pair set? What if, for example, the robot chose a which-box experiment with objects not wholly in a single box? A partial object was never reported.

Without free will, the not-conscious robot must use some "mechanical" choice procedure. Investigating, the experimenter finds, for example, that it flips a coin. Heads, a which-box experiment; tails interference. The experimenter is troubled by the mysterious correlation between the landing of the coin and what was presumably actually in a particular box-pair set.

To avoid that inexplicable correlation, the experimenter replaces the robot's coin flipping with the one choice method she is most sure is not correlated with the contents of a box-pair set: her own free choice. Pushing a button telling the robot which experiment to do, she establishes what she would by doing the experiment directly, that by her conscious free choice she can establish either of two contradictory physical situations. In the end, the robot argument establishes nothing.

5. CONCLUSION

Extending the implications of quantum mechanics beyond the microscopic realm admittedly leads to ridiculous-seeming conclusions. Nevertheless, the experimental results are undisputed, and quantum theory is the most basic and most battle-tested theory in all of science.

The embarrassing intrusion of the conscious observer into physics can be mitigated by focusing on observation in quantum theory, the collapse (or decoherence) of the wavefunction. However, the inescapable assumption of free choice by the experimenter displays the intrusion of the observer in the quantum-theory-neutral quantum experiment, logically prior to the quantum theory.

The intrusion was less disturbing when confined to never-directly-observed microscopic objects. However, the vast no-man's-land that once separated the microscopic and the macroscopic realms, allowing a tacit acceptance of this view, has been invaded by technology.

Bell's theorem, and the experiments it stimulated, seems to rule out a resolution of the quantum measurement problem by the existence of an underlying structure, somehow involving only properties localized in quantum objects. An overarching structure, somehow involving conscious free will, seems required (Squires 1991).

Acknowledgments: Parts of this article have been taken from the second edition of our book, Quantum Enigma: Physics Encounters Consciousness, with the permission of Oxford University Press.

References

Aspect A., Dalibard J., Roger G. (1982). Experimental Test Of Bell Inequalities Using Time-Varying Analyzers. Physical Review Letters 49, 1804-1807.

Bell, J.S. (1964). On the Problem of Hidden Variables in Quantum Mechanics. Physics 1, 195.

Bell, J. S. 1981. Bertlmann's Socks and the Nature of Reality. Journal de physique 42, 41.

Clauser, J.F. (2010). deBroglie Wave Interference of Small Rocks and Live Viruses. In: Cohen, R.S., Horne, M., Stachel, J. (Eds.), Experimental Metaphysics: Quantum Mechanical Studies for Abner Shimony, Volume One. Kluwer Academic, Dordrecht, The Netherlands, pp. 1-12.

Einstein, A., Podolsky, B., Rosen, N. (1935). Can quantum-mechanical description of physical reality be considered complete? Physical Review 47, 777-780.

Elitzur, A. Dolev, S., Kolenda, A. (Eds), (2006). Quo Vadis Quantum Mechanics. Springer, Berlin.

Feynman, R.P., Leighton, R.B., Sands, M. (2006). The Feynman Lectures on Physics, Vol. III. Addison-Wesley, Reading, MA.

Freedman, S.J., Clauser, J. F. (1972). Experimental Test Of Local Hidden-Variable Theories. Physical Review Letters 28, 938-941.

Greenstein, G., Zajonc, A.G. (1997). The Quantum Challenge: Modern Research on the Foundations of Quantum Mechanics. Jones and Bartlett, Sudbury, MA.

Griffiths, D.J. (1995). Introduction to Quantum Mechanics. Prentice Hall, Englewood Cliffs, NJ.

Haroche, S., Raimond, J-M. (2006). Exploring the Quantum: Atoms, Cavities, and Photons. Oxford University Press, Oxford.

Heisenberg, W. (1971) Physics and Beyond: Encounters and Conversations. Harper & Row, New York.

Jacques, V., Wu, E., Grosshans, F., Treussart, F., Grangier, P., Aspect, A., Roch, J-F. (2007). Experimental realization of Wheeler's delayed-choice gedanken experiment. Science 315, 966-968.

Penrose, R. (2006). Foreword. In: Elitzur, A. Dolev, S., Kolenda, A. (Eds), Quo Vadis Quantum Mechanics. Springer, Berlin, pp. v-viii.

Rosenblum, B., Kuttner, F. (2002) The Observer in the Quantum Experiment. Foundations of Physics 32, 1273-1293.

Squires, E.J. (1991). One Mind Or Many - A Note On The Everett Interpretation Of Quantum-Theory. Synthese, 89, 283-286.

Squires, E.J. (1993). Quantum-Theory And The Relation Between The Conscious Mind And The Physical World. Synthese, 97, 109-123.

Squires, E.J. (1994). Quantum Theory And The Need For Consciousness. Journal of Consciousness Studies 1, 201-204.

Stapp, H.P. (1972). The Copenhagen Interpretation. American Journal of Physics 40, 1098-1115.

Wheeler (1980). Delayed Choice Experiments and the Bohr-Einstein Dialog. In: The American Philosophical Society and the Royal Society: Papers Read at a Meeting June 5, 1980. American Philosophical Society, Philadelphia, PA.

48. Does Quantum Mechanics Require A Conscious Observer?
Michael Nauenberg

Physics Dept. University of Califonia Santa Cruz, CA, USA

Abstract

The view that the implementation of the principles of quantum mechanics requires a conscious observer is based on misconceptions that are described in this article.

KEY WORDS: Quantum Physics, Wave function, Observer, Consciousness

The notion that the interpretation of quantum mechanics requires a conscious observer is rooted, I believe, in a basic misunderstanding of the meaning of a) the quantum wavefunction ψ, and b) the quantum measurement process. This misunderstanding originated with the work of John von Neumann (1932) on the foundations of quantum mechanics, and afterwards it was spread by some prominent physicists like Eugene Wigner (1984); by now it has acquired a life of its own, giving rise to endless discussions on this subject, as shown by the articles in the Journal of Cosmology (see volumes 3 and 14).

Quantum mechanics is a statistical theory that determines the probabilities for the outcome of a physical process when its initial state has been determined. A fundamental quantity in this theory is the wavefunction ψ which is a complex function that depends on the variables of the system under consideration. The absolute square of this function, ψ^2, gives the probability to find the system in one of its possible quantum states. Early pioneers in the development of quantum mechanics like Niels Bohr (1958) assumed, however, that the measurement devices behave according to the laws of classical mechanics, but von Neumann pointed out, quite correctly, that such devices also must satisfy the principles of quantum mechanics. Hence, the wavefunction describing this device becomes entangled with the wavefunction of the object that is being measured, and the superposition of these entangled wavefunctions continues to evolve in accordance with the equations of quantum mechanics. This analysis leads to the notorious von Neumann chain, where the measuring devices are left forever in an indefinite superposition of quantum states. It is postulated that this chain can be broken, ultimately, only by the mind of a conscious observer.

Forty five years ago I wrote an article on this subject with John Bell who became, after von Neumann, the foremost contributor to the foundations of quantum mechanics, where we presented, tongue in cheek, the von Neumann paradox as a dilemma:

> The experiment may be said to start with the printed proposal and to end with the issue of the report. The laboratory, the experimenter, the administration, and the editorial staff of the Physical Review are all just part of the instrumentation. The incorporation of (presumably) conscious experimenters and editors into the equipment raises a very intriguing question... If the interference is destroyed, then the Schrodinger equation is incorrect for systems containing consciousness. If the interference is not destroyed, the quantum mechanical description is revealed as not wrong but certainly incomplete (Bell and Nauenberg, 1966).

We added the remark that "we emphasize not only that our view is that of a minority, but also that current interest in such questions is small. The typical physicist feels that they have been long answered, and that he will fully understand just how, if ever he can spare twenty minutes to think about it." Now the situation has changed dramatically, and interest in a possible role of consciousness in quantum mechanics has become widespread. But Bell, who died in 1990 , believed in the second alternative to the von Neumann dilemma, remarking that :

> I think the experimental facts which are usually offered to show that we must bring the observer into quantum theory do not compel us to adopt that conclusion (Davies and Brown, 1986).

Actually, by now it is understood by most physicists that von Neumann's dilemma arises because he had simplified the measuring device to a system with only a few degrees of freedom, e.g. a pointer with only two states (see Appendix). Instead, a measuring device must have an exponentially large number of degrees of freedom in order to record, more or less permanently, the outcome of a measurement. This recording takes place by a time irreversible process. The occurrence of such processes in Nature already mystified 19th century scientists, who argued that this feature implied a failure in the basic laws of classical physics, because these laws are time reversible. Ludwig Boltzmann resolved this paradox by taking into account the large number of degrees of freedom of a macroscopic system, which implied that to a very high degree of probability such a system evolved with a unique direction in time. Such an irreversibility property is also valid for quantum systems, and it constitutes the physical basis for the second law of thermodynamics, where the arrow of time is related to the increase of entropy of the system.

Another misconception is the assumption that the wavefunction ψ describing the state of a system in quantum mechanics behaves like a physical object. For example, the authors of a recent book discussing quantum mechanics and consciousness claim that

> In quantum theory there is no atom in addition to the wavefunction of the atom. This is so crucial that we say it again in other words. The atom's wave-functions and the atom are the same thing; "the wave function of the atom" is a synonym for "the atom". Since the wavefunction ψ is synonymous with the atom itself, the atom is simultaneously in both boxes. The point of that last paragraph is hard to accept. That is why we keep repeating it (Rosenblum and Kuttner, 2006).

If the wavefunction ψ is a physical object like an atom, then the proponents of this flawed concept must require the existence of a mechanism that lies outside the principles governing the time evolution of the wavefunction ψ in order to account for the so-called "collapse" of the wavefunction after a measurement has been performed. But the wavefunction ψ is not a physical object like, for example, an atom which has an observable mass, charge and spin as well as internal degrees of freedom. Instead, ψ is an abstract mathematical function that contains all the statistical information that an observer can obtain from measurements of a given system. In this case there isn't any mystery that its mathematical form must change abruptly after a measurement has been performed. For further details on this subject, see (Nauenberg, 2007) and (van Kampen, 2008). The surprising fact that mathematical abstractions can explain and predict real physical phenomena has been emphazised by Wigner (Wigner 1960), who wrote:

> The miracle of appropriateness of the language of mathematics for the formulation of the laws of physics is a wonderful gift which we neither undestand nor deserve.

I conclude with a few quotations, that are relevant to the topic addressed here, by some of the most prominent physicists in the second half of the 20th century.

Richard P. Feynman (Nobel Prize, 1965):

> Nature does not know what you are looking at, and she behaves the way she is going to behave whether you bother to take down the data or not (Feynman et al., 1965).

Murray Gellmann (Nobel Prize, 1969):

> The universe presumably couldn't care less whether human beings evolved on some obscure planet to study its history; it goes on obeying the quantum mechanical laws of physics irrespective of observation by physicists (Rosenblum and Kuttner 2006, 156).

Anthony J. Leggett (Nobel Prize 2003):

> It may be somewhat dangerous to explain something one does not understand very well [the quantum measurement process] by invoking something [consciousness] one does not understand at all! (Leggett, 1991).

John A. Wheeler:

> Caution: "Consciousness" has nothing whatsover to do with the quantum process. We are dealing with an event that makes itself known by an irreversible act of amplification, by an indelible record, an act of registration. Does that record subsequently enter into the "consciousness" of some person, some animal or some computer? Is that the first step into translating the measurement into "meaning" meaning regarded as "the joint product of all the evidence that is available to those who communicate." Then that is a separate part of the story, important but not to be confused with "quantum phenomena." (Wheeler, 1983).

John S. Bell:

> From some popular presentations the general public could get the impression that the very existence of the cosmos depends on our being here to observe the observables. I do not know that this is wrong. I am inclined to hope that we are indeed that important. But I see no evidence that it is so in the success of contemporary quantum theory.

So I think that it is not right to tell the public that a central role for conscious mind is integrated into modern atomic physics. Or that `information' is the real stuff of physical theory. It seems to me irresponsible to suggest that technical features of contemporary theory were anticipated by the saints of ancient religions... by introspection.

The only 'observer' which is essential in orthodox practical quantum theory is the inanimate apparatus which amplifies the microscopic events to macroscopic consequences. Of course this apparatus, in laboratory experiments, is chosen and adjusted by the experiments. In this sense the outcomes of experiments are indeed dependent on the mental process of the experimenters! But once the apparatus is in place, and functioning untouched, it is a matter of complete indifference - according to ordinary quantum mechanics - whether the experimenters stay around to watch, or delegate such 'observing' to computers, (Bell, 1984).

Nico van Kampem:

> Whoever endows ψ with more meaning than is needed for computing observable phenomena is responsible for the consequences. (van Kampen, 1988).

Appendix. Schrodinger's Cat: This cat story is notorious. It requires one to accept the notion that a cat, which can be in innumerable different biological states, can be represented by a two component wavefunction ψ, a bit of nonsense that Erwin Schrodinger, one of the original inventors of quantum mechanics, himself originated. One of the two components represents a live cat, and the other a dead cat. The cat is enclosed in a box containing a bottle filled with cyanide that opens when a radioactive nucleus in the box decays. Thus, this fictitious cat is a measuring device that is supposed to determine whether the nucleus has decayed or not when the box is opened. But according to the principles of quantum mechanics formulated by von Neumann, such a cat ought to be in a superposition of life and dead cat states, yet nobody has ever observed such a cat. Instead, it is expected that a movie camera - a real measuring device - that is also installed in the box containing the cat, would record a cat that is alive until the unpredictable moment that the radioactive nucleus decays, opening the bottle containing the cyanide that kills the cat. For obvious reasons such a gruesome experiment has never been performed. It is claimed that Schrodinger never accepted the statistical significance of his celebrated wavefunction.

References

Bohr, N. (1958). Quantum Physics and Philosophy, Causality and Complementarity in Essays 1958/1962 on Atomic Physics and Human Knowledge. Vintage Books

Bell, J. S., Nauenberg, M. (1966). The moral aspects of quantum me- chanics, in Preludes in Theoretical Physics, edited by A. De Shalit, Herman Feschbach, and Leon van Hove (North Holland, Amsterdam), pp. 279-286. Reprinted in J.S. Bell Speakable and Unspeakable in Quantum Mechanics (Cambridge Univ. Press 1987) p. 22

Bell, J. S. (1987). Introductory remarks at Naples-Amal meeting, May 7, 1984. In: Bell, J.S. Speakable and Unspeakable in Quantum Mechanics. Cambridge Univ. Press, p.170,

Davies, P.C.W., Brown, J.T. (1986). Ghost in the Atom. Cambridge Univ. Press, Interview with J. Bell, pp. 47-48

Feynman, R.P., Leighton, R.B., Sands,M. (1965). The Feynman lectures on Physics vol. 3 (Addison Wesley, Reading) 3-7

Leggett, A. (1991) Reflections on the Quantum Paradox, In: Quantum Implications, Routledge, London, p. 94

Nauenberg, M. (2007). Critique of Quantum Enigma: Physics encounters Consciousness, Foundations of Physics 37, 1612-162

Rosenblum, B and Kuttner, F. (2006). Quantum Enigma, Physics Encounters Consciousness. Oxford Univ. Press, p. 106

van Kampen, N. G. (1988). Ten theorems about quantum mechanical measurements Physica A 153, 97 .

van Kampen, N.G. (2008) The Scandal in Quantum Mechanics, American Journal of Physics 76, 989

von Neumann, J. (1932) Measurement and Reversibility, Chapters V and VI of Mathematische Grundlagen der Quantemmechanik, translated into English by R.T. Mayer, Mathematical Foundations of Quantum Mechanics, Princeton Univ. Press, Princeton (1955) pp. 347-445

Wigner, E. (1984). Review of the Quantum-Mechanical Measurement Problem, Science, Computers and the Information Onslaught, eds. D.M. Kerr et al.. Academic Press, New York, pp. 63-82 Reprinted in "The Collected Works of Eugene Paul Wigner", Part B, vol. 6, Springer-Verlag, Berlin, p. 240

Wheeler, J. A. (1983). Law without law. In: Quantum Theory and Measurement, edited by Wheeler, J.A. and Zurek, W.H., Princeton Univ. Press, Princeton, p. 196.

Wigner, E. (1960) The Unreasonable Effectiveness of Mathematics, Communications in Pure and Applied Mathematics 13, 1-14.

49. Quantum Physics, Advanced Waves and Consciousness

Antonella Vannini Ph.D[1], and Ulisse Di Corpo[2]

[1]Lungotevere degli Artigiani 32 – 00153 ROME - ITALY
[2]Lungotevere degli Artigiani 32 – 00153 ROME - ITALY

Abstract

An essential component of the Copenhagen Interpretation of quantum mechanics is Schrödinger's wave equation. According to this interpretation, consciousness, through the exercise of observation, forces the wave function to collapse into a particle. Schrödinger's wave equation is not relativistically invariant and when the relativistically invariant wave equation (Klein-Gordon's equation) is taken into account, there is no collapse of the wave function and no justification for consciousness as a prerequisite to reality. Klein-Gordon's wave equation depends on a square root and yields two solutions: retarded waves which move forwards in time and advanced waves which move backwards in time. Advanced waves were considered to be unacceptable since they contradict the law of causality, according to which causes always precede effects. However, while studying the mathematical properties of Klein-Gordon's equation, the mathematician Luigi Fantappiè noted that retarded waves are governed by the law of entropy (from Greek en=diverge, tropos=tendency), whereas advanced waves are governed by a law opposite to entropy which leads to concentration of energy, differentiation, complexity, order and growth of structures. Fantappiè named this law syntropy (syn=converge, tropos=tendency) and noted that its properties coincide with the qualities of living systems, arriving in this way at the conclusion that life and consciousness are a consequence of advanced waves (Fantappiè, 1942).

KEY WORDS: Consciousness, Advanced waves, Syntropy, Feeling of Life, Free Will, Quantum Physics

1. Introduction

The Copenhagen Interpretation of quantum mechanics was formulated by Niels Bohr and Werner Heisenberg in 1927 during a joint work in Copenhagen, and explains the dual nature of matter (wave/particle) in the following way:

- Electrons leave the electronic cannon as particles.
- They dissolve into waves of superposed probabilities, in a superposition of states.
- The waves go through both slits, in the double slit experiment, and interfere, creating a new state of superposition.
- The observation screen, performing a measurement, forces the waves to collapse into particles, in a well defined point of the screen.
- Electrons start again to dissolve into waves, just after the measurement.

An essential component of the Copenhagen Interpretation is Schrödinger's wave equation, reinterpreted as the probability that the electron (or any other quantum mechanical entity) is found in a specific place. According to the Copenhagen Interpretation, consciousness, through the exercise of observation, forces the wave function to collapse into a particle. This interpretation states that the existence of the electron in one of the two slits, independently from observation, does not have any real meaning. Electrons seem to exist only when they are observed. Reality is therefore created, at least in part, by the observer.

In the paper Quantum Models of Consciousness it is argued that quantum models of consciousness can be divided in three main categories (Vannini, 2008):

1. models which assume that consciousness creates reality and that consciousness is a prerequisite of reality;
2. models which link consciousness to the probabilistic properties of quantum mechanics;
3. models which attribute consciousness to a principle of order of quantum mechanics.

Considering the criteria of scientific falsification and of biological compatibility Vannini (2008) notes that:

- Quantum models of consciousness which belong to the first category show a tendency towards mysticism. All these models start from the Copenhagen Interpretation of quantum mechanics and assume that consciousness itself determines reality. These models try to describe reality as a consequence of panpsychism, and assume that consciousness is an immanent property which precedes the formation of reality. The concept of panpsychism is explicitly used by most of the authors of this category. These assumptions cannot be falsified.

- Quantum models of consciousness which belong to the second category consider consciousness to be linked to a realm, for example that of the Planck's constant, which cannot be observed by modern science and which is impossible to falsify or test using experiments.
- Quantum models of consciousness which belong to the third group attribute consciousness to principles of order which have been already discovered and used for physical applications (laser, superconductors, etc.). The order principles on which most of these models are based require extreme physical conditions such as, for example, absolute zero temperatures (-273 C°). These models do not meet the criteria of biological compatibility.

Vannini concludes that only the models which originate from the Klein-Gordon equation, which unites Schrödinger's wave equation (quantum mechanics) with special relativity and are not pure quantum mechanical models, survive the selection of scientific falsification and biological compatibility.

2. Klein-Gordon's Equation

In 1925 the physicists Oskar Klein and Walter Gordon formulated a probability equation which could be used in quantum mechanics and was relativistically invariant. In 1926 Schrödinger simplified Klein-Gordon's equation in his famous wave equation (ψ) in which only the positive solution of Klein-Gordon's equation was considered, and which treats time in an essentially classical way with a well defined before and after the collapse of the wave function. In 1927 Klein and Gordon formulated again their equation (2) as a combination of Schrödinger's wave equation (quantum mechanics) and the energy/momentum/mass equation of special relativity (1).

$$E^2 = c^2 p^2 + m^2 c^4 \qquad \text{1)}$$

Energy/momentum/mass equation

Where E is the Energy of the object, m the mass, p the momentum and c the constant of the speed of light. This equation simplifies in the famous $E=mc^2$ when the momentum is equal to zero (p=0).

$$E\psi = \sqrt{p^2 + m^2}\psi \qquad \text{2)}$$

Klein-Gordon's wave equation

Klein-Gordon's wave equation depends on a square root and yields two solutions: the positive solution describes waves which diverge from the past to the future (retarded waves); the negative solution describes waves which diverge from the future to the past (advanced waves). The negative solution introduces in science final causes and teleological tendencies. Consequently, it was considered to be unacceptable.

In 1928 Paul Dirac tried to get rid of the unwanted negative solution by applying the energy/momentum/mass equation to the study of electrons, turning them into relativistic objects. But, also in this case, the dual solution emerged in the form of electrons (e^-) and antiparticles (e^+). The antiparticle of the electron, initially named neg-electron, was experimentally observed in 1932 by Carl Anderson in cosmic rays and named positron. Anderson became the first person who proved empirically the existence of the negative energy solution; the negative solution was no longer an impossible mathematical absurdity, but it was an empirical evidence. Dirac's equation predicts a universe made of matter which moves forwards in time and antimatter which moves backwards in time. The negative solution of Dirac's equation caused emotional distress among physicists. For example Heisenberg wrote to Pauli: "The saddest chapter of modern physics is and remains the Dirac theory" (Heisenberg, 1928); "I regard the Dirac theory ... as learned trash which no one can take seriously" (Heisenberg, 1934). In order to solve this situation, Dirac used Pauli's principle, according to which two electrons cannot share the same state, to suggest that all states of negative energy are occupied, thereby forbidding any interaction between positive and negative states of matter. This ocean of negative energy which occupies all positive states is called the Dirac sea.

It is important to note that it appears to be impossible to test the existence of advanced waves in a laboratory of physics:

- According to Fantappiè, advanced waves do not obey classical causation, therefore they cannot be studied with experiments which obey the classical experimental method (Fantappiè, 1942).

- According to Wheeler's and Feynman's electrodynamics, emitters coincide with retarded fields, which propagate into the future, while absorbers coincide with advanced fields, which propagate backward in time. This time-symmetric model leads to predictions identical with those of conventional electrodynamics. For this reason it is impossible to distinguish between timesymmetric results and conventional results (Wheeler & Feynman, 1949).

- In his Transactional Interpretations of Quantum Mechanics, Cramer states that "Nature, in a very subtle way, may be engaging in backwards-in-time handshaking. But the use of this mechanism is not available to experimental investigators even at the microscopic level. The completed transaction erases all advanced effects, so that no advanced wave signalling is possible. The future can affect the past only very indirectly, by offering possibilities for transactions" (Cramer, 1986).

3. The Law of syntropy

At the end of 1941, the mathematician Luigi Fantappiè was working on the equations of relativistic and quantum physics when he noted that the dual solution of the Klein-Gordon equation explains two symmetrical laws:

- $+E\psi$ (retarded waves) describes waves diverging from causes located in the past, governed by the law of entropy;
- $-E\psi$ (advanced waves) describes waves converging towards causes located in the future and governed by the law of syntropy.

According to Fantappiè the main properties of retarded and advanced waves are:
1. Retarded waves:
   a. Causality: diverging waves exist as a consequence of causes located in the past.
   b. Entropy: diverging waves tend towards the dissipation of energy (heat death).
2. Advanced waves:
   a. Retrocausality: converging waves exist as a consequence of causes located in the future.
   b. Syntropy:
      - converging waves concentrate matter and energy in smaller spaces (ie this principle is well described by the large quantities of energy accumulated by living systems of the past and now available in the form of coal, petrol and gases).
      - Entropy diminishes. Entropic phenomena are governed by the second law of thermodynamics according to which a system tends towards homogeneity and disorder. The inversion of the time arrow also inverts the second law of thermodynamics, so that a reduction in entropy and an increase in differentiation are observed.
      - Final causes, attractors, which absorb converging waves are observed. From these final causes syntropic systems originate.
      - Because syntropy leads to the concentration of matter and energy, and this concentration cannot be indefinite, entropic processes are needed to compensate syntropic concentration. These processes take the form of the exchange of matter and energy

with the environment. For example metabolism is divided into:
- anabolism (syntropy) which includes all the processes which transform simple structures into complex structures, for example nutritive elements into bio-molecules, with the absorption of energy.
- catabolism (entropy) which includes all the processes which transform higher level structures into lower level structures, with the release of energy.

Fantappiè noted that the properties of syntropy coincide with the qualities of living systems: finality, differentiation, order and organization.

Other authors suggested the existence of the law of syntropy associated to living systems. For example:

- Albert Szent-Gyorgyi (Nobel prize 1937 and discoverer of vitamin C) underlined that "One major difference between amoebas and humans is the increase in complexity, which presupposes the existence of a mechanism which is capable of contrasting the second law of thermodynamics. In other words a force must exist which is capable of contrasting the universal tendency of matter towards chaos, and of energy towards heat death. Life processes continuously show a decrease in entropy and an increase in inner complexity, and often also in the complexity of the environment, in direct opposition with the law of entropy." In the 1970s Szent-Gyorgyi concluded that in living systems there was wide evidence of the existence of the law of syntropy, even though he never managed to infer it from the laws of physics. While entropy is a universal law which leads towards the disintegration of all types of organization, syntropy is the opposite law which attracts living systems towards forms of organization which are always more complex and harmonic (Szent-Gyorgyi, 1977). The main problem, according to Szent-Gyorgyi, is that "a profound difference between organic and inorganic systems can be observed ... as a man of science I cannot believe that the laws of physics lose their validity at the surface of our skin. The law of entropy does not govern living systems." Szent-Gyorgyi dedicated the last years of his life to the study of syntropy and its conflict with the law of entropy (Szent-Gyorgyi, 1977).

- Erwin Schrödinger talks about the concept of negative entropy. He was looking for the nutrient which is hidden in our food, and which defends us from heat death. Why do we need to eat biological food; why can we not feed directly on the chemical elements of matter? Schrödinger answers this question by saying that what we feed on is not matter but neg-entropy, which we absorb through the metabolic process (Schrödinger, 1944).

- Ilya Prigogine, winner in 1977 of the Nobel prize for chemistry, introduced in his book "The New Alliance", a new type of thermodynamics, the "thermodynamics of dissipative systems", typical of living systems. Prigogine stated that this new type of thermodynamics cannot be reduced to dynamics or thermodynamics (Prigogine, 1979).

- Hermann Haken, one of the fathers of the laser, introduced a level that he named "ordinator", which he used to explain the principles of orders typical of living systems (Haken, 1983).

4. Experiments

According to the Copenhagen Interpretation no advance effects should be possible, since time flows from the past to the future. On the contrary Fantappiè's syntropy model suggests that life and consciousness are a consequence of advanced waves (Fantappiè, 1942) and should therefore show anticipatory reactions. Is it possible to devise experiments in order to test which of the two models is correct?

In 1981 Di Corpo extended Fantappiè's hypothesis suggesting that structures which support vital functions, such as the autonomic nervous system (ANS), should show anticipatory reactions since they need to acquire syntropy. Consequently, if the Advanced Waves Interpretation is correct the parameters of ANS, such as heart rate and skin conductance, should react before stimuli (Di Corpo, 1981, 2007; Vannini & Di Corpo, 2008, 2009, 2010), on the contrary if the Copenhagen Interpretation is correct no reactions before stimuli should be observed.

Since 1997, anticipatory pre-stimuli reactions in the parameters of the autonomic nervous system have been reported in several studies, for example:

1. The first experimental study was produced by Radin in 1997 and monitored heart rate, skin conductance and fingertip blood volume in subjects who were shown for five seconds a blank screen and for three seconds a randomly selected calm or emotional picture. Radin found significant differences, in these autonomic parameters, preceding the exposure to emotional versus calm pictures. In 1997 Bierman replicated Radin's results confirming the anticipatory reaction of skin conductance to emotional versus calm stimuli and in 2003 Spottiswoode and May, of the Cognitive Science Laboratory, replicated Bierman and Radin's experiments performing controls in order to exclude artifacts and alternative explanations. Results showed an increase in skin conductance 2-3 seconds before emotional stimuli are presented ($p=0.0005$). Similar results have been obtained by other authors, using parameters of the autonomic nervous system (McDonough et al., 2002), (McCraty et al., 2004), (May Paulinyi & Vassy, 2005) and (Radin, 2005).

2. In the article "Heart Rate Differences between Targets and Nontargets in Intuitive Tasks", Tressoldi describes two experiments which show anticipatory heart rate reactions (Tressoldi et al., 2005). Trials were divided in 3 phases: in the presentation phase 4 pictures were shown and heart rate data was collected; in the choice phase pictures were presented simultaneously and the subject was asked to guess the picture which the computer would select; in the target phase the computer selected randomly one of the four pictures (target) and showed it on the monitor. In the first experiment a heart rate difference of 0.59 HR, measured in phase 1 during the presentation of target and non target pictures, was obtained (t = 2.42, p=0.015), in the second experiment the heart rate difference was 0.57 HR (t = 3.4, p=0.001).

3. Daryl Bem, psychology professor at the Cornell University, studies retrocausality using well known experimental designs in a "time-reverse" pattern. In his 2011 article "Feeling the Future: Experimental Evidence for Anomalous Retroactive Influence on Cognition and Affect" Bem describes 9 well-established psychological effects in which the usual sequence of events was reversed, so that the individual's responses were obtained before rather than after the stimulus events occurred. For example in a typical priming experiment the subject is asked to judge if the image is positive (pleasant) or negative (unpleasant), pressing a button as quickly as possible. The response time (RT) is registered. Just before the image a "positive" or "negative" word is briefly shown. This word is named "prime". Subjects tend to respond more quickly when the prime is congruent with the following image (both positive or negative), whereas the reaction times become longer when they are not congruent (one is positive and the other one is negative).

In retro-priming experiments Bem used IAPS (International Affective Picture System) emotional pictures. Results show the classical priming effect with reaction times faster when the prime is congruent with the image. Considering all 9 experiments, conducted on a sample of more than 1,000 students, the retrocausal effect size is $p = 1.34 \times 10^{-11}$.

4. In the article "Collapse of the wave function?" Vannini and Di Corpo describe 4 experiments which gradually control different types of artefacts and show a statistical significance of prestimuli heart rate effects of $p=1/10^{27}$ (Vannini & Di Corpo, 2010).

5. How Can These Results Be Interpreted?

Anticipatory pre-stimuli reactions seem to be incompatible with the Copenhagen Interpretation, since Schrödinger's wave equation treats time in an essentially classical way and rejects the possibility of pre-stimuli reactions (effects before causes). Dick Bierman tried to overcome this limit of the Copenhagen Interpretation with his CIRTS model (Consciousness Induced Restoration of Time Symmetry), presented at the PA 2008 conference (Bierman, 2008). This model states that almost all formalisms in physics are time-symmetric. Nevertheless the Copenhagen Interpretation of quantum mechanics, which postulates the collapse of the wave function, introduces a break of time symmetry at the point of collapse. The assumption of CIRTS is that the brain, when it is sustained by consciousness, is such a special system that it partially restores time-symmetry and therefore allows advanced waves to occur. The time symmetry restoring condition is not the brain per se but the brain sustained by consciousness. The restoration of time symmetry is suggested to be proportional to the brain volume involved in consciousness. CIRTS considers consciousness to be a pre-requisite of reality with special properties which restore time-symmetry. However, in CIRTS the rationale behind consciousness is missing and its special properties seem to arise from nothing. Contrary to Bierman's model, Luigi Fantappiè's syntropy model and Chris King's quantumtransactions model describe consciousness as a consequence of the properties of advanced waves: – Fantappiè states that, according to the converging properties of advanced waves, living systems are energy and information absorbers and that the "feeling of life" can be described as a consequence of these converging and absorbing properties of advanced waves. On the contrary it would be difficult to justify the feeling of life as a consequence of diverging and emitting properties which characterize retarded waves. The equivalence "feeling of life = advanced waves" leads to the conclusion that systems based on the retarded solution, as for example machines and computers, would never show the "feeling of life" independently from their complexity, whereas systems based on the advanced solution, as for example life itself, should always have a "feeling of life", independently from their complexity.

According to King, the constant interaction between information coming from the past and information coming from the future would place life in front of bifurcations. This constant antagonism between past and future would force life into a state of free will and consciousness. Consequently consciousness would be a property of all living structures: each cell and biological process would be forced to choose between information coming from the past and information coming from the future (King, 1996). This constant state of choice would be common to all levels of life and would give form to chaotic behaviour on which the conscious brain would feed. King (1996) states that "The chaotic processes which are observed in the neuronal system can be the result of behaviour which is apparently random and probabilistic, since they are non local in space and time. This would allow neuronal networks to connect in a subquantum way with non local situations and explain why behaviour results in being non deterministic and non computational."

The followings are some of the fundamental differences between Bierman's CIRTS model and Fantappiè's syntropy model:

1. Fantappiè focused on the Klein-Gordon's equation and excluded other time-symmetric equations, such as the electromagnetic wave equation. The rational of this choice is that at the quantum level time would be unitary (past, present and future would coexist) whereas at the macro-level time flows forward and advanced waves would be impossible. This conclusion was reached considering the mathematical properties of retarded waves which obey classical causation and propagate from the past to the future, and of advanced waves which obey final causation and propagate from the future to the past. Fantappiè noted that in diverging systems, such as our expanding universe, entropy prevails forcing time to flow forwards and forbidding advanced solutions. On the contrary in converging systems, such as black holes, syntropy prevails, time flows backwards and retarded solutions would be impossible; whereas in systems balanced between diverging and converging forces, such as atoms, time would be unitary, past, present and future would coexist and both advanced and retarded waves would be possible. In the CIRTS model Bierman considers advanced solutions possible also at the macro level, without taking into account the restrictions posed by the law of entropy.
2. Fantappiè argued that, as a consequence of the fact that advanced waves exist at the quantum level, living systems need a way to "extract" advanced waves from the quantum level in order to sustain living functions and contrast the destructive effects of entropy. Fantappiè found this mechanism in water, in the hydrogen bridge, a bond among the hydrogen atom and two electrons, found by Maurice Huggins in 1920, which allows to explains the anomalous properties of water (Ball, 1999). The hydrogen bridge makes water totally different from other liquids, mainly by increasing its cohesive forces (syntropy) and this would be the reason why water is so essential to life, since it allows the flow of advanced waves from the micro to the macro level. Consequently, in the syntropy model advanced waves are not associated to the brain, but are considered a fundamental property of all living systems. On the contrary, CIRTS suggests that advanced waves are mediated by consciousness and therefore should be a consequence of conscious brain activities. Bierman produces evidence in experiments conducted with meditators, but this evidence can be easily read as an increase of the role of the autonomic nervous system during meditation, and not as a consequence of consciousness. It is well known that, while meditating, subjects often experience a state of trance as a consequence of the fact that the aim is usually that of "turning off the mind".

3. In the CIRTS model consciousness is a pre-requisite of reality. In the syntropy model the feeling of life is a consequence of the cohesive and unitary properties of advanced waves. According to the syntropy model, any form of life has a feeling of life. Consequently we would have a feeling of life also when no brain activity is observed. This would explain why all forms of life, even the most simple ones, show anticipatory reactions (Rosen, 1985) and why, for example, patients during surgery in a state of anesthetic-induced unconsciousness tend to defend themselves and subjects with no brain activity react and defend themselves when their organs are removed for transplant. According to the syntropy model, the feeling of life does not reside in the brain; however, the brain provides memory which allows us to remember and reason regarding our conscious experiences.
4. CIRTS associates pre-stimuli reactions to coherence whereas the syntropy model associates pre-stimuli reactions to feelings and emotions. Coherence is a concept which is quite difficult to measure, whereas emotions can be easily measured using the parameters of the autonomic nervous system. Nevertheless Bierman introduces a formula in order to justify why pre-stimuli reactions are lower than post-stimuli reactions. In this formula the volume of the brain affected by coherence is divided by the total volume of the brain. The example reported by Bierman, relative to skin conductance, seems to support this formula. However, when using heart rate measurements pre-stimuli reactions and post-stimuli reactions tend to have the same size of effect. Even though effects vary greatly among subjects and generalization seems not to be appropriate, HR data contradicts Bierman's formula.

6. Conclusion

According to the syntropy model the dual manifestation of the quantum world in the form of waves and particles is not the consequence of the collapse of the wave equation, but the consequence of the dual causality at the quantum level: retarded waves, past causality, and advanced waves, future causality (Cramer, 1986). The advanced waves model does not need the collapse of the wave function and, consequently, does not need a time-symmetry restoration system. Advanced waves would explain not only the dual manifestation particle/waves, but also non-locality and entanglement (De Beauregard, 1977). On the contrary the CIRTS model finds its justification within the Copenhagen Interpretation of quantum mechanics and requires the collapse of the wave function.

The Copenhagen Interpretation was formulated in 1927 and can be considered the expression of the Zeitgeist, "the spirit of the time", since it reflects the idea of men as semi-Gods who, through the exercise of consciousness, can create reality. When Erwin Schrödinger discovered how Heisenberg and Bohr had used his wave equation, with ideological and mystical implications which provided powers of creation to consciousness, he commented: "I don't like it, and I am sorry I ever had anything to do with it" (Schrödinger, 1944).

References

Anderson C.D. (1932). The apparent existence of easily deflectable positives, Science, 76:238. Ball P. (1999). H_2O A Biography of Water, Weidenfeld & Nicolson, London.

Bem D.J. (2010). Feeling the Future: Experimental Evidence for Anomalous Retroactive Influence on Cognition and Affect. Journal of Personality and Social Psychology (in press), DOI: 10.1037/a0021524, http://dbem.ws/FeelingFuture.pdf

Bierman D.J., Radin D.I. (1998). Conscious and anomalous non-conscious emotional processes: A reversal of the arrow time? Toward a Science of Consciousness, Tucson III. MIT Press, 367-386.

Bierman D.J. (2008). Consciousness Induced Restoration of Time-Symmetry (CIRTS), a Psychophysical Theoretical Perspective, Proceedings of the PA Convention, Winchester, England, 13-17 August 2008.

Cramer J.G. (1986). The Transactional Interpretation of Quantum Mechanics, Reviews of Modern Physics, 58: 647-688.

De Beauregard C. (1977). Time Symmetry and the Einstein Paradox, Il Nuovo Cimento, 42(B).

Di Corpo U. (1981), Un nuovo approccio strutturale ai fondamenti della psicologia. Ipotesi teoriche ed esperimenti. Dissertation, University of Rome "La Sapienza".

Di Corpo U. (2007). The conflict between entropy and syntropy: the vital needs model. SSE Proceedings, Norway, 132-138.

Dirac P. (1928). The Quantum Theory of the Electron, Proc. Royal Society, London, 117: 610-624; 118: 351-361.

Fantappiè L. (1942). Sull'interpretazione dei potenziali anticipati della meccanica ondulatoria e su un principio di finalità che ne discende. Rend. Acc. D'Italia, 4(7).

Haken H. (1983). Synergetics, an Introduction: Nonequilibrium Phase Transition and Self-Organization in Physics, Chemistry, and Biology, Springer-Verlag.

Heisenberg W. (1928). Letter to W. Pauli, PC, May 3, 1: 443.

Heisenberg W. (1934). Letter to W. Pauli, PC, February 8, 2: 279.

King C.C. (1996). Quantum Mechanics, Chaos and the Conscious Brain. Journal of Mind and Behavior, 18: 155-170.

May E.C., Paulinyi T., Vassy Z. (2005). Anomalous Anticipatory Skin Conductance Response to Acoustic Stimuli: Experimental Results and Speculation about a Mechanism, The Journal of Alternative and Complementary Medicine, 11(4): 695-702.

McCratly R., Atkinson M., Bradely R.T. (2004). Electrophysiological Evidence of Intuition: Part 1, Journal of Alternative and Complementary Medicine, 10(1): 133-143.

McDonough B.E., Dons N.S., Warren C.A. (2002). Differential event-related potentials to targets and decoys in a guessing task, in Journal for Scientific Exploration, 16: 187-206.

Prigogine I. and Stengers I. (1979). La Nouvelle Alliance, Gallimard.

Radin D.I. (1997). Unconscious perception of future emotions: An experiment in presentiment. Journal of Scientific Exploration, 11(2): 163-180.

Radin D.I., Schlitz M. J. (2005). Gut feelings, intuition, and emotions: An exploratory study, Journal of Alternative and Complementary Medicine, 11(4): 85-91.

Rosen R. (1985). Anticipatory Systems, Pergamon Press, USA.

Schrödinger E. (1944). What is Life? The Physical Aspects of the Living Cell, Cambridge University Press.

Szent-Gyorgyi A. (1977). Drive in Living Matter to Perfect Itself, Synthesis, 1(1): 14-26.

Spottiswoode P., May E. (2003). Skin Conductance Prestimulus Response: Analyses, Artifacts and a Pilot Study, Journal of Scientific Exploration, 17(4): 617-641.

Tressoldi P. E., Martinelli M., Massaccesi S., Sartori L. (2005). Heart Rate Differences between Targets and Nontargets in Intuitive Tasks, Human Physiology, 31(6): 646–650.

Vannini A. (2005). Entropy and Syntropy. From Mechanical to Life Science, NeuroQuantology, 3(2): 88-110.

Vannini A. (2008). Quantum Models of Consciousness, Quantum Biosystems, 2: 165-184.

Vannini A. and Di Corpo U. (2008). Retrocausality and the Healing Power of Love, NeuroQuantology, 6(3): 291-296.

Vannini A. and Di Corpo U. (2009). A Retrocausal Model of Life, in Filters and Reflections. Perspective on Reality, ICRL Press, Princeton, NJ, USA, 231-244.

Vannini A. and Di Corpo U. (2010). Collapse of the wave function? Pre-stimuli heart rate reactions. NeuroQuantology, 8(4): 550-563.

Wheeler J.A. and Feynman R. (1949). Classical Electrodynamics in Terms of Direct Interparticle Action. Reviews of Modern Physics, 21: 425-433.

50. Consciousness and the Quantum
Don N. Page, Ph.D.

Theoretical Physics Institute, Department of Physics,
University of Alberta, Edmonton, Alberta, Canada

ABSTRACT

Sensible Quantum Mechanics or Mindless Sensationalism is a framework for relating consciousness to a quantum universe. It states that each conscious perception has a measure that is given by the expectation value of a corresponding quantum "awareness operator" in a fixed quantum state of the universe. The measures can be interpreted as frequency-type probabilities for a large set of perceptions that all actually exist with varying degrees of reality, so detailed theories within this framework are testable. The measures are not propensities for potentialities to be actualized, so there is nothing indeterministic in this framework, and no free will in the incompatibilistic sense. As conscious perceptions are determined by the awareness operators and the quantum state, they are epiphenomena. No fundamental relation is postulated between different perceptions (each being the entirety of a single conscious experience and thus not in direct contact with any other), so SQM or MS, a variant of Everett's "many-worlds" framework, is a "many-perceptions" framework but not a "many-minds" framework.

Keywords: Consciousness, quantum, universe, physics, perceptions, observations, cosmology, sensible quantum mechanics, mindless sensationalism, measure, probability, mathematics, reality, theories, simplicity, elegance, precision, time, awareness, experience, free will, epiphenomenon, epiphenomenalism

Consciousness and the Quantum

For hundreds of years, physicists have sought to find and understand at least one theory that will give a good description and explanation of our universe. It is typically preferred that such a theory have simple principles, an elegant form, and yet make precise statements. For these purposes, mathematical theories are often the ideal.

On the other hand, physics is usually regarded as necessarily resting upon observations (in contrast to, say, pure mathematics, which in principle can be divorced from experience, though in practice it too is usually based on observed patterns). However, observations themselves seem less simple, elegant, and precise than what physicists would often want for their theories. I am of the opinion that this is one of the causes for the tendency to regard the simple, elegant, and precise elements of theories of physics as more real than the apparently complex, sometimes ugly, and usually imprecise observations that we pay lip service to as the foundation of physics. Whatever we cannot understand in terms of the simple, elegant, precise elements of our mathematical theories, we tend to dismiss as less real.

In particular, I personally regard my own first-person subjective experience of consciousness as overwhelming evidence of its existence, but its apparent complexity, inelegance, and imprecision often seems to lead many physicists to dismiss it as less real than, say, elementary particles, spacetime, or quantum states. I am not at all decrying simplicity, elegance, and precision, which I do take to be extremely important, but I also do not wish to dismiss such a central feature of our experience as consciousness.

In fact, as a physicist, I take the observations that are considered to be the foundation of physics to be most simply and fundamentally conscious perceptions. This viewpoint raises the question of how consciousness is related to the structures that are more common in current theories of physics. In particular, the most fundamental structure of many of the current theories of physics is the quantum framework. Although we cannot be certain that our universe really is quantum, such a hypothesis helps explain so much of both our present theories and our observations that I shall take it as one of my central working hypotheses, along with the hypothesis of the existence of conscious perceptions.

In view of our desire to formulate the most simple, elegant, precise theories possible, I am not be content with leaving these two fundamental hypotheses unrelated but instead want to integrate them together. How might that be done? Here I wish to describe and explore a framework I have developed for relating consciousness and the quantum, which I have called Sensible Quantum Mechanics (Page, 1995) or Mindless Sensationalism (Page, 2003). I should emphasize that this is so far just a framework, since the details to make it into a proper precise theory are not yet known. But even just the framework itself has various consequences that may be explored.

Because Sensible Quantum Mechanics builds upon quantum theory, I should first say what I regard that pillar to be. I regard the essence of quantum theory to be a C^*-algebra of quantum operators and a quantum state that gives expectation values to each operator.

I cannot go into precise details here, but let me give a crude description of these elements. Quantum operators are mathematical entities that may be adjointed, multiplied by complex numbers, added together, or multiplied together to give other operators. For example, if A and B are two operators and c is a complex number (so c = a + ib with a and b real numbers and i the square root of −1), then the adjoints $A^\dagger$ and $B^\dagger$, cA, cB, A + B, AB, and BA are also quantum operators. (It is not assumed that AB = BA; the order of two operators that are multiplied together generally matters.) An operator A might represent the position of a particle, and B might represent its momentum, but for the general structure we do not need to assign specific meanings to the operators.

A quantum state may be regarded as a positive linear functional σ on the quantum operators, a rule for assigning a complex number to each quantum operator that is called its expectation value in that quantum state. For example, the expectation value of the operator A in the quantum state σ, denoted σ[A], represents a particular complex number associated with that operator. This rule is required to have the form that σ[cA]= cσ[A] for any operator A and complex number c, so the expectation value of the operator cA that is the complex number c multiplied by the operator A is simply c times the expectation value of A. The rule is also required to give σ[A + B]= σ[A]+ σ[B] and σ[I] = 1, where I is the identity operator such that IA = AI = A for each quantum operator A, as well as to make σ[$A^\dagger$A] a nonnegative real number.

Quantum theory is often regarded as having other basic elements, but I shall regard them as either being part of this basic formalism or as not really being a necessary part of quantum theory. For example, often one talks about the time evolution of a quantum state, but one can reformulate this into the Heisenberg picture in which the quantum state stays fixed and the operators change with time, and then one can re-interpret the time dependence of each operator as representing a whole family of operators, each labeled by a time parameter in addition to other labels of what the operators are. The dynamics of the operators in the usual approach would expressed in terms of the algebra of all the operators in all the families labeled by the time parameter and by the other labels. In this view, there would be nothing fundamentally special about time; it would just be one of many labels for the operators.

The other pillar of Sensible Quantum Mechanics is consciousness. I shall assume here that there is a countable discrete set M of all possible conscious experiences or perceptions p. By a "conscious experience," I mean all that one is consciously aware of or consciously experiencing at once. Lockwood (2003) has called this a "phenomenal perspective" or "maximal experience" or "conscious state." It could also be expressed as a total "raw feel" that one has at once.

Because I regard the basic conscious entities to be the conscious experiences themselves, which might crudely be called sensations if one does not restrict the meaning of this word to be the conscious responses only to external stimuli, and because I doubt that these conscious experiences are arranged in any strictly defined sequences that one might define to be minds if they did exist, my framework has sensations without minds and hence may be labeled Mindless Sensationalism (Page, 2003). In this way the framework of Mindless Sensationalism proposed here is a particular manifestation of Hume's ideas (Hume, 1988), that "what we call a mind, is nothing but a heap or collection of different perceptions, united together by certain relations, and suppos'd, tho' falsely, to be endow'd with a perfect simplicity and identity" (p. 207), and that the self is "nothing but a bundle or collection of different perceptions" (p. 252). As he explains in the Appendix (p. 634), "When I turn my reflexion on myself, I never can perceive this self without some one or more perceptions; nor can I ever perceive any thing but the perceptions. 'Tis the composition of these, therefore, which forms the self." (Here I should note that what Hume calls a perception may be only one component of the "phenomenal perspective" or "maximal experience" (Lockwood, 1989) that I have been calling a perception or conscious experience p, so one of my p's can include "one or more perceptions" in Hume's sense.)

I should also emphasize that by a conscious experience, I mean the phenomenal, first-person, "internal" subjective experience, and not the unconscious "external" physical processes in the brain that accompany these subjective phenomena. In his first chapter, Chalmers (1996) gives an excellent discussion of the distinction between the former, which he calls the phenomenal concept of mind, and the latter, which he calls the psychological concept of mind. In his language, what I mean by a conscious experience (and by other approximate synonyms that I might use, such as perception or sensation or awareness) is the phenomenal concept, and not the psychological one.

The next idea is that not all possible conscious perceptions p occur equally, but that there is a normalized measure $w(p)$ associated with each one (a non-negative real number which sums to unity when one adds up the values for all the p's in the full set M). This measure in some sense gives the level or degree of reality that the conscious perception p has. Perceptions with large measures have high degrees of reality, whereas perceptions with very low measures have tiny degrees of reality and effectively can be ignored. One can also say that the weight $w(p)$ is analogous to the probability for the conscious experience p, but it is not to be interpreted as the probability for the bare existence of p, since any conscious experience p exists (is actually experienced) if its weight is positive, $w(p) > 0$. Rather, $w(p)$ is to be interpreted as being proportional to the probability of getting this particular experience if a random selection were made.

Because the specification of the conscious experience p completely determines its content and how it is experienced (how it feels), the weight $w(p)$ has absolutely no effect on that—there is absolutely no way within the experience to sense anything directly of what the weight is. A toothache within a particular conscious experience p is precisely as painful an experience no matter what w(p) is. It is just that an experience with a greater w(p) has a greater degree of existence and is more likely in the sense of being more probably chosen by a random selection using the weights w(p).

Finally, Sensible Quantum Mechanics assumes the connection between consciousness and the quantum in the form that for each conscious perception p, there is an associated quantum "awareness operator" A(p), and that the measure for the conscious perception p is the expectation value, in the quantum state of the universe, of the corresponding experience operator, w(p)= σ[E(p)].

One can summarize this by saying that Sensible Quantum Mechanics or Mindless Sensationalism is given by the following three basic postulates or axioms (Page, 1995, 2003):

Quantum World Axiom: The "quantum world" Q is completely described by an appropriate C^*-algebra of operators O and by a suitable state σ (a positive linear functional of the operators) giving the expectation value σ[O] of each operator O.

Conscious World Axiom: The "conscious world" M, the countable discrete set of all conscious experiences or perceptions p, has a fundamental normalized measure w(p) for each perception p.

Psycho-Physical Parallelism Axiom: The measure w(p) for each conscious experience p is given by the expectation value of a corresponding quantum "awareness operator" A(p) in the state σ of the quantum world, w(p)= σ[A(p)].

One might note that in comparison with the more general assumptions of (Page, 1995, 2003), here for simplicity I am making the more restrictive hypothesis that the set M of all conscious perceptions p is a countable discrete set. I am also assuming that the measure is normalized, $\sum_p w(p) = 1$.

The Psycho-Physical Parallelism Axiom is the simplest way I know to connect the quantum world with the conscious world. One could easily imagine more complicated connections, such as having w(p) be a nonlinear function of the expectation values, say m(p), of a positive "experience operator" E(p) depending in the p (Page, 1995, 2003). Instead, my Psycho-Physical Parallelism Axiom restricts the function to be linear in the expectation values. In short, I am proposing that the psychophysical parallelism is linear.

Of course, the Psycho-Physical Parallelism Axiom, like the Quantum World Axiom, is here also deliberately vague as to the form of the awareness operators A(p), because I do not have a detailed theory of consciousness, but only a framework for fitting it with quantum theory. My suggestion is that a theory of consciousness that is not inconsistent with bare quantum theory should be formulated within this framework. I am also suspicious of any present detailed theory that purports to say precisely under what conditions in the quantum world consciousness occurs, since it seems that we simply don't know yet. I feel that present detailed theories may be analogous to the cargo cults of the South Pacific after World War II, in which an incorrect theory was adopted, that aircraft with goods would land simply if airfields and towers were built.

Since all conscious perceptions p with $w(p) > 0$ really occur in the framework of Sensible Quantum Mechanics or Mindless Sensationalism, it is completely deterministic if the quantum state and the awareness operators $A(p)$ are determined: there are no random or truly probabilistic elements in SQM or MS. Neither is there any free will in the incompatibilist sense, and consciousness may be viewed as an epiphenomenon (Page, 1995, 2003). Nevertheless, because the framework has normalized measures $w(p)$ for conscious perceptions, these can be interpreted as probabilities for the perceptions, given the theory. In particular, one can interpret the measure $w_i(p_j)$ that a theory T_i assigns to one's particular perception p_j as the likelihood of the theory. Then if one assigns different SQM or MS theories prior probabilities $P(T_i)$, one can use Bayes' theorem to calculate the posterior probability of the theory, given the observation or conscious perception p_j, as $P(T_i|p_j) = P(T_i)w_i(p_j) / \Sigma_k P(T_k)w_k(p_j)$. In this way different SQM or MS theories are testable.

A major problem at the frontier of theoretical cosmology is essentially to develop one or more theories that give the measures $w(p)$ for conscious perceptions, except that most theorists are hesitant to focus on conscious perceptions and hence ask for the probability of an observation O_j given a theory T_i, $P(O_j|T_i)$. It is usually left rather vague what is supposed to constitute an observation. For me the most fundamental entities that can be identified with observations are conscious perceptions, so I would take $P(O_j|T_i)$ to be $w_i(p_j)$, the normalized measure that theory T_i assigns to the conscious perception p_j. In Sensible Quantum Mechanics, a theory T_i would assign an awareness operator $A_j = A(p_j)$ to each conscious perception and give a quantum state σi so that $P(O_j|T_i) = w_i(p_j) = σi[A_j]$, the expectation value of the awareness operator in the quantum state. (Here, for compactness, I do not explicitly display the dependence of the A_j operators on the theory T_i, but different theories can differ not only in their quantum states but also in their awareness operators.)

For theorists hesitant to identify observations with conscious perceptions, they may still wish to say that the probability of an observation O_j given a theory T_i is $P(O_j|T_i) = σi[A_j]$. In this generalized view, A_j is simply the operator in the theory T_i whose expectation value in the quantum state given by that theory gives the probability of the observation O_j.

Quantum theories of this generalized form (whether or not an observation is taken to be a conscious perception) can be considered to have three parts: (1) the algebra of the full set of quantum operators, (2) the quantum state σi, and (3) the particular operators A_j for each observation (or for each conscious perception in Sensible Quantum Mechanics or Mindless Sensationalism). Part

(1) includes the dynamical laws of physics, which historically have often been naïvely called a 'Theory of Everything' or TOE, though it certainly is not. Part (2) includes the boundary conditions that specify which solution of the dynamical laws describes our actual universe, but even Part (1) augmented with Part (2) is not sufficient. Part (3) includes the rules for extracting the probabilities of observations from the quantum state.

The logical independence of Part (3) is becoming widely recognized with the measure problem of cosmology (Linde, 1986; Garcia-Bellido et al., 1994; Vilenkin, 1995a,b; Guth, 2000; Tegmark 2005; Aguirre, 2007; De Simone et al., 2008; Linde & Noorbala, 2010; Bousso et al., 2010). If a universe had a definite number N_j of occurrences (each occurrence with the same degree of reality) of each kind of observation O_j, and a finite total number of occurrences $N = \sum_k N_k$, it would be natural to say that the probability of the observation O_j is the fraction of the number of its occurrences, $P(O_j) = N_j/N$. However, theories of eternal inflation (Linde, 1986; Garcia-Bellido et al., 1994; Vilenkin, 1995a,b; Guth, 2000; Tegmark 2005; Aguirre, 2007; De Simone et al., 2008; Linde & Noorbala, 2010; Bousso et al., 2010) suggest that the universe may have expanded to become infinitely large, in which case most of the numbers N_j of occurrences are infinite, and the ratio N_j/N is ambiguous. Therefore, a lot of work has gone into different proposed schemes for regularizing the infinities.

I have pointed out that even in finite universes, quantum uncertainties in the numbers of occurrences also leads to ambiguities in the probabilities of observations (Page, 2008, 2009a,b,c, 2010). In particular, I have shown that Born's rule does not work in the sense that the operators A_j cannot be projection operators, so that one must choose other operators, and the ambiguity of that choice is the measure problem. The ambiguity occurs even for finite universes, but it is particularly severe for infinite universes. So whether or not the operators A_j whose expectation values give the the probabilities of observations are interpreted to be awareness operators in Sensible Quantum Mechanics or Mindless Sensationalism (in which the observations are conscious perceptions), it is now recognized that there is the challenge of finding these operators, in addition to the challenge of finding the dynamics or algebra of all operators and the quantum state.

In conclusion, I am proposing that Sensible Quantum Mechanics or Mindless Sensationalism is the best framework we have for understanding the connection between consciousness and the quantum universe. Of course, the framework would only become a complete theory once one had the set of all conscious experiences, the awareness operators, and the quantum state of the universe.

This research was supported in part by the Natural Sciences and Engineering Research Council of Canada.

Bibliography

Aguirre, A. (2007) On making predictions in a multiverse: Conundrums, dangers, and coincidences. In: Carr, B. J. (Ed.), Universe or Multiverse?, Cambridge University Press, Cambridge, UK, pp. 367-386 [arXiv:astro-ph/0506519] <http://arxiv.org/abs/astro-ph/0506519>.

Bousso, R.,Freivogel, B.,Leichenauer, S., Rosenhaus, V. (2010). Geometric origin of coincidences and hierarchies in the landscape. arXiv:1012.2869 [hep-th] <http://arxiv.org/abs/arXiv:1012.2869>.

Chalmers, D. J. (1996). The Conscious Mind: In Search of a Fundamental Theory. Oxford University Press, New York, USA.

Garcia-Bellido, J., Linde, A. D., Linde, D. A. (1994). Fluctuations of the gravitational constant in the inflationary Brans-Dicke cosmology. Physical Review D, 50, 730-750 [arXiv:astro-ph/9312039] <http://arxiv.org/abs/astro-ph/9312039>.

Guth, A. H. (2000). Inflation and eternal inflation. Physics Reports, 333, 555

574 (2000) [arXiv:astro-ph/0002156] <http://arxiv.org/abs/astro-ph/0002156>. Hume, D. (1988). A Treatise of Human Nature. Clarendon Press, Oxford, UK. Linde, A. D. (1986). Eternally existing self-reproducing chaotic inflationary uni

verse. Physics Letters B, 175, 395-400. Linde, A., Noorbala, M. (2010). Measure problem for eternal and non-eternal inflation. Journal of Cosmology and Astroparticle Physics, 1009, 008 [arXiv:1006.2170

[hep-th]] <http://arxiv.org/abs/arXiv:1006.2170>.

Lockwood, M. (1989). Mind, Brain and the Quantum: The Compound 'I.' Basil Blackwell Press, Oxford, UK.

Lockwood, M. (2003) Consciousness and the quantum world: Putting qualia on the map. In Smith, Q., Jokic, A.(Eds.), Consciousness: New Philosophical Perspectives, Oxford University Press, Oxford, pp. 447-467.

Page, D. N. (1995). Sensible quantum mechanics: Are only perceptions probabilistic? arXiv:quant-ph/9506010 <http://arxiv.org/abs/quant-ph/9506010>.

Page, D. N. (2003) Mindless sensationalism: A quantum framework for consciousness. In Smith, Q., Jokic, A.(Eds.), Consciousness: New Philosophical Perspectives, Oxford University Press, Oxford, pp. 468-506 [arXiv:quant-ph/0108039] <http://arxiv.org/abs/quant-ph/0108039>.

Page, D. N. (2008). Cosmological measures without volume weighting. Journal of Cosmology and Astroparticle Physics, 0810, 025 [arXiv:0808.0351 [hep-th]] <http://arxiv.org/abs/arXiv:0808.0351>.

Page, D. N. (2009a). Insufficiency of the quantum state for deducing observational probabilities. Physics Letters B, 678, 41-44 [arXiv:0808.0722 [hep-th]] <http://arxiv.org/abs/arXiv:0808.0722>.

Page, D. N. (2009b). The Born rule fails in cosmology. Journal of Cosmology and

Astroparticle Physics, 0907, 008 [arXiv:0903.4888 [hep-th]] <http://arxiv.org/abs/arXiv:0903.4888>. Page, D. N. (2009c). Born again. arXiv:0907.4152 [hep-th] <http://arxiv.org/abs/arXiv:0907.4152>. Page, D. N. (2010). Born's rule is insufficient in a large universe. arXiv:1003.2419

[hep-th] <http://arxiv.org/abs/arXiv:1003.2419>.

De Simone, A., Guth, A. H.,Salem, M. P., Vilenkin, A. (2008). Predicting the cosmological constant with the scale-factor cutoff measure. Physical Review D, 78, 063520 [arXiv:0805.2173 [hep-th]] <http://arxiv.org/abs/arXiv:0805.2173>.

Tegmark, M. (2005). What does inflation really predict? Journal of Cosmology and Astroparticle Physics, 0504, 001 [arXiv:astro-ph/0410281] <http://arxiv.org/abs/astro-ph/0410281>.

Vilenkin, A. (1995a). Predictions from quantum cosmology. Physical Review Letters, 74, 846-849 [arXiv:gr-qc/9406010] <http://arxiv.org/abs/gr-qc/9406010>. Vilenkin, A. (1995b). Making predictions in eternally inflating universe. Physical Review D, 52, 3365-3374 [arXiv:gr-qc/9505031] <http://arxiv.org/abs/gr-qc/9505031>.

51. Quantum Reality and Mind
Henry P. Stapp, Ph.D.

Lawrence Berkeley Laboratory, University of California, Berkeley, California

Abstract

Two fundamental questions are addressed within the framework of orthodox quantum mechanics. The first is the duality-nonduality conflict arising from the fact that our scientific description of nature has two disparate parts: an empirical component and a theoretical component. The second question is the possibility of meaningful free will in a quantum world concordant with the principle of sufficient reason, which asserts that nothing happens without a sufficient reason. The two issues are resolved by an examination of the conceptual and mathematical structure of orthodox quantum mechanics, without appealing to abstract philosophical analysis or intuitive sentiments.

Key Words: Quantum Reality, Mind, Mind-Matter, Free Will, Duality, Mental monism

1. Introduction

The first purpose of this article is to explain the nature of the connection between mind and matter and how orthodox quantum mechanics is both dualistic and nondualistic: it is dualistic on a pragmatic, operational level, but is nondualistic on a deeper ontological level.

The second purpose is to reconcile a meaningful concept of human freedom with the principle of sufficient reason; with the principle that nothing happens without a sufficient reason.

To lay a framework for discussing these two issues I shall begin by describing some contrasting ideas about the nature of reality advanced by three towering intellectual figures, Rene Descartes (1596-1650), Isaac Newton (1642-1727), and William James (1842-1910).

René Descartes conceived nature to be divided into two parts: a mental part and a physical part. The mental part, which he called "res cogitans", contains our thoughts, ideas, and feeling, whereas the physical part, called "res extensa", is defined here to be those aspects of nature that we can describe by assigning mathematical properties to space-time points. Examples of physical aspects of our understanding of nature are trajectories of physical particles, the electric field $E(x,t)$, and the quantum mechanical field (operator) $A(x,t)$. Descartes allowed the mental and physical aspects to interact with each other, but only for those physical parts that are located inside human brains. This is the classic Cartesian notion of duality.

Isaac Newton built the foundations of "modern physics" upon the ideas of Descartes, Galileo, and Kepler. The astronomical observations of Tycho Brahe led to Kepler's three laws of planetary motion. These laws, coupled to Galileo's association of gravity with acceleration, led directly to Newton's inverse square law of gravitational attraction, and his general laws of motion. Newton extended these dynamical ideas, with tremendous success, down to the scale of terrestrial motions, to the tides and falling apples etc.. He also conjectured an extension down to the level of the atoms. According to that hypothesis, the entire physically described universe, from the largest objects to the smallest ones, would be bound by the precept of physical determinism, which is the notion that a complete description of the values of all physically described variables at any one time determines with certainty the values of all physically described variables at any later time. This idea of universal physical determinism is a basic precept of the development of Newtonian dynamics into what is called "classical physics".

1a. The Omission of the Phenomenal Aspects of Nature

The dynamical laws of classical physics are formulated wholly in terms of physically described variables: in terms of the quantities that Descartes identified as elements of "res extensa". Descartes' complementary psychologically described things, the elements of his "res cogitans", were left completely out: there is, in the causal dynamics of classical physics, no hint of their existence. Thus there is not now, nor can there ever be, any rational way to explain, strictly on the basis of the dynamical precepts of classical physics, either the existence of, or any causal consequence of, the experientially described aspects of nature. Yet these experiential aspects of nature are all that we actually know.

This troublesome point was abundantly clear already at the outset:

Newton: "…to determine by what modes or actions light produceth in our minds the phantasm of colour is not so easie."

Leibniz: "Moreover, it must be confessed that perception and that which depends upon it are inexplicable on mechanical grounds, that is to say, by means of figures and motions."

Classical physics, by omitting all reference to the mental realities, produces a logical disconnect between the physically described properties represented in that theory and the mental realities by which we come to know them. The theory allows the mental realities to know about the physical aspects of nature, yet be unable to affect them in any way. Our mental aspects are thereby reduced to "Detached Observers", and Descartes' duality collapses, insofar as the causally closed physical universe is concerned, to a physics-based nonduality; to a physical monism, or physicalism. Each of us rejects in actual practice the classical-physics claim that our conscious thoughts and efforts can have no affects on our physical actions. We build our lives, and our political, judicial, economic, social, and religious institutions, upon the apparently incessantly reconfirmed belief that, under normal wakeful conditions, a person's intentional mental effort can influence his physical actions.

William James, writing in 1892, challenged, on rational grounds, this classical-physics-based claim of the impotence of our minds. At the end of his book Psychology: The Briefer Course he reminded his readers that "the natural science assumptions with which we started are provisional and revisable things". That was a prescient observation! Eight years later Max Planck discovered a new constant of nature that signaled a failure of the precepts of classical physics, and by 1926 the precepts of its successor, quantum mechanics, were firmly in place.

The most radical shift wrought by quantum mechanics was the explicit introduction of mind into the basic conceptual structure. Human experience was elevated from the role of 'a detached observer' to that of 'the fundamental element of interest':

Bohr: "In our description of nature the purpose is not to disclose the real essence of phenomena but only to track down as far as possible relations between the multifold aspects of our experience." (Bohr, 1934, p.18).

Bohr: "The sole aim [of quantum mechanics] is the comprehension of observations...(Bohr, 1958, p.90).

Bohr: The task of science is both to extend the range of our experience and reduce it to order (Bohr, 1934, p.1)

Heisenberg: "The conception of the objective reality of the elementary particles has evaporated not into the cloud of some new reality concept, but into the transparent clarity of a mathematics that represents no longer the behaviour of the particles but our knowledge of this behavior" (Heisenberg, 1958a, p. 95).

This general shift in perspective was associated with a recasting of physics from a set of mathematical connections between physically described aspects of nature, into set of practical rules that, eschewing ontological commitments, predicted correlations between various experiential realities on the basis of their postulated dynamical links to certain physically describable aspects of our theoretical understanding of nature.

In view of this fundamental re-entry of mind into basic physics, it is nigh on incomprehensible that so few philosophers and non-physicist scientists entertain today, more than eight decades after the downfall of classical physics, the idea that the physicalist conception of nature, based on the invalidated classical physical theory, might be profoundly wrong in ways highly relevant to the mind-matter problem.

Philosophers are often called upon to defend highly counter-intuitive and apparently absurd positions. But to brand as an illusion, and accordingly discount, the supremely successful conceptual foundation of our lives---the idea that our conscious efforts can influence our physical actions---on the basis of its conflict with a known-to-be-false theory of nature that leaves out all that we really know, is a travesty against reason, particularly in view of the fact that the empirically valid replacement of that empirically invalid classical theory is specifically about the details of the connection between our consciously chosen intentional actions and the experiential feedbacks that such actions engender.

1b. Von Neumann's Dualistic Quantum Mechanics.

The logician and mathematician John von Neumann (1955/1932) formalized quantum mechanics in a way that allowed it to be interpreted as a dualistic theory of reality in which mental realities interact in specified causal ways with physically described human brains. This orthodox quantum ontology is in essential accord with the dualistic ideas of Descartes.

An objection often raised against Cartesian dualism is couched as the query: How can ontologically distinct aspects of nature ever interact? Must not nature consist ultimately of one fundamental kind of stuff in order for its varied components to be able to cohere.

This objection leads to a key question: What is the ontological character of the physical aspect of quantum mechanics?

2. The Ontological Character of the Physical Aspect of the Orthodox (von Neumann-Heisenberg) Quantum Mechanics.

The physical aspect of the quantum mechanical conception of nature is represented by the quantum state. This state is physical, in the defined sense that we can describe it by assigning mathematical properties to space-time points. But this physical aspect is does not have the ontological character of a material substance, in the sense in which the physical world of Newtonian (or classical) physics is made of material substance: it does not always evolve in a continuous manner, but is subject to abrupt "quantum jumps", sometimes called "collapses of the wave function".

Heisenberg (1958b) couched his understanding of the ontological character of the reality lying behind the successful quantum rules in terms of the Aristotelian concepts of "potentia" and "actual". The quantum state does not have the ontological character of an "actual" thing. It has, rather, the ontological character of "potentia": of a set of "objective tendencies for actual events to happen". An actual event, in the von Neumann-Heisenberg orthodox ontology, is "The discontinuous change in the probability function [that] takes place with the act of registration...in the mind of the observer". (Heisenberg, 1958b, p. 55)

The point here is that in orthodox (von Neumann-Heisenberg) quantum mechanics the physical aspect is represented by the quantum state, and this state has the ontological character of potentia---of objective tendencies for actual events to happen. As such, it is more mind-like than matter-like in character. It involves not only stored information about the past, but also objective tendencies pertaining to events that have not yet happened. It involves projections into the future, elements akin to imagined ideas of what might come to pass. The physical aspects of quantum mechanics are, in these ways, more like mental things than like material things.

Furthermore, a quantum state represents probabilities. Probabilities are not matter-like. They are mathematical connections that exist outside the actual realities to which they pertain. They involve mind-like computations and evaluations: weights assigned by a mental or mind-like process.

Quantum mechanics is therefore dualistic in one sense, namely the pragmatic sense. It involves, operationally, on the one hand, aspects of nature that are described in physical terms, and, on the other hand, also aspects of nature that are described in psychological terms. And these two parts interact in human brains in accordance with laws specified by the theory. In these ways orthodox quantum mechanics is completely concordant with the defining characteristics of Cartesian dualism.

Yet, in stark contrast to classical mechanics, in which the physically described aspect is matter-like, the physically described aspect of quantum mechanics is mind-like! Thus both parts of the quantum Cartesian duality are ontologically mind-like.

In short, orthodox quantum mechanics is Cartesian dualistic at the pragmatic/operational level, but mentalistic on the ontological level.

This conclusion that nature is fundamentally mind-like is hardly new. But it arises here not from some deep philosophical analysis, or religious insight, but directly from an examination of the causal structure of our basic scientific theory.

3. Natural Process, Sufficient Reason, and Human Freedom.

There are two fundamentally different ways to cope with the demands of the theory of relativity.

The first is the classical-physics-based Einsteinian idea of a Block Universe: a universe in which the entire future is laid out beforehand, with our experiences being mere perspectives on this preordained reality, viewed from particular vantage points in spacetime.

A second way to accommodate, rationally, the demands of special relativity is the quantum-physics-based Unfolding Universe, in which facts and truths become fixed and definite in the orderly way allowed by relativistic quantum field theory (Tomonaga, 1946; Schwinger 1951; Stapp, 2007, p. 92) with our experiences occurring in step with the coming into being of definite facts and truths.

I subscribe to the latter idea: to the idea of an unfolding reality in which each experienced increment of knowledge is associated with an actual event in which certain facts or truths become fixed and definite.

I also subscribe to the idea that this unfolding conforms to the principle of sufficient reason, which asserts that no fact or truth can simply "pop out of the blue", with no sufficient reason to be what it turns out to be.

An important question, then, is whether such a concordance with the principle of sufficient reason precludes the possibility of meaningful human freedom. Is human freedom an illusion, in the sense that every action that a person makes was fixed with certainty already at the birth of the universe?

Laplace's classical argument for the "certainty of the future" states (in condensed form):

"For a sufficiently powerful computing intellect that at a certain moment knew all the laws and all the positions, nothing would be uncertain, and the future, just like the past, would be present before its eyes."

This view argues for "certainty about the future"; for a certainty existing at an earlier moment on the basis of information existing at that earlier moment. It contemplates:

1. A computing intellect existing outside or beyond nature itself, able to "go" in thought where the actual evolving universe has not yet gone.

2. Invariant causal laws.

But nothing really exists outside the whole of nature itself! Thus nature itself must make its own laws/habits. And these habits could themselves evolve. Even if there is a sufficient reason, within the evolving reality, for every change in the laws, and every generation of a fact, it is not evident that any intellect standing outside the evolving reality itself could compute, on the basis of what exists at a certain moment, all that is yet to come. For the evolution of reason-based reasons may be intrinsically less computable than the evolution of the mathematically formulated physically described properties of the classical-physics approximation to the actual laws of nature. Reason encompasses computable mathematics, but computable mathematics may not encompass reason. Reason is a category of explanation more encompassing than mathematical computation.

The laws of classical mechanics are cast in a particular kind of mathematical form that allows, in principle, a "mathematical computation" performed at an early time to predict with certainty the state of the universe at any later time. But this is very special feature of classical mechanics. It is far from obvious that the---definitely nonclassical---real world must exhibit this peculiar 'computability" feature.

The notion that a reason-based unfolding of the actual world is computable is an extrapolation from the classical-physics approximation that is far too dubious to provide the basis of a compelling argument that, in a mind-based quantum universe evolving in accordance with the principles of both quantum mechanics and sufficient reason, the outcomes of human choices are certain prior to their actual occurrence. It is far from being proved that, in a universe of that kind, the exact movement of the computer key that I am now pressing was certain already at the birth of the universe. "Reasons" could lack the fantastic computability properties that the physically described features of classical physics enjoy. But in that case our present reason-based human choices need not have been fixed with certainty at the birth of the universe. Within orthodox quantum mechanics meaningful human freedom need not be an illusion.

4. Reason-Based Dynamics Versus Physical-Description-Based Dynamics.

The arguments given above rest heavily upon the contrast between the reason-based dynamics of the unfolding universe described by quantum mechanics and the physical-description-based dynamics of the block universe described by classical mechanics. In this final section I shall pinpoint the technical features of orthodox quantum mechanics that underlie this fundamental difference between these two theories.

Von Neumann created a formulation of quantum mechanics in which all the physical aspects of nature are represented by the evolving quantum mechanical state (density matrix) of the universe. Each subsystem of this physically described universe is represented by a quantum state obtained by performing a certain averaging procedure on the state of the whole universe.

Each experience of a person is associated with an "actual event". This event reduces, in a mathematically specified way, the prior quantum state of this person's brain---and consequently the quantum state of the entire universe---to a new "reduced" state. The reduction is achieved by the removal of all components of the state of this person's brain that are incompatible with the increment of knowledge associated with the experience. The needed mapping between "experiential increments in a person's knowledge" and "reductions of the quantum mechanical state of that person's brain" can be understood as being naturally created by trial and error learning of the experienced correlations between intentional efforts and the experiential feedbacks that these efforts tend to produce.

A key feature of von Neumann's dynamics is that it has two distinct kinds of mind-brain interaction. Von Neumann calls the first of these two processes "process 1". It corresponds to a choice of a probing action by the person, regarded as an agent or experimenter. The second kind of mind-brain interaction was called by Dirac "a choice on the part of nature". It specifies nature's response to the probing action selected by a logically preceding process 1 action. Von Neumann uses the name "process 2" to denote the physical evolution that occurs between the mind-brain (collapse) interactions. I therefore use the name "process 3" to denote the reduction/collapse process associated with nature's response to the process 1 probing action.

The mathematical form of process 1 differs from that of process 3. This mathematical difference causes these two processes to have different properties. In particular, process-1 actions have only local effects, in the sense that the dependence of the predictions of quantum mechanics upon a process-1 action itself, without a specification of the response, is confined (in the relativistic version) to the forward light-cone of the region in which the process-1 physical action occurs: the empirically observable effects of a process-1 action never propagate faster than the speed of light. On the other hand, nature's response (process 3) to a localized process-1 action can have observable statistical effects in a faraway contemporaneous region. The no-faster-than-light property of the empirically observable effects of any process-1 action is what justifies the word "relativistic" in relativistic quantum theory, even though the underlying mathematical description involves abrupt process-1-dependent faster-than-light transfers of information in connection with nature's response to the process-1 action.

Process 2 is a generalization of the causal process in classical mechanics, and, like it, is deterministic: the state of the universe at any earlier time completely determines what it will evolve into at any later time, insofar as no process 1 or process 3 event intervenes. But if no process 1 or process 3 event intervenes then the process 2 evolution would take the initial "big bang" state of the universe into a gigantic smear in which, for example, the moon would be smeared out over the night sky, and the mountains, and the cities, and we ourselves, would all be continuously spread out in space.

It is the process 1 and process 3 actions that, in the orthodox ontology, keep the universe in line with human experiences. On the other hand, the von Neumann ontology certainly does not exclude the possibility that non-human-based analogs of the human-based process 1 and follow-up process 3 actions also exist. Rather, it explains why the existence of reduction processes associated with other macroscopic agents would be almost impossible to detect empirically. These features of the von Neumann ontology justify focusing our attention here on the human involvement with nature.

Process 1, unlike process 2, is not constrained by any known law. In actual practice our choices of our probing actions appear to us to be based on reasons. We open the drawer in order to find the knife, in order to cut the steak, in order to eat the steak, in order to satisfy our hunger. Whilst all of this chain of reasons would, within the deterministic framework of classical physics, need to be, in principle, explainable in mathematical ways based upon the physical description of the universe, there is no such requirement in orthodox quantum mechanics: the sufficient reasons could be "reasons"; reasons involving the experiential dimension of reality, rather than being fully determined within the physical dimension. And these reasons could be, at each individual moment of experience, sufficient to determine the associated process 1 choice, without those choices having been mathematically computable from the state of the universe at earlier times.

The process 3 selection on the part of nature, unlike the process 1 choice, is not completely unconstrained: it is constrained by a statistical condition. According to the principle of sufficient reason, the process 3 choice must also be, in principle, determined by a sufficient reason. But, as emphasized above, nature's choice is nonlocal in character. Thus the reason for a process 3 choice need not be located at or near the place where the associated process 1 action occurs. Yet, as was clear already in classical statistical mechanics, there is an á priori statistical rule: equal volumes of phase space are equally likely. The (trace-based) statistical rule of quantum mechanics is essentially the quantum mechanical analog of this á priori statistical rule. The quantum statistical rule is therefore the natural statistical representation of the effect of a reason-based choice that is physically far removed from its empirical process 3 manifestation.

In closing, it is worth considering the argument of some physicalist philosophers that the replacement of classical mechanics---upon which physicalism is based---by quantum mechanics is not relevant to the resolution of the mind-matter problem for the following reason: that replacement has no bearing on the underlying problem of human freedom. The argument is that the essential difference between the two theories is (merely) that the determinism of classical mechanics is disrupted by the randomness of quantum mechanics, but that an introduction of randomness into the dynamics in no way rescues the notion of meaningful human freedom: a random choice is no better than a deterministic choice as an expression of meaningful human freedom.

This physicalist argument flounders on the fact that the element of quantum randomness enters quantum mechanics only via process 3, which delivers nature's choice. Man's choice enters via process 1, which is the logical predecessor to nature's process 3 "random" choice. In orthodox quantum mechanics, no elements of quantum randomness enter into man's choice. Nor is man's choice fixed by the deterministic aspect of quantum mechanics: that aspect enters only via process 2. Von Neumann's process 1 human choice is, in this very specific sense, "free": it is von Neumann's representation of Bohr's "free choice of experimental arrangement for which the quantum mechanical formalism offers the appropriate latitude" (Bohr 1958. p.51). Human choices enter orthodox quantum mechanics in a way not determined by a combination of the deterministic and random elements represented in the theory.

References

Bohr, N. (1934). Atomic Physics and the Description of Nature. Cambridge University Press, Cambridge, UK.

Bohr, N. (1958). Atomic Physics and Human Knowledge. Wiley, New York, US.

Heisenberg, W. (1958a). The representation of ature in contemporary physics. Daedalus, 87 (summer), 95-108.

Heisenberg, W. (1958b). Physics and Philosophy. Harper, New York, US.

James. W. (1892). Psychology: The Briefer Course. In: William James: Writings 1879-1899. Library of America (1992), New York, US.

Schwinger, J. (1951). Theory of Quantized Fields 1. Physical Review, 82, 914-927.

Stapp, H.P. (2007). Mindful Universe: Quantum Mechanics and the Participating Observer. Springer, Berlin.

Tomonaga, S. (1946). On a relativistically invariant formulation of the quantum theory of wave fields. Progress of Theoretical Physics, 1, 27-42.

Von Neumann, J. (1955/1932). Mathematical Foundations of Quantum Mechanics. Princeton University Press, Princeton New Jersey, US. (Translation of the German original: Mathematische Grundlagen der Quantenmechanik, Springer, Berlin, 1932.)